6th International Conference of Science, Technology, and Interdisciplinary Research (IC-STAR 2020)

IOP Conference Series: Materials Science and Engineering
Volume 1173

Bandar Lampung, Indonesia
20 - 21 October 2020

Editors:

Irza Sukmana
Khairuddin
Seri Rahayu Ya'akub
Fethma M. Nor
Taufiq Mulyanto
Muhammad Hanif

ISBN: 978-1-7138-4633-8
ISSN: 1757-8981

Printed from e-media with permission by:

Curran Associates, Inc.
57 Morehouse Lane
Red Hook, NY 12571

Some format issues inherent in the e-media version may also appear in this print version.

Institute of Physics
Dirac House, Temple Back
Bristol BS1 6BE UK

Phone: 44 1 17 929 7481
Fax: 44 1 17 920 0979

techtracking@iop.org

Additional copies of this publication are available from:

Curran Associates, Inc.
57 Morehouse Lane
Red Hook, NY 12571 USA
Phone: 845-758-0400
Fax: 845-758-2633
Email: curran@proceedings.com
Web: www.proceedings.com

TABLE OF CONTENTS

Preface

It has been a tremendous experience to organize the 6th International Conference of Science, Technology, and Interdisciplinary Research (IC-STAR 2020). The annual international conference is held in conjunction with other conferences throughout series of event spanning from October to December 2020. The scope of IC-STAR 2020 includes chemical and biotechnology, civil and environmental, computer science and information technology, electrical and communications, materials sciences, industrial engineering, mechanical and aerospace, geodetic and geophysics, applied mathematics and physic sciences, and also the interdisciplinary research. IC-STAR conference series intend to integrate engineering and life sciences fields in order to provide a cross section roadmap from basic research, technological and developments, processes development, and their applications and integrity in the real-world usage.

The event's series started with a virtual event which is hosted at Universitas Lampung, in Bandar Lampung, Indonesia, on 20 – 21 October 2020. The conference featured Keynote Speakers list of Prof. Ahmad Saudi Samosir (Universitas Lampung, Indonesia), Prof. Rizalman Mamat (Universiti Malaysia Pahang, Malaysia), Prof. Sunny J. Kalayathankal (K. E. College, India), Prof. Naoya Kasai (Yokohama National University, Japan) and Prof. Gavin Brown (Auckland University, New Zealand). During the event, there are about 50 authors presenting their papers in oral presentation.

The series proceeded with another virtual event hosted at Institut Teknologi Bandung, in Bandung, Indonesia, on 25-26 November 2021 by ICASA-MATRIX. Keynote Speakers for the conference were Prof. Koji Shimoyama (Tohoku University, Japan), Prof. Erry YT Adesta (International Islamic University of Malaysia, Malaysia), Prof. Tatacipta Dirgantara (Institut Teknologi Bandung, Indonesia) and Prof. Bambang Kismono Hadi (Institut Teknologi Bandung, Indonesia). There were about 40 oral presentations during the event.

The series concluded with a virtual event jointly hosted and supported by Universiti Teknologi Malaysia, Curtin University Malaysia, Universiti Teknologi Brunei, and Universitas Lampung, on 8 – 9 December 2021. The Keynote Speakers were Prof. Hadi Nur (Universiti Teknologi Malaysia, Malaysia), Prof. Ramesh Singh (Universiti Teknologi Brunei, Brunei Darussalam), Dr. Rino Rakhmata Mukti (Institut Teknologi Bandung, Indonesia), Prof. Mohammed Rafiq Abdul Kadir (Universiti Teknologi Malaysia, Malaysia), Prof. M. Akbar Rhamdhani (Swinburne University, Australia) and Prof. Ahad Ali (Lawrence Technological University, USA). There were about 80 oral presentations in the event.

We appreciate all distinguished speakers and all authors who presented during the conference series. We also acknowledge the effort of all committees, including all reviewers, for their contribution to the conference series and in making this proceeding. Thank you and hope to see you in future events of the conference series.

Publication and Proceeding Editors,
Irza Sukmana (Universitas Lampung, Indonesia)
Khairuddin (Universitas Lampung, Indonesia)
Seri Rahayu Ya'akub (Universiti Teknologi Brunei, Brunei Darussalam)
Fethma M Nor (Curtin University, Malaysia)
Taufiq Mulyanto (Institut Teknologi Bandung, Indonesia)
Muhammad Hanif (Universitas Lampung, Indonesia)

Peer review declaration

All conference organisers/editors are required to declare details about their peer review. Therefore, please provide the following information:

- **Type of peer review:** Single-blind
  The IC-STAR 2020 Committee use a single-blind review process where the identity of author is made known to the reviewers, but the authors do not know who reviewed their article.

  For each paper, the Editors will do first screening by determining (a) the suitability with scope, (b) similarity index, using official iThenticate (Turnitin) that should be max 20% similarity, (c) presentation style, language and formatting, and (d) overall technical correctness.

  Once the paper passes the screening, the Editors will assign at least 2 reviewers. Once comments and recommendation from reviewers are collected, the Editors will decide acceptance, revision, or rejection. For revision, the authors will be asked to address the reviewer's comments along with the revised manuscript.

- **Conference submission management system:** EasyCair management system
- **Number of submissions received:** 153
- **Number of submissions sent for review:** 99
- **Number of submissions accepted:** 73
- **Acceptance Rate (Number of Submissions Accepted / Number of Submissions Received X 100):** 47,71%
- **Average number of reviews per paper:** 1,5
- **Total number of reviewers involved:** 50
- **Any additional info on review process:** The IC-STAR 2020 committee used similarity detection software for each submitted full paper contribution. For clearance to the next stage of review process, the maximum similarity index is at 25%. Full papers with higher similarity index were individually checked for possible international plagiarism or duplication. When there is clear indication of international plagiarism or duplication, the particular paper is rejected. Otherwise, the paper is cleared to the next stage of review process.

- **Contact person for queries:**
  Dr. Irza Sukmana, Faculty of Engineering, Universitas Lampung
  Email: icstar@eng.unila.ac.id

Characteristics of medium-low rank coal blending on performance and efficiency steam power plant

Hendri[1,*], R Nurhasanah[1], and A F Lubis[1]

[1] Department of Mechanical Engineering, Faculty Technology and Energy Business, Institut Teknologi PLN, Jalan Lingkar Luar Barat Duri Kosambi Jakarta 11750, Indonesia

*Email: hendri@itpln.ac.id

Abstract. Blending medium rank coal (MRC) with low-rank coal (LRC) is carried out to meet the caloric content which follows the boiler specifications. Negative effects of blending with LRC are environmental, specific fuel consumption (SFC), and efficiency issues. The purpose of this research is the optimum blending composition of LRC and MRC coal, which can be used by power plants. The research methodology used is to simulate coal blending with MRC percentage in the MRC-LRC from 0, 20%, 40%, 50%, 60%, 80%, and 100%. The type of coal used is categorized in MRC1, MRC2, LRC1, LRC2, LRC3. The combination used is MRC1-LRC1, MRC1-LRC2, and MRC2-LRC3. From the analysis obtained, if the MRC percentage in MRC-LRC blending increases, losses will decrease mass flow will decrease, SFC will decrease, and boiler efficiency will increase. Blending MRC2-LRC3 has the lowest losses, low mass flow with an average of 298.78 ton/h, low SFC with an average of 0.466 kg/kWh, and boiler efficiency between 84.85% to 85.71 %. Blending MRC2-LRC3 is most recommended for use in combustion plant with a minimum MRC2 percentage of 50%.

Keywords: coal blending, medium rank coal, low rank coal, boiler efficiency

1. Introduction

At present the need for fuel generation power plant in the word 36 % comes from coal fossil fuels [1]. Some countries in the world, most still use coal fossil fuels, such as South Africa 93%, Poland 92%, China 79%, and Australia 77% [2]. In general, power plant is designed with coal fuel which has a specific composition and heating value. Coal will be blending to meet boiler specifications. The caloric value of coal mining product in Indonesia is generally of low quality anthracites 0.3 %, bituminous 13%, sub-bituminous 26.7%, and lignite 58.7% [3]. Blending coal quality is basically a combination of all coal parameter, in order to get high boiler efficiency [4]. Optimization of coal blending depends on the method and equipment, stockyard size, coal quality distribution, coal management quality control system, and sustainable coal procurement inventory system [5]. In conducting coal blending there are several reasons that are reducing electricity production cost, guarantee of electricity supply for consumers, according to specifications of power plant, regulation on emissions and pollutant [6]. Optimization of coal blending can be increase coal quality diversity, stability of electricity production. Coal blending also helps solve coal transportations problems [7].

From the results of research it has been obtained, coal blending Cibabuyun Garut, Adaro South Kalimantan, Bayah Lebak Banten still produce high emissions, emission pollutant volatile matter is high

[8]. The results of the optimization of blending low rank and high rank coal with a ratio of 80:20 can reduce electricity productions cost to 342 IDR/kWh [4]. Other indicators of combustion from coal blending can be seen from the heat value, volatile matter, melting point, ash content, and high sulphur levels will be cause corrosion [9], and high heating value an impact on boiler efficiency [10–12]. Calculation of boiler efficiency with the heat loos method helps to identity the sources of heat losses [10]. Environmental issues that are always associated with combustion in plants ace Carbon Dioxide (CO2), Nitrogen dioxide (NOx), Sulphur Dioxide (Sox) emissions. The percentage of unburned carbon decreased the impact is NOx emissions level increase if that percentage of sub-bituminous coal rises and the percentage of unburnt carbon rises. So that increase in NOx can be estimated from unburnt carbon levels [13–16]. Indonesia Power Company UP Surayala (UPS) is power plant with coal fuel a capacity of 3400 MW, coal received has a different heating value, can be grouped type of medium rank (sub-bituminous) and low rank. So that the caloric value of the coal fed to the boiler meets requirements, the coal will be blending. This study was conducted to determine the effect of medium rank and low rank coal blending percentage on boiler efficiency.

2. Methodology

In this research the coal blending uses method by Canterper, and approach taken by Jianbo Li, et.all [16], coal blending is based on proximate analysis, ultimate analysis as shows in table 1, data obtained from UPS. In table1, MRC is medium rank coal, and LRC is low-rank coal. While the blending composition of percentage used in this research 0% MRC, 20% MRC, 40% MRC, 50% MRC, 60% MRC, 80% MRC dan 100% MRC. The boiler efficiency method is calculated according to ASME PTC 4-1 standard and heat loss approach [10], i.e.

$$\eta = 100 - L_1 - L_2 - L_3 - L_4 - L_5 - L_6 - L_7 \quad (1)$$

where, L_1: losses form dry exhaust gas
L_2: losses moisture from burning hydrogen
L_3: losses moisture in coal fuel
L_4: losses moisture in the air
L_5: losses form imperfect combustion
L_6: losses from radiation and convection are assumed is 0.19 % [10].
L_7: losses from carbon does not burn are assumes is 1 %.

Same advantages using the heat loss method include more easily calculate losses by measuring the coal mass flow rate, easy to check losses for are controlled and uncontrolled losses, boiler efficiency can be increased by knowing the type of losses [17].

Table 1. Proximate dan Ultimate

Parameter	MRC1	MRC2	LRC1	LRC2	LRC3
Proximate analysis (wt. %)					
Moisture	24.23	24.44	28.49	26.15	28.16
Ash content	5.09	5.21	4.92	2.22	4.73
Fixed carbon	37.02	35.90	33.71	34.28	34.04
Volatile matter	33.66	34.45	32.88	37.55	33.07
Ultimate analysis (wt. %)					
Carbon	53.76	53.09	50.55	53.85	51.97
Hydrogen	3.75	3.72	3.54	3.48	3.71
Oxygen	11.20	11.63	11.38	13.51	10.37
Nitrogen	1.09	1.08	0.77	0.71	0.78
Sulphur	0.87	0.84	0.34	0.18	0.28
HHV (kcal/kg)	5109	5041	4888	4821	4920

3. Results and Discussions

3.1. Proximate, ultimate, and calorific value analysis

From the proximate and ultimate test results that need attention are ash content, nitrogen, sulphur content can cause corrosion [9]. On the hand the company has set standards for ash content nitrogen and sulphur content. Effect of ash content on boiler will reduce the flame in heat exchanger, increase excess air, increase in exhaust gas [18]. Company has implemented a maximum standard of ash content in coal is 5% by wight. Based on Figure 1, MRC1-LRC1, MR1-LRC2 and MRC2-LRC3 coal blending meets the maximum ash content limit requirement of 5%, if the percentage of MRC content increase, the ash content in coal also rises, this results of this research are corresponding to the results of the study [11]. MRC1-LRC1 and MRC2-LRC3 coal blending is recommended for combustion in boiler.

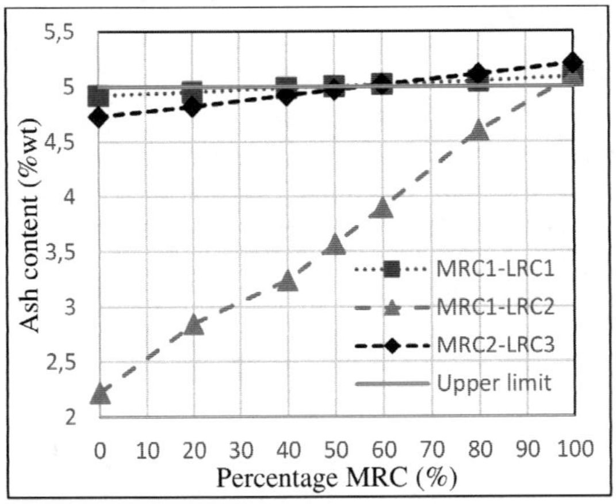

Figure 1. Effect Percentage MRC-LRC to ach content

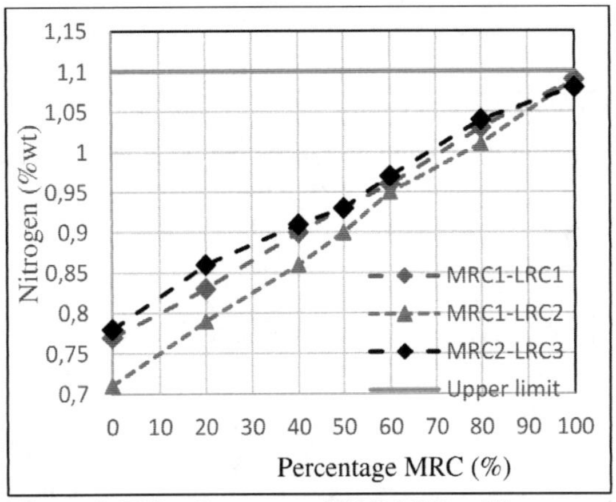

Figure 2. Effect Percentage MRC-LRC to Nitrogen content

Figure 2 shows the amount of nitrogen in coal based on analysis proximate test; company has determined that the nitrogen in coal is maximum 1.11 % by weight. Nitrogen in combustion will potentially produce NOx exhaust emissions, the higher the nitrogen in coal, the higher the potential NOx emissions. This results will be relevant to the results of research form [13,15,19]. From Figure 2, it can also be seen that if the percentage MRC increase, the weight nitrogen in coal also rises, this relevant to study [11], where the MRC-LRC percentage has a linier relationship with nitrogen in coal.

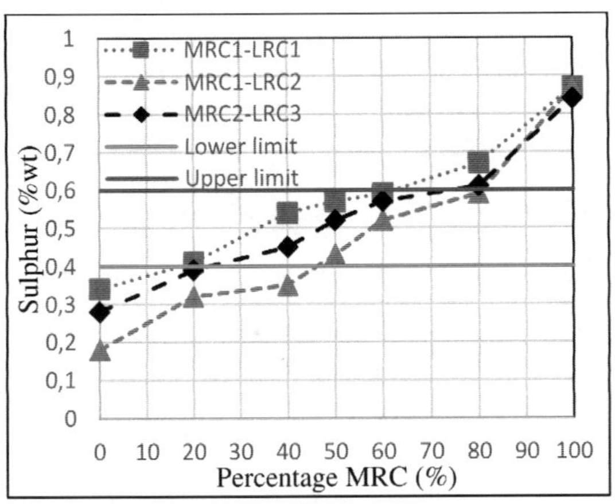

Figure 3. Effect Percentage MRC-LRC on sulphur

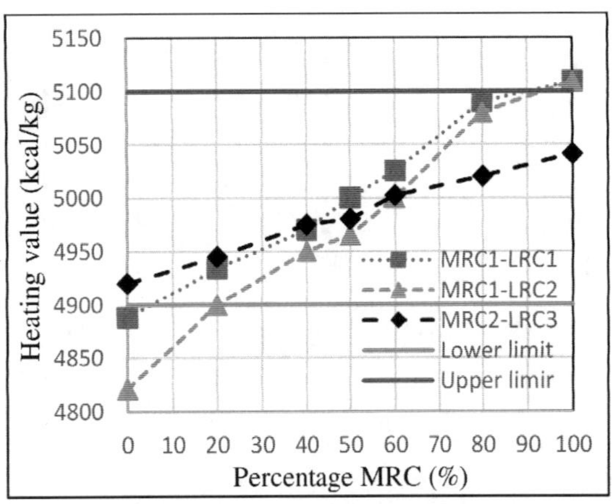

Figure 4. Effect Percentage MRC-LRC on Heating Value

From Figure 3, shows the relationship between percentage of MRC on sulphur, if percentage MRC rises, the sulphur content also rises. Sulphur in coal will cause corrosion in boiler pipes [9], then the company set the standard sulphur in coal between 0.4% to 0.6% by weight. Coal blending with a percentage of 20:80 to 60:40 meets the sulphur content standard in coal. If percentage of MRC in mixture rises, the sulphur content in coal also rises, this has an impact on increasing Sox emission levels. Results of this research are relevant to results of research from [7,11].

Figure 4 shows that the relationship of proportion of MRC-LRS to caloric value of the coal blending. The company has set a permissible caloric between 4900 kcal to 5100 kcal/kg. Based on Figure 4, the percentage of at least 20% MRC meets the caloric value coal requirement to be fed the boiler. Each increase percentage of MRC, caloric value in coal also increase, meaning that caloric value has linier relationship with increase percentage MCR in coal blending. The results of this study are relevant to results of study [11,20] which states that if percentage of MRC in coal blending rises and carbon content in mixture rises, then the HHV coal also rises.

3.2. Analysis Losses

Based on the results of data processing, that three losses that have percentage of more than 1%, that is L_1, L_2, and L_3. If the percentage of losses from unburned carbon increases, the efficiency of boiler decreases, and each increase of carbon burned rises between 5%-8%, the efficiency of the boiler will be decrease by 2% [21].

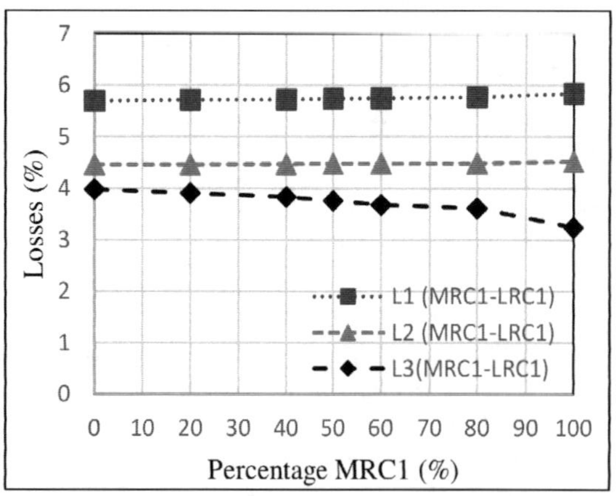

Figure 5. Effect Percentage MRC1-LRC1 on Losses

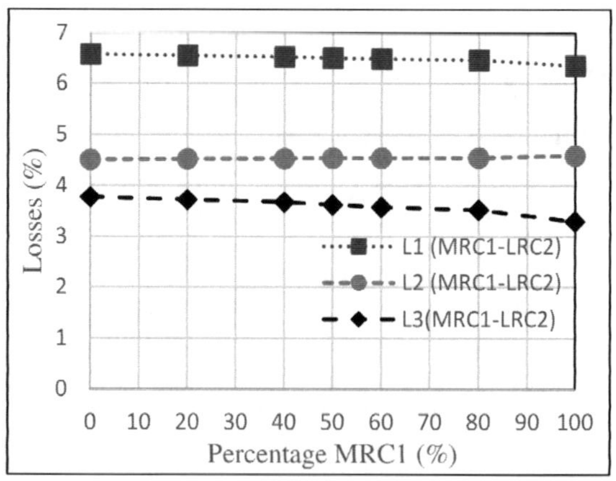

Figure 6. Effect Percentage MRC1-LRC2 on Losses

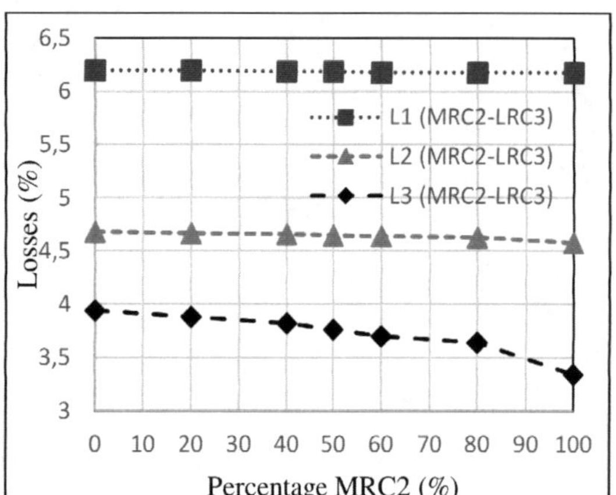

Figure 7. Effect percentage MRC2-LRC3 on losses

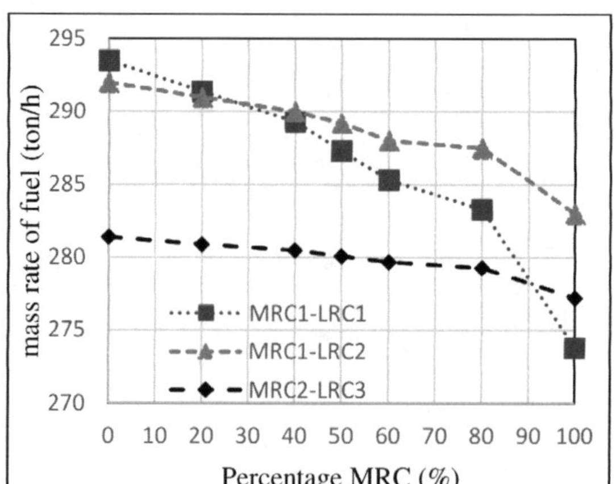

Figure 8. Effect percentage MRC-LRC on mass rate of fuel

In Figure 5, shows influence of percentage of MRC1-LRC1 on losses. With the simple linier regression method, gradient for $L_1$is 0.0007, average of losses 5.74 %, gradient for L_2 is 0.0003 and average of losses 4.48%, and gradient for L_3 is -0.0055 with average of losses 3.372 %. From the gradient difference its can be said the percentage of blending MRC10LRC1 does not effect, on the amount of dry gas exhaust losses, and moisture in hydrogen combustion process, but the influence on the moisture losses in coal.

Form Figure 6, average L1 and L2 losses for MRC1-LRC2 blending are, respectively is 6.49% and 4.54%, and from Figure 7, average L1 and L2 losses for MR2-LR3 blending are, respectively is 61.18% and 4.64%. The L1 and L2 losses does not have a linier relationship with the MRC-LRC percentage increase. While the L3 average losses for MRC1-LRC2 and MRC2-LRC3 are respectively 3.59% and 3.73%. L3 losses have linier relationship with MRC-LRC where each gradient is -0.017 and -0.005, the meaning that if percentage of MRC-LRC in blending rises, losses due to moisture in coal will be reduced. The results of this research are relevant to the results of the study [4, 22], which states that if the quality of coal in blending increase, then the losses from combustion moisture will decrease.

3.3. *Mass flow, SFC and Boiler Efficiency Analysis*

Efficiency is basically the ratio of energy out compared to the energy input in a system. Boiler efficiency is ratio of boiler output production compared to the input rate[18]. Boiler efficiency can be improved through combustion optimization by increasing carbon and caloric value in coal blending [13]. On the other hand by reduction the rate mass of coal fed to boiler will reduce the cost electricity production [4]. Figure 8, coal mass rate has an inverse linier relationship with MRC-LRC percentage. This means that if percentage of MRC in coal blending increase, the mass of coal will go down. Reduction coal mass flow rate is relevant to results the study [10,11,22]. With the linier regression approach, gradient of MRC1-LRC1 is -0.0821, gradient MRC1-LRC2 is -0.0821 and gradient MRC2-LRC3 is -0.38 with average mass flow 297.87 ton/h, its means the use of MRC2-LRC3 coal blending has impact on the use coal mass flow [4].

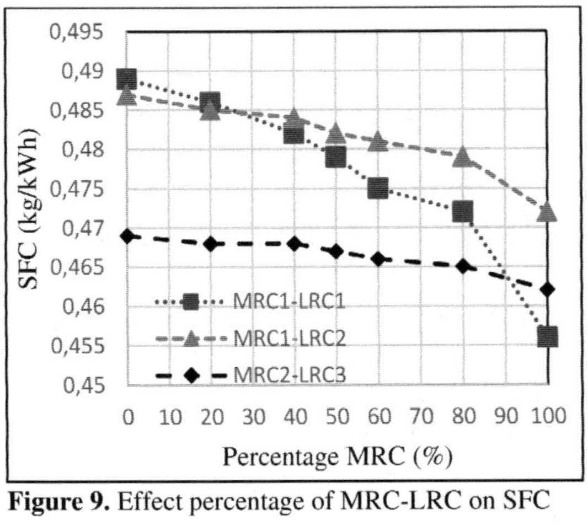

Figure 9. Effect percentage of MRC-LRC on SFC

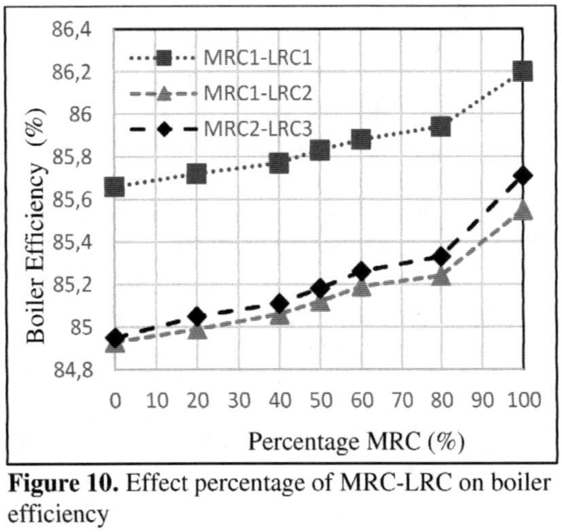

Figure 10. Effect percentage of MRC-LRC on boiler efficiency

Specific fuel consumption (SFC) is the amount of mass fuel flow need to produce power per hour for one hour. Figure 9, if percentage of MRC in coal blending increase, SFC decrease. The lowest average SFC is a mixture MRC2-LRC3 which is 0.4664 kg/kWh. This mean that fuel demand for coal mass flow decrease, if there is an increase in the heating value, which relevant to the results [10]. The reduction in SFC will have an impact on the reduction in electricity production costs, relevant to results of study [4]. This it is recommended to use MRC2-LRC3 coal blending to feed to the boiler.

Figure 10, that boiler efficiency has a linier relationship on increase MRC percentage in MRC-LRC blending, and correlation is positive. This means that if percentage of MRC in blending increase, the boiler efficiency will also increase. Boiler efficiency increase, if coal quality increase, this is relevant to results of study [10,22]. The minimum boiler efficiency MRC1-LRC2 is 84.95%, and maximum MRC1-LRC1 is 76.21 %%. With the linier regression approach, greatest gradient is the MRC-LRC3 pair, increasing efficiency it will have an impact the use of coal mass flow, so it will reduce the decrease in production costs. The recommended coal blending is MRC2-LRC3 pair with a composition of least 50% MRC2.

4. Conclusion

Based on the research results, it can be concluded that if the MRC percentage in MRC-LRC blending increases, losses will decrease mass flow will decrease, SFC will decrease, and boiler efficiency will increase. Blending MRC2-LRC3 has the lowest losses, low mass flow with an average of 298.78 ton/h, low SFC with an average of 0.466 kg/kWh, and boiler efficiency between 84.85% to 85.71 %. Blending MRC2-LRC3 is most recommended for use in combustion plant with a minimum MRC2 percentage of 50%.

Acknowledgement

This research was supported by Grant "Penelitian Dosen Pemula" (PDP) of the Ministry of Research and Technology and BRIN for the 2020 budget year.

Reference

[1] A. I. Anukam, S. N. Mamphweli, P. S. Mabizela, and E. L. Meyer, 2016, *J. Chem.*, **3** 1–8
[2] M. Huda, Gandhi K. Hudaya, N. S. Ningrum, and Suganal, 2012, *Teknol. Miner. dan Batubara*, **8** 152–163
[3] Amiral Aziz and Andi Rinaldi Hasan, 2015, *Energi Lingkung.*, **11**(1) 1–6.
[4] H. H. Santoso, W. Kurniawan, H. Setiawan, and T. R. Biyanto, 2016, *Int. J. Eng. Sci. Invent*, **5(8)** 25–30
[5] M. Yorukoglu, 2017, *Madencilik*, **56(3)** 109–116
[6] L. Sloss, 2014, July. IEA Clean Coal Centre
[7] M. Zaid, M. A. Wahid, M. Mailah, M. A. Mazlan, and A. Saat, 2019, *AIP Conf. Proc.*, 2062
[8] Widodo, D. Fatimah, and L. M. Estiaty, 2018, *IOP Conf. Ser. Earth Environ. Sci.*, **118** 1
[9] J. Cai, 2019, *AIP Conf. Proc.*, July, **2122**
[10] T. Y. Mun, T. Z. Tumsa, U. Lee, and W. Yang, 2016, *Energy*, **115** 954–962
[11] S. Santhosh Raaj, et.all, M. Muthukrishnan, S. Krishnamoorthy, S. Arumugam, and N. Anantharaman, 2016, *Appl. Therm. Eng.*, **102** 9–16
[12] M. C. Barma, R. Saidur, S. M. A. Rahman, A. Allouhi, B. A. Akash, and S. M. Sait, 2017, *Renew. Sustain. Energy Rev.*, **79** 970–983
[13] B. H. Lee, E. G. Eddings, and C. H. Jeon, 2012, **26 (11)** 6803–6814
[14] J. Zhang, Q. Wang, Y. Wei, and L. Zhang, 2015, *Energy and Fuels*, **29 (2)** 1196–1209
[15] C. Moon, Y. Sung, S. Eom, and G. Choi, 2014, *Exp. Therm. Fluid Sci.*, **62 (1)** 99–108
[16] J. Li, M. Zhu, Z. Zhang, K. Zhang, G. Shen, and D. Zhang, 2017, *Fuel*, **195** 131–142
[17] M. K. Bora and S. Nakkeeran, 2014, *Int. J. Adv. Res.*, **2(5)** 561–574
[18] M. S. Bhatt, 2006, *Energy sources*, **28** 25–41
[19] A. S. Jatoi, 2017, *J. Appl. Emergi Sci.*, **7** 34–39
[20] A. V. Akkaya, 2008, *Fuel Process. Technol.*, **90 (2)** 165–170
[21] L. Chao, L. Ke, W. Yongzhen, M. Zhitong, and G. Yulie, 2017, *Energy Procedia*, **105** 3045–50
[22] A. M. Lahijani, E. E. Supeni, and F. Kalantari, 2018, *J. Ind. Pollutan Control*, **34 (1)** 1825–1832

Comparison of the winglet aerodynamic performance in unmanned aerial vehicle at low Reynolds number

S H S Putro[1*], B J Pitoyo[1], N Pambudiyatno[2], Sutardi[3], and W A Widodo[3]

[1] Department of Aircraft Engineering, Politeknik Penerbangan Surabaya, Jemur Andayani I/73 Wonocolo, Surabaya, 60236, Indonesia

[2] Department of Air Navigation and Telecommunications Engineering, Politeknik Penerbangan Surabaya, Jemur Andayani I/73 Wonocolo, Surabaya, 60236, Indonesia

[3] Department of Mechanical Engineering Faculty of Industrial Technology, Institut Teknologi Sepuluh Nopember Surabaya, Jalan Arief Rahman Hakim, Surabaya 60111 Indonesia

* Email: hudzaifahsetyo@gmail.com

Abstract. The Aerodynamic performance of an object is very dependent on the flow of fluid through it and the geometric shape of the object. The use of Unmanned Aerial Vehicle (UAV) demands high aerodynamic performance because it strongly supports the cruising range and fuel or battery used. By reducing drag and increasing the lift optimally, the resulting lift to drag ratio will get maximum results. This study compares the use of winglets in UAVs at low Reynolds numbers. This is very important because generally UAVs are used at low Reynolds numbers. The freestream used is Re = 2.34 x 10^4 on several low to high angle of attack. The type of airfoil used is Eppler 562 which is equipped with a winglet. Winglets used are simple winglets, blended winglets, and wingtip fence. The geometry was analyzed using numerical simulations namely computational fluid dynamic (CFD). The turbulent model used in this study is K-ω SST on Ansys 19.1. From this study it was found that the use of a wingtip fence produced the best aerodynamic performance compared to other configurations. In addition, the use of a wingtip fence can delay stall and separation better than simple and blended winglets.

Keywords: lift to drag ratio, unmanned aerial vehicle, blended winglet, simple winglet, wingtip fence

1. Introduction

Winglets have become a significant requirement in aircraft operations. At present, it can be ascertained that all new designs for commercial aircraft will use winglets on their wings. It is a common agreement in various journals and magazines that the first generation of the winglet design was designed by NASA's Richard Whitcomb in 1976. The design is claimed to have increased efficiency by 2.5-3% in fuel economy compared to without winglets. The second generation has been installed on the Boeing 737 Classics, 737 NG, 757, and 767 where it is claimed to save 4-6% fuel. This is followed by the third generation installed on the 737 MAX which is claimed to have saved 1-2% compared to the second generation. In the Airbus design, the use of Sharklets is claimed to have reduced wingtip vortices and reduced total drag [1]. Other researchers have also discussed and compared winglets on aircraft, as presented by Jahanmiri [2] and Scholz [3].

The use of other flying objects seeks to continue to improve its performance through the use of winglets. Flying objects that are similar to airplanes include the Unmanned Aerial Vehicle (UAV). Its use has been widely used in traffic monitoring, agriculture, volcano monitoring, pollution monitoring, delivery, military, and others [4]. UAV utilization is very dependent on aerodynamic aspects, although the power source also plays a major role in increasing the range.

During its development, the UAV was not only in the form of a fixed-wing and rotary wing like an airplane but developed into a UAV convertiplane and tail-sitter UAV. This development makes the types and forms of UAV more varied [5].

Belferhat et al examined the effect of cant angle β variation on the blended winglet, which was modified at the leading edge[6]. This research was conducted on an airfoil wing NACA 0012 with β = 0°, 55°, 65°, and 75°. The velocity used in the wind tunnel are 20, 30, and 40 meter/second (Re = 2.35 x 10^5, 2.84 x 10^5 and 3.03 x 10^5). The angles of incidence used are + 0°, + 5°, + 10°, and + 19°. From the results of this study, it was found that β = 55° resulted in better aerodynamic performance than other cant angle variations.

Munoz et al conducted a study on fixed-wing UAV[7]. The airfoil used on the wing is NACA 23015 with an Eppler 387 winglet airfoil. Freestream velocity used in the wind tunnel is 30 meter/second (Re = 3.5 x 10^5). From this research, it was found that with wingtip blowing using coanda jet and multi winglet can reduce induced drag. However, for commercial use, it is not recommended because it uses a very large amount of energy. Multi winglets show better results although slightly compared to plain wings. The use of multi-winglets is highly recommended, especially for light aircraft using low velocity.

Xin et al [8] examined the performance of the bionic wing design in UAVs using winglets. The bionic referred to imitates the wing shape of a seagull. The bionic wing is in the form of a plain wing compared to the plain wing which is equipped with a winglet. The use of bionic wings and winglets can increase the pressure distribution on the wing, reduce the drag coefficient, and reduce the separation point on the wing. With a decrease in the drag coefficient, the lift to drag ratio of the wings also increases. Besides, it can be seen that with the winglets, the downwash angle of attack and the downwash velocity can be reduced.

The above studies have provided further inspiration to compare the aerodynamic performance of several types of winglets that can be used in UAVs. The choice of UAV is emphasized because it uses a relatively smaller velocity. This study uses a rectangular or straight wing as the basis for the use of other wing types. The output of this research is the drag coefficient, lift coefficient, lift to drag ratio, and also some wing visualizations that can show the evolution of the fluid flow around it.

2. Materials and Methods

Computational Fluids Dynamic numerical simulation in this study using Ansys 19.1. K-ω SST was used as a turbulent model with a freestream velocity of 10 meter/second (Re = 2.34 x 10^4). This numerical simulation refers to Hariyadi's research [9–11]. Before analyzing aerodynamic performance and numerical visualization, an independent grid is carried out to see the extent of the efficiency of each mesh that will be used. Several types of mesh used can be seen in table 1. Table 1 shows the condition of the model used for running Ansys 19.1.

Table 1 shows the boundary conditions and properties of the model and the flow that passes through the Eppler 562 wing airfoil. Referring to the Kontogianis research [12] on y+ and Anderson [13] regarding the drag coefficient, it is found that Mesh B in table 2 is used in this study. Figure 1 shows the boundary conditions that refer to Mulvany's research [14]. Figure 2 shows the model used in the study.

Table 1. Description of the Model Condition

Description	Model Condition
Model	3D, Unsteady/Transient
	Plain wing
	Plain wing with simple winglet
	Plain wing with blended
	Plain wing with forward wingtip fence
	Plain wing with rearward wingtip fence
Airfoil	Eppler 562
Fluid	Air
Fluids Properties	Density 1.225 kg/m3
	Viscosity 0.000017894 kg/m-s
Boundary condition	Wing Wall
	Outlet Outflow
	Inlet Velocity Inlet
	Wall Wall
Reynolds Number	Re = 2,34 x 10^4

Table 2. Grid analysis model E562 three dimensions without winglets

Meshing Type	Number of Cells	C_D	y^+	Skewness Average
Meshing A	768.081	0,883	1,4	0,347
Meshing B	569.313	0,902	0,8	0,343
Meshing C	469.682	0,858	2,1	0,346
Meshing D	353.120	0,921	2,1	0,334
Meshing E	335.582	0,933	2,8	0,351

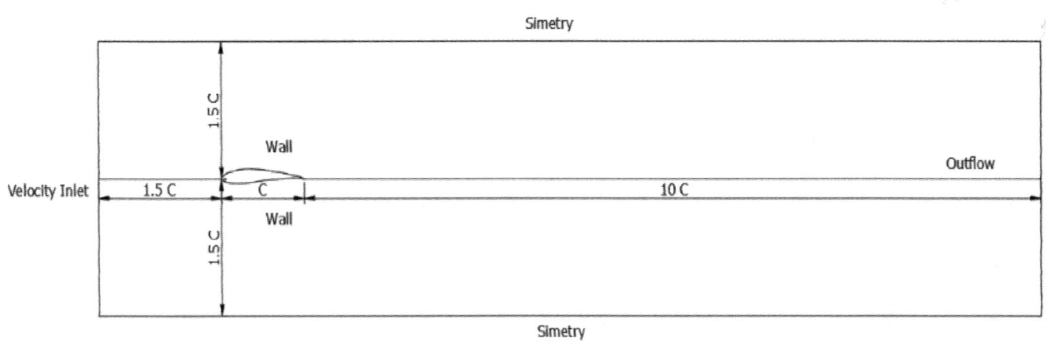

Figure 1. Simulation domain and research boundary conditions [14]

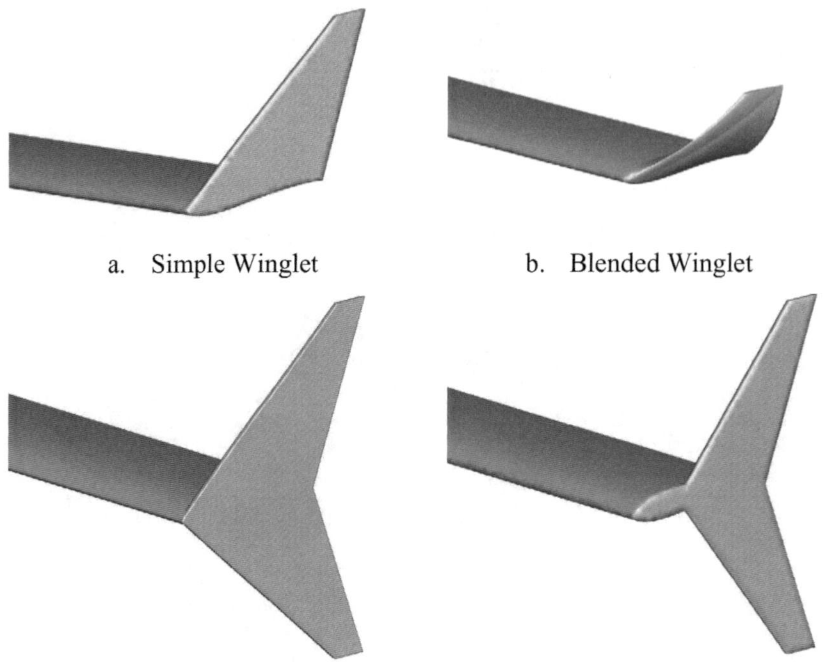

a. Simple Winglet b. Blended Winglet

c. Forward Wingtip Fence d. Rearward Wingtip Fence

Figure 2. The model used in the simulation

3. Result and Discussion

In the numerical simulation output results, the total drag coefficient, lift coefficient, and lift to drag ratio are obtained. Besides, various kinds of visualizations can be generated that can support the lift and drag output. Although the definitions and distribution of drag types differ in some literature, for example in the Kundu [15] and Panagiotou [16], the Ansys 19.1 output results refer to viscous drag and pressure drag. The results of this study show the results of viscous drag and pressure drag.

3.1. Drag Coefficient, Lift Coefficient, and Lift to Drag Ratio

Figure 3 shows the results of the drag coefficient of the plain wing, the plain wing equipped with the blended winglet, simple winglet, rearward wingtip fence, and forward wingtip fence. Figure 3 shows that the addition of a blended and simple winglet increases the total drag compared to the plain wing. In this figure it is also shown that the forward and rearward wingtip fence produces a smaller total drag than the plain wing.

In general, it is shown that as the angle of attack increases, the total drag also increases. The results obtained in this study have higher drag than Turanoguz's research [17,18] which uses Re = 2.356×10^6. This is possible because in the Turanoguz study using a higher Reynolds number so that the ability of the flow to resist viscous drag is higher than this study. which uses Re = 2.34×10^4.

To see the contribution of viscous drag and pressure drag, it can be explained in figures 4 and 5. Figure 4 shows that the role of viscous drag in the forward and rearward wingtip fence is quite dominant. This is because the surfaces of the forward and rearward wingtip fences are wider than the plain, simple, and blended winglets. Figure 5 shows the effect of drag pressure on the overall drag. The frontal area or cross-sectional area of the winglet configuration to the direction of flow will determine how much flow resistance the geometric configuration accepts. Therefore, the amount of drag pressure is not determined by the side view area of the wing because the flow is assumed to come from the direction of the X-axis of the flow.

IOP Conf. Series: Materials Science and Engineering 1173 (2021) 012002 doi:10.1088/1757-899X/1173/1/012002

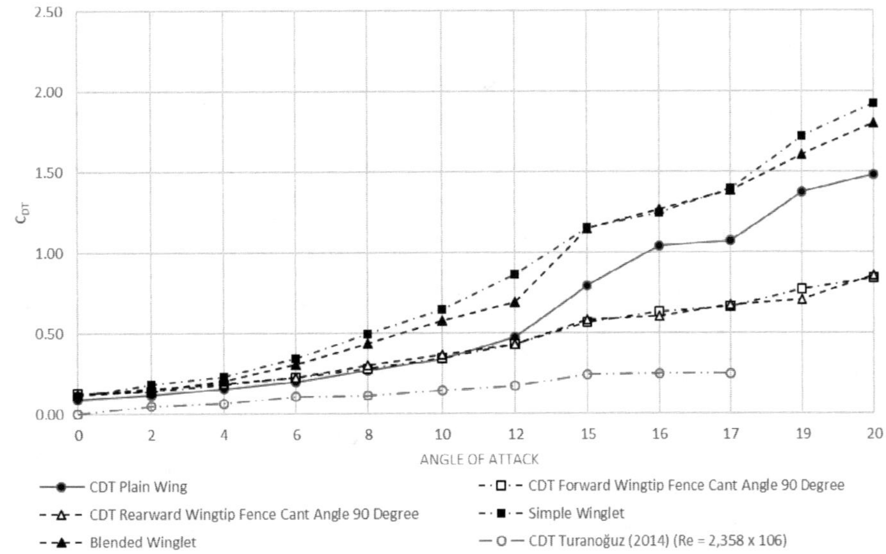

Figure 3. Total Drag Coefficient Comparison of the Research Model

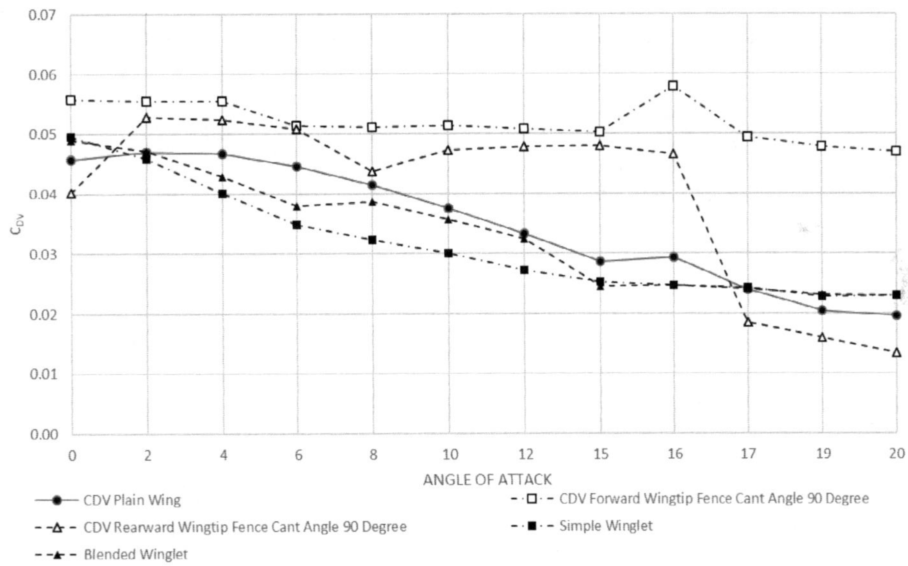

Figure 4. Viscous Drag Coefficient Comparison of the Research Model

Figure 6 shows the lift coefficient of the plain wing, the plain wing equipped with the blended winglet, simple winglet, rearward wingtip fence and forward wingtip fence. The lift coefficient produced by Turanoguz's study [17,18] is higher than this study. Again, this is because the Reynolds number used is higher than this research. The thing of concern is the ability of the winglet to delay the occurrence of a stall compared to the plain wing.

The stall point on the plain wing occurs at an angle of attack of 12°, as in Turanoguz's research. Likewise, the plain wing which is equipped with a simple winglet. The plain wing which is equipped with a blended winglet shows that the stall occurs at an angle of attack of 15°, while the rearward wingtip fence occurs at 16°. The forward wingtip fence shows the best result by showing the stall occurred at 17° angle of attack.

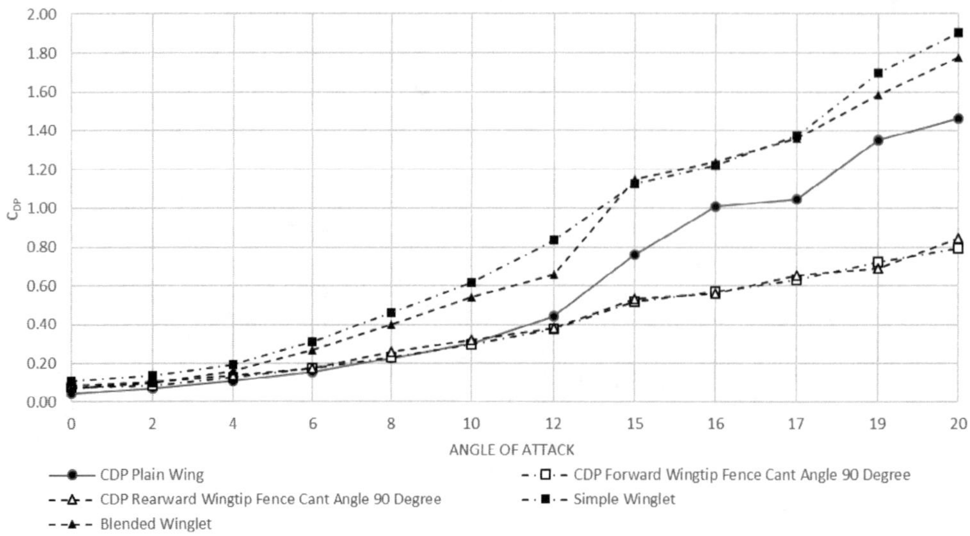

Figure 5. Pressure Drag Coefficient Comparison of the Research Model

From the results of the drag coefficient and the lift coefficient, the lift to drag ratio is obtained as shown in figure 7. Plain wings equipped with blended and simple winglets produce lower performance than plain wings. The rearward wingtip fence produces better performance than the plain wing from an angle of attack of 10°, while the forward wingtip fence is better than the plain wing from the angle of attack of 6°. The important thing to note is that the aerodynamic performance for each velocity is different from the other velocity. In this study using the minimum velocity used by the UAV. This refers to the research of Panagiotou [16,19–21], Turanoguz [17,18], and Kontogianis [12,22].

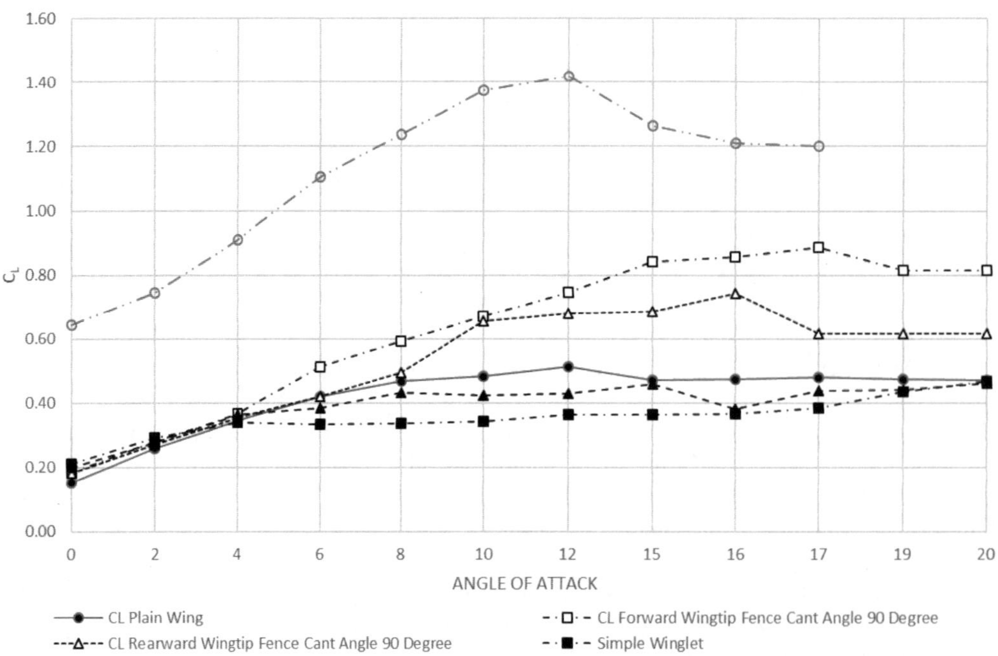

Figure 6. Lift Coefficient Comparison of the Research Model

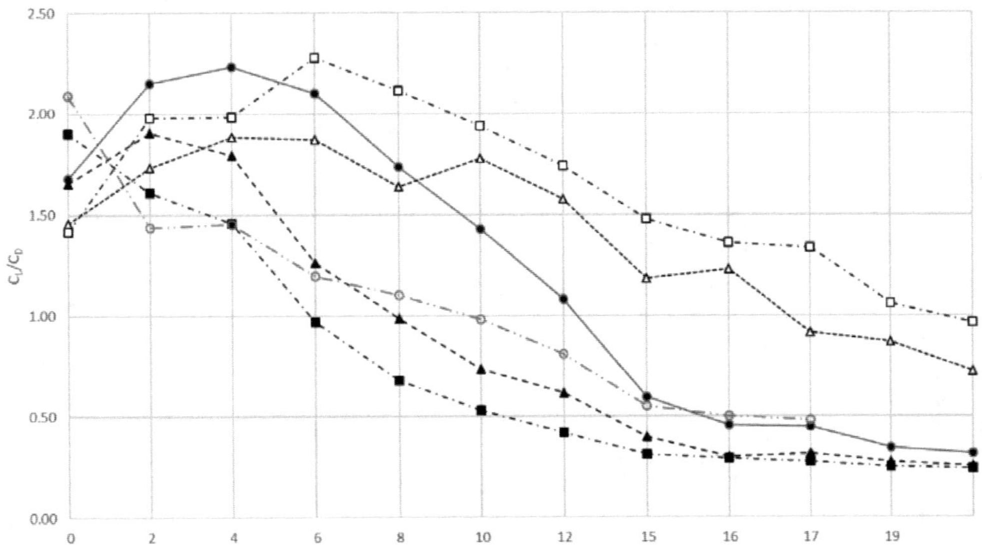

Figure 7. Lift to Drag Ratio Comparison of the Research Model

3.2. Pressure Contour

Figure 8 shows the comparison of the evolution of the pressure contour on the upper side of the plain wing, the plain wing equipped with the blended winglet, the simple winglet, the rearward wingtip fence, and the forward wingtip fence at $\alpha = 15^o$ and $\alpha = 17^o$. The plain wing shows that the wingtip tip has lower pressure, both at $\alpha = 15^o$ and $\alpha = 17^o$. The low-pressure area tends to expand to the surrounding area as the angle of attack increases, although the value is slightly higher than the wingtip area. The blended winglet shows that the pressure with a low value has shifted to the winglet. It can be seen that the blended winglet serves as an extension of the wing. The pressure value in the wingtip area is slightly higher than that of the plain wing.

The simple winglet shows that the pressure contour has a lower value and the area is wider than the plain and blended winglets. In the area on the inside of the winglet, there is an area of pressure that is higher than the surrounding pressure. This requires further discussion because it is thought to be an area with a more complex secondary flow. The forward wingtip fence shows the wingtip area with higher pressure compared to other configurations. Although the existing geometry configuration is similar to the simple winglet, the area on the side of the fence does not show pressure with a higher value than the other areas. In the rearward wingtip fence $\alpha = 15^o$, it shows signs that pressure with a low value is starting to form at the end of the leading edge. Although the area is not as large as the simple winglet, the area with this low value is quite clear and broad compared to other configurations. At $\alpha = 17^o$, it can be seen that the area with this low value extends towards the wingspan and is connected to the wing root. This area has a low value and is the most extensive compared to other configurations. The area that is formed in the simple winglet is not visible on the inside of the fence. As with the simple winglet, the area with a lower area is interesting to discuss further because it involves secondary flow which is broad and complex.

From the winglet configuration used in this study, it can be seen that there is an effect of flow leakage at the end of the wingtip from the leading edge. An area of pressure with a low value is an indication of a leakage flow from the lower side to the upper side. The simple winglet and rearward wingtip fence are configurations that have a wider area with a low value. The shape and geometry of the winglets greatly influence the extent to which they can withstand the leaps and leaks of the fluid flow.

IOP Conf. Series: Materials Science and Engineering 1173 (2021) 012002 doi:10.1088/1757-899X/1173/1/012002

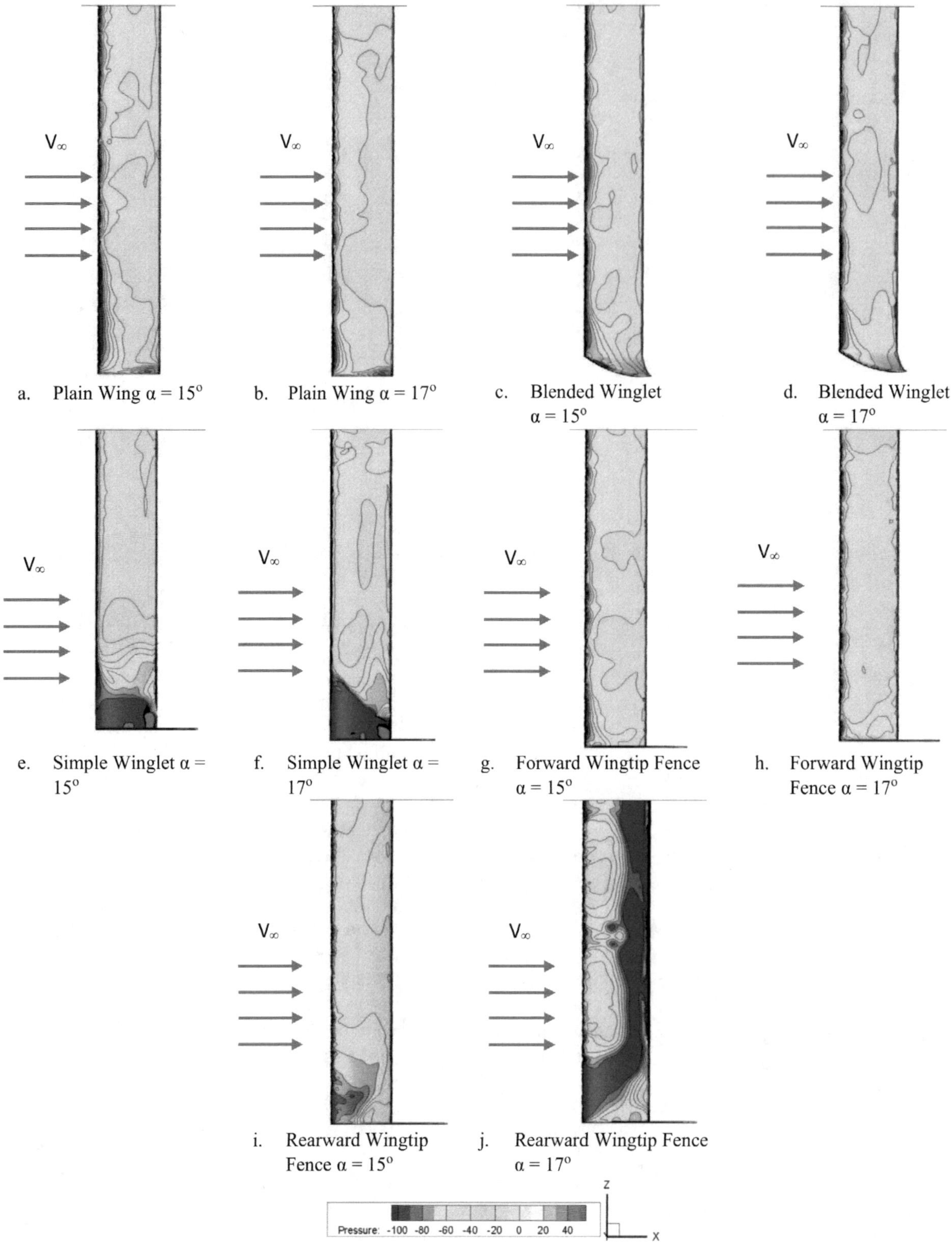

Figure 8. Pressure Contours Comparison on the Upper Side of the Research Model at $\alpha = 15^{o}$; $\alpha = 17^{o}$

4. Conclusion

In this study, the comparison for several types of winglets have been carried out. From the research results above, there are several things that need to be considered. The viscous drag and pressure drag play an important role in the formation of total drag. With the addition of a winglet to the plain wing, the addition of the forward and rearward wingtip fence results in higher viscous drag compared to other configurations. In this case, the blended and simple winglet produces greater value than other configurations. Also, the forward and rearward wingtip fences provide better aerodynamic performance than others. This can be seen from the resulting lift to drag ratio. The rearward wingtip fence produces better performance than the plain wing from an angle of attack of 10°, while the forward wingtip fence is better than the plain wing from the angle of attack of 6°. While the leap of fluid flow from the lower surface to the upper surface cannot be completely eliminated but can be reduced as much as possible. Furthermore, the shape and geometry of the leading edge on the wingtip greatly affect the ability of the winglet to withstand the leap of fluid flow. Forward wingtip fence and blended winglet indicate a relatively smaller low-pressure area. This suggests that both configurations are better able to withstand flow leakage at the wingtip. The results of this study require a follow-up study, mainly on the secondary flow around the winglet and its effect on the flow towards the mid span. The flow dynamics around the winglets and their effects on the flow in the mid span require further study and a more detailed discussion.

References

[1] Aircraft Operations 2017 The benefits of winglets and performance enhancing kits *Airline Cost Conference* 48 5
[2] Jahanmiri M 2014 Aircraft Drag Reduction : An Overview Aircraft Drag Reduction : An Overview Division of Dynamics
[3] Scholz D 2018 Definition and discussion of the intrinsic efficiency of winglets *INCAS Bulletin* 10 117–34
[4] Boukoberine M N, Zhou Z and Benbouzid M 2019 A critical review on unmanned aerial vehicles power supply and energy management: Solutions, strategies, and prospects *Applied Energy* 255 113823
[5] Saeed A S, Younes A B, Cai C and Cai G 2018 A survey of hybrid Unmanned Aerial Vehicles *Progress in Aerospace Sciences* 98 91–105
[6] Belferhat S, Meftah S M A, Yahiaoui T and Imine B 2013 Aerodynamic optimization of a winglet design *EPJ Web of Conferences* 45 18–21
[7] Céron-Muñoz H D, Cosin R, Coimbra R F F, Correa L G N and Catalano F M 2013 Experimental investigation of wing-tip devices on the reduction of induced drag *Journal of Aircraft* 50 441–9
[8] Xin H, Shuo Q, Luo K and Liang W 2015 Unmanned Aerial Vehicle (UVA) Bionic Wing Design and Performance Analysis 2031–4
[9] Setyo Hariyadi S P, Sutardi, Widodo W A and Mustaghfirin M A 2018 Aerodynamics analisys of the wingtip fence effect on UAV wing *International Review of Mechanical Engineering* 12 837–46
[10] Setyo Hariyadi S P, Sutardi and Widodo W A 2018 Numerical study of flow characteristics around wing airfoil Eppler 562 with variations of rearward wingtip fence *AIP Conference Proceedings* 1983
[11] Hariyadi S S P, Sutardi and Widodo W A 2018 Drag reduction analysis of wing airfoil E562 with forward wingtip fence at cant angle variations of 75°and 90° *AIP Conference Proceedings* vol 2001 (American Institute of Physics Inc.)
[12] Kontogiannis S G, Mazarakos D E and Kostopoulos V 2016 ATLAS IV wing aerodynamic design: From conceptual approach to detailed optimization *Aerospace Science and Technology* 56 135–47

[13] Anderson J D and Anderson Jr J D 1995 *Computational Fluid Dynamics The Basics with Applications*
[14] Mulvany N, Chen L, Tu J and Anderson B 2004 Steady-State Evaluation of Two-Equation RANS (Reynolds-Averaged Navier-Stokes) Turbulence Models for High-Reynolds Number Hydrodynamic Flow Simulations *Department of Defence, Australian Government* 1–54
[15] Ajoy Kumar Kundu *Aircraft Design* (Cambridge Aerospace Series)
[16] Panagiotou P and Yakinthos K 2019 Aerodynamic efficiency and performance enhancement of fixed-wing UAVs *Aerospace Science and Technology* 1 105575
[17] Turanoguz E 2014 Design of a Medium Range Tactical
[18] Turanoguz E and Alemdaroglu N 2015 Design of a medium range tactical UAV and improvement of its performance by using winglets *2015 International Conference on Unmanned Aircraft Systems, ICUAS 2015* 1074–83
[19] Panagiotou P, Kaparos P, Salpingidou C and Yakinthos K 2016 Aerodynamic design of a MALE UAV *Aerospace Science and Technology* 50 127–38
[20] Panagiotou P, Kaparos P and Yakinthos K 2014 Winglet design and optimization for a MALE UAV using CFD *Aerospace Science and Technology* 39 190–205
[21] Panagiotou P, Salpingidou C, Kaparos P and Yakinthos K 2015 a Cfd-Aided Design Procedure , Performance Estimation and Optimization Study of a Male Uav
[22] Kontogiannis S G and Ekaterinaris J A 2013 Design, performance evaluation and optimization of a UAV *Aerospace Science and Technology* 29 339–50

Structure from Motion (SfM) method to characterize fluvial sedimentology in Way Semaka river in Lampung province, Indonesia

R C Wibowo[1,*], B S Mulyatno[1], O Dewanto[1] and M Sarkowi[1]

[1] Department of Geophysical Engineering, Universitas Lampung, Jalan Prof. Soemantri Brojonegoro No.1, Bandar Lampung 35145, Indonesia

*Email: rahmat.caturwibowo@eng.unila.ac.id

Abstract. Structure from motion (SfM) has been applied recently in fluvial sedimentology. This application is due to the availability of many low-cost unmanned aerial vehicles/drones, which can help overcome challenging terrain, provide efficient and reproducible and high-accuracy topographic images and data. The current study describes the application of SfM to build a geomorphological model and estimate the surface water velocity of the Way Semaka River in the Bandar Negeri Suoh (BNS) region in the West Lampung area. Way Semaka is a river type with an extensive meandering system and stable extension tectonic regime relatively. River morphology approximately 96.62 m width and 4 m depth in straight (relatively) area, whereas in channel area approximately 171.22 m width and 5 m depth. The velocity range between 0.39 – 1.56 m/s based on image analysis proven by current meter measurements onsite with an RMS error of 0.25. Manual geomorphic unit level 2 analysis revealed that 9% of the coverage consisted of basins for the channel's zone, 35% of the coverage consisted of convexity, 9% consisted of planar features, and 37% of the coverage consisted of the transition zone.

Keywords: SfM, sedimentology, fluvial, Way Semaka, BNS

1. Introduction

One of the geological footprints that has important information about natural resources found on earth such as hydrocarbon reservoirs, mineral resources, water resources, and flood risk management [1]. Geometry and heterogeneity are examples of depositional models that are needed and essential for economic and social aspects. Several problems remain in question today, mainly how fluvial deposits can be used to reconstruct geological history. One of the essential parts of a fluvial system is a channel that connects all river networks and provides information about the dynamics of sedimentation and the basins from which the sediments originate [2].

Structure-from-Motion (SfM) with multi-view stereo (MVS), is a combination of advances in computer vision and conventional photogrammetry used for topographic survey techniques. This method can produce a dense three-dimensional (3D) point cloud with high resolution of an object or surface at an affordable operating cost. SfM, while it is only applied to geoscience applications that have relatively transformative effects on scientific disciplines [3], provides a very economical and spatially fast 3D survey over a wide area. The density and level of point accuracy comparable to other survey methods can be seen in Table 1. Several pragmatic reasons for using this SfM method, namely the data obtained in the form of data resulting from unmanned aerial vehicle (UAV) shooting, which UAV for now has developed very rapidly and is easy to use even for ordinary consumers. Thus, the image data obtained from the UAV has made the SfM photogrammetric method more attractive than conventional photogrammetry; for example, [4] developed a complex photogrammetric concept orthorectifying photos obtained from drone surveys.

Table 1. Relationship between the SfM photogrammetric characteristics and area, density, and accuracy [5].

	Without SfM	**With SfM**
Spatial extent (km^2)	TS, dGPS: <1.0 TLS: <5.0 AP: <50 ALS, MBES:<100	Ground-based platforms: 0.01 to 1.0 Airborne platform: <5.0
Spatial density (pts/m^2)	TS, dGPS: <5.0 AP: <10 ALS, MBES: <10 TLS: <10,000	1 to 10,000
Point acquisition rate (pts/hr)	TS: 10^2 dGPS: 10^3 AP, MBES: 10^4 ALS, TLS: 10^6	Millions
Point accuracy (m)	TS: <0.001 dGPS: <0.005 TLS, MBES: <0.05 ALS: <0.2 AP: <0.5	0.01 to 0.2

However, SfM can generate more than ordinary topographical contour because the extracted orthophoto motion is obtained from a full 3D color point cloud. One of the newest methods in the field of geoscience that utilizes SfM has considered the extraction of other information besides the primary fluvio-morphological data and the development of the SfM method in the fluvial environment, for example, is topographic mapping and surface water velocity.

For applied fluvio-morphological investigations in the modern era, sophisticated technology has been presented, which provides an excellent opportunity to obtain more accurate and detailed topographic contours, mapping automation, and quantitative modeling of river shape and activity. Field mapping and morphometric analysis provide basic templates for various toolkits for integrative river science. Morphometric analysis is often used as a classification exercise to identify specific features, process zones, and “river landscape” parts of continuous data, although its analysis can produce continuous derivative outputs.

River flow monitoring is a significant thing that the development of research and management of river science. An essential component in calculating the flow of a river is velocity. This can be obtained using in situ velocity meters such as ultrasonic meters, current meters, and acoustic Doppler current filters (ADCP). The use of terrestrial and aerial radar and camera sensors is an example of several long-range flow monitoring techniques that have been introduced in recent years. Image velocity is a river flow monitoring technique using a camera. Large-Scale Particle Image Velocimetry (LSPIV), Fujita in 1974 [6] developed the first image velocity technique in outdoor environments to apply the classic Particle Image Velocimetry (PIV) technique for fluvial (large-scale) environment. The application of image velocity has been successfully used in various river flow monitoring applications [7-8].

This study aims to introduce SfM and a detailed description of the methods used; make an initial sedimentological analysis about the geometry and internal architecture of fluvial sedimentology, including surface water flow velocity. The focus here is to outline a practical workflow that earth scientists and practitioners interested in deploying SfM can apply to geomorphological research.

2. Materials and Methods

2.1. Image's acquisition, processing, and accuracy

Photos are the core data in carrying out the SfM reconstruction; thus, digital cameras are the primary tools used to collect data. Photos used as input significantly affect the quality of the resulting model; this applies to other photogrammetry techniques. The photos obtained during the acquisition can improve accuracy and reduce time; they can also improve processing efficiency. Turner et al. [9] showed that absolute spatial accuracy <1 m can be achieved with coordinated photo images. The addition of ground control points (GCP) can be added during processing on cameras that do not have geotagging capabilities.

Determination of specific and objective locations is the optimal strategy in obtaining the required number of photos and the level of overlap. The overlap between adjacent photos> 60% has resulted in a reasonably good model [10]. The function of the area size and the amount of overlap between images is the definition of the total number of photos required. The main thing in object reconstruction must be visible in at least three photographs for the SfM algorithm to find individual points, and this is a general rule. Fewer photos may cause cracks, holes, or distortion in the SfM model compared to more photos. However, the large number of photos can result in extended processing times, and the resulting file is large, making it challenging to manipulate after processing.

Before applying the UAV and SfM techniques to the Way Semaka Sub-watershed, the test was carried out regarding the absolute accuracy and relative reconstruction of the open field. The UAV-SfM technique has a mean horizontal and vertical error of 2-5 m measured using GCP when reconstructing it in absolute space. The measurement comparison between the object of the SfM reconstruction and the direct measurement (the object whose size is known) becomes a test of relative accuracy. This relative test did not produce a statistically significant difference, indicating that the relative accuracy rate was very high in the SfM reconstruction. The results of this test indicate that the UAV-SfM technique can be applied to map medium-scale fluvio-morphology (1-100 meters) in the Way Semaka Sub-watershed.

2.2. Surface flow water velocity

2.2.1. Current meters

The ideal flow meter should respond instantaneously and consistently to any change in water velocity. There are different types of flow meters available, and they are grouped into three main categories: electromagnetic current meters, mechanical current meters, and Acoustic Doppler speed meters. In this research, to measure the surface flow water velocity is using the mechanical current meter.

2.2.2. PIVlab

Thielicke and Stamhuis developed PIVlab software as a Graphical User Interface (GUI) in MATLAB, which functions to analyze particle image velocity [7] and can be used for LSPIV applications [8]. For ease of use, tutorials and software can be downloaded for free from the internet. Image analysis in PIVlab is carried out on extracted georeferenced images. The analysis stage is carried out in a sequence of continuous frames. The multipass window deformation is applied with two interrogation paths: (i) 64 to 32 px^2 and (ii) 32 to 16 px^2. The chaotic height vector and the contrasting riparian vector were removed during the final image processing. The assessment lasts 5 minutes, referring to the power of the computer and the selection of the number of frames.

3. Results and Discussions

The first series of measurements were carried out through equipment installed on the river surface. The river surface has the advantage of following the surface flow velocity. After three consecutive measurements carried out for 5 minutes, the mean surface velocity was determined to be 0.34 m/s. The second measurement and so on can be seen in table 2.

Table 2. GCP locations and surface water velocity locations include surface water velocity data from the flow meter & PIVlab [7].

GCP	Lat	Long	Velocity current meter (m/s)	Velocity PIVlab (m/s)	RMS error
1	-5.203340985	104.276977	0.34	0.57	0.25
2	-5.240679011	104.298941	0.62	0.77	0.26
3	-5.204309011	104.277563	1.45	1.49	0.27
4	-5.212021032	104.282493	0.57	0.68	0.29
5	-5.212545991	104.282622	1.33	1.56	0.31
6	-5.21352198	104.28365	0.89	0.98	0.32
7	-5.219899016	104.287472	0.42	0.71	0.35
8	-5.220987992	104.288718	0.71	1.35	0.37
9	-5.232144976	104.296531	0.36	0.39	0.05
10	-5.23926096	104.297974	0.74	0.68	0.06

The use of PIVlab in analyzing water flow velocity is described in Figure 1. The yellow arrow is a velocity vector which is calculated as an average indicating direction and magnitude. The pattern formed at each location is a vector. No tracer could be seen above the constructed barrier, which should have resulted in a velocity pattern. There are several reasons for this absence, such as (a) carrying out the measurement process when no leaves or twigs are floating in the river, (b) the river is muddy, which results in difficulty distinguishing floating particles; and (c) the flight is carried out at an altitude of 80 m, which resulted in the absence of an adequate record of particle size. The measured distance must be predefined to convert the frame pixels to their actual size. The width of the point bar known when measured at the location becomes a reference. Statistics and graphs can be obtained using the PIVlab based on the cross-section or designed area.

The middle of the river has a higher speed; this is due to the high depth and the absence of obstacles. Meanwhile, the two lanes on the left and right have slow velocity due to rocks and vegetation. The middle point bar yields a very slow velocity, as expected. This is because there is no sufficient turbulence pattern to get the velocity field. The middle of the river has a speed > 1.4 m/s, which is more significant when compared to the left side, which is 17 m away, and the right side is 23 m from the bank because there are no significant obstacles. Two locations have very low velocities, at a distance of 3–5 m and 29–39 m from the left side of the cutting edge, because of vegetation and rocks.

Figure 1 illustrates that the Way Semaka Sub-watershed has pockets of discontinuous floodplain, confirming a partially limited valley arrangement. Sometimes, the boundary is the boundary of the valley itself (i.e., the hillside). The channel has a reasonably tortuous plan shape, partly due to transitions between the margins and partly from the bends themselves. The groove has limited capacity to adjust laterally unless there is sufficient width of the valley floor. This range is characterized by a partially delimited valley with moderate sinuosity, a discontinuous floodplain controlled by a valley.

Manual geomorphic unit level 2 analysis [11] revealed that for the zone within the channel, 9% of the coverage consisted of basins, 35% of the coverage consisted of convexity, 9% consisted of planar features, and 37% of the coverage consisted of the transition zone (Figure 1). Larger basins, more temporary sediment storage, a small number of planar units, and many transitions between individual units indicate a fairly complex habitat.

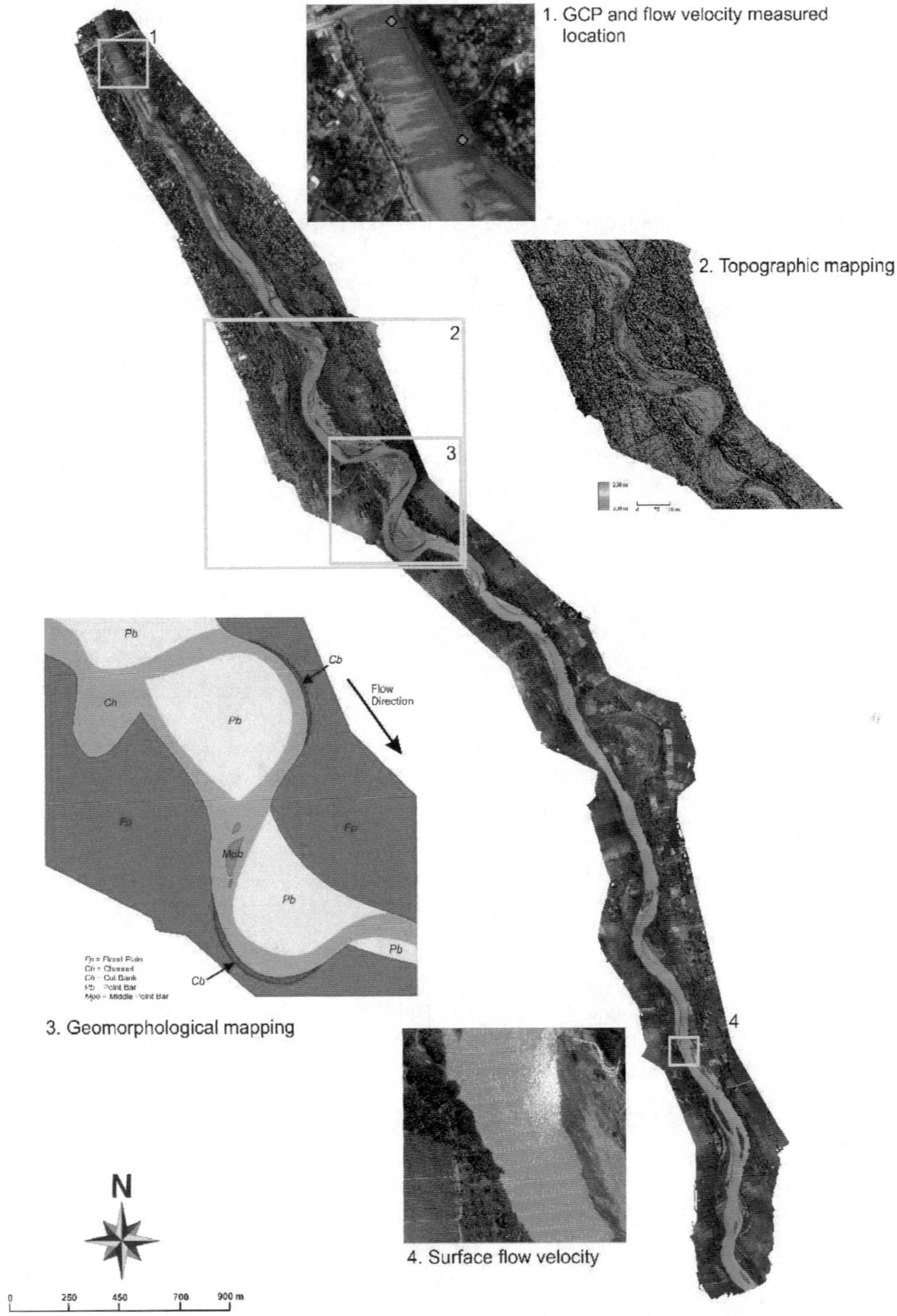

Figure 1. SfM applications to detect fluvio-morphological characteristics of the Way Semaka sub-watershed.

The large-scale (> 1 km) fluvio-morphological characterization that has been carried out requires cost efficiency, especially the amount of GCP usage. Recent studies also suggest that using a suitable GCP saves costs and produces a suitable model [12-13]. Developments in UAV technology such as more optimal battery performance, ease of data retrieval, and software that demands less CPU make UAV-based frameworks a sensible choice for robust and accurate (> 1 km) wide area fluvio-morphological characterization.

Comparing the proposed framework performance with geomorphological environmental classification techniques becomes the following work. Remote sensing methods are needed to overcome the shortcomings of UAV images related to resolution and spatial coverage. However, some aspects of fluvio-morphological characterization can be obtained from UAV images. For example, the high-resolution image textures required for SfM implementations can also interfere with implementing frameworks over large areas. If the image texture is disturbed, SfM will not detect a feature match between the overlapping images and fail to produce an orthomosaic image at the measured location. So further research is needed to address these issues before the framework is adopted for monitoring large areas.

The RGB images collected for this study allow a qualitative assessment of fluvio-morphological features (such as topographic mapping and calculations of surface water velocity) but do not facilitate quantitative depth estimation. Some authors have highlighted that RGB images present a limited radiometric resolution which precludes restoration of topography in darker parts (e.g., shadows and deep water) [14]. Such problems can affect the level of accuracy in all calculations that are reflected in the error value.

4. Conclusion

The aerial photograph data and current meter measurements can describe the fluvio-morphological conditions in the Way Semaka Sub-watershed. The test results give confidence that the UAV-SfM technique is suitable for testing the Way Semaka Sub-watershed features on a medium scale (1-100 meters). The speed range is between 0.39 - 1.56 m/s based on image analysis and proven by measuring the current meter at the location with an RMS error of 0.25. Manual geomorphic unit level 2 analysis shows that for the groove zone, 9% coverage consists of basins, 35% coverage consists of convexities, 9% planar features, and 37% coverage consists of transition zones.

References

[1] Wibowo R C, Sarkowi M, Setiawan A F, Yudamson A, Asrafil, Kurniawan M and Arifianto I 2020 Flash flood hazard areas assessment in Bandar Negeri Suoh (BNS) region using an index based approaches and analytical hierarchy process *J. Phys. Conf. Ser.* **1434**

[2] Sambrook Smith G H, Nicholas A P, Best J L, Bull J M, Dixon S J, Goodbred S, Sarker M H and Vardy M E 2019 The sedimentology of river confluences *Sedimentology* **66** 391–407

[3] Niedzielski T 2018 Applications of Unmanned Aerial Vehicles in Geosciences: Introduction *Pure Appl. Geophys.* **175** 3141–4

[4] Javernick L, Brasington J and Caruso B 2014 Modeling the topography of shallow braided rivers using Structure-from-Motion photogrammetry *Geomorphology* **213** 166–82

[5] Carrivick J L and Smith M W 2019 Fluvial and aquatic applications of Structure from Motion photogrammetry and unmanned aerial vehicle/drone technology *Wiley Interdiscip. Rev. Water* **6** e1328

[6] Fujita I, Muste M and Kruger A 1998 Large-scale particle image velocimetry for flow analysis in hydraulic engineering applications *J. Hydraul. Res.* **36** 397–414

[7] Thielicke W and Stamhuis E J 2014 PIVlab – Towards User-friendly, Affordable and Accurate Digital Particle Image Velocimetry in MATLAB *J. Open Res. Softw.* **2**

[8] Naves J, Anta J, Puertas J, Regueiro-Picallo M and Suárez J 2019 Using a 2D shallow water

model to assess Large-Scale Particle Image Velocimetry (LSPIV) and Structure from Motion (SfM) techniques in a street-scale urban drainage physical model *J. Hydrol.* **575** 54–65
[9] Turner D, Lucieer A and Wallace L 2014 Direct georeferencing of ultrahigh-resolution UAV imagery *IEEE Trans. Geosci. Remote Sens.* **52** 2738–45
[10] Chesley J T, Leier A L, White S and Torres R 2017 Using unmanned aerial vehicles and structure-from-motion photogrammetry to characterize sedimentary outcrops: An example from the Morrison Formation, Utah, USA *Sediment. Geol.* **354** 1–8
[11] Wheaton J M, Fryirs K A, Brierley G, Bangen S G, Bouwes N and O'Brien G 2015 Geomorphic mapping and taxonomy of fluvial landforms *Geomorphology* **248** 273–95
[12] Cucchiaro S, Fallu D J, Zhang H, Walsh K, Oost K Van, Brown A G and Tarolli P 2020 Multiplatform-SfM and TLS data fusion for monitoring agricultural terraces in complex topographic and landcover conditions *Remote Sens.* **12**
[13] Agüera-Vega F, Carvajal-Ramírez F and Martínez-Carricondo P 2017 Assessment of photogrammetric mapping accuracy based on variation ground control points number using unmanned aerial vehicle *Meas. J. Int. Meas. Confed.* **98** 221–7
[14] Woodget A S, Austrums R, Maddock I P and Habit E 2017 Drones and digital photogrammetry: from classifications to continuums for monitoring river habitat and hydromorphology *Wiley Interdiscip. Rev. Water* **4** e1222

Optimization, modelling and simulation to scale-up the production of rice bran extract and defatted rice bran

G Fitriyano[1,*], N H Fithriyah[2], R A Nugrahani[2], A B Syamsudin[1], M Kosasih[2]

[1]Department of Chemical Engineering, Faculty of Engineering, Universitas Muhammadiyah Jakarta, Indonesia

[2] Magister's Engineering Program, Department of Chemical Engineering, Faculty of Engineering, Universitas Muhammadiyah Jakarta, Indonesia

*Email: gema.fitriyano@ftumj.ac.id

Abstract. Rice bran is the source of rice bran oil, but much of it is used as livestock feed without being extracted for oil. Some food industries utilize the stabilized rice bran as a source of fiber, protein, and cooking oil. It contains lipase that catalyzes triglyceride hydrolyzation into glycerol and fatty acids, which leads to rancidity. A stabilization process must therefore be introduced to increase oxidation resistance in rice bran and oil hydrolysis. Defatted rice bran still contains protein, fiber, vitamins, and minerals. There has not been much research on integrated production process of crude rice bran oil and defatted rice bran. This research aimed to study the procedures to optimize, model, and simulate, generate Process Engineering Flow Diagram (PEFD). The phases in this study included laboratory-scale extraction with rice bran feed capacities of 10g, 25g, 100g, dan 500g; optimization using mathematical computations, simulations of mass-balance scale up and generation of PEFD from the pilot plant of 5kg rice bran capacity. The results of this research, including optimization, modelling, and simulation showed that to increase production scale from a number of raw materials of 57.690g to 5kg in a single batch required 5kg rice bran and 23.121kg n-hexane and produced 0.671kg oil and 4.121kg defatted rice bran, 13.212kg of reusable n-hexane and this study also provided a PEFD of the rice bran oil production process for a 5 kg rice bran capacity.

Keywords: Flow Diagram, Optimization, Modelling, Rice bran, Simulation

1. Introduction

Rice cultivation in Indonesia amounts high productivity. According to BPS and BPPT, rice harvest area in the country in 2018 covered an area of 10.90 million hectares, and produced 56.54 million metric tons of dry un-milled grain [1]. When this number is converted into milled grain, the number turns into 32,42 million tons of milled rice grain. Rice bran is a significant part of rice milling by-product, because it contributes to 10% of un-milled rice grain gross. Rice bran is a source of vitamins, minerals, essential fatty acids, and other sterol food fiber, as well as γ-oryzanol [2].

Currently, rice bran is directly used mostly for livestock feed without any treatment, such as oil extraction. Some practitioners in the beverage product already use stabilized rice bran as a source of fiber and proteins, incorporate rice bran oil in their products [3]. An important limiting factor of rice bran oil utilization in the food industry is its short shelf life. Lipase, present in rice bran oil, catalyzes the hydrolyses of triglycerides into glycerol and fatty acids, which leads to rancidity [4].

Therefore, utilization of rice bran oil requires stabilization of rice bran to increase oxidation resistance on the rice bran and its oil. Rice bran that has undergone oil extraction is called defatted rice bran, which still contains proteins and other nutrients, such as fiber, vitamins, and minerals. Defatted rice bran can be incorporated in food products, like biscuits [5].

Defatted rice bran is a low-value agroindustry waste and is commonly used in the feed industry. It contains 97.37mg/ 100g of γ-aminobutyric acid [6]. Research on oil extraction from rice bran in laboratory scale has been frequently conducted. In a study did a large-scale pilot of 900-kg capacity of rice bran [7].

On another research conducted a study on modelling, flow-sheeting, and economic analyses on the extraction of rice bran oil from a combination of batch supercritical fluid extraction (SFE) and counter current (CC) processes. In this study, they used SFE to separate the free fatty acid (FFA) fraction from the extracted oil. However, there was no further treatment of the defatted rice bran [8].

This study aimed to develop an integrated procedure to optimize, generate a model, and simulate, as well as produce a Process Engineering Flow Diagram (PEFD) of the crude rice bran oil and defatted rice bran processing at the scale of 5-kg rice bran.

2. Materials and Methods

There were three phases involved in this study. The first phase was laboratory-scale extractions to obtain oil and defatted rice bran at the capacities of 10g, 25g, 100g, and 500g of rice bran. The second phase was to optimize the production by conducting mathematical computations on the data obtained from the previous laboratory-scale trials. The last phase was to simulate balanced-mass scale up and produce PEFD of the pilot plant with a 5-kg capacity.

2.1. Rice bran oil extraction

This process was performed on a laboratory-scale with varying initial capacities of 10g, 25g, 100g, and 500g rice bran. The material used is rice bran from rice mills that are randomly selected around the border area between Jakarta and West Java, Indonesia and n-hexane (98% purity, Merck). The equipment consisted of glassware (Pyrex), vacuum pump (Yhzkb), condenser (400 mm, liebig), ultrasonic extractor (Wisd^{tm} WUC-D06H), centrifuge (Gemmy industral corp, PLC-03) and pneumatic press machine (ATS). The working procedure is presented in Figure 1.

2.2. Optimization of Data with Golden Section Method

The optimization process performed was a maximization using the golden section method. This procedure is used to solve non-linear programming problems with one variable in the form of maximization or minimization $f(x)$ with the error value of $a \leq \alpha \leq d$. [9]. Golden Section is a numerical optimization technique that find the maximum or minimum function of a single variable at an interval [10]. This method essentially narrows the area (α) that may produce optimal values of an objective function iteratively. Using this method, the optimal value from a laboratory phase can be more easily achieved [11].

The first step to determine the optimal point was conducted by deciding the simulation interval based on the laboratory trial data, which contains the maximum peak. This step was applied to both trials, which were the determinations of optimal conditions for obtaining oil and defatted rice bran.

The presentation of the resulting data is in the graphic form based on the non-linear formula from the golden section method. This process was assisted by Quick Basic™ software.

IOP Conf. Series: Materials Science and Engineering 1173 (2021) 012004 doi:10.1088/1757-899X/1173/1/012004

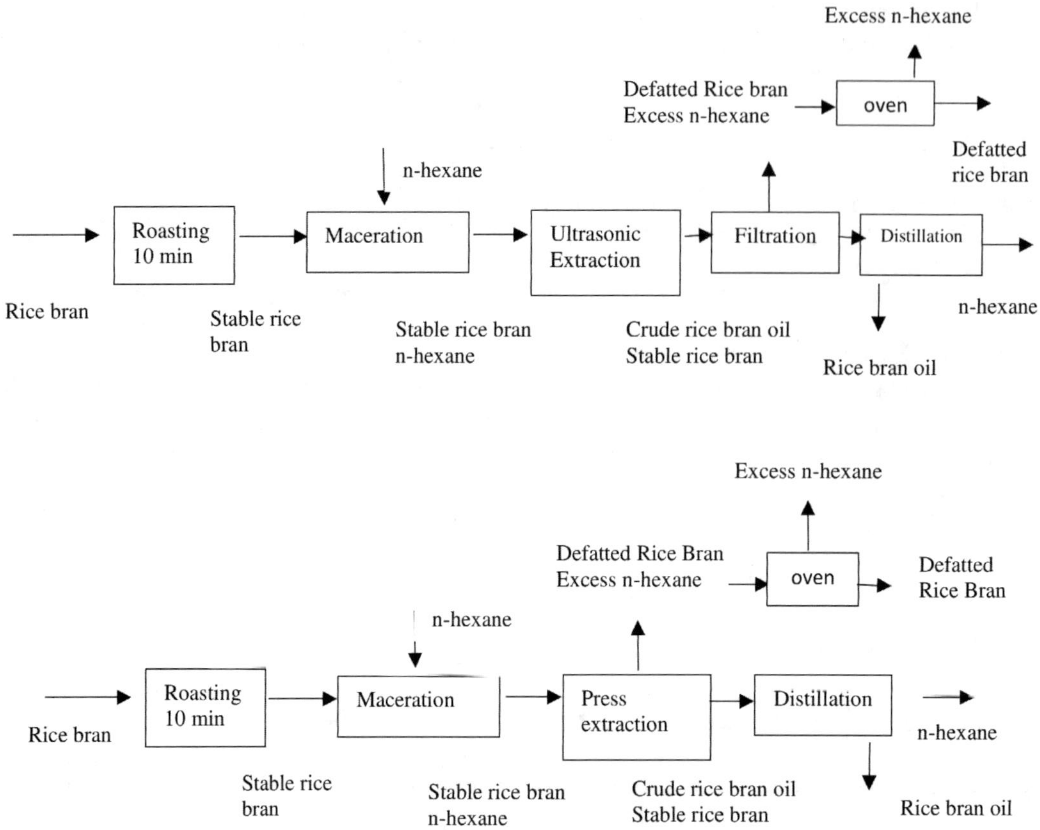

Figure 1. Block diagram of oil extraction and defatted rice bran processes

2.3. Scale up simulation

Using the data based on the optimal production of rice bran oil, scale up production at 5-kg capacity was devised. The computational process was performed with MS Excel^{TM}. The phases of the study are presented on Figure 2.

2.4. Process Engineering Flow Diagram

The next step was employing the results of Mass Balance calculations to generate a Process Engineering Flow Diagram. Process Flow Diagram (PFD) is a flow of major equipment and controls for a production process. The major equipment is usually in the form of columns, reactors, heat exchangers, pumps, agitators, filters, etc.

The information contained in the PFD is the equipment number and a brief description, material and energy balance , equipment size, material, flow direction [12].

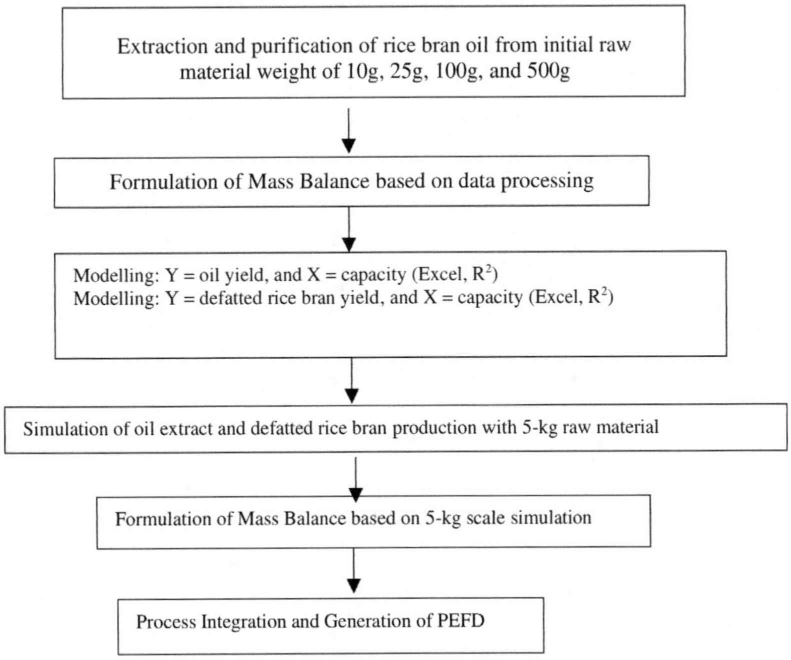

Figure 2. Trial stages in optimization, modelling, and scale-up production simulation of rice bran oil and defatted rice bran.

3. Results and Discussions

Results and discussion are presented on each phase of the study. The analyses included the percent yields of rice bran oil and defatted rice bran from different initial masses of raw material, optimal capacities in productions of rice bran oil and defatted rice bran, and the mass balance of production with 5-kg initial mass of rice bran.

3.1. Results of laboratory-phase of rice bran oil extraction

The results of rice bran oil extraction and defatted rice bran at laboratory-scale at various capacities of initial rice bran weight of 10g, 25g, 100g, and 500g are shown in Table 1.

Table 1. Results of rice bran extraction from different initial capacities of raw material

No	Capacity	Treatment step	Dry Rice Bran (g)	N-hexane (ml)	Oil (g)	Defatted Rice Bran (g)
1	10 g, n-hexane 70 ml by ultrasonic	Roasting	8.9			
		Maceration Ultrasonic	8.9	70		
		Filtration	8.9	70		
		Distillation		30	0.64	
		Oven	8.9	40	0.64	3.85
2	25g N-hexane 175 ml by ultrasonic	Roasting	22.62			
		Maceration Ultrasonic	22.62	175		
		Filtration	22.62	175	2.49	
		Distillation		100	2.49	
		Oven	22.62	75		20.13

Cont'd. Table 1

3	100 g N-hexane 700 ml by ultrasonic	Roasting	95.4			
		Maceration	95.4	700		
		Ultrasonic				
		Filtration	95.4	700		
		Distillation		400	8.6	
		Oven	95.4	300	8.6	89.57
4	500 g N-hexane 3.5 L by manual press	Roasting	450			
		Maceration	450	3500		
		Press				
		Filtration	450	3500	30.96	
		Distillation		2000	30.96	
		Oven	450	1500		400

3.2. Results of data optimization with golden section method

The optimization process was initiated by deciding the intervals among data from the laboratory trials. The first modelling was on the optimization of rice bran oil extraction and used the result data to plot a capacity vs oil yield curve. This graph shows the optimal peak occurring between 10g and 100g (Figure 3). This optimization resulted in the optimal point of 57.69 g capacity with a yield of 14.49%.

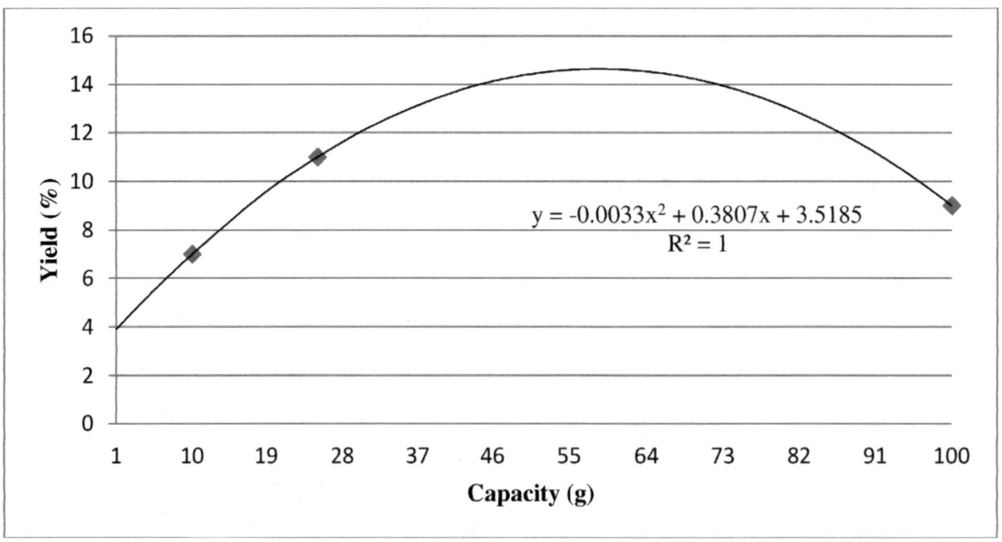

Figure 3. Correlation curve between capacity (g) and oil yield (%)

Based on Figure 3, the determination of capacity that would produce the optimal yield was conducted with golden section method using Quick Basic™ software. The second model was on the defatted rice bran by generating a curve on capacity vs defatted rice bran yield. This exercise produced the optimal point in between 25g and 500g as shown in Figure 4.

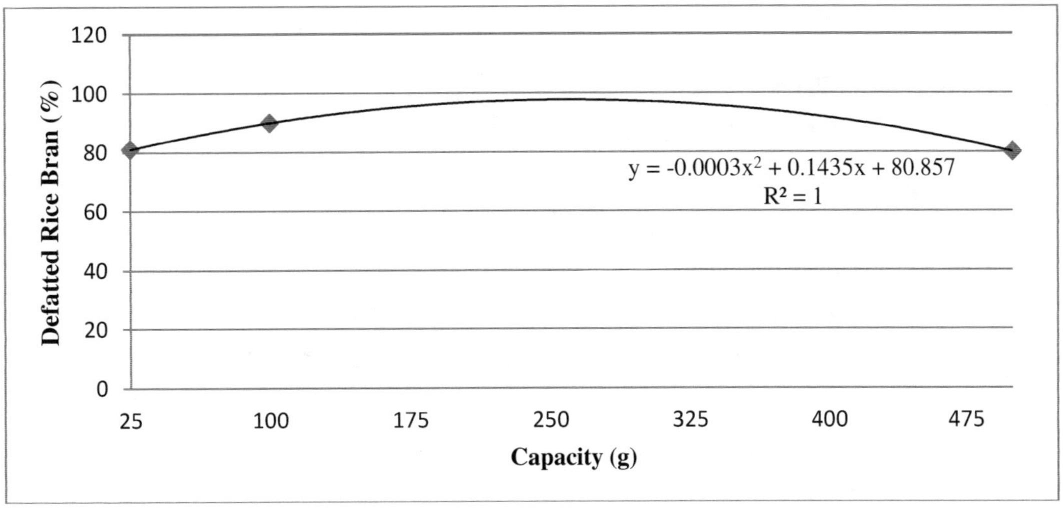

Figure 4. Correlation curve between capacity (g) and defatted rice bran yield (%)

Based on Figure 4, the determination of capacity that would produce the optimal yield was conducted with golden section method using Quick BasicTM software. This optimization analysis produced an optimal point at 239.19-gr capacity with a defatted rice bran yield of 98.01%. The optimization results were compared and 57.69-gr capacity was favoured because it generated the highest yield of oil extract of 14.49% and acceptable defatted rice bran yield of 88.18%, as shown in Table 2.

Table 2. Yield comparison between production capacities of 57g and 239g

No	Capacity (gr)	Oil Yield (%)	Defatted Rice Bran Yield (%)
1	57.69	14.49	88.13
2	239.19	8.30	98.01

3.3. Scale-up simulation results

The scale-up simulation of 5-kg capacity of raw rice bran (5 - 10% water content) generated a Mass balance that referred to the optimal production capacity of 57.69 gram. The calculation results are presented in Table 3.

Table 3. The mass balance results from 5-kg capacity simulation with various equipment

Maceration Tank

Material	F1	F2	F3
Rice bran (g)		4631.245	4631.245
N-hexane (g)	23121		23121
Oil (g)			
Defatted rice bran (g)			
Amount	23121	4631.245	27752.25
Total	27752.25		27752.25

Cont'd. Table 3

Filter Press

Material	**F3**	**F4**	**F5**
Rice bran (g)	4631.245	4121	
N-hexane (g)	23121	9909	13212
Oil (g)			671
Defatted rice bran (g)			
Amount	35674	17376	18298
Total	35674	35674	

Oven

Material	**F4**	**F6**	**F7**
Rice bran (g)	4121		
N-hexane (g)	9909	9909	
Oil (g)			
Defatted rice bran (g)			4121
Amount	14031	9909	4121
Total	14031	14031	

Distillation Column

Material	**F5**	**F8**	**F9**
Rice bran (g)			
N-hexane (g)	13212	13212	
Oil (g)	671		671
Defatted rice bran (g)			
Amount	13883	13212	671
Total	13883	13883	

The next step was employing the results of mass balance calculations to generate a process engineering flow diagram which is shown in Figure 5.

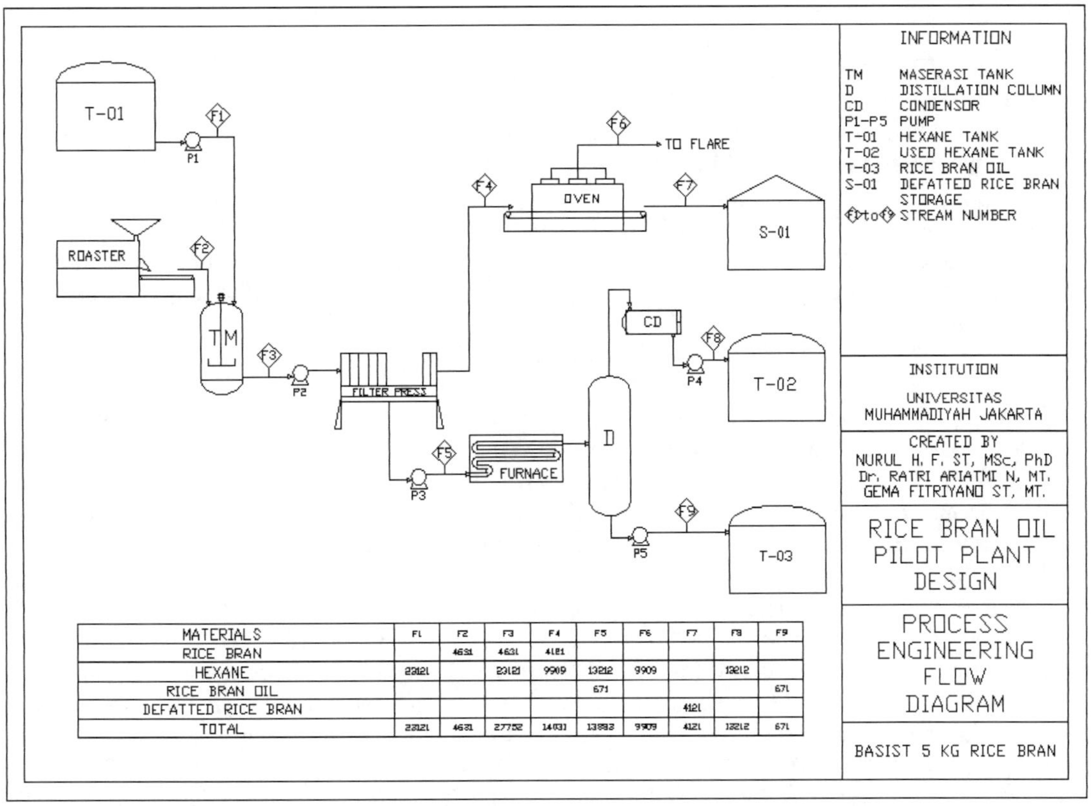

MATERIALS	F1	F2	F3	F4	F5	F6	F7	F8	F9
RICE BRAN		4631	4631	4121					
HEXANE	23121		23121	9909	13212	9909		13212	
RICE BRAN OIL					671				671
DEFATTED RICE BRAN							4121		
TOTAL	23121	4631	27752	14031	13883	9909	4121	13212	671

Figure 5. Process Engineering Flow Diagram (PEFD) for a Rice Bran Oil Plant Design

4. Conclusions

This study produced several conclusions. The first is that ultrasonic extraction method yields more than manual press. The second is that the result of optimization process using golden section method is 14.49% weight yield of rice bran oil from an initial weight of 57.69 grams. The third is drawn from the production of rice bran oil and defatted rice bran that has been scaled up from 57.69 grams to 5 kg. Based on the production simulation, when the initial capacity is scaled up to 5 kg of raw rice bran, the process requires 23.121 kg of n-Hexane and produces 0.671 kg of rice bran oil and 4.121 kg of defatted rice bran, as well as 13.212 kg of reusable n-Hexane. and also provided a PEFD of the rice bran oil production process for a 5 kg rice bran capacity. For further research, it can be directed to simulate an increase in pilot plant scale production capacity and it is necessary to study the compatibility factor of equipment to raw materials and products characteristics.

Acknowledgments

The authors would like to express a profound gratitude for the grant awarded by the Ministry of Research, Technology, and Higher Education of the Republic of Indonesia, and The authors would like to express a gratitude for laboratory as a place for this research held at Department of Chemical Engineering, Faculty of Engineering, Universitas Muhammadiyah Jakarta and also The authors would like to express a gratitude for administration process support by LPPM- Universitas Muhammadiyah Jakarta.

References

[1] BPS & BPPT, "RINGKASAN EKSEKUTIF LUAS PANEN DAN PRODUKSI BERAS DI INDONESIA 2018," 2018.

[2] M. Sardarodiyan and E. A. Salehi, "Bioactive phytochemicals in Rice Bran: processing and functional properties," *Int. J. PharmTech Res. CODEN IJPRIF*, vol. 9, no. 6, pp. 401–408, 2016.

[3] U. Issara and S. Rawdkuen, "Rice bran: A potential of main ingredient in healthy beverage," *Int. Food Res. J.*, vol. 23, no. 6, pp. 2306–2318, 2016.

[4] O. I. Oluremi, A. O. Solomon, and A. A. Saheed, "Fatty acids, metal composition and physico-chemical parameters of Igbemo Ekiti rice bran oil," *J. Environ. Chem. Ecotoxicol.*, vol. 5, no. 3, pp. 39–46, 2013.

[5] R. A. Nugrahani, N. H. Fithriyah, and Nelfiyanti, "No Title," in *IOP Conference Series: Materials Science and Engineering 543*, 2019, pp. 1–7.

[6] D. Wongwaiwech, M. Weerawatanakorn, S. Tharatha, and C.-T. Ho, "Comparative study on amount of nutraceuticals in by-products from solvent and cold pressing methods of rice bran oil processing," *J. food drug Anal.*, vol. 27, no. 1, pp. 71–82, 2019.

[7] R. Schramm, G. Gonzalez, N. Walker, and M. Lima, "Pilot Scale Mill Characterization and Evaluation of Rice Bran Oil Concentration of Jazzman Rice," *Am. J. Plant Sci.*, vol. 6, pp. 1885–1892, 2015.

[8] L. Danielskia, C. Zetzla, H. Henseb, and G. Brunner, "No TitA process line for the production of raffinated rice oil from rice bran.le," *J. Supercrit. Fluids*, vol. 34, pp. 133–141, 2005.

[9] D. A. Mahkya, H. Yasin, and A. Mukid, M, "Aplikasi Metode Golden Section untuk Optimasi Parameter Pada Metode Exponential Smoothing," *J. Goussian*, vol. 3, no. 4, pp. 605–614, 2014.

[10] S. K. Chakraborty and G. Panda, "Golden section search over hyper-rectangle: A Direct search method," *Int. J. Math. Oper. Res.*, vol. 8, no. 3, p. 279, 2016.

[11] F. Akbari, A. Setyanto, and F. W. Wibowo, "Optimasi Parameter Pemulusan Algoritma Brown Menggunakan Metode Golden Section Untuk Prediksi Data Tren Positif dan Negatif," *J. RESTI (Rekayasa Sist. dan Teknol. Informasi)*, vol. 2, no. 1, pp. 307 – 314, 2018.

[12] S. Hall, *Rules of Thumb for Chemical Engineers*, Fifth Edit. Elsevier Inc., 2012.

Optimization of shell and tube heat exchanger using the water cycle algorithm

M F I M Hanafi, A Bahreininejad*, and N Uddin
Faculty of Engineering, Universiti Teknologi Brunei, Gadong, Brunei

* Corresponding author: ardeshir.bahreininejad@utb.edu.bn

Abstract. The shell and tube heat exchangers (STHEs) are used to transfer the thermal energy between two or more media. Due to its importance for conserving energy, numerous studies of heat exchanger design optimization have been carried out. The traditional design approaches are usually time-consuming, and may not guarantee to produce an economically optimal solution. A complete optimization procedure was conducted to find the minimum cost of the heat exchanger system. Heat exchanger design optimization was carried out using a recently proposed optimization algorithm called the water cycle algorithm (WCA). Three design variables are considered consisting of the shell inside diameter, tube outer diameter, and baffle spacing. The total annual costs which include the investment costs and the operating costs will be considered as the objective function.

Keyword: heat exchanger, thermal energy, water cycle algorithm, baffle spacing.

1. Introduction

A heat exchanger is a mechanical system used to transfer heat between two fluids stream at different temperatures, hot and cold streams, which are separated by a solid wall. The applications of heat exchangers are normally found in petrochemical industries as well as waste heat recovery, power generation, refrigeration, and air conditioning [1]. Among the different types of heat exchangers, shell and tube heat exchangers (STHEs) are the most widely used heat exchanger and have a wide range of operating pressures and temperatures. A typical STHE is shown in Figure 1.

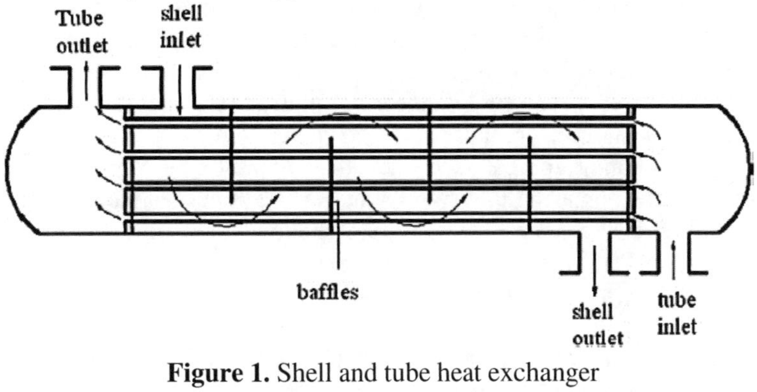

Figure 1. Shell and tube heat exchanger

The design of STHE includes several geometric and operating variables to find a heat exchanger geometry that can satisfy the heat duty requirement with a particular set of design constraints. Normally a geometric layout of the heat exchanger is chosen as a reference at first with a fixed allowable pressure drop value. Then, based on the design specifications and several assumptions of thermodynamic and mechanical parameters, the values of the design variable for the heat exchanger are defined [1, 2].

As the STHEs play an important role in conserving energy, several techniques of design optimization problem by considering different objective functions have been proposed such as mixed-integer nonlinear programming, genetic algorithm, simulated annealing, imperialist competitive algorithm, particle swarm optimization, and so on. Amongst the mentioned optimization technique, genetic algorithm (GA) received the most attention in the design optimization of STHEs for the past few years. Caputo et al. [3] proposed an optimization of heat exchanger using the GA from an economical perspective. They minimized the total cost of the heat exchanger which includes the total discounted operating expenditures and the capital investment. Three different case studies were carried out to validate the capability of the suggested method. A reduction in the total costs higher than 50% was obtained from the examined cases.

Guo et al. [4] conducted a study where they used GA along with entropy generation minimization to optimize the design of STHE. The modified entropy number in this case is the dimensionless entropy generation rate, is set as the objective function. Multiple design variables were considered. The results showed a significant increase in heat exchanger effectiveness and also a reduction in pumping power. Selbas et al. [5] utilized GA by considering the use of a logarithmic mean temperature difference (LMTD) method which is used to identify the dimensions of the STHE characteristics. They included the maintenance cost and the capital recovery factor in their proposed cost estimation.

Patel and Rao [6] used particle swarm optimization (PSO) for the design optimization of STHE in terms of the economic point of view with the minimization of the total annual cost as its objective function. For the design variables, the outer tube diameter, shell internal tube diameter, and the baffle spacing was considered. The results of using PSO were then compared with the other four different case studies of using GA to show the effectiveness of the two methods.

A constructal theory, which was suggested by Bejan [7], has been used as an optimization design method for heat exchanger. For a system such as STHE, the constructal theory suggests that a design (geometry, flow pattern, configuration) with higher durability, stability, and conservation can be accomplished by increasing the access level of heat and working fluids that flow through it. Yang et al. [8, 9] performed an optimization design approach for STHE based on the constructal theory, with all the design parameters following the Tubular Exchanger Manufacturers Association (TEMA) standards. The objective function is the total cost of STHE by using genetic algorithm method.

Azad et al. [10] applied GA that is based on a constructal theory for the optimization design to reduce the total cost of the exchanger. It was shown that the overall heat transfer coefficient of the STHE increases by using the constructal theory and hence reducing the cost of manufacturing the heat transfer surface. Moreover, the constructal theory also helps to reduce the operational energy cost of the pump which was used to overcome the pressure loss. The results show a reduction of costs by more than 50%.

Another optimization technique which was applied for the cost minimization of STHE is based on the imperialist competitive algorithm (ICA) [11]. This algorithm is based on the human's socio-political evolution that has been used in various engineering applications such as data clustering, thermal engineering, and industrial engineering. It has been shown that ICA has a better rate of convergence and also finding the global minima compared to the other evolutionary algorithm. Hadidi et al. [12] performed a design optimization using ICA by varying several important parameters of the STHE. The ICA method provides a faster solution to the design problem and able to explore many other different solutions of good quality. With the results showing a saving in operating cost as high as 94% and reduction of the total cost up to 53%, it proves that ICA method is more efficient than the traditional GA method.

Later, Hadidi et al. [13] also applied the biogeography-based optimization (BBO) algorithm to heat exchanger design optimization. BBO method was first proposed by Simon [14], a search algorithm based on the geographical distribution of biological organisms. BBO algorithm mainly operates based on mutation and migration and has a similar way of sharing data between solutions similar to PSO and GA methods. The results produced by BBO method showed a slight improvement over the results given by the ICA with saving in operating cost as high as 96% and reduction of the total cost up to 56.1% [13].

Fettaka et al. [15] performed a multi-objective optimization, also called Pareto optimization, of the pumping power and heat transfer of STHE. The multi-objective optimization offers several Pareto-optimal solutions without degrading the values of the other objective function. In the research, they considered nine design variables which were the number of tube passes, tube layout pattern, baffle cut, baffle spacing, shell-to-baffle diametrical clearance, tube-to-baffle diametrical clearance, tube outer diameter, tube length, and tube wall thickness. The optimization was done via the fast and elitist non-dominated sorting genetic algorithm (NSGA-II). The results showed that multi-objective optimization offers a Pareto front with a larger extent of optimal design variables.

Turgut et al. [16] used an improved intelligent tuned harmony search (I-ITHS) algorithm, an improved form of the harmony search algorithm, for optimizing the total cost of STHE. Four design variables were considered such as the shell diameter, tube outer diameter, baffle spacing, and the number of tube passes. Results showed that I-ITHS algorithm can be applied to STHE design problem.

Asadi et al. [17] optimized the design of STHE using a cuckoo-search-algorithm (CSA). The objective function is the total annual cost of the heat exchanger. The obtained results by using CSA indicated the possibility of energy saving by 77%. A different kind of algorithm was used for the optimization of STHE by Mohanty [18] namely the firefly algorithm (FA) which was inspired by the nature of the flashing light of fireflies. Considering the total annual cost which includes the operating cost and investment cost as the objective function, a reduction in the operating cost by 77% and the total cost by 29% were obtained. Mohanty also performed a different algorithm called the gravitational search algorithm (GSA). The gravitational search algorithm is a heuristic type of search algorithm which is derived from the law of gravity and mass interactions. The GSA methodology was utilized to find the ideal geometric configuration by considering the complexity of the STHE design and its applications. The results obtained were then compared with several other types of the algorithm used in STHE [19]. In this study, heat exchanger design optimization was carried out by performing water cyle algorithm (WCA) simulations.

2. Shell and tube heat exchange model

2.1. Heat transfer coefficient

Depending on the system of the flow, the tube side heat transfer coefficient h_t can be computed from the following correlation [2, 3]:

$$h_t = \frac{k_t}{d_i}\left[3.657 + \frac{0.0677\left(Re_t Pr_t\left(\frac{d_i}{L}\right)\right)^{1.33}}{1+0.1Pr_t\left(Re_t\left(\frac{d_i}{L}\right)\right)^{0.3}}\right], Re_t \leq 2300 \quad (1)$$

$$h_t = \frac{k_t}{d_i}\left[\frac{\left(\frac{f_t}{8}\right)(Re_t-1000)Pr_t}{1+12.7\left(\frac{f_t}{8}\right)^{\frac{1}{2}}\left(Pr_t^{\frac{2}{3}}-1\right)}\left(1+\frac{d_i}{L}\right)^{0.67}\right], 2300 \leq Re_t \leq 10{,}000 \quad (2)$$

$$h_t = \frac{k_t}{d_i}0.027{Re_t}^{0.8}{Pr_t}^{\frac{1}{3}}\left(\frac{\mu_t}{\mu_{wt}}\right)^{0.14}, Re_t \geq 10{,}000 \quad (3)$$

Nomenclature

D_s	shell inside diameter	h_s	shell side convective coefficient
d_o	tube outside diameter	U	overall heat transfer coefficient
B	baffles spacing	A	heat exchanger surface area
N_t	number of tubes	L	tubes length
S_t	tube pitch	ΔP_t	tube side pressure drop
v_t	tube side fluid velocity	f_s	shell side friction coefficient
Re_t	tube side Reynolds number	ΔP_s	shell-side flow pressure drop
f_t	tube side friction coefficient	C_i	capital investment
Pr_t	tube side Prandtl number	C_o	annual operating cost
D_e	equivalent shell diameter	C_{od}	total discounted operating cost
A_s	0.0137	C_{tot}	total annual cost
Re_s	Shell side Reynolds number	h_s	shell side convective coefficient
Pr_s	shell side Prandtl number	U	overall heat transfer coefficient
v_s	shell-side fluid velocity	ΔP_s	shell side pressure drop
h_t	tube side convective coefficient	$LMTD$	logarithmic mean temperature difference
T_{hi}	hot fluid inlet temperature	T_{ho}	hot fluid outlet temperature
T_{ci}	cold fluid inlet temperature	T_{co}	cold fluid outlet temperature

where k_t and f_t are the tube side thermal conductivity and the Darcy friction factor respectively. The tube inner diameter d_i is calculated as:

$$d_i = 0.8 d_o \tag{4}$$

The Darcy friction factor can be calculated by [3, 20],

$$f_t = (1.82 \log_{10} Re_t - 1.64)^{-2} \tag{5}$$

The Prandtl number of the tube side Pr_t is given as:

$$Pr_t = \frac{\mu_t c_{pt}}{k_t} \tag{6}$$

The Reynolds number of the tube side Re_t is given as:

$$Re_t = \frac{\rho_t V_t d_i}{\mu_t} \tag{7}$$

The tube side flow velocity can be calculated using [21]:

$$V_t = \frac{m_t}{\left(\frac{\pi d_i^2}{4}\right)\rho_t}\left(\frac{n}{N_t}\right) \tag{8}$$

where m_t is the tube side mass flow rate, n is the number of tube passes and N_t is the number of tubes which can be determined from [1, 3]:

$$N_t = K_1 \left(\frac{D_s}{d_o}\right)^{n_1} \tag{9}$$

The D_s is the shell inside diameter, d_o is the tube outer diameter, K_1 and n_1 are the numerical coefficient depending on the number of passes and the flow arrangement. These coefficients are presented in Table 1.

Table 1. K_1 and n_1 coefficients for $S_t = 1.25 d_o$ [2].

No. of passes	Triangular tube pitch		Square tube pitch	
	K_1	n_1	K_1	n_1
1	0.319	2.142	0.215	2.207
2	0.249	2.207	0.156	2.291
4	0.175	2.285	0.158	2.263
6	0.0743	2.499	0.0402	2.617
8	0.0365	2.675	0.0331	2.643

The shell side heat transfer coefficient h_s is calculated using Kern's formulation for segmental baffle STHE [21]:

$$h_s = \frac{k_s}{D_e} 0.36 Re_s^{0.55} Pr_s^{\frac{1}{3}} \left(\frac{\mu_s}{\mu_{ws}}\right)^{0.14} \quad (10)$$

where D_e is the shell equivalent diameter which can be calculated by [2, 21];

$$D_e = \frac{4\left[S_t^2 - \left(\frac{\pi d_o^2}{4}\right)\right]}{\pi d_o}; \quad \text{(for square pitch)} \quad (11)$$

$$D_e = \frac{4\left[0.43 S_t^2 - 0.5\left(\frac{\pi d_o^2}{4}\right)\right]}{0.5\pi d_o}; \quad \text{(for triangular pitch)} \quad (12)$$

where S_t is the tube pitch as illustrated in Figure 2.

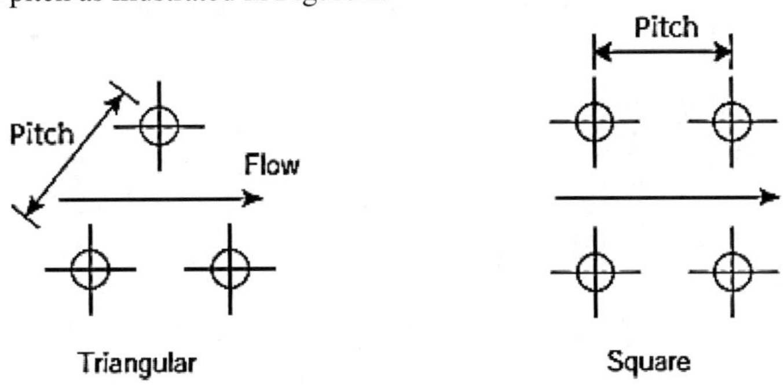

Figure 2. Tube arrangements

The Prandtl number for shell side Pr_s is given as:

$$Pr_s = \frac{\mu_s c_{ps}}{k_s} \quad (13)$$

The Reynolds number for shell side Re_s is given as:

$$Re_s = \frac{\rho_s V_s D_e}{\mu_s} = \frac{m_s D_e}{A_s \mu_s} \quad (14)$$

The shell side flow velocity V_s can be calculated by taking into consideration the shell side cross-section area normal to flow direction A_s:

$$V_s = \frac{m_s}{\rho_s A_s} \quad (15)$$

$$A_s = \frac{D_s \cdot B \cdot Cl}{S_t} \quad (16)$$

where m_s is the shell side mass flow rate, B is the baffle spacing and Cl is the shell side clearance given as:

$$Cl = S_t - d_o \quad (17)$$

After finding the heat transfer coefficient for both shell and tube side, the overall heat transfer coefficient U can be determined by considering the fouling resistance R_f on both sides [2]:

$$U = \frac{1}{\left(\frac{1}{h_s}\right) + R_{fs} + \left(\frac{d_o}{d_i}\right)\left(R_{ft} + \frac{1}{h_t}\right)} \quad (18)$$

The fouling resistances are determined from the literature data based on the operating temperatures and the type of fluids. Considering the flow between the adjacent baffles as crossflow, the logarithmic mean temperature difference $LMTD$ can be calculated by:

$$LMTD = \frac{(T_{hi} - T_{co}) - (T_{ho} - T_{ci})}{\ln\left(\frac{(T_{hi} - T_{co})}{(T_{ho} - T_{ci})}\right)} \quad (19)$$

Based on the flow configuration involved, the temperature difference correction factor F is introduced where it is a function of dimensionless temperature ratio for most flow configuration of interest [1, 2].

$$F = \frac{(R^2+1)^{\frac{1}{2}}}{R-1}\left[\frac{\ln\left(\frac{1-P}{1-PR}\right)}{\ln\left(\frac{2-P\left(R+1-(R^2+1)^{\frac{1}{2}}\right)}{2-P\left(R+1+(R^2+1)^{\frac{1}{2}}\right)}\right)}\right] \quad (20)$$

where R is the correction coefficient and P is the efficiency and can be found by the following equations:

$$R = \frac{(T_{hi}-T_{ho})}{(T_{co}-T_{ci})} \quad (21)$$

$$P = \frac{(T_{co}-T_{ci})}{(T_{hi}-T_{ci})} \quad (22)$$

Taking into consideration the overall heat transfer coefficient, *LMTD,* and the correction factor, the heat exchanger surface area A is calculated by:

$$A = \frac{Q}{U \cdot F \cdot LMTD} \quad (23)$$

where Q is the heat transfer rate for sensible heat transfer expressed as:

$$Q = m_h c_{ph}(T_{hi} - T_{ho}) = m_c c_{pc}(T_{co} - T_{ci}) \quad (24)$$

The necessary tube length L can be determined based on the heat exchanger surface area:

$$L = \frac{A}{\pi d_o N_t} \quad (25)$$

2.2. Pressure drops

In many heat exchangers, heat transfer is closely related to the pressure drop in terms of economical and physical. The pressure-drop is the fluid pressure available to pump the fluid through the exchanger. For a heat exchanger that is going to be designed, under a constant heat capacity, increasing the flow velocity will also increase the heat transfer coefficient. In this matter, a compact with lower dimension heat exchanger can be made with lower investment cost.

However, high flow velocity results in the high-pressure drop which in return requires a more powerful pump and this mean additional running cost. Hence, the pressure drop must be taken into account along with the heat transfer when designing a heat exchanger to obtain the best solution for the system [6, 21]. The tube side pressure drop is computed as the summation of the distributed pressure drop along the tubes length and the concentrated pressure losses in elbows and in the inlet and outlet nozzles [21]:

$$\Delta P_t = \Delta P_{tube,length} + \Delta P_{tube,elbow} = \frac{\rho_t {V_t}^2}{2}\left(\frac{Lf_t}{d_i} + p\right)n \quad (26)$$

where p is a constant and has different values given in the literature. Kern [21] assumed the value of $p = 4$ while Sinnott [2] assumed the value of $p = 2.5$. Whereas the pressure-drop of the shell side can be calculated by:

$$\Delta P_s = f_s\left(\frac{\rho_s {V_s}^2}{2}\right)\left(\frac{L}{B}\right)\left(\frac{D_s}{D_e}\right) \quad (27)$$

$$f_s = 2b_o {Re_s}^{-0.15} \quad (28)$$

where f_s is the friction factor and $b_o = 0.72$ valid for $Re_s < 40{,}000$.

2.3. Costs estimations

The total annual cost C_{tot} is assessed as the objective function by considering the capital investment C_i, the annual operating cost C_o, the energy cost C_e and the total discounted operating cost C_{od} [3]:

$$C_{tot} = C_i + C_{od} \quad (29)$$

The capital investment is evaluated as a function of the heat exchanger surface area using Hall's correlation [23]:

$$C_i = a_1 + a_2 A^{a_3} \tag{30}$$

where $a_1 = 8000$, $a_2 = 259.2$ and $a_3 = 0.93$ for heat exchanger made of stainless steel for both shells and tubes.

The total discounted operating cost is influenced by the pumping power P to deal with the frictional losses and can be calculated as:

$$C_{od} = \sum_{j=1}^{ny} \frac{C_o}{(1+i)^j} \tag{31}$$

$$C_o = PC_eH \tag{32}$$

$$P = \frac{1}{\eta}\left(\frac{m_t}{\rho_t}\Delta P_t + \frac{m_s}{\rho_s}\Delta P_s\right) \tag{33}$$

where the pumping efficiency η is taken as 80%.
From all of the above calculations, the total cost of the heat exchanger can be determined from Eq. (29).

3. Water cycle algorithm (WCA)

The WCA is a population-based metaheuristic algorithm based on the observation of the water cycle and how rivers and streams flow downhill towards the sea in the real world [24, 25, 26]. Rivers and streams are formed whenever water moves downhill from one location to another. Most rivers are formed at higher altitudes where snow or glaciers melt. Streams and rivers eventually end up to a sea. Water is also allowed to evaporate into the atmosphere to generate clouds which then condenses in the colder atmosphere, releasing the water back to the earth in the form of rain or precipitation.

The initial population generated randomly is called raindrops. Hence, a raindrop represents an individual or single solution comprising a number of design/decision variables. The best individual (best raindrop) is chosen as a sea to represent an optimal solution to a given problem. Further details of the WCA are given in [24, 25, 26].

4. Results and discussions

4.1. Case study

The case study was investigated in [3, 6, 18, 19] which was originally taken from [21]. The considered heat exchanger is having a heat duty of 1.44 MW. According to the original specification, the fluid flowing through the shell side is hot kerosene while the fluid flowing through the tube side is cold crude oil. The heat exchanger is a single shell pass and four tube passes exchanger with a square pitch arrangement.

The physical properties and the operating parameters for both shell and tube side fluids are summarized in Table 2.

Table 2. Physical properties for the case study [3].

Process parameters	Case study	
	Shell side: kerosene	Tube side: crude oil
Mass flow rate, m (kg/s)	5.52	18.80
Inlet temperature, T_i (°C)	199.0	37.8
Outlet temperature, T_o (°C)	93.3	76.7
Specific heat, c_p (m^2K/W)	2.47	2.05
Density, ρ (kg/m^3)	850.0	995.0
Viscosity at tube wall, μ (Pa. s)	0.0004	0.00358
Viscosity at core flow, μ_w (Pa. s)	0.00036	0.00213
Thermal conductivity, k (W/m. K)	0.13	0.13
Fouling resistance, R_f (m^2K/W)	0.00061	0.00061

The upper and lower bounds for the optimization design variables used in this work are similar to the ones considered in [3]. The shell internal diameter D_s ranging between 0.1 m to 1.5 m, tube outer diameter d_o ranging from 0.01 m to 0.051 m and baffle spacing B ranging from 0.05 m to 0.5 m were considered. Likewise, all the values used for discounted operating costs calculation are energy cost C_e = 0.12 €/kW h, ny = 10 years, annual discount rate j = 10% and an annual work hours H = 7000 h/year, similar with the other researchers.

4.2. Discussions

The simulations for WCA were run 20 times by setting the number of populations as 20 and the number of maximum iterations as 20. This is because the objective function was found to be converged to its minimum value less than 12 generations as shown in Figure 3.

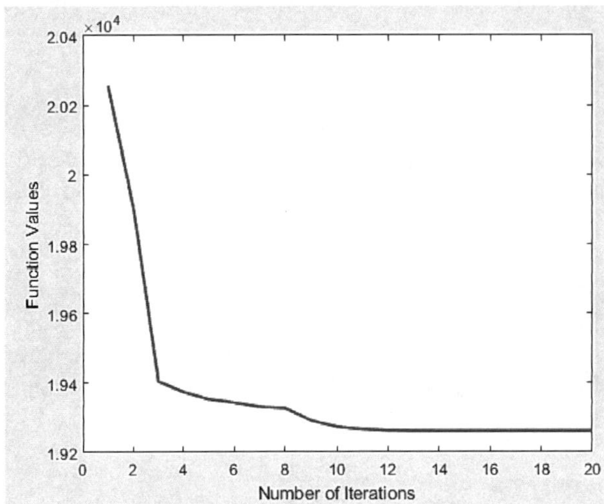

Figure 3. Convergence of WCA

The results obtained were compared with the original design reported in [21] and other optimization methods reported in the literature [3, 6, 18, 19]. The outcomes are presented in Table 3.

The results show WCA offers a reduction in the tube length compared to the original design and GA while the tube length given by WCA is longer than the lengths given by PSO, FA, and GSA. However, the number of tubes given by PSO, FA, and GSA is more than the number given by WCA. The heat transfer coefficients for both shell and tube sides given by WCA were found to be significantly lower than the ones given by other methods. Thus, a considerably lower overall transfer coefficient was observed. The heat exchanger area given by WCA is lowered by 29.2%, 10%, 15.9%, and 13.4% compared to the original design, GA, FA, and GSA respectively. Consequently, the investment cost given by WCA is lower than the other optimization methods except for PSO.

The tube side velocity causes the pressure drop on the tube side to be significantly higher than the values obtained from other optimization methods. The increase in the tube length also contributes to the increase in the pressure drop. On the contrary, the small shell side velocity results in a small pressure drop on the shell side as compared to the values obtained from other optimization methods. Based on the pressure drops on both sides, comparing with the other methods, the annual operating cost given by WCA is lower by 68.3%, 6.13%, and 21% for the original design, GA and PSO respectively while it is much higher than FA and GSA. Thus, making the total discounted annual operating cost to be lower compared to GA and PSO.

Table 3. Optimal STHE design using different optimization methods.

Parameter	Original design [21]	GA [3]	PSO [6]	FA [18]	GSA [19]	WCA (Present work)
D_s (m)	0.539	0.63	0.59	0.7276	0.62	0.623
d_o (m)	0.025	0.02	0.015	0.01575	0.015	0.02
B (m)	0.127	0.12	0.1112	0.1054	0.11	0.135
N_t	158	391	646	924	718	400
S_t (m)	0.031	0.025	0.0187	0.01968	0.0187	0.0244
v_t (m/s)	1.44	0.87	0.93	0.677	0.75	0.99
Re_t	8,227	4,068	3,283	2,408	3,102	4,281
f_t	0.033	0.041	0.044	0.049	0.046	0.040
Pr_t	55.2	55.2	55.2	55.2	55.2	56.5
D_e (m)	0.025	0.019	0.0149	0.0156	0.0148	0.0193
A_s (m^2)	0.0137	0.0148	0.0131	-	-	0.01687
Re_s	25,281	18,327	15,844	14,448	15,004	15,814
Pr_s	7.5	7.5	7.5	7.5	7.5	7.6
v_s (m/s)	0.47	0.43	0.495	0.4	0.476	0.385
h_t (W/m^2 K)	619	1,168	1205	1,262	1,488	584
h_s (W/m^2 K)	920	1,034	1,288	1,156	1,512	985
U (W/m^2 K)	317	376	409.3	347.6	348	221
A (m^2)	61.5	52.9	47.5	56.6	54.98	47.6
L (m)	4.88	2.153	1.56	1.64	1.317	1.941
ΔP_t (Pa)	49,245	14,009	16,926	9374	8449	17,458
f_s	0.315	0.331	0.337	0.3422	0.34	0.3377
ΔP_s (Pa)	24,909	15,717	21,745	12,768	17,962	9,844
C_i (€)	19,007	17,599	16,707	18,202	17,639	16,720
C_o (€/year)	1,304	440	523.3	210.2	290.1	413
C_{od} (€)	8,012	2,704	3,215.6	1,231	1,642	2,541
C_{tot} (€)	27,020	20,303	19,922.6	19,433	19,281	**19,261**

Therefore, based on the total investment cost and annual discounted operating cost, a reduction of total cost by 28.7% is associated with the original design and saving of 5.1%, 3.3%, and 0.9% by using GA, PSO, and FA optimization method respectively. The comparisons of the total costs for each optimization methods are shown in Figure 4.

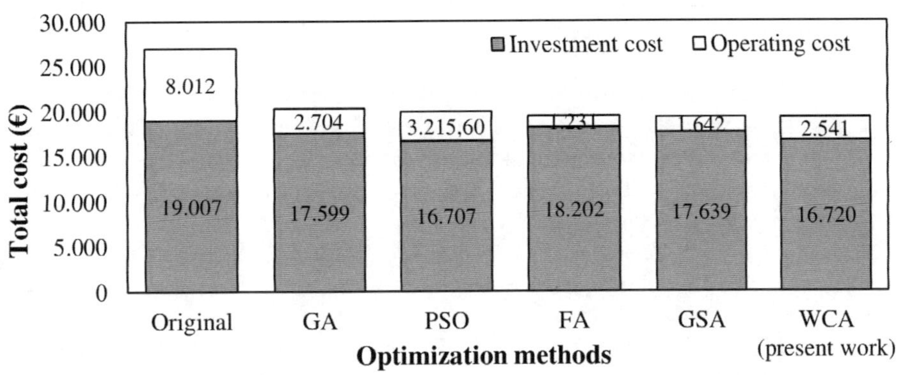

Figure 4. Comparison of the total cost for each optimization methods.

5. Conclusions

This paper examined the application of a recently proposed optimization method called the water cycle algorithm (WCA) to shell and tube heat exchanger optimization to lower the cost of a heat exchanger. The objective function considered is the total annual cost which includes the total investment cost and operating cost. Comparing the results obtained from the WCA, against the results obtained from other optimizers reported in published literature shows that WCA produced better results in terms of a reduction of investment cost as much as 12% and in operating cost of 68.3%, causing a decrease in total cost up to 28.7%. This illustrates the efficiency of the WCA in finding better solutions. Furthermore, WCA obtained the best values in 12 generations since WCA uses more efficient mechanisms to search for terms of exploration and exploitation.

References

[1] Shah, R. K., Sekulic, D. P., (2003). *Fundamentals of Heat Exchanger Design.* New Jersey: John Wiley & Sons. Retrieved from I.I.T-Delhi.

[2] Sinnott, R. K., (2005). *Coulson & Richardson's Chemical Engineering* (Vol. 6). Oxford, UK: Butterworth-Heinemann.

[3] Caputo, A. C., Pelagagge, P. M., Salini, P., (2008). Heat exchanger design based on economic optimization. *Applied Thermal Engineering, 28*, 1151-1159.

[4] Guo, J., Cheng, L., Xu, M., (2009). Optimization design of shell-and-tube heat exchanger by entropy generation minimization and genetic algorithm. *Applied Thermal Engineering, 29*, 2954-2960.

[5] Selbas, R., Kızılkan, O., Reppich, M., (2006). A new design approach for shell and-tube heat exchangers using genetic algorithms from economic point of view. *Chemical Engineering and Processing, 45*, 268-275.

[6] Patel, V., Rao, R., (2010). Design optimization of shell-and-tube heat exchanger using particle swarm optimization technique. *Applied Thermal Engineering, 30*, 1417-1425.

[7] Bejan, A., 1997. *Advanced Engineering Thermodynamics* (2nd ed.). New York: John Wiley & Sons.

[8] Yang, J., Fan, A., Liu, W., Jacobi, A. M., (2014). Optimization of shell-and-tube heat exchangers conforming to TEMA standards with designs motivated by constructal theory. *Energy Conversion and Management, 78*, 468-476.

[9] Yang, J., Oh, S.-R., Liu, W., (2014). Optimization of shell-and-tube heat exchangers using a general design approach motivated by constructal theory. *International Journal of Heat and Mass Transfer, 7877*, 1144-1154.

[10] Azad, A. V., Amidpour, M., (2011). Economic optimization of shell and tube heat exchanger based on constructal theory. *Energy, 36*, 1087-1096.

[11] Atashpaz-Gargari, E., Lucas, C., (2007). Imperialist competitive algorithm: An algorithm for optimization inspired by imperialistic competition. *IEEE Congress on Evolutionary Computation.* Singapore: IEEE.

[12] Hadidi, A., Hadidi, M., Nazari, A., (2013). A new design approach for shell-and-tube heat exchangers using imperialist competitive algorithm (ICA) from economic point of view. *Energy Conversion and Management, 67*, 66-74.

[13] Hadidi, A., Nazari, A., (2013). Design and economic optimization of shell-and-tube heat exchangers using biogeography-based (BBO) algorithm. *Applied Thermal Engineering, 51*, 1263-1272.

[14] Simon, D., (2008). Biogeography-based optimization. *IEEE Transactions on Evolutionary Computation, 12*(6), 702-713.

[15] Fettaka, S., Thibault, J., Gupta, Y., (2013). Design of shell-and-tube heat exchangers using multiobjective optimization. *International Journal of Heat and Mass Transfer, 60*, 343-354.

[16] Turgut, O. E., Turgut, M. S., Coban, M. T., (2014). Design and economic investigation of shell and tube heat exchangers using Improved Intelligent Tuned Harmony Search algorithm. *Ain Shams Engineering Journal, 5*, 1215-1231.

[17] Asadi, M., Song, Y., Sunden, B., Xie, G., (2014). Economic optimization design of shell-and-tube heat exchangers by a cuckoo-search-algorithm. *Applied Thermal Engineering, 73*, 1032-1040.

[18] Mohanty, D. K., (2016). Application of firefly algorithm for design optimization of a shell and tube heat exchanger from economic point of view. *International Journal of Thermal Sciences, 102*, 228-238.

[19] Mohanty, D. K., (2016). Gravitational search algorithm for economic optimization design of a shell and tube heat exchanger. *Applied Thermal Engineering, 107*, 184-193.

[20] Incropera, F. P., Dewitt, D. P., Bergman, T. L., Lavine, A. S., (2016). *Fundamentals of Heat and Mass Transfer* (6th ed.). John Wiley & Sons.

[21] Kern, D. Q., (1950). *Process Heat Transfer.* McGraw-Hill.

[22] Peters, M. S., Timmerhaus, K. D., (1991). *Plant Design and Economics for Chemical Engineers.* New York: McGraw-Hill.

[23] Taal, M., Bulatov, I., Klemes, J., Stehlik, P., (2003). Cost estimation and energy price forecasts for economic evaluation of retrofit projects. *Applied Thermal Engineering, 23*, 1819-1835.

[24] Eskandar, H., Sadollah, A., Bahreininejad, A., Hamdi, M., (2012). Water cycle algorithm – A novel metaheuristic optimization method for solving constrained engineering optimization problems. *Computers & Structures, 110-111*, 151-166.

[25] Sadollah, A., Eskandar, H., Bahreininejad, A., Kim, J. H., (2015). Water cycle algorithm with evaporation rate for solving constrained and unconstrained optimization problems. *Applied Soft Computing, 30*, 58-71.

[26] Bahreininejad, A., (2019). Improving the performance of water cycle algorithm using augmented Lagrangian method. *Advances in Engineering Software, 132*, 55-64.

[27] David, S., (1993). *The water cycle.* New York: Thomson Learning.

Classification of *Civet* and *Canephora* coffee using Support-Vector Machines (SVM) algorithm based on order-1 feature extraction

R Z H Suyoto, M Komarudin*, G F Nama, and T Yulianti

Informatics Engineering Study Program, Department of Electrical Engineering, Faculty of Engineering, Universitas Lampung, Jl. Prof. Sumantri Brojonegoro No.1 Bandar Lampung 35145, Indonesia

*Email: m.komarudin@eng.unila.ac.id

Abstract. Identification of coffee beans types is a process carried out in determining the category of civet and robusta coffee. The selection of civet coffee manually by the community in determining the variety of coffee beans. Manual selection makes the process slow. There is no certainty whether the selected coffee stated original civet coffee or a mixture with Robusta coffee. It needs to be built a system in identifying the types of coffee beans. In this study, the classification technology used for the selection process of Civet and Robusta coffee beans. This research uses first-order feature extraction with the Support-Vector Machines (SVM) algorithm, which aims to recognize civet and robusta coffee beans' patterns using texture analysis on grayscale images and feature extraction. In the study, the authors used the Rapid Application Development approach to build the selection system. The data used in this study were 120 images, consisting of 110 training data and 10 test data. The accuracy in identifying the types of coffee beans by using this method is 87.27%.

Keywords: Civet coffee, robusta coffee, classification, SVM

1. Introduction

Lampung Province is the second-largest producer of Robusta coffee exports, with production in 2013-2017 of 110.12 thousand tons of coffee per year. In 2015 coffee production was concentrated in five districts. West Lampung Regency with the highest production reached 52.65 thousand tons or 47.81% of the total Robusta coffee production in Lampung Province. Other districts namely, Tanggamus Regency with a production yield of 26.92% or 29.64 tons, North Lampung Regency with production results 9.41% or 10.37 tons, Way Kanan Regency with a production yield of 8.38% or 9.23 tons, Pesisir Barat Regency with a production yield of 4.06% or production reaching 4.47 tons. Center for Agricultural Data and Information Systems - Ministry of Agriculture 2017 [1].

Civet Coffee is a delicious taste of coffee beans that are loved by coffee lovers or coffee connoisseurs in the world because coffee produced from imperfect digestion by civet civets makes the most expensive coffee on the world market [2]. Unique coffee whose coffee beans pass through the fermentation process in wet processing can improve the body (thick taste) and Milky (fat taste) [3]. Civet itself is a wild animal that consumes good quality fruits in the forest or farmer's garden, including coffee fruit. Civet

has a very sensitive sense of smell, so clever to choose which coffee fruit is the best and has been cooked perfectly so that the quality Civet coffee produced is of high quality.
Because coffee that is eaten by civet coffee is quality coffee making civet coffee known as the most expensive and distinctive coffee (unique) to date Civet coffee production is still in very limited quantities. Civet coffee itself has a physical color of dark skin and brown (enzymatic browning), has epidermis and tends to be darker [4].

Civet coffee prices are expensive and limited production, leading to fraud by unscrupulous by mixing civet coffee with other types of coffee such as Robusta. The fraud that occurred in the person gave rise to the idea to create an application that could distinguish between Civet coffee and Robusta coffee. In this study distinguish between civet and robusta coffee by classifying the image of civet and robusta coffee beans, general the stages in the image classification process are, inserting images, converting RGB images to grayscale, extracting features, training, testing and measuring the accuracy of images. classification aims to classify the object into certain classes based on the value of attributes associated with the object being observed.

This study aims to identify the image of coffee beans using texture analysis on grayscale images and color feature extraction in color images. Grayscale image elements have elements of mean, entropy, variance, skewness, and kurtosis. The result of feature contraction will be the input value for SVM to classify the type of coffee beans used is the image of coffee beans consisting of 120 coffee beans consisting of 60 types of robusta coffee beans and 60 types of civet beans.

2. Literature review

2.1. Previous research
Sri Citra, conducted research with use image processing techniques for coffee bean processing. Classifying coffee beans into four classes based on the image area, height, width, and parameters. However, due to the time of manual verification, it was not very thorough, so from the results of the program around 81% success [5]. Wahyu aji, researched the identification of coffee fruit maturity using artificial neural networks to study patterns and identify coffee maturity using color features in the fruit. Based on the results of the test produces the best image with 100% accuracy without error in every test performed [6]. Other research on applying the Rapid Application Development (RAD) methodology in developing training application systems, using RAD enables rapid implementation in real environmental systems. RAD's research, it proved to be the right method for this work project because it served the purpose of rapid implementations for small systems [7]. Zhibin Liau, in his research on evaluating the ability of cooperative income villages by using SVM algorithm to evaluate Supplier Cooperative Design Ability (SCDA), introduces 2 specific gravity factors and sample weight factors. As a result, the accuracy and ability of SVM have a strong classification function and good learning so it has broad application prospects in the SCDA section [8]. Beynon-Davies, in his research on integrating Rapid Application Development (RAD), discusses the development of systems using RAD on business and uncertainty and Participatory Design (PD). RAD and PD systems are reviewed in the area of possible cross-fertilization, which relates to aspects of RAD approach research, specifically scenario concepts, and detailed design for work on RAD [9]. Yun Lin, in his research on the classification of texts based on the SVM and K-Nearest Neighbor (KNN) algorithm, combines the SVM and KNN algorithm. As a result, the classification of SVM and KNN can obtain better accuracy from the grouping of SVM and KNN and can improve computing that is very small [10].

2.2. Feature extraction
Feature extraction is a method of characterization based on image histogram characteristics. The histogram shows the probability value of the appearance of the degree of gray pixel values in an image. Values that exist in the resulting histogram, can be calculated with several parameters of first-order characteristics, including Mean, Skewness, Variant, Kurtosis, and Entropy [11].

Based on the values in the resulting histogram, several characteristic parameters can be calculated, including the mean, variance, skewness, kurtosis, and entropy, namely:

1. Mean Feature (μ) shows the size of the dispersion of an image

$$\mu = \sum_{n=0}^{N} f_n \, P(f_n) \tag{1}$$

Explanation:
f_n = gray intensity value
$P(f_n)$ = histogram value

2. Variance Feature (σ2) shows the variation of elements in the histogram of an image

$$\sigma^2 = \sum_{n=0}^{N} (f_n - \mu)^2 \, P(f_n) \tag{2}$$

Formula Description:
f_n = gray intensity value
μ = mean value
$P(f_n)$ = histogram value

3. Skewness Feature (a3) indicates the relative skewness level of the histogram curve of an image.

$$\alpha_3 = \frac{1}{\sigma^3} \sum_{n=0}^{N} (f_n - \mu)^3 P(f_n) \tag{3}$$

Formula Description:
a3 = standard deviation of gray intensity values
f_n = gray intensity value
μ = mean value
P(f_n) = histogram value

4. Characteristics of Kurtosis (a4) shows the relative degree of shock of the histogram curve of an image

$$\alpha_4 = \frac{1}{\sigma^4} \sum_{n=0}^{N} (f_n - \mu)^4 P(f_n) - 3 \tag{4}$$

Formula Description:
α 4 = standard deviation of gray intensity values
f_n = gray intensity value
M = mean value
$P(f_n)$ = histogram value

5. Entropy (H) Characteristics Indicates the size of the irregular shape of an image

$$\mathrm{H} = \sum_{n=0}^{N} P(f_n) .^2 \log P(f_n) \tag{5}$$

Formula Description:
f_n = gray intensity value
$P(f_n)$ = histogram value

2.3. Support Vector Machine (SVM)

Support Vector Machine one of the methods in supervised learning that is usually used for classification (such as Support Vector Classification) and regression (Support Vector Regression). Related to the number of decision boundaries cannot be modeled with a form or linear equation, one must model a decision boundary as a non-linear function [12]. SVM is used to find the best hyperplane by maximizing the distance between classes. The hyperplane is a function that can be used for separating classes. In 2-D functions used for class, classifications are referred to as lines whereas, functions used for class classifications in 3-D are called planes similarly, whereas functions used for classifications in classrooms of higher dimensions are called hyperplanes. In SVM the outermost data object that is closest to the hyperplane is called a support vector. Objects called support vectors are the most difficult to classify due to positions that almost overlap with other classes Related to the number of decision boundaries, it cannot be modeled with a form or linear equation, it must model a decision boundary as a non-linear function.

3. Research methods

The research method used in the implementation and execution of this thesis is the Rapid Application Development (RAD) method, which is a model of an international software development process, especially for short processing times [13]. The research flow is in Figure 1.

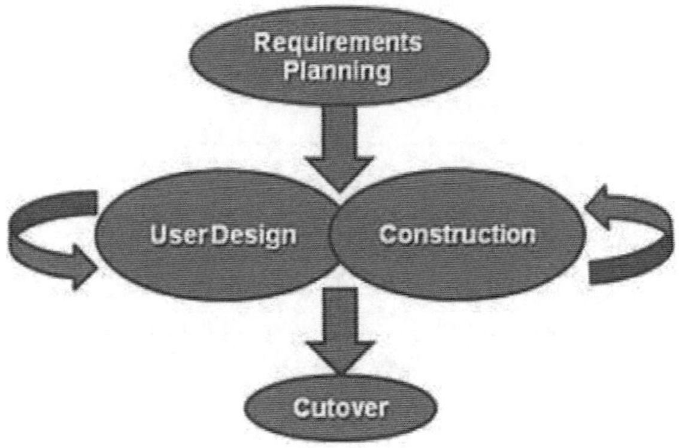

Figure 1. RAD Research stage.

The stages of the RAD method:

1. Requirements planning or Need Planning.
   At this stage, the authors design the needs and planning carried out to determine the goals and expectations for research as well as current and potential problems that need to be addressed. In this stage the writer collected the materials and made observations on the needs of the civet and robusta coffee bean samples, the methods used and the MATLAB application compared to the information analyzed to obtain system specifications.
2. User Design or the Design Process
   At this stage, the authors carry out the design process in the form of modeling the required data based on application modeling and defining its attributes and their relationships with other data. This stage the author uses ERD and LRS for database modeling so that it can know what attributes are needed and how the relation of the data, as well as making the application mockup process to be used. At this stage the authors also made changes during the process, starting from working on the system, in the first experiment the authors used GLCM feature extraction, but during the research work, the authors experienced changes in GLCM feature extraction into first-order feature extraction due to the value of features that could implement research.
   Furthermore, in the classification method used, the previous author used the KNN classification, however, there was a change to the SVM classification because SVM classification uses the separating elements between classes using hyperplane lines, and the mockup display on the Matlab used. Furthermore, the authors use the use case as a process identification, use case definition, and activity diagrams as modeling the application process. This stage is done to get the best design of an application system.
3. Construction (construction) and Implementation.
   At this stage, it breaks down into preparations for a fast workmanship framework, the development of programs and applications, coding in systems, units, integration, and system testing. At the time of the research, the author made the application using the MATLAB 2017b GUI, then coding the system until the application was successfully made so that it could be implemented on the system. In making and implementing changes occur when the author does the work process such as adding features to the Matlab GUI display that was not previously displayed then it is displayed in the GUI,

and the display successfully identifies from coffee beans that were not displayed before, the author displays the results of the identification of the application system.

4. Cutover
   At this stage, the application was made successfully. Through this stage, the writer also implements the application functions that have been defined about defining the data[14]. After a successful application, the program will be tested whether there is an error or not before it is published. The most important thing is user involvement is needed so that the system can be developed if something goes wrong and gives satisfaction to the user, other than that the old system does not need to be run in parallel with the new system or the system that has been successfully published.

4. Results and discussion

4.1. Implementation

This study uses a comparison of training (training data) and testing (test data) of 70:30 because the typical image research uses the ratio of 70:30. The training data used amounted to 110 images and the test data amounted to 10 images. The image used as training data is taken from datasets with sequence numbers 1 to 55 in each class, and images used as test data are taken from datasets with sequence numbers 1 to 5 in each class. The following is the display to enter the image.

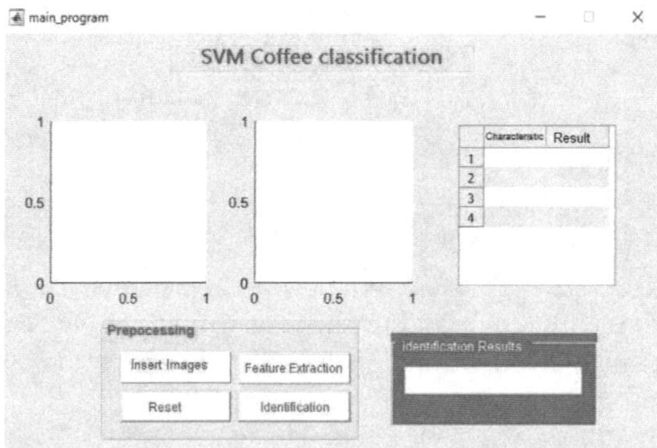

Figure 2. Display of the Matlab GUI application.

Based on the image is the display after pressing the "Enter image" then the image conversion will exit from the RGB Image to Grayscale Image. In pre-processing stages are carried out as is done. Next to the Grayscale image is a table that contains the conversion of images or the value of an image and the features that are in it.

Table 1. The result of the data training feature extraction value

Sample	Mean	Ent	Var	Skew	Kurto
Civet 1	119	6.62	4318	0.29	-1.72
Civet 2	115	6.64	4441	0.34	-1.70
Civet 3	108	6.68	5142	0.28	-1.69
Civet 4	124	6.50	4942	0.06	-1.81
Civet 5	113	6.87	4457	0.21	-1.68
Robusta 1	124	6.95	4395	0.08	-1.67
Robusta 2	114	6.69	3761	0.24	-1.50
Robusta 3	112	6.61	3746	0.25	-1.70
Robusta 4	118	6.84	3311	0.32	-1.42
Robusta 5	111	6.89	3045	0.39	-1.34

Table 3 is the result of feature extraction testing using the first-order method which included 5 sample data from each type of coffee from a total of 110 sample data types. The results of feature extraction can be seen from the output values such as mean, entropy, variance, skewness and kurtosis. The above value is used as a reference to classify 110 test sample data in accordance with the standard values.

4.2. Testing and identification

In the testing stage, new samples are used that have been distinguished from training data samples or (training) consisting of 5 images or image objects of each type of coffee bean. In the application, the test is carried out to see whether the application has been able to read or identify an image in the training data that has produced output in accordance with the expected pattern. The following results identify the types of coffee beans as Table 2.

Table 2. Application identification results

Sample	Input	Output	Information
1	Civet	Civet	True
2	Civet	Civet	True
3	Civet	Civet	True
4	Civet	Civet	True
5	Civet	Civet	True
1	Robusta	Robusta	True
2	Robusta	Robusta	True
3	Robusta	Robusta	True
4	Robusta	Robusta	True
5	Robusta	Robusta	True

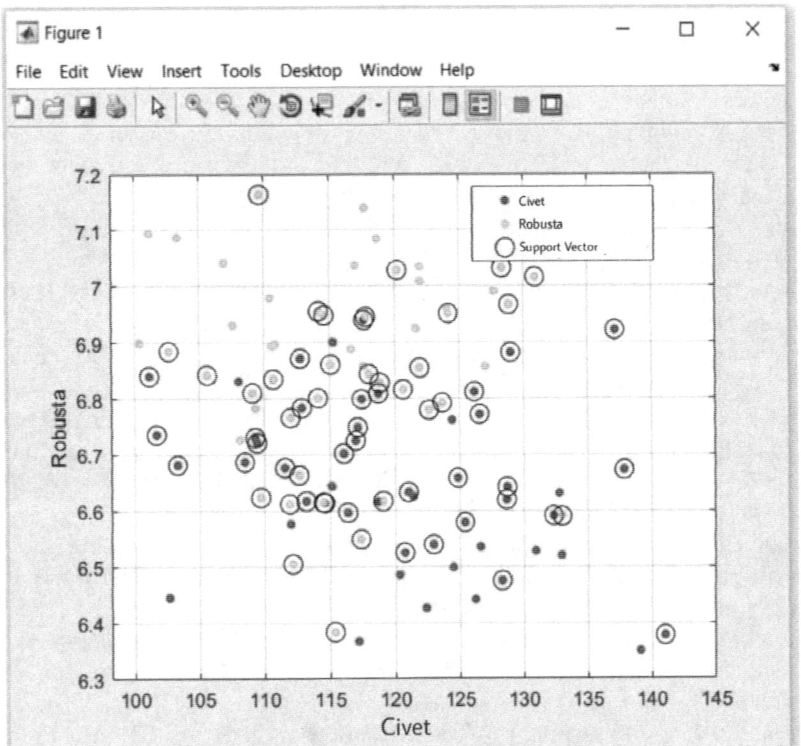

Figure 3. Support Vector Machine (SVM) classification.

Based on Table 3 is a truth table 10 of 110 sample data types of coffee beans. In the test, there were 96 coffee beans declared correct and stated 14 coffee beans wrong because there was an error in this classification into an SVM algorithm having an accuracy rate of 87.27%.

Figure 3 shows the classification formed by a support vector machine or SVM. Distinguishing between types of civet coffee and robusta coffee with red color as civet coffee and robusta coffee in blue and circle indicates a vector machine that divides it using 110 training data samples.

4.3. Cutover

In addition to conducting their testing or functional, the author conducts testing of the community to conduct testing. The author tests the community or user. In testing conducted on the system using training data that is tested by first the user runs the program, then the user enters an image, then the user performs the feature extraction and the user identifies. In testing the user program successfully identified by the program.

5. Conclusions

Based on the results of trials that have been carried out in this study, it is proven that the SVM algorithm can be implemented to classify the types of coffee beans. At an accuracy of 87.27% by successfully recognizing 96 coffee beans out of 110 and not recognizing 14 beans, then for subsequent research, it can be developed with other algorithms and methods. Coffee set image data types need to be added so that the level of accuracy can be increased. The author approves the process of introducing types of coffee beans using an android application that is far more easily accessed by the public.

Acknowledgements

The authors would like to thank Department of Electrical Engineering Universitas Lampung.

References

[1] Suwandi, Outlook 2017 Agricultural Commodities, Coffee Plantation Sub Sector. Center for Agricultural Data and Information Systems General Secretariat - Ministry of Agriculture 2017, 2017.

[2] B. Ikhwan, "Enchantment of Kopi Luwak," Export News, Ministry of Trade, pp. 1–20, 2013.

[3] J. Towaha, A. Aunillah, E. H. Purwanto, and H. Supriadi, "Effects of Elevation and Processing on Chemical Content and Taste of Lampung Robusta Coffee," J. Tanam. Ind. and Refresher, vol. 1, no. 1, p. 57, 2019, doi: 10.21082 / jtidp.v1n1.2014.p57-62.

[4] Yusianto, "Kopi Luwak and testing its authenticity," Pus. Researcher. Coffee and Cocoa Indones., pp. 1–44, 2014.

[5] S. C. Madi yuliana, "Deciding Coffee Beans by Using Image Processing [Thesis]," Inst. pertan. Bogor, 2010.

[6] W. Aji Pulungan, "Identification of Coffee Fruit Maturity Using Artificial Vector Neutral Neural Networks," University of Lampung, 2019.

[7] M. Daud, A. Bakar, and H. Rusli, "Implementing Rapid Application Development (RAD) methodology in developing practical training application systems," Proc. 2010 Int. Symp. Inf. Technol. - Syst. Dev Appl. Knowl. Soc. ITSim '10, vol. 3, pp. 1664–1667, 2010, doi: 10.1109 / ITSIM.2010.5561634.

[8] Z. Liu and L. Bai, "Evaluating the supplier cooperative design ability using a novel support vector machine algorithm," Proc. 2008 12th Int. Conf. Comput Support. Coop Work Des. CSCWD, vol. 2, pp. 986–89, 2008, doi: 10.1109 / CSCWD.2008.4537113.

[9] P. Beynon-Davies and S. Holmes, "Integrating rapid application development and participatory design," IEE Proc. Softw., Vol. 145, no. 4, pp. 105–112, 1998, doi: 10.1049 / ip-sen: 19982196.

[10] Y. Lin and J. Wang, "Research on text classification based on SVM-KNN," Proc. IEEE Int. Conf. Softw. Eng. Serv. Sci ICSESS, pp. 842–844, 2014, doi: 10.1109 / ICSESS.2014.6933697.

[11] A. Fadlil, Pattern Recognition Lecture Module, no. 002. Yogyakarta: Ahmad Dahlan University, 2012.
[12] J. Wira, Introduction to the Concept of Machine Learning and Deep Learning Edition 1.3, vol. 1.3. Tokyo, Japan: Wiragotama, 2019.
[13] O. Irnawati and et al, "Rapid Application Development (RAD) Method in the Design of Inventory Website of PT. Sarana Abadi Makmur Bersama (S.A.M.B), "vol. 6, pp. 12–18, 2018.
[14] A. Ismail and A. Aziz, "ITourism travel buddy mobile application," Int. Conf. Next Gener. Mob Appl. Serv. Technol., Pp. 82–87, 2016, doi: 10.1109 / NGMAST.2016.22.

Geophysical approach for assessment of seawater intrusion in the coastal aquifer of Bandar Lampung, Indonesia

Rustadi[1,2,*], I G B Darmawan[2], N Haerudin[2], Suharno[2], and A Setiawan [1]

[1] Department of Environtment Science, Universitas Lampung, Jalan Prof. Soemantri Brojonegoro No.1, Bandar Lampung 35145, Indonesia

[2] Department of Geophysical Engineering, Universitas Lampung, Jalan Prof. Soemantri Brojonegoro No.1, Bandar Lampung 35145, Indonesia

*Email: rustadi_2007@yahoo.com

Abstract. Overexploitation in 1980-2000 caused a lot of groundwater loss in the aquifer coastal of Bandar Lampung. Groundwater has been degraded with increasing salinity due to the intrusion of seawater. In 2013, the intrusion extended 400 m from the coastline. Subsurface geological mapping is carried out to examine alluvial formation's material constituents and the ability to transfer seawater to land. Three well explorations combined with three geoelectric tomography and 51 points of gravity. The gravity method is used to map structures that can connect coastal aquifers and land aquifers. Meanwhile, geoelectric tomography to map the presence of aquifers and low resistivity zones by seawater intrusion. The coastline has relatively shallow aquifers at a depth of 2 - 10 m in a sand layer with coral inserts, while the lower part is composed of silt and clay. The aquifer has experienced a decrease in resistivity of fewer than 10 ohms.m appears in the three circular patterns, caused by seawater. In addition to the permeable zone of shallow aquifers, there is a northwest-southeast trending structure that can channel seawater to land aquifers.

Keywords: Coastal aquifer, seawater intrusion, geophysics, Bandar Lampung

1. Introduction

Since the Dutch colonial period, the Bandar Lampung coast has been the gateway to commerce between regions. The VOC relied on Teluk Betung to collect and transport the spices obtained in Lampung Province. As an economic center for a long time, it has been able to trigger regional development. Developing the manufacturing industry, tourism, and settlements with a population density of 8050/km^2.

The main component of clean water sources on the coast of Bandar Lampung is groundwater. Overexploitation in 1980 - 2010 has caused degradation in coastal aquifers and increased sodium and chloride, caused by seawater's intrusion. In 2013, the distribution zone reached a radius of 500 m from the coastline [1]. Seawater intrusion in coastal areas is a process of infiltrating saltwater from the sea into groundwater on land. In natural conditions, groundwater will flow continuously into the sea. When the balance is disturbed, the intrusion could be extended inland. The problem of groundwater contamination by seawater has become a global problem [2-4].

Increasing groundwater salinity on the coast of Bandar Lampung requires regular monitoring and is supported by integrated data. However, there are various problems, and the research focuses on mapping aquifer layers and faults that can affect the interaction between seawater and freshwater on land. Geoelectric and gravity methods are used to obtain initial information on coastal aquifers [5] and structures [6]. Geoelectric mapping is often performed to delineate aquifers [7-9] and evaluate zones that have been exposed by seawater intrusion [10-12]. Meanwhile, the gravity method is quite useful in interpreting the thickness of the sediment layer and the possible presence of structures [13-15].

2. Coastal geology in Bandar Lampung

The geology of the coastal of Bandar Lampung is composed of sedimentation from land runoff and ocean wave circulation to produce coastal alluvial deposits (Qa). This formation has a limited distribution and forms inconsistent contacts with the Quaternary sedimentary rock formations, Young Mountain Rock Formation (Qhv) and the Lampung Formation (QTI) and Tertiary sedimentary rocks, the Campang Formation (TPOC), and the Tarahan Formation (TPOT). The geological setting of the Bandar Lampung coast and its surroundings is shown in Figure 1.

Sedimentary rock formations and volcanic products cover bedrock in the form of metamorphic rocks consisting of migmatite, schist, genes, alabaster, and quartzite, which are included in the Gunung Kasih Complex (Pzg). Gunung Kasih Complex is in pre-tertiary age, and in some places, it has a relatively shallow position [16]. The Bandar Lampung region has a complex geological setting that is influenced by tectonic pressure, which results in an increase in magmatic through the bedrock and forming intrusions and lenses of igneous rock. The tectonic process forms various faults in the direction of the subduction path. Various faults can act as permeable zones that allow interactions between the saline water to move to extend inland.

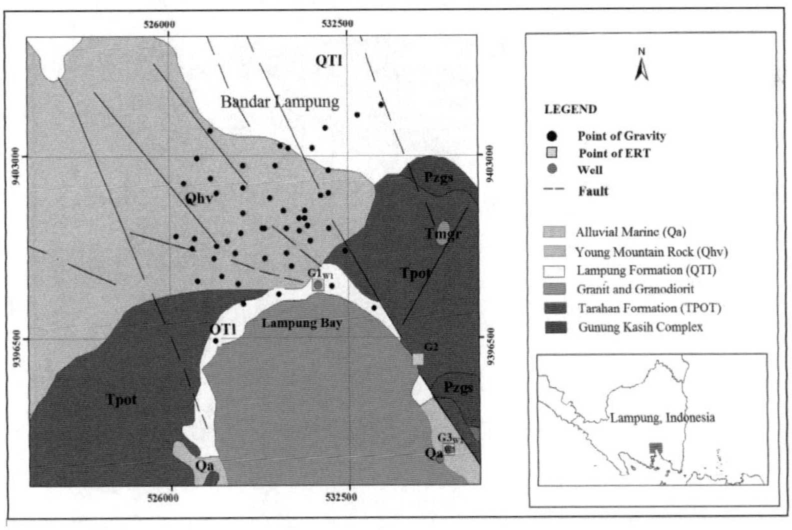

Figure 1. Geology map, well exploration, measure point of gravity, and line ERT of the study area.

3. Materials and methods

There are three exploration wells in coastal alluvial deposits (Qa) intending to determine soil mechanics' constituent lithology and characterization. Imaging to map the conductive zone related to aquifers' presence and seawater intrusion distribution is carried out through 3 lines of electrical resistivity tomography (ERT). Measurements using the ARES instrument, with the Wenner-Schlumberger configuration, electrode spacing 6 m. Other sediment formations and faults were analyzed through gravity measurements to map the alluvial coastal sediment layer's thickness. Measurements using a LaCoste and Romberg gravity meter with points scattered randomly. Each data point from the well is represented by a red dot (W), the electrical resistivity tomography (ERT) is symbolized by green rectangular (G), and a thick black dot represents the gravity measuring point.

4. Results and discussions

4.1. Gravity

The anomaly Bouguer from gravity data is shown in Figure 2, and the anomaly value ranges from 32 - 60 mgal. Low anomalies correlate with young formations, especially coastal alluvials in the south. There are indications of basins in coastal alluvials with anomalies ranging from 34 - 38 mgal, as shown on Fig. 2.

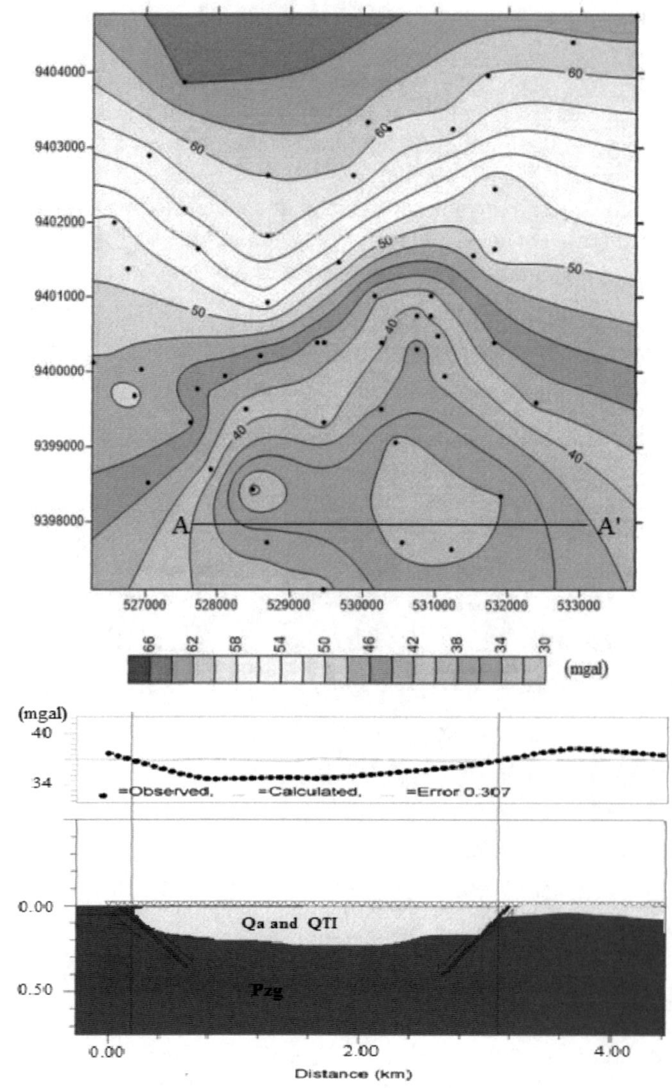

Figure 2. Map of the Bouguer anomaly with contour intervals of 2 mgal.

The contour pattern of the Bouger anomaly in the Lampung Bay area is in line with the existence of the fault structure in the area. The low anomaly in the south is oriented towards the northwest-southeast. The direction of this anomaly is also in line with the direction of the dominant structure. An incision profile was carried out on the A-A' path to identify this dominant structure's effect. Qualitative interpretation through 2.5 D modelling on slice A-A' to obtain subsurface images of the basin on the coast of Bandar Lampung.

The modelling results are shown in Figure 2 show that the two main structures are trending northwest-southeast. This structure is a normal fault structure with opposite slopes to form a sag basin. This structural pattern confirms the presence of the basin from the Bouguer anomaly contour. This basin area is interpreted as a zone of the potential groundwater aquifer.

4.2. Electrical resistivity tomography

The measurement data of three ERT lines in G-1, G-2, and G-3 is processed using Res2DInv software. The resulting image is shown in Figures 3, 4, and 5.

Tomographic inversion results is presented in Figure 3 show a low resistivity value at low depth. These results indicate that groundwater has been mixed with seawater in shallow aquifers. However, this pattern still does not show a seawater intrusion pattern, but rather a mixture of groundwater in coastal areas. This condition is reinforced by observations in the field of saltwater conditions in the coastal area of Lampung Bay. The majority of residents no longer use groundwater wells and instead use clean water by buying. Meanwhile, in the port area (Panjang Port), the resistivity tomography inversion results also showed the same pattern as the Lampung Bay (Figure 4).

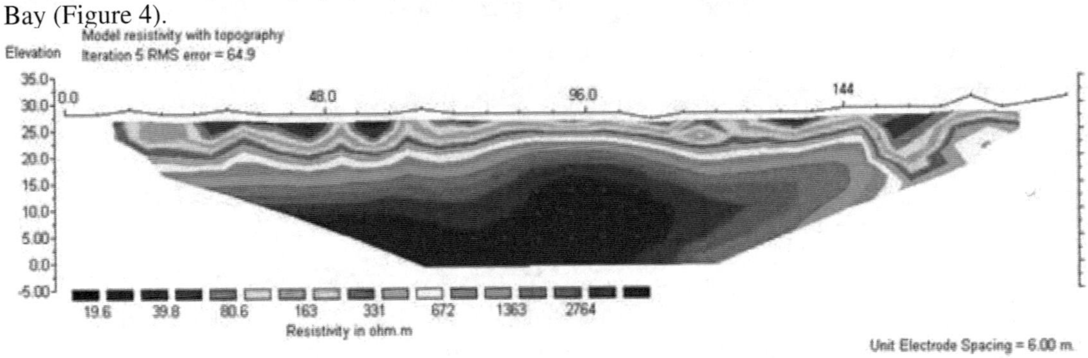

Figure 3. Inverted electrical resistivity models for Wenner-Schlumberger measurements on ERT line 1.

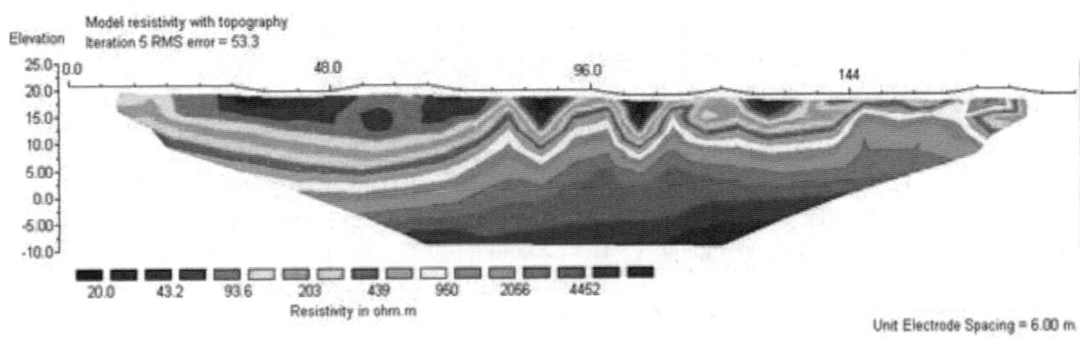

Figure 4. Inverted electrical resistivity models for Wenner-Schlumberger measurements on ERT line 2.

The absence of a seawater intrusion resistivity pattern in this measurement indicates that seawater contamination only occurs in shallow groundwater aquifers, specifically in coastal areas. This condition is seen at low resistivity values with a thickness of up to 10 m at the beginning of the measurement path. The low resistivity value indicates the effect of mixed saltwater on groundwater. Two resistivity patterns appeared that formed a trench in the middle of the measurement path with a low resistivity value <10 Ohm.m. This pattern is interpreted as a trench or shallow aquifer pathway that may have been formed due to the Lampung-Panjang Fault's existence. This low resistivity pattern also indicates the presence of an accretion structure filled with saline water.

Although the ERT line 1 and line 2 measurement paths have not shown a seawater intrusion pattern, exciting results are shown by the third line. This route is located in the industrial area and the population, namely Srengsem Village. This area is the location for massive groundwater utilization. Several problems have emerged regarding the need for clean water in this area. The low resistivity pattern indicating the seawater intrusion zone is shown in Figure 5.

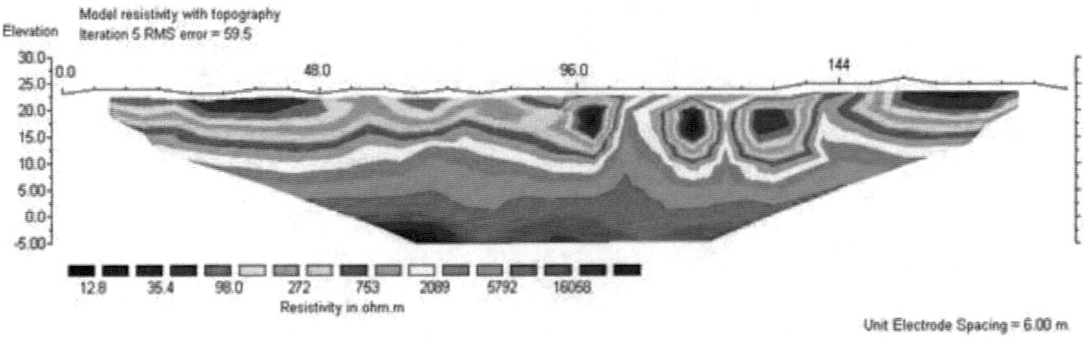

Figure 5. Inverted electrical resistivity models for Wenner-Schlumberger measurements on ERT line 3.

The seawater intrusion channel that appears in the middle of the track is composed of three circular patterns. This pattern has a low resistivity value of fewer than 10 ohms. The seawater path is estimated to be in the geological structure gap that develops in the region. This seawater intrusion pattern can occur due to the massive use of groundwater to disturb the aquifer's stability and the soil.

4.3. Well exploration

The constituent materials in the coastal alluvial formation produce variations in lithology and thickness in the three wells. The drilling results' interpretation is shown in Figure 6 and site class testing from the SPT compression test, the layers are still in the soft group up to a depth of 12 m in well-1, 26 m in well-2, and 32 m in well-3.

The first well (Well-1) layer is dominated by clay at a depth of 2 - 25 meters below sea level. Meanwhile, the sand layer with coral inserts is in the second layer with a thickness of 2 meters, overlain by a clay layer above it. The thickness of the clay layer in well-1 indicates an impermeable layer that holds the aquifer above it so that seawater intrusion cannot penetrate the clay layer. However, surface aquifers that are not too thick are likely to be contaminated by seawater. The sand layer's density with coral inserts shows that the shallow marine still influences forming this rock layer to deep marine depositional environment.

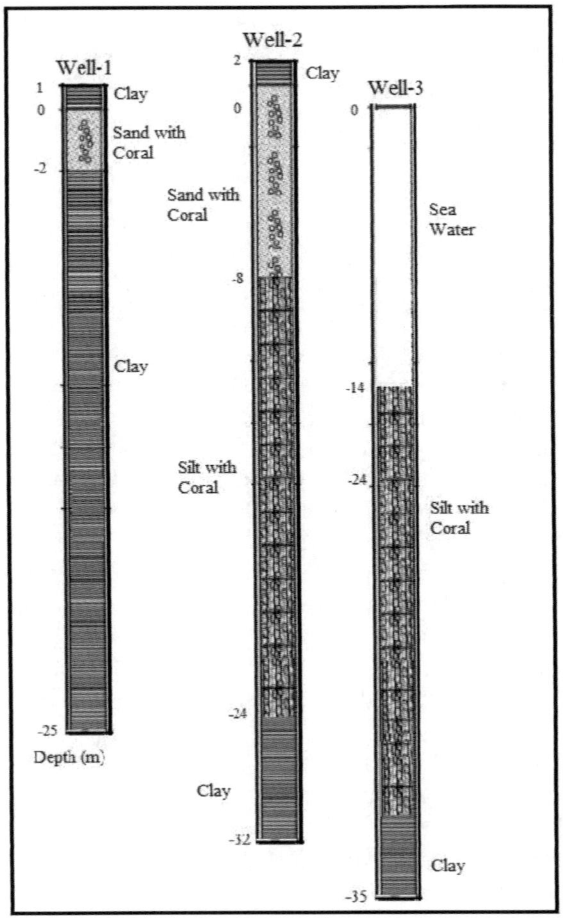

Figure 6. Lithology section on three explorations well

The location of well-2 and well-3 is in the same location, namely Srengsem Village, Panjang District. Well-3 is closer to the coast than well-2. Well-3 starts with a seawater depth of up to 14 m until it meets a silt layer with coral and clay insert at the bottom. This layer indicates a depositional environment in the deep sea to the shallow sea transition area. In line with well-2, the presence of a silt layer with coral and clay insertion at the bottom indicates a similar depositional environment in well-3. However, in well-2, there is an aquifer layer of groundwater in the form of coral sand with a thickness of approximately 6 m. This layer has the potential to act as a groundwater pathway that can be contaminated by seawater.

5. Conclusions

The results of the geophysical approach to assessing the intrusion zone of seawater in the Bandar Lampung coastal aquifer have shown a mixed pattern of groundwater and seawater. The gravity method has succeeded in identifying the groundwater aquifer's basin area in Lampung Bay controlled by two normal fault structures. One of the three ERT measurement paths successfully detected a seawater intrusion pattern in the Srengsem Village area, namely the ERT line 3. The resistivity value is relatively low (<10 ohms), appears in the three circular patterns, and is interpreted as a seawater intrusion zone. This finding is also reinforced from drill data, especially in well-2, which shows the groundwater aquifer layer is in a sand layer with coral inserts.

Acknowledgements
Acknowledgments were delivered to the Universitas Lampung, especially the Department of Geophysical Engineering and all colleges who have helped collect data, processing, and model, until completing this paper.

References
[1] Syafriadi, Zaenudin A, Kusumastuti D I, and Suharno 2013 Penggunaan metode geolistrik untuk pemodelan intrusi air laut di daerah pesisir Kota Bandar Lampung *Jurnal Teori dan Aplikasi Fisika* **01**(02)
[2] Loáiciga H A, Pingel T J, Garcia E S 2012 Sea water intrusion by sea-level rise: Scenarios for the 21st century *Ground Water* **50**(1):37–47
[3] Das S, Maity P K, Das R 2018 Remedial Measures for Saline Water Ingression in Coastal Aquifers of South West Bengal in India *MOJ Ecol Environ Sci* **3**(1)
[4] Perera M D N D, Ranasinghe T K G P, Piyadasa R U K, Jayasinghe G Y 2018 Risk of seawater intrusion on coastal community of Bentota river basin Sri Lanka. *Procedia Eng* **212** 699–706 https://doi.org/10.1016/j.proeng.2018.01.090
[5] Pratama W R 2019 Aplikasi Metode Geolistrik Resistivitas Konfigurasi Wenner-Schlumberger Untuk Mengidentifikasi Litologi Batuan Bawah Permukaan dan Fluida Panas Bumi Way Ratai di Area Manifestasi Padok di Kecamatan Padang Cermin Kabupaten Pesawaran Provinsi Lampung *J. Geofis. Eksplor.* **5** 30–44
[6] Wibowo R C, Mulyatno B S, Romosi M and Zaenudin A 2018 Determination of slip surface area using geoelectric, MASW, and soil mechanics data in cimuncang village, west Java *EAGE-HAGI 1st Asia Pacific Meet. Near Surf. Geosci. Eng.* 0–4
[7] El_Hameed A G A, El-Shayeb H M, El-Araby N A, Hegab M G 2017 Integrated geoelectrical and hydrogeological studies on Wadi Qena, Egypt *NRIAG J Astron Geophys* **6**(1):218–29 http://dx.doi.org/10.1016/j.nrjag.2017.03.003
[8] Mohamaden M I I, Abuo Shagar S, Allah G A 2009 Geoelectrical survey for groundwater exploration at the Asyuit governorate, Nile Valley, Egypt *J King Abdulaziz Univ Mar Sci.* **20**(1):91–108
[9] Sultan S A, Essa K S A T, Khalil M H, El-Nahry A E H, Galal A N H 2017 Evaluation of groundwater potentiality survey in south Ataqa-northwestern part of Gulf of Suez by using resistivity data and site-selection modeling *NRIAG J Astron Geophys* **6**(1):230–43 http://dx.doi.org/10.1016/j.nrjag.2017.02.002
[10] Bouderbala A, Remini B, Hamoudi A S 2016 Geoelectrical investigation of saline water intrusion into freshwater aquifers: A case study of Nador coastal aquifer, Tipaza, Algeria. *Geofis Int.* **55**(4):239–53
[11] Alfarrah N, Walraevens K 2018 Groundwater overexploitation and seawater intrusion in coastal areas of arid and semi-arid regions *Water (Switzerland)* **10**(2)
[12] Sathish S, Elango L, Rajesh R, Sarma V S 2011 Application of Three Dimensional Electrical Resistivity Tomography to Identify Seawater Intrusion **4**(I):21–8
[13] Al Farajat M 2009 Characterization of a coastal aquifer basin using gravity and resistivity methods: A case study from Aqaba in Jordan *Acta Geophys* **57**(2):454–75
[14] Pool D R 2008 The utility of gravity and water-level monitoring at alluvial aquifer wells in southern Arizona *Geophysics* **73**(6)
[15] Delinom R M 2009 Structural geology controls on groundwater flow: Lembang fault case study, West Java, Indonesia *Hydrogeol J.* **17**(4):**10** 11–23
[16] Darmawan I G B, Setijadji L D and Wintolo J 2015 Geology and Geothermal System in Rajabasa Volcano South Lampung Regency, Indonesia (Approach to Field Observations, Water Geochemistry and Magnetic Methods) *Proceedings World Geothermal Congress* (Melbourne, Australia).

IOP Conf. Series: Materials Science and Engineering 1173 (2021) 012008 doi:10.1088/1757-899X/1173/1/012008

Reservoir properties prediction using seismic inversion and geostatistical integration

O Dewanto[1], B S Mulyatno[1], P R Ordas[2], and R C Wibowo[1,*]

[1] Department of Geophysical Engineering, Faculty of Engineering, Universitas Lampung, Jalan Prof. Sumantri Brojonegoro No. 1 Bandar Lampung 35145, Indonesia

[2] Department of Geological Engineering, Universitas Padjajaran, Jalan Raya Bandung Sumedang Km 21, Jawa Barat, Indonesia

*Email: rahmat.caturwibowo@eng.unila.ac.id

Abstract. Exploration and production in the sand reservoir still have their challenges due to the low porosity and permeability characteristics. This study was conducted to analyse the characteristics of a tight sand reservoir based on a log property distribution map, using acoustic impedance inversion and multi-attribute analysis. Stepwise regression multi-attribute analysis is a method that uses the best attributes to predict the target log by going through a trial and error process. Choosing the right seismic attributes can provide a better depiction of the target zone. This research was conducted to obtain a map of subsurface geological structures, acoustic impedance volumes. Then, we performed a multi-attribute analysis to obtain a prediction of volume log properties such as pseudo gamma-ray, density and porosity using the stepwise regression method. The results of acoustic impedance seismic inversion and stepwise regression multi-attribute analysis shows that the reservoir is a gas with tight sand lithology, which has a range of acoustic impedance values of 15,000 ((ft/s)*(g/cc)) up to 30,000 ((ft/s)*(g/cc)), porosity of 12% to 24%, and the distribution of Sw of 8-13%. The density and porosity maps obtained from the multi-attributes analysis can help in the long-term exploration and production stages. Its aims are to improve primary recovery and tertiary recovery, understanding the stratigraphic traps, and continuity of reservoir layers.

Keywords: qualitative and quantitative analysis, porosity, saturation water, petrophysics

1. Introduction

One of the reservoir characteristics that can be measured by geophysical methods is porosity [1]. Porosity modeling can be obtained through acoustic impedance seismic inversion (IA), using both 2D and 3D seismic data [2]. Seismic inversion IA is a method that can estimate subsurface physical properties in the form of IA values through seismic data as input and well data as control [3]. Seismic data is highly influenced by the bandwidth with a thin layer below the thickness of the tuning resolution which cannot be separated properly. While the well data has high resolution, so it can display the rock layers in detail. The integration of the two can be used to solve the problem of thin layer under tuning thickness, depending on the quality and completeness of the data [4].

The multi-attribute seismic approach can be used to predict well log property data and seismic volume. including porosity [5]. The porosity obtained through the multi-attribute approach is closer to the actual subsurface conditions than using the one-attribute approach (AI) [6]. The objective of a multi-attribute transformation is a linear or nonlinear transformation between the attribute subset and the target log value. The selected subset is determined by the step wise regression process, namely by looking for a larger subset of attributes. The conventional extension of the cross-plot involves the use of convolutional operators to resolve the frequency difference between the target log and the seismic data [7].

AI distribution maps and log property distribution maps (porosity, density, etc.) can be used for reservoir characteristic analysis and reservoir prospect zone location determination. The data generated from the AI inversion process and multi-attribute seismic will later be used to create a porosity distribution in the research area so that it can be used to analyze reservoir characteristics.

2. Methods

2.1. Data

We use seismic data in the form of 3D Post-Stack Time Migration (PSTM) data, which is equipped with an inline of 300 lines running from east-west and an xline that runs from north-south as many as 800 lines, and there are data from 4 wells, namely the ZR-1, ZR-2, ZR-3, and ZR-4 well. This well log data will assist in identifying hydrocarbon zones and lithological characters in the study area. In addition, there is check-shot data in the form of depth with a travel time of Two-Way Time (TWT). In addition, there is also marker data that contains information on the depth of the layers of a rock. In data processing, there are several processes carried out including qualitative and quantitative interpretation, well seismic tie, horizon interpretation, fault interpretation, time structure map creation, time structure map conversion to depth, isopach map creation, initial modeling (initial model), pre-analysis. inversion, AI distribution map, and multi attribute analysis to get seismic volume.

2.2. Geostatistical inversion

Russell (1988) [3] divides the inversion seismic method into two groups, namely pre-stack inversion and post-stack inversion. This research will discuss post-stack inversion related to amplitude inversion, in which this inversion consists of several algorithms such as bandlimited (recursive) inversion, model-based inversion, and sparse-spike inversion.

We tried to increase the conventional inversion seismic constraints such as predicting log AI properties. Attributes in seismic data have a statistical relationship with well log data such as porosity logs [8-11]. So that multi-attribute seismic analysis is also called a term that includes all geostatistical methods that use more than one attribute to predict several physical properties of the earth. Bivariate geo-statistics is clearly the simplest subset of multivariate techniques and standard co-kriging techniques, otherwise known as multivariate geo-statistics. However, in this study we will use a geostatistical method that uses more than two variables. Even so, there are many different methods included in this heading. There are three main subcategories [7]:

a) The expansion of co-kriging includes more than one secondary attribute to predict the main parameter.
b) A method that uses a covariance matrix to predict the parameters of the linear weighted number of input attributes.
c) A method that uses artificial neural networks or nonlinear optimization techniques to combine attributes into the desired parameter estimation.

In this study, the multi-attribute method will be used as in the second point.

3. Results and discussion

3.1. Reservoir zone potential

In the ZR-1 well, the top reservoir layer is at a depth of 4188 ft, and the base reservoir is at a depth of 4425 ft, so that the reservoir thickness indicated to contain hydrocarbons in the ZR-1 well is about 237 ft or about 72 m. Based on the resistivity value in the LLD log, the very high resistivity value in the reservoir layer, with a value of more than 80 ohm-m, indicates that the type of hydrocarbon fluid in the ZR-1 well is a gas.

In the ZR-2 well, the top reservoir layer is at a depth of 4225 ft, while the base reservoir is at a depth of 4540 ft, so that the reservoir thickness in the ZR-2 well is about 315 ft or about 96 m. The type of hydrocarbon fluid in the ZR-2 well is also indicated as gas, because of its high resistivity value, which is more than 80 ohm-m. In the ZR-3 well, the top reservoir layer is at a depth of 3978 ft, while the base reservoir is at a depth of 4258 ft, so that the reservoir thickness in the ZR-2 well is about 280 ft or about 85 meters. The resistance value of the reservoir layer in the ZR-3 well is very high, more than 80 ohm-m, and some layers even reach 2000 ohm-m. This high resistivity value indicates that the type of hydrocarbon in the reservoir in the ZR-3 well is gas. In the ZR-4 well, no reservoir layers were found. Based on the Gamma-Ray (*GR*) log, there are only a few thin layers that have good permeability which are indicated by low *GR* values, namely at the depths of 4473-4486 ft, 4840-4852 ft, 5016-5034 ft and 5168-5199 ft. Although there are several permeable layers, in the ZR-4 well, no cross-section (separation) was found between the RHOB and NPHI logs. Based on the resistivity value in the LLD log, there was no layer that had a high contrast resistance value in a layer that had good permeability, so it can be concluded that in the ZR-4 well there was no prospect layer containing hydrocarbons.

3.2. Sensitivity analysis

Sensitivity analysis is used to obtain the lithological distribution and characteristics of the reservoir or zone of interest. The lithological separation is based on the results of the cross-plot data between p-impedance, gamma ray and porosity. The p-impedance value we get from the multiplication of the p-wave and the density. The cross-plot result of this sensitivity analysis can see the separation between the interest zone which is sandstone and shale lithology. Based on the results of the cross-plot that has been carried out, the separation of sand and shale is considered sensitive because it is able to separate the boundary between sand and shale in the target zone.

From the cross plot between acoustic impedance logs, neutron porosity and gamma ray, it can be seen that high acoustic impedance values are associated with relatively low porosity values, because the lithology of this reservoir is tight sand. In the four wells, the results of the cross-plot analysis were divided into two zones, namely tight sand and shale. The tight sand zone is characterized by high impedance values, low gamma rays, and relatively low porosity. As well as the shale zone has a low impedance value, high gamma ray, and relatively high porosity. The results of this analysis can make it easier to determine the character and reservoir model of the research area. From the results of this cross plot, it is found that the acoustic impedance limit for tight sand in the ZR-1 well and the ZR-2 well is greater than 30,000 ((ft/s)*(g/cc)), while the ZR-3 well and ZR-4 well are more than 27,500 ((ft/s)*(g/cc)).

3.3. Initial model

Initial modeling is performed as a basic model for seismic inversion. This initial model determines whether the inversion result is good or not. The initial model will be used as a control in inversion. If the model results are good, then inversion can be continued. However, if the model does not match the subsurface model, then the model is made again by changing the inversion parameters, such as wavelet, well tie and picking horizon. Based on the results of the model obtained, the reservoir zone has an acoustic impedance value of 15,000 - 35,000 ((ft/s)*(g/cc)).

Table 1 shows the correlation values of the inversion results of the wells ZR-1, ZR-2, ZR-3 and ZR-4. This analysis is carried out to see whether the results of our initial model are correct or not. If the error is small and the correlation between the acoustic impedance value of the log, the initial model and the inversion result are close to 1 then the inversion can be performed. However, if the error is large and the correlation is small, then we have to create a new initial model by updating the wavelet, well tie or picking horizon results.

Table 1. The Correlation Value of ZR-1, ZR-2, ZR-3 and ZR-4 wells inversion results

No	Well	Correlation
1	ZR-1	0.981
2	ZR-2	0.984
3	ZR-3	0.995
4	ZR-4	0.975

The results of the inversion analysis using the model-based method. This analysis was carried out several times until a parameter that was deemed good enough to be applied to inversion was obtained. Analysis can be said to be good if the correlation value is high and the error rate is low. Based on the results of the inversion analysis, the correlation value between the acoustic impedance values of the wells, models and inversions is very good for the four wells, namely the average correlation is 0.983. This value is declared good so that inversion can be done.

3.4. Inversion result analysis

After the initial model has been formed and the error is small, the AI inversion is then carried out according to the inversion AI model that has been made. Acoustic impedance inversion is done by using model-based method after obtaining the best initial model. Inversion is carried out at the boundary of the top reservoir horizon zone to the base reservoir.

Based on Figure 1, the inversion results produce a cross section of the acoustic impedance model and are able to distinguish acoustic impedance zones based on the color and scale of the acoustic impedance values. The zone with the lowest impedance is shown in green and the zone with the highest impedance is shown in purple.

To see the suitability of the inversion AI value with the AI value from the log, a comparison is made on the cross section of the AI inversion result from the log, which can be seen in Figure 1. Based on this comparison, it shows that there is a match between the inversion AI value and the log AI value. For the four wells used, when the AI value from the log shows a low value which is marked in yellow to red in the cross section of the inversion results it shows a low value, as well as a high AI value, so it can be concluded that the inversion results made are very low.

In general, the results of Acoustic Impedance inversion show low values in the target zone, namely the top reservoir and the base reservoir. Where in the targeted sandstone reservoirs in the ZR-1, ZR-2 and ZR-3 wells, the acoustic impedance is in the range of values 15,000 ((ft/s)*(g/cc)) - 30,000 ((ft/s)*(g/cc)) indicated by yellow to red. The low AI value in the reservoir zone is in accordance with the previous sensitivity analysis, that the AI value which is less than 30,000 ((ft/s)*(g/cc)) is a zone containing gas, while the ZR-4 well has a high acoustic impedance value which is more than 30,000 ((ft/s)*(g/cc)). The high AI value in the ZR-4 well is in accordance with the previous petrophysical analysis, that this well does not contain hydrocarbons, so the AI value is relatively high because the lithology in the research area is tight sand.

After inversion is carried out, slicing is carried out in the reservoir zone to see the distribution of AI. The AI slice is carried out at a time of 0 + 50 millisecond below the horizontal top reservoir as shown in Figure 1. It can be seen that the AI distribution value in the area ranges from 8000-30000 ((ft/s)*(g/cc)). However, for the productive reservoir areas, namely in the ZR-1, ZR-2, and ZR-3 wells, there is a low AI value, namely 8000-20000 ((ft/s)*(g/cc)), while for the area in the ZR-4 well has an AI value higher than the productive area, because the well does not contain hydrocarbons.

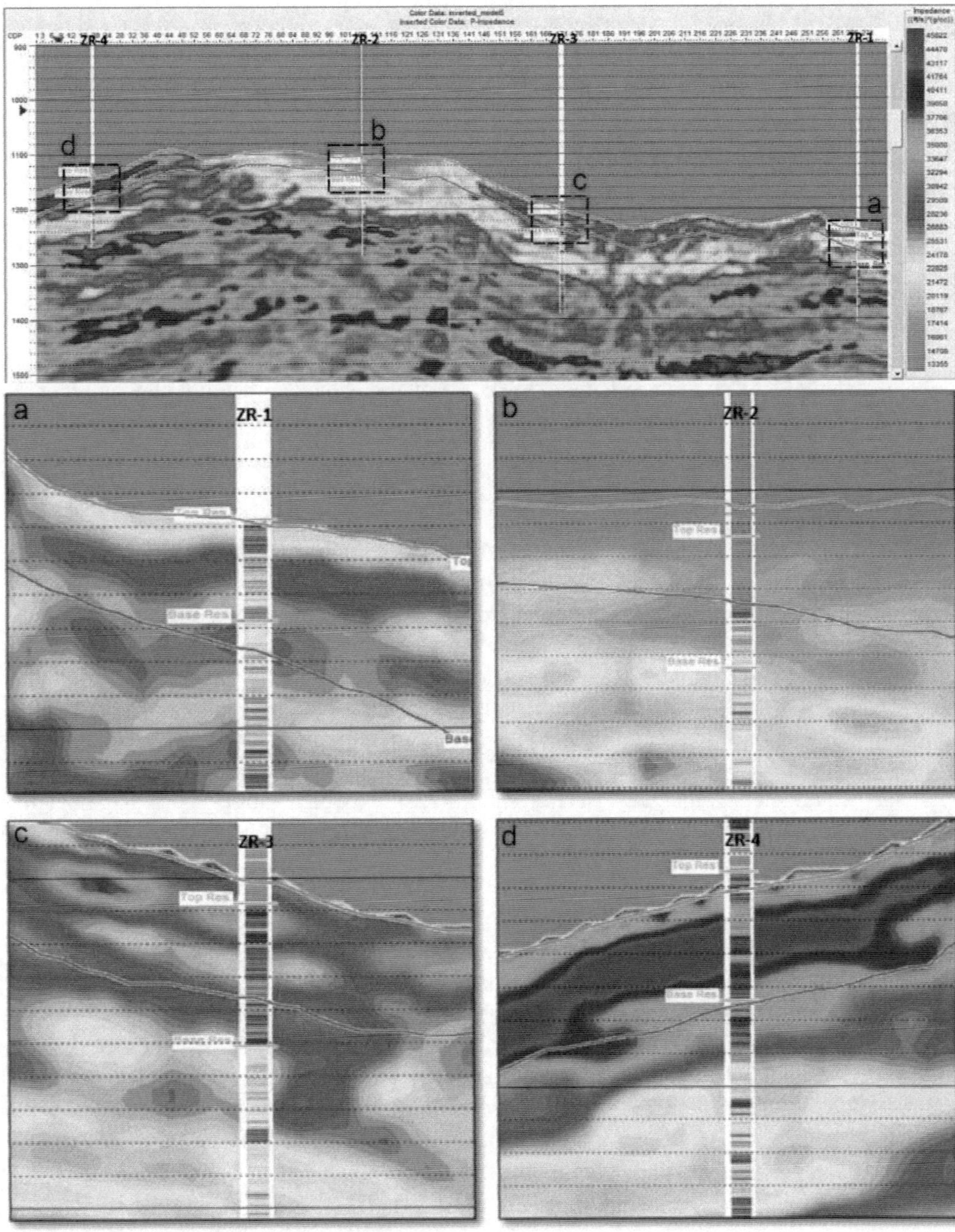

Figure 1. The results of the inversion on the seismic section using a model-based method with details on each well (ZR-1 to ZR-4).

3.5. Properties distribution map

The map of the distribution of porosity is obtained from the equation of the relationship between density and effective porosity in petrophysical analysis. Based on the map of the distribution of porosity in the figure, it can be seen that the value of the porosity in the prospect zone is high, which is around 12-24%.

The map of the distribution of porosity is obtained from the equation of the relationship between density and effective porosity in petrophysical analysis. The effective porosity and density values of the petrophysical analysis were plotted to obtain the equation of the relationship between porosity and density. The equation of the relationship between density and porosity obtained in the study area is as shown in Figure 2. From the relationship graph above, it is obtained the equation y = -0.392x + 1.054 and it can be seen that the correlation value is 0.905, this indicates a strong correlation between porosity and density. After obtaining this equation, a porosity distribution map has made. Based on the porosity distribution map in the figure high value prospect zone, which is around 12-24%.

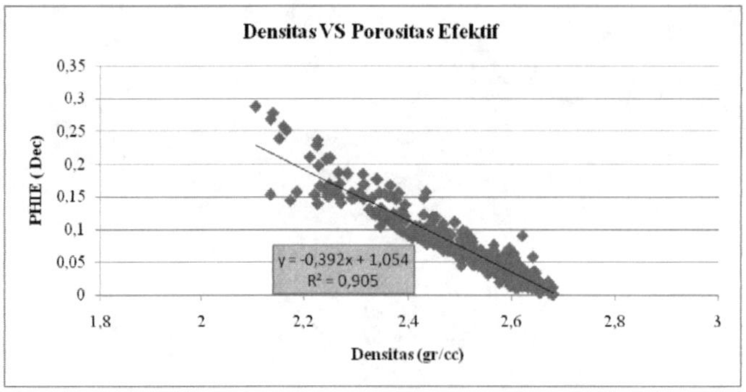

Figure 2. Density and porosity relationship.

The water saturation (*Sw*) value can be obtained from the division between the constant value and the effective porosity. This constant value is obtained from the division between water saturation and effective porosity obtained from log processing. Then the average value of the constant is taken. To get the water saturation value that will be distributed on the map, this constant value is divided by the porosity value resulting from the decrease in AI inversion. Based on the calculation results, the constant value is 0.017. Based on the *Sw* distribution map above, it can be seen that in the prospect area marked with ZR-1, ZR-2 and ZR-3 wells, they have a small *Sw* value, which ranges from 8-13%. This small value indicates that the area is very prospective for containing hydrocarbons, because it has a low water content.

4. Conclusions

The results of the petrophysical analysis show that the ZR-1, ZR-2 and ZR-3 wells have a target zone containing gas hydrocarbons, while the ZR-4 well does not have a prospect zone. The area around the wells ZR-1, ZR-2 and ZR-3 is a prospect area containing gaseous hydrocarbons, which is characterized by the distribution of AI (acoustic impedance) of 15,000 (ft/s)*(g/cc) - 30,000 (ft/s)*(g/cc), the distribution of porosity value is 12-24%, and the distribution of *Sw* is 8-13%.

References

[1] Riel P van 2000 The past, present, and future of quantitative reservoir characterization *Lead. Edge* **19** 878–81
[2] Wibowo R C and Mulyatno B S 2012 Karakterisasi Reservoar Menggunakan Metode Inversi Impedansi Akustik dan Neural Network Pada Lapangan " ICL " Cekungan Sumatera Selatan *PROCEEDINGS PIT HAGI 2012* (Palembang: HAGI) pp 2–4
[3] Russell B H 1988 *Introduction to Seismic Inversion Methods* (Tulsa: Society of Exploration Geophysicists)
[4] Wibowo R C, Sarkowi M, Mulyatno B S, Dewanto O, Zaenudin A, Amijaya D H and Aspari A A 2020 Thinned coal distribution modeling based on integrated geological and geophysical data : Case study CBM resources in Central Palembang Sub-Basin *2nd International Conference on Earth Science, Mineral, and Energy* vol 070011 (Yogyakarta: AIP Publishing) pp 1–9

[5] Hampson D P, Schuelke J S and Quirein J A 2001 Use of multiattribute transforms to predict log properties from seismic data *Geophysics* **66** 220–36

[6] Setiawan M H F, Asy'ari M R, Wibowo R C, Amijaya D H and Aspari A A 2015 Parasequence Concepts, Problems And Solutions in CBM Exploration Using Seismic Data Case Study: Muara Enim Formation, South Sumatra Basin *INDONESIAN PETROLEUM ASSOCIATION Thirty-Ninth Annual Convention & Exhibition* (Jakarta: INDONESIAN PETROLEUM ASSOCIATION) p 10

[7] Russel B, Hampson D, Schuelke J and Quirein J 1997 Multiattribute seismic analysis *Lead. Edge* 5

[8] Hampson D P, Schuelke J S and Quirein J A 2001 Use of multiattribute transforms to predict log properties from seismic data *GEOPHYSICS* **66** 220–36

[9] Dinata O, Mulyanto B S, Ramadian R and R D A 2020 Prediksi Porositas Menggunakan Metode Neural Network Pada Lapangan Ozza, Cekungan Sumatra Tengah *J. Geofis. Eksplor.* **6** 77–86

[10] Pratama E and Mulyatno B S 2020 Analisis Reservoar Migas (Sandstone) Menggunakan Multiatribut Seismik Pada Lapangan Tg12, Cekungan Barito, Kalimantan Selatan *J. Geofis. Eksplor.* **5** 3–14

[11] Ramdhani E, Dewanto O, Karyanto K and Yulianto N 2020 Perhitungan Cadangan Hidrokarbon Formasi Talang Akar Menggunakan Analisis Petrofisika Dan Seismik Inversi AI Dengan Pendekatan Map Algebra Pada Lapangan Bisma, Cekungan Sumatera Selatan *J. Geofis. Eksplor.* **4** 3–14

The effect analysis of wind speed variation to the horizontal axis wind turbine design with Q-blade

A S Samosir[1,*], **and A Riszal**[2]

[1] Department of Electrical Engineering, Universitas Lampung, Jalan Prof. Soemantri Brojonegoro No.1. Bandar Lampung 35145, Indonesia.

[2] Department of Mechanical Engineering, Universitas Lampung, Jalan Prof. Soemantri Brojonegoro No.1. Bandar Lampung 35145, Indonesia.

*Email: saudi.ahmad@gmail.com

Abstract. Wind energy is renewable energy that environmentally friendly. One of the utilizations of wind energy is conversing the wind turbine. The wind turbine is an energy conversion device to convert wind energy into mechanical energy and convert it into electricity through the generator. The size of the power produced by wind turbine depends on the wind speed and the type of turbine that is used. In this research, the researchers design the wind turbine with wind speed variations. The purpose of the designing and the analysing the wind speed is to design a low wind turbine using number of blade 5 model. The objective of this study was to design a wind turbine with variations in wind speed of 2 m/s, 3 m/s, 4 m/s, 5 m/s and 6 m/s using number of blade 5 model wind turbine horizontal axis, rotor diameter of 1.4 m and this rotor used NACA 4412 air foil type. The method that is used in this study is Blade Element Momentum (BEM) and Q-Blade software for designing the simulation. The results of this study, the researchers found there is an increase in power as the wind speed increases.

Keywords: wind turbine, Q-blade software, Blade Element Momentum (BEM)

1. Introduction

Energy consumption using fossil energy sources is increasing every year in the world. Therefore, the environment pollution is huge due to the used of fossil energy. On the other hand, the renewable energy is quite potential in Indonesia. The use of renewable energy has an important role in reducing the use of fossil energy and it may help Indonesia to avoid energy crisis. One of the most potentials in Indonesia is wind energy. Renewable energy in Indonesia can be maximumly utilized for energy needs to support sustainable development and human needs in the energy sector [1]. The Utilization of wind energy conversion can be used by using a horizontal axis wind turbine. Unfortunately, in real conditions not all regions in Indonesia have the ideal wind potential. In this planning, a wind turbine design with low wind potential will be made, therefore, the design uses a number of blades of 5 to increase the output power. The researcher may also vary several speeds in order to predict the output power according to the real wind speed.

IOP Conf. Series: Materials Science and Engineering 1173 (2021) 012009 doi:10.1088/1757-899X/1173/1/012009

This research is using Q-blade software. Numerical analysis research with Q-blade software and using the BEM method, airfoil DU86-084, E387, SD2030 and SG6041, were selected for the design of wind turbine rotor blades with airfoil type E387 in the tip speed ratio (TSR) design with the highest power coefficient [2]. BEM theory is used to design and analyse the 300 watt wind turbine with design procedures codified in MATLAB to simplify routine. Airfoil used by air foil SG6040 and SG6041 was selected as hub and tip airfoil assumption a wind speed of 5m/s. The analysis showed a maximum annual energy production (AEP) of 538 kWh and a maximum torque of 15 Nm [3]. The design of a three-blade wind turbine using the 4418-airfoil type using Q-blade software using the BEM method, as a stage in the future. More aspects are also needed to be considered such as CFD analysis and stress analysis of the material. The performance analysis shows that the design of wind turbine blades can obtain a relatively high-power coefficient of 0.45 [4]. In this research, four airfoils NACA 4412, SG6043, SD7062 and S833 have been selected and investigated in Q-Blade. The results showed that the CP power coefficient of NACA 4412 on the tip speed ratio was superior to the other three airfoil [5].

2. Methods

This research was using HAWT (Horizontal Axis Wind Turbine) model with blade 5 number. The blade design uses Airfoil NACA 4412 with a rotor diameter of 1.4 m. The researcher was varying the wind speed 2-6 m/s and variations of Rpm 50, 100, 150, 200, 250 and 300 using Q-blade software and the BEM (Blade Element Momentum) method. BEM is used to predict wind turbine blades to increase performance with principle of calculating the aerodynamic force on the blade element.

2.1 Blade Element Momentum Method

Blade Element Theory has two methods. First, this theory uses momentum theory or disk actuator theory with a mathematical model. Second, Blade Element Theory is used to calculate the aerodynamics of the blade. The theoretical blade element is frequently used to calculate the blade elements on each airfoil [6]. This equation can be iterated as a solver. Iteration uses two variables, named axial and radial in BEM Theory. Then the variables can be defined as follows [7].

$$\alpha = \frac{1}{\frac{4FSin^2 \varphi}{(\sigma\, C_N)} + 1} \tag{1}$$

$$\alpha' = \frac{1}{\frac{4F\, Sin\, \varphi\, Cos\, \varphi}{(\sigma\, C_T)} - 1} \tag{2}$$

where is, φ is inflow angel, C_T and C_N is tangential and normal force coefficient and σ is solidity [8]. It can be expressed as:

$$\alpha = \frac{cB}{2\pi r} \tag{3}$$

c is the chord length, B is the number of blades, and r is the disk radius.
The tangential C_T and C_N is force coefficient in the wind turbine:

$$C_T = C_L\, Sin\, \varphi - C_D\, Cos\, \varphi \tag{4}$$

$$C_N = C_L\, Cos\, \varphi + C_D\, Sin\, \varphi \tag{5}$$

IOP Conf. Series: Materials Science and Engineering 1173 (2021) 012009 doi:10.1088/1757-899X/1173/1/012009

2.2 Q-blade Airfoil

The representation of Q-blade Airfoil and rotor blade design is presented on Fig. 1 dan Fig. 2 below.

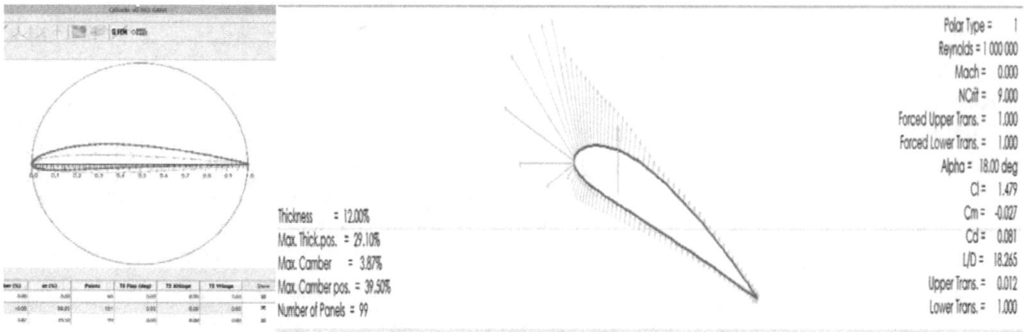

Figure 1. Airfoil pressure profile NACA 4412.

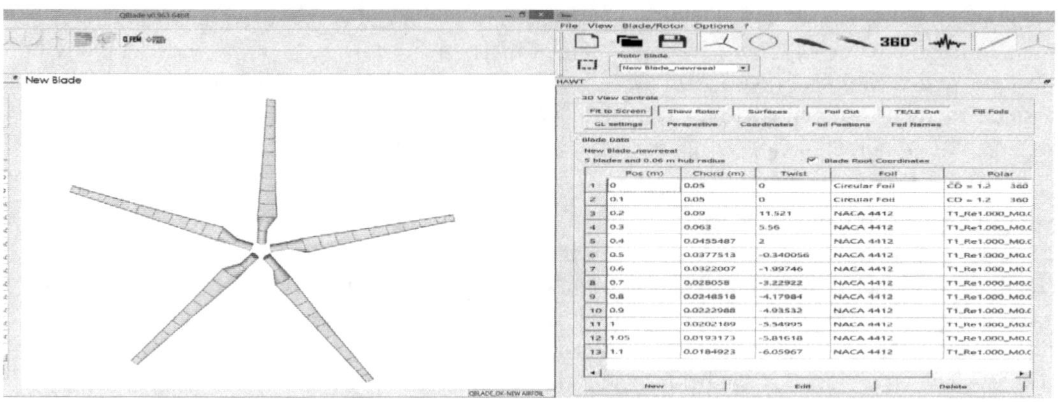

Figure 2. Rotor blade design.

3. Results and Discussions

This research discussed about the analysis of wind turbine design with five blades. This research will be carried out by varying the wind speed of the TSR (Tip Speed Ratio) to determine the wind turbine output power. In this study also carried out variations in the RPM of wind speed.

In figure 3, it shows the relationship between wind speed of 2 m/s to 6 m/s with Tip Speed Ratio, it can be seeing the maximum TSR value is at the TSR value of 7 and the maximum wind speed of 6 m/s is due to the determinants of TSR such as wind speed and rotation on the rotor, so that with the increasing turbine rotation then the resulting TSR will increase like the Figure 3.Based on all wind speed tests, the maximum TSR value at a maximum speed of 6 m/s with five blades is as big as at TSR 7 with an output power value of 250 Watt while for the lowest value at a wind speed of 2 m/s is generated, namely TSR 7 with a value of 10 Watt. Therefore, it can be seen that the variation in wind speed greatly affects the TSR value generated by the turbine.

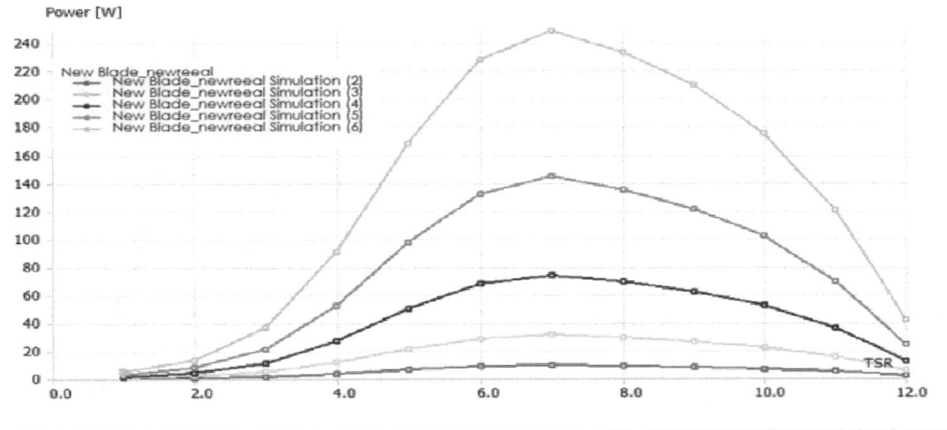

Figure 3. Effect of wind speed variations on TSR.

Figure 4 shows the effect of TSR on Power Coefficient (Cp) at a wind speed of 6 m/s. It can be seen that the overall value of the maximum power coefficient on TSR 7 is 0.45. for TSR 1 to TSR 6 indicates an increase in CP.

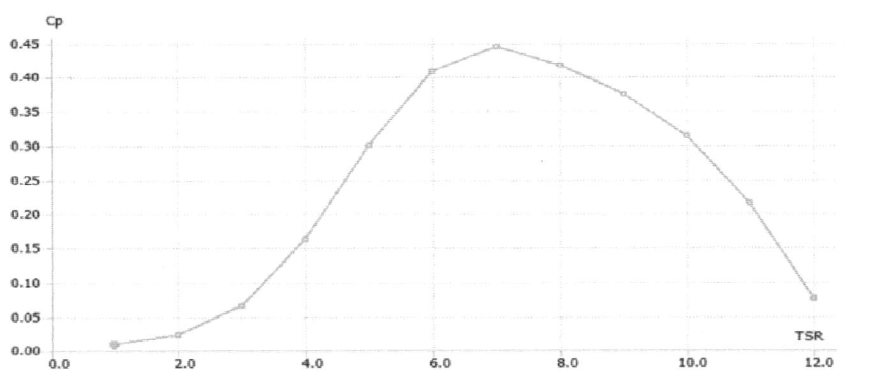

Figure 4. Effect of wind speed 6 m/s on TSR.

However, at TSR 8 to TSR 12 there is a decrease, this is because at low TSR the flow angle through the blade is in accordance with the conditions of the turbine design being made, while at high TSR, the flow angle through the blade is quite large and when the flow passes through the blade, flow spacing occurs. which causes the rotor to experience a decrease in lift force and an increase in drag force. A decrease in lift and an increase in drag forces the torque in the rotor to decrease so that the output power decreases.

Further evaluation is to examine the variation of wind speed toward variation of rotational speed which presented on Fig. 5.

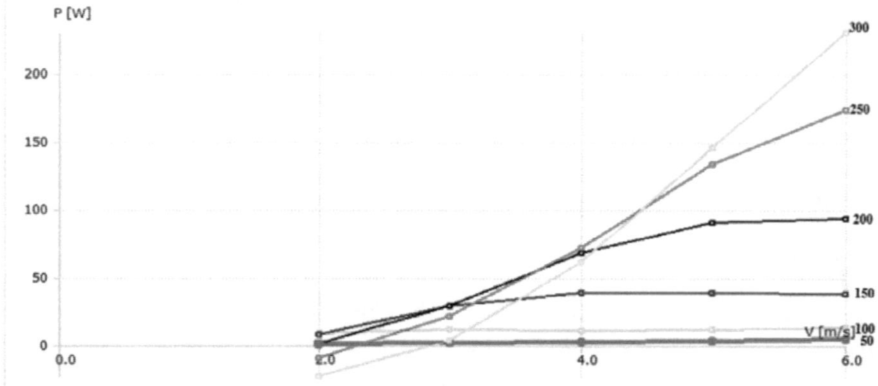

Figure 5. Variation of wind speed toward variation of rotational speed (1/min).

Figure 5 shows that the wind speed is directly proportional to the resulting rotational speed. The greater the wind speed, the greater the turbine rotation is produced, the greater the wind energy, the greater the energy that the turbine converts into increased rotation. From the graph above, it can be seen that at 300 rpm the power is 247 Watt with a maximum wind speed of 6 m/s. while for the minimum rotation at 50 rpm that occurs at wind speed 6 m/s with output power 9 Watt.

4. Conclusions

The results of five blades wind turbine simulation, based on variations in wind speed and variations in rotational speed, shows that the higher the wind speed was resulting high output power according to the design. Moreover, the power coefficient is influenced by the speed of the wind in and out. The greater the power coefficient value, the greater the wind which is converted depending on the TSR value.

References

[1] Daryanto Y, Yohanes F. A. dan F. Hasim 2005 BPPT Tangerang
[2] Priyanka R. Dhurpate and Kailashnat B. Sutar 2016 Imperial Journal of Interdiciplinary Research (IJIR) **1426** 1430
[3] Naveen Prakash Noronha, M. Krishna 2020 International Journal of Advance Science and Technology **8877** 8885
[4] Abdul Goffar Al Mubarok 2018 International Journal of Scientific & Engineering Research **1018** 1021
[5] Ali Said, Mazharul Islam, Mohiuddin A.K.M, Moumen Idres 2019 International Journal of Recent Technology and Engineering (IJRTE) **153** 157
[6] G. Ingram 2011 Durham University **5** 21
[7] V.Esfahanian, A. Salavati Pour, et all 2013 J.Wind.Eng.Ind.Aerodyn **35**
[8] J. F. Manwell, J. G. Mcgowan, A. L. Rogers.,2009 Wiley. **131** 677

Coal velocity and proximate analysis relationship using Multiple Linear Regression

S Erfani, F M S Siregar, A Zaenudin, Rustadi, I B S Yogi, and R C Wibowo*

Department of Geophysical Engineering, Faculty of Engineering, Universitas Lampung, Jalan Prof. Dr. Soemantri Brodjonegoro No. 1 Bandar Lampung 35145, Indonesia

*Email: rahmat.caturwibowo@eng.unila.ac.id

Abstract. Coal properties such a velocity (*Vp*) are essential to building a lateral distribution of coal seam using seismic data. The experimental determination of velocity analysis is sophisticated, long time consumed, and expensive. On the contrary, statistical approaches such as linear regression can be run rapidly. The study's two main objectives were to develop models for coal velocity using well log data variables (density and natural Gamma-Ray) and found the relationship between velocity with proximate analysis results. Multiple linear regression (MLR) methods were applied to estimate Vp's relationship between estimated and proximate analysis. By conducting cross-validation, the prediction analysis of the models has been tested by using R^2. The result showed that between *Vp* estimated versus Vp log have R^2 0.80 and *Vp* estimated versus proximate analysis that reflected have R^2 of 0.52. Correlations can estimate the relationship between *Vp* and proximate analysis, then applied that correlation to distributed in seismic volume to obtain coal seam characteristic.

Keywords: velocity, coal, multiple linear regression, proximate, prediction

1. Introduction

The use of geophysical parameters to explore coal basins and delineate coal seams and geological structures that affect coal deposits is now widely practiced. Since the oil crisis in 1973, geophysical methods have increasingly been used to identify coal deposits by estimating coal targets that have economic value [1]. Several techniques have been used in oil and gas exploration and adapted for coal exploration purposes. This study uses the geophysical method, namely geophysical logging, to determine the relationship between its characteristics and the coal seam's proximate value. Coal as a lithology responds well to the most geophysical method in that its physical properties contrast with those of other lithologies commonly found in coal-bearing sequences. In general, coal has a lower density, a lower seismic velocity, a lower magnetic susceptibility, a higher electrical resistivity, and low radioactivity compared with surrounding rocks in typical coal-bearing sequences [1]. Density measurements of rocks are not usually measured in-situ but in the laboratory on some small outcrop or drill core samples. Singh and Dubey conducted previous research (2000) [2] where the value of the P wave was used to determine the effect of minerals from sandstones where the higher the P-wave, the values of uniaxial compressive strength (UCS), Young's Modulus (*E*), and density as well. However, the value of P-wave velocity (*Vp*) is inversely proportional [3]. In the research carried out, taking one of the equations is looking for the *Vp* value to be used in estimating the relationship with proximate analyses. This research aimed to approach to predict the quality of the distribution of coal seams in a coalfield.

Geophysical logging is the measurement of the variation with depth of particular physical properties of surrounding rocks with geophysical measuring tools (probes) located in boreholes. Measurement are made by lowering a probe attached to the end of a cable to the bottom of the borehole and then raising the probe back out of the borehole at a constant rate to record the geology. In coal exploration, density logs are useful in identifying coal because coal has a lower density value than other lithologies such as shale and sandstone [4].

2. Methods

One of the methods that can be applied in predicting the distribution of coal quality is statistical approaches. This approach has been carried out by many researchers, one of them is Singh and Dubey (2000) [2], where they conducted experiments by calculating the value of the P wave, which is used to determine the effect of minerals from sandstones where the higher the P-wave, the uniaxial strength values, Compressive (UCS), Young's modulus (*E*). From some of the resulting equations, this study refers to the research that has been done by taking one of the equations, namely looking for the value of *Vp*. Where the empirical equation is given is:

$$V_p = 0.0009d^{1.0104} \quad (1)$$

The *Vp* value of the density log data is based on the well log measurement results to find the *Vp* value from the sonic log based on multiple regressions and will be correlated with the sonic log of the well log results. *Vp* values are correlated with proximate data to determine the relationship between these values and existing concepts. The empirical formula given can be used to determine the quality distribution of the wells measured (Table 1).

The regression calculations carried out sequentially is to find the correlation value between the Sonic logs calculated from the empirical formula with the velocity well log (*Vp* cal vs *Vp* log) to validate the *Vp* calculated (*Vp* cal), then correlating the Sonic logs calculated from the proximate data, namely *Vp* cal vs Ash, *Vp* cal vs Inherent Moisture, *Vp* cal vs Volatile Matter and *Vp* cal vs Fixed Carbon.

3. Results and Discussions

3.1. Calculated Vp calculated vs. Vp well log

The calculation and graph in Figure 1 obtained an excellent relationship between the two regressed variables, which indicates that the two variables have a directly proportional relationship. This means that the velocity calculation (*Vp* cal) results are as good as the well log results with a correlation of 80%.

3.2 Ash vs. Vp calculated

Ash is a residual material that remains during the coal combustion process with specific conditions where the main elements are oxides and sulfates. Based on the calculation results (Figure 2), the relationship between Ash vs. *Vp* is inversely proportional, which means it explains that the higher the Ash content, the lower the velocity value and vice versa. This is not following the concept. The higher the value of ash content, the higher the *Vp* response. An example is lignite coal, which has a high ash content value tends to have a high sonic response compared to the *Vp* value of anthracite coal.

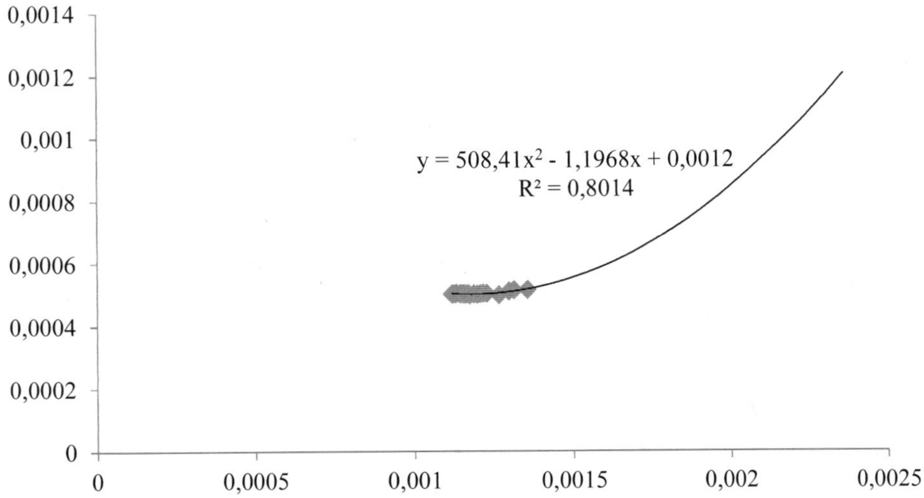

Figure 1. Relation of *Vp* calculated vs. *Vp* well log.

Table 1. Coal data (velocity, density, and proximate analysis) were used in this study

Sample Number	Density core (g/cm^3)	Density (g/cm^3)	*Vp* (m/s)	*Vp* (m/s)	Ash (%)	IM (%)	VM (%)	FC (%)
1	1.2	1.30	0.00050	0.0012	2.6	14.4	39.4	45.0
2	1.2	1.40	0.00050	0.0013	2.6	14.8	39.8	46.6
3	1.17	1.29	0.00050	0.0012	1.3	13.8	39.4	42.4
4	1.23	1.33	0.00050	0.0012	5.0	13.4	41.9	43.8
5	1.25	1.28	0.00050	0.0012	9.8	12.3	40.6	36.9
6	1.18	1.27	0.00050	0.0011	2.9	12.6	40	46.9
7	1.18	1.25	0.00050	0.0011	3.2	13.0	42.6	43.9
8	1.21	1.33	0.00050	0.0012	3.8	11.8	41.2	43.5
9	1.21	1.31	0.00050	0.0012	3.4	10.2	38.8	32.9
10	1.26	1.30	0.00050	0.0012	6.6	11.6	39.8	43.3
11	1.24	1.29	0.00050	0.0012	6.6	7.7	41.1	41.5
12	1.29	1.28	0.00050	0.0012	14.3	8.2	43.6	41.3
13	1.21	1.27	0.00050	0.0011	2.3	10.8	43.6	38.0
14	1.27	1.24	0.00050	0.0011	2.1	11.4	40.7	37.1
15	1.24	1.24	0.00050	0.0011	3.8	11.5	41.6	42.8
16	1.3	1.25	0.00050	0.0011	9.3	9.0	42	41.8
17	1.28	1.50	0.00052	0.0014	7.7	15.48	39.2	33.93
18	1.32	1.36	0.00050	0.0012	2.9	14.42	28.36	39.8
19	1.46	1.45	0.00051	0.0013	1.4	15.98	39.04	37.26
20	1.28	1.43	0.00051	0.0013	3.4	16.34	39.18	39.0
21	1.3	1.35	0.00050	0.0012	4.6	16.4	38.74	41.43

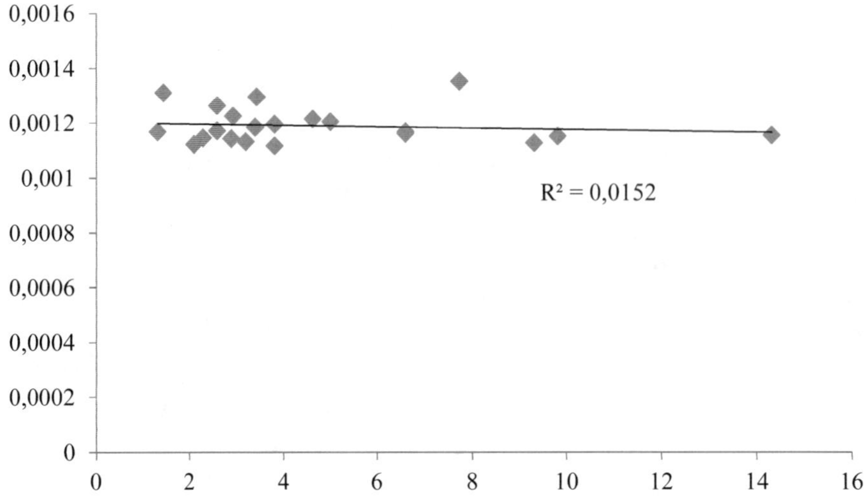

Figure 2. Relation of Ash vs. *Vp* calculated.

3.3. Inherent Moisture vs. Vp calculated

Inherent Moisture (IM) is the amount of water contained in coal microspores. In other words, the value of IM is not affected by the amount of external water, such as formation water. IM values generally get lower as coal ranks rise. The IM value should directly proportional to the velocity value, namely, the higher the IM value, the higher the velocity value. This correlation is shown in Figure 3.

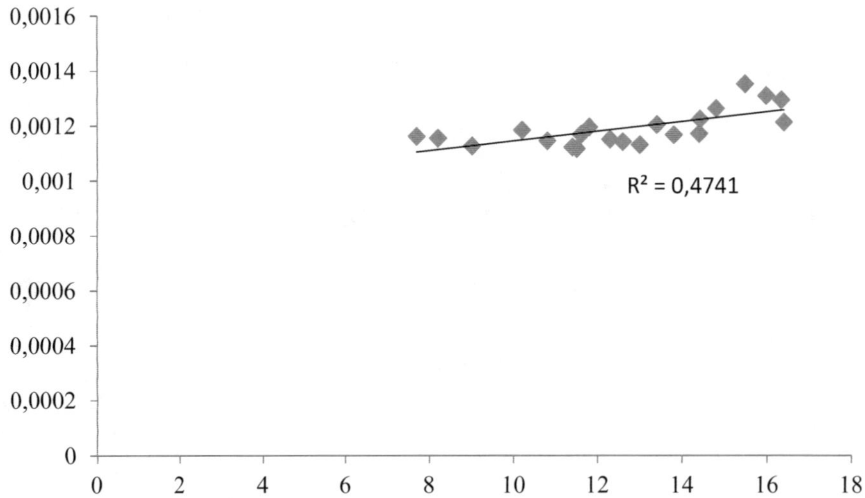

Figure 3. Relation of Inherent Moisture vs. *Vp* calculated.

3.4 Volatile Matter vs. Vp calculated

As determined by standard test methods (i.e., ASTM D-3175; ISO 562), the volatile matter is the percentage of volatile product/material, i.e., water vapor that is released during the coal heating process from the graphic showed in Figure 4. The higher the Volatile value, the higher the product/material contained in a coal sample. Velocity is directly proportional to the Volatile Matter value. The higher the velocity response, the higher the VM value should be.

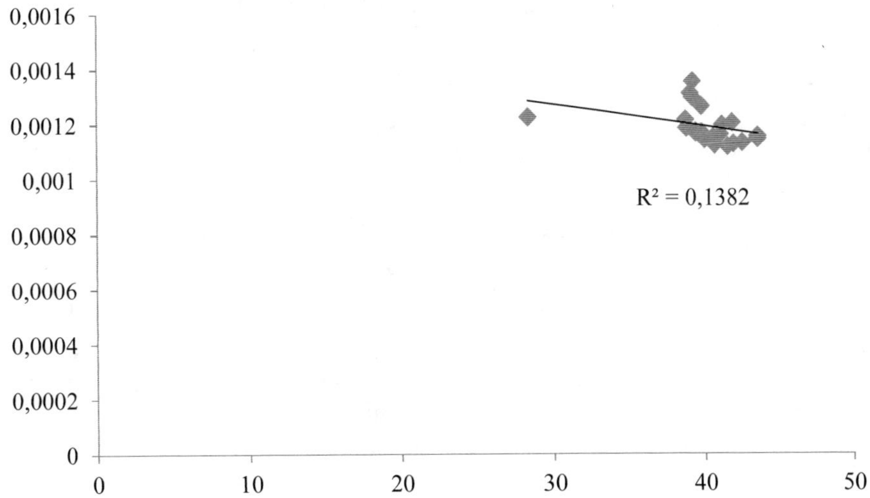

Figure 4. Relation of Volatile Matter vs. *Vp* calculated.

3.5 *Fixed Carbon vs. Vp calculated*

Fixed carbon is the remaining material after determining the test results of moisture, volatile matter, and ash [4]. From the graph in Figure 5, we got the higher the coal rank, the higher the Fixed Carbon value, and the lower velocity log response. The results of the calculations show conformity to the concept where the correlation is inversely proportional.

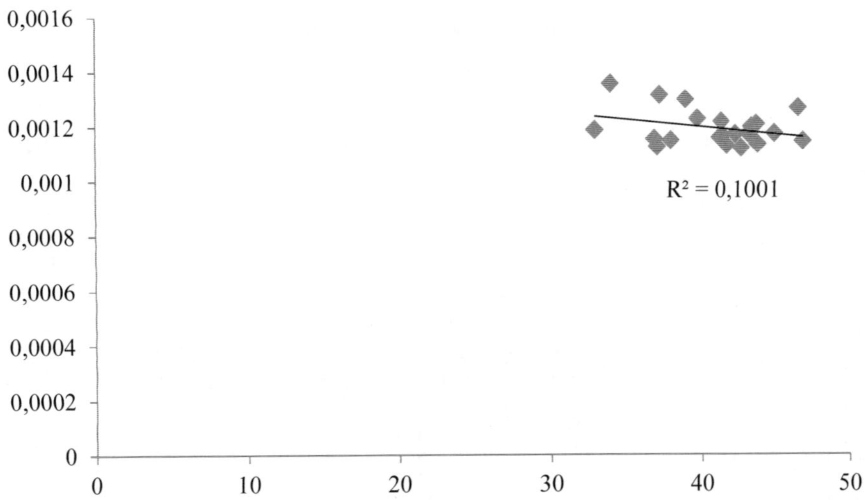

Figure 5. Relation of Fixed Carbon vs. *Vp* calculated.

The multiple regression analysis results with the sonic value input data as the dependent variable (y) with proximate values such as ash, inherent moisture, volatile matter, and fixed carbon as the independent variable (*x*) as shown in Table 2.

Table 2. Multiple linear regression analysis results between velocity (*Vp* calculated) and proximate analysis.

	Coefficients (10^{-5})	Standard Error (10^{-5})	P-value	R^2
Intercept	112.590	27.507	0.001	
X 1	1.049	0.498	0.051	
X 2	2.307	0.668	0.003	0.520
X 3	-0.126	0.518	0.811	
X 4	-0.541	0.373	0.166	

The correlation degree R2 is 0.523 or 52.3%, where it could be said that the relationship between sonic responses to proximate values has a strong enough relationship and influences each other. The standard error of 0.00000062 indicates that the average data observed is excellent. It is close to the regression line. The P-value obtained from the analysis shows that the sonic variable (intercept = 0.00) and (x2 = 0.00) shows a statistically significant variable. This variable is the content of Inherent Moisture (IM) statistically has a significant relationship with sonic value. Meanwhile, the proximate variables ash content, volatile matter and fixed carbon have a p-value of x1 = 0.005, x3 = 0.81 and x4 = 0.17 indicates these variables are not statistically significant.

4. Conclusions

Based on calculation results and multiple regression analysis, the empirical formula is sufficiently feasible to apply to coal seams. There are two correlation calculations following the concept of coal, namely the correlation between the *Vp* calculated from the empirical formula with the *Vp* log with a correlation degree of 0.80 indicates that *Vp* log computation with *Vp* log has almost no difference if applied especially if there is no log *Vp* data available, we can approach to calculate *Vp* by ourselves. The multiple linear regression analysis correlations are 0.52 (50%) that multiple regression analysis can be used as an alternative to log density in interpreting coal quality. It is hoped that the relationship between this approach's results can be applied to other cases with many data to estimate the distribution of coal quality. So it can be concluded that the log velocity value can be used to distribute quality properties in the coal seam.

Acknowledgements

We want to thank the Engineering Faculty, the University of Lampung, for funding this research activity through the DIPA Faculty of Engineering program.

References

[1] Thomas L 2013 *Coal Geology* (Oxford: John Wiley & Sons, Ltd Registered)

[2] Singh T N and Dubey R K 2000 Study of transmission velocity of primary wave (P-wave) in coal measure sandstone *J. Sci. Ind. Res. (India).* **59** 482–6

[3] Hutami H Y, Fitriyani F, Priniarti T L and Handoyo H 2020 Analisis Model Kecepatan Gelombang-P pada Coal-Seam Gas Studi Kasus Cekungan Sumatera Selatan, Indonesia *J. Geofis. Eksplor.* **6** 113–20

[4] Wibowo R C, Sarkowi M, Mulyatno B S, Dewanto O, Zaenudin A, Amijaya D H and Aspari A A 2020 Thinned coal distribution modeling based on integrated geological and geophysical data : Case study CBM resources in Central Palembang Sub-Basin *2nd International Conference on Earth Science, Mineral, and Energy* vol 070011 (Yogyakarta: AIP Publishing) pp 1–9

Investigation of conical draft tube of Kaplan turbine with spiral liner and without spiral liner using CFD simulation

Sirojuddin[*], I Ekayana, B C H Nugroho, H R Aziz, A Kholil, and S Harahap

Department of Mechanical Engineering, Faculty of Engineering, State University of Jakarta, Jalan Rawamangun Muka, Jakarta 13220, Indonesia

*Email: sirojuddin@unj.ac.id

Abstract. The draft tube is a component in the Kaplan turbine to create positive suction pressure using kinetic energy that is not used when leaving the turbine runner and increasing turbine power efficiency. This research aimed to find out the highest efficiency of variant geometries of a conical draft tube with spiral liner and without the spiral liner. The design geometry was processed using 2D software and 3D using Inventor and simulated in CFD software. The inlet water discharge was 0.1 m^3s^{-1} with head 5 m and theoretical turbine power 4.9 kW. Based on the simulation results showed that the spiral liner draft tube variant was the best power and efficiency.

Keywords: draft tube, efficiency, Kaplan turbine, kinetic energy, spiral liner

1. Introduction

Kaplan turbines are known as axial flow reaction turbines that use large amounts of water and low head [1]. The draft tube is one of the significant components in the reaction turbine. Without this, the pressure at the outlet can drop due to a lack of water, which can affect plant efficiency and may even fail to produce the desired power [2]. Investigation of the effect of the inlet nozzle shape on the performance and internal flow of cross-flow turbines for hydropower by modifying the diffuse length and angle using CFD shows the results that the narrow and convergent shape of the inlet nozzle gives a better effect on turbine performance [3]. Another simulation of the conical draft tube with different lengths and diffuser angles by calculating the head loss, head recovery coefficient using CFD and ANSYS CFX 13.0.

The results showed that the maximum efficiency was achieved at a diffuser angle of 5 degrees with L / D3 = 19 [4]. Design Optimization of the elbow draft tube model with multiple diffuser angles for the same or increased diffuser angle to resolve the problem of draft tube design and reduce prototyping costs using CFD and analysis in ANSYS 14.0 [5]. In this paper, the CFD analysis of the Mixed Elbow draft has been performed, and results for the same are compared with experimental reading. From results were found within limits and to compare the results of experiments and CFD with a simple elbow draft tube. [6]. The efficiency and losses of draft tube are calculated from the pressure and velocity distribution to determine the effect of geometric parameters on the draft tube performance to obtain the best results. The best performance is achieved with a height ratio close to 2.24 and a length ratio L / D1 of 6.0 [7].

In another study investigated the hydraulic performance of the Francis turbine at a 15 MW hydroelectric power plant. The design parameters used were the thickness of the guide vane, three types of draft tubes, and the characteristics of the cord using three-dimensional numerical methods with SST turbulent models. The vortex and flow field results are depicted using different design parameters and operating conditions [8]. An experimental study of draft tubes in crossflow turbines found that the correct shape and size of the draft tubes can have a positive effect on improving turbine performance at each turbine head. The result showed that the 300 mm draft tube variant has the highest output [9]. In the present work, there were three types of conical draft tubes with a length of 22.45 m, 30.4 m and 38.4m and were modelled using a diffuser angle of 4°. The results showed that the draft tube with a length of 30.4 m is the most suitable where the efficiency is 22.7% [10].

The objective of this study was to investigate draft tube with spiral liner and without spiral liner against the efficiency of power generated by the Kaplan turbine.

2. Materials and Methods

This study consisted of 3 variations of the draft tube with the same height and diffuse angle but the number of different spiral liner, namely DT-1 without spiral liner, DT-2 with 6 spiral liner, and DT-3 with 12 spiral liners. The parameters and dimensions of the turbine used are 0.1 $m^3.s^{-1}$ discharge and 5 m head and can see in Table 1. The 2D geometry using AutoCAD, 3D model using Inventor software, and water flow using flow simulation of CFD software. The boundary conditions can see in Table 2. The results of the pressure and velocity can see in Figure 1 and Figure 2. Theoretical power is calculated by the formula:

$$P_{th} = \dot{m}.g.H = \rho.g.Q.H \quad (1)$$

Where Pth is theoretical power (Watt), Q is water discharge (m^3s^{-1}), ρ is density of water ($kg.m^{-3}$) and H is head (m).

Table 1. Boundary condition

Item	Unit
Operation pressure	101325 Pa
Gravitation	9.81 m/s^2(Y - Axis)
System unit	SI
Type of analysis	Internal
Cavitation	No
Reference axis	Y
Fluid	Water (O)
Type of flow	Turbulence and laminer
Roughness	0.01
Wall thermal condition	Adiabatic wall
Temperature	293.2 K
Turbulence model	2%
Turbulence length	0.003 m
Turbulence energy	1 J/kg
Turbulance dissipation	1 W/kg
Rotation	Assumption stop for a moment

Turbine power (P_d) is obtained using the horizontal drag force approach (F_{dh}) multiplied by tangential velocity (U_1). Axial or vertical drag force (F_{dv}) values are obtained from CFD analysis on the runner blade. Drag force and power are calculated by the following formula:

IOP Conf. Series: Materials Science and Engineering 1173 (2021) 012011 doi:10.1088/1757-899X/1173/1/012011

$$F_{dh} = \frac{F_{dv}}{\tan\theta} \tag{2}$$

$$P_d = F_{dh}.U_1 \tag{3}$$

The calculation with drag force approach formulation on the blade surface is done because the calculation by velocity triangle cannot predict the force due to blade shape geometry. The result of efficiency can be seen in Table 3.

$$\eta_h = \frac{P_d}{P_{th}} x100\% \tag{4}$$

$$\eta_t = \eta_h.\eta_m \tag{5}$$

Where the parameter η_h is hydraulic efficiency, η_m is mechanical efficiency = 95%, and η_t is turbine efficiency.

3. Results and discussion

From the CFD simulation, are obtained the velocity and pressure contour as shown in figure 1 and figure 2, while the flow analysis to determine the power generated and efficiency in Table 2 and Table 3.

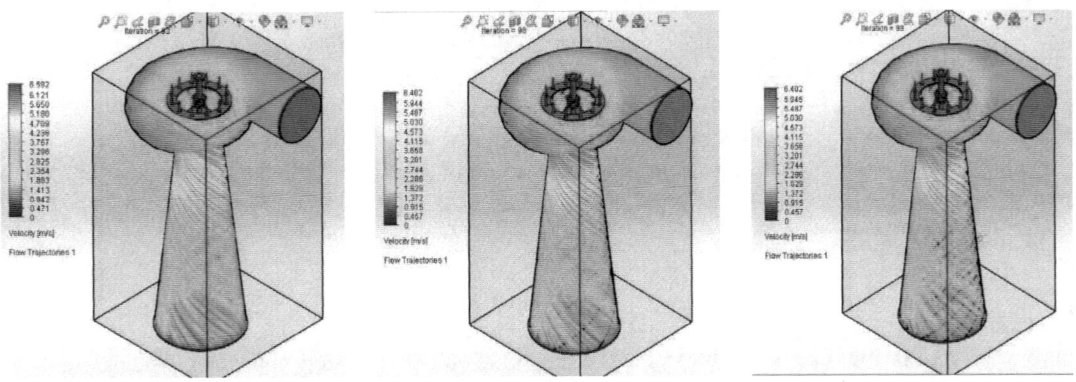

(a) (b) (c)

Figure 1. Velocity contour of draft tube (a) DT-1, (b) DT-2, (c) DT-3

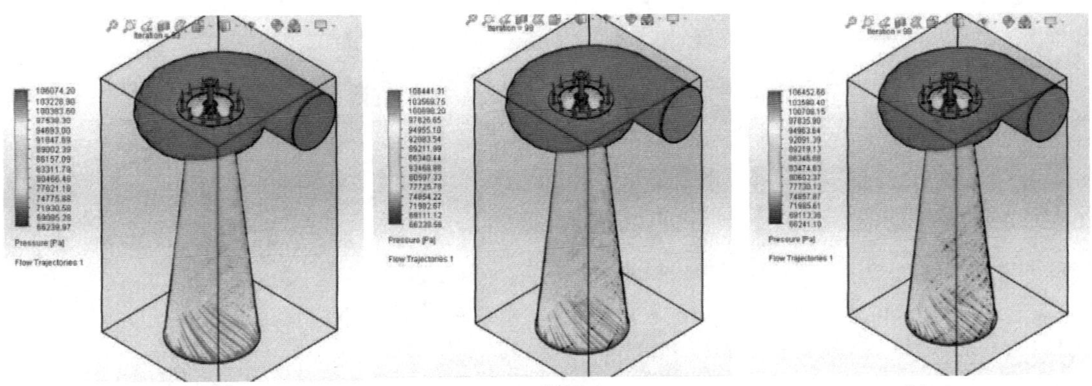

(a) (b) (c)

Figure 2. Pressure contour of draft tube (a) DT-1, (b) DT-2, (c) DT-3.

From figure 1, DT-1 has the bluest flow contour it means DT-1 has the highest average velocity. For DT - 2 and DT-3 it keeps going down. DT-1 has 1,885 $m.s^{-1}$, DT-2 has 1,880 $m.s^{-1}$, and the last one DT-3 has the lowest velocity with 1,862 $m.s^{-1}$. Then from figure 2, DT-3 has the highest pressure with an almost red contour on the flow simulation. DT-1 has the lowest pressure under DT-2 and DT-3. DT-3 has pressure 95552.08 Pa, DT-2 has 95529.52 Pa, and the last one DT-1 has 95362.93 Pa. It can be seen that the highest pressure has the lowest velocity, it keeps going down.

Table 2. Drag force and power of the draft tube

Draft Tube Variant	DT -1	DT -2	DT -3
Fd (N)	328.963	341.28	349.407
Fdh (N)	375.4	461.2	472.2
Pd (Watt)	4570.5	4741.1	4854.2

Table 3. Turbine efficiency

Draft Tube Variant	Pth (Watt)	η_h (%)	η_t (%)
DT -1	4905	93.2	88.5
DT -2	4905	96	91.2
DT 3	4905	98	93.1

From Table 2 and Table 3, it can be seen the value of drag force and power of the draft tube of each variant. The DT–1 variant has a drag force of 328.963 N with a power of 4570.5 Watt, and a turbine efficiency of 88.5%. The DT–2 variant has a drag force of 341.28 N with a power of 4741.1 Watt, and 91.2% turbine efficiency. The DT–3 variant has a drag force of 249.407 N with a power of 4854.2 Watt, and a turbine efficiency of 93.1%.

4. Conclusions

From the results and discussion above can be found that using the spiral liner draft tube can increase the efficiency of the Kaplan turbine. The draft tube without spiral liner variant DT-1 produces an efficiency of 88.5%, the lowest efficiency. Variant DT-2 with 6 spiral liners produces an efficiency of 91.2% and variant DT-3 with 12 spiral liners produce 93.1%. It means that the best and the highest efficiency can be reached with 12 spiral liners.

Acknowlegements

Authors wishing to acknowledge to all my colleagues and mechanical engineering laboratory UNJ that already supported this research.

References

[1] R S R Gorla and A A Khan 2003 *Turbomachinery Design and Theory* (Amsterdam Marcel Dekker,Inc)

[2] Wadhibasme S J and et al 2016 *Int. J. of Sci Eng. and Tech. Research* **5** 3

[3] Choi Y D and Son S W 2012 *J. of the Korean Society of Marine Eng.* **36** 351

[4] Khare R, Prasad V and Verma M, 2012 *Int. J. of Advances in Eng. Sci. and Tech.* **2** 21

[5] Gupta J and Sahun S 2017 *Int. Research J. of Advanced Eng.and Sci.* **2** 144

[6] Pal C, Gulshan T, Shashi S, Nisbat A and Chandan M 2018 *Int. J.for Research in Eng. Application and Management* **04** 15

[7] Prashad V, Khare R and Chincholikar A 2010 *Int. Group for Hydraulic Efficiency Measurement* 252

[8] Joen J H, Byeon S S and Kim Y J 2013 *IOP. Series: Materials Sci.and Eng* **52** 5

[9] Reddy H, Seshadri V and Kothari D P 1996 *Energy Sources*, **18** 143

[10] Nema U and Rajvaidya R 2017 *Int. J. of Eng. Sci. and Research Tech.* ***6*** *2*

Spectrophotometric analysis of caffeine in local product of Arabica: observed at different roasted temperatures

Misto, K Alawiyah, L Rohman, Supriyadi, Mutmainnah and E Purwandari*

Department of Physics, Faculty of Mathematics and Natural Sciences, University of Jember, Jalan Kalimantan 37 Jember 68121, Indonesia

* Email: endhah.fmipa@unej.ac.id

Abstract. Indonesia's geographical location supports the existence of various coffee commodities. As a plantation commodity, coffee has a high economic value. Coffee has a lot of substance that is beneficial to the body, one of is caffeine that serve to increase the stamina. Caffeine is naturally present in many types of plants as secondary metabolism. This research was done to know the content of caffeine in a local Arabica pure coffee with their variation of roasted temperatures. Coffee roasting is a stage of processing coffee beans into ready to grind, which can produce variations in the taste and content of coffee. The roasting process uses a roasting machine that automatically adjusts the roast temperature. Variations of roasted coffee are ground and used as a solution to determine the caffeine content. The determination of the coffee content is carried out using the spectrophotometric method, which measures the absorbance of caffeine in solution when it is passed by electromagnetic waves in the UV-Vis spectrum. The test results showed that the highest caffeine content was on samples of pure Arabica coffee with a roasted temperature of 195°C to 215°C with five levels of coffee roast. Caffeine content in sequential 3.217% and 2.597%.

Keywords: Arabica coffee, Caffeine, UV-Vis spectrophotometer

1. Introduction

The coffee is one of Indonesia's plantation crops originating from Africa. In Indonesia, the existence of this plant has a very good impact on the country's economic growth [1]. One of the regions in East Java Province producing the second largest coffee commodity is Jember [2]. The type of coffee dominating plantations in Jember is Arabica coffee. Coffee grown at different heights and types of soil will produce different aromas and substances. Different qualities will also be obtained when the coffee beans are processed with certain roasting temperature variations [3]. The coffee content that has the most significant function is caffeine, which is believed to boost stamina for the body [4]. As an organic compound, caffeine contains nitrogen with a dual cyclic structure in the methylxanthine group (heterocyclic alkaloids) crystalline powder.

The chemical name of caffeine is 1,3,7-trimethyl xanthine, with the chemical formula C8H10N4O2 is shown in Figure 1 [5]. The purpose of this study was to determine the caffeine content of Arabica coffee, which is produced by the plantation in Jember. The caffeine content of Jember Arabica coffee was studied to identify the qualities of local Jember products analysed in various roasting conditions. The existence of pure Arabica coffee began to be modified by the addition of maize as to coffee blend to obtain an economical profit by making the final product less expensive [6]. Most people have modified pure Arabica coffee products with maize ingredients to reduce the price of coffee which is still too expensive [7]. In this study, we will also analyse how the effect of maize with a composition of 10% and 20% of the total coffee on its caffeine content.

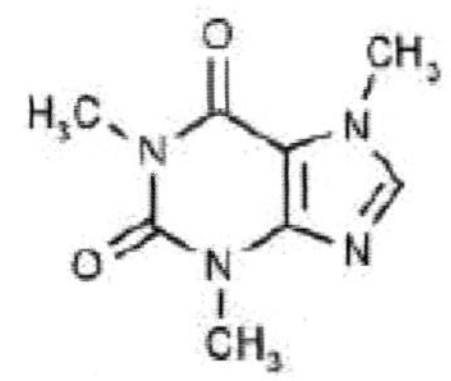

Figure 1. Chemical formula of caffeine [5]

Solvents such as Chloroform, Methyl Chloride, Ethanol, Acetone and Ethyl acetate are commonly used for the solvent extraction of caffeine. Chloroform ($CHCl_3$) is a compound capable of dissolving caffeine and is assisted by $CaCO_3$ (lime). In the chloroform extraction process, caffeine can dissolve 94.53% [8]. The tool used to identify the caffeine content in Arabica coffee with variations in roasting temperature is the UV-Vis Spectrophotometer [9]. The UV-Vis spectrophotometry method, as shown in Figure 2, works based on the interactions between monochromatic rays and light sources. The result of the interaction in the form of energy absorbed by a specific value can cause the electron to be excited from the ground state to a higher energy state.

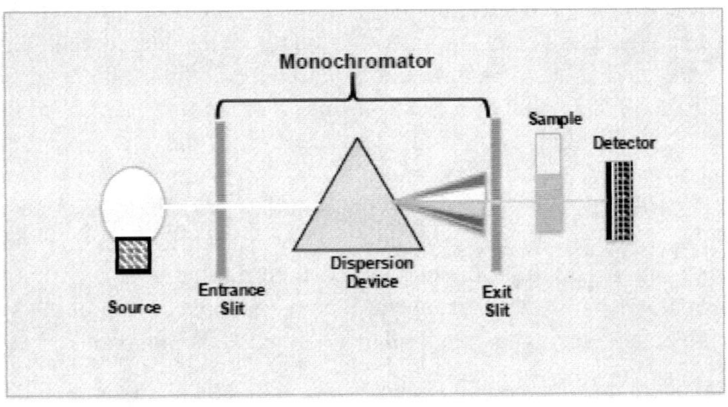

Figure 2. Schematic diagram of Spektrofotometer UV-Vis [10]

Quantitatively, the rays that are absorbed by matter can be arranged according to the Lambert-Beer law. This law states that the amount of visible light absorbed or transmitted by a solution is an exponential function of the concentration of the substance and the thickness of the solution. The Lambert-Beer law has the following equation [11],

$$T = \frac{I_t}{I_0} = I_0{}^{-acl} \tag{1}$$

$$\log(T) = \log\frac{I_0}{I_t} = -acl \tag{2}$$

$$\log\left(\frac{1}{T}\right) = \log\frac{I_0}{I_t} = acl = A \tag{3}$$

2. Materials and Methods

The material used was a local coffee Arabica pure and mixed Arabica-maize. We used $CaCO_3$ and $CHCl_3$ as standard caffeine and the solvents of caffeine. The instrument used for analysing caffeine absorbance is a UV-Vis Spectrophotometer with the wavelength range 250-300 nm. The five samples of Arabica coffee were prepared using a roaster, which is equipped with a roasting temperature controller. They were set at 194°C, 215°C, 225°C, 230°C and 240°C respectively to represent the variation types of roasting process.

2.1. Manufacture of standard solution caffeine

A standard solution of 250 mg of caffeine is put into a beaker glass and then added 90°C-95°C distilled water. This solution is diluted again in a 100 mL volumetric flask. This standard of caffeine produces a concentration of 2500 ppm. This solution was piped with a size of 1 ml, 0.8 ml, 0.6 ml, 0.4 ml, and 0.2 ml and then put in a 50 ml volumetric flask. The concentration results after being homogenized are 50 ppm, 40 ppm, 30 ppm, 20 ppm, and 10 ppm, respectively.

2.2. Determination of the maximum absorbance wavelength

Absorbance measurements were carried out in standard caffeine solutions with 10 ppm, 20 ppm, 30 ppm, 40 ppm, and 50 ppm in the wavelength range of 250-300 nm. The results of the wavelength scanning showed that the maximum absorbance for all samples was 273 nm. This wavelength was applied then to observe the caffeine content of various Arabica coffee roasted in various temperature.

2.3. Sample Preparation of pure Arabica and adulteration of 10% and 20% maize

Caffeine extraction was proposed to separate the caffeine from coffee grounds. A 2-grams sample of coffee was put into a beaker, and 150 ml of hot water was added. The coffee solution obtained is then filtered through a funnel and filter paper into the Erlenmeyer. A total of 1.5 g of Calcium Carbonate ($CaCO_3$) was added to the filtrate that had been put into a separating funnel, and then extracted three times with the addition of 25 ml of chloroform. The function of the addition of chloroform is to bind caffeine compounds to form caffeine extraction.

The extraction of caffeine at the bottom of the solution will then be evaporated with a rotary evaporator so that the chloroform can evaporate completely. The caffeine extract from each solvent-free coffee sample was put into a 100 ml volumetric flask, then diluted with distilled water. Besides pure Arabica coffee, this research will also analyze the type of Arabica coffee in the form of a mixture of pure Arabica and maize. The amount of maize that is mixed is as much as 10% and 20% of the total mixed coffee made [12].

2.4. Determination of the caffeine content

Calculation of caffeine content in ppm was obtained through linear regression from the standard caffeine concentration and the maximum absorbance. The conversion of ppm units into mg/g was calculated using the following equation [9].

$$\text{Caffeine content} = \frac{\text{Concentration (X)} \times \text{The weight of extract} \times \text{Fp}}{\text{Sample weight}} \qquad (4)$$

Where X is the concentration in mg/ml, the weight of the extract is the caffeine extract from each solvent-free coffee sample which is diluted with 100 ml distilled water and measured in litres, Fp is the dilution factor that is carried out several times dilution, and the sample weight is the sample weight. Ground coffee that has not been brewed in grams. The caffeine content then compared with Standart Nasional Indonesia (SNI).

3. Result and Discussion

The absorbance of caffeine standard solution with a concentration of 50-10 ppm at a wavelength of 250-300 nm is shown in Figure 3. The wavelength with the maximum absorbance occurs at a wavelength of 273 nm.

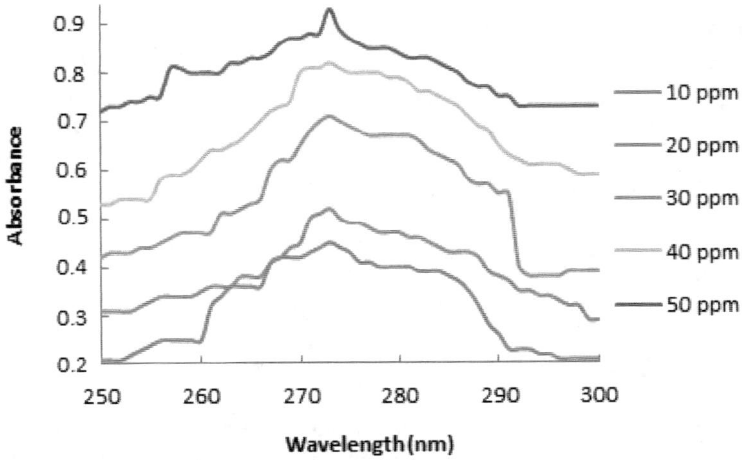

Figure 3. The absorbance of caffeine standards solution on the variation of the concentration in the range of the wavelength spectrum of 250 – 300 nm.

The absorbance measurement of the caffeine standard was carried out three times, so that the mean absorbance value and its uncertainty value are shown in Table 1.

Table 1. The maximum absorbance of caffeine standart using Spectrophotometer UV-Vis.

No.	Caffeine Consentrarion (ppm)	Maximum Absorbance
1	10	0.437 ± 0.008
2	20	0.517 ± 0.004
3	30	0.707 ± 0.004
4	40	0.810 ± 0.007
5	50	0.937 ± 0.008

Based on this table, an increase in the concentration of caffeine solution causes the maximum light intensity absorbed in $\lambda = 273$ also increase. Calibration curves for caffeine concentration and maximum absorbance produce a linear regression equation used to determine caffeine content (refer to Fig. 4).

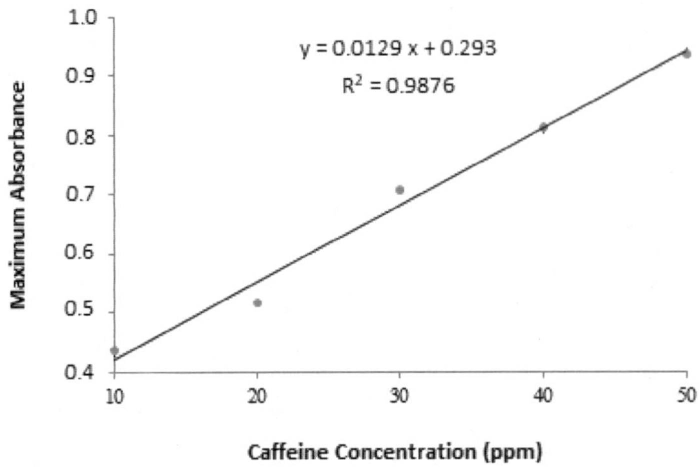

Figure 4. Calibration curve of caffeine standard solution of caffeine.

The calibration curve used to calculate the caffeine content of coffee powder sample. Furthermore, the measurement value of the maximum absorbance on ground coffee pure Arabica and a mixture of Arabica-maize 10% and 20%. Table 2 has shown the maximum absorbance of each coffee.

Table 2. The measurement of maximum absorbance of ground coffee pure Arabica at a wavelength of λ = 273 nm.

Coffee	Roasting Temperature	Absorbance Maximum
Pure arabica	195°C	1.123 ± 0.004
	215°C	0.963 ± 0.011
	225°C	0.637 ± 0.008
	230°C	0.473 ± 0.004
	240°C	0.387 ± 0.004
Mixture arabica-maize 10%	195°C	0.707 ± 0.008
	215°C	0.640 ± 0.007
	225°C	0.427 ± 0.029
	230°C	0.373 ± 0.004
	240°C	0.347 ± 0.004
Mixture arabica-maize 20%	195°C	0.517 ± 0.004
	215°C	0.467 ± 0.004
	225°C	0.327 ± 0.004
	230°C	0.237 ± 0.004
	240°C	0.300 ± 0.000

According to Table 2, the higher the roasting temperature used, the smaller the absorbance value measured. The decrease in maximum absorbance when the roasting temperature is increased is closely related to caffeine (caffeine content) in pure coffee. The high roasting temperature, past its melting point, causes the atomic bonds in caffeine to change. The melting point of caffeine is at a temperature of 228°C [5]. This property influences the change of caffeine absorbance to electromagnetic waves applied. We have also calculated the caffeine content on a 2 mg serving three times of coffee consumption per day and presented by Table 3.

Table 3. The Total caffeine content of coffee consumed three times a day with the weight of the sample every dish as much as 2 g

Coffee	Roasting Temperature	Caffeine (mg/day)
Pure Arabica	195°C	193.0
	215°C	155.8
	225°C	79.9
	230°C	41.9
	240°C	21.7
Mix Arabica/Maize with Maize content of 10 wt %	195°C	96.1
	215°C	80.6
	225°C	31.0
	230°C	18.6
	240°C	12.4
Mix Arabica/Maize with Maize content of 20 wt %	195°C	51.9
	215°C	40.3
	225°C	7.8
	230°C	4.7
	240°C	1.6

When referring to the FDA (Food Drug Administration) standards mentioned by Liska [13], the dose of caffeine consumed per day is 100-200 mg/day. Also, according to SNI 01-7152-2006, the threshold limit for caffeine in food and beverages is 150 mg/day [14]. Based on the fifteen samples that have been tested, according to both the FDA and SNI, this type of local Arabica coffee in Jember at 195°C roasting temperature and 215°C in pure coffee exceeds the limit of caffeine consumption for each serving three times a day. Arabica coffee provides acidic, sharp aromas and little bitter. The number of caffeine as its substance makes it popular to be consumed. Coffee has various benefits for the body, that it can be useful as an antioxidant and also for stimulating brain performance [15]. However, consuming a substance in excess amounts will also have a negative impact on the body.

4. Conclusion

The caffeine content of Arabica coffee in Jember plantations has been identified based on variations in roasting temperatures. It was obtained based on the results of the analysis of the maximum absorbance changes at each roasting temperature variation. Increasing the roasting temperature causes the maximum absorbance value at a wavelength of 273 nm to decrease. A decrease in the maximum absorbance indicates a decrease in the caffeine content in coffee. Local Arabica coffee in the plantation area in Jember has caffeine content according to SNI for daily consumption, except for the pure Arabica types, which are roasted at 195°C until 215°C. The addition of maize as a mixing ingredient of up to 20% still shows sufficient caffeine levels for daily consumption.

Acknowledgments

This research was funded by the University of Jember under community based research project. The authors were thanks to University of Jember for funding the research by Hibah Keris 2020.

Reference

[1] Rahardjo P 2012 *Panduan Budidaya dan Pengolahan Kopi Arabika dan Robusta* (Jakarta: Penebar Swadaya)

[2] Aji J M M, Ebban B K, Dewi C H S 2017 Employment absorptive capacity of smallholder coffee Plantations and sustainable coffee development in Jember district *Proceedings of The International Conference* of FoSSA 151-162

[3] Danarti and Najiyati 2001 *Budidaya Kopi dan Pengolahan Pasca Panen* (Jakarta: Penebar Swadaya)

[4] Clarke 1987 Coffee Technology Elsevier *Applied Science* 2

[5] Mumin A Md, Akhter F K, Abdein Z Md, Hossain Z Md 2006 *Malaysian Journal of Chemistry* **8** 045-051

[6] Daniel D, Lopes FS, dos Santos, V.B.; do Lago CL 2018 Detection of coffee adulteration with soybean and maize by capillary electrophoresis-tandem mass spectrometry. *Food Chem* **243** 305–310

[7] Ini baru media 2018 *Kopi Jagung dan Masyarakat Melarat Zaman Kolonial* Media Grup [https://www.inibaru.id/coffreak/begini-cara-membedakan-kopi-jagung-dan-kopi-murni]

[8] Sunarto and K Roosenda 2016 *Jurnal Kimia Dasar* **4**

[9] Rismawati S 2019 *Identifikasi Kandungan Kafein dan Warna RGB pada Kopi dengan Variasi Sangrai* (Jember: Universitas Jember)

[10] Misto, T Mulyono, B Eko, E Agustina 2019 *Jurnal Food Science and Technology* **7** 57-67

[11] Sumarni A R 2016 *Jurnal Ilmu Teknosains* **2** 2

[12] Aini N 2018 *Pola Absorbansi Kopi Bubuk Menggunakan Spektrofotometer UV-Vis*. (Jember: Universitas Jember)

[13] Liska 2004 *Drugs and The Body with Implication for Society* 7 Ed. (New Jersey: Pearson)

[14] Badan Standarisasi Nasional 2006 SNI 01-7152-2006 *Bahan Tambahan Pangan* (Jakarta: Badan Standarisasi Indonesia)

[15] A. Farida E R R and A C Kumoro 2013 *Jurnal Teknol Kimia* **2** 2-70

Reservoir properties modelling using multi-attribute seismic analysis in south Sumatra basin, Indonesia

B S Mulyatno[1], F S Parameswari[1], N Hikmah[2], O Dewanto[1] and R C Wibowo[1,*]

[1] Department of Geophysical Engineering, Faculty of Engineering, Universitas Lampung, Jalan Prof. Dr. Soemantri Brodjonegoro No. 1 Bandar Lampung 35145, Indonesia

[2] PT Pertamina EP Asset 2 Prabumulih, Jalan Jend. Sudirman Prabumulih, South Sumatera

*Email: rahmat.caturwibowo@eng.unila.ac.id

Abstract. Seismic inversion and multi-attribute analysis method are used in this research to characterize the reservoir properties. Where both of these methods will be resulted an acoustic impedance volume and log property volumes. The property log volumes were created from density, porosity, Gamma-Ray (*GR*) and Water Saturation (*Sw*) log. There are 4 wells that are used in this research, they are EM-31, EM-32, EM-33 and EM-34 and a 3D seismic data post-stack time migration. Based on a map of acoustic impedance distribution, the reservoir target has an acoustic impedance value of 6.000 ((m/s)*(g/cc)) to 8.000 ((m/s)*(g/cc)). Furthermore, based on the map of the distribution of the log property, the reservoir has a density value with a range of 2.3 g/cc up to 2.4 g/cc, porosity value > 10%, the *GR* value of less than 100 API and *Sw* 50% to 55%. Thus, it can be concluded that reservoir has a lithology of sandstones, has a fairly good porosity and contains oil fluids.

Keywords: multi-attribute seismic, reservoir, south Sumatra basin, modelling

1. Introduction

The high demand for oil and gas needs is to encourage gas companies to take an exploration stage in discovering new reserves. Many studies and research have been conducted in recognizing the reservoir character. To perform characteristic reservoir there are several methods that can be used. They are seismic inversion method and multi-attribute analysis method.

According to [1] seismic inversion method is an inversion technique in the form of a geological condition to the back (invers modelling), this method provides the results of the underground geological appearance, so as to identify the character and distribution of the reservoir on the target area in the form of the interpretation of geological, lithology, and fluid.

Multi-attribute analysis is a method of analysing the relationship between well data and seismic data at the location of wells in performing the volume estimation of log properties on the overall seismic volume. This method is a method which uses many attributes to predict some of the log properties of which are porosity [2-4].

In this research, the characteristic analysis of reservoir is done based on acoustic impedance distribution map and the map of the log properties that will be conducted analysis on the location of lead zone reservoir. The data generated in this study which was obtained from the Acoustic Impedance (AI) inversion process and multi-attribute analysis can be used to create a log property distribution in the research area that will be able to be known characteristic of the location indicated as reservoir containing hydrocarbons in it.

2. Regional Geology

The basin of South Sumatra is a basin bordered by the Tiga Puluh mountains to the northwest separating the South Sumatra basin and the Central Sumatran Basin, the Sunda display is in the northeast, Barisan and the Sumatran Fault in the southwest, and Lampung High in the southeast. Figure 1 shows the column stratigraphy of the southern Sumatran basin, it appears that the Baturaja formation above from the Talang Akar formation, the Baturaja formation itself consists of limestone and has a varying thickness of about 60 – 75 mm, while the Talang Akar formation itself is above the Lahat formation and Baturaja formation, consisting of sandstone that comes from the delta plain, silt, quartz with coaly carbonaceous intercalated claystone and in some places there is a conglomerate. The thickness of the Talang Akar formation around 460-610m [5].

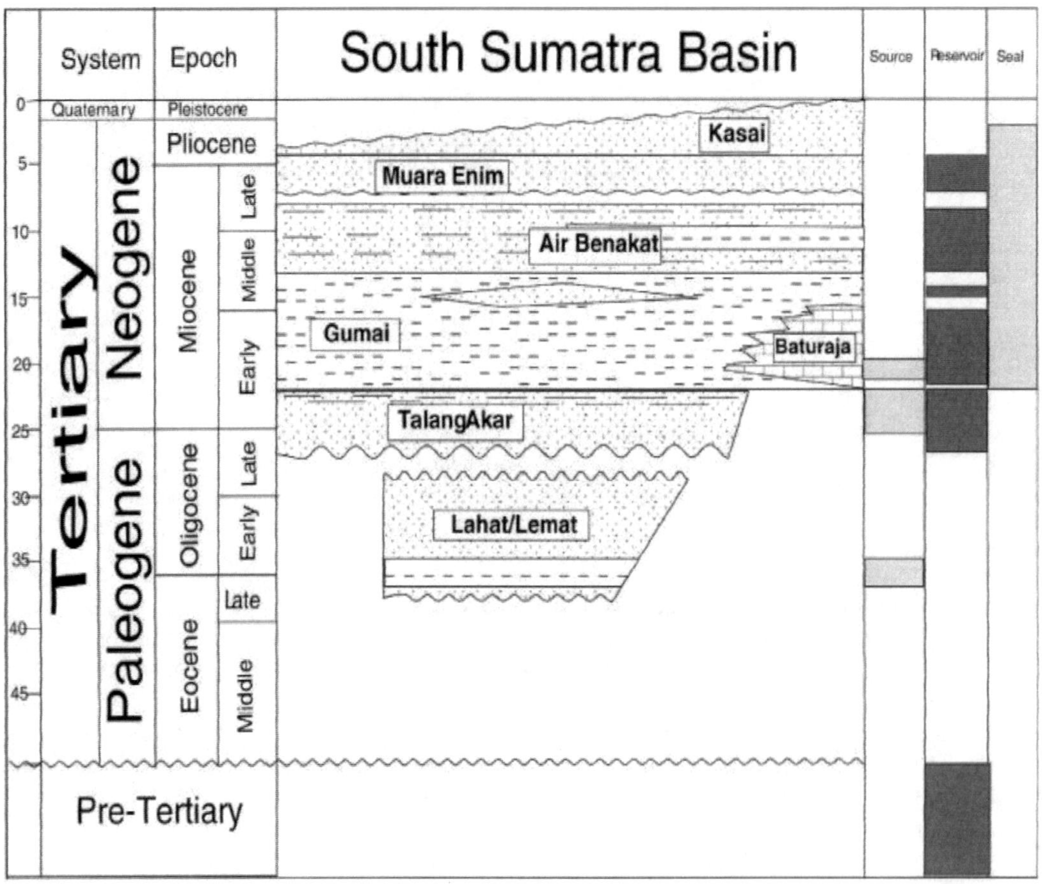

Figure 1. Stratigraphy column south Sumatera basin [5].

3. Materials and Methods

3.1. AI inversion

The AI inversion is done by analysing the data sensitivity of the well, where this stage serves to determine the value of the acoustic impedance as to what indicates the hydrocarbon lead zone in the research area. This analysis of sensitivity is done by creating a cross-plot between the log AI and other logs. Therefore, the cross-plot results will be known as the AI value range that shows the lead zone of the Reservoir and create the initials model (initial model). Then after the initials model is obtained, then perform a pre-inversion analysis. Pre-inversion analysis is done for simulation or control to perform inversion stage, that to know the best inversion results to be obtained. Once considered pre-inversion results are already good. Next, apply the inverse result of all well at the seismic volume as a whole by applying the parameters at the pre-inversion stage[6].

3.2. Multi-attribute analysis

Multi-attribute analysis is an analysis of the relationship between seismic data and well data that will be used to perform a volume estimation of log properties on the overall seismic data volume [7-9]. Multi-attribute analysis uses multiple attributes in predicting multiple log properties from the log data used, in conducting multi-attribute analysis of used seismic data, the log data being targeted, internal seismic attribute data as well as seismic external attribute data. In this study, the log target is density log, porosity log, Gamma Ray (*GR*) log, and Log saturation water (*Sw*). The initial stage is to perform a single attribute analysis, which will be cross plot next to each attribute with the target log property. Then, perform multi-attribute analysis by using the step-wise regression method, in this step channelled the determination of many good attributes is used based on the validation error values.

4. Results and Discussion

The prospect zone of hydrocarbons in the well data has a low gamma ray value, and there are crosses or separations between low density (RHOB) and low NPHI. If viewed from Based on qualitative analysis, it can be concluded that the coating with a good indication of hydrocarbon is the layer of BRF, A0 and A and A2. However, due to the layers of A0 and A2 is a thin layer, the research is focused on the layers of BRF and A.

This Sensitivity analysis is done before doing an inversion of the inverse, which from the analysis result of the data logs This will provide an information about the relationship log AI data with other data logs. From the sensitivity analysis results that have been done before, it is noted that the target zone has an acoustic impedance value of 6000 (m/s)*(g/cc) to 800 (m/s)*(g/cc), density value 2.4 g/cc to 2.3 g/cc, porosity value > 10%, *GR* value < 100 API as well as *Sw* value < 70%. Good inversions can be seen quantitatively by looking at the parameters of the RMS error, the correlation coefficient as well as the relatively minor errors. According to table 1 that the correlation value that has been produced has a strong quality according to Schober et al. (2018) [10] which is above 0.9 which is followed by an RMS error value of 400 to 480 and has a relative error value less than 0.5. Where this indicates that the pre-inversion analysis performed is reasonably good, which further applies the pre-inversion results to inversion the overall seismic data volume.

Multi-attribute analysis carried out as many as 4 log targets, namely density, porosity, *GR*, and *Sw* (Figure 2). This multi-attribute analysis uses the step wise regression and probabilistic neural network (PNN) methods. The step wise regression method is a method that includes both internal and external attributes that serve to produce a model capable of matching the target log. While the PNN method is a method that performs analyses with non-linear methods. The parameter that should be observed in conducting multi-attribute analysis is the number of attributes to be used, the validation error value and the resulting correlation value. As for the research the initial attributes are used as much as 10 attributes. After, obtaining the attribute to be used, the cross-plot stage is performed to find out the relationship of the original density log and the predicted density log of the results from the multi-attribute analysis.

Then, subsequent results are multi-attributes by using the about regression method and the PNN method compared to each other. From the results of the analysis process generated multi-attribute log prediction of density logs, porosity, *GR*, and *Sw*. Further, applied to the seismic data volume. Then, the slicing is done on the BRF and A layers and generating a spread of property log.

The characteristic analysis of reservoirs is performed to determine the nature and distribution and character of reservoir in seismic data relating to log data. In conducting analysis, the characteristic reservoirs are carried out based on a map of the log AI distribution, the map of the density log distribution, the map of *GR* log distribution, and the map of the *Sw* log spread on the zone considered as the lead zone. Based on a log sensitivity analysis that has been done that the target zone has an AI value of 6000 (m/s)*(g/cc) to 8000 (m/s)*(g/cc). Where the wells of EM-31, EM-32, EM-33 and EM-34 showed an acoustic impedance value of 6000 (m/s)*(g/cc) on the BRF layer. On A layer indicates that the EM-32 well has an acoustic impedance value of 8000 (m/s) * (g/cc), for the wells of EM-31, EM-33, and EM-34 has an acoustic impedance value of 7250 (m/s) * (g/cc). So that it can be concluded based on the acoustic impedance value map that the lead zone has an acoustic impedance value of 6,000 (m/s)*(g/cc) to 8,000 (m/s)*(g/cc). On the map of the BRF layer that the density value of the lead zone consists of the EM-31, EM-32, EM-33 and EM-34 amounted to 2.3 g/cc. Whereas, on a map of the layer A density value in the lead zone of 2.4 g/cc.

Table 1. Quantitative Pre-Inversion Analysis

Well	RMS *Error*	Coefficient correlation	*Relative Error*
EM-31	465.367	0.986	0.265
EM-32	450.527	0.902	0.437
EM-33	411.481	0.978	0.214
EM-34	477.147	0.993	0.258

On The map of the porosity of the wells of EM-31, EM-32, EM-33 and EM-34 that porosity value is above 10%. Hence, it can be concluded that the porosity owned by the lead zone has a quality that is according to that Wei (2016) [11]. Based on the *GR* distribution map that the prospect Zone is a zone that has a moderate to moderate radioactive value, which is correlated with the density value that the lithology in the lead zone is the lithology of sandstones. Based on a map of the SW range, That the *Sw* value in the lead zone on the BRF layer for the EM-31 wells, the EM-33, the EM-34 indicates a *Sw* value of 0.55 or 55% while the EM-32 well indicates a *Sw* value of 0.5 or 50%. At layer A *Sw* value at the well EM-31, EM-32, EM-33, and EM-34 show a value of 50%. As it is known according to Wibowo (2020) [12], that the value of $Sw < 0.25$ is oil and gas fluid, *Sw* 0.25 – 0.75 is oil fluid, and > 0.75 is water fluid. It can be concluded that the *Sw* value in the lead zone is a zone that has fluid in the form of oil. However, judging from each map, the zone that will be used as the target zone has not been clearly visible, making it difficult to determine the location of the lead zones. Thus, a method can be demonstrated that the location of the lead zone is targeted in accordance with the value of acoustic impedance, density, porosity, *GR* and *Sw*, where the method is spatial analysis method, which combines five maps into one.

Spatial Analysis method is conducted by classifying acoustic impedance distribution map, density map, porosity map, *GR* map and *Sw* map which are classified as 0 and 1. Where 0 is declared as an area that is not a prospect and 1 is declared as the lead area. A zone declared as the hydrocarbon prospect Zone is an eligible zone having an acoustic impedance value of 6,000 (m/s)*(g/cc) to 8,000 (m/s)*(g/cc), the density value is 2.3 g/cc-2.4 g/cc. The value of porosity > 10%, *GR* value < 100 API and the average *Sw* value 55%. From the analysis that has been done on the lead zone distribution map of the area BRF

IOP Conf. Series: Materials Science and Engineering 1173 (2021) 012013 doi:10.1088/1757-899X/1173/1/012013

that indicates hydrocarbons are located in the north to the east of the map and on layer an area that indicates hydrocarbons are located in the north to the northeast of the map.

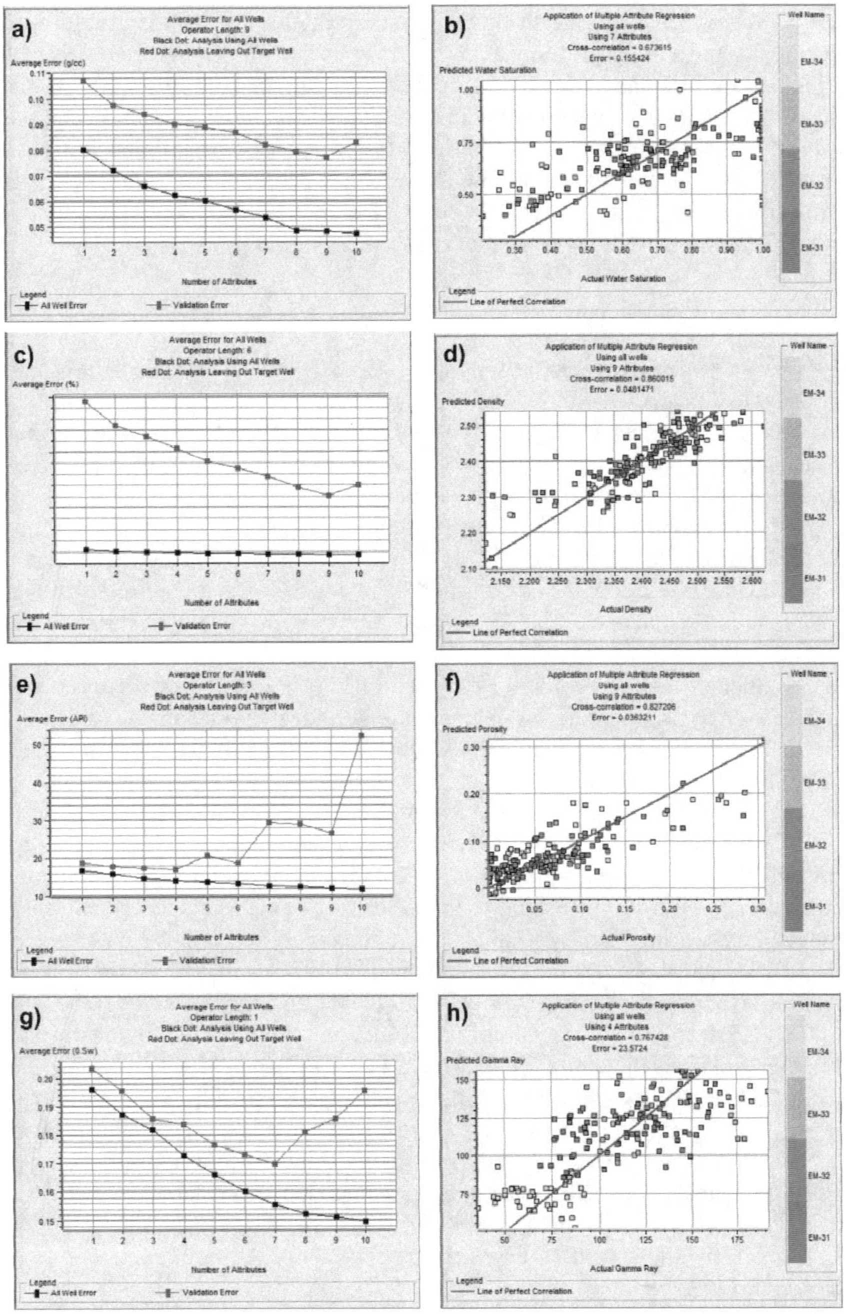

Figure 2. Cross-plot total of attributes used with average error in the analysis of the multi-attribute density (a & b), porosity (c & d), GR (e & f), Sw log (g & h).

5. Conclusion

The result of the volume of AI inversion and the volume of the Log property that has been made have good quality, indicated by the average correlation value in the output of AI of 0.959, multi-attribute analysis of density log of 0.744, multi-attribute analysis of the logs of 0.797, a multi-attribute analysis of the *GR* log of 0.732 and a multi-attribute analysis 0.744 of *Sw*. It is known that the spread of property logs is performed on the BRF and a layer which are the target zones in this study. Based on acoustic impedance distribution map and log property distribution map, that in the lead zone has an acoustic impedance value of 6,000 (m/s)*(g/cc) to 8,000 (m/s)*(g/cc), density of 2.3 g/cc to 2.4 g/cc, porosity > 10%, *GR* < 100API and *Sw* average 55%. The analysis has been done on the layer of BRF area that indicates hydrocarbons are in the north to east of the map and on layer an area that indicates hydrocarbons are in the north to the northeast of the map.

References

[1] Russell B H 1988 *Introduction to Seismic Inversion Methods* (Tulsa: Society of Exploration Geophysicists)
[2] Basin M, Marin W, Aldana M, Bolívar U S and Sierra J 2014 Seismic multi-attribute integration for the characterization of a gas reservoir, Lower Magdalena Basin, Colombia *SEG Denver 2014 Annual Meeting* (Denver: SEG) pp 2693–7
[3] Wibowo R C, Sarkowi M, Mulyatno B S, Dewanto O, Zaenudin A, Amijaya D H and Aspari A A 2020 Thinned coal distribution modeling based on integrated geological and geophysical data : Case study CBM resources in Central Palembang Sub-Basin *2nd International Conference on Earth Science, Mineral, and Energy* vol 070011 (Yogyakarta: AIP Publishing) pp 1–9
[4] Colpaert A, Mienert J, Dvorkin J and Henriksen L B 2005 Combining seismic stratigraphy, multi attribute analysis and rock physical modeling to identify carbonate facies and karst along an Upper Palaeozoic carbonate Platform, Barents Sea *SEG Houston 2005 Annual Meeting* (Houston: SEG) pp 724–7
[5] Wibowo R C 2013 "Unconventional Reservoir" Shale Gas Potential Based On Source Rock Analysis In Sumatran Back Arc Basin *Proceedings of International Conference on Geological Engineering Geological Engineering Department, Engineering Faculty, Gadjah Mada University* ed N I Setiawan, W Budianta and A Idrus (Yogyakarta: Geological Engineering Department, Engineering Faculty, Gadjah Mada University) pp 151–63
[6] Hansen H D, Magnus I, Edvardsen A and Hansen E 1997 Seismic inversion for reservoir characterization and well planning in the Snorre Field *Lead. Edge* **16** 269–74
[7] Hampson D P, Schuelke J S and Quirein J A 2001 Use of multiattribute transforms to predict log properties from seismic data *GEOPHYSICS* **66** 220–36
[8] Wibowo R C, Ariska S and Dewanto O 2020 Inversi Geostatistik Menggunakan Analisa Multi-Atribut Stepwise Regression Untuk Karakterisasi Reservoir *Ris. Geol. dan Pertamb.* **30** 187
[9] Febridon M N, Mulyatno B S and Wijaksono E 2020 Analisis Sifat Fisis Pada Reservoar Batupasir Menggunakan Metode Seismik Inversi Impedansi Akustik (Ai) Dan Multiatribut Pada Lapangan "Mnf" Cekungan Bonaparte *J. Geofis. Eksplor.* **4** 3–14
[10] Schober P and Schwarte L A 2018 Correlation coefficients: Appropriate use and interpretation *Anesth. Analg.* **126** 1763–8
[11] Wei W, Zhu X, Meng Y, Xiao L, Xue M and Wang J 2016 Porosity model and its application in tight gas sandstone reservoir in the southern part of West Depression, Liaohe Basin, China *J. Pet. Sci. Eng.* **141** 24–37
[12] Wibowo R C, Arlinsky D, Ariska S, Wiranatanegara B W and Riyadi P 2020 Gas Saturated Sandstone Reservoir Modeling Using Bayesian Stochastic Seismic Inversion *J. Geosci. Eng. Environ. Technol.* **05** 25–31

Synthesis and characterization exopolysaccharide from algae *Spirulina* sp. using technique sol-gel as adsorbent Pb(II) ion

R A Kausar[1,*], Buhani[1,*], Suharso[1], and A Setiawan[1]

[1]Departement of Chemistry, Faculty of Mathematics and Natural Sciences, University of Lampung, Jalan Prof. Soemantri Brojonegoro No.1, Bandar Lampung, 35145, Indonesia.

*E-mail: radhoalkausar@gmail.com; buhani_s@yahoo.co.id

Abstract. In this study, a modification of exopolysaccharide (EPS) from *Spirulina* sp. algae biomass with silica (HES) and exopolysaccharide from *Spirulina* sp. algae biomass with silica coated by magnetite particles (HESM) has been done successfully. A series of experiments with the batch method were conducted to test the ability of HESM adsorption of Pb(II) ions in solution. Identification of functional groups of HESM adsorbents was studied using infrared spectrometer (IR) while surface morphological analysis and composition of the constituent elements of HESM were performed using scanning electron microscope-energy (SEM). Concentration of Pb(II) ion in the adsorption process were analysed by inductively coupled plasma-atomic emission spectrometry (ICP-AES Adsorption of Pb(II) ions with HES and HESM is optimum at pH 7 and concentration at 25 ppm with the adsorption capacity on HES at 3.85 mg g^{-1} while on HESM at 272.63 mg g^{-1}, contact time of 15 min and it tends to follow second orde pseudo kinetic model and Langmuir adsorption isotherm model.

Keywords: Sol-gel, exopolysaccharide, adsorption, silica coated magnetite.

1. Introduction

Heavy metal pollution to the environment is closely related to the activities carried out by humans. Much of the heavy metal pollution is generated from agricultural activities such as making fertilizers and applying insecticides containing heavy metals, industrial factory activities, as well as ceramics and metal smelting activities. Heavy metal contamination activity which is carried out continuously every day will result in the accumulation of heavy metals in large quantities [1]. Heavy metals are usually present in the form of ions, either in the form of single ions or pair ions in the water [2].

Dangerous metal ions such as Cd, Pb, Zn, Hg, Cu, and Fe are elements that have high molecular weight types [3]. To reduce heavy metal contamination, many techniques have been used including coagulation, complexation, ion exchange and adsorption methods. Of all these methods, the adsorption method is the method most widely used to reduce the impact of heavy metal pollution [4]. This adsorption method has several advantages including, the process is very simple, the cost used is quite cheap, and it is environmentally friendly [5].

The adsorbents that are usual used in the adsorption process are activated carbon, silica gel, alumina, and zeolite. At this time starting to develop the use of alternative adsorbents derived from nature because it is more economical. One example of such adsorbent is algae biomass. Several types of algae have received attention because of their high ability to adsorb ionic ion bound to biomass and the possibility of reusing biomass as a biosorbent that can be used for wastewater treatment in liquid form [6].

However, algae biomass has several disadvantages, low density, easily damaged due to degradation by other microorganisms and technical difficult to use in columns to be used as adsorbent [7]. This weakness is carried out by immobilization of algae biomass with a silica gel matrix through a sol-gel process which to homogenize the solution but does not damage its structure so that the immobilization process of algae biomass in the silica matrix is expected to maintain the activity of functional groups present in the adsorbent and increase ion-adsorption capacity. Metal ions, especially in heavy metals. One example of an adsorbent that can be used well in handling heavy metals is exopolysaccharides. Exopolysaccharides play a role in biosorption [8]. Exopolysaccharides can adsorb heavy metals because exopolysaccharides are negatively charged [9].

Exopolysaccharides are mobile because they are generally composed of organic materials such as monosaccharides and disaccharides [10]. The application of exopolysaccharides as heavy metal biosorbents has been carried out and the results showed that exopolysaccharides are good media for absorbing metal ions and have a high adsorption capacity for wastewater treatment containing lead, nickel, and zinc ions [11]. exopolysaccharides are water soluble. Therefore, in this research, an immobilization process was carried out with a silica gel matrix through the sol-gel process. The sol-gel technique is a colloidal suspension of silica particles that is glued to a solid form based on hydrolyzed molecular precursors, which are mostly metal alkoxides and semi metal. The application of this sol gel technique in the exopolysaccharide immobilization process will produce adsorbents that are homogeneous, well mixed in multi-component systems, the size, shape and properties of the particles can be controlled [13].

In this study, the synthesis of exopolysaccharide hybrids from the algae cultivation of *Spirulina* sp. with a silica-magnetite matrix and its application as an adsorbent of Pb(II) ion in solution. Adsorption data were analyzed to determine the kinetic model and adsorption isotherm of Pb(II) ion which was absorbed by the adsorbent. The exopolysaccharide-silica-magnetite hybrid material from the algae cultivated by *Spirulina* sp. effectively absorbs Pb(II) ions in solution.

2. Materials and Methods

2.1. Isolation of Exopolysaccharide

Microalgae *Spirulina* sp. ready to harvest separated by centrifuge at 8.500 rpm at 4°C for 5 min, then the precipitate or gel was obtained from Spirulina sp. and the filtrate. Twice the volume of the filtrate, ethanol was added and left to stand for 4 h. Then it is deposited using a centrifuge at a speed of 8.500 rpm and a temperature of 4°C for 5 min. Sediment of *Spirulina* sp. and the exopolysaccharide formed was washed using distilled water and dried using frezee dry. Biomass of *Sprirulina* sp. and the dry exopolysaccharide was characterized using an IR spectrometer.

2.2. Shynthesis of HESM

Solution A, as much as 5 mL of TEOS mixed with 2.5 mL of distilled water is put into a plastic glass, then added with magnetite with concentrations 50 mg, then add a few drops of 1 M HCl to pH 2. Stir with a stirrer magnet until the solution is homogeneous. Solution B, consisting of 5 mL of ethanol plus an optimum exopolysacride of 0.1 gram silica and stirred with a magnetic stirrer. The two solutions are mixed to form a gel. The wet gel that is formed is left to stand for 24 hours then washed with ethanol and distilled water to pH 7, dried using freeze dry, then crushed until smooth. The resulting material is used for the adsorption test.

2.3. Adsorption kinetics

The optimum dose of HES was included in 5 Erlenmeyers. Then 25 mL of 100 ppm Pb (II) ion solution with optimum pH was added to each Erlenmeyer. Solution was shaken using a stirrer with time variations from 5, 15, 30, 60, and 120 min and the filtrate was analyzed by ICP-AES.

2.4. Adsorption isotherm

An amount of 50 mg of HES and HESM adsorbents with optimum concentration was put into 5 different plastic containers. Then added 20 ml of solution with different concentrations (10, 25, 30, 100, 200, and 300 mg L^{-1}) with optimum stirring time. The optimum pH was made using citrate buffer solution with pH range of 3.0-6.0 for the acidic pH and the phosphate buffer solution with pH range of 7.0-9.0 for the alkaline pH. After stirring, the adsorbent and solution were separated using centrifugation. The concentrations of ion Pb(II) in the adsorption process were analyzed inductively coupled plasma-atomic emission spectrometry (ICP-AES).

3. Results and discussion

3.1. Characterization of adsorbent by IR spectrometer

From the synthesis that has been done in this practical work. A functional group analysis was performed on the materials of EPS, HES, and HESM which were varied by magnetite using an *IR* spectrophotometer. The spectra IR of the adsorbent material will be shown in (Figure 1) to identify the functional groups of the adsorbent material before and after being modified with magnetite. Spectra *IR* shows several peaks that the functional groups present in the sample both belong to the material and functional groups that are possible from impurities that cannot be cleaned completely. Characterization using an IR spectrophotometer shows a success from the synthesis that has been carried out.

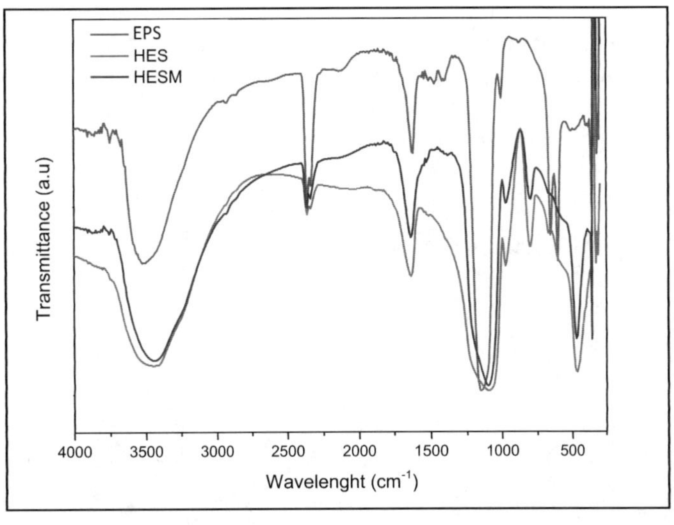

Figure 1. Spectra *IR* of EPS, HES, and HESM from algae *Spirulina* sp.

The IR spectra of EPS can be observed that the presence of a sharp peak in the absorption band 1149.57 cm^{-1} shows the characteristics of the polysaccharide, there is stretch vibration absorption from O–H and is strengthened by the presence of an absorption band at 1002.98 cm^{-1} which is a stretching vibration C–O (Figure 1). The absorption area of 1620.21 cm^{-1} shows asymmetric carbonyl. It was also found in the absorption of 1473.62 cm^{-1} of an asymmetric carbonyl. *IR* (HES) spectra in it can be observed that a sharp peak in the absorption band 1103.28 cm^{-1} shows the characteristics of the polysaccharide, then there is a change in the absorption band in the exopolysacride after being modified with silica gel especially in the absorption band at a frequency of around 470.63 cm^{-1}, it is the bending vibration absorption of O atoms from the siloxane functional group (Si–O–Si). The presence of siloxane functional groups is strengthened by the appearance of absorption bands in the area of 794.67 cm^{-1} which is the presence of Si–O (Si–O–Si) symmetric stretching vibrations.

In the absorption band around 3448.72 cm^{-1} is the stretching vibration of (–OH) from the silanol (Si–OH) group. In the area of wave number 2931.8 cm^{-1}, there is an absorption band originating from the (C–H) stretching vibration of (–CH_2) aliphatic. The frequency of the wave number 1635.64 cm^{-1}, it is an absorption that shows the OH vibration attached to the carbonyl group (–C=O) originating from the exopolysacride. Spectra *IR* (HESM) can be observed that there is a sharp peak in the absorption band 1103.57 cm^{-1} shows the characteristics of polysaccharides, then there is a change in the absorption band in the exopolysacride after being modified with silica gel with magnetite variations, especially in the absorption band at a frequency of around 470.63 cm^{-1}, which is the bending vibration absorption of O atoms from the siloxane (Si–O–Si) functional group.

The presence of siloxane functional groups is strengthened by the appearance of absorption bands in the area of 794.67 cm^{-1} which is the presence of Si–O (Si–O–Si) symmetric stretching vibrations. In the absorption band around 3448.72 cm^{-1} is the stretching vibration of –OH from the silanol (Si–OH) group. In the area of wave number 2931.8 cm^{-1}, there is an absorption band originating from the (C–H) stretching vibration of (–CH_2) aliphatic. At the frequency of the wave number 1635.64 cm^{-1} is an absorption that shows the vibration of (O–H) which is attached to the carbonyl group (–C=O) which comes from the exopolysacride.

3.2. Characterization of adsorbent by SEM

The results of analysis by SEM to determine the morphology of HES and HESM (Figure 2). Can be seen that there are morphological differences between HES and HESM, the morphology of HES is more homogeneous because of the presence of gelatinous silica which acts as a supporting matrix for EPS. Silica gel is an inorganic solid which is used to enrich the surface of exopolysaccharide solids as adsorbent. Silica gel was chosen as the support matrix because it has a large surface area and active sites such as silanol (–Si–OH) and siloxane (Si–O–Si) which can bind chemically with the functional groups found in exopolysaccharides.

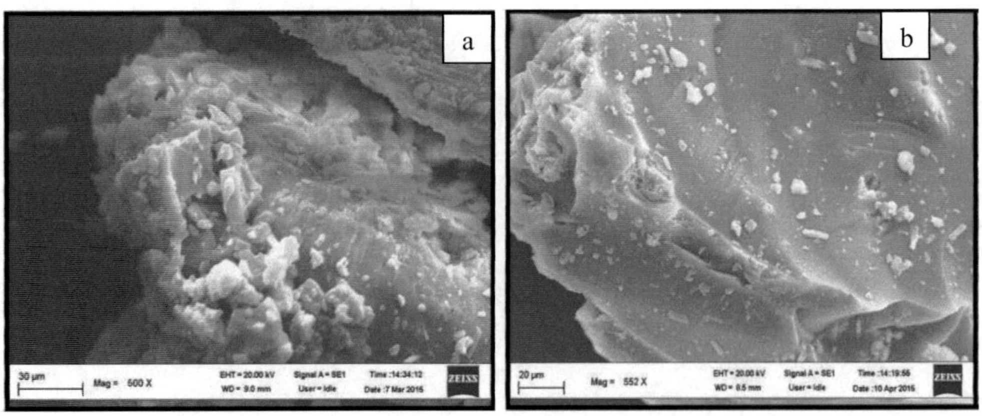

Figure 2. Scanning Electron Microscope (SEM) results (a) HES with 500x magnification and (b) HESM with 500x magnification.

From the results of this analysis it is in accordance with Figure 2. The largest element contained in HESM is Si. The results of HES and HESM characterization by SEM are compatible with the *IR* spectrophotometer characterization. In Figure 2, the different SEM results between HES and HESM are shown, in HESM there is Si obtained from the synthesis of exopolysaccharides with silica gel, indicating a more homogeneous surface morphology. Silica gel is an inorganic solid which is used to enrich the surface of exopolysaccharide solids as adsorbent.

3.3. Adsorption

3.3.1. Adsorption kinetics

The adsorption kinetics of metal ions in solution with the adsorbent can be determined by determining the adsorption rate. The adsorption rate between the adsorbent against metal ions plays a role in determining the optimum conditions for the interaction of metal ions from the solution and its equilibrium state to the adsorbent. The data in (Figure 3) were analyzed using the first-order pseudo and second-order pseudo equations with the kinetics of Pb(II) ion adsorption kinetics parameters. The results of the analysis show that the pseudo second order rate constant (k_2) of Pb(II) ions in HESM and HESM against Pb(II) ions is the greatest, this identify that HESM against Pb(II) ions has a greater adsorption rate than others. The data in this study shows that the HES and HESM adsorbents in this study tend to follow second-order pseudo kinetics because the R^2 values for HES and HESM for Pb(II) ions tend to be around 0.99 (Table 1).

Table 1. Second-order kinetic parameters on HES and HESM with respect to Pb(II) ion.

Adsorbent	*qe* (g mg^{-1})	K_2 (g mg^{-1})	R^2
HES	10.97	3.48	0.99
HESM	2.91 × 10^{-8}	0.34 × 10^{-8}	1

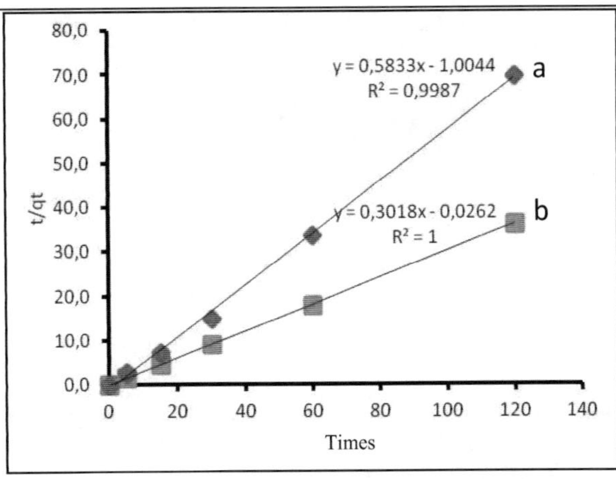

Figure 3. Kinetics of pseudo second-order Pb(II) ion on a. HES and b. HESM.

3.3.2. Adsorption isotherm

Adsorption isotherm can be used to determine the interaction between the amount of adsorbed substance (adsorbate) and the amount of adsorbent (adsorbent) and the optimum ability achieved by the adsorbent to adsorb metal ions.

The adsorption isotherm is a very important parameter in adsorption because it plays a role in determining the maximum conditions for optimal adsorption [15]. In this study, two forms of adsorption isotherm equation were used, namely the Langmuir and Freundlich adsorption isotherms to analyze the Pb(II) ion solution, presented on Table 2.

Table 2. Langmuir and Freundlich adsorption isotherm parameters.

Adsorbent	Langmuir			Freundlich		
	n_m (mg g^{-1})	K_L (L mol^{-1})	R^2	n (mol g^{-1})	K_F (mg g^{-1}) (L mg^{-1})$^{1/n}$	R^2
HES	3.85	39.39	0.86	1.098	2.55	0.63
HESM	272.63	99.99	0.96	1.101	1.15	0.77

In Figure 4 it can be seen that the adsorption isotherm model more closely follows the Langmuir adsorption isotherm model with an R^2 value of 0.9631 which is greater than the R^2 value in the Freundlich model. The adsorption capacity of the HESM adsorbent against Pb(II) was greater than the adsorption capacity of the HES adsorbent against Pb(II). The higher adsorption capacity indicates that the Pb(II) ion is more capable of making the site active and reliable. In the Langmuir isotherm, it explains that the adsorbent surface, namely HESM, has an active site which is proportional to the surface area of the adsorbent. Langmuir and Freundlich adsorption parameters including adsorption capacity and equilibrium constant seen in Table 2.

From the data analysis of adsorbent using Langmuir isotherm model which can be seen in Table 2, the adsorption capacities of HES and HESM for Pb(II) ions are 3.85 and 272.63 mg g^{-1}, respectively, K_L values (adsorption equilibrium constant) of 39.39 and 99.99 L mol^{-1}, respectively. The greater the K_L value, the greater the absorption affinity by the bio-sorbent can be seen in Figure 4.

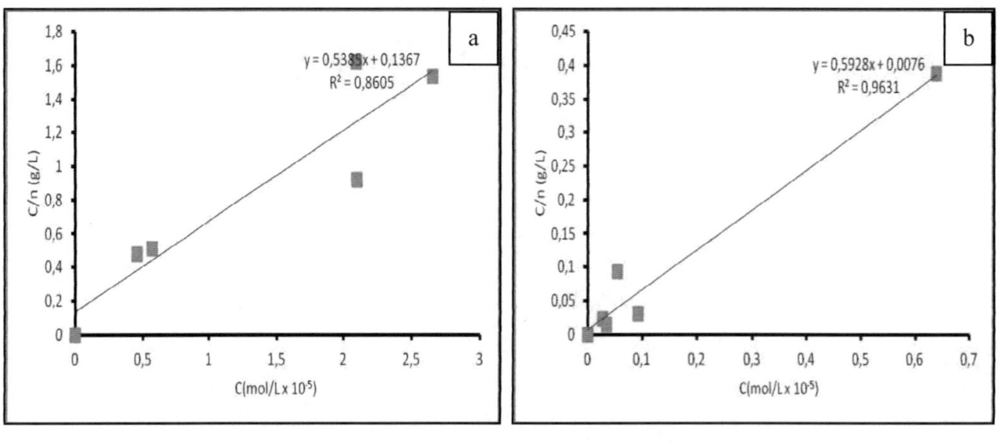

Figure 4. Adsorption isotherm curve of (a) Pb(II) ion in HES and (b) Pb(II) ion in HESM Langmuir model

Langmuir adsorption isotherm is an adsorption process that occurs because of the interaction between the adsorbate and the active sites on the adsorbent surface. The adsorption of metal ions by the adsorbent in this study tends to follow the Langmuir isotherm, so that the adsorption capacity can be determined directly using the Langmuir isotherm based on the slope value [16].

4. Conclusions

Adsorption of Pb(II) ion with HES and HESM is optimum at pH 7 and concentration at 25 ppm with the adsorption capacity on HES at 3.85 mg g^{-1} while on HESM at 272.63 mg g^{-1}, contact time of 15 min and it tends to follow second order pseudo kinetic model and Langmuir adsorption isotherm model. The HESM are effective adsorbents for removing heavy metals specifically Pb(II) ion in solution.

Acknowledgements

The authors would like to thank the Directorate of Research and Community Service, Directorate General for Research and Development, Ministry of Research, Technology and Higher Education of the Republic of Indonesia who have funded this research in accordance with contract number: 179/SP2H/AMD/LT/DRPM/2020.

References

[1]Kausar R A, Buhani & Suharso 2019 IOP Conference. Materials Science and Engineering. **857**.
[2]Seshadri S, Bishop P L,& Agha A M 1994 *Waste Manage.* **14** 127–137.
[3]Sen S & Demirer GN 2003 *J. Environt. Eng.* **129** 595-601.
[4]Buhani, Hariyanti F, Suharso, Rinawati, & Sumadi 2019 *Desalin Water Treat.* **142** 331–340.
[5]Buhani, Suharso, & Sumadi 2010 *Desalination.* **259** 140-146.
[6]Gupta SS& Bhattacharyya K G 2006 *J. Colloid. Interf. Sci.* **295** 21-32.
[7]Buhani, Suharso, Luziana F, Rilyanti M, & Sumadi 2019 *Desalin Water Treat.* **171** 281–293.
[8]Gupta V K, Nayak A, & Agarwal S 2015 *Environ. Eng. Res.* **20** 1–18.
[9]Celekli A, Atmaca, M Y,& Bozkurt H 2010 *J. Hazard. Mater.* **173** 123–129.
[10]Morillo P J A, García R R, & Quesada T 2008 *J.* of Microbiology & Biotechnology. **24** 2699–2704.
[11]Wei W, Wang Q, Li A, Yang J, Ma F, Pi S, & Wu D. 2016. *J. Scientific Reports*, **6** 1.
[12]Peng Q, Liu Y, Zeng G, Xu W, Yang C, & Zhang *J* 2010 *J. Hazard.* Mater. **177** 676-682.
[13]Liu Y, Zeng Y, Xu W,Yang C, & Zhang J 2010 *J. Hazard. Mater.* **177** 676-682.
[14]Buhani, Musrifatun, Pratama D S, Suharso, & Rinawati 2017 *Asian J. Chem.* **29** 2734-2739.
[15]Entezari MH & Bastami T R 2013*Mater. Res. Bull.* **48** 3149–3156.
[16]Buhani, Herasari D, Suharso, & Yuwono S. D 2017 *Orient. J. Chem.* **33** 418-429.

Investigation and modelling of the flood control system in the Aerotropolis of Yogyakarta International Airport

S Fadilah[1*], Istiarto[1], and D Legono[1]

[1] Department of Civil and Environtmental Engineering, Faculty of Engineering, Universitas Gadjah Mada, Jl. Grafika No. 2 Kampus UGM, Yogyakarta 55281, Indonesia

*Email: shofwatul.fadilah@mail.ugm.ac.id

Abstract. The New Yogyakarta International Airport has become one of Indonesia's National strategic projects and potentially support the development of the Aerotropolis area around the airport. Located on the southern coast of Yogyakarta with rivers at the airport's upstream, flooding can be a threat to the Aerotropolis area and the Yogyakarta International Airport. This research aims to investigate and model the flood review of the existing design of the flood control system of the Aerotropolis area and find out the possible option of better flood control in the area. The one-dimensional Hydrologic Engineering Center-River Analysis System (HEC-RAS) 4.1 was used to simulate the flood and its control in the Aerotropolis area. Simulation of flood control options includes river normalization, optimization of flood gates, retarding basin, and flood pumps. The present study proposed adjustment of performing in the use of a flood control alternative. Alternative flood control at each flood-prone area was adjusted to the level of flooding risk in the area. In conclusion, the use of retarding basin and flood pumps highly recommended to prevent flooding at the Aerotropolis area of Yogyakarta International Airport.

Keywords: Aerotropolis, flooding, investigation, modelling, flood control, HEC-RAS

1. Introduction

The New Yogyakarta International Airport has become one of Indonesia's National strategic projects [1]. It will be the central hub for international flights to and from Yogyakarta [2] and potentially support the development of the Aerotropolis area around the airport [3]. Aerotropolis is a metropolitan sub-region whose infrastructure, land use, and economy are centered on the airport [4]. As a new urban development paradigm of the 21st century, Aerotropolis has increasingly become a portal to national and regional economic growth [5-7].

Located at the southern coast of Yogyakarta, in the Temon district of Kulon Progo regency, with rivers at the airport's upstream, flooding can be a threat to the Aerotropolis area and the Yogyakarta International Airport [8,9]. Floods is one of a natural disaster frequently occurs there. Based on the data in the regional Disaster management Agency (BPBD) Kulon Progo, Temon district has a high flood hazard index value [10]. The flood is an overflow of water that submerges the land that is usually dry and inundated in many ways [11]. On 17th March 2019, a flood happened in the Temon sub-district and caused 344 people to evacuate. The New Yogyakarta International Airport has affected the condition, physical capacity, and morphology of the airport's rivers.

Based on these conditions, the investigation and the existing design review are necessary to find out the possible option of better flood control in the area. Due to its complexity and risk, the flood system modeling in the Aerotropolis area needs to do. Many software has been developed to predict future floods and possible affected areas. One of the most commonly used software for hydraulic flood modeling is one-dimensional Hydrologic Engineering Center-River Analysis System (HEC-RAS).

There have been several previous studies related to flood inundation in the Kulon Progo area, both individually and by institutions. In 2018, PT. Angkasa Pura researched flood inundation simulation using MIKE 21 and HEC-RAS software. The alternative flood control proposed includes normalization, construction of river dike, and the use of four retarding basins, but retarding basins have not been simulated with HEC-RAS [9]. Aditya K et al. (2019) analyzed the land-use changes in the Temon sub-districts after YIA construction with the study area is in the West Carik channel, the Sindutan channel, the Poncol channel and the Lawang channel [12].

Alternative flood control proposed is river normalization. In 2019, PT. Virama Karya In this work, PT Virama Karya simulated the flood occurrence using HEC-RAS. The simulation was modelled by dividing into three major systems, namely the Carik Barat System, the Ledeng System and the Carik Timur System. For existing conditions, the system simulated using steady flow. While for the redesign, the system simulated using unsteady flow analysis [8]. In 2020, Akhyar Fahrizal et al. analyzed the effect of land-use changes on the runoff potential in the Serang watershed, Kulon Progo Regency, which is the Yogyakarta International Airport [13]. In the same year of 2020, Festi and Novi analyzed the estuary conditions for flooding occurrence in the Serang River, Kulon Progo Regency. This study focused on the behaviour of the Serang River estuary using HEC-HMS and HEC-RAS to determine the effectiveness of the existing section of the estuary [14].

In this research, flood control was simulated using HEC-RAS 4.1 software with unsteady flow analysis for a whole of drainage and river systems. Simulation of flood control options includes river normalization, optimization of flood gates, retarding basin, and flood pumps. The present study proposed adjustment of performing. Alternative flood control at each flood-prone area was adjusted to the level of flooding risk in the area. This research was more detailed than previous studies, so that the result of the simulation will be better than others.

2. Study of Literature

2.1. Hydraulic flood modelling

Flood hydraulic modelling is a mathematical model used to evaluate the main elements of fluids flow to predict flooding. The model can simulate the 1D, 2D, or 3D forms depending on the user's needs. Flood hydraulic modelling was used to determine the ability of the river to drain flood discharge [15]. In hydraulic flood modelling, the model analyzes the input data such as boundary conditions, channel geometry, forces, and other channel parameters into output data, including flow velocity, flow discharge, water surface level, flow pressure, and others [16], so that the modelling results highly dependent on input data [17].

The nature of flooding causes the flood modelling simulation cannot be modelled precisely as in reality but just close to it [18]. Even though a model has been calibrated to get optimal results, it does not guarantee that the results are the same as a real condition [17]. Many software has been developed to simplify the calculation of flood hydraulics such as HEC-RAS, LISFLOOD, MIKE, and others. Those have their advantages and disadvantages. One of the well-known and widely used software is HEC-RAS[19].

2.2. Overview of HEC-RAS software

HEC-RAS is a free software designed by M. Gary W. Brunner in 1995 and was developed at the Hydrologic Engineering Center (HEC) to perform hydraulic behaviour of flood in one-dimensional (1D), two-dimensional (2D) or a combined of 1D and 2D calculation for permanent (steady) or non-permanent flow (unsteady) both in the natural and constructed channels [20,21]. HEC-RAS also enables users to perform quasi unsteady and full unsteady flow sediment transport-mobile bed modelling, water temperature analysis, and generalized water quality modelling (nutrient fate and transport) [20].

The choice of steady or unsteady flow analysis simulation depends on the analysis's objectives; for example, to simulate the flood routing in a river, a non-permanent flow simulation is needed. HEC-RAS applies conservation laws to the basis of its hydraulic calculations. Conservation laws divide into three are mass conservation law, momentum conservation law, and energy conservation law [22]. The mass conservation law in open-channel flow states that the volume of liquid entering a flow system will be proportional to that comes out as in equation 1 [20,23-24].

$$\frac{dA_T}{dt} + \frac{dQ}{dx} = q_i \quad (1)$$

Where dA_T/dt is the rate of storage change, dQ/dx is the flow rate, and q_i is lateral inflow (m^3). The momentum conservation law is expressed as the momentum equation. The momentum conservation law in a control volume is equal to the amount of momentum flux entering and leaving the control volume plus the sum of forces acting on the control volume that causes a change momentum on the control volume, defined as equation 2 [25].

$$\frac{\partial Q}{\partial t} + \frac{d(Qv)}{dx} + gA\left(\frac{\partial y}{\partial x} + S_f\right) = 0 \quad (2)$$

The law of energy conservation in the hydraulic system is stated according to the Bernoulli equation, as equation 3. Bernoulli's law states that the amount of water energy from each flow through a channel cross-section is equal to the sum of elevation head, pressure head, and velocity head.

$$H = z + d\ cos\theta + \frac{v^2}{2g} \quad (3)$$

Where H is the total head (m), z is the elevation of the channel bottom (m), θ is slope channel (rad,^{0}), d is depth (m), v is the velocity (ms^{-1}), g is the gravitational force (ms^{-2}), and ⟨ is coefficient of kinetic correction. If the channel slope is small, then the value of θ can be assumed to be 0, so that the value of d is proportional to the value of h measured vertically. Therefore, the Bernoulli equation changes to equation 4.

$$H = z + h + \frac{v^2}{2g} \quad (4)$$

If the energy in is equal to the sum of the energy out and friction energy, equation 4 can be written as equations 5 and 6.

$$H_{in} = H_{out} + H_f \quad (5)$$

$$z + h + \frac{v^2}{2g} = z + h + \frac{v^2}{2g} + H_f \quad (6)$$

The open channel flow velocity in HEC-RAS is calculated using one of the empirical formula, a widely used and very versatile formula in water resources, the manning equation. Robert Manning first presented the manning formula in 1889. The manning equation was originally developed for uniform channels with a uniform steady state. However, in reality, no flow can be uniform and stable so that the manning equation can ultimately be applied to unsteady flow with non-prismatic channels. Equation 5 is a widespread and useful form of the Manning equation [26].

$$V = \frac{1}{n} R^{\frac{2}{3}} S^{\frac{1}{2}} \qquad (7)$$

Where V is the flow rate (m^3s^{-1}), n is the Manning's roughness coefficient (unitless), R is the hydraulic radius (m), and S is slope energy gradient (unitless). Manning's roughness coefficient value is a function of the wall material. According to [27], n value of manning of the channel is classified into several classes, as in Table 1.

Table. 1. Manning's roughness coefficient value [27]

No	Type of channel and description	*n* Manning's value		
		Minimum	Normal	Maximum
1	Concrete			
	-Culvert, straight and free of debris	0,010	0,011	0,013
	-Culvert with bends, connections, and some debris	0,011	0,013	0,014
	-Finished	0,011	0,012	0,014
	-Sewer with manholes, inlet, etc., straight	0,013	0,015	0,017
2	Earth, straight and uniform			
	-Clean recently completed	0,016	0,018	0,02
	-Clean, after weathering	0,018	0,022	0,025
	-Gravel, uniform section, clean	0,022	0,025	0,03
	-with short grass, few weeds	0,022	0,027	0,033
3	Natural streams			
	-Clean, straight, full stage, no rifts or deep pools	0,025	0,03	0,033
	-clean, winding, some pools and shoals	0,033	0,04	0,045
	-Sluggish reaches, weedy, deep pools	0,05	0,07	0,08

3. Materials and Methods

3.1. Study Area

The research location is in Temon District, Kulon Progo Regency, Special Region of Yogyakarta Province. Based on the topography, the research location area was dominated by gardens, rice fields, and residential areas. It is located between the Bogowonto River and the Serang River. During the rainy season, the two rivers affect the airport drainage area's flow system and its surroundings. The figure of the research location shown in Figure 1.

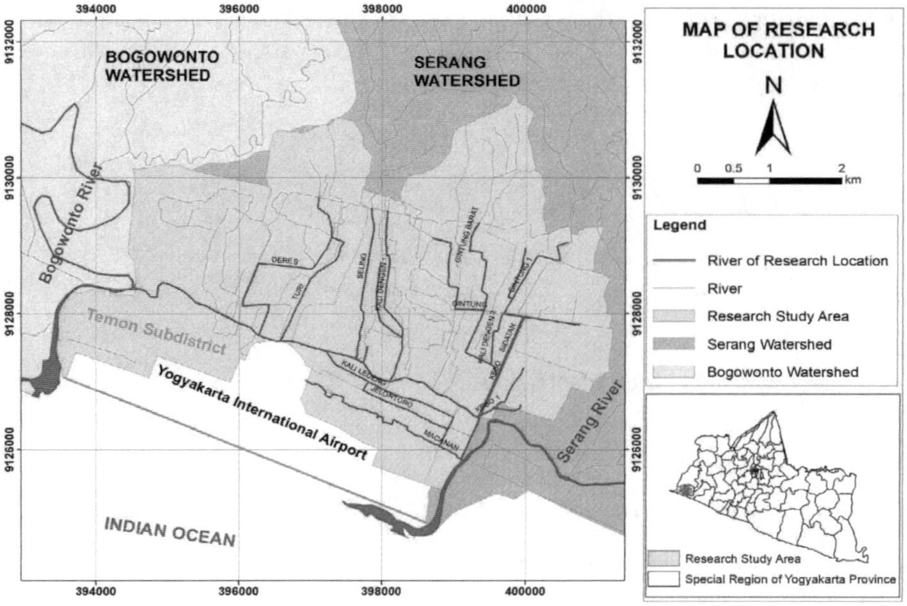

Figure 1. Research location

3.2. Data Resources and Analysis

Primary data in this study were obtained from the investigation at the research location. Those data are including flood information data, documentation of photographs at the research location, and inventory of buildings along the river/channel. Flood information data is obtained from the investigation at the research location, through interviews with stakeholders and residents affected by flooding in the Aerotropolis Yogyakarta International Airport area. Flood information data will be used as a calibration so that it becomes a reference in modelling flood control using HEC-RAS 4.1.

Secondary data required in this study were obtained from related agencies and previous studies. This research uses secondary data consisting of Map of the RBI, river profile data, tidal data, and hydrologic analysis data.

The analysis of this research was a hydraulic analysis using HEC-RAS software with unsteady flow analysis. The primary and secondary data were used as input data for simulation using the HEC-RAS. The simulation was divided into two steps, both the existing simulation and the simulation with flood control. In the existing simulation, the flood system was simulated to determine which locations were affected by flooding and needed to apply for flood control. The simulation with flood controls was performed by trying various flood control options, including river normalization, optimization of flood gates, retarding basin, and flood pumps. The present study proposed adjustment of performing. Alternative flood control at each flood-prone area was adjusted to the level of flooding risk in the area.

4. Result and Discussions

4.1. Evaluation of existing channel capacity

Existing channels system was simulated using one-dimensional HEC-RAS with an unsteady flow analysis. The unsteady flow analysis was appropriate to use in this hydraulic analysis because the type of simulation is flood routing. Based on investigation, most of the channel modelled is still a natural channel and there is a tidal influence from the sea in the downstream of both Bogowonto river and Serang river. There are two boundary conditions used in the simulation using HEC-RAS are stage hydrograph and flow hydrograph. At the downstream boundary condition, a stage hydrograph was applied with tidal data. A flow hydrograph was applied at the upstream boundary conditions with a discharge resulting from the hydrological analysis. The water level conditions strongly influence floods in the Aerotropolis area of Yogyakarta International Airport downstream of the Serang and Bogowonto Rivers. The downstream of those rivers causes backwater.

The stage and flow hydrograph conditions downstream of the Bogowonto and Serang rivers in Figure 2. In the Bogowonto River, the stage hydrograph adjusted to the tide observations. At the beginning of the simulation, the flow hydrograph increases until the end of the simulation. On the contrary, the height of the tides' level is decreasing. In the Serang river, the tide level is constant at 1 m. The sea level in the Serang river was kept constant at a maximum of 1 m because there is a tourism object in that area.

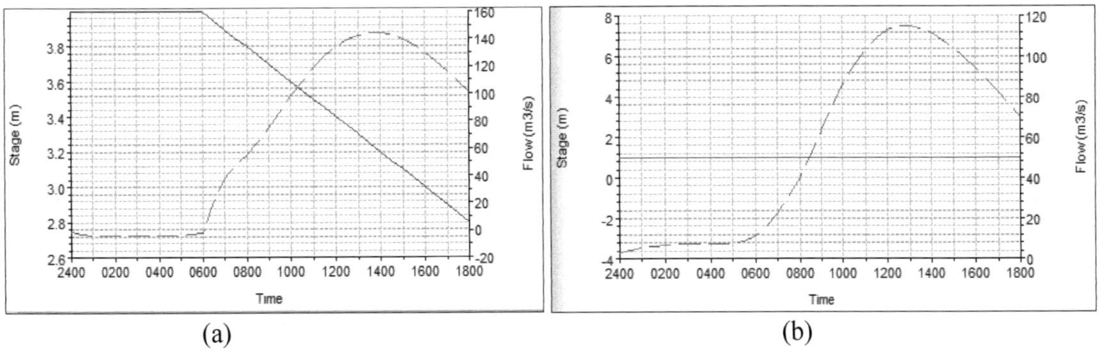

Figure 2. Stage and flow hydrograph condition at downstream (a) Bogowonto river (b) Serang river

The simulation result of the existing conditions, as shown in Figure 3, explains that several channels are overflowing, causing flood inundation around the area. The overflow channel was represented as the red line symbol, while the result of HEC-RAS simulation shown in Figure 4. Based on investigation, there were water structures existing in the existing conditions, such as gate, culvert, bridge, and weir. All existing floodgates cannot function optimally, and some were broken. It is highly recommended to repair the floodgates so that the floodgate's function in preventing flooding can operate optimally. Each water structure has different hydraulic characteristics which will have a direct influence on hydraulics along the open channel, such as decreasing the flow rate, increasing the water surface level, and others.

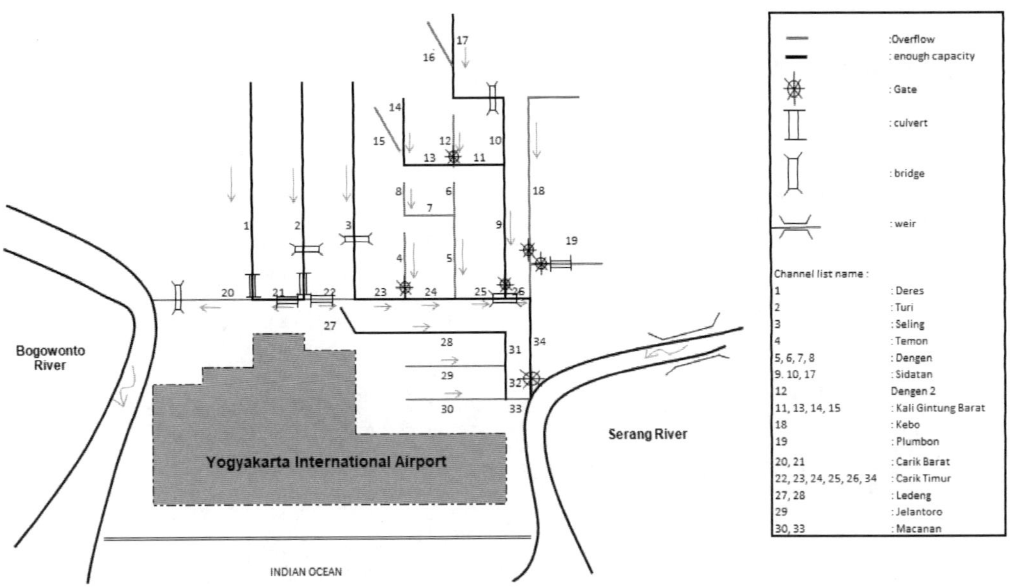

Figure 3. Existing scheme of the channel system

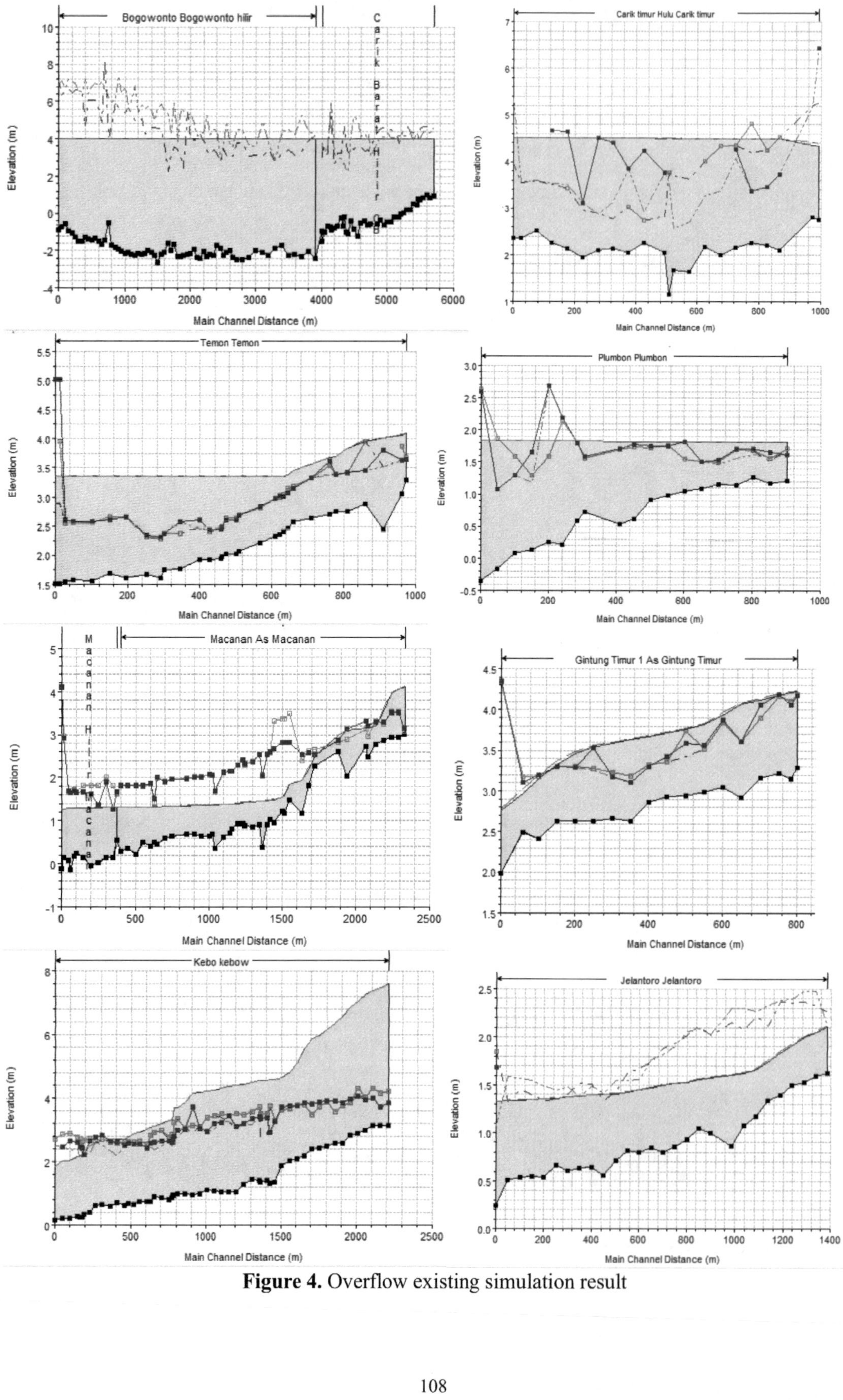

Figure 4. Overflow existing simulation result

4.2. Evaluation of redesign channel system

Simulation of the redesign channel system is a simulation of floods with channel conditions different from existing conditions. Overflow channels in the existing simulation were normalized with the new dimensions in each path as needed, as shown in Figure 5 with the green line. Normalization aims to normalize or restore drainage conditions as before, carried out when the channel dimensions are not uniform, or there is a narrowing in several sections, which results in a reduction in capacity and inhibits river flow rates [28].

The floodgate in the redesign channel system was fully functionalized, as shown in figure 4, with the blue colour gate symbol. A new floodgate was applied in the junction between Carik Timur channel (number 22) and Ledeng channel (number 27), as shown in Figure 5 with the red colour gate symbol. It was carried out to reconnected both channels. It makes different from previous studies.

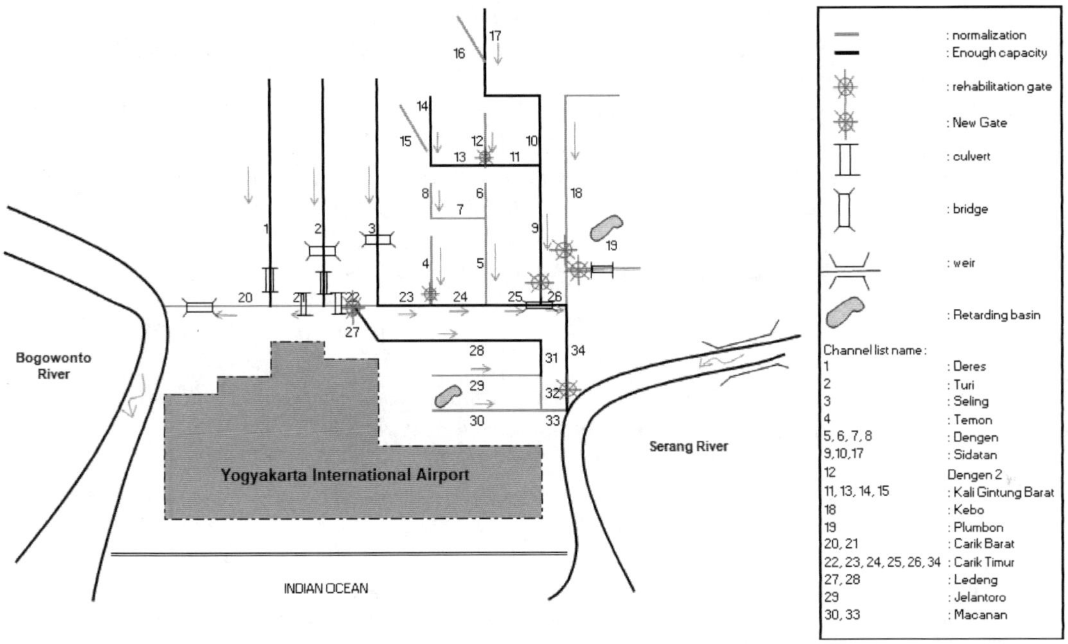

Figure 5. Scheme of the redesign channel system

The result of the simulation with the river flow conditions that had been normalized and completed by the floodgates shows that there are still flooded areas, especially in channel number 18 and number 19 (Plumbon area) and in channel number 29 and number 30 (Jelantoro area). Therefore, as a solution, a retarding basin was built and modelled in the location area between channel number 18 and number 19 and in the area between channel number 29 and 30. The planning of retarding basin was equipped with flood pumps. Pumps are greatly optimum in overcoming flood problems, especially in areas downstream of river flows such as estuaries [29]. The pump will release floods in the retention pond/reservoir out to the channel or sea. In this simulation, the pump only operates when the water level is at a certain level, according to the pump's elevation-discharge function.

5. Conclusions

Based on the simulation results, it can be concluding that Aerotropolis area needs some flood handling. The first is the normalization in some channels to handle the floods around the Aerotropolis drainage. The second is the existing floodgate needs to repair and need a new flood gate between channel number 22 and number 27 to connect the Carik Timur channel to the Ledeng channel. The third is the use of retarding basin and flood pumps highly recommended to prevent flooding at the Aerotropolis area of Yogyakarta International Airport. For further research, the flood simulation can be re-simulated using different discharge from the updated hydrological analysis with predictive land use change information.

Acknowledgment
The authors gratefully acknowledge of Indonesia Endowment Fund for Education (LPDP) for funding this research. The authors also thank the assistance of related agencies for the data collection.

References

[1] Minister of National Development Planning 2015 Attachment to Presidential Regulation of the Republik of Indonesia Number 2 Year 2015 about the Medium-Term National Development Plan (RPJMN) 2015-2019 3rd Book Regional development agenda (Indonesia : Minister of National Development Planning)
[2] Kadarisman M 2019 Policy Implementations of Yogyakarta International Airport Development (Yogyakarta : Universitas Gadjah Mada)
[3] Edita E P 2019 *At What Cost, to Whom ? An Analysis of Social and Environtmental Impacts of New Yogyakarta International Airport (NYIA) project, Indonesia* (Sweden : Lund University Centre for Sustainability Studies)
[4] Kasarda J D 2019 *The Wiley Blackwell Encyclopedia of Urban and Regional Studies* ed A Orum (USA : John Wiley & Sons Ltd)
[5] Liou J J H, Hsu C C and Li C S J 2018 *Technological and Economic Development of Economy* **24** 1080—1103
[6] Wang Y, Chou C C and Yeo G T 2013 *The asian journal of shipping and logistics* **29** 395-414
[7] Charles M B, Barnes P, Ryan N and Clayton J 2007 *J. Futures* **39** 1009-1028.
[8] KSO PT. Virama Karya (Persero) Semarang –PT. Indra Karya (Persero) Engineering Div. II 2019 *Laporan Hidrologi, Hidrolika, dan Hidro-Oceanografi detail Desain Pengamanan Pantai dan Pengendalian Banjir Sungai Bogowonto dan Sungai Serang.*(Yogyakarta: KSO PT. Virama Karya (Persero) Semarang –PT. Indra Karya (Persero) Engineering Div. II) [In Bahasa Indonesia]
[9] Angkasa Pura 2019 *Laporan Kajian Potensi Banjir di Kawasan Sekitar Bandara* (Yogyakarta: Angkasa Pura) [In Bahasa Indonesia]
[10] Regional Disaster Management Agency 2013 *Peta Bahaya Banjir di Kabupaten Kulon Progo, Daerah Istimewa Yogyakarta*. [In Bahasa Indonesia]. Available at https://bpbd.Kulon Progo.go.id accesed on March 20,2020
[11] Doswell III C A 2003 Flooding (USA: Elsevier Science Ltd.)
[12] Aditya K 2019 *Penanganan Dampak Genangan Akibat Mega Proyek Yogyakarta International Airport* [research] (Yogyakarta, Universitas Gadjah Mada) [In Bahasa Indonesia]
[13] Akhyar F, Joko S, and Istiarto 2020 *Kajian Pengaruh Perubahan Penggunaan Lahan Terhadap Potensi Limpasan di DAS Serang Kabupaten Kulon Progo* [Research] (Yogyakarta: Universitas Gadjah Mada) [In Bahasa Indonesia]
[14] Festi W P and Novi A S P 2020 *Prosiding Webminar Nasional Teknik Sipil 2020* (Surakarta: Engineering Faculty of Muhammadiyah Surakarta University) ISSN: 2459-9727
[15] Wahyuningtyas A, Pahlevari J E, Darsono S dan Budeny H 2017 *Jurnal Karya Teknik Sipil* **6** 161—171
[16] Novak P, Guinot V, Jeffrey A and Reeve D E 2010 *Hydraulic Modelling-An Introduction* (London and USA: Spoon Press)
[17] Lim N J 2018 Modelling, Mapping and Visualisation of Flood Inundation Uncertainties (Sweden: Gävle University Press)

[18] Teng J, Jakeman A J, Vaze J, Croke B F W, Dutta D and Kim S *Environtmental Modelling and Software* **90** 201-216
[19] Quiroga V M, Kure S, Udo K and Mano A 2016 *RIBAGUA-revista Iberoamericana del Aqua* **3** 25—33
[20] US Army Corps of Engineers 2016 *HEC-RAS River Analysis System : Hydraulic Reference Manual Version 5.0* (Davis: Hydrologic Engineering Center of US Army Corps of Engineers)
[21] Dimitriadis P, Tegos A, Oikonomou A, Pagana V, Koukouvinos A, Mamassis N, Koutsoyiannis D and Efstratiadis A 2016 *Journal of Hydrology* **534** 478—492
[22] Jobson H E and Froehlich D C 1988 *Basic Hydraulic Principles Of Open-Channel Flow* (Virginia: U.S. Geological Survey)
[23] Chaudhry M H 2008 *Open-Channel Flow Second Edition* (Columbia: Springer)
[24] Akan A O 2006 *Open Channel Hydraulics* (Oxford : Butterworth-Heinemann)
[25] French R H 1985 *Open-Channel Hydraulics* (New York: McGraw-Hill Book Company)
[26] Ball J, Babister M, Nathan R, Weeks W, Weinmann E, Retallick M, Testoni I *Australian Rainfall and Runoff: A Guide to Flood Estimation* Book 6 (Australia: Commonwealth of Australia [Geoscience Australia]) pp 18—27
[27] Chow V T 1958 *Open-Channel Hydraulics* (New York: McGraw Hill Book Company, Inc.) pp 110-114
[28] Evy H and M. Soemantoro 2015 *Jurnal Teknik Sipil* **1** 71—80
[29] J. J. Lian, K. Xu and C. Ma 2013 *Hydrology Earth System Sciences* **17** 679-689

The effects of Ozonation and ultrasonic cavitation on batik wastewater treatment with coagulation-flocculation as pre-treatment

I N J Siahaan[1], A E Saputra[1], and E F Karamah[1*]

[1]Department of Chemical Engineering, Faculty of Engineering, Universitas Indonesia, Kampus UI Depok 16424, Indonesia

*Email: eva.fathul@ui.ac.id

Abstract. The batik industry produces large volumes of wastewater with high organic content. Dyes in batik wastewater are difficult to degrade by biological processes. Therefore, further processing is needed to reduce pollutants before being released into the environment. This study aims to observe the performance of the ultrasonic cavitation, ozonation, and combination of ozonation/ultrasonic cavitation to treat batik wastewater, which was first given a coagulation-flocculation pre-treatment with Polyaluminium Chloride (PAC) coagulants. The variations applied were ultrasonic intensity (20%, 30%, and 60%) and pH of wastewater (4, 7, and 10). In this study, it was found that the combination of the ozonation/ultrasonic cavitation method with a pH of 4 and ultrasonic intensity of 20% produced the best performance, namely COD, TSS, and color (Pt-Co) removal of 77.02%, 95.15%, and 94.88%. The ultrasonic intensity and initial pH of wastewater played an important role in the performance of the treatment process. In the ultrasonic cavitation method with an ultrasonic intensity of 20%, the highest percentage of COD, TSS, and color (Pt-Co) removal was 65.59%, 91.51%, and 93.41%. The ozonation method in acidic conditions (pH 4) produced a better performance with COD, TSS, and color (Pt-Co) removal of 70.51%, 94.35%, and 96.10% for 60 minutes.

Keywords: Batik wastewater, coagulation, flocculation, ozonation, ultrasonic cavitation

1. Introduction

The process of producing batik requires large amounts of water, which causes large amounts of wastewater production. In Indonesia, the volume of water needed for batik processing is around 25 million m^3 per year [1]. Nearly 85% of the water used ends up being wastewater that has the characteristic of cloudy and thick color, along with high temperature, pH, and organic content. The major contaminant in the batik industry wastewater comes from synthetic dyes. Synthetic dyes have a stable composition, are difficult to decompose, and prevent sunlight from entering the water, which can disrupt the balance of the aquatic ecosystem.

The coagulation process aims to destabilize colloidal particles by adding a coagulant and stirring at high speed. The addition of coagulants intends to bind colloidal particles in wastewater so that it can form the flocs. Meanwhile, flocculation is a process of forming flocs on slow stirring to increase the fusion of particles or agglomeration so it can easily deposit. Coagulation-flocculation is a cost-effective and efficient process of removing color, so it is prospective to applicate in batik wastewater treatment. However, the coagulation-flocculation produces wastewater that does not meet the quality standards of wastewater disposed to the environment. Therefore, further processing is required.

Ultrasonic cavitation and ozonation are advanced oxidation processes that can increase wastewater biodegradability. Ozonation can increase biodegradability by oxidizing organic compounds directly and changing the organic chemical structure of wastewater. Meanwhile, ultrasonic cavitation is a technology that uses high-frequency ultrasonic waves to break down complex organic compounds into simpler compounds. The combination of ozonation and ultrasonic cavitation methods can reduce the concentration of organic pollutants and increase biodegradability in batik wastewater by producing a total organic carbon (TOC) removal of 85% [2] and decolorization of dyes up to 99.2% [3].

In this study, a pre-treatment of batik wastewater using coagulation-flocculation was intended to remove the suspended or colloidal contaminants contained in the batik wastewater. The coagulant used is Polyaluminium Chloride (PAC) with the chemical formula of $Al_{12}Cl_{12}(OH)_{24}$. The reason for choosing PAC is because commonly used in wastewater treatment processes, works over a wide pH range, and has low corrosivity. Further, the methods of ultrasonic cavitation (U.C), ozonation (O), and a combination of ozonation and ultrasonic cavitation (O/U.C) were carried out to treat batik wastewater.

2. Materials and Methods

The duration of the batik wastewater treatment process in each method was 60 minutes. The independent variables used in this study were the ultrasonic wave intensity (20%, 30%, and 60%) for the ultrasonic cavitation and the initial pH of batik wastewater (4, 7, and 10) for the ozonation and combined method of ozonation/ultrasonic cavitation. The evaluation carried out on the performance of batik wastewater treatment was based on several parameters, namely COD, TSS, color (Pt-Co), and TOC.

The main equipment used in this study were Ultrasonic Processor E-Chrom Tech UP-800, Ozonator Aquasuper AS-002 with an average ozone production of 398.25 mg/hour at a flow rate of 6 L/minute, Nocchi centrifugal pump, and 5 Litres reservoir tank. The scheme of the research equipment is shown in Figure 1.

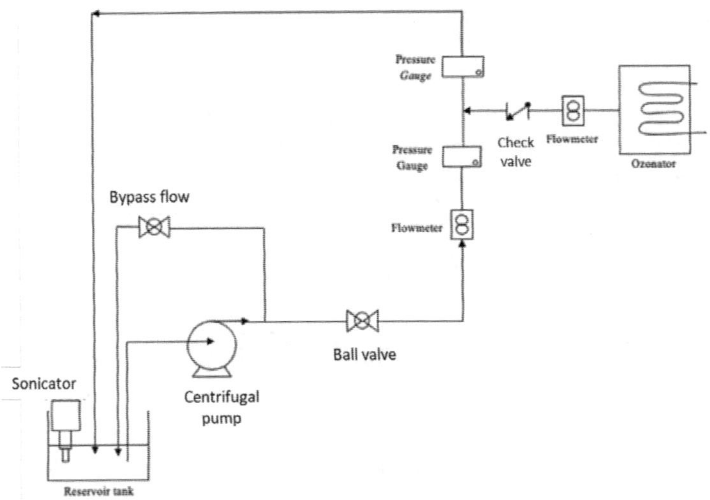

Figure 1. Research Equipment Scheme.

The materials used in this study were batik wastewater, 500 ppm PAC coagulant, 0.1 M potassium iodide (KI), 2 M sulfuric acid (H_2SO_4), 0.05 M sodium thiosulfate ($Na_2S_2O_3$), 2% starch indicator, 0.001

N potassium permanganate ($KMnO_4$), sodium hydroxide (NaOH), and water. All chemicals were supplied from Merck in the pro analysis class (p.a.), except sodium hydroxide (technical grade).

3. Results and Discussions

3.1 Characterization of Batik Wastewater

Initial characterization of batik wastewater aims to identify the amount of pollutant load on the wastewater so it can be used as a reference to measure the effectiveness of the wastewater treatment process. At first, batik wastewater went through a filtering process with a stainless-steel filter. The purpose of filtering was to reduce the contaminant on the wastewater. Furthermore, dilution was carried out with a ratio of one to four to minimize the formation of foam that interfere with the process. Table 1 shows the results of the initial characterization of batik wastewater.

Table 1. Comparison of the Results of Initial Characterization of Batik Wastewater.

Parameter	Unit	Raw Wastewater	After Filtering and Dilution	Adjustment of pH 4	Government Standard
pH	–	10.08	9.97	4	6-9
COD	mg/L	29,509	5,428	4,282	150
TSS	mg/L	1,525	285	1,990	50
Pt-Co	mg/L	26,756	4,270	28,280	-

In Table 1, it can be seen that the filtering and dilution treatments provided significant results for reducing levels of COD, TSS, and color (Pt-Co). The filtering process removed large particles by the filter media, while dilution caused the particles contained in the wastewater to be more dissolved. Before entering the next process, the pH adjustment was given to the wastewater. Adjusting pH to 4 caused a decrease in COD and an increase in TSS and color. This was due to the formation of flocs which came from the formation of salt due to the addition of acidic compounds to the alkaline wastewater. The formation of flocs triggers an increase in suspended solids and makes the color cloudy. Previous studies had shown that pH adjustment in batik wastewater with the addition of acid could reduce COD with a removal percentage of up to 79% [4].

3.2 Performance of Coagulation-Flocculation Method with Jar Test

In this study, the coagulation-flocculation method aims to reduce the pollutant load on the wastewater, especially colloidal particles in batik wastewater. The coagulation-flocculation process was carried out using a jar test with the addition of PAC coagulant. The comparison between the parameters of batik wastewater before and after the coagulation-flocculation process is shown in Table 2.

Table 2. Results of Parameter Testing on Batik Wastewater on Before and After Coagulation-Flocculation.

Parameter	Unit	Before Coagulation-Flocculation	After Coagulation-Flocculation
pH	–	4	4.12
COD	mg/L	4,282	1,019
TSS	mg/L	1,990	145
Pt-Co	mg/L	28,280	1,078

Table 2 shows that there was a significant reduction in COD, TSS, and Pt-Co. COD reduction occurred because of the high concentration of colloids after the adjustment of pH 4 in the batik wastewater. It increased the suspended solids content and turbidity. The increase in colloid concentration in the wastewater caused the collision frequency of the destabilized particles to increases so that the formation of flocs more easily occurred. Moreover, the addition of a coagulant that is a cation to wastewater that has negative electric charges causes the formation of micro flocs and eventually forms particles of larger size (macro flocs).

The PAC coagulant underwent a hydrolysis reaction that results in insoluble aluminum polyhydroxide that settled and formed a large volume of flocs. The flocs formed are then removed the color and colloidal particles by adsorption onto/within the formed metal hydroxide [5]. Thus, there was a reduction in TSS and color (Pt-Co) in batik wastewater.

3.3 Hydroxyl Radical Quantification

Hydroxyl radical quantification test was carried out by circulating water for 60 minutes. The sample obtained was then tested by the permanganometric titration method. The use of water as a substitute for batik industrial wastewater was intended to prevent the formation of complex molecules that could interfere with the hydroxyl radical quantification. The results of the quantification of hydroxyl radicals can be seen in Figure 2.

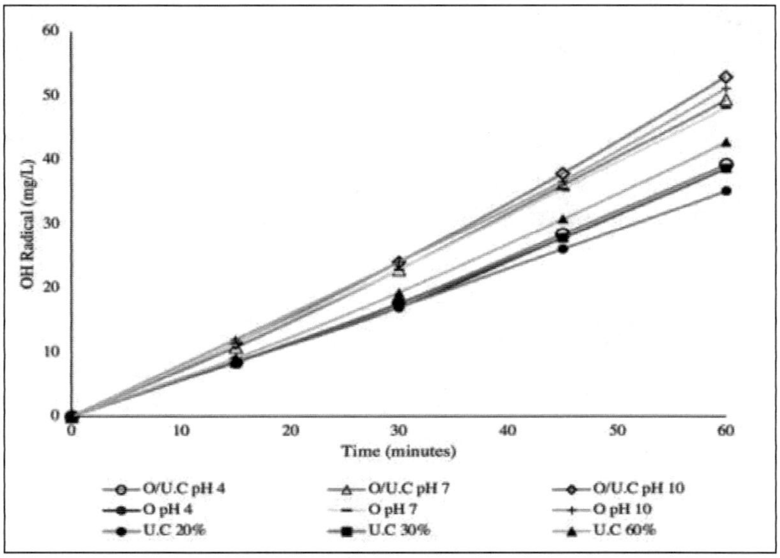

Figure 2. Hydroxyl Radical Profile towards Time at Various Ultrasonic Wave Intensities (20%, 30%, and 60%) and pH (4, 7, and 10) (Condition: Initial Temperature 25°C)

Based on Figure 2, the highest number of hydroxyl radicals was produced by the combination of ozonation/ultrasonic cavitation with a pH of 10, which was 53 mg/L in the 60 minutes. Meanwhile, the least amount of hydroxyl radicals was produced by the ultrasonic cavitation with an ultrasonic wave intensity of 20%, which was 35 mg/L.

The combination of ozonation/ultrasonic cavitation with a pH of 10 produced the highest number of hydroxyl radicals because the formation of the largest oxidizing compounds occurred in alkaline conditions. In alkaline solutions, hydroxide ions (OH^-) were more abundant. The hydroxide ion acted as an initiator in the decomposition reaction of ozone to hydroxyl radicals [6]. Radical compounds formed between ozone and hydroxyl ions would react over with ozone to produce more hydroxyl radicals. Equations 1 and 2 show the decomposition reaction of ozone to hydroxyl radicals.

$$O_3 + OH^- \rightarrow HO_2^- + O_2 \quad (1)$$

$$O_3 + HO_2^- \rightarrow \bullet OH + O_2^{\bullet -} + O_2 \quad (2)$$

3.4 Ultrasonic Cavitation on Batik Wastewater Treatment

In the ultrasonic cavitation, the circulated wastewater had been given a pH adjustment so that it became pH 4. The application of acidic pH to azo dye treatment by the ultrasonic cavitation resulted in a higher degradation rate than alkaline conditions [7]. Figure 3 shows the percentage of parameter removal at 60 minutes for each of the independent variables.

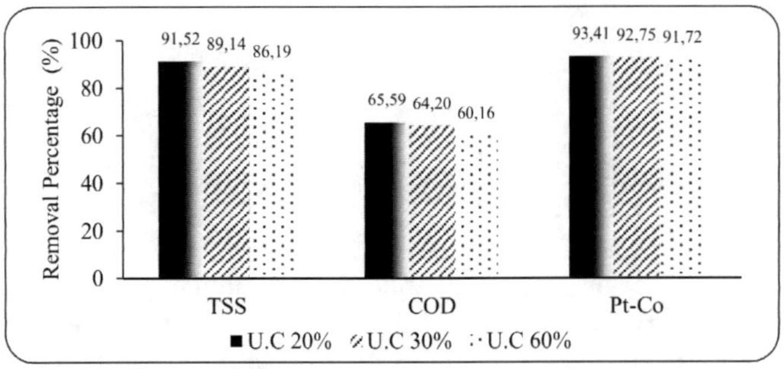

Figure 3. The Comparison of the Results of TSS, COD, and Pt-Co Removal in Batik Wastewater with Ultrasonic Cavitation Method with Ultrasonic Wave Intensities of 20%, 30%, and 60% (Condition: pH = 4, Processing Time = 60 Minutes).

The results presented on Fig. 3 shows that the ultrasonic cavitation with an ultrasonic intensity of 20% produced the highest percentage removal of TSS, COD, and Pt-Co. The high intensity of ultrasonic waves triggered a condition in which the solution is saturated with oxygen and peroxyl so that it formed hydrogen peroxide, as shown in equations 3 and 4 [8].

$$O_2 + \bullet H \rightarrow \bullet HO_2 \tag{3}$$
$$\bullet HO_2 + \bullet HO_2 \rightarrow H_2O_2 \tag{4}$$

Hydrogen peroxide has a lower oxidation potential than ozone (2.07 V) and hydroxyl radical (2.8 V), which is 1.78 V. The lower oxidation potential can reduce the efficiency of the ultrasonic cavitation.

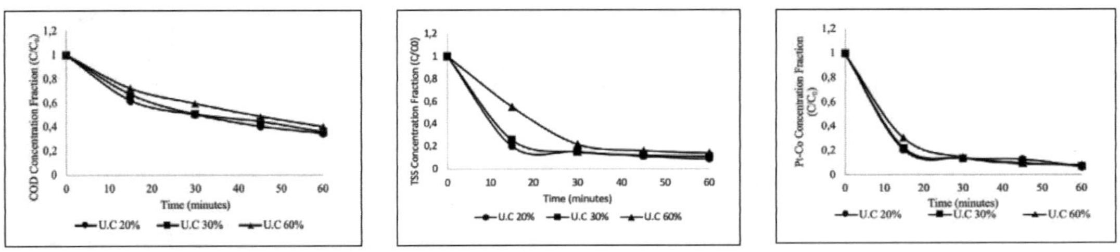

Figure 4. Concentration Fraction a) COD, b) TSS, and c) Pt-Co towards Time at Various Ultrasonic Wave Intensities (20%, 30%, and 60%) (Condition: Initial Temperature 25°C, pH 4).

Based on Figure 4, the concentration fractions of COD, TSS, and Pt-Co tended to decrease with the length of processing time. The highest removal rate occurred in the ultrasonic cavitation with 20% ultrasonic intensity, with the removal of COD from 887 mg/L to 305 mg/L, removal of TSS from 165 mg/L to 14 mg/L, and removal of Pt-Co from 2,172 mg/L to 143 mg/L.

Then, followed by removal using the ultrasonic wave intensities of 30% and 60%. In the ultrasonic cavitation, hydroxyl radicals, hydrogen, and other radicals (H_2O_2, $\cdot HO_2^-$, $\cdot O_2$) are produced by pyrolysis. These radical compounds, especially hydroxyl radicals, attack organic compounds in the bulk liquid and break down suspended solids in batik wastewater into soluble particles so that the concentration of TSS and Pt-Co decreases. Color removal can also occur due to processing carried out at pH 4. Acidic conditions accelerate the decolorization rate and this acceleration is caused by hydrogenation of negative charges, such as SO_3^-, in an acidic environment ($SO_3^- \rightarrow HSO_3^-$) [7].

3.5 Ozonation on Batik Wastewater Treatment

The process of wastewater treatment using the ozonation method was carried out with various pH (4, 7, and 10) to evaluate the performance of the ozonation in acidic, neutral, and alkaline conditions. Figure 5 shows the comparison of parameter removal produced by each variation.

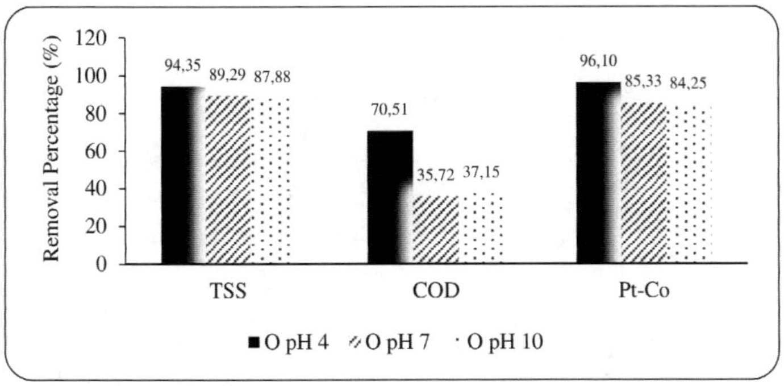

Figure 5. The Comparison of the Result of TSS, COD, and Pt-Co Removal in Batik Wastewater with the Ozonation at pH 4, 7, and 10 (Processing Time = 60 Minutes).

Figure 5 shows that the largest percentage of parameter removal occurred in ozonation with a pH of 4. This was due to the use of auxiliary substances of soda ash (sodium carbonate) in the wax removal process in batik production [9]. The presence of carbonate ions in wastewater could be a hydroxyl radical scavenger or an inhibitor of OH. Carbonate is the largest OH scavenger that inhibited ozone decomposition [10]. Furthermore, the carbonate ion concentration increases with increasing pH [11]. At alkaline pH, carbonate consumed OH and inhibited the formation of HO_2^-, $\bullet O_2^-$, $\bullet O_3^-$, or other intermediates that can accelerate the decomposition of O_3 [12]. Thus, the presence of carbonate slowed down the decomposition process of ozone into hydroxyl radicals, thereby increasing ozone stability.

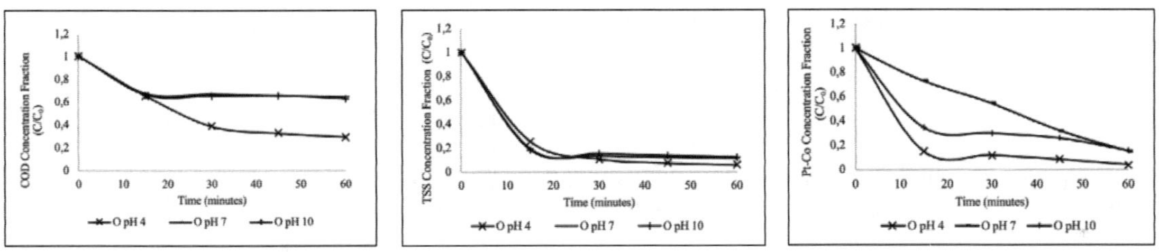

Figure 6. Concentration Fraction a) COD, b) TSS, and c) Pt-Co towards Time at Various pH (4, 7, and 10) with the Ozonation Method (Condition: Initial Temperature 25°C).

Based on Figure 6, the concentration of COD, TSS, and Pt-Co tends to decrease with the length of processing time. The highest removal of COD occurred during ozonation with pH 4, which reduced COD levels from 916 mg/L to 270 mg/L, TSS levels from 177 mg/L to 10 mg/L, and Pt-Co levels from 1,060 mg/L to 41 mg/L. Then followed by the ozonation method with pH 7 and pH 10. Under acidic conditions, O_3 molecules tend to react directly by breaking unsaturated bonds in azo compounds contained in the wastewater, such as cleavage of –N=N– and –C–N– bonds [13]. Therefore, ozone can decompose organic compounds in batik wastewater and reduce COD levels.

O_3 reacts with the suspended solids in the wastewater by forming flocs that absorb colloids in the wastewater [14]. O_3 does not degrade suspended solids but acts as a coagulant agent which helps form flocs from suspended solids. Thus, the flocs formed will cause a decrease in TSS in the system [15]. In the ozonation, the color decreases due to the oxidative cleavage of chromophores (coloring compounds). Furthermore, the rate of dye decolorization in the ozonation was faster due to the electrophilic reaction between O_3 and the dye parent compound as well as the oxidation of the intermediate compound [7].

3.6. Combination of Ozonation/Ultrasonic Cavitation on Batik Wastewater Treatment

The treatment of wastewater with a combination of ozonation/ultrasonic cavitation used an ozone flow rate of 6 L/minute and an ultrasonic wave intensity of 20%. Figure 7 shows the results of the comparison of the largest TSS, COD, and Pt-Co removal percentages generated by each variation.

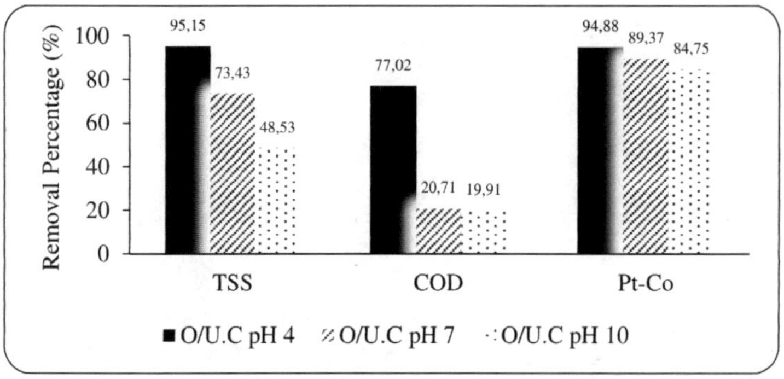

Figure 7. Comparison of TSS, COD, and Pt-Co Removal in Batik Wastewater with Combination Method of Ozonation and Ultrasonic Cavitation at pH 4, 7, and 10 (Processing time = 60 minutes).

Based on Figure 7, the condition with pH 4 resulted in the highest percentage reduction in TSS, COD, and Pt-Co, then followed by pH 7 and pH 10. At higher pH, the degradation rate did not increase by radical reactions due to dissociation in •OH to an oxygen anion radical ($O^{\bullet -}$), which caused a reduction in •OH. The presence of inhibitors such as carbonates and bicarbonates play a role in inhibiting the decomposition of O_3 to •OH.

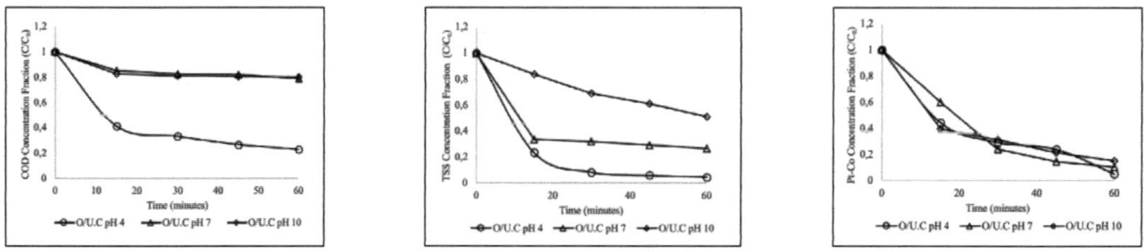

Figure 8. Concentration Fraction a) COD, b) TSS, and c) Pt-Co towards Time at Various pH (4, 7, and 10) with the Combined Ultrasonic Ozonation/Cavitation (Condition: Initial Temperature 25°C).

The combination method of ozonation and ultrasonic cavitation at pH 4 resulted in the highest removal of COD, TSS, and Pt-Co, namely removal of COD levels from 1,328 mg/L to 305 mg/L, removal of TSS levels from 165 mg/L to 8 mg/L, and removal of Pt-Co levels from 1,862 mg/L to 95 mg/L. The application of ultrasonic cavitation was able to increase the mass transfer of O_3 to the liquid phase thereby displacing the consumed O_3 and leading to a higher percentage of COD removal [16]. Besides, the O_3 molecule acts as an oxidizing agent for compounds that are more dominant than other active compounds under acidic conditions.

Therefore, the content of aromatic compounds in batik wastewater can be directly attacked by O_3 resulting in a higher COD removal. On the other hand, the low removal of COD at neutral and alkaline pH conditions can be caused by an increase in the foam produced during waste treatment. The foam causes O_3 to be trapped in it so that the decomposition and utilization of O_3 is reduced. O_3 produced in the combination of the two methods will react with the suspended solid organic fraction to convert it into dissolved organic matter [16]. The reduction of suspended solids contained in the wastewater will result in reduced color in the wastewater.

3.7. Ozonation and Ultrasonic Cavitation Method on TOC Removal of Batik Wastewater

In this study, the TOC levels test was carried out at the beginning and the end of the process at optimum conditions. The optimum condition chosen was the highest COD removal percentage from each process variation. The optimum condition for all methods is shown by the processing of batik wastewater with a pH of 4. The intensity of the ultrasonic waves selected for the TOC test using the ultrasonic cavitation is an ultrasonic intensity of 20%.

Total Organic Carbon (TOC) removal shows the amount of organic carbon that is completely mineralized into simple inorganic compounds, such as CO_2 and H_2O. The percentage of TOC removal for each method is presented in Figure 9.

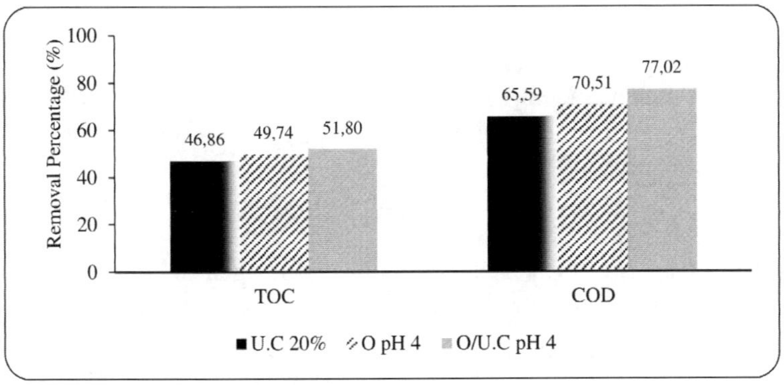

Figure 9. Comparison of TOC and COD removal on Ultrasonic Cavitation, Ozonation, Combination of Ozonation/Ultrasonic Cavitation at the Optimal Conditions (Processing Time = 60 Minutes).

Based on the research conducted, the most effective method for TOC removal is the combination method of ultrasonic cavitation and ozonation which can reduce TOC levels by 447 mg/L. The combination of ozonation/ultrasonic cavitation increases the mineralization of organic compounds in batik wastewater into harmless inorganic compounds. This is because the mass transfer of ozone in solution increases in the presence of ultrasonic irradiation. The microbubbles produced by ultrasonic cavitation cause a large part of the O_3 to enter the liquid phase or react at the gas-liquid interface [2]. The occurrence increases the contact area between the ozone in the gas phase and the liquid phase so that the mass transfer of O_3 into the wastewater will be greater. Thus, more ozone reacts with organic compounds.

There is a correlation between COD and TOC values. By definition, COD is the amount of oxygen needed to oxidize organic compounds and TOC is the amount of organic carbon that is oxidized. Based on Figure 9, the TOC removal for each method is lower than the COD removal. This is following the theory where the value of TOC reduction will always be smaller than the value of COD reduction because the TOC value shows how much the mineralized product is CO_2 and H_2O.

4. Conclusion

Based on the research conducted, it was found that the combination of ozonation/ultrasonic cavitation resulted in better removal of COD, TSS, and TOC than the ozonation and ultrasonic cavitation processes applied separately. In the ultrasonic cavitation, the lower ultrasonic wave intensity caused higher oxidation of organic compounds. Furthermore, the lower initial pH of batik wastewater in the ozonation method and the combination of ozonation/ultrasonic cavitation caused higher oxidation of organic compounds. To identify organic compounds contained in the wastewater as well as compounds that were produced by advanced oxidation processes, further testing with LC/MS is required.

Acknowledgement

The authors would like to thank Publikasi Terindeks Internasional (PUTI) Prosiding Tahun Anggaran 2020 No. NKB-1102/UN2.RST/HKP.05.00/2020 from Universitas Indonesia for providing financial support for this research.

References

[1] Indrayani L 2018 Pengolahan limbah cair industri batik sebagai salah satu percontohan ipal batik yogyakarta *Ecotrophic* **12** 174

[2] Song S, Xia M, He Z, Ying H, Lu B and Chen J 2007 Degradation of p-nitrotoluene in aqueous solution by ozonation combined with sonolysis *J. Hazard. Mater.* **144** 532–7

[3] Shen Y, Xu Q, Wei R, Ma J and Wang Y 2017 Mechanism and dynamic study of reactive red X-3B dye degradation by ultrasonic-assisted ozone oxidation process *Ultrason Sonochem* **38** 681–92

[4] Birgani P M, Ranjbar N, Abdullah R C, Wong K T, Lee G, Ibrahim S, et al 2016 An efficient and economical treatment for batik textile wastewater containing high levels of silicate and organic pollutants using a sequential process of acidification, magnesium oxide, and palm shell-based activated carbon application *J. Environ. Manage.* **184** 229–39

[5] Gaceva G B, Bužarovska A and Dimzoski, B. 2008 Discoloration of synthetic dyeing wastewater using polyaluminium chloride *G. U. J. Sci.* **21** 123–8

[6] Tizaoui C, Karodia N and Aburowais M 2009 Kinetic study of the manganese-based catalytic hydrogen peroxide oxidation of a persistent azo-dye *J. Chem. Technol. Biotechnol.* **85** 234–242

[7] Ince N H, Tezcanli-Guyer G 2004 Impacts of pH and molecular structure on ultrasonic degradation of azo dyes *Ultrasonics* **42** 591–6

[8] Ince N H, Tezcanli G, Belen R K and Apikyan I G 2009 Ultrasound as a catalyzer of aqueous reaction systems: the state of the art and environmental applications *Appl. Catal. B-Environ.* **29** 167–76

[9] Maghfiroh, Sari N 2017 Studi komparasi efektivitas penggunaan beberapa senyawa natrium sebagai zat bantu dalam proses pelepasan malam pada batik tulis *Jurnal Unikal* **31** 8

[10] Beltran F J 2004 *Ozone Reaction Kinetics fot Water and Wastewater System* (New York: Lewis Publisher)

[11] Lower S K 1996 Carbonate Equilibria in Natural Waters (Burnaby: Simon Fraser University)

[12] Khuntia S, Majumder S K and Ghosh P 2015 Quantitative prediction of generation of hydroxyl radicals from ozone microbubbles Chem. Eng. Res. Des 98 231-9

[13] Zheng Q, Dai Y and Han X 2016 Decolorization of azo dye C.I. Reactive Black 5 by ozonation in aqueous solution: influencing factors, degradation products, reaction pathway and toxicity assessment *Water Sci. Technol.* **73** 1500–10

[14] Isyuniarto I and Andrianto A 2009 Pengaruh waktu ozonasi terhadap penurunan kadar BOD, COD, TSS, dan fosfat pada limbah cair rumah sakit Ganendra **12** 45–9

[15] Karamah E F, Primasto A R, Najeges R R and Bismo S 2019 Treatment of tofu industry's wastewater using combination of ozonation and hydrodynamic cavitations method with venturi injector *J. Phys.: Conf. Ser.* **1198** 7

[16] Alfonso-Muniozguren P, Bohari M H, Sicilia A, Avignone-Rossa C, Bussemaker M, Saroj D, et al 2020 Tertiary treatment of real abattoir wastewater using combined acoustic cavitation and ozonation *Ultrason. Sonochem.* **64** 104986

The effect of Ozonation on the chemical structure of microplastics

A N Fitri[1], D Amelia[1], and E F Karamah[1, *]

[1]Department of Chemical Engineering, Faculty of Engineering, Universitas Indonesia, Jalan Fuad Hasan, Depok, 16424, Indonesia

*Email: eva.karamah@yahoo.com

Abstract. Microplastics are not susceptible to microbial degradation, thus often end up in aquatic ecosystem. Moreover, microplastics cause danger to aquatic biota and human. Effective technological solutions to degrade microplastics in wastewater treatment plants are desirable. Biodegradation is one of the most applied techniques, but takes a long time. This study evaluates the ozonation as pre-treatment to transform the chemical structure of polyethylene microplastics into more susceptible to biodegradation. The process was done by using a corona-discharge ozonator and batch reactor, varied pH value (6, 7, 8, 10, 12), ozone flowrate (1, 3, 5 L/min), and contact duration (1, 2, 3 hours). The study was begun by quantification of OH radicals and dissolved ozone, ozonation of microplastics, evaluation of ozonation by gravimetric weight loss including the change of microplastics chemical structure through FT-IR (Fourier Transform Infrared). The results revealed chemical structure changes of polyethylene after ozonation confirmed by the appearance of carbonyl bonds and the loss of weight. The optimum operating condition appeared at pH 12 with 5 L/min ozone flowrate, resulted in 0.0482% weight loss and carbonyl bond intensity reached 104.556% after 3 hours ozonation.

Keywords: Biodegradation, microplastics, ozone, ozonation, polyethylene

1. Introduction

Nowadays plastics have transformed human daily life. The high rising production of plastics in line with the waste generation Often plastics waste ends up in aquatic ecosystems. It has been revealed that Indonesia is the second country with the highest amount of plastics waste in the oceans [1]. Plastic debris may be broken down and become microplastics (MPs) in the size range of 0.001 - 5 mm [2]. Microplastics originate from two sources, primary and secondary. Primary MPs are intentionally manufactured, usually for toothpaste, exfoliating facial scrubs, resin pellets, etc. Secondary MPs are formed from the breakdown of larger plastics, come from household appliances, plastic bags, etc. Microplastics are not easily removed owing to the recalcitrance to microbial degradation. Moreover, microplastics have the potential to transfer chemicals and endanger aquatic biota and human, resulting from ingestion.

One of the solutions to reduce the microplastics contaminants in aquatic ecosystems is through Wastewater treatment plants (WWTPs). However, the technologies in WWTPs have not been able to completely remove microplastics [3]. Biodegradation is the most widely applied treatment for organic compounds waste. However, it takes longer time than physical or chemical methods. Therefore, to increase the efficiency of biodegradation, pre-treatment of abiotic degradation is necessary [4]. Ozonation as pre-teratment allows the reaction not merely between microplastics and ozone (direct reaction), however reaction between microplastics with OH radicals as a result of ozone decomposition in water also occurs (indirect reaction). Tian, *et al.* (2017) revealed ozonized polystyrene (β-14C) before incubation with *Penicillium variabile*~~,~~ degraded microplastics higher than the non-ozonized [5]. The result suggested ozone pre-treatment ability to change the polymer chemical structure into shorter chains, thus more susceptible to biodegradation [6].

In the present study, we evaluated the potential of ozonation pre-treatment to transform the chemical structure of polyethylene (PE) microplastics into more susceptible to biodegradation. PE was known as the most commonly found non-biodegradable solid waste [7]. Pratama obtained the optimal reduction of oil content was at the highest ozone dose, namely 0.53 g/hour and the highest pH, namely pH 8, suggesting the effect of pH and ozon dosage variation [8]. This study was varied on the initial pH value (6, 7, 8, 10, 12), ozone flowrate (1, 3, 5 L/min), and contact duration among ozone and microplastics (1, 2, 3 hours). Result parameters were seen based on the gravimetric weight loss, associated with the changes in the chemical structure which will be analyzed by FT-IR (Fourier Transform Infrared), and pH changes.

2. Materials and Methods

2.1 Microplastics

Microplastics used were commercial polyethylene powder with white color and a size of 20 mesh. The microplastic was weight for 0.5000 g and mixed with 200 mL aquadest.

2.2 Preparation Stage

The preparation stage was divided into the ozonator productivity test using Iodometric titration method, ozone flow rate calibration, hydroxyl (OH) radical quantification through permanganate titration method, and ozone solubility test in water. The ozone dissolved was carried out using Merck Ozone Reagent Test Kit No. 100607 and the UV VIS Spectrophotometer.

2.3 Microplastics Ozonation

Process was done in batch system. The equipment used at this phase was ozonator HYPRO GJ-8002K and Erlenmeyer Pyrex 250 mL as batch reactor. The equipment was arranged according the following scheme.

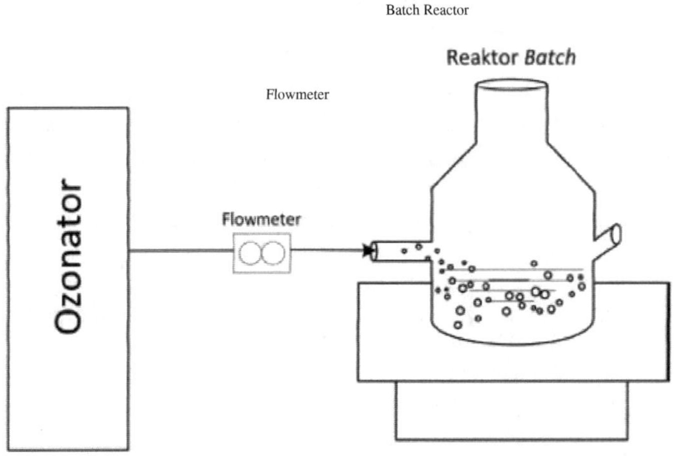

Figure 1. Schematic tool of the ozonation process

2.4 Determination of Weight Loss of PE Microplastics

The procedure was validated based on research by Auta, *et al.* (2018) [9]. The ozonized microplastics in aquadest were filtered by Whatman grade 41 filter paper and dried in a hot-air oven BIOBASE BOV-T50F at 105 °C for 4 hours. The residual polymer weight measured to determine the weight loss using the AND GR-200 analytical balance with a readability of 0.0001 g. The weight loss of the PE microplastics in percentage was determined using Eq. 1:

$$\%W = \frac{W_0 - W_x}{W_0}\ x\ 100\% \tag{1}$$

where W_0 is the initial weight of the microplastics (g) and W_x is the weight of microplastics after ozonation (g).

2.5 Fourier transform infrared (FTIR) analysis of Microplastics

The changes in the chemical structure of the microplastics were analyzed using Nicolet iS 5 FT-IR (Fourier Transform Infrared) Spectrometer in the frequency range of 4000–450 cm^{-1}.

2.6 pH Change Measurement

pH after ozonation was a key factor of microorganisms' activity in the next biodegradation process [8]. The pH change was calculated by subtracting the final pH of the ozonized microplastic solution with initial condition using pH meter ST 20. The result then was analysed to recommend the optimal condition since most of the bacteria living in soil or aquatic ecosystems grow and work optimally in a pH range of 6-8 [10].

3. Result and Discussion

3.1 Preparation Stage

The ozonator HYPRO GJ-8002K has an average of 262.4 mg/h productivity. Due to the incremental of flow rate, the production rate of ozone increased, namely 64.8 mg/h, 118.8 mg/h, and 248.4 for 1, 3, 5 L/min respectively. Furthermore, the longer contact duration resulted lower productivity. The ozonation procedure was firstly varied by initial pH solution during 1, 2, 3 hours with constant flowrate of 3 L/min.

Then the ozonized microplastics were observed by final pH condition to determine the constant value of pH in ozone flowrate variation. pH 12 was selected due to the lowest pH changes in the range of base. The quantification of OH radical increased along with the high pH, ozone flowrate, and contact duration. As for dissolved ozone concentration tended to be unstable. However, on average, the concentration in alkaline was less than acidic and neutral conditions. Higher ozone flowrate resulted greater dissolved ozone in the solution.

3.2 Weight loss of PE microplastics

The weight loss of the PE microplastics results are presented in Fig 2.

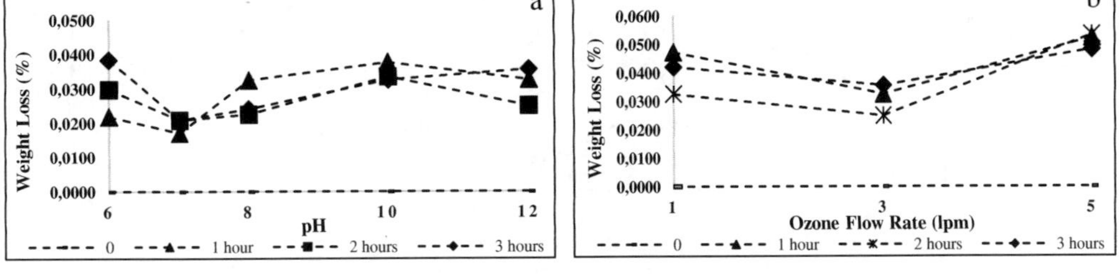

Figure 2. Effect of pH (a) and ozone flowrate (b) on weight loss of microplastics

The graphics showed the effect of pH, ozone flowrate, and contact duration. As pH value increased, the weight loss of microplastics increased as a result of higher number of OH radicals. Likewise, higher ozone flowrate and longer contact duration resulted higher weight loss percentage. Even so, fluctuated results occurred due to the decreasing of ozonator productivity and the possibility for sample to contact with air during gravimetric procedure.

In different study, Auta, *et al.* (2018) obtained the highest weight loss reached 6.40% in granular polypropylene after UV pre-treatment and 40 days of incubation with *Rhodococcus* sp. strain 36 [9]. Middleton, *et al.* (2013) showed ozonation process reduce the weight of polymers up to 0.30% after 28 days of ozone injection [11]. In the present study, the highest weight loss of PE microplastics was 0.53%. The lower yields compared to previous studies were due to differences in the variation in the duration of contact with the degrading agent.

The weight loss was a result of polymer molecules attacked by ozone and the loss of certain properties, hence hinting at the physical breakdown and degradation of polymers [12]. FT-IR analysed the presence of missing groups. In all samples varied by pH, absorption peaks were disappeared at 1059.73; 1135.61; 1182.29; and 1212.96 cm-1, were attributed to the C-O group. The S = O and C = C groups assigned 1400.05 and 1635.61 cm^{-1} were also disappeared. At flowrates of 1 and 3 L / minute, the presence of missing C-O groups was marked at 1059.73; 1135.61; 1212.96 cm^{-1}. The S = O and C = C groups assigned 1400.05 and 1635.61 cm^{-1} were also disappeared in all samples varied by flowrate.

In addition to the FT-IR, the cause of weight loss has observed by the Olympus CX41 microscope with the finding of cavity in microplastics surface. Tian, *et al.* (2015) supported the results of the rough and hollow of Polystyrene through SEM (Scanning Electron Microscopy), followed by the reduction of molecular weight using GPC (Gel Permeation Chromatography), indicates chain cutting [13] and polymer degradation after ozonation and biodegradation.

3.3 Fourier transform infrared (FTIR) analysis of Microplastics

The infrared bands of polyethylene were the CH_2 asymmetric and symmetric appear at 2917 and 2852 cm^{-1}, respectively. CH_2 scissoring vibration at 1468 cm^{-1} and rocking vibration at 718 cm^{-1} [14]. The reactions of ozone and polyethylene explained by Kefeli, *et al.* (1971) [15]:

$$RH + O_3 \rightarrow RO\bullet_2 + OH\bullet \quad (2)$$
$$RO\bullet_2 + RH \rightarrow ROOH + R\bullet \quad (3)$$
$$R\bullet + O_2 \rightarrow RO\bullet_2 \quad (4)$$
$$ROOH + O_3 \rightarrow RO\bullet_2 + O_2 + OH\bullet \quad (5)$$
$$RO\bullet_2 \rightarrow product + r\bullet \quad (6)$$
$$2r\bullet \rightarrow product \quad (7)$$

Based on the above reaction, the peroxide radical ($RO\bullet_2$) was generated as a result of the interaction of polyethylene with ozone then led to the formation of carbonyl, carboxyl, and hydro peroxide. Below FT-IR spectrum was presented based on the position of new carbonyl, suggested by Tian, *et al.* (2015) as the evidence of polymer degradation. The carbonyl (C = O) were identified at wavenumbers 1650-1818 cm^{-1}. Figure 3 shows the effect of pH on the appearance of carbonyl groups after 3 hours ozonation.

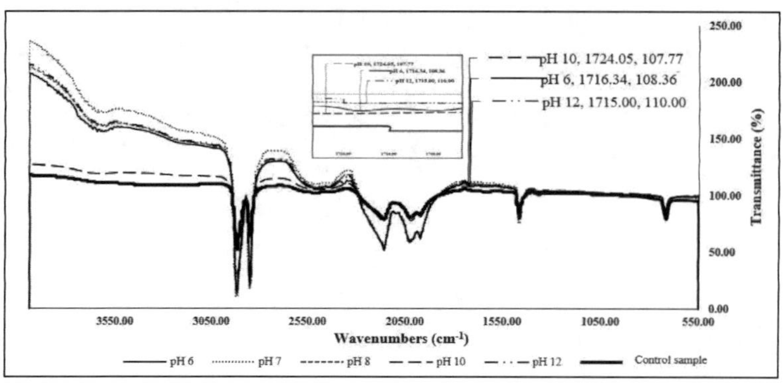

Figure 3. Effect of pH on the appearance of carbonyl groups in the FT-IR spectrum (Flowrate = 3 L/min; contact duration = 3 hours)

The FT-IR spectra showed the presence of a carbonyl absorption band at pH 6, 10, and 12, specifically ketones, aldehyde, and carboxylic acid, respectively. pH 10 as the second highest number of OH radical and dissolved ozone, had the highest carbonyl intensity with 107.77% of transmittance. The addition of a new group was formed at 3200 - 3700 cm^{-1}, was assigned to the OH stretching band on samples varied by pH. Although pH 7 and 8 had no carbonyl found, evidence of PE changes could be seen from the new O-H bands. Compared with control sample, the transmittance of the CH_2 group stretching at pH 6, 7, 8, and 12 decreased, indicates the increase of CH_2 intensity, corresponding with Vimala & Mathew [16]. On the contrary, CH_2 stretching vibration at pH 10 was higher than control sample. This likely due to the formation of carbonyl bonds with the highest intensity. Figure 4 shows the effect of ozone flowrate on the appearance of carbonyl groups after 3 hours ozonation.

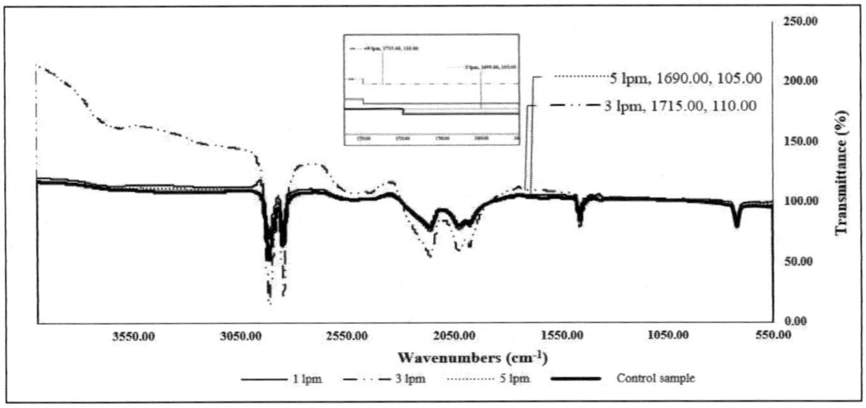

Figure 4. Effect of flowrate on the appearance of carbonyl groups in the FT-IR spectrum (pH = 12, contact duration = 3 hours)

The carbonyl group was formed in flowrate of 3 and 5 L/min, specifically ketones and aldehyde, respectively. The flowrate of 5 L/min with second highest number of OH radicals and highest dissolved ozone has the greatest carbonyl intensity. Even so, the flowrate of 1 L/min with no carbonyl found, has the highest C-H groups reduction. Samples with 5 L/min had the most frequent O-H stretching groups appearance, followed by 3 L/min and 1 L/min. Another new group is C = C (alkene) at 1 and 5 L/min. Figure 5 shows the effect of contact duration on the appearance of carbonyl groups at pH 12 and 5 L/min ozone flowrate.

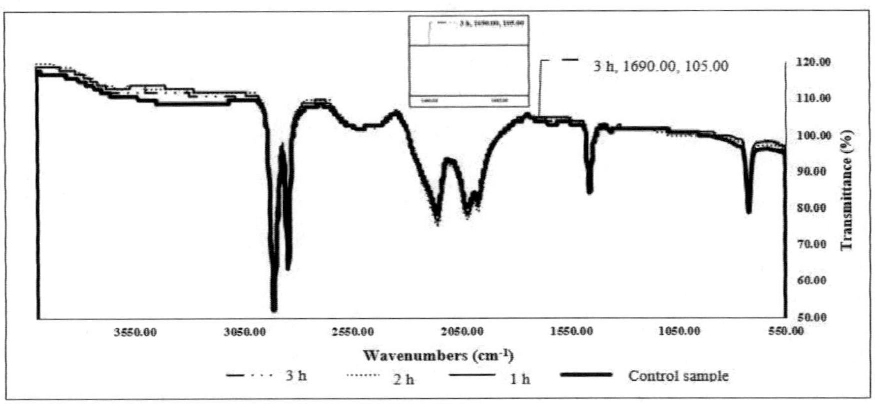

Figure 5. Effect of contact duration on the appearance of carbonyl groups in the FT-IR spectrum (pH = 12, flowrate = 5 L/min)

Carbonyl groups were only formed after 3 hours ozonation. Furthermore, the intensity of the CH_2 rocking, scissoring, and stretching increased with the longer contact duration.

3.4 pH Changes Measurement

Based on research, the ozonation process resulted in a decrease of pH, indicated the intermediate product formed. All samples in the study obtained acidic conditions, except sample at pH 12. ΔpH at pH 12 was the smallest in the range 0.21-2.63 after 3 hours ozonation. The highest ΔpH was 7.10 which occurred at pH 10. Therefore, based on the pH final condition value and high carbonyl intensity, the optimal condition concluded at pH 12 with a flowrate of 5 L/min during 3 hours ozonation. In this condition, if the ozone level to be enlarged, the final pH possibly obtained are just in the range 6-8, namely the optimum pH range for bacteria to live in biodegradation process.

4. Conclusion

The ozonation process has been shown to showed changes of the chemical structure of polyethylene microplastics along with the influence on pH, ozone flow rate and contact duration among ozone and microplastics. Result study obtained the escalation in the weight loss and the structural changes of polyethylene microplastics using FT-IR, confirmed by the disappearance of certain groups and carbonyl groups (ketone, aldehyde, alcohol, and carboxylic acid) appearance. These changes pointed to the ability of the polymer fragment resulted from ozonation to be more accessible to biodegradation process. Based on the pH final condition and high carbonyl intensity formed, the optimal conditions occurred at pH 12 with a flowrate of 5 L/min during 3 hours ozonation. This condition was obtained pH range 6-8, namely the optimum pH range for bacteria to live in biodegradation process.

Acknowledgement

The authors would like to thank PUTI 2020 No. NKB-1102/UN2.RST/HKP.05.00/2020 from Universitas Indonesia for providing financial support for this research.

References

[1] Jambeck J, Geyer R, Wilcox C, Siegler T, Perryman M, Andrady A, Narayan R and Law K 2015 *Science* **347** 768–71
[2] Shim W J, Hong S H and Eo S 2018 *Microplastic Contamination in Aquatic Environments* 1–26
[3] Talvitie J, Mikola A, Koistinen A and Setälä O 2017 *Water Research* **123** 401–7
[4] Onodera K, Mukumoto H, Katsuyaya Y, Saiganji A, Tani Y 2001 *Polym Degrad Stabil* **72** 323–327
[5] Tian L, Kolvenbach B, Corvini N, Wang S, Tavanaie N, Wang L and Ji R 2017 *New Biotechnology* **38** 101–5
[6] Cortez S, Teixeira P, Oliveira R and Mota M 2011 *Journal of Environmental Management* **92** 749–755
[7] Harshvardhan K and Jha B 2013 *Marine Pollution Bulletin* **77** 100–6
[8] Pratama F A M 2013 *Tesis Fakultas Teknik Universitas Indonesia*
[9] Auta H, Emenike C, Jayanthi B and Fauziah S 2018 *Marine Pollution Bulletin* **127**15-21
[10] Yani R B 2016 *Skripsi Universitas Andalas*
[11] Middleton J, Burks B, Wells T, Setters A M, Jasiuk I, Predecki P, Hoffman J and Kumosa M 2013 *Polymer Degradation and Stability* **98** 436-45
[12] NIIR Board of Consultants & Engineers 2006 *Asia Pacific Business Press, Delhi*
[13] Raddadi N and Fava F 2019 *Science of The Total Environment*
[14] Bonifazi G and Serranti S 2010 *The Open Waste Management Journal* **3** 35–45

[15] Kefeli A A, Razumovskii S D and Zaikov G Y 1971 *U.S.S.R* **13** 904–11
[16] Vimala P P and Mathew L 2016 *Procedia Technology* **24** 232–239.

Batik wastewater treatment by the hydrodynamic cavitation and Ozonation with coagulation-flocculation pretreatment

A E Saputra[1,*], I N J Siahaan[1], and E F Karamah[1]

[1] Department of Chemical Engineering, Faculty of Engineering, Universitas Indonesia, Kampus Baru UI Depok 16424, Jawa Barat, Indonesia

*Email: andreasexasaputra1@gmail.com

Abstract. Wastewater discharged by Batik's industry still contains a high concentration of dyes and pollutants, thus can contaminate the water's ecosystem. Because of that, in this study, Batik wastewater was treated by the hydrodynamic cavitation, ozonation, and combination of those two. By using PAC, a coagulation-flocculation-based pretreatment technic was conducted first to increase the effectiveness of the main wastewater treatment process. Then, variations in flow rate (2 L/min, 4 L/min, and 6 L/min) and initial pH of wastewater (4, 7, and 10) were evaluated to analyze its effect on the pH changing and the degradation of TSS, COD, color (Pt-Co), and TOC. The best result obtained from this research was by the application of the combination technic, which can eliminate 95.19%; 78.85%; 96.42%; and 60.56% of TSS, COD, color (Pt-Co), and TOC, respectively after 60 minutes treatment.

Keywords: Batik Wastewater, Coagulation, Hydrodynamic Cavitation, Ozonation

1. Introduction

Industrial Ministry of Indonesia highlights that in the past few years Batik industry has developed rapidly, based on the revenue generated from the export. It directly gives positive contributions to the betterment of the nation's economy and the community's life quality. Nonetheless, the negative impact in the increasing amount of wastewater generated from the production process should be taken into account. Batik wastewater mostly contains dyes that are rich in the content of unsaturated organic compounds like aromatic hydrocarbon, phenol, nitrous hydrocarbon, and their derivatives [1].

Prior to being discharged, those hazardous dyes must be treated first in order not to cause water pollution. To handle that, generally, Batik industry applies the coagulation-flocculation process combined with the filtration technic. The ease of operation and the use of a cheap-in-price and easy-to-find chemical have favored the application of this process. However, this process has some shortcomings like the insignificant decolorization percentage (only 50-80%) and also low COD degradation (30-70%) which further means the use of polishing technologies is mandatory in order to comply with the government regulation about the wastewater discharge standard [2].

Ozonation by using ozone can be a promising polishing technology. As a strong oxidant, ozone can react directly and decompose easily to form hydroxyl radical which further can degrade the organic pollutant and dissolved metal within the wastewater. Ozone is also a well-known chemical for decolorizing, and deodorizing the wastewater. Moreover, the major advantages of ozonation are an environmental-friendly technology, requiring the relatively small-sized area to operate, not a time-consuming process, not using the other chemicals to generate some oxidative compounds, and not producing hazardous byproducts for the environment [3]. By considering that ozone is a hardly dissolved chemical and reacted selectively with the other species in water, thus the combination with hydrodynamic cavitation was conducted in this research.

Hydrodynamic cavitation can in general be defined as the phenomena of the formation, growth, and subsequent collapse of microbubbles or cavities produced by pressure variations which are obtained using the geometry of the system creating velocity variation. These phenomena can be created by passing in the fluid through a constriction such as a venturi injector, cavities generator unit used in this research. The combination of hydrodynamic cavitation and ozonation aims for increasing the wastewater degradation effectivity due to the enhancement of ozone molecules mass transfer from the gas phase to the bulk solution [4].

The objective of the present work is to evaluate the effect of the combination of the hydrodynamic cavitation and ozonation as a polishing technology for Batik wastewater. Variations in flow rate and initial pH of wastewater were conducted. Furthermore, to evaluate the performance of the process, some parameters such as Total Suspended Solid (TSS), Chemical Oxygen Demand (COD), color (Pt-Co), and Total Organic Carbon (TOC) were tested.

2. Materials and Methods

2.1 Materials

Batik wastewater was obtained from a local Batik Plant in Tegal, Central Java.

2.2 Experimental Setup

The experimental setup was consisted of a venturi injector-based hydrodynamic cavitation reactor and an ozonator, illustrated in Figure 1. The venturi injector used was Venturi Mazzei type 384. This injector was equipped with a nozzle as an inlet for ozone gas generated by the ozonator which enabled direct contact between cavities and ozone. Ozonator used was a commercial-type Aquasuper AS-002 with the ozone production rate was 398.25 mg/h.

2.3 Methods

The initial step of wastewater treatment was conducted by the coagulation-flocculation pretreatment by utilizing the Poly Aluminium Chloride (PAC, molecular formula: $Al_{12}Cl_{12}(OH)_{24}$) coagulant as presented on Fig. 1.

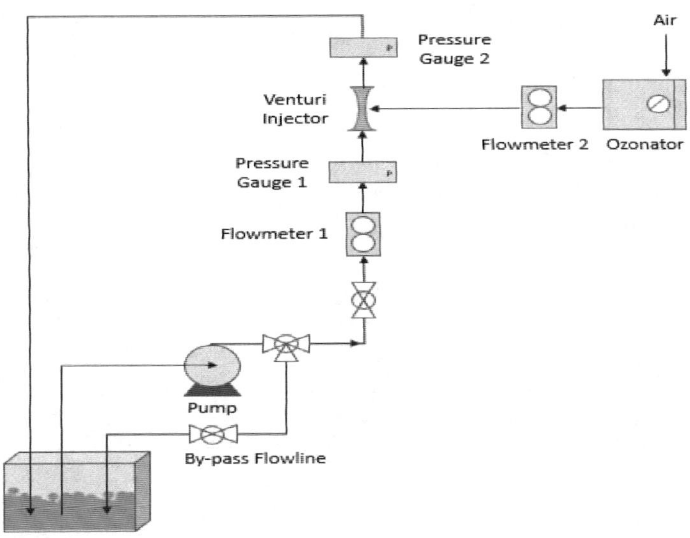

Figure 1. Experimental setup

By understanding that the coagulant works optimum in the certain range of pH, thus before being added, the pH of Batik wastewater should be adjusted to 4 in accordance with the result obtained by other by adding the sulphuric acid (H2SO4 p.a.) from Merck [5]. Then, the PAC solution 500 ppm was added into 500 mL of Batik wastewater. By using the jar tester Velp Scientifica JLT6 unit, firstly the wastewater was stirred quickly with the velocity of 200 rpm for 4 minutes and then stirred slowly (40 rpm) for 10 minutes. To make the colloidal particles settled, let the wastewater rested for 30 minutes and after that filtered it by using the filter paper. Batik wastewater was further treated by the cavitation- and ozonation-based polishing technologies.

3. Results

3.1 Effect of Coagulation-Flocculation Pretreatment

The pH adjustment and coagulation-flocculation pretreatment processes have changed the characteristic of Batik wastewater whose changes are depicted as follow.

Table 1. The comparison of Batik wastewater characteristics after pH adjustment and coagulation-flocculation

Parameter	Unit	Batik Wastewater Treatments			Standard Discharge Limit
		Initial Characteristic	After pH Adjustment to 4	After Coagulation-Flocculation	
pH	–	10,08	4	4,12	6 – 9
TSS	mg/L	1525	1990	145	50
COD	mg/L	29509	4282	1019	150
Pt-Co	mg/L	26756	28280	1078	–

Based on Table 1, the pH adjustment process has increased the TSS and Pt-Co concentration. This result indicated that when sulphuric acid was added into the naturally-alkaline wastewater, flocs (which were salts) were produced as a product of the neutralization reaction. Those macroflocs could not dissolve perfectly in the wastewater, hence the concentration of the suspended solid was higher after the addition of acid. Furthermore, this increased the turbidity of wastewater, which made the percentage of light penetrate the water became lower (Pt-Co concentration was higher).

Meanwhile, the pollutant concentration has decreased when the coagulation-flocculation process was applied on wastewater. This was happening because PAC as a coagulant could destabilize colloidal particles in wastewater which initiated the formation of huge and heavy flocs, thus settled due to the effect of gravitational force. Filtration was then applied to separate those flocs from the wastewater, leaving the wastewater with a lower concentration of TSS, COD, and Pt-Co.

3.2 Batik Wastewater Polishing Technology

3.2.1 Hydrodynamic Cavitation-based Polishing Technology

For each circulation flow rate, the following is the result of Batik wastewater treatment with the hydrodynamic cavitation-based polishing technology at 60 minutes.

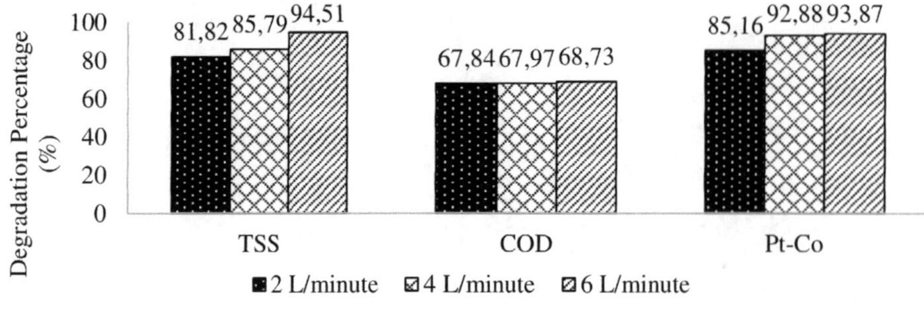

Figure 2. Effect of Wastewater Circulation Flow Rate on the Degradation Percentage of TSS, COD, and Pt-Co by the Hydrodynamic Cavitation Process (Initial pH = 4, Processing Time = 60 minutes)

Figure 2 shows that the optimum degradation of TSS, COD, and Pt-Co was by the application of circulation flow rate 6 L/min, followed by 4 L/min and 2 L/min. Higher flow rate meant higher fluid velocity passing through the constriction of a venturi injector than the inlet fluid velocity, which further indicated a higher pressure drop happening on that constriction area. The fluid pressure fell lower below its saturated vapor pressure indicating more gases would be formed. These gases were actually cavities. More fraction of cavities collapsed, thus more hydroxyl radicals would be produced. The higher concentration of hydroxyl radicals, consequently the degradation of organic pollutants in wastewater would be more massive which could be directed to the reduction of TSS, COD, and Pt-Co concentration. This finding was actually appropriate with the result obtained by Dhanke and Wagh [6] and Randhavane and Khambete [7].

3.2.2 Ozonation-based Polishing Technology

For each initial pH of wastewater, the following is the result of Batik wastewater treatment with the ozonation-based polishing technology at 60 minutes.

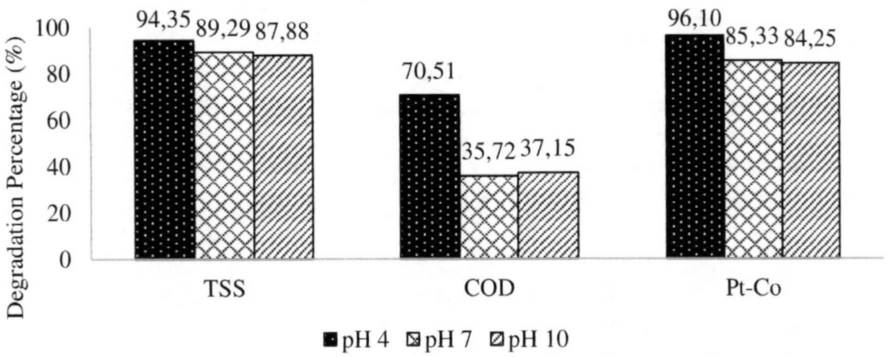

Figure 3. Effect of Wastewater Initial pH on the Degradation Percentage of TSS, COD, and Pt-Co by the Ozonation Process (Wastewater Flow Rate = 6 L/min, Ozone Flow Rate = 6 L/min, Processing Time = 60 minutes)

Figure 3 shows that the optimum degradation of TSS, COD, and Pt-Co was by the application of acidic conditions (initial pH of wastewater = 4) followed by the neutral and alkaline conditions, respectively. This result indicated that the degradation of the organic pollutant was favored by the direct and selective reaction between ozone molecules with the unsaturated hydrocarbon compounds (contained the bonds of C=C, C=N, N=N, etc) through the Criegee Mechanism to form peroxide hydrogen (H_2O_2) which later initiated the non-selective radical mechanism [8].

Naturally, Batik wastewater is alkaline in pH and contains carbonate ions in a huge number. These strong alkali ions come from the crystal or powder of sodium carbonate (Na_2CO_3) which in the Batik production process are the important chemical to accelerate the integration of dyes into the fiber of Batik, to adjust the pH, and to improve dyes' purity. Within wastewater, these chemicals were dissociated to form carbonate ions which were the hydroxyl radical scavenger agent or the strong inhibitor for the hydroxyl radical, due to the fact that they could consume hydroxyl radical resulted by the ozone decomposition through the reaction mechanism as follow [8,9].

$$CO_3^{2-} + HO\bullet \rightarrow OH^- + CO_3^{-}\bullet \quad (1)$$

Khuntia (2015) also stated that the carbonate ions could inhibit the formation of these ions including HO_2^-, $\bullet O_3^-$, O_2^-, or the other intermediates which were the catalyst for the ozone decomposition reaction [10]. This inhibition would stabilize the ozone in its molecular form, which consequently initiated the selective reaction with the organic pollutant.

3.2.3 The Combination of Hydrodynamic Cavitation and Ozonation-based Polishing Technology

For each initial pH of wastewater, the following is the result of Batik wastewater treatment with the combination of hydrodynamic cavitation and ozonation-based polishing technology at 60 minutes.

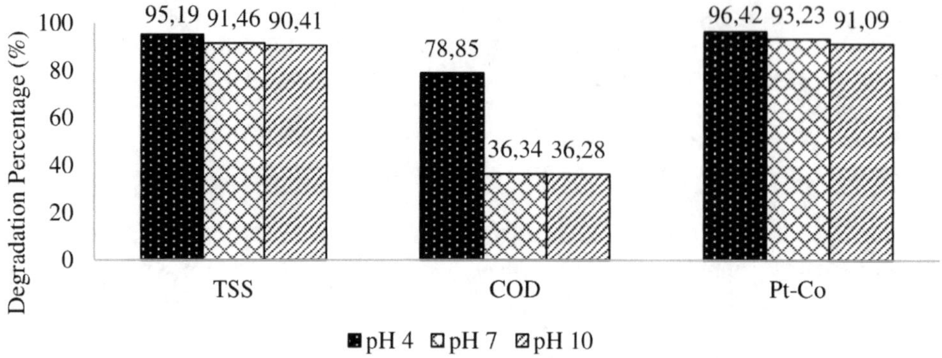

Figure 4. Effect of Wastewater Initial pH on the Degradation Percentage of TSS, COD, and Pt-Co by the Combination Process (Wastewater Flow Rate = 6 L/min, Ozone Flow Rate = 6 L/min, Processing Time = 60 minutes)

Figure 4 shows that the optimum degradation of TSS, COD, and Pt-Co was by the application of acidic conditions (initial pH of wastewater = 4) followed by the neutral and alkaline conditions, respectively. This finding indicated that the direct oxidation of dyes by the ozone molecule would be dominant to happen than the indirect oxidation reaction initiated by the hydroxyl radicals. Saharan, *et al*. (2011) explained the reasons behind that finding.

In the acidic condition, a huge amount of SO_3^- ions which could be found within dyes would be protonated. As a result, these dyes were present in a molecular state, became more hydrophobic, thus more likely to be located or preferred to be at the gas-water interface of the collapsing cavities. In that region, the dyes would be reacted directly with the hydroxyl radical, consequently increasing the decolorization percentage. Meanwhile, in the condition where pH > 7, dyes would be ionized, became more hydrophilic, thus more likely to be located at the bulk liquid.

Due to the hydroxyl radical recombination reaction in the cavities, hence the fraction of radicals that could diffuse into the bulk liquid region would be insignificant (generally around 10%), which further minimizing the dyes' degradation reaction with the hydroxyl radical and reducing the decolorization percentage [11]. This result was actually supported by Barik and Gogate [12] and also Pradhan and Gogate [13].

3.3 The Performance Comparison among Several Polishing Technologies

Below is the comparison of the TSS, COD, and Pt-Co degradation percentages for each optimum condition in each process.

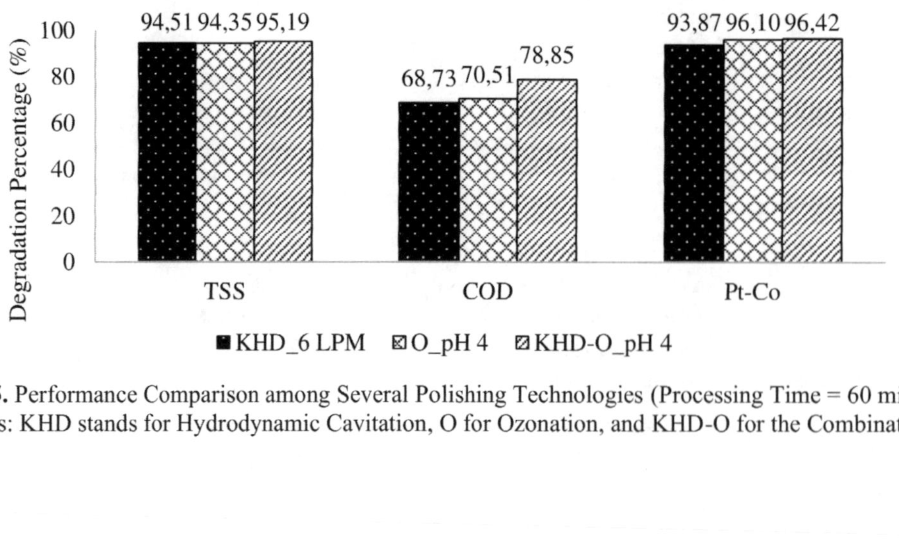

Figure 5. Performance Comparison among Several Polishing Technologies (Processing Time = 60 minutes). Notes: KHD stands for Hydrodynamic Cavitation, O for Ozonation, and KHD-O for the Combination

Figure 5 shows that the treatment by using the combination-based polishing technology gave the best result in TSS, COD, and Pt-Co degradation compared with the single cavitation or ozonation. This occurred because of three factors. First, the mechanical action of the hydrodynamic cavitation enhanced the dissolution of ozone and the generation of additional hydroxyl radicals, leading to a faster reaction rate with the organic substrate. Second, the hydrodynamic cavitation improved the mass transfer of ozone, which resulted in more ozone entering the liquid phase or reacting on the gas-water interface. Third, aeration of ozone might increase the turbulence of the aqueous solution and enabled more related substances to migrate from the collapsing cavities into the bulk of the solution [14].

3.4 Effect of Polishing Technologies on TOC Degradation

Total Organic Carbon (TOC) is a parameter to measure the mineralization rate of an oxidation process. Chang, *et al.* (2006) supported by Saharan, *et al.* (2011) stated that the mineralization is a reaction phenomenon to convert the organic compounds into the simple and nontoxic inorganic compounds [11,15]. Higher TOC degradation level means the higher conversion of organic compounds into the inorganic compounds, such as CO_2 and H_2O. In this part, TOC degradation is measured only at the optimum condition of each polishing technology, and the results can be observed as follow.

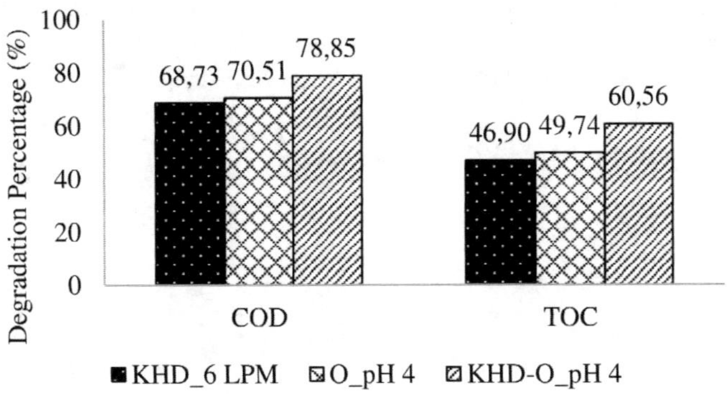

Figure 6. Effect of Polishing Technologies at Its Optimum Condition on TOC Degradation of Batik Wastewater (Processing Time = 60 minutes)

Figure 6 shows that the combination-based polishing technology gave the best result for TOC degradation. This was caused by the fact that the oxidation of organic compounds initiated by more oxidative agents (ozone and hydroxyl radical) in number and the enhancement of ozone mass transfer occurred in the solution which further could accelerate the mineralization and make it right on target than the single cavitation or ozonation.

4. Conclusion

It can be summarized that to treat the Batik wastewater which has pre-treated first by the coagulation-flocculation process, the combination of hydrodynamic cavitation and ozonation-based polishing technology gave the best result in degradation of TSS, COD, Pt-Co, and TOC up to 95.19%; 78.85%; 96.42%; and 60.56% respectively. Also, in the application of the hydrodynamic cavitation-based technology, a higher wastewater circulation flow rate gives higher degradation percentages in the parameters of TSS, COD, Pt-Co, and TOC. The operation of ozonation- and combination-based technology to treat the Batik wastewater was favored under acidic conditions and was very much dependent on the pH of wastewater, state (whether molecular or ionic), and nature (hydrophobic or hydrophilic) of the pollutants.

Acknowledgment

The authors would like to thank Program Publikasi Terindeks Internasional (PUTI) Prosiding Tahun Anggaran 2020 No. NKB-1102/UN2.RST/HKP.05.00/2020 from Universitas Indonesia for providing financial support for this research.

References

[1] Indrayani L and Rahman N 2018 *Jurnal Rekayasa Proses* **12** 41-50
[2] Verma A K, Dash R R and Bhunia P 2012 *Journal of Environmental Management* **93** 154-168
[3] Von Gunten U 2003 *Water Research* **39** 3189-98
[4] Gogate P R and Patil P N 2015 *Ultrason. Sonochem.* **25** 60-69
[5] Purnawan I, Febriasari A, Setyaputra B, Yolandini T T, Windriyo M J, Karamah E F and Kartohardjono S 2020 *IOP Conf. Ser.: Earth and Environmental Science* **442** 012003
[6] Dhanke P B and Wagh S M 2020 *Emerging Contaminants* **6** 20-32
[7] Randhavane S B and Khambete A K 2018 *KSCE Journal of Civil Engineering* **22** 2219-25
[8] Coca M, Pena M and Gonzalez G 2005 *Chemosphere* **60** 1408-15
[9] Ratpukdi T, Siripattanakul S and Khan E 2010 *Water Research* **44** 3531-43
[10] Khuntia S, Majumder S K and Ghosh P 2015 *Chemical Engineering Research and Design* **98** 231-9
[11] Saharan V K, Badve M P and Pandit A B 2011 *Chem. Eng. J.* **178** 100-7
[12] Barik A J and Gogate P R 2016 *Ultrason. Sonochem.* **30** 70–78
[13] Pradhan A A and Gogate P R 2010 *Chem. Eng. J.* **156** 77–82
[14] He Z, Lin L, Song S, Xia M, Xu L, Ying H and Chen J 2008 *Separation and Purification Technology* **62** 376-81
[15] Chang M C, Shu H Y and Yu H H 2006 *Journal of Hazardous Materials* **138** 574-81

Production of palm shell based activated carbon by two stage phosphoric acid impregnation and physical activation

Yuliusman[1*] and T M Afifah[1]

[1] Department of Chemical Engineering, Faculty of Engineering, Universitas Indonesia, Depok 16424, Indonesia

*E-mail: usman@che.ui.id

Abstract. Activated carbon (AC) as a promising gas adsorbent has been widely used in various applications. One of the raw materials that can be utilized is palm shell waste. Palm shell has great potential in terms of quantity and composition as raw material for AC. Palm shell waste production in Indonesia is classified as high and increasing every year. The high amount of carbon and low ash in the palm shell can produce good porous carbon. In this study, The Palm shell was processed through preparation, carbonization, and activation. After being dried and reduced in size, the palm shell was immersed in phosphoric acid (H_3PO_4) for 24 hours and continued with carbonization at 350°C for 30 minutes. Then, it was second chemically activated by H_3PO_4 with activator mass variations and physically activated by N_2 gas at 600°C for 1 hour. AC with the best characteristics was produced by a mass ratio of 2:1 in the second chemical activation. The iodine number and surface area obtained are 1164 mg/g and 1158 m^2/g. Furthermore, Pores have formed perfectly on the surface of carbon and the carbon content of AC has reached 76.9%.

Keywords: Activated carbon, palm shell, phosphoric acid

1. Introduction

Activated carbon (AC) has the largest gas adsorption capacity (n_{Gibbs}) than other adsorbents like metal oxides and natural zeolite [1]. Due to its high surface area, its adsorption capacity is also high. This potential makes the research of AC keep developing. AC can be synthesized from biomass waste, such as palm shells. Palm shells has a high carbon content, a low amount of ash, is inexpensive, and is not easily damaged during storage. The compositions of palm shell are 49.5% carbon, 0.9% ash, 7.23% moisture, and 73.67% volatile matter [2-3]. Besides, its availability in Indonesia is very abundant. Therefore, palm shell has the potency as a precursor of AC.

The combination of chemical activation and physical activation can produce AC. Phosphoric acid (H_3PO_4) has several advantages as a chemical activator. H_3PO_4 can produce AC with high surface area, maximum adsorption capacity, high yield, and low ash [4]. Meanwhile, N_2 as a physical activator is preferable to CO_2 because it produces a higher surface area [5]. The amount of chemical activation carried out can affect the characteristics of AC produced. The method of producing AC with two-stage of chemical activations using KOH, namely before and after pyrolysis, can increase 31% iodine number compared with one chemical activation after pyrolysis [6].

The objective of the present work is to evaluate and analyze the production of palm shell based activated carbon by two stage phosphoric acid impregnation and physical activation. Variation in impregnation ratio of H_3PO_4 was conducted. The activated carbon produced was evaluated by its overall yield, Iodine number, surface area, Scanning electron microscopy (SEM) image, and Energy dispersive X-Ray spectroscopy (EDS) result.

2. Materials and Methods

2.1 Preparation of palm shell activated carbon

Palm shell was washed in distilled water to remove dirt. Then, it was oven dried at 105 °C for 3h to evaporate moisture and crushed to obtain smaller size. Palm shell was immersed in 65%b/b H_3PO_4 for 24 hours with an impregnation ratio 3:1 at ambient temperature. Then, the soaked palm shell was carbonized at 350 °C for 30 minutes in a muffle furnace with a heating rate of 10 °C/min. After that, it went through the second chemical activated by 65%b/b H_3PO_4 with impregnation ratio 1:1, 2:1, and 3:1 for 24 hours at ambient temperature. Next, the activated sample went through physical activation by N_2 gas at 600 °C for 1 hour in a tubular furnace. The produced AC was washed with hot distilled water until it has a neutral pH. Finally, The AC was dried in the oven at 105 °C for 1 hour and can be stored in desiccator.

2.2 Characterization Process

The characterizations were done by calculating overall yield, Iodine number, and surface area, Scanning electron microscopy (SEM), and Energy dispersive X-Ray spectroscopy (EDS). The iodine number analysis was calculated as an approach for the surface area calculation. Meanwhile, SEM showed the image of pores formed and EDS provided chemical elements contained in the produced AC.

The overall yield of the production of AC was calculated using equation 1.

$$Yield(\%) = \frac{mass\ of\ activated\ carbon\ (g)}{mass\ of\ palm\ shell\ (g)} \times 100 \quad (1)$$

The iodine number (IN) and surface area (SA) of activated are obtained by following ASTM D4607-94 guidelines. The formulas used in this test are in equations 2 and 3.

$$IN = \frac{(V_B - V_F) \times N_T \times f_p \times 126.93}{m} \quad (2)$$

$$SA = \frac{IN \times 10^{-3}}{M_1} \times N \times \omega_1 \quad (3)$$

Where V_B is volume of iodine solution (ml), V_F is volume of $Na_2S_2O_3$ (ml), N_T is concentration of $Na_2S_2O_3$ (M), f_p is the dilution factor, m is AC mass (g), M_1 is Iodine molar mass (126.92 g/mol), N is Avogadro constant (6.023×10^{23} mol^{-1}), and ω_1 is Iodine surface area (0.2096×10^{-18} m^2).

3. Results and Discussions

3.1 Overall yield the production of activated carbon

The yield range obtained in this study was 48.6% - 56.1%. (figure 1). The increase in the mass ratio of H_3PO_4 in the second stage of phosphoric impregnation was inversely proportional to the yield produced. This was because excessive activators can cause undesired oxidation and can reduce yield [7]. The yield value in this study was not much different compared to previous studies using H_3PO_4. The yield obtained from date pits (*Phoenix dactylifera*) as raw material was 47.59% [8] and 50% for palm shell [2].

IOP Conf. Series: Materials Science and Engineering 1173 (2021) 012019 doi:10.1088/1757-899X/1173/1/012019

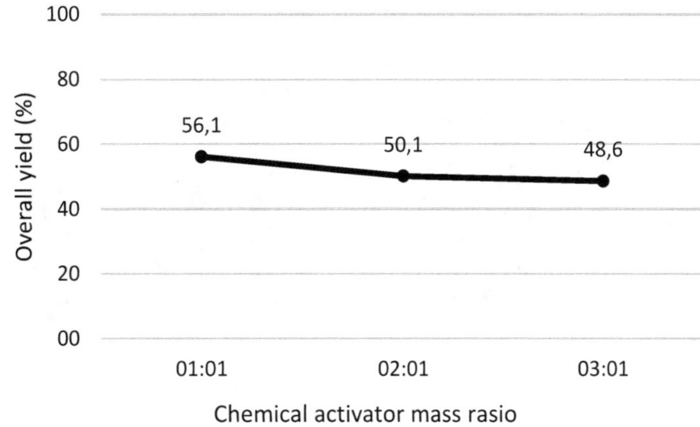

Figure 1. Effect of activator mass ratio on overall yield.

3.2 Iodine number and surface area

Brunauer-Emmett-Teller (BET) method is generally used as an analysis to determine the surface area. However, some studies have stated that iodine number analysis could be done to calculate surface area. This number correlates with the surface area owned by AC. Before carrying out the iodine number test, the sample must be washed first and confirmed that it was neutral. If the sample is acidic, the polyphosphate left in the pores of the AC can reduce the surface area. Figure 2 shows the iodine number and surface area from the carbonization to the second chemical activation with variations in the activator mass ratio.

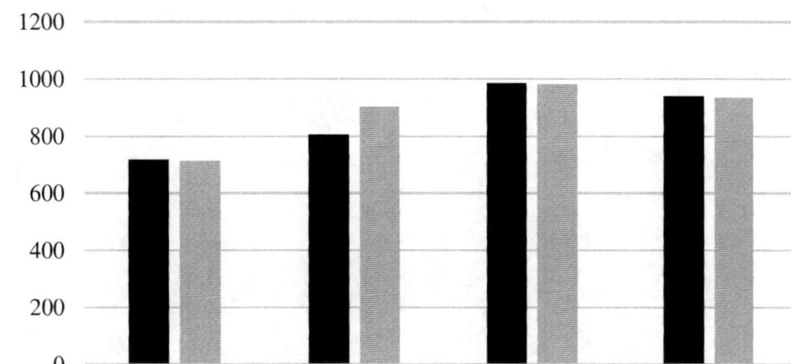

	Carbonization	2nd chem. act. 1:1	2nd chem. act. 2:1	2nd chem. act. 3:1
■ Iodine num. (mg/g)	717	806	985	940
■ Surface area (m2/g)	713	902	980	935

Figure 2. The results of iodine adsorption analysis from carbonization to impregnation ratio variation on second chemical activation.

In Figure 2, The second chemical activation was successfully proven to be able to increase the surface area and iodine number. This was because H_3PO_4 was directly reacted to carbon. Hence, new pores were produced. The highest surface area obtained was 980 m^2 / g which belongs to the impregnation ratio 2:1. Next, the iodine number and surface area at the physical activation stage are shown in Figure 3.

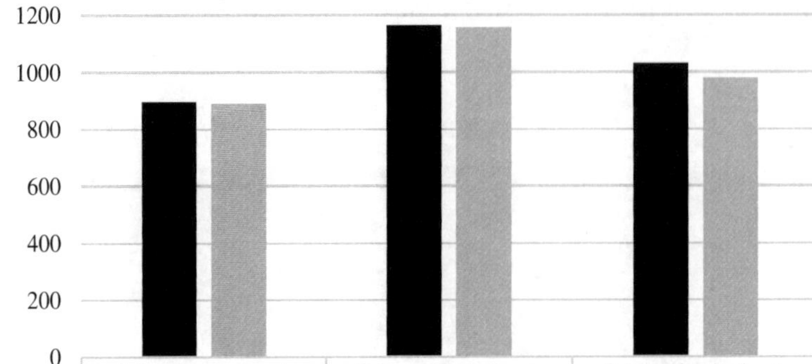

	1:1 + physical act.	2:1 + physical act.	3:1 + physical act.
■ Iodine num. (mg/g)	896	1164	1030
■ Surface area (m2/g)	891	1158	980

Figure 3. The results of iodine adsorption analysis after physical activation.

Figure 3 shows the highest surface area obtained of AC which was activated by a second chemical impregnation at a ratio of 2: 1 and continued by physical activation. The surface area value is 1158 m^2 / g. This surface area was in the surface area range of commercial AC, which was between 800-1500 m^2 / g [9]. Furthermore, the surface area decreased if the activator mass ratio is increased to 3:1. This was because excessive impregnation can form a phosphorus layer that covers active carbon particles and reduced contact with nitrogen gas on physical activation [7].

3.3 Scanning electron microscopy (SEM)

SEM is a characterization carried out to see the structure and pores of AC formed. Samples that were characterized are palm shell which only through the stage one of phosphoric impregnation (figure 4) and the produced AC (figure 5).

Figure 4. SEM analysis results for sample after first chemical activation.

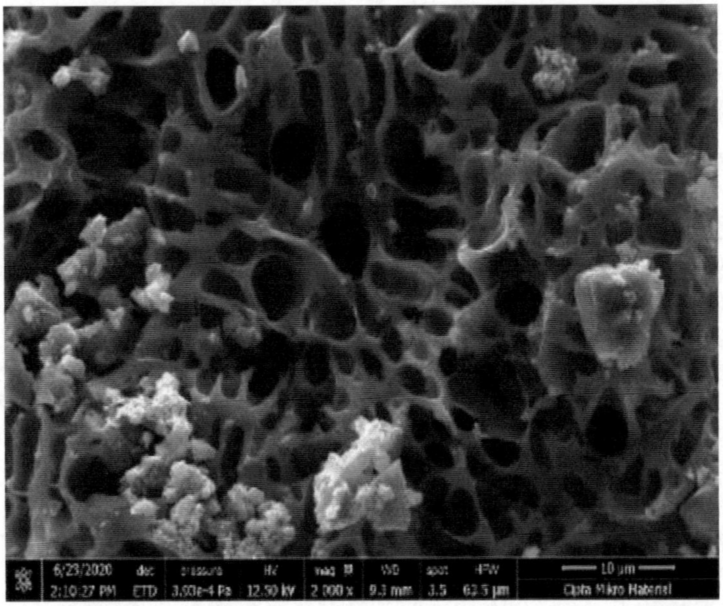

Figure 5. SEM analysis results for the produced AC.

Based on figure 4, the first chemically activated palm shell did not have pores yet. This was because the thermal decomposition has not been applied. Hence, the pore formation mechanism has not yet begun to occur. Whereas in figure 5, the AC produced in this study wasporous carbon. The use of H_3PO_4 as a chemical activator generated perfect pores formation. Pores on the outer surface of AC formed because of the evaporation of phosphoric acid during carbonization and physical activation. So, it left the cavity that was previously occupied by phosphoric acid. The number of pores shown in the pictures above is in line with the high surface area of AC.

3.4 Energy dispersive X-Ray spectroscopy (EDS)

EDS is a method of identifying elements or chemical elements contained in the test sample. This was done to determine the effect of activation and impregnation on the chemical composition contained in AC. The instrument used in EDS analysis was FEI type QUANTA 650. The following results of EDS characterization of AC samples are shown in Table 1.

Table 1. EDS analysis.

Element	Weight percentage (%)
Carbon (C)	76.9
Oxygen (O)	15.6
Phosphorus (P)	6.7
Silicone (Si)	0.4
Alumunium (Al)	0.4

Table 1 showed that AC has the dominant element C which is 76.9%. This AC has met the standard of commercial AC in SNI 06-7370-1995, which is at least 65% of the total mass. The element P comes from H_3PO_4 which was used in chemical activation. The content of phosphorus in AC tends to be high. Meanwhile, AC was activated by H_3PO_4 in previous research [10] has 1.59% of phosphorus. This difference was due by the rudimentary neutralization of AC. Other elements contained on AC are Si and Al. These elements come from palm shell ash that is still attached to the surface of AC [11].

4. Conclusion

Palm shell waste can be converted to a promising AC because of the high surface area and overall yield obtained in this study. The combination two-stage of phosphoric impregnation and physical activation was proven can increase the surface area to 1158 m^2/g for the ratio mass chemical activator 2:1. Meanwhile, the range of overall yield obtained is 48.6% - 56.1%. The SEM analysis showed that the AC was porous and the EDX showed that the carbon content has met the standard of commercial AC.

References

[1] Yuliusman, Purwanto W W and Nugroho Y S 2013 Pemilihan adsorben untuk penjerapan karbon monoksida menggunakan model adsorpsi isotermis langmuir *Jurnal Reaktor* **14** 223-33

[2] Lim W C , Srinivasakannan C and Balasubramanian N 2010 Activated of palm shells by phosphoric acid impregnation for high yielding activated carbon. *J. Analytical and Applied Pyrolysis* **88** 181-6

[3] Sumathi S, Bhatia S, Lee K T and Mohamed A R 2009 Optimization of microporous palm shell activated carbon production for flue gas desulphurization: experimental and statistical studies. *Bioresource tech.* **100** 1614

[4] Thomas B N and George S C, 2015 Production of activated carbon from natural resources. *Trends in Green Chemistry* **1** 1

[5] Yuliusman and Afdhol M 2018 Carbon monoxide and methane adsorption of crude oil refinery using activated carbon from palm shells as biosorbent *IOP Conf. Ser.: Mater. Sci Eng.* **316** 012016

[6] Jamilatun S, Setyawan M, Salamah S, Purnama D A A and Putri R U M 2015 Pembuatan arang aktif dari tempurung kelapa dengan aktivasi sebelum dan sesudah pirolisis *Seminar Nasional Sains dan Teknologi* (Jakarta: Universitas Muhammadiyah Jakarta)

[7] Ching L W 2011 *Preparation of porous carbon from palm shell using phosphoric acid activation: process optimisation, adsorption reagent recovery, and reuseability* (Bandar Sunway: Monash University)

[8] Devi A S, Kalavathy M H and Miranda L R 2015 Optimization of the process parameters for the preparation of activated carbon from low cost phoenix dactylifera using response surface methodology *Austin Chem Eng* **2** 1021

[9] Bansal R C and Goyal M 2005 *Activated carbon adsorption* (New York: CRC Press)

[10] Yuliusman, Ayu M P, Hanafi A and Nafisah A 2020 Activated carbon preparation from durian peel wastes using chemical and physical activation *AIP Conference Proceedings* **2230** 030020

[11] Nizamuddin S, Shrestha S, Athar S, Ali B S and Siddiqui M A 2016 A critical analysis on palm kernel shell from oil palm industry as a feedstock for solid char production *Reviews in Chemical Engineering* **32** 1

Influence of groundnut shell ash on compressive and tensile strengths of concrete

Samsunan[1,*], I Y Salena[1], Rosiana[2] and Ruslaini[2]

[1] Department of Civil Engineering, Faculty of Engineering, Teuku Umar University, Alue Peunyareng street, Ujong Tanoh Darat, Meureubo, Aceh Barat, Aceh 23615, Indonesia

[2] Students of Civil Engineering, Faculty of Engineering, Teuku Umar University, Alue Peunyareng street, Ujong Tanoh Darat, Meureubo, Aceh Barat, Aceh 23615, Indonesia

*Email: samsunan@utu.ac.id

Abstract. Research objectives are to determine the effect of groundnut shell ash used as partial cement replacement on concrete compressive strength. Groundnut shell ash obtained from the combustion results and passed sieve No. 200. Design mix is 21.7 MPa, with different variation percentage of groundnut shell ash in the order of 0%, 5%, 7.5%, 10% and 12.5% as partial replacement of cement for mixes. Total of 60 concrete specimens with 150 mm in diameter and height of 300 mm and tested for compression and tensile strength in 14 and 28 days. The result according to variations percentages in compression strength test on average at the age of 14 respectively were 18.495; 19.250; 20.382; 22.647 and 20.005 MPa. Whereas 28 days respectively were 18.495; 19.250; 20.382; 21.703 and 22.741 MPa. The results in tensile strength at the age of 14 respectively were 2.218; 2.029; 1.746; 1.557 and 1.463 MPa and 28-day age concrete respectively were 1.934; 2.029; 2.123; 2.029 and 1.840 MPa. The highest compressive and tensile strength in days 14 were 22.647 and 2.218, and in days, 28 were 22.741 and 2.123. It was observe that the strength has increased and groundnut shell ash seems promising material in concrete production.

Keywords: Compressive strength, Groundnut shell ash, tensile strength and normal concrete

1. Introduction

Groundnut (Arachis hypogea) is widely cultivated in Indonesia, but groundnut shells only become waste [1]. The Government of West Aceh District in 2019 developed 4,000 hectares of groundnut plants with a total harvest of 1.25 tons per hectare [2]. The results of the chemical properties test in Baristan [3] show that the peanut shell contains groundnut SiO_2 (19,06%), CaO (7,53%), K_2O (5,55%), Al_2O_3 (4,6%), ferrous oxide (1,59%), MgO (0,18%), Na_2O (2,71%). Lakshmi and Sagar [4] state that the silica content in groundnut shells, so that it can be used as a building material for concrete construction and used as an alternative for conducting research as a partial substitute for cement. Groundnut shell ash is composed of ferrous oxide, aluminium oxide, magnesium oxide, calcium oxide, sodium oxide, potassium oxide and silica so that it has the potential as agent concrete.

Portland cement is an inorganic binder, which is mix with water as a hydraulic binder. Tjokrodimuljo [5] states that Portland cement combines fine aggregate and coarse aggregate so that it hardens into concrete. The raw material components for Portland cement are; Limestone (CaO) = 60 - 67%; Silica sand (SiO2) = 17-25%; Alumina (Al2O3) = 0.3 - 0.8%; Magnesia (MgO) = 0.3 - 0.8% and sulfur (SO3) = 0.3 - 0.8%. SNI 15-2049-2015 defines Portland cement as hydraulic cement produced by grinding Portland cement slag mainly consisting of calcium silicate, which is hydraulic in nature and milled together with additional materials in the form of one or more crystalline forms of calcium sulphate compounds and may be add with other additives [6]. The use of groundnut shell ash as a partial substitute for cement in concrete can reduce costs and prevent environmental pollution. It is also expected to be able to cover the cavities or pores between the aggregate particles, by filling the cavities it will increase the area of the concrete so that the ability of the concrete to withstand loads can be perfect.

2. Materials and Methods

This research was conduct at the Laboratory of the Department of Civil Engineering, Teuku Umar University. Concrete forming materials consist of coarse aggregate, fine aggregate, cement, water and groundnut shell ash (AKKT). This research begins with material preparation, equipment preparation, processing of peanut shells, checking the physical properties of the aggregate, planning the concrete mix (mix design), working on the concrete mix, testing fresh concrete (slump test), making the test object, treating the test object, and testing.

2.1. Materials

The materials used are coarse aggregate / gravel, fine aggregate / sand, cement, water and groundnut shell ash. Coarse aggregate in the form of crushed stone (split) from one of the stone crushers at Nagan Raya district. The sand used comes from the Krueng Meurubo River, of Aceh Barat district. Groundnut shells come from Pante Ceureumen, Aceh Barat district. Coarse aggregate and fine aggregate are for physical properties checked to meet the quality of the material according to ASTM standards. Examination of aggregates is in the form of sieve analysis, bulk density, specific gravity and absorption. The cement has checked for the integrity of the package and only seen visually for the lump and smoothness of the cement. Water is not be checked because it meets the water quality requirements for concrete mixtures. The ash from the peanut shells checked for chemical elements according to the chemical content in cement.

Table 1. Chemical Composition of Groundnut Shells Ash

Number	**Composition**	**Percentage**
1	Silica oxide (SiO)	19.06 %
2	Potassium oxide (K_2O)	7.53 %
3	Calcium oxide (CaO)	5.55 %
4	Aluminium oxide (Al_2O_3)	4.6 %
5	Sodium oxide (Na_2O)	2.71 %
6	Ferro oxide (FeO)	1.59 %
7	Magnesium oxide (MgO)	0.18 %

Table 1 indicates that the groundnut shell ash contains a lot of silica oxide and Calcium Oxide also which is mostly contained in cement.

2.2. Method

This research uses normal concrete with concrete quality f'c = 21.7 MPa. Planning is based on the method of comparing the weight and volume of the concrete forming material The percentage of groundnut shell ash (GSA) used as a partial substitute for cement is 0%, 5.0%, 7.5%, 10.0% and 12.5%.

The test is at 14 days and 28 days of age in the form of a compressive test and tensile test. The number of test specimens is 30 cylindrical Ø15 cm, H = 30 cm with variations in the percentage of groundnut shell ash is 3 pieces.

Table 2. Research design

Specimens code	Proportion of groundnut shell ash	Number of specimen by Testing method and age of testing			
		Compressive test		Tensile test	
		14 day	28 day	14 day	28 day
GSA0	0 %	3	3	3	3
GSA 5	5.0 %	3	3	3	3
GSA 7,5	7.5%	3	3	3	3
GSA 10	10.0 %	3	3	3	3
GSA 12,5	12.5%	3	3	3	3
Summary of age's testing		15	15	15	15
Summary of method		30		30	
Summary of specimen		60			

The tests are carrying out after the test object is first remove from the soaking tub the day before the test is carry out after the treatment process. Then let stand for 24 hours until the test object is surface dry, then weighed to get the weight per specimen. Compressive tests and split tensile tests were carrying out at the age of 14 and 28 days using a compression tested machine.

3. Results and Discussions

3.1. Result of materials properties

3.1.1. Physical aggregate properties

The results of the examination of the physical properties of coarse and fine aggregates are in the form of bulk density, specific gravity saturated surface dry, specific gravity oven dry, absorption and fineness modulus (FM) of aggregate, as shown in the table 3.

Table 3. Result of materials properties

Material Checked		Bulk density	Specific Gravity		Absorption	Fineness Modulus (FM)
			Sat. Surface Dry	Oven Dry		
Unit		(Kg/l)	(Kg/l)	(Kg/l)	(%)	(%)
Coarse Aggregate	Result	1.98	2.663	2.602	2.348	6.695
	Requirement	1.6 – 1.9	1.60 – 3.20	1.60 – 3.20	0.2 – 4.0	5.5 – 8.5
	Reference	ASTM	ASTM	ASTM	ASTM	ASTM
Fine Aggregate	Result	1.817	2.473	2.426	1.956	3.049
	Requirement	>1.445	1.60 – 3.20	1.60 – 3.20	0.2 – 2.0	2.2 – 3.1
	Reference	Orchard	ASTM	ASTM	ASTM	ASTM

The results of physical properties checked of the aggregate show that all material checks have met the requirements. Only the results of the examination of the bulk density coarse aggregate are higher than required.

3.1.2. Result of slump test

The results of the slump test indicate that the use of groundnut shell ash as a substitute for part of the cement in the concrete mixture can reduce the slump test value, as shown in Figure 1 below.

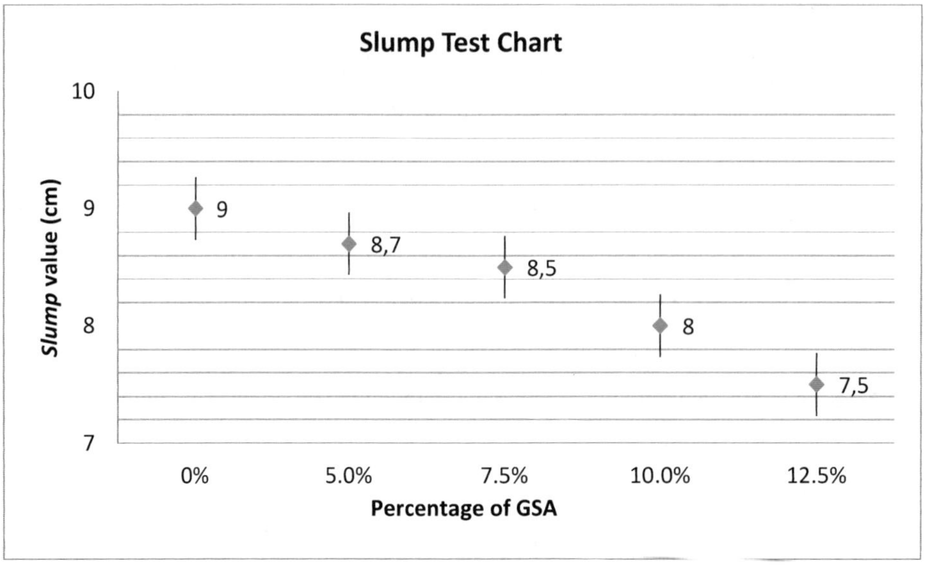

Figure 1. Value of slump test chart

Fig. 1 shows that the more the percentage of groundnut shell ash, the thicker the concrete, but it is still within the planning boundary between than 7.5 - 10 cm.

3.1.3. Weight of specimen

The substitution of groundnut shell ash (GSA) for cement in concrete affects the density of the concrete as shown in Figure 2 below.

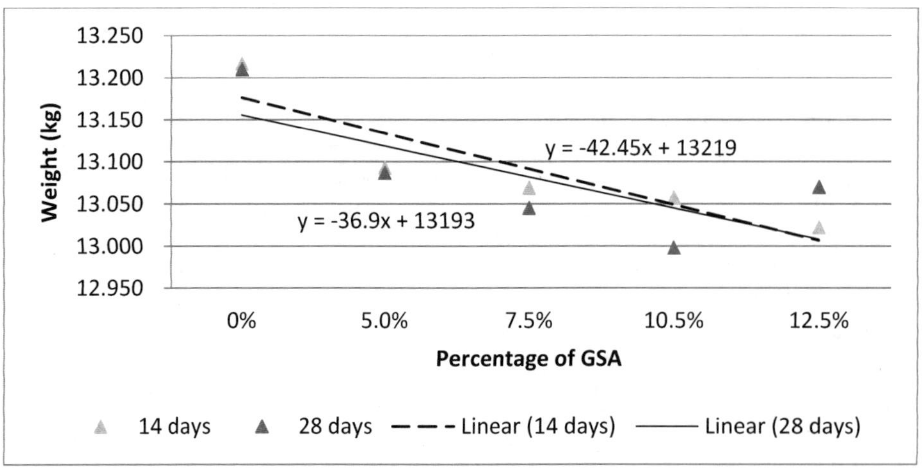

Figure 2. Weight of sample

From the figure 2 above, it shows that the more addition of groundnut shell ash causes the concrete to become lighter. Based on the age of the concrete, the test at the age of 28 days is lighter than the concrete at the age of 14 days.

3.1.4. Result of compression test

The results of the concrete compressive test indicate that the use of groundnut shell ash in the concrete mixture can increase the compressive strength of the concrete. The results of the concrete compression test of 14 days and 28 days of age are in Figure 3 below.

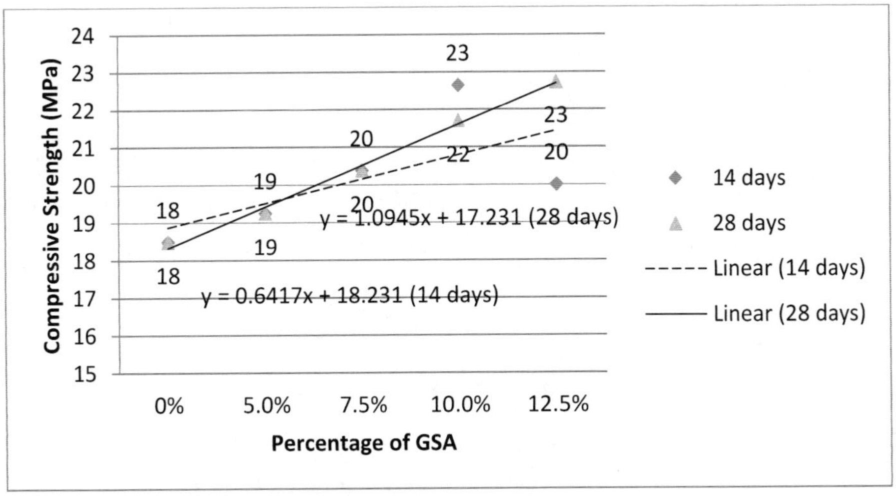

Figure 3. Compressive strength test

Figure 3 above, shows that the concrete test of age 14 shows an increase in the compressive strength of the concrete, with a composition of GSA content up to 10%, but decreases at 12.5%. The maximum compressive strength is obtained at the composition of 10% GSA, with value of linear regression of $y = 0.6417x + 18.231$. Whereas in the 28 day age test, the greater the use of GSA, the higher the results of the concrete compressive test with a linear regression of $y = 1.0945x + 17,231$. In this study, the maximum concrete compressive strength test results occurred at a composition of 12.5%.

3.1.5. Result of split tensile test

The results of the split tensile test at the age of 14 days and 28 days with the percentage of peanut shell ash aged 14 days at a variation of 0%; 5.0%; 7.5%; 10.0% and 12.5% by weight of cement, respectively (MPa) is 1.580; 1.895; 1.679; 1.356 and 1.329 MPa. The results of the split tensile test aged 28 days were 1.761; 1.891; 1.895 and 1.840 MPa. More details can be seen in Table 4 the increase or decrease in the tensile strength of concrete at the age of 14 and 28 days as follows:

Table 4. Result of tensile test

Age's	Tensile strength (MPa)				
	0%	5.0%	7.5%	10.0%	12.5%
14 days	1.580	1.895	1.679	1.356	1.329
28 days	1.749	1.761	1.891	1.895	1.840

The results of the split tensile test at the age of 14 days showed that the use of soil skin ash reduced the tensile strength of the concrete. The optimum tensile strength of 14 days at 5.0% is 1.895 MPa and the lowest is at 12.5% which is 1.329 MPa. While the results of the split tensile test at the age of 28 days have increased with the optimum value at 10.0% variation, namely 1.895 MPa and the lowest at the percentage of 5.0% at 1.761 MPa. The results of the comparison of the tensile strength of concrete at the age of 14 and 28 days shown in Figure 4 below.

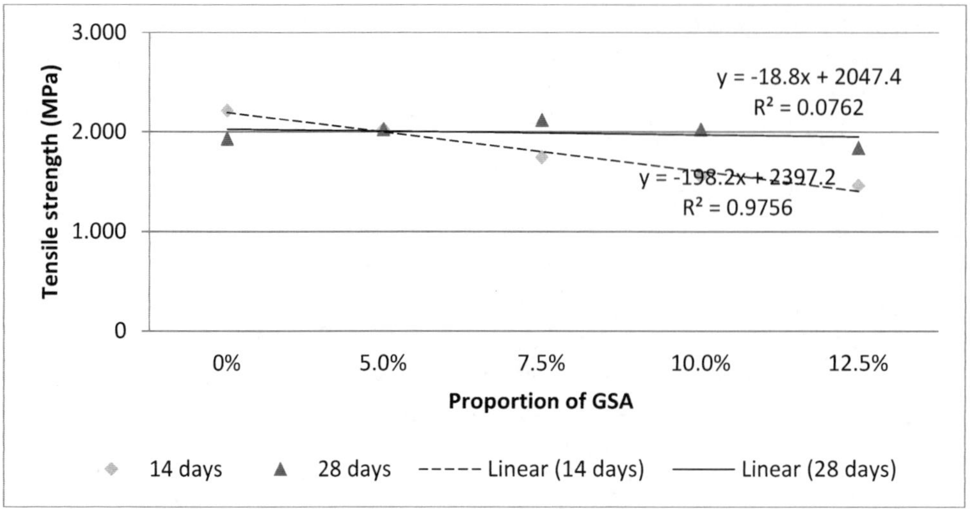

Figure 4. Tensile strength test

From Figure 4 shown that the split tensile test at age 14 decreased the split tensile strength due to the use of GSA based on the equation $y = -0.1982x + 2.3972$ with a value of $R^2 = 0.9756$. Meanwhile, in the 28-day test, there was also a decrease in split tensile strength with the equation $y = -0.0188x + 2.0474$, where the value of $R^2 = 0.0762$.

3.2. Discussion

The effect of substitution of GSA as a partial substitute for cement can affect the performance of concrete and the strength of the concrete. The use of GSA can cause the concrete to become thicker by decreasing the slump test value. The use of GSA also affects the density of concrete, where the more GSA composition, the lighter the concrete. This occurs because the density of GSA is much lighter than the density of cement. So that it can use be as an alternative use in mixing lightweight concrete. Based on the age of concrete, the weight of concrete at 28 days is lighter than that of 14 days. The longer the life of the concrete, the lighter it will be.

The results of the concrete compressive test show that the use of GSA can increase the compressive strength of concrete. The 28-day-old concrete test showed a more significant increase in the compressive strength of the concrete compared to the 14-day.The use of GSA in the split tensile test was not able to increase the results of the split tensile test. In the 14 days test, the split tensile strength decreased due to the substitution of GSA in the concrete mixture. At 28 days of age testing, there was an increase in the tensile strength of the split to 10% composition, but at larger compositions, the split tensile strength decreased. This can be show based on the linear regression value which is negative, as in a good agreement with others [5,6].

4. Conclusions

The results of the slump test show a decrease in the slump value, the higher the percentage of peanut shell ash the higher the water absorption in the concrete mixture. The weight of the concrete is lighter by substituting GSA as a partial substitute for cement in the concrete mixture. The results show that partial substitution of GSA in the concrete mixture can increase the compressive strength of concrete. The split tensile test results are not able to increase the split tensile strength with the use of GSA in the concrete mixture.

Acknowledgements

We thank you for the participation of all parties, especially the Civil Engineering Laboratory of Teuku Umar University. We thank too for the leaders of the civil engineering department, and the leaders of the engineering faculty at Teuku Umar University.

References

[1] Sumarno 1986, *Teknik Budidaya Kacang Tanah*, Penerbit Sinar Baru, Bandung.
[2] Safrizal., 2019, *https://aceh. Antara news. Com/berita/97370/aceh–barat–kembangkan–4000–hektare–tanaman–kacang-tanah*, Aceh Barat.
[3] Baristan., 2019, *Komposisi Kimia Abu Kulit Kacang Tanah*, Balai Riset dan Standarisasi Kementerian Perindustrian, Banda Aceh.
[4] Lakshmi, N.V., dan Sagar, P.S., 2017, *Study on Partial Replacement of Groundnut Shell Ash With Cement, Challenge Journal of Concrete Research Letters 8 (3) (2017) 84–90*, GMR Institute of Technology, Razam, Andhra Pradesh 532127, India.
[5] Tjokrodimuljo,K., 1996, *Teknologi Beton*, Biro Penerbit Universitas Gadja Mada, Yogyakarta.
[6] Badan Standarisasi Nasional, 2015, SNI 2049-2015, *Semen Portland*, Badan Standarisasi Nasional, Jakarta.

Hydrological modelling of Air Bengkulu river watershed in Indonesia using SUH and HEC-HMS models

G Gunawan*

Department of Civil Engineering, Faculty of Engineering, Bengkulu Univertsity, Jalan W.R Supratman No. 1 Bengkulu 22107, Indonesia

*Email: g.gunawan@unib.ac.id

Abstract. The purpose of this research is to develop a synthetic unit hydrograph model to determine the most suitable model for predicting peak discharge in the Air Bengkulu Watershed. Rainfall data were obtained from the Bengkulu Province Meteorology, Climatology, Geophysics Agency, from 2000 to 2019 year. The Synthetic Unit Hydrograph (SUH) method used to estimate the discharge in the Air Bengkulu River Watershed is the synthetic unit hydrograph, namely Snyder, ITB-1 and ITB-2, in addition to that also used HEC-HMS software. The estimation results of peak discharge and peak time for each type of hydrograph in the Air Bengkulu watershed are: Peak Discharge of Snyder 1,302.76 m^3 / s with Peak Time 6,596 hours, Peak Discharge of ITB-1 2,395.50 m^3 / s with Peak Time of 4,705 hours, Peak Discharge of ITB-2 2,165.857 m^3 / s with Peak Time of 6,341 hours. It can be concluded that the synthetic unit hydrograph has been able to estimate the peak discharge based on the time sequence of a rain event, so there is a relationship between peak time and peak discharge. The peak and peak times obtained from the output of each method are different, so more in-depth research is needed if this model is to be used as a basis for decision making.

Keywords: synthetic unit hydrograph, Air Bengkulu, ITB-1, ITB-2, Snyder, HEC-HMS

1. Introduction

Decision making in water resources management requires information about the discharge value [1]. The required discharge is design discharge or peak discharge [2-4]. There are various methods to get the peak discharge value [5-8]. Each method requires different data. In general, the peak discharge value calculated using different methods will result in different values, although the difference is not too big.

In general, the choice of method is based on the availability of data on the location of the activity plan, so that adequate knowledge of the location of the planned activity, the process of rainfall-runoff relationship is very important to estimate the amount of discharge generated in a particular catchment area.

Traditional techniques for planning flood estimation include rational methods, empirical methods, flood frequency methods, unit hydrograph techniques, and watershed modelling [1]. Unit hydrograph techniques and watershed models can be used to estimate the planned flood hydrograph in addition to the magnitude of the planned flood peak. The hydrological response of a watershed to precipitation can be estimated using hydrological modelling. The choice of model depends on the characteristics of the watershed and the purpose of carrying out the hydrological prediction.

The first researcher to introduce the concept of synthetic unit hydrograph (SUH) was Sherman in 1932 [1-4]. The next researchers who developed HSS were Snyder (1938) [9], Nakayasu (1940), Gray (1961) and Soil Conservation Service (1972). The development of synthetic unit hydrographs in Indonesia was carried out by Harto (1985) [5], Natakusumah (1990) [6], Lasidi, et al. (2003) [7], and Limantara (2009) [8].

SUH can be used to estimate peak discharge (Qp), peak time (Tp) in a watershed that does not have an automatic rainfall gauge and an automatic water level meter. SUH has been widely used by hydrologists to estimate flood designs in data-poor watersheds, the output of this method is able to provide an estimated discharge value that is close to the measured discharge value. In Indonesia, synthetic unit hydrograph analysis methods commonly used include the Snyder-Alexeyev, Snyder-SCS, Nakayasu and GAMA-1, GAMA-2, ITB-1, ITB-2 methods and others [3].

The Air Bengkulu Watershed is one of the watersheds in Indonesia that often experiences flooding due to overflowing rivers, namely the Air Bengkulu River. In the Air Bengkulu Watershed, there is no complete automatic water level measurement tool so that the prediction of flood discharge can only be done using SUH.

However, each SUH provides unequal discharge forecast results, so that the dimensions of the planned flood control building design are also not the same. The SUH used in the discharge estimation must be in accordance with the flood discharge value that occurs at the project location. The purpose of this study is to develop a SUH model to determine the most suitable SUH model for predicting peak discharge in the Air Bengkulu Watershed.

2. Methods

2.1. Research site

Geographically, the location or position of the Air Bengkulu watershed is between 5 ° 40'2 "south latitude to 100 ° 40" 104 "east longitude. This Air Bengkulu River flows through seven districts along the river flow. The research location is presented in Figure 1

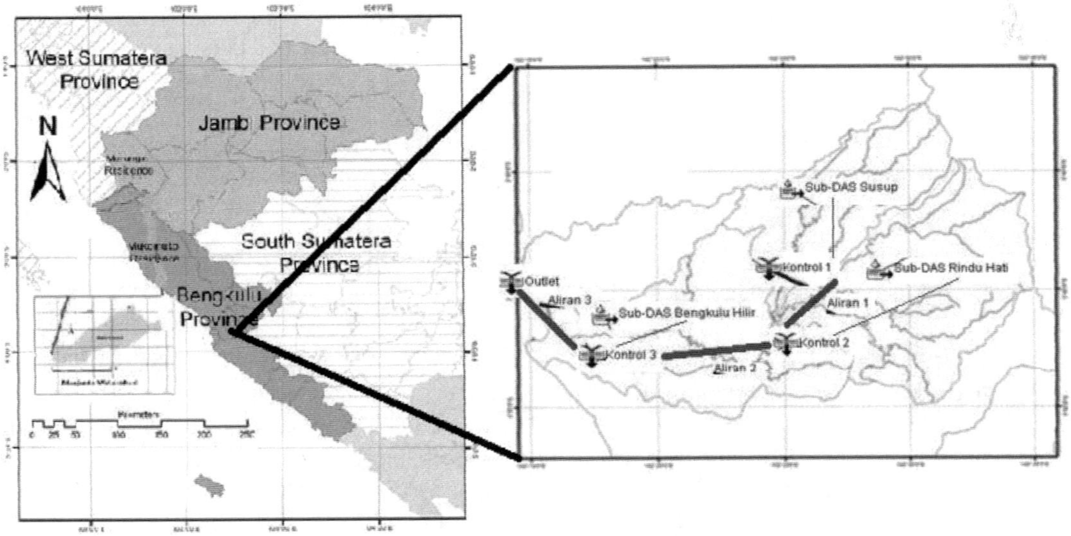

Figure 1. Research site.

2.2. Data collection

Rainfall data were obtained from the Meteorology, Climatology, Geophysics Agency of Bengkulu Province. Rainfall data is presented in Table 1

Table 1. Rainfall data

No	Year	Rainfall Station			$R_{theissen}$
		Tanjung jaya (mm)	Bajak (mm)	Taba Mutung (mm)	(mm)
1	2000	105.40	122.20	98.70	107.531
2	2001	200.10	115.00	69.00	119.367
3	2002	136.20	122.20	98.70	116.167
4	2003	180.70	94.00	79.30	112.080
5	2004	166.00	110.30	105.80	124.011
6	2005	218.50	105.50	94.50	132.523
7	2006	137.60	450.00	112.80	219.507
8	2007	107.00	109.20	114.10	110.660
9	2008	117.50	153.70	77.70	111.343
10	2009	151.20	94.60	163.70	139.753
11	2010	122.20	95.20	98.20	104.042
12	2011	95.00	100.00	90.80	94.699
13	2012	223.50	53.50	71.20	108.667
14	2013	80.20	28.00	103.50	74.632
15	2014	131.00	120.00	87.70	109.396
16	2015	113.00	200.00	161.60	159.333
17	2016	198.00	240.40	129.90	181.684
18	2017	175.00	100.00	126.70	132.344
19	2018	186.00	126.00	61.90	115.659
20	2019	130.80	177.00	131.10	144.594
Mean		148.745	135.84	103.85	125.900

Source: BMKG Agency, Bengkulu Province, 2020

2.3. Synthetic Unit Hydrograph (SUH)

The SUH method has been widely used by hydrologists to estimate flood plans in watersheds that do not have automatic water level recording devices. The SUH method is able to provide an adequate estimate of the peak discharge value and peak time. In Indonesia, the commonly used SUH methods include the Snyder-Alexeyev, Snyder-SCS, Nakayasu and GAMA-1, GAMA-2, ITB-1, and ITB-2 methods. The choice of this method is based on the consideration that the amount of data required is simpler than other methods.

In this study, the synthetic unit hydrograph tested to estimate the discharge in the Bengkulu watershed is Snyder, ITB-1 and ITB-2. The equations used for each synthetic unit hydrograph are presented in Table 2.

Table 2. Synthetic Unit Hydrograph (SUH) Equation

Parameters	Synthetic Unit Hydrograph (SUH) Equation		
	Snyder-Alexeyev	ITB-1	ITB-2
Physical Input	A = Area of the watershed L = Length of the longest river Lc = Length of river to watershed center	A = Area of the watershed L = Length of river	A = Area of the watershed L = Length of river
Non Physical Input	R = Rainfall Unit Tr = Standard rain duration Cp = Peak Discharge Coef (0.59-0.66) Ct = Coef Time (1-1.2) n = 0.2-0.3	R = Rainfall Unit Tr = Standard rain duration Ct = Coef Time (1-3)	R = Rainfall Unit Tr = Standard rain duration Ct = Coef Time (1-3)
Peak Discharge	Qp = (0.275 Cp A) / Tp Cp = Coef Discharge (For calibration)	Qp = R / (3,6 Tp) A_DAS / A_SUH ASUH = 1.69	Qp = R / (3,6 Tp) A_DAS / A_SUH ASUH = 1.454
Time Lag tp	tp = Ct (L. Lc) n Cp = Coef Time (For calibration)	tp = Ct 0.8122 L0.6 Cp = Coef Time	tp = Ct 0.8122 L0.6 Cp = Coef Time
Effective rain Peak time (t_p)	$t_e = t_p / 5,5$ te > Tr ⟶ Tp = tp + 0.25 (Tr – te) te < Tr ⟶ Tp = tp + 0.50 Tr	t_e = t_p / 5,5 te> Tr Tp = tp + 0.25 (Tr - te) te <Tr Tp = tp + 0.50 Tr	t_e = t_p / 5,5 te> Tr Tp = tp + 0.25 (Tr - te) te <Tr Tp = tp + 0.50 Tr
Time Base	Tb = 5.0 (Tp + Tr/2)	Tb = 10 Tp	Tb = 10 Tp
Recession Coefficient	Not stated explicitly but follows the shape of the SUH curve	Not stated explicitly but follows the shape of the SUH curve	Not stated explicitly but follows the shape of the SUH curve
Curve Properties	A single curve changes and turns into a watershed	A single curve cannot represent a watershed	Double curve can't imagine watershed

2.4. Hydrologic Engineering Centre-Hydrologic Modeling System (HEC-HMS)

The procedures used in developing the HEC-HMS model are as follows: first, identifying the characteristics of the Air Bengkulu watershed.

The method used in the calculation is the SCS method with the criteria used for modeling are presented in the Table 3

Table 3. The parameters used in HEC-HMS Model

Sub-Code	Area (km^2)	SCS Curve Number		Baseflow (m^3/s)	
Rindu Hati	192,08	Initial Abraction (mm) Curve Number	20,75 71	Initial Discharge	1,61
Susup	98,9	Initial Abraction (mm) Curve Number	20,27 71	Initial Discharge	1,61
Bengkulu Hilir	224,02	Initial Abraction (mm) Curve Number	20,27 71	Initial Discharge	1,92

Simulation of rainfall to surface runoff using HEC HMS in each sub-watershed requires several model components, namely: Rain models, Loss models, Transform models, and Baseflow models [10], [11].

3. Results and Discussions

The SUH used to estimate the discharge in the Air Bengkulu Watershed are Snyder, ITB-1 and ITB-2. The parameters used as input to estimate each SUH models are presented in Table 4.

Table 4. The parameters used to estimate each synthetic unit hydrograph

Parameters	Synthetic Unit Hydrograph (SUH)		
	Snyder-Alexeyev	ITB-1	ITB-2
Physical Input	$A = 194.64\ km^2$ L = 27.36 km Lc = 13.68 km	$A = 184\ km^2$ L = 34.2 km	$A = 515\ km^2$ L = 40 km
Non Physical Input	R = 0,988 Tr = 1 hour Cp = 1.2 Ct = 1 n = 0.042	R = 1 Tr = 1 Ct = 1	R = 1 Tr = 1 Ct = 1
Peak Discharge	$Qp = 10.894\ m^2/s$ Cp = 1.2	$Qp = 4.855\ m^3/s$ $A_{SUH} = 1.57$	$Qp = 16.41\ m^3/s$ $A_{SUH} = 1.45$
Time Lag (tp)	tp = 5.915 hours Cp = 1.2	tp = 6.705 hours Cp = 1.2	tp = 7.429 hours Cp = 1.2
Effective rain Peak time (T_p)	$t_e = 1.075$ hours Tp = 5.896 hours	te = 1.23 hours Tp = 6.705 hours	te = 1.35 hours te > Tr Tp = 7.341 hours
Time Base	Tb = 31.98 hours	Tb = 67.05 hours	Tb = 73.41 hours
Recession Coefficient	Not stated explicitly but follows the shape of the SUH curve	Not stated explicitly but follows the shape of the SUH curve	Not stated explicitly but follows the shape of the SUH curve
Curve Properties	A single curve changes and turns into a watershed	A single curve cannot represent a watershed	Double curve can't imagine watershed
Curve Shape	$Qt = 4.829\ m^3/s$ Where: $\lambda = 0.330$ a = 0.530	$Qt = 0.741\ m^3/s$	$(0 \leq t \leq 1)$ Qt = 10.93

3.1. The synthetic unit hydrograph in the Air Bengkulu Watershed.

The estimation results using the synthetic unit hydrograph Snyder, ITB-1 and ITB-2. After the hydrograph form is obtained, the flood discharge due to the planned rain is calculated with the superposition. In practice, the hydrograph superposition process can be calculated in tabular form as it is easy to find in various reference books. In this research is used the 6 hour rain distribution. As an indicator of the accuracy of the calculation results, the principle of mass conservation is used, namely the volume of rain the effective fall in the watershed must be equal to the volume of flood hydrograph produced. The total Runoff height calculated for the three hydrographs is very close to the total rainfall height given as input. The final superposition result in the form of SUH is shown in Figure 2.

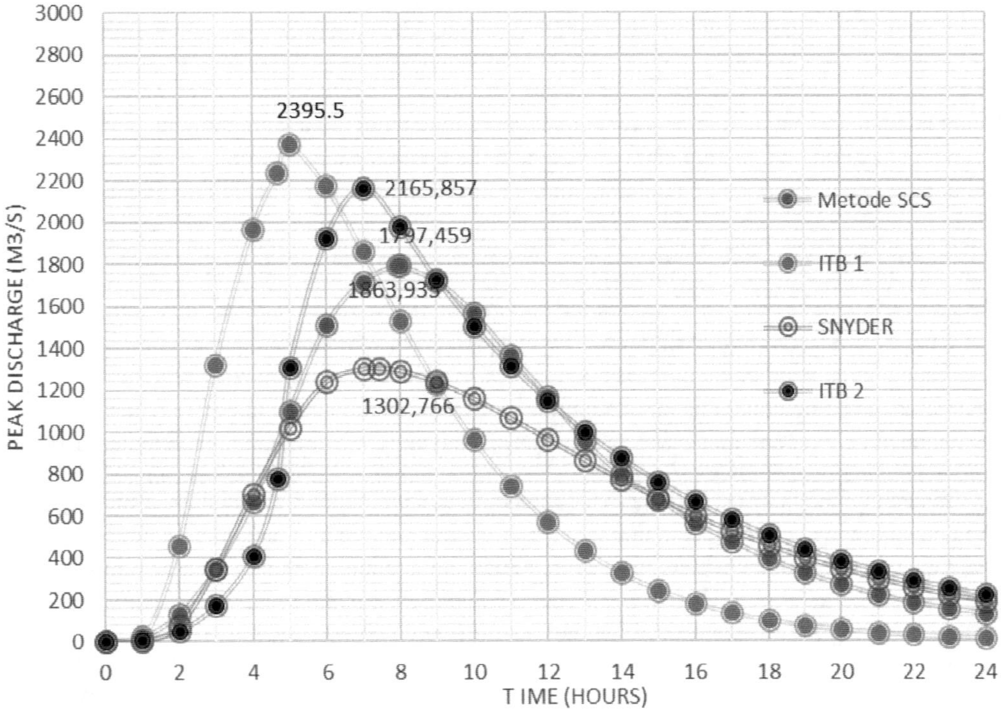

Figure 2. The synthetic unit hydrograph in the Air Bengkulu Watershed.

The final superposition result estimation of the peak discharge of Air Bengkulu Watershed using 3 Synthetic Unit Hydrograph methods namely Snyder-Alexevey, ITB-1 and ITB-2 are shown in Table 5

Table 5. The estimation results of the peak discharge for each synthetic unit hydrograph

Parameters	**Synthetic Unit Hydrograph (SUH)**		
	Snyder-Alexeyev	ITB-1	ITB-2
Peak Discharge	Qp = 1.302,766 m^3/s	Qp = 2395,5m^3/s	Qp = 2.165,857 m^3/s
	Cp = 1.2	A_{SUH} = 1.57	A_{SUH} = 1.45
Time Lag (tp)	tp = 5.915 hours	tp = 6.705 hours	tp = 7.429 hours
	Cp = 1.2	Cp = 1.2	Cp = 1.2
Peak Time (T_P)	Tp = 6.596 hours	Tp = 4.705 hours	te > Tr Tp = 6.341 hours

3.2. The hydrograph in the Air Bengkulu Watershed from HEC-HMS

The hydrograph form of the output of the HEC-HMS model developed in the Air Bengkulu Watershedis shown in Figure 3.

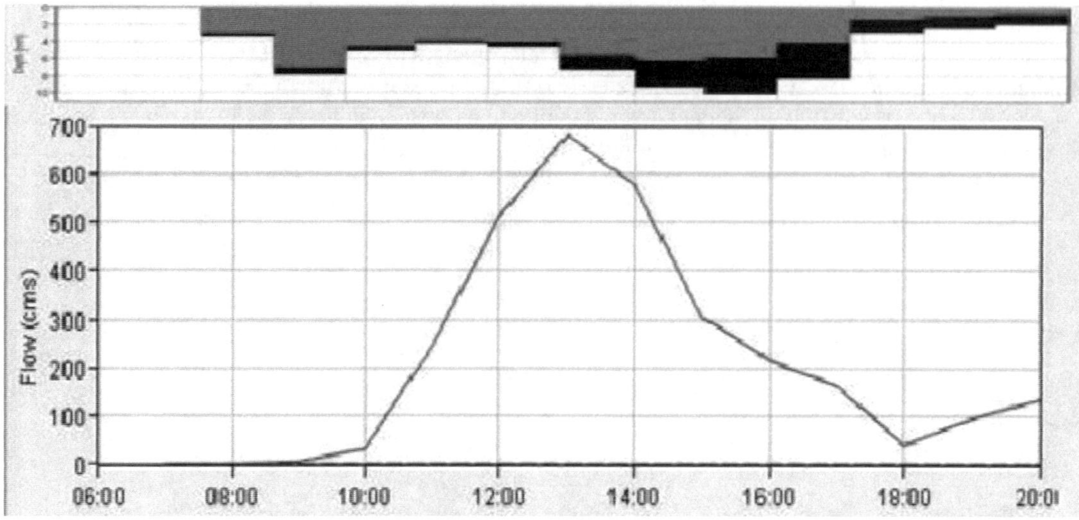

Figure 3. Hydrograph in the Air Bengkulu River watershed using HEC-HMS.

The flow obtained from the HEC-HMS output is 690 cms (cubic meters per second) with the time the peak occurred at 13.00, as presented on Fig. 4. This discharge value is much smaller when compared to the three synthetic unit hydrograph models (Nakayasu, ITB-1 and ITB-2) developed at the research location. The difference in peak discharge values that occurs between SUH and HEC-HMS is caused by various factors in the process of estimating the relationship between rain to flow.

One of the factors is the difference in rainfall data input. SUH uses maximum daily rainfall data, namely from rainfall data from 2000 to 2019 year. HEC-HMS uses maximum daily rainfall data during floods in the Air Bengkulu Watershed. In addition, the SUH method used in this study did not take into account the land cover and soil type factors, while the HEC-HMS took these two factors into account. Land cover and soil type affect the surface runoff value, if the land cover is predominantly forest and the soil type has a high permeability value, then the surface flow that occurs in the watershed will be smaller.

4. Conclusions

Based on the results of synthetic hydrograph modelling that has been carried out in the watershed using Snyder, ITB-1 and ITB-2 and HEC-HMS synthetic hydrographs, it can be concluded that the synthetic unit hydrograph has been able to estimate the peak discharge based on the time sequence of a rain event, so that the relationship between peak time and peak discharge. However, the value of the peak discharge and peak time obtained from the output of each method is different, so that more in-depth research is needed if this model is to be used as a basis for decision making.

References

[1] Safarina, A.B. 2011. *Clusterization of Synthetic Unit Hydrograph Methods Based on Watershed Characteristics*, International Journal of Civil & Environmental Engineering IJCEE-IJENS Vol: 11 No: 06.

[2] Salami, A. W. et al. 2009. *Evaluation of synthetic unit hydrograph methods for the development of design storm hydrographs for Rivers in South-West, Nigeria*, Journal of American Science 2009;5(4):23-32

[3] Chow VT, Maidment DR and Mays. LW. 1998. *Applied Hydrology*: McGraw – Hill Publishing company, New York.
[4] Singh, P.K., Mishra, S.K. and Jain, M.K., 2014. *A review of the synthetic unit hydrograph: from the empirical UH to advanced geomorphological methods Hydrological,* Sciences Journal – Journal des Sciences Hydrologiques, 59 (2) 2014, http://dx.doi.org/10.1080/
[5] Sri Harto, 1985, Hidrograf Satuan Sintetik Gama I, Badan Penerbit Departemen Pekerjaan Umum, Jakarta.
[6] Natakusumah, D.K., 2009. *Prosedure Umum Penentuan Hidrograf Satuan Sintetis Untuk Perhitungan Hidrograph Banjir Rencana*. Seminar Nasional Teknik Sumber Daya Air Peran Masyarakat, Pemerintah dan Swasta sebagai Jejaring dalam Mitigasi Bahaya Banjir Bandung, 11 Agustus 2009
[7] Lasidi, E dan Anwar, N., 2003., Hidrograf Satuan Sintetik Aβγ (Hss- Abg), Prosiding Seminar PIT XX HATHI, 20-21 Oktober
[8] Limantara, L., M., 2009., Hidrograf Satuan Sintetik Limantara (Studi Kasus Di Sebagian Das Di Indonesia) Jurnal Rekayasa Sipil / Volume 3, No.3– 2009 Issn 1978 – 5658
[9] Ramirez. J,. A. 2000. *Prediction and Modeling of Flood Hydrology and Hydraulics*. Chapter 11 of Inland Flood Hazards: Human, Riparian and Aquatic Communities. Edited by Ellen Wohl; Cambridge University Press.
[10] Kalita, D, N. 2008. *A Study of Basin Response Using HEC-HMS and Subzone Reports of Cwc* Conference Paper, · August 2008 DOI: 10.13140/2.1.1470.2724
[11] Fleming, M. and Brauer,T. (2016). *Hydrologic Modelling System HEC-HMS Quick Start Guide*. [online] Version 4.2. US Army Corps of Engineers, Institute of water Resources.

Partial discharge type detection and identification based on its sources

R W Putra[1], H H Sinaga[1,*], N Purwasih[1], D Permata[1], Y Yuniati[1], and H B H Sitorus[2]

[1] Department of Electrical Engineering, Universitas Lampung, Jalan Prof. Soemantri Brojonegoro No.1, Bandar Lampung 35145, Indonesia

[2] Department of Electrical Engineering, Universitas Pembangunan Nasional Veteran Jakarta, Jalan RS Fatmawati, Cilandak, Kota Jakarta Selatan, DKI Jakarta 12450, Indonesia

*Email: herman.h.sinaga@eng.unila.ac.id

Abstract. Partial discharge (PD) is a symptom of initial damage to high voltage equipment insulators which if left for a long period will cause total damage to high voltage equipment. This study aims to detect and identify the type of PD based on the source of its discharge so that it can be useful in terms of monitoring and maintenance of high voltage equipment. In this research, the Hilbert fractal antenna sensor is used in the detection process of surface, cavity, and corona PD with different input voltage variables that successfully produce a total of 600 PD signal data on the oscilloscope. To reduce noise on the PD signal, the denoising process is done by utilizing the sym4 wavelet feature found in the MATLAB software. The denoising process generates new data so that the research data becomes 600 original PD signal data and 600 denoising PD signal data. With a statistical approach, all PD signal data is extracted successfully into the mean, skewness, kurtosis, and standard deviation parameters which are useful as input for the PD type identification process. From each of the PD signal statistical data, 450 data are used in the training data process and 150 data are used in the data testing process. The PD type identification process is performed using a back propagation neural network with a mean square error (MSE) level of 0.01. The identification results show that back propagation neural networks are able to identify PD types based on statistical input accurately. In addition, the denoising process also affects the accuracy of the identification results of the PD type that is 95.33% for the original discharge signal to 97.33% for the denoising signal.

Keywords: partial discharge (PD), PD identification, Hilbert fractal antenna, back propagation

1. Introduction

Partial Discharge (PD) is a phenomenon of discharge which results in the deterioration of insulating material on high voltage equipment. This situation sometimes occurs in power transformers due to the electric field pressure that exceeds the threshold value of the insulator (inception voltage). If the PD event is left for a long period of time, there will be a decrease in the quality of the insulating material or even cause a breakdown event. Therefore, the detection and identification of PD signal characteristics is very important so that the risk of damage to high voltage equipment can be avoided [1].

In general, PD detection methods can be grouped into two types, namely conventional and non-conventional. The conventional method is a PD detection method that refers to the IEC 60270 standard, while the non-conventional method is a method that does not follow the standard rules according to IEC 60270. Basically, this method can be divided into several types namely the Acoustic, Optical, and Ultra High Frequency methods (UHF) [2].

Some previous PD detection studies have actually been carried out. Researchers [3,4] have successfully analyzed PD from detection results based on magnitude, pattern, and time of occurrence of PD using the IEC 60270 method. PD detection using the optical fiber method has also been carried out by [5]. In this study, a peak-plate arrangement was used to produce PD sources, then conventional methods and fiber optic methods were used to detect and find out the location of PD. In addition, PD detection studies using acoustic methods have also been successfully carried out by other researchers [6-8] who are known to be very effective acoustic methods when used in liquid insulators such as transformer oil.

2. PD detection and identification

2.1. Partial Discharge type

Partial Discharge (PD) usaually can be grouped into three types, namely discharge that occurs on the surface (surface discharge), cavities (void discharge), and corona (corona discharge). Surface discharge is a phenomenon of discharge that occurs on the surface of insulating material. Void discharge is a discharge phenomenon in the inner cavity of the insulator. Air cavities can be produced due to production defects in the insulating material. Corona discharge is an eruption that occurs due to the acceleration of ionization events under the pressure of an electric field on sharp metal material in the open air.

2.2. Hilbert fractal antenna for PD detection

In PD detection, a sensor that can work detects electromagnetic waves at high frequencies is required. One of the sensors that meet the criteria is the Hilbert fractal antenna sensor [9]. Based on research done by [9], this sensor is rated effective for detecting electromagnetic waves which work at a frequency range of 300 MHz to 1 GHz [9]. This frequency range is at the UHF frequency range [1, 6, 10]. In addition to this research, it was proposed to use the Hilbert fractal 4th order antenna sensor because the size can be made very small and have a large spectrum frequency bandwidth [11].

In accordance with its name, the Hilbert fractal antenna sensor is a micro-strip sensor type with continuous fractal curve consisting of 1st order up to 4th order. The larger the fractals order, the greater the gain is generated on the sensor [12]. The fractal order can be seen in the Figure 1.

In its use, the performance of this sensor is influenced by several factors namely such as surface area (L), segment Length (d), number of segments (S), and curve order (n). In general, these parameters can be calculated using some of the following equations [12].

$$d = \frac{L}{2^n - 1} \tag{1}$$

$$S = (2^{2n} - 1)d \tag{2}$$

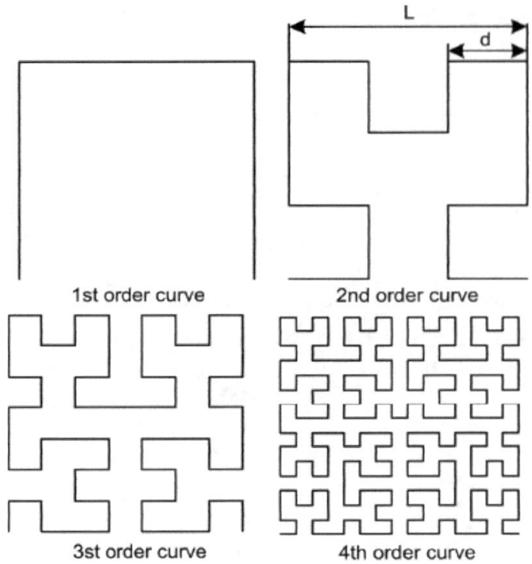

Figure 1. Hilbert Fractal Curve 1st up to 4th order [12].

The shape of the Hilbert fractal antenna sensor used in this study can be seen in Figure 2 below.

Figure 2. Microstrip Hilbert fractal 4th order antenna.

In designing and manufacturing Hilbert sensor, IE3D software is used to determine S-parameter of the sensor. The return loss and VSWR simulation results of Hilbert sensor 4th order with dimensions 5 cm x 5 cm using IE3D software can be seen in Figure 3

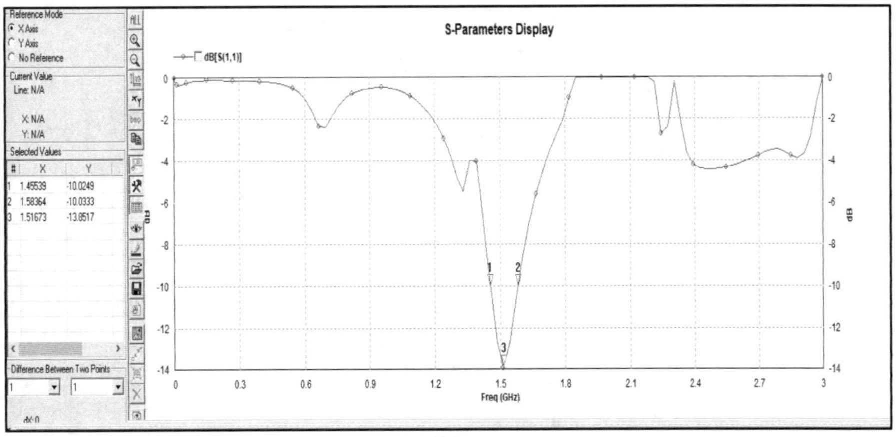

(a)

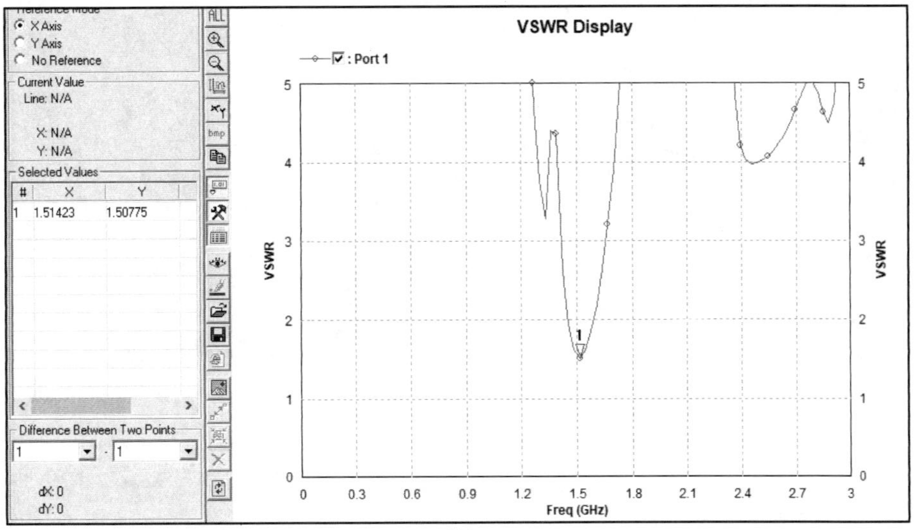

(b)

Figure 3. Hilbert sensor S-parameter simulation using IE3D software (a) return loss (b) VSWR.

2.3. PD setup test

PD detection tests are carried out to produce PD signals from several types of discharge sources. In this research, a Hilbert fractal antenna sensor is used with the UHF method in detecting the discharge event that occurred. The PD detection test circuit can be seen in Figure 4.

220V / 22kV test transformer is used to raise the high voltage AC to produce a source of discharge. In the test circuit there is also a voltage divider circuit that functions to measure the output value on the transformer. In the test model the Hilbert fractal antenna is placed to detect the discharge signal which is then displayed on the oscilloscope.

In the PD detection test to determine the signal characteristics of the three types of PD consisting of corona, surface, and cavity, in the test circuit the voltage input variables and three different types of insulators are used as in Figure 5.

IOP Conf. Series: Materials Science and Engineering 1173 (2021) 012022 doi:10.1088/1757-899X/1173/1/012022

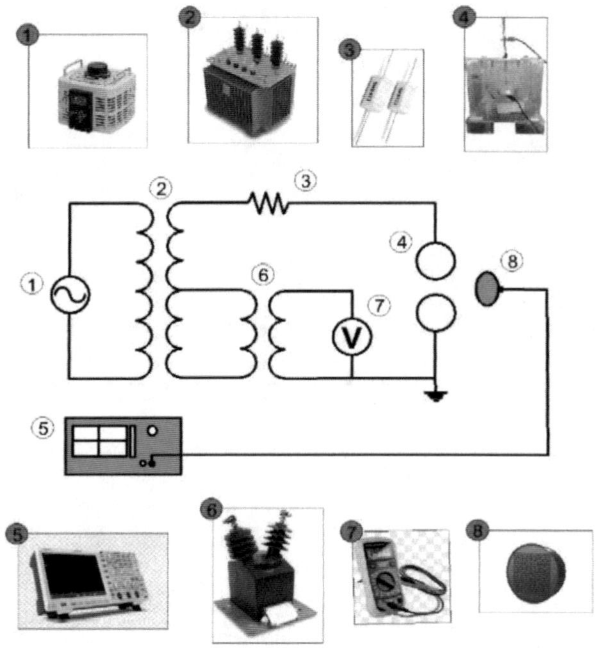

1. AC Voltage Regulator
2. Transformer 220V/22kV
3. Resistor 156 kΩ
4. Test Kit
5. Oscilloscope
6. Potential Transformer
7. Multimeter
8. Sensor Hilbert

Figure 4. PD setup test.

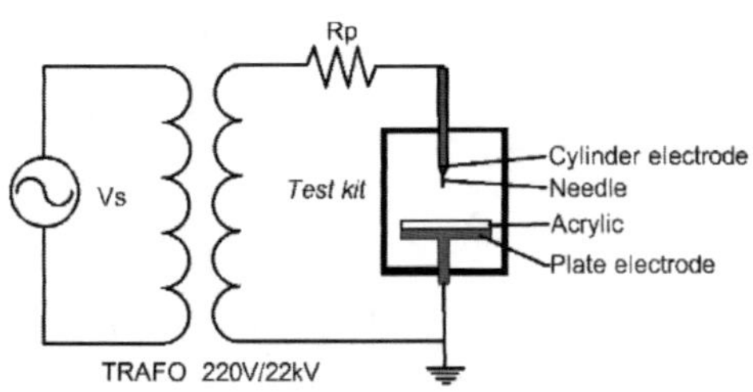

(a)

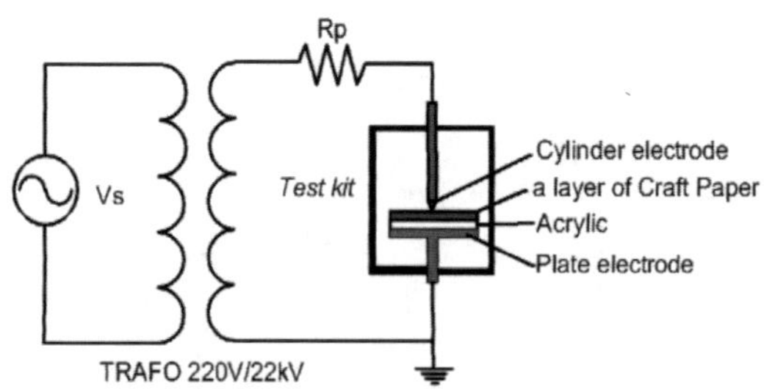

(b)

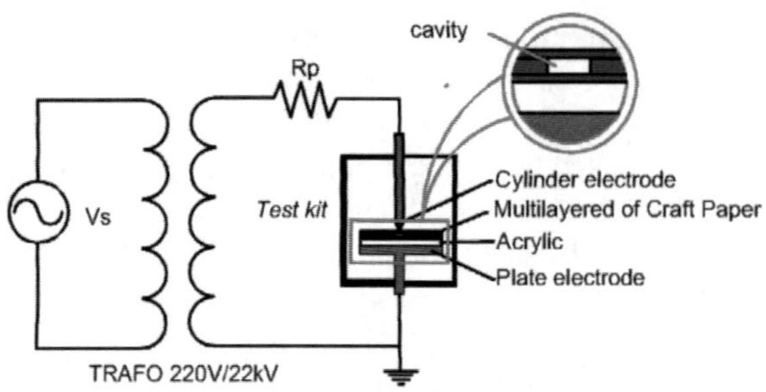

(c)

Figure 5. PD detection test circuit with various isolators (a) corona (b) surface (c) cavity.

In Figure 5a, the tip of the rod electrode, a needle is added and the two electrodes are separated by 5 cm. It is intended that the air acts as an insulator to produce corona type PD. The variable input voltage used ranges from 10-16 kV. In Figure 5b, a craft paper with a thickness of 0.2 cm is used as a test isolator to produce surface type PD. The variable input voltage used ranges from 0-10 kV. In Figure 5c, a layered craft paper with a 0.2 cm diameter hole in the middle is used to produce a cavity type PD source. The variable input voltage used ranges from 0-10 kV.

2.4. Wavelet denoising tool

In PD detection, sometimes the signal of the decay captured by the sensor is very small [5]. Unwanted signal interference (noise) is sometimes the main cause of the matter. Radio, television and telecommunications signals, as well as thermal noise are examples of the types of noise that can affect PD detection results. Therefore, to increase the sensitivity of PD detection, the amount of noise needs to be minimized [5]. There are many methods used in the process of denoising the signal, one of which is using Multivariate Wavelet Denoising. This method is useful for separating the noise signal with the original signal. As for Multivariate Wavelet Denoising using regression models as follows [13].

$$X(t) = f(t) + \varepsilon(t), t = 1 \quad (3)$$

Where X (t) is a observed signal, *f* (t) is the denoised signal and the ε (T) is a spatial correlated noise signal.

In this research, 600 samples of the original signal data are then denoised using sym4-wavelet level 5 found in the MATLAB toolbox. The results of the PD signal denoising process can be seen in Figure 6 below.

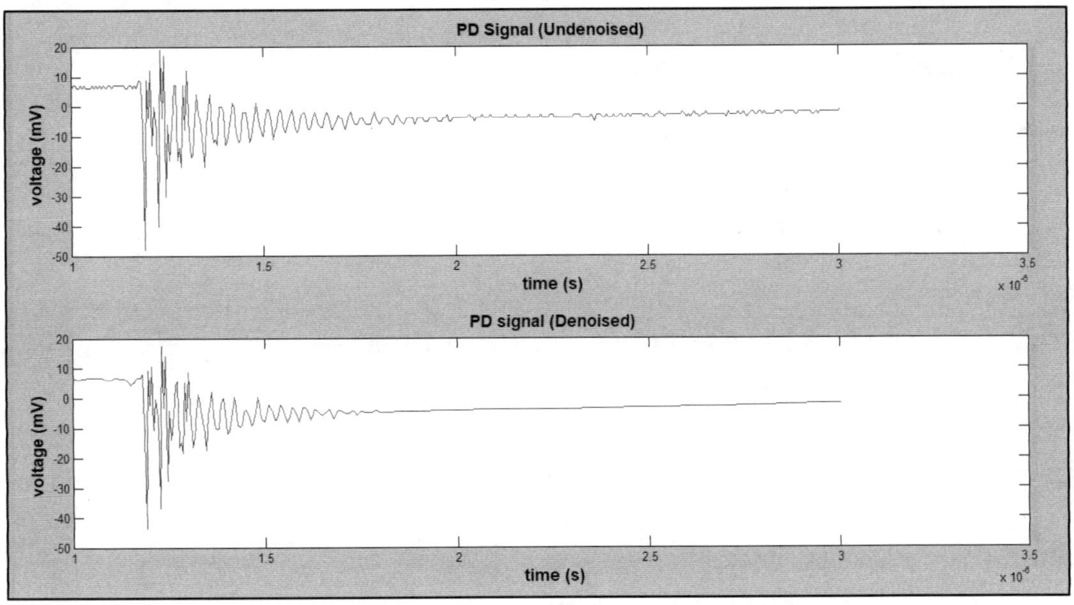

Figure 6. PD signal denoised using wavelet.

The process of denoising the signal does not always provide maximum results, there is still a possible loss of the original signal from the denoising process [13]. Therefore, it is necessary to review and consider first whether the process of denoising against the original signal is profitable or otherwise detrimental [13].

2.5. PD extraction and identification

The original PD signal and the denoising results are then extracted into statistical parameters consisting of mean, standard deviation, skewness and kurtosis. These parameters will be used as input to the neural network to identify the type of PD.

The PD type identification process is carried out using an artificial neural network (ANN) with back propagation type. Based on statistical parameter data, PD can be identified based on the source of discharge.

3. Results and Discussions

In each test, 200 sample discharge signals have been taken from the oscilloscope, bringing the total sample to 600 discharge signals. The average magnitude of the type of PD can be seen in Table 1 as follows.

Table 1. PD magnitude test results.

Type of PD	The Lowest Magnitude (mV)	The Highest Magnitude (mV)
Corona	124,76	294,47
Surface	19,26	36,46
Void	39,63	112,62

In Table 1 it can be seen that the magnitude of the corona type PD is greater than the surface and cavity type PD. The magnitude of the magnitude value at the corona discharge is thought to be due to the electromagnetic field moving from the needle electrode to the free electrode plate without any obstructions so that a large magnitude is generated. Meanwhile, for PD cavity type which has a magnitude greater than the surface type, this is presumably due to the air cavity that is inside the test isolator. The air cavity will bear a higher voltage in the discharge testing process compared to solid insulators due to differences in the permittivity coefficient between solid and gas insulators.

The denoising process to reduce the noise signal contained in the original signal begins by entering the original PD signal in the MATLAB software. A total of 600 samples of the original signal data are then denoised using sym4-wavelet level 5 found in the MATLAB toolbox.

The original signal and denoising results are then extracted into statistical parameters which can be seen in the following Table 2 and Table 3.

Table 2. Statistical data of undenoised PD signal

Statistical Parameter	Type of PD (mV)		
	Corona	Surface	Void
Mean	207,46	29,09	60,35
Skewness	-0,31	-1,41	-0,76
Kurtosis	1,75	11,47	5,81
Deviation Standard	16,03	3,48	10,73

Table 3. Statistical data of denoised pd signal

Statistical Parameter	Type of PD(mV)		
	Corona	Surface	Void
Mean	207,46	29,08	60,29
Skewness	-0,280	-1,16	-0,68
Kurtosis	1,60	9,74	5,49
Deviation Standard	15,93	3,30	10,48

PD type identification using ANN back propagation type consists of the training process (train dataset) and testing (test dataset). 450 of the 600 PD signal statistical data are used as input in the training process (train dataset). The training process (train dataset) is completed when the mean square error (MSE) of 0.01 is reached. This is shown by the training data performance graph in Figure 7.

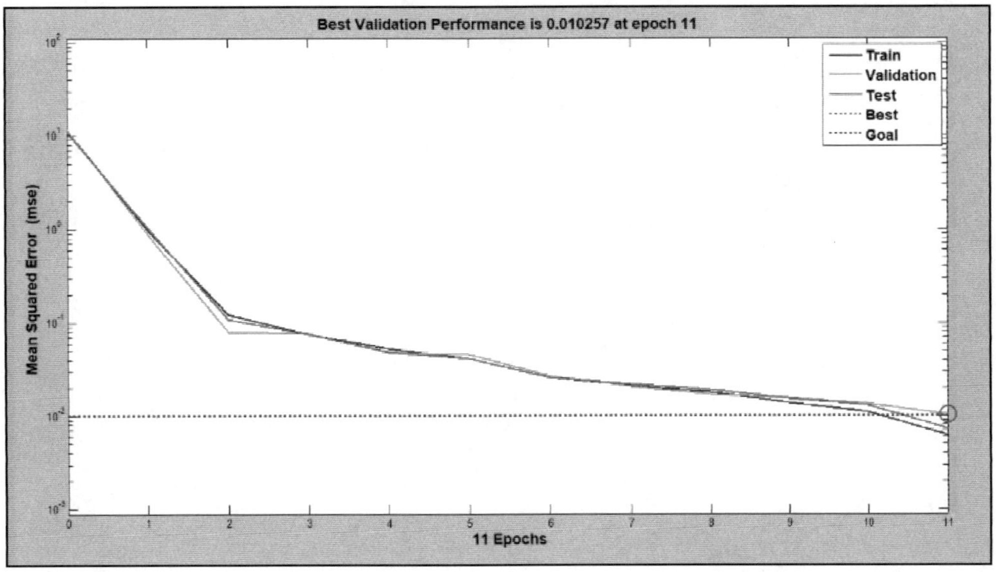

Figure 7. Mean square error (MSE) graph of training data (train dataset) ANN back propagation type.

The PD type identification process is then continued through testing (test dataset) using the remaining 150 PD statistical data. The graph of the results of identification of type PD by ANN back propagation type can be seen in Figure 8.

Based on Figure 8 that shows the test results (test dataset) identification of PD type based on the source of discharge, it can be seen that the x-axis is 150 sequential PD static data and the y-axis is the target target value. The red graph shows the test target that must be achieved at the PD type identification output value, while the blue graph is the result of PD type identification output. In the testing process of 150 PD signal statistical data, it can be seen that the first data up to the 50th data are corona type discharge signal data marked as target 1, 51st to 100th data are surface type discharge signal data (surface discharge) marked as target 2, and the 101st to 150th data are void discharge discharges marked as target 3. Prior to the denoising process as shown in Figure 8a, there are several identification outputs the type of PD that is not in accordance with the test target is proven by the output graph that does not coincide with the test target graph.

However, after a denoising process on the PD signal is shown in Figure 8b, errors from the neural network in identifying the type of PD can be minimized. The error (result) identification of the type of PD by ANN allegedly can be caused by high levels of noise in the PD signal data. In addition, errors in the process of identifying the type of PD sometimes originate from ANN itself or can also be caused by errors in the detection of PD. To minimize the level of error (ANN) in identifying the type of PD caused by noise, can be overcome by denoising the PD signal. On the other hand, to overcome errors (errors) originating from ANN itself, can be minimized by increasing the amount of PD training data. If the error (error) identification results caused by inaccurate detection of PD can be minimized by increasing the precision of the measuring instrument.

Overall, the process of detecting and identifying the type of PD in this study has been successfully carried out with fairly accurate results. If comparing the output value with the PD target data, then the percentage of ANN accuracy can be calculated in identifying the type of PD. The results of the ANN accuracy comparison test in identifying the type of PD between the original discharge signal and the signal from the denoising process can be seen in Table 4.

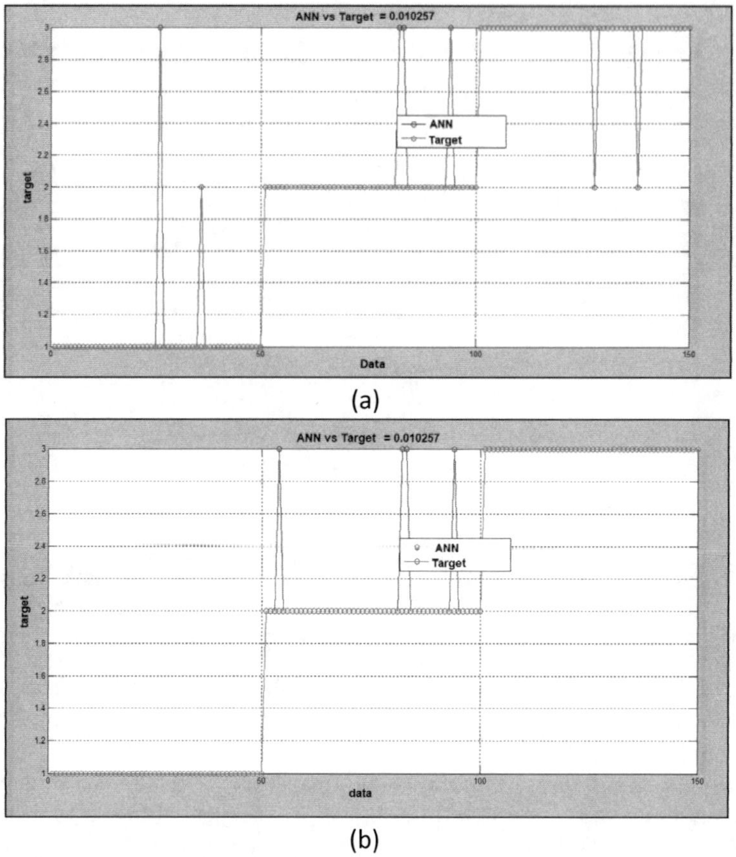

Figure 8. Graphs of PD type identification results (a) Before the denoising process (b) After the denoising process.

Table 4. The results of the ann accuracy level

PD type	Sample number	Correct identificate	
		Original signals	Denoised signals
Corona	50	48	50
Surface	50	47	46
Void	50	48	50
Accuracy (%)		95,33%	97,33%

Based on Table 4, the corona type PD consisting of the original discharge signal and the denoising signal has the difference in the number of true samples in the identification process. Corona original signal identification produces 48 correct samples from a total of 50 samples, while for the signal from the denoising process, produces 50 correct samples from a total of 50 samples in the identification process. At the surface type discharge, the number of true samples in the original signal is 47 samples from a total of 50 samples and after the denoising process the number of true samples drops to 46 samples.

At the discharge of the cavity type obtained the number of true samples on the original signal as many as 48 samples from a total of 50 samples and after the denoising process carried out the number of true samples increased to 50 samples. For PD type identification results based on Table 4, it can be analyzed that the denoising process has little effect on corona and cavity type discharge signals in improving the accuracy of identification of PD type by ANN, whereas for surface type discharge signals, denoising process reduces the level of accuracy of PD type identification by ANN. This is presumably because the denoising process can sometimes eliminate the characteristics of the PD signal or it may also be due to inaccuracy in the detection of PD. Based on the results of the identification of the type of PD in this study, obtained a large percentage of the level of accuracy of the identification of the PD signal by ANN that is equal to 95.33% for the original signal and 97.33% for the denoising signal.

4. Conclusions

PD detection using the Hilbert fractal antenna sensor has been successfully carried out by each producing 200 surface discharge, void discharge, and corona discharge signals. In this study, the process of identifying the introduction of PD types based on the source of discharge has been successfully carried out. In data extraction, statistic parameters consisting of mean, standard deviation, skewness, and kurtosis are generated from discharge signal data. This is done to reduce the size of the data that is too large, but is still recognized by ANN. The denoising process on the discharge signal does not always provide an effective result at the signal sealing level by ANN because there is still the possibility of losing signal characteristics due to the denoising process.

Acknowledgements

The authors would like to express their gratitude to Kemenristek-DIKTI Indonesia for supporting this research. This research was supported by Kemenristek-DIKTI Indonesia with the Fundamental Research Grant, contract number: 857/UN26.21/PN/1999.

References

[1] Bartnikas R 2003 Partial Discharge , Their Mechanism, Detection and Measurement *J.IEEE Trans. on Dielectrics and Insulation* **9** 763

[2] Garnacho F and Trasmonte I 2008 On-Site Measurement s experiences in insulation condition for medium and high viltage cables *CIGRE,* **D1.201**

[3] Phung B T 1997 Computer-based Partial Discharge Detection and Characterisation," *PhD Thesis Department of Electrical Power Engineering, University Of New South Wales, Sydney,* 1997.

[4] Sinaga H H, Sitorus H B H, Permata D, Yuniati Y and Vraja R B 2020 Denoising of partial discharge waveforms using multivariate wavelet method, *IOP Conference Series: Materials Science and Engineering* 857(1)

[5] Contin A, Cavallini A, Montanari G C, Pasini G and Puletti F 2000 Artificial Intelligence Methodology for Separation and Classification of Partial Discharge Signals *Annual Report Conference on Electrical Insulation and Dielectric Phenomena* 522-526

[6] Cavallini A, Conti M, Contin A and Montanari G C 2003 Advanced PD Inference in On-Field Measurements II. Identification of Defects in Solid Insulation Systems *J. IEEE Transactions on Dielectrics and Electrical Insulation* **10** 528-538

[7] Miller R K, Shu F, Nunez A and Ternowcheck S 2003 Advances in Accoustic Emission Testing for Detecting , Location and Assessing Electrical and Thermal Faults *EPRI 2003: Substation Equipment Diagnosis Conf*

[9] Sinaga H H, Sitorus H B H, Permata D and Soedjarwanto N 2019 Fractal Hilbert Sensor to Detect Partial Discharge on Transformer, *Journal of Engineering and Scientific Research* (JESR), **1**(2) 94-100

[10] Sinaga H H, Phung B T and Blackburn T R 2012 UHF sensor array for partial discharge location in transformers, Proceedings of 2012 *IEEE International Conference on Condition Monitoring and Diagnosis* 979–982
[8] Muhr M, Strehl T, Gulski E, Feser K, Gockenbach E and Hauschild W 2006 Sensor and Sensing Used For Non Conventional PD detection *Cigre* D1-102_2006
[11] Wang Y, Wang Z and Li J 2013 UHF Moore Fractal Antennas for Online GIS PD Detaection *J. IEEE Antennas and Wireless Propagation Latters* **16** 852-855
[12] Li J, Jiang T, Wang C and Cheng C 2012 Optimization of UHF Hilbert Antenna for Partial Discharge Detection of Transformers *J. IEEE Trans Antennas and Propagation* **60** 2536-2540
[13] Aminghafari M, Cheze N and Poggi J M 2006 Multivariate de-noising using wavelets and principal component analysis, *Computational Statistics & Data Analysis*, **50** 2381-2398

Performance comparison of single and double pass PV/T solar collectors integrated with rectangular plate fin absorber

Amrizal*, Y P Hidayat, and M Irsyad

Mechanical Engineering Department, University of Lampung, Jln. Prof. Soemantri Brodjonegoro No.1, Bandar Lampung 35145, Indonesia

*Email: amrizal@eng.unila.ac.id

Abstract. Solar panel can act as a medium for transforming sunlight into electrical energy. Nevertheless, not all of the sunlight received on this device can be converted into electrical energy and the remainder will be dissipated into thermal energy. If a solar panel is continuously exposed to the sun, so its temperature will increase and thereby reduce its electrical efficiency. To deal with the issue, it may combine a PV panel with a thermal collector which is called a hybrid PV/T collector. Furthermore, in the present study the effects of fluid flow direction and mass flow rate on the thermal and electrical efficiency of the PV/T collectors are extensively analyzed. Single and double pass collectors with rectangular plate fin absorber were attached to backside of the PV panel. They were tested using a solar simulator with different flow rate ranging between 0.01 and 0.05 kg/s of the working fluid. The results show that the use of double pass PV/T air collector increase its performance compared with the single pass one. The highest values obtained from this type collector are found to be 73.23% in the thermal efficiency and 10.16% in the electrical efficiency with the working fluid mass flow rate of 0.048 kg/s.

Keywords: single, double pass, PV/T collectors, rectangular fin absorber

1. Introduction

PV/T collector is built by joining a PV cell and a thermal collector that produce thermal energy and electrical energy at the same moment. However, the electrical efficiency of PV/T collector decreases if the surface temperature of PV cell increases [1]. In general, a working fluid in the form of liquid and air can act as a coolant as reported in a comprehensive review by other [2]. The working fluid aims to absorb the heat waste of the collectors and reduce its surface temperature. In this type of the collector, the implementation of fins led an enhancement in the area of heat transfer and could be a better way to increase the thermal efficiency. Several researchers have conducted studies on the different geometry fins as reported in several literatures [3-6].

Furthermore, it has been also investigated that solar PV/T air collectors using v-grooved fins as a thermal absorber [7]. Air is circulated along the upper groove and returned in the counter flow direction along the lower groove. The results show that the use of an extended surface appearing as v-grooved fin give a significant impact on increasing the performance of the collectors. This is because of the working fluid absorb more waste heat through the fins attached backside the PV panel. Several other studies related to the improvement of collector performance, the effect of working fluid, fluid flow geometry and configuration have also been reported in the literatures. Since the air has lower heat capacity compared with water, so it is important to develop the performance of PV/T air collector as observed by other [8]

In addition, several other researchers investigated the performance of solar PV/T collectors for both single [9] and double passes [10-13] which were applied in various fin geometries. Meanwhile, a comparison research between the single and double passes of PV/T air collectors was carried out by others [13,14]. The results reported that better performance was obtained from the double pass design. Furthermore, the thermal efficiency of various PV/T air collectors associated with the double pass design is better than others because of uniform cooling for the surface temperature of PV panel, so enhance the electrical performance as investigated and reported by other [15]. Further study assumed that thermal efficiency of a trapezoidal corrugated plate thermal air collector [16]. This type collector was tested under the climate of Konya region (Turkey). There is the difference between outlet and inlet fluid temperatures up to 9°C for the trapezoidal corrugated plate thermal air collector compared with the traditional thermal collector.

Based on the previous references reported from researchers as mentioned earlier, the impact on PV/T air collector performance may give various characteristics depend on several parameters. This makes the development of PV/T air collectors should be interesting to be explored in term of variations in geometries, materials and various test locations. In the present study, the performance of the single and double passes of PV/T air collectors combined with rectangular plate fin absorber is investigated. The inlet fluid temperature as a working fluid is based on the tropical climate of Lampung, Indonesia. Further, this new design of the proposed PV/T collector uses fins as absorber with a specific geometry. The material of fin used is aluminum which has good thermal conductivity and low density. Then, the two collectors were built in the form of single and double passes for flow arrangement of the working fluid. Furthermore, the influence of the mass flow rate of the working fluid on the performance of the two collectors is elaborated in the present work.

2. Design of PV/T collector

A photograph of the test installation is given in Figure 1. Solar simulator is the main component for simulating solar radiation energy. This device consists of 12 halogen lamps with a power of 300 W. Other's equipment are installed to measure data associated with radiation, temperature, fluid mass flow rate, current and voltage of the PV panel. Furthermore, the type of PV panel collector is polycrystalline PV 100Wp with dimensions of 1020 x 670 x 30 mm. The specifications of the PV panel tested are also presented in Table 1. Meanwhile, for circulating the working fluid along the rectangular plate fin, a blower is used.

Figure 1. Photograph of experimental setup.

Table 1. PV panel specifications (Polycristaline)

P_{max}	100 W
V_{mp}	17.6 V
I_{mp}	5.69 A
V_{oc}	2.85 A
I_{sc}	3.04 A
NOCT	45°C±2°C

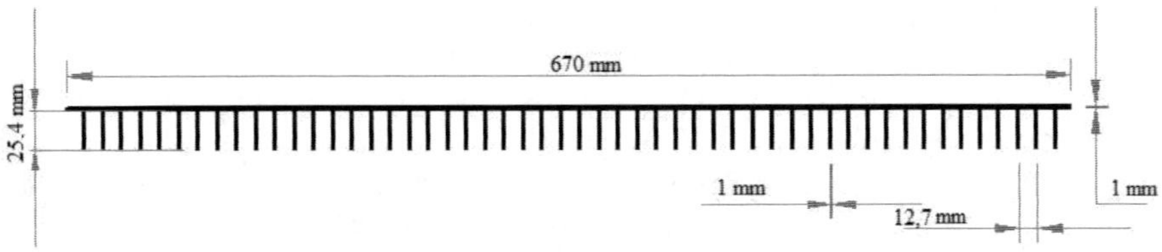

Figure 2. Dimensions of rectangular plate fin absorber.

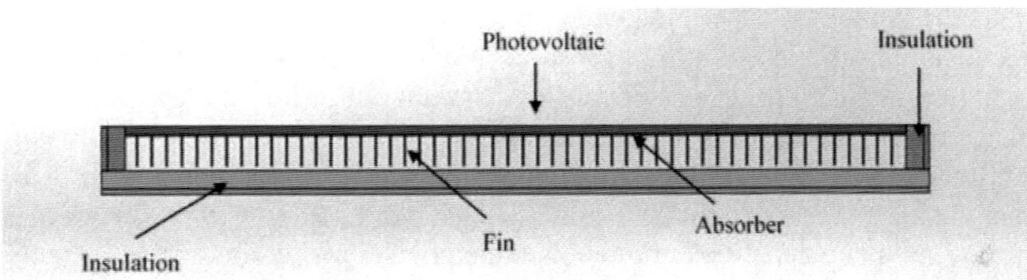

Figure 3. Front view of single pass collector

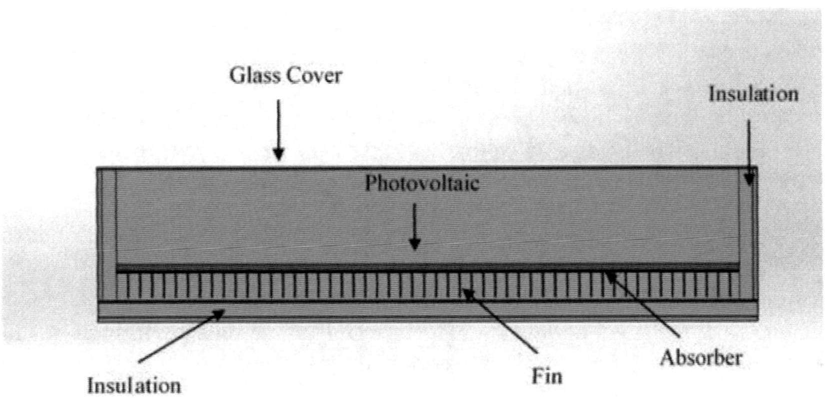

Figure 4. Front view of double pass collector.

Figure 5. Side view and fluid flow direction of double pass collector

The PV/T air collector performed in the present work was combined by attaching the solar thermal collector to backside the PV cell surface. The dimensions of rectangular plate fin absorber fit the available surface of PV collector as seen in Figure 2. Fin height-space ratio setting is found to be 2:1. The differences between the two types of collectors are illustrated in detail in Figures 3 and 4, respectively. Besides, Figure 5 shows the side view of double pass collector which is developed from a single pass collector, where the outlet fluid of the single pass one is circulated back to the top of the PV collector surface.

3. Materials and Methods

The two types collectors were tested indoor by several halogen lamps as a substitute for solar radiation energy. The radiation intensity used was 860 W/m^2. In data collection, the collectors were exposed in direction perpendicular to radiation. The experiments were carried out in Bandar Lampung where is the capital of the Lampung Province (Indonesia). Bandar Lampung region has a typical tropical climate which is characterized as dry and rainy seasons. Air as a working fluid in this work acts as a cooling medium for rectangular plate fin absorber of PV/T collectors. It was varied in the range of 0.012, 0.024, 0.036 and 0.048 (kg/s), respectively, for the mass flow rate values. Furthermore, the performances of the two collectors were investigated and analyzed as described in the following sections.

3.1. Experimental procedures

A data acquisition system recorded every 10 s interval was implemented in the present study. The experimental data of the two collectors were taken such as inlet, outlet and ambient fluid temperatures, incident radiation, voltage and electrical current. Besides, the fluid temperatures and radiation intensity were measured using K-type Thermocouples and Solar Power Meter SPM 1116SD, respectively. The working fluid was regulated by using a valve for the *mass flow rate*. A power meter and electrical loading resistors were then installed to obtain the electrical data of the two collectors.

3.2. Characterization of PV/T solar collector

The thermal performance of the PV/T collectors was characterized referring to the working fluid and the surface temperature of the PV collectors. Furthermore, the experiments were also to determine the electrical efficiency of the two collectors. Moreover, the thermal performance in the form of instantaneous efficiency is associated with next equation:

$$\eta_{th} = \frac{\dot{m}\, C_p(T_{fo}-T_{fi})}{A\, G} \quad (1)$$

where the thermal efficiency is represented by η_{th} (%). Further, the specific heat of the working fluid and the mass flow rate are symbolized by C_p (W/m^{2}C) and $\dot{m}$ (kg/s), respectively. Then, the inlet and oulet of fluid temperatures, the absorber area and the intensity of radiation are given by T_{fi} (°C), T_{fo} (°C), A (m^2) and G (W/m^2), respectively.

Furthermore, testing electrical performance can use the following formula:

$$\eta_{el} = \frac{I\,V}{A\,G} \tag{2}$$

where the parameters of equation (2) namely electric current and voltage are represented by I(A) and V(V), then area of the PV surface and the intensity of radiation are given by A (m^2) and G (W/m^2), respectively.

4. Results and Discussions

The two types of the PV/T collector have been tested using several halogen lamps as a substitute for solar radiation. The experiments were carried out with a radiation intensity of 860 W/m^2 and various flow rate ranging between 0.01 and 0.05 kg/s of the working fluid. The ambient temperature in this work was implemented under the Bandar Lampung climatic (latitude 5°27′S and longitude 105°16′E). The experiment results are then analyzed and compared between the two collectors as described in the following paragraphs.

4.1. PV/T collector thermal performance

The results obtained from the experiments are presented in the form the temperature distribution and the thermal efficiency for the two collectors as illustrated in the Figures 6 and 7.

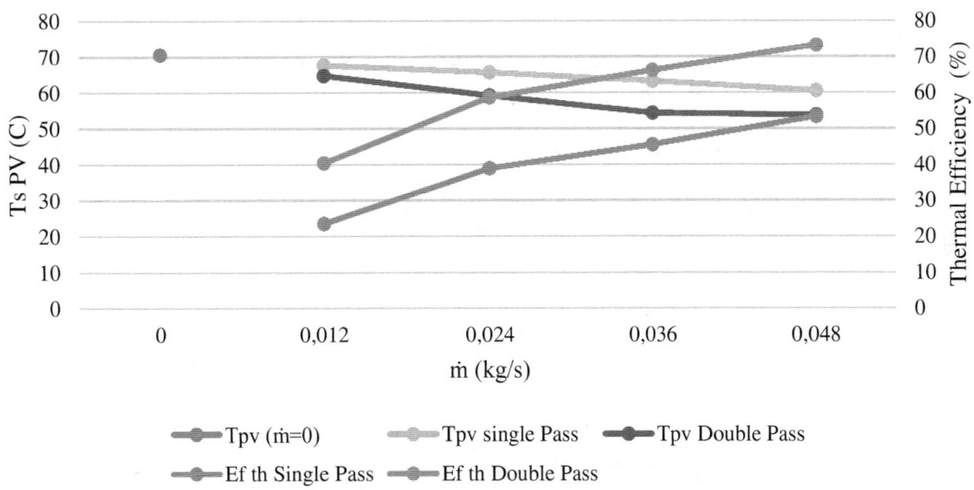

Figure 6. Thermal efficiency and PV surface temperature in accordance with mass flow rate of working fluid *without* ambient air speed (V = 0 m/s).

Figures 6 and 7 show the relationship between thermal efficiency and surface temperature of the two collectors associated with the mass flow rate of the working fluid. During the experimental tests, the two collectors were also performed *with* (velocity of 1 m/s) and *without* ambient air speed moving over the collector surface which was circulated by a fan. As for the cooling process, air as a working fluid supported by a blower was circulated through the rectangular plate fin absorber.

In general, from Figures 6 and 7 it is explained that the greater the working fluid mass flow rate, the temperature of PV surface will decrease and then the thermal performance enhances for the two types of the collectors. This is because the greater the working fluid mass flow rate of the collectors, the greater the waste *heat* absorbed by flowing air and move it to the outlet section, causing decreased the PV panel surface temperature. For this context, there is the difference in the temperature of PV surface for the double pass one (see the T_{PV} *double pass* curve) in Figure 7.

This condition is indicated by sharper curve than the others which decrease *the* temperature of PV surface around 10°C. The advantage of the double pass one, the outlet fluid of the first channel is circulated back to the top of the PV collector surface (second channel) may give the working fluid longer time to contact with the surface. Consequently, it can absorb more waste heat to improve the performance of the collectors in comparison with just a single pass one. Thermal efficiency is able to reach 73.23% at the working fluid mass flow rate of 0.048 kg/s for double pass one.

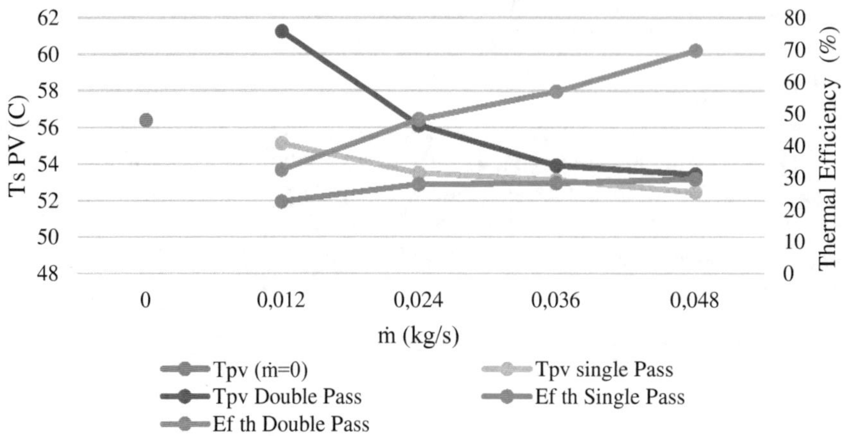

Figure 7. PV surface temperature and thermal efficiency in accordance with mass flow rate of working fluid *with* ambient air speed (V = 1 m/s).

4.2. PV/T collector electrical performance

To find out the electrical performance, the test procedure is similar to that of the thermal performance testing. The data obtained from experiments are then calculated and presented in the form of electrical efficiency and surface temperature of the collectors.

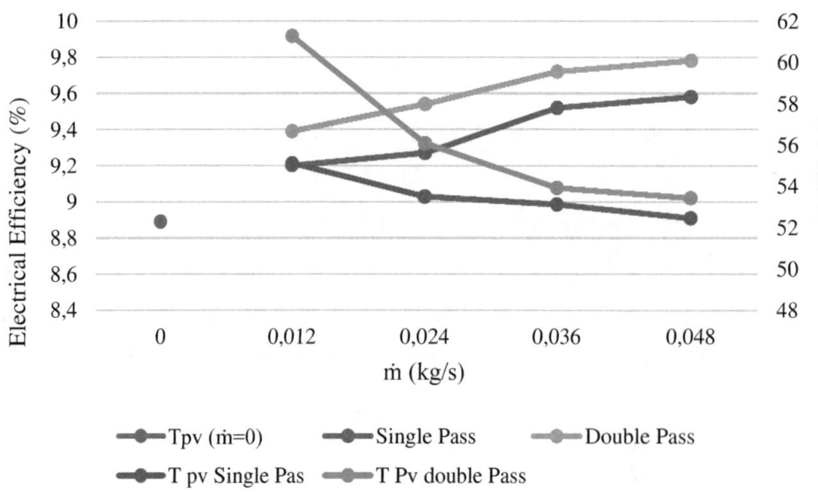

Figure 8. PV surface temperature and electrical efficiency in accordance with the mass flow rate of fluids *without* wind speed.

Figures 8 and 9 show impact of various flow rate of the working fluid on surface temperature and electrical efficiency of the two different types of PV/T collectors *with* and *without* ambient air speed.

Similar to that of the thermal performance characteristics, increasing the mass flow rate also enhance the performance of the PV/T collectors associated with the electrical efficiency and reduce the surface temperature for both types of the collectors. With the use of ambient air speed in the experimental tests, the surface temperature of the PV panel is lower than *the* results obtained from *without* ambient air speed. This is because in the testing *with* the ambient air speed, a convection heat transfer process occurs on the PV panel surface to ambient fluid flow. The waste heat of the PV panel surface will be absorbed by the ambient air moving over the collector surface.

That condition may be advantageous because a decrease in the surface temperature. It will enhance the PV/T collector performance related to the electrical efficiency. The graph of the impact of the mass flow rate for double pass collector as seen in Figure 9 provide the highest electrical efficiency value of 10.16% in the working fluid mass flow rate of 0.048 kg/s.

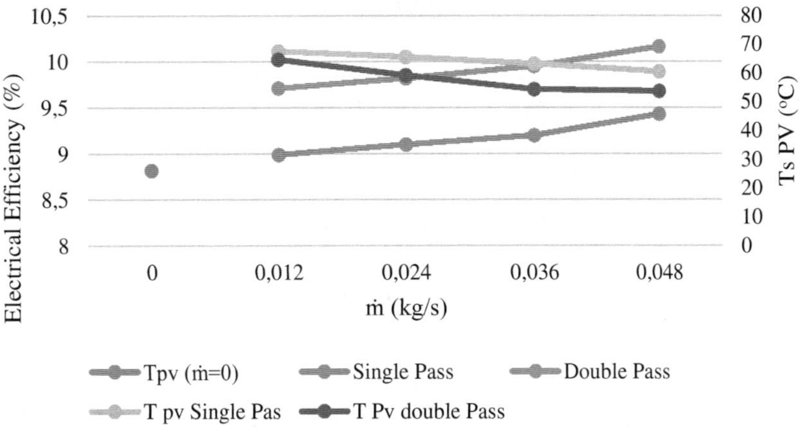

Figure 9. PV surface temperature and electrical efficiency in accordance with mass flow rate of fluid *with* wind speed.

5. Conclusions

The results have been obtained from this study indicate that thermal and electrical efficiency of the PV/T air collectors are affected by geometries, working fluid mass flow rate, flow arrangement and test locations (weather and climate). With fin height-space ratio setting (2:1) and different flow rate ranging between 0.01 and 0.05 kg/s of the working fluid and inlet fluid temperature (T_{fi}) under the tropical climate of Lampung (Indonesia), there is a different response of the two types of collectors. In general, the double pass PV/T collector has better thermal and electrical efficiencies than the single pass one. The highest values obtained from the double pass collector with the working fluid mass flow rate of 0.048 kg/s are found to be 73.23% in thermal efficiency and 10.16% in the electrical efficiency, respectively. Meanwhile, there is also an increase in electrical efficiency of 0.89% for double pass collector *with* the use of ambient air speed. The results of this study indicate that the use of double pass PV/T collector may have the potential to be developed in the tropical climate.

Acknowledgements

The authors acknowledge the support of University of Lampung in providing research facilities.

References

[1] S. Natarajan, T. Mallick, M. Katz and S. Weingaertner 2011 *Numerical investigations of solar cell temperature for photovoltaic concentrator system with and without passive cooling arrangements* International Journal of Thermal Sciences Vol. 50, pp. 2514-2521,

[2] Sourav Diwania, Sanjay Agrawal, Anwar S. Siddiqui, Sonveer Singh 2020 *Photovoltaic–thermal (PV/T) technology: a comprehensive review on applications and its advancement* International Journal of Energy and Environmental Engineering 11:33–54

[3] Tonui, J.K., Tripanagnostopoulos, Y 2008 *Performance improvement of PV/T solar collectors with natural air flow operation* Sol. Energy 82(1), 1–12

[4] Mojumdera, J.C., Chonga, W.T., Onga, H.C., Leong, K.Y., Al Mamoon, A. 2016 *An experimental investigation on performance analysis of air type PV thermal collector system integrated with cooling fins design.* Energy Build. 130, 272–285

[5] Hussain, F., Othman, M.Y., Yatim, B., Ruslan, H., Sopian, K., Anuar, Z., Khairuddin, S. 2012 *Comparison study of air base photovoltaic/thermal (PV/T) collector with different design of heat exchanger.* WREF 1, 189–194

[6] M. Irsyad, M. Akmal, M. D. Susila, Amrizal and Amrul 2020 *Heat Transfer Characteristics of Paraffin in Staggered Fins Heat Exchanger for the Cooling Process to Room Cooling Applications*, IOP Conf. Series: Earth and Environmental Science 520

[7] Othman, M.Y., Ruslan, H., Sopian, K., Jin, G.L. 2009 *Performance study of photovoltaic-thermal (PV/T) solar collector with V-grooved absorber plate* Sains Malays. 38, 537–541

[8] Niccolò Aste Giancarlo, Chiesa FrancescoVerri 2008 *Design, Development and Performance Monitoring of a Photovoltaic-Thermal (PVT) Air Collector*, Renewable Energy 33, , 914-927

[9] Jin, G.L., Ibrahim, A., Chean, Y.K., Daghigh, R., Ruslan, H., Mat, S., Othman, M.Y., Ibrahim, K., Zaharim, A., Sopian, K. 2010 *Evaluation of single-pass photovoltaic-thermal air collector with rectangle tunnel absorber.* Am. J. Appl. Sci. 7, 277–282

[10] Amin Elsafi and P. Gandhidasan 2015. *Performance of a Photovoltaic or Thermal Double-Pass Solar Air Heater with Different Fin Configurations.* Journal of Clean Energy Technologies JOCET 2015 Vol.3(1): 28-33, ISSN: 1793-821X

[11] Mohd. Yusof Othman, Baharudin Yatim, Kamaruzzaman Sopian, Mohd. Nazari Abu Bakara 2007 *Performance studies on a finned double-pass photovoltaic-thermal (PV/T) solar collector*, Desalination, Volume 209, Issues 1–3, Pages 43-49

[12] Marc A. Rosen, Rakesh Kuma 2011. *Performance of a Photovoltaic/ Thermal Solar Air Heater: Effect of Vertical Fin on a Double Pass System.* International Journal of Energy and Enviromental Engineering ISSN: 9163, Vol.2 / No.4 (pp.1-12)

[13] Kumar, R., Rosen, M.A. 2011 Performance of a double pass PV/T solar air heater with and without fins. Appl. Therm. Eng. 31, 1402–1410

[14] K. Sopian, K. S. Yigit, H. T. Liu, S. Kaka and T. N. Veziroglu 1996 *Performance Analysis of Photovoltaic Thermal Air Heaters,* Energy Conversion And Management Vol. 37, no. 11, pp. 1657-1670

[15] Slimani, M.E.A., Amirat, M., Kurucz, I., Bahria, S., Hamidat, A., Chouch, W.B. *A detailed thermal-electrical model of three photovoltaic/thermal (PV/T) hybrid air collectors and photovoltaic module: comparative study under Algiers climate condition.* Energy Convers. Manag. 133, 458–476

[16] Selçuk Darici & Anil Kilic, 2020. Comparative study on the performances of solar air collectors with trapezoidal corrugated and flat absorber plates, Heat and Mass Transfer, Springer-Verlag GmbH

Optimization of X-ray parameter monitor wireless system based on internet of things

A B Triani, S Budiyanto*, A Adriansyah, and M Alaydrus

Department of Electrical Engineering, Faculty of Engineering, Universitas Mercu Buana, Jakarta, Indonesia

* Email : sbudiyanto@mercubuana.ac.id

Abstract. The purpose of this research is to develop a non-invasive X-ray acceptance test device with an internet of things (IoT) based on monitoring system as an interface connection. This experiment uses an application with IoT technology as a graphical user interface for the X-ray acceptance test prototype. The experiment is done by measuring the value of the kilovolt and millisecond on the General X-ray machine. The data result will be compared with the traceable X-ray calibrator as reference value. The overall average percentage of error result is 99.26% and the percentage of accuracy is 0.74%. It is generally concluded that this prototype experiment has good results and can be used as a X-ray acceptance test equipment.

Keywords: X-ray acceptance test, Internet of Things (IoT), development software

1. Introduction

One of the diagnostic techniques in the healthcare industry is X-rays. The generation of X-ray radiation depends on the kilovolt X-ray tube and the exposure time. Diagnostic X-rays use a range of radiation to analyze the patient's body. If the received radiation dose exceeds the threshold, the patient's health can be threatened. This situation can be happen due to a malfunction of the X-ray machine. To improve the performance of the X-ray machine, it is necessary to assess the tube voltage and the exposure time to ensure that the values are not exceed the specified threshold. [1-3].

The acceptance test for diagnostic and interventional radiology x-ray is a series of testing activities to ensure the X-ray machine is in reliable condition (passes the parameter test value) [4]. The aim of a quality assurance program is to assist a radio-diagnostic facility in consistently obtaining adequate radiological information with a minimum of dose and cost [5]. An integrated part of a quality assurance program is quality control ascertaining quality by measurements and other procedures [6]. Radiology service standards state that a diagnostic radiology room must have radiation protection for staff, patients and the environment [7]. The radiology room requires an X-ray acceptance test at a specific location to prevent unwanted gaps in X-ray radiation.

This research tries to develop technology in the field of radiology by utilizing internet of things (IoT) technology. IoT is used as an intermediary for the results of X-ray radiation acceptance tests that can cover a large area. The research was conducted by designing a monitoring test system for the optimization of kilovolt and millisecond values on X-ray machine, in the form of X-ray accpentance test prototype and a commercial Android operating system application.

2. Materials and Methods

2.1. Internet of Things
The Internet of Things (IoT) is a new paradigm in modern wireless telecommunications which has a broad concept in digitizing the physical world. IoT connects smart devices to an exponential network with addressing scheme that can interact with each other and work together between IP addresses to achive common goals. IoT uses machine language which aims to facilitate communication between software and hardware [8].

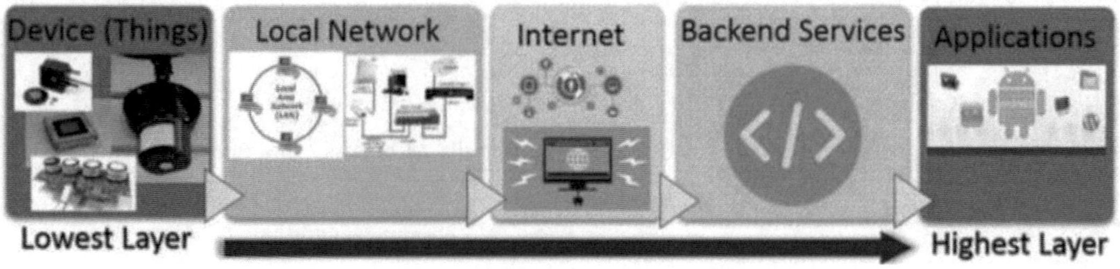

Figure 1. IoT architectute [9]

This research refers to the IoT technology which functions as monitoring X-ray acceptance test result. from the IoT architecture in Figure 1, the first layer is a sensor that collects data from X-rays. The second layer is a wireless network for device connectivity with prototypes. The IoT gateways manage data traffic between various networks and protocols. The information then goes to back-end services which involve various servers and processors such as the IoT cloud, to perform different analytics and process large amounts of data in real time which are the third and fourth layers respectively. The top layer is the application layer which is the user interface as the display of data [9].

2.2. Components
Much of the works on x-ray detectors has concentrated on devices especially for the manufacture of x-ray suitability test kits. The development of solid state-based electronic x-ray detectors such as photodiode detectors evaluates fast and real-time parameters in radiological imaging. Since there are many commercial photodiodes with different characteristics, there is interest in using low cost photodiodes with good radiation sensitivity [10,11].This study used a low-cost x-ray imaging tool based on the commercialization of a BPW-34 silicon PIN photodiode. The photodiode BPW34 takes advantage of the x-ray sensitivity that has been filtered by aluminum to detect and convert it into an analog voltage. The analog voltage is processed on the microcontroller to get kV results with the interpolation linear equation and the calculation of exposure time [12,13].

Arduino is a commercial 8-bit microcontroller development board with a USB programming interface to connect to a computer and additional connection sockets to external electronics. The ESP8266 WiFi board is a SOC with integrated TCP/IP protocol stack that can give any secondary microcontroller access to a WiFi network. Arduino microcontroller processes the data that obtained from the photodiode and sends the data to the backend service database with the ESP8266 WiFi module connection with the local internet network [14,15].

2.3. *Operating system developement software*
The application functioned as a graphical user interface on the test result monitor. It is created using the commercial development application Android MIT App Inventor 2, which functions to form the Android Package Kit (APK) format. This APK format will be used for installation applications on the commercial operating system Android.

MIT App Inventor is an online platform designed to teach computational thinking concepts through blocks-based programming tool which simplifies building Android applications. Android application development on MIT App Inventor 2, is made using the block editor that has a function to display the processed results from the webview. The webview displays stored data back-end database that processed from the microcontroller [16,17]. The MIT App Inventor 2 Android software developement can be seen in Figure 2 and the result of display can be seen in Figure 3.

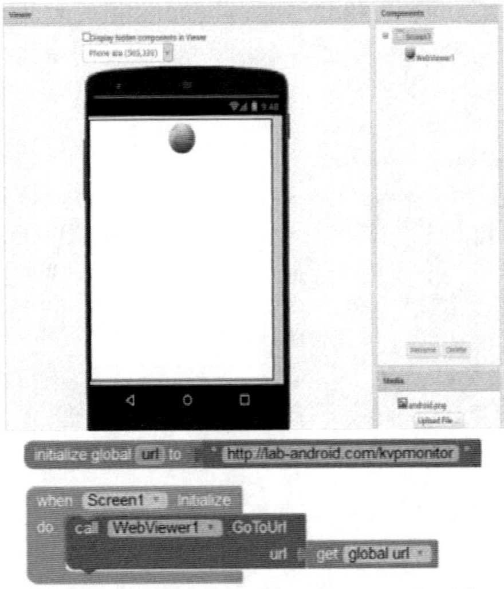

Figure 2. MIT app inventor android software development.

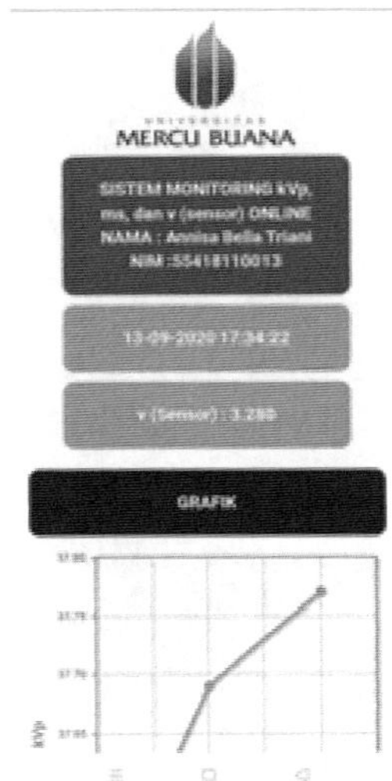

Figure 3. Result of display.

The radiology room has two rooms, namely the operator room and the examination room. In general, X-ray acceptance test is carried out in that both rooms. This is because the examination room has a wall covered with an X-ray anti radiation shield protection which is used for the safety of X-ray radiation exposure, so that it is not exposed to other rooms especially the operator room.

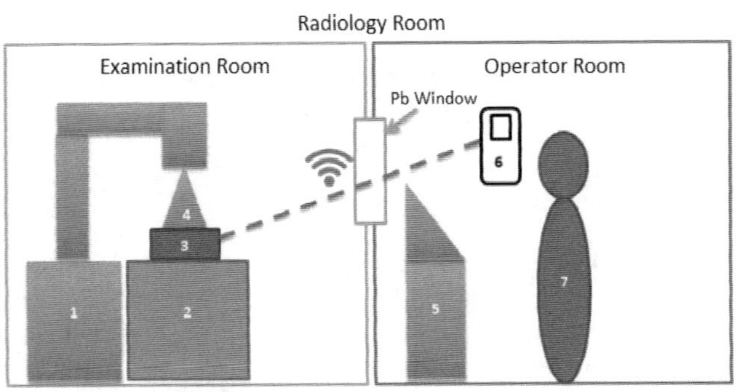

Figure 4. Implementation ilustration.

On the Figure 4, implementation of experiment start from setting the parameter values on the x-ray control panel (5). After setting parameter values, the next step is pressing the X-ray radiation exposure button. The X-ray machine (1) will emit an X-ray (4) on a prototype (3) located on the patient's table (2). After the X-ray irradiation is complete, the results of the parameter values analysis are received by the user (7) through the monitoring application on the smart device with an internet connection (6).

2.4. Block diagram

The X-rays radiation emitted by X-ray tube and will be detected by the two photodiodes that have been coated with an aluminum foil x-ray filter. The X-ray radiation will be converted into an analog voltage and transferred to a ratio circuit. The ratio circuit has a functions to obtain the value of the transmission ratio between the two photodiodes detector. The analog voltage will also amplified by ratio circuit and transferred to the ADC module which functions to convert the analog voltage into digital voltage. The ADC1115 commercial module also functions to obtain an accurate level of digital voltage readings on the microcontroller. The digital voltage that has been processed by the ADC115 module will enter the microcontroller for processing the interpolation linear equation calculation and the calculation of the voltage emission time.

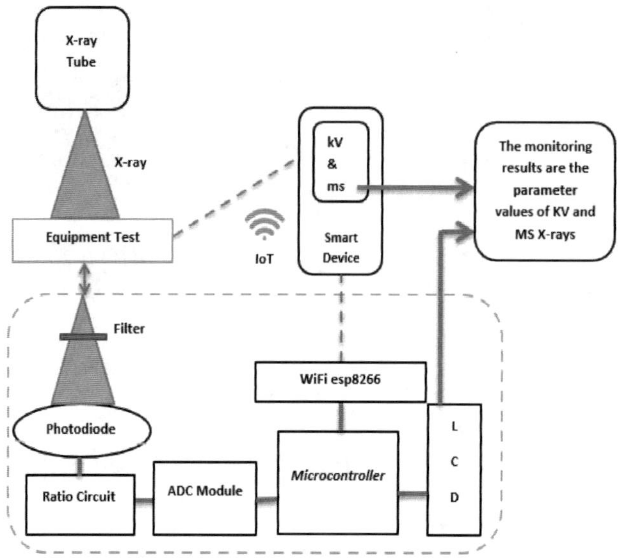

Figure 5. Block diagram.

The processing results will be displayed on the LCD screen and stored on the backend cloud. The ESP8266 module will connect to a local network with the SSID name and WiFi password according to the microcontroller program script. After the ESP8266 module is connected to the local network, the resulting data will be transferred using the local internet network to the backend cloud. The results on the backend cloud will display the test results to the Android application. An overview of the block diagram can be seen in Figure 5.

3. Results and Discussions

The overall analysis results will be compared with an acceptable X-ray calibrator. The average values generated by the calibrator becomes a reference for getting the percentage error and accuracy from the experimental results data. Data analysis refers to a guidlines testing and calibration of medical devices accordingly an establish aninstitutional local standard.

3.1. Results and monitor analysis with IoT

The results of the user interface in Figure 6, show a delay in sending data (pending) of 3-5 seconds from the reference calibrator data transmission. The reason is, data transmission depends on the quality of service from the internet network provider operator. data transmission also affects the distance of the WiFi signal source which is limited by the distance between the control room and the examination room which is coated with anti radiation protection.

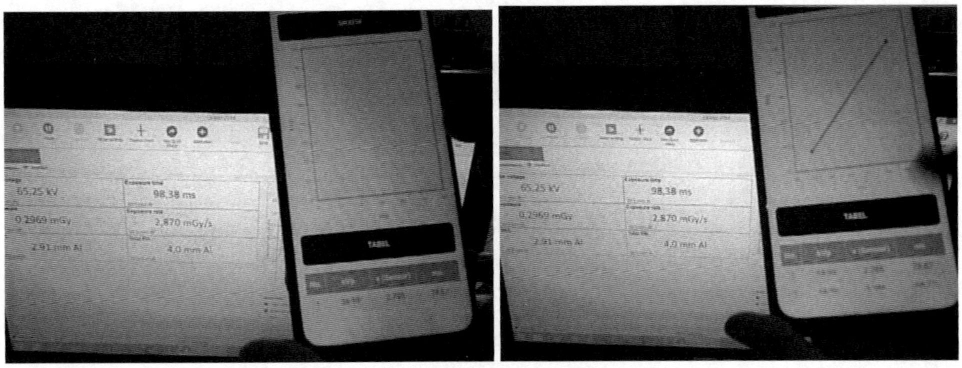

Figure 6. Results monitor with IoT Network.

3.2. Results and analysis of retrieval data

The percentage error and accuracy of the average voltage and time for three measurements with a parameter setting points: 100 mA with 40 ms for voltage measurement and 100 mA with 50 kV for time measurement, can be seen in Table 1 and Table 2.

Table 1. Results and analysis of retrieval of kV

Setting kV	Average Calibrator	Average Experiment	Persentage Error (to Calibrator)	Persentage accuracy
45	45.12	44.71	0.91	99.09
50	49.90	49.61	0.58	99.42
55	54.61	54.30	0.56	99.44
60	60.06	59.77	0.48	99.52
65	65.25	64.96	0.44	99.56

Table 2. Results and analysis of retrieval of ms

Setting ms	Average Calibrator	Average Experiment	Persentage Error (to Calibrator)	Persentage accuracy
40	38.70	37.85	2.18	97.82
50	48.74	48.21	1.09	98.91
63	61.54	61.93	0.64	99.36
80	78.35	78.25	0.13	99.87
100	97.90	98.25	0.36	99.64

The error percentage used to determine the error of the kV and ms values results and the accuracy percentage used to determine the closeness of the measurement results to the reference calibrator value. The average of percentage error voltage value is 99.41% and for accuracy is 0.59%. The average of percentage error time value is 99.12% and for accuracy is 0.88%. The overall average percentage of error result is 99.26% and the percentage of accuracy is 0.74%.

4. Conclusions

Although the delivery of data results is delayed (pending), the result data can be monitored in a broad scope. Smart devices that support the commercial Android operating system, can access applications for monitor test results with the connection to local internet network. this experiment has sucessfully monitor and give a good result as for the X-ray acceptance test. The analysis result is still in the requirements for all equipment performance in accordance an establish aninstitutional local standard.

Acknowledgements
This research was supported by grant of Penelitian Tesis Magister, Ministry of Research and Technology, National Research and Innovation Agency of Indonesia, contract number: 064.ADD/LL3/PG/2020.

References
[1] Jasim A K, Hulugalle and Haider 2017. A Quality Control Test for General X-ray Machine. *World Scientific News*. **90** 11-30
[2] Tilak R and Shanthi 2014. *The International Journal Of Engineering And Science*. **3** 69-73
[3] Pepenene and Jozela 2015. *European Journal of Medical Physics*. **31** S1-s22 S18
[4] Bapeten 2018 *PERKA: No. 2 Uji Kesesuaian Pesawat Sinar-X Radiologi Diagnostik Dan Intervensional* (Jakarta: Badan Pengawas Tenaga Nuklir)
[5] Kemenkes 2020 *Keputusan Kementerian Kesehatan Nomor : HK.02.02/V/0412/2020 Pedoman Pengujian Dan Kalibrasi Alat Kesehatan* (Jakarta: Kementerian Kesehatan RI)
[6] Jermain E and Petal S 2018. *Journal of Medical Diagnostic Methods*. **7** 1-8
[7] Kemenkes RI 2008 *Keputusan Mentri Kesehatan RI: No. 1014/MENKES/SK/XI/2008 Standar Pelayanan Radiologi Diagnostic Disarana Pelayanan Kesehatan* (Jakarta: Kementerian Kesehatan RI)
[8] Setiyo B, Arissetyanto N, Beny N and Fadli S 2017. IP over Radio: a Performance Evaluation for IoTSystem with Various Data Transmission Technique *Advanced Science Letters, American Scientific Publishers*. **23** pp 5581-5583
[9] Shivanjali K and Michael T 2020 *Internet of Things: An Overview* vol **1118** (Singapore: Springer) pp 71-79
[10] Mohammad N, Kune Y S and Afshin M 2016. *Review of Scientific Instruments*. **87** 073502 1-6
[11] Zhao, Lioliou and Barnett 2018. *Nuclear Inst. and Methods in Physics Research*. **887** 138–143
[12] Emirhan, Bayrak, BarlasYücel, Yücel and Ozben 2016. *Nuclear Inst. and Methods in Physics Research*. **819** 1–5
[13] Hanifah R F, Zaipul R, Djoko S 2019. *International Journal of Electrical and Computer Engineering*. **9** 2328-2335
[14] Patnaikuni 2017. *International Journal of Advanced Research in Computer Science*. **8** 2350-2352
[15] Setiyo B, Galang P N, Ahmad F and Fajar R 2020. *Sinergi*. Sensor Selection Comparison Between Fuzzy Topsis Algorithm And Simple Additive Weighting Algorithm Inautomatic Infuse Monitoring System Application pp 207-211
[16] Ghassan B and Saman Z 2015 *Teaching Cyber-Physical Systems Using Mit App Inventor 2* (Istanbul, Turkey) pp 1209-1216
[17] Evan W P, Michael T and Farzeen H 2019 *MIT App Inventor: Objectives, Design, and Development* (Cambridge: Massachusetts Institute of Technology) pp 31-49

Concentration profile and water level effect on the evaporation in the spent nuclear fuel storage pool using a diffusion approach

T Sundari[1,2,*] and A Wahid[1]

[1] Sustainable Energy Research Group, Chemical Engineering Department, Faculty of Engineering, Universitas Indonesia, Kampus UI Depok, 16424, Indonesia

[2] Center for Radioactive Waste Technology, National Nuclear Energy Agency of Indonesia (BATAN), Gd. 50 Kawasan Puspiptek Serpong, Tangerang Selatan 15310, Indonesia

*Email: titikbatan@gmail.com

Abstract. In the spent nuclear fuel storage pool (SFSP) of GA Siwabessy Multipurpose Reactor (GAS-MPR), there are three status of pool water level with unknown effect of water level on evaporation rate. This study aims to determine the effect of pool water level on the evaporation rate in the SFSP, so that it can be a consideration in minimizing the evaporation rate. A prediction of water evaporation rate can be explained as a diffusion model through a stagnant gas film. A numerical method, 4th order Runge-Kutta was used to describe the concentration profile of water and air over a SFSP column to the top of the pool. The effect of the water level can be explained then. The various distance of the water to the pool top at constant water temperature and constant air temperature shows a result that the evaporation rate is decreasing while the distance of the water to the pool top is increasing. The investigation results conclude that the closer of the liquid-water surface to the gas-air flow, increases the evaporation rate. This approach can be applied in a SFSP that has an abundant amount of water, the water temperature remained constant by operation of the water-cooling system, air temperature, pressure, and relative humidity remained constant as well by the operation of ventilation and air conditioning (VAC) system. Those constant conditions were stated as normal operation in SFSP of GAS-MPR. In the case of abnormal conditions such as the VAC not operating or the water-cooling system not operating, there will be dynamics of pool water and air temperatures which require further modeling research.

Keywords: concentration profile, evaporation, spent nuclear fuel, water level effect.

1. Introduction

Spent Nuclear Fuel Storage Pool (SFSP) is a storage pool designed to temporarily store spent nuclear fuel for a certain period to reduce radioactive decay heat before being stored in a dry storage facility [1]. GA Siwabessy Multipurpose Reactor (GAS-MPR) spent nuclear fuel storage pool is rectangle in shape with 14 m length, 5 m width, and -9.1 m depth [2]. Spent nuclear fuel is stored in demineralized water as deep as 3.6 meters from the top of the fuel so that the water can provide sufficient radiation shielding to workers and the environment. The Safety Analysis Report (SAR) of the SFSP explains three-step of the safety warning system, that is Normal Condition (NC), Safety System Setting (SSS), and Safety Limit (SL). The water temperature is classified as a normal condition when it is no more than 35°C, as safety system setting when it is around 36-40°C, and as safety limit when it reaches 67°C. The water level is classified as a normal condition when it has a minimum of 4.8 m from the pool bottom, as safety system setting when it is less than 4.8 m but still more than 4.6 m, and as safety limit when it starts going through 4.2 m. In every three steps of this safety warning system, there is not a description of how the water pool behaves especially for evaporation occurrences due to water diffusion into the air.

The study of the evaporation rate has been conducted by many researchers in some applications such as swimming pools, water storage, tanks, and nuclear spent fuel storage pools. Carrier [3,4], Lovell et al. [5], Shah [6,7], Boelter et al. [8], Lecoq et al. [9], Asdrubali [10], Blazquez et al. [11] investigated factors that influence the evaporation rate of swimming pool with and without ventilation and air conditioning (VAC).

For water evaporation in SFSP, some investigation is well presented by Shah [12], Hugo [8,13], Brewster [14] in various cases such as a normal condition and Station Blackout (SBO). However, the concentration profile and the effect of water level have never been explained. In this paper, the concentration profile and the effect of water level on the water evaporation rate were investigated as a diffusion approach using numerical methods and being compared with the analytical methods.

2. Methodology

This study was conducted with real SFSP design data from GAS-MPR under normal operating conditions and steady-state processes. The air temperature and humidity were maintained constantly due to VAC operation, air pressure is maintained at 759.44 mmHg. To maintain the water level and replace the loss of pool water due to evaporation, demineralized water was supplied as make-up water to the SFSP. In the investigation of the water level effect, the water temperature was maintained constantly by circulating pool water through a cooling system.

The concentration profile of water and air mixture as a mass transfer process or evaporation were studied using numeric methods and compared with the analytical method. A simple description of a spent fuel storage pool and its associated systems is shown in Figure 1.

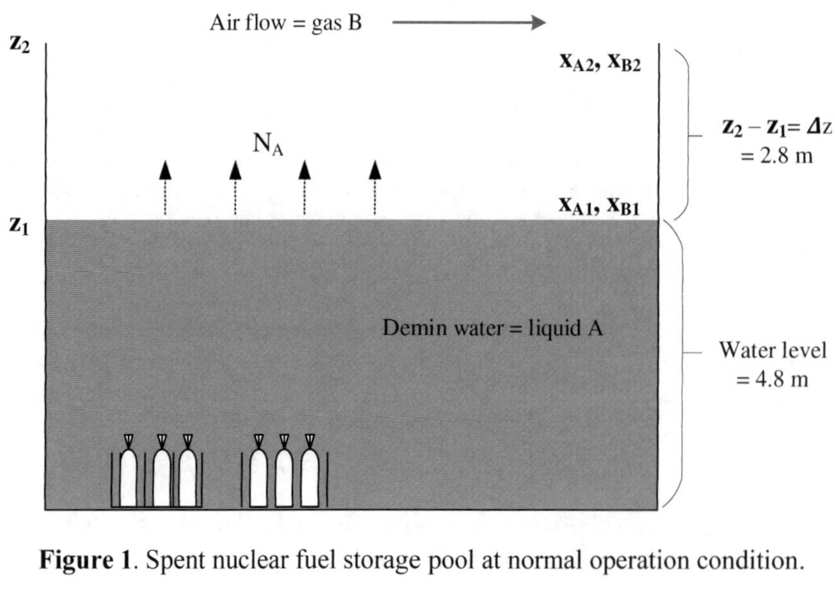

Figure 1. Spent nuclear fuel storage pool at normal operation condition.

Figure 1 shows the diffusion system in which pool water (liquid A) is evaporating into room air (gas B). This study adopted the theory of evaporation as a mass transfer diffusion of liquid surface to stagnant gas film by Bird et al. [15]. The combined flux consists of molecular flux and convective flux. The number of moles of pool water (liquid A) that go through a unit area in unit time is defined as N_A. The unit area which is the cross-sectional area of the pond filled with pool water is fixed in space. The distance of the pond water surface to the upper end of the pond construction is determined as the z-coordinate. Accordingly, the relation of the molar flux to the concentration gradient in z-coordinate is presented on the equation below [15].

$$\underbrace{N_{A,z}}_{\text{combined flux}} = \underbrace{-c\,D_{AB}\frac{dx_A}{dz}}_{\text{molecular flux}} + \underbrace{x_A(N_{A,z} + N_{B,z})}_{\text{convective flux}} \tag{1}$$

In this case, there were some systems that maintains the water level at z = z_1, as it is stated in the Safety Analysis Report (SAR) that the water level at normal operation is 4.8 m. Right at the pool water-room air interface, the concentration of pool water in vapor phase is expressed as mole fraction X_{A1}. X_{A1} is the vapor pressure of pool water divided by the total pressure in the system, P_A^{vap}/P.

A gas mixture pool water-room air flows slowly past the top of the pool to maintain the mole fraction of pool water at X_{A2} for $z = z_2$. The pool water system was kept at constant temperature by operation of water-cooling system to remove the spent nuclear fuel heat generation in the bottom of the pool. The vapor velocity at the pool wall and in the center of the pool were assumed to be uniform. By assuming the evaporation process attained in a steady-state system, there was a net motion of pool water (A) away from the interface, and the species of room air (B) was stationary. With $N_{B,z} = 0$, solving for $N_{A,z}$, the rate of mass transfer at the pool surface along z-coordinate is defined as Eq.2[15].

$$N_{A,z} = \frac{-c\,D_{AB}}{(1 - x_A)}\frac{dx_A}{dz} \tag{2}$$

Mass balance at steady state over the increment Δz, states that water vapor entering is equal to the leaving ones. Accordingly, in cross-sectional area S of the pool column, the mass balance was defined as Eq.3.

$$SN_{A,z}\big|_z - SN_{A,z}\big|_{z+\Delta z} = 0 \tag{3}$$

Division by SΔz and taking the limit as Δz→0 [15] gives:

$$\frac{-dN_{A,z}}{dz} = 0 \tag{4}$$

Substitution of Eq.2 into Eq.4 gives:

$$\frac{d}{dz}\left(\frac{c\,D_{AB}}{1 - x_A}\frac{dx_A}{dz}\right) = 0 \tag{5}$$

The gas phase of pool water and room air were assumed to be ideal gases. The solubility of air in pool water is negligible. For the ideal gas mixture, $p = cRT$, so that c must be constant at constant pressure and temperature. Gases D_{AB} is very nearly independent of the composition. Therefore, cD_{AB} can be moved to the left of the derivative operation and be as follows,

$$\frac{d}{dz}\left(\frac{1}{1 - x_A}\frac{dx_A}{dz}\right) = 0 \tag{6}$$

The concentration profile as moles fraction of pool water (x_A) expressed in Eq.6 and it is a second-order differential equation. The boundary conditions were:

Boundary condition 1: At $z = z_1$, $x_A = x_{A1}$

$z_1 = 0$

x_{A1} is the vapor pressure of A divided by the total pressure = P_A^{vap}/P.

Boundary condition 2: At $z = z_2$, $x_A = x_{A2}$

$z_2 = 280$ cm

$x_{A2} = 0$

As the mole fraction in a mixture, the relation of x_A and x_B is $x_A + x_B = 1$. The pressure, P, in the pool atmosphere, is defined as the pressure in a normal operating system at 759.44 mmHg. For the vapor pressure of water, P_A^{vap}, which is a function of temperature, it uses data in Table 1 [16].

Table 1. Vapor pressure of water

Temperature [°C]	Pressure [mmHg]
26	25.2
35	42.2
40	55.3
67	205

2.1. Concentration profile using 4th Runge-Kutta numeric method

The second-order differential equation (Eq.6) was solved using the fourth-order Runge-Kutta (RK4) method with the two boundary conditions. Firstly, the second-order differential equation Eq.6 was simplified into a first-order differential equation as Eq.7 and Eq.8 as follows:

$$\frac{dx_A}{dz} = x_A' \tag{7}$$

$$\frac{d}{dz}\left(\frac{1}{1-x_A}\frac{dx_A{}'}{dz}\right) = 0 \tag{8}$$

The first-order differential equation (Eq.8) was solved then as follows:

$$\frac{1}{1-x_A}\frac{dx_A{}'}{dz} + x_A{}'\frac{d}{dz}\left(\frac{1}{1-x_A}\right) = 0$$

$$\frac{dx_A{}'}{dz} = -(1-x_A)\,x_A{}'\frac{d}{dz}\left(\frac{1}{1-x_A}\right) = 0$$

$$\frac{dx_A{}'}{dz} = -(1-x_A)\,x_A{}'\left(\frac{\frac{d(1-x_A)}{dz}}{(1-x_A)^2}\right) = 0$$

$$\frac{dx_A{}'}{dz} = \frac{(1-x_A)}{(1-x_A)^2}\,x_A{}'\,\frac{dx_A}{dz} = 0$$

$$\frac{dx_A{}'}{dz} = \frac{1}{(1-x_A)}\,(x_A{}')^2 = 0 \tag{9}$$

The Eq.7 and Eq.9 then were solved using fourth-order Runge-Kutta (RK4) in Spreadsheet, to get the mole fraction of water (x_A) and mole fraction of air (x_B) in each increment of $\boldsymbol{\Delta}$z. The calculation takes increment, $\boldsymbol{\Delta}$z = 2.8 cm as the distance from the water surface to the top of the pool (280 cm) divided by 100 points. The calculation of error between the model of RK4 and the analytical method was calculated then.

2.2. Concentration profile using analytical method

The analytical solution for the problem as Eq.6 and the two boundary conditions was described by Bird et al. [15] for the concentration profile of water (A) as follows:

$$\frac{(1-x_A)}{(1-x_{A1})} = \left(\frac{1-x_{A2}}{1-x_{A1}}\right)^{\frac{(z-z1)}{(z2-z1)}} \quad (10)$$

Or the concentration profile of air (B) as follows:

$$\left(\frac{x_B}{x_{B1}}\right) = \left(\frac{x_{B2}}{x_{B1}}\right)^{\frac{(z-z1)}{(z2-z1)}} \quad (11)$$

2.3. Water evaporation rate in a stagnant air/gas film

The rate of mass transfer at the water-air interface, that is the evaporation rate, was obtained from Eq.2 as follows:

$$N_{A,z} = \frac{-c\, D_{AB}}{(1-x_A)}\frac{dx_A}{dz} = \frac{c\, D_{AB}}{x_A}\frac{dx_B}{dz} = \frac{c\, D_{AB}}{z_2 - z_1}\ln\left(\frac{x_{B2}}{x_{B1}}\right) \quad (12)$$

Where c = P/RT, P is the total pressure, R is the gas constant, T is the air temperature and D_{AB} is diffusion constant of water vapor into air that is a function of temperature. The values of diffusion constant of water vapor into the air at temperature 26, 35, 40, and 67°C can be obtained from linearization from reference data in Table 2 [17].

Table 2. Diffusion of water vapor into air.

Temperature, T_w [°C]	Diffusion constant, D_{AB} [cm^2/sec]
20	0.246
30	0.260
40	0.275
60	0.305
70	0.321

The evaporation rate was calculated then in the mass of water transferred per time using Eq.13.

$$W = N_{A,z}.M_A.S \quad (13)$$

Where W is the mass of water evaporate (g/hr then will be converted to kg/hr), M_A is the mass molecule of water (18.02 g/mole), and S is the pool water surface exposed to evaporation (cm^2).

3. Results and Discussions

In this study, the concentration profile and rate of mass transfer were calculated in various water temperatures and also in various water level in the next. First, 26°C is the water temperature in the reality at that time (August 2020), 35°C is the water temperature at the maximum limits of normal operating conditions criteria, 40°C is the water temperature specified as the safety system setting, and 67°C is the water temperature specified as the safety limit.

3.1. Concentration profile

In this section, the profile of water vapor concentration and room air was modeled first to determine the effect of pool water temperature. The In the liquid-vapor interface maintained at a fixed position, the diffusion model of pool water (A) through a stagnant room air (B) in a steady-state system is shown in Figures 2 to 5.

Figures 2 to 5 show how the concentration profile varies with different water temperature variations. The red-lines are a mole fraction of water and the green-lines are a mole fraction of air. Both of them were distributed in a pool column as z-coordinate from z_1 (pool water surface) to z_2 (top of the pool).

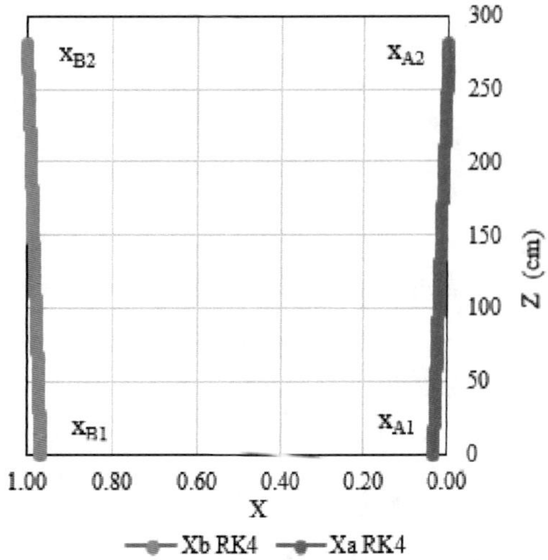

Figure 2. Concentration profile in z-coordinate using 4th order Runge-Kutta at $Tw = 26°C$.

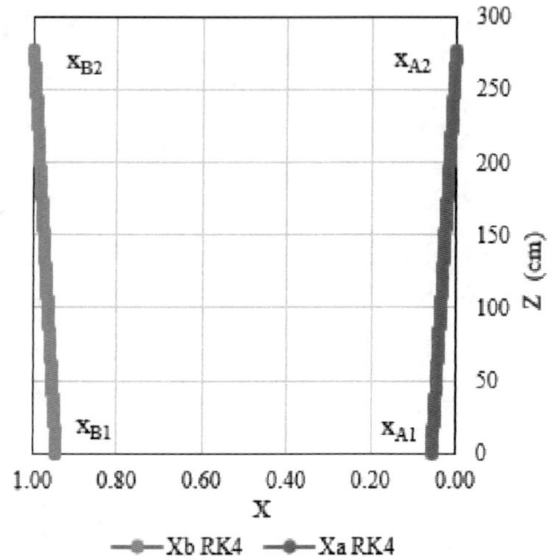

Figure 3. Concentration profile in z-coordinate using 4th order Runge-Kutta at $Tw = 35°C$.

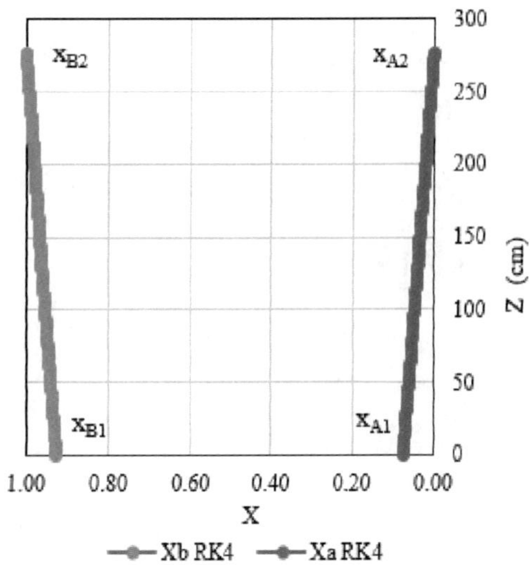

Figure 4. Concentration profile in z-coordinate using 4th order Runge-Kutta at $Tw = 40°C$.

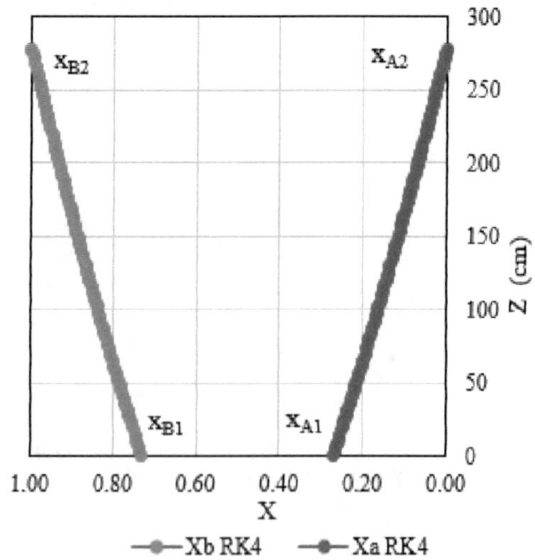

Figure 5. Concentration profile in z-coordinate using 4th order Runge-Kutta at $Tw = 67°C$.

With the same Δz (z_2-z_1), various water temperatures were resulting in differences in initial mole fraction (x_{A1} and x_{B1}). This is because different water temperatures caused different water vapor pressure and x_{A1} is the water vapor pressure divided by the total pressure. Increasing water temperature increased the water vapor pressure then increased the mole fraction of water. The final mole fraction ($x_{A2} = 0$ and $x_{B2}=1$) in various water temperatures is the same since it is the boundary conditions. This case demonstrates that the water level, water temperature, air temperature, air pressure, and air humidity were maintained at its value (constant).

The water evaporated that passed the top of the pool (at z_2, z = 280 cm) was removed by airflow (gas B) of the pool top continuously to maintain the same boundary conditions. From the calculation of the error between the RK4 numerical method and the analytical method, the error for each water temperature of 26, 35, 40, and 67 is 0.017%, 0.059%, 0.130%, and 1.610% respectively.

3.2. Water evaporation rate in various water level

When a mixture has concentration differences at two points, there will be a migration of molecules from the high to the low concentration region. The water molecules migrate from z_1 to z_2 reaching the top of the pool. Assuming the air humidity was maintained at the same value by removing the humid air, there will be a steady-state water mass transfer from the pool water surface that is called water evaporation. The rate of evaporation is the rate of mass transfer at the liquid water-air gas interface. The effect of water temperature on the water evaporation rate is shown in Figure 6.

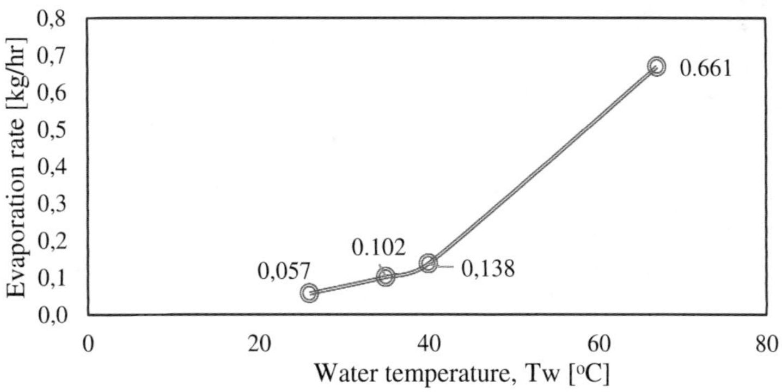

Figure 6. Evaporation rate as a rate of mass transfer in various water temperature.

Figure 6 shows that the increase in water temperature will increase the evaporation rate. Water temperature at 26°C, 35°C, and 40°C show a low increasing gradient. However, there is a significant increase in the evaporation rate from water temperature at 40°C to 67°C. This is because the water vapor pressure was also increasing significantly from 55.3 to 205 mmHg. The amount of water evaporated in pool water should be replaced by make-up water (demineralized water) to maintain the water level. The production capacity of the demineralized water system was about 0.8 m^3/hr or about 800 kg/hr. As the limit of safe operation, the water temperature at 67°C caused the evaporation reaching 0.661 kg/hr.

According to the Safety Analysis Report (SAR) of the spent nuclear fuel storage pool, there are three-step conditions for level warning system: (1) normal operating conditions; (2) safety system setting; and (3) safety limit, that are shown in Table 3.

Table 3. Steps conditions for water level.

Step conditions	Water level [cm]	Distance from water level to pool top, Δz [cm]
Normal operation	480	280
Safety system setting	460	300
Safety limit	420	340

The concentration profile and rate of mass transfer (evaporation rate) were calculated in each step conditions above at constant water temperature (for 67°C) to predict the effect of the water level towards evaporation that occurs. In the same way, the concentration profile was calculated using the 4th order Runge-Kutta in various Δz. The concentration profiles are shown in Figure 7 and the evaporation rate predictions are shown in Figure 8.

The concentration profile for various Δz in Figure 7 shows that the initial mole fractions are at the same point, but the mole fraction ended at a different point because of the different distance from the water surface to the pool top. The mole fraction in each step of increments changed gradually for each x_B and x_A in every Δz. The average error between the model and the analytical calculation for each Δz 280, 300, and 340 cm is 1.61%, 1.60%, and 1.58%. The prediction of evaporation rates for various Δz at t_{wo} = 67°C in Figure 8 shows that the evaporation rate is decreasing while Δz is increasing. This explains that the closer the water surface A to the gas flow B, the greater the evaporation rate that occurs.

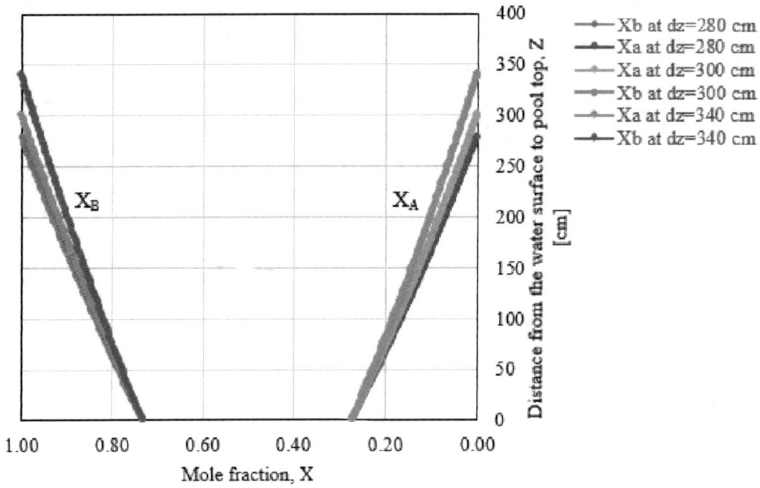

Figure 7. Concentration profile for water temperature 67°C in various Δz.

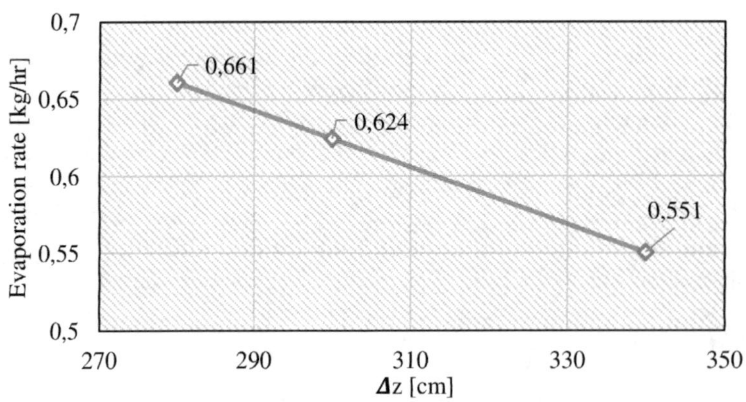

Figure 8. Water evaporation rate for water temperature 67°C in various Δz.

However, this approach only applies when all parameter conditions such as water temperature, air temperature, pressure, and humidity are remained constant and the amount of water is abundant. Those constant conditions and abundant water were stated as normal operation in actual SFSP of GAS-MPR. The amount of water is kept abundant within normal conditions to provide sufficient radiation shielding.

The water temperature is remained constant by operation of the water-cooling system. The air temperature, pressure, and relative humidity are remained constant as well by the operation of VAC system. In the case of abnormal conditions such as the VAC not operating or the water-cooling system not operating, there will be dynamics of pool water and air temperatures which require further modeling research temperature.

4. Conclusions

A numerical method is used to show the concentration profile of water and air over a spent nuclear fuel storage pool. This paper demonstrates the water level, water temperature, air temperature, air pressure, and air humidity maintained at its value (constant) then the water evaporated, passed the top of the pool, and removed by the airflow of the pool top continuously to maintain the same boundary conditions. A prediction of water evaporation rate for a variable distance of the water to the pool top at constant water temperature and constant air temperature shows that the evaporation rate is decreasing while the distance of the water to the pool top is increasing. The closer a liquid-water surface to the gas-air flow, the greater the evaporation rate that occurs. It can be applied in a spent nuclear storage pool which has an abundant amount of water, the water temperature is remaining constant by operation of the water-cooling system, the air temperature, pressure, and relative humidity are remaining constant by operation of the VAC system. This circumstance is approaching the spent nuclear storage pool at normal operation.

Acknowledgments

The authors would like to acknowledge the Ministry of Research and Technology of the Republic of Indonesia (KEMENRISTEK/BRIN) for financial support under the "Beasiswa Saintek 2019", and also support from the Center for Radioactive Waste Technology (CRWT), National Nuclear energy Agency of Indonesia (BATAN) where the real designs were taken for modelling.

References

[1] Ratiko 2017 *Optimasi Sistem Ventilasi dan Tata Udara Instalasi Penyimpanan Sementara Bahan Bakar Nuklir Bekas dari Reaktor Nuklir Riset GA Siwabessy* (Depok: Universitas Indonesia) p 8
[2] Sundari T, Kusuma M, Budiyono, Puspito M, Parjono, Aji D, Santoso I, Ismarwanti S 2019 *J. Phys. Conf. Ser.* **1198** 2
[3] Carrier W 1918 *ASHVE Trans.* **24** 25–50
[4] Hugo B and. Kinsel W 2014 *Int. Nucl. Saf. J.*, **3** 50–56
[5] Lovell D, Rickerby T, Vandereydt B, Do L, Wang X 2019 *Build. Environ.* **160** June
[6] Shah M 2002 *Int. J. HVAC&R Res.*, **8** 125–131
[7] Shah M 2003 *Energy Build.* **35** 707–713
[8] Hugo B 2015 *Modeling Evaporation from Spent Nuclear Fuel Storage Pools* (Washington: Washington State University)
[9] Lecoq L, Flick D, and Laguerre O 2017 *Int. J. Therm. Sci.*, **111** 450–462
[10] Asdrubali F 2009 *Energy Build.* **41** 311–319
[11] Blázquez J, Maestre I, Gallero F, and Gómez P 2017 *Energy Build.* **149** 133–141
[12] Shah M 2012 *Energy Build.* **49** 306–309
[13] Shah M 2014 *ASHRAE Transactions* **120** 1–15
[14] Brewster M 2018 *Int. J. Heat Mass Transf.* **116** 703–714
[15] Bird R, Stewart W, and Lightfoot E 2002 *Transport Phenomena*, 2nd ed. (New York: John Wiley & Sons Inc.) p 545-548
[16] http://intro.chem.okstate.edu/1515SP01/Database/VPWater.html. [Accessed: 12-Sep-2020]
[17] Rumble J 2018 *CRC Handbook of Chemistry and Physics*, 99th ed. (Ohio: CRC Press) p 6-228.

Study of the potential utilization of local Lampung Province resources in development of dental implant bioceramics

Y Hendronursito[1,2], I Sukmana[2,*], A Y E Risano[2]

[1] Research Unit of Mineral Processing – Indonesian Institute of Sciences
Jl. Ir. Sutami Km. 15 Tanjung Bintang, Lampung Selatan, Indonesia – 35361
[2] Mechanical Engineering Department, Engineering Faculty, University of Lampung
Jl. Prof. Soemantri Brojonegoro No.1 Bandar Lampung 35143Telp/Fax. 0721-704947

* E-mail: irza.sukmana@eng.unila.ac.id

Abstract. The increasing of dental implants needs to be supported by local materials' development as a form of independence for a nation. With its rich natural resources, Indonesia has the potential to be used as a dental implant material. Dental implant materials can be in bioceramic materials consisting of bioactive, bioresorbable, and bioinert. One of the bioactive materials of concern today is glass-ceramics from basalt rock. Meanwhile, the hydroxyapatite material sourced from limestone is the most widely used bioresorbable type. These materials, basalt, and limestone are very much found in Indonesia, including Lampung Province. However, the use of these two materials as bioceramic materials is still very little. Several methods of treating basalt as glass-ceramics and limestone as Hydroxyapatite are described in this article review. The utilization of local potential as a source of dental implant material will provide support for the independence of a nation in the welfare of its people.

Keywords: bioceramics, basal rock, glass-ceramics, dental implant

1. Introduction

The need for dental implants to replace missing teeth and improve tooth structure to support craniofacial reconstruction and orthodontic treatment is increasing. Three main types of synthetic biomaterials for dental implants are ceramics and carbon, metals and alloys, and polymer [1]. Metal and alloy materials are the primary choices because of their high strength. However, in its use, artificial bone made of metal needs careful attention. Corrosion, biofilm development, and hypersensitivity reactions are the most significant risk factor for metallic materials. However, titanium and its alloys are still widely used in osseointegration dental implants. There are still many confirmed cases of titanium hypersensitivity reactions [2].

Therefore, nowadays, medical devices made of metal have switched to ceramic or polymer materials because of the many innovations and excellent biocompatibility of these materials [3]. Bioceramics has been a lot conducted research and clinical professionalism of teeth with new and exciting substances for more than two decades. Various bioceramic formulations have been understood because of the chemical similarity to human bone. Bioceramics are often used as coating materials to improve the biocompatibility of metal implants. Serve as a resorbable lattice to rebuild the replaced tissue. Bioceramics have good thermal and chemical stability, wear-resistance, and high strength [4].

Bioceramics are non-toxic and durable materials. It can interact with the surrounding tissue, are biodegradable, dissolve, or resorbable. Sugar and protein can bind with ceramics. For example, Blood vessels can penetrate some ceramic prosthetics and bone materials, eventually replacing them. Bioceramics is a unique biomedical material.

1.1. History of bioceramics development

The first bioceramics research was carried out, namely the "Paris" plaster (CaSO4, H2O). This plaster was first used to repair bone damage in 1892, which Dressman published. Furthermore, in 1920 the use of tricalcium phosphate was successfully used. The development of bioceramic materials began in the 1960s with Hulbert's work. It continued to be developed until the 1970s and 1980s [5]. In 1977 a study was carried out regarding the very low coefficient of friction between zirconia and alumina. History shows that 400 thousand hip joint femoral heads made of zirconia have been used from 1985 to 2001. The use of bioceramic materials in dental implants is classified as shown in Figure 1 [6].

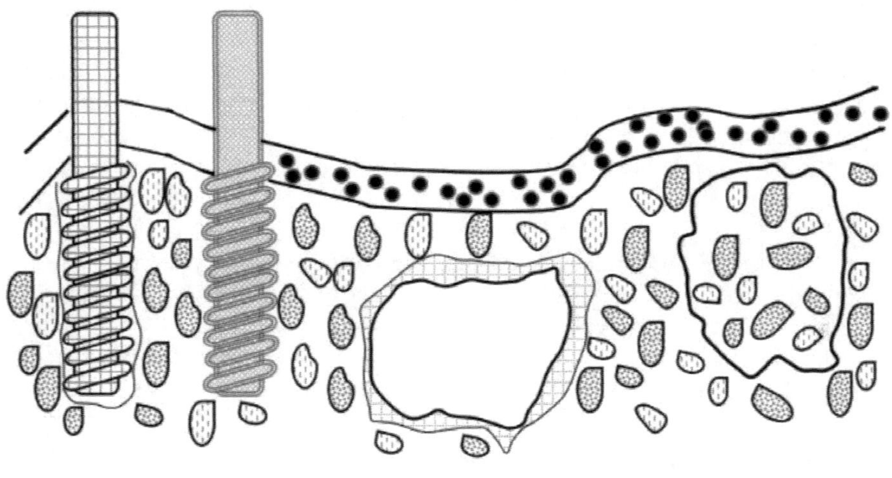

Figure 1. Classification of bioceramics according to their bioactivity [6]

There are four types of body response to the implanted matter, which allow the attachment of the material to the muscles: (a). bioinert (alumina dental implant); (b). bioactive (hydroxyapatite); (c). surface-active (bioglass); and (d). bioresorbable (tri-calcium phosphate/ $Ca_3(PO_4)_2$). The main groups of bioceramics are divided according to the tissue reaction to the implant surface. Bioactive materials include bioglass and glass ceramics. Bioresorbable materials such as calcium phosphate and hydroxyapatite (HAp). Bioinert include alumina, zirconia, and carbon.

Bioinert has a high level of stability in vivo, high mechanical properties, and when implanted in living bone tissue, the material fuses according to the osteogenetic contact pattern. Bioactive ceramic is osteoconduction and can bind chemically with living bone. In general, the mechanical strength of bioactive ceramics is lower than that of bioinert ceramics [7]. Meanwhile, bioresorbable is a ceramic material that can degrade when it is replaced by regenerating tissue [8].

Several types of bioactive and glass-ceramics have been developed. Glass-ceramics with different functions, such as superior mechanical properties and quick setting capabilities. Glass which is the most widely used research in the implant field, is mainly based on silica (SiO_2) which may contain several other critical phases. The use of silicate ceramics ($SiO_2 - Na_2O - CaO$) with a silica content of 65% or more. Bioglass 45S5 with a composition of 45% SiO_2, 24.5% CaO, 24.5% NaO has been used since 1971 but is brittle and brittle. The addition of 6% P2O5 provides a high strength increase.

Produced glass-ceramics named AW glass-ceramics were reported. These glass-ceramics were containing oxy-fluor-apatite $Ca_{10}(PO_4)_6$ (OH, F_2) and wollastonite (CaO.SiO_2) in a MgO-CaO-SiO_2 glass matrix [9]. These glass ceramics spontaneously bind to living bones without the formation of surrounding fibrous tissue. The development of machine-capable bioactive glass with appetite and phlogopite phases has been carried out. Used clinically as a vertebra makean [6].

1.2. Basalt as Glass-Ceramics Material

In the early 1980s, glass-ceramics from basalt rock were developed. The method of forming glass-ceramics from basalt rock was patented since 1977. It is described as making glass-ceramic from basalt rock melted and enriched with CaO, MgO, and SiO_2 to reach a specific ratio. The melted basalt is then cooled and thermally forms rigid glass-ceramics crystals. Other studies studying the effects of basalt on the bodies of living things have also been reviewed. They experimented on the development of asbestos and basalts fiber on mice for six months [10]. In the case of using asbestos fiber in mice, one-third of the mice died, and at a dose of 2.7 g/kg of asbestos, all mice died. In the case of basalt fiber, all mice survived at doses up to 10 g/kg. Likewise, the experiment explained in his investigation that basalt showed a reaction that was not harmful to the human body [11].

Local basalt material from Mount Kopaonik as a glass-ceramic material was done. The phenomenon of glass crystallization formation was studied using X-ray (X-RD) phase analysis, optical microscopy, and several other techniques. In this study, heat treatment experiments were carried out and their influence in controlling the microstructure and properties of the product. The heat treatment used is by melting the basalt rock in a ladle over a temperature range of 1,250 - 1,300 °C. Crystallization by reheating at 950, 1,000, and 1,050 °C for 3 to 8 hours. The results of the phase formation observations obtained Diopside CaMg $(SiO_3)_2$ and Hypersthene ((Mg, Fe,) SiO_3) as crystalline phases. The size of the crystals formed ranged from 8 to 480 μm. Microhardness from 6.5 to 7.5, compressive strength between 2,000 - 6,300 kg/cm2 and wear resistance between 0.1 to 0.2 g/cm^2 [12].

Due to its tough nature and good biocompatibility, glass ceramics from basalt rock are being developed. In its use, it is often made in powder form and used as a coating or as a reinforcing filler in bioceramic composites. Glass-ceramic made from basalt rock with heat treatment in various temperatures of 800, 900, and 1,000 °C for 60, 120, 180, and 240 minutes [13]. Glass-ceramics is used as a coating material for AISI 1040 using the plasma spray technique. The use of the plasma spray method aims to form crystalline oxide-based ceramics suitable for glass formation. The formation of amorphous glass structures is required to make glass-ceramic before crystallization heat treatment. From the DSC test, it is known that the glass transition temperature (Tg) is in an endothermic condition of around 804 oC and two exothermic peaks at 841 °C and 880 °C indicate crystal formation. The appearance of two crystallization peaks on the DSC curve implies that are two different crystal phases were formed during the heat treatment. From the XRD test, it was found that the heat treatment was 800, 900, and 1,000 °C for 2 hours to form a crystalline phase of augite [(CaFeMg) -SiO_3], aluminian-diopside [Ca(Mg,Al) $(Si,Al)_2O_6$], and diopside [$Ca(Mg_{0.15}Fe_{0.85})$ $(SiO_3)_2$]. This diopside-augite phase provides superior wear resistance and chemical resistance. The presence of Fe oxidation explained the glass-ceramics made of basalt rocks do not require a nucleant agent. Fe oxidized to Fe_3O_4, which acts as a nucleating agent and a site for crystal growth. The X-Ray Defragment analysis showed that the higher the crystallization temperature caused, the higher the diopside-augite peak. The heat treatment temperature leads to the formation of a large number of crystal phases in the glass.

The use of glass-ceramic basalt as a reinforcing filler in HAp composites against the possible dangers of use in the human body has been carried out in several studies on the safety and resistance of the body to implanted artificial materials. Even the use of basalt in the medical world has been widely used as a traded material. Service-disabled veteran-owned small business (SDVOSB) produces advanced materials. One of which is basalt. States that basalt products do not cause toxic reactions to water or air are not flammable and explosive. When in contact with chemicals, basalt does not produce chemical reactions that can damage health and the environment [14]. The physical and chemical stability of powdered basalt (BS) to the biological environment is complete. In this research, composites from HAp powder add 5 and 10% wt of basalt to HAp are safe. The powder is then stirred and pressed, and then the sintering process is carried out. The selected sintering temperatures are 700, 900, and 1,200 °C.

The biological liquid environment was 0.9% NaCl solution at pH 6.7; Ringer's solution (NaCl; and KCl; and $CaCl_2$); Ringer – Locke's solution (NaCl; $NaHCO_3CaCl_2$; KCl; and glucose). By increasing the BS content by up to 10% and the anneal temperature to 900 °C, the X-phase lines which are part of the BS become clearer. The interaction of the HAp composite system with 5 and 10% BS was almost the same. It has been shown that the HAp + 10% BS at sinter temperature 900 oC composite system is a promising material for reconstructive surgery, given its strength properties. The hardness of 0.86 GPa Hv; modulus of elasticity 23.7 GPa, close to natural bone properties. From these studies, that BS is safe for the living environment of organisms [15]. The research about physical and mechanical tests of basalt powder reinforced HAp composites showed that the addition of 5% and 10% basalt powder as a filler increased the physical and mechanical properties of HAp products. At low temperatures, the β-TCP phase is formed. The β-TCP phase builder is considered to be more biocompatible than the α-TCP phase. The possibility of adjusting the HAP - β-TCP ratio is very promising for producing materials with bio-resorbing properties. The highest compressive strength that the HAp-BS composite can accept was obtained from the addition of 5% basalt to HAp, while 10% basalt strength decreased [16]. The manufacture and development of glass-ceramic materials from raw materials can improve mechanical and physical properties. Besides, this method is a significant segment of the story of the modern world [15,16].

1.3. *Hydroxiapatite (HAp)*

This time, significant progress has been made in the use of dental implants to replace missing teeth and damage. Calcium phosphate-associated hydroxyapatite (HAp) is the primary inorganic material in human teeth. The natural components of teeth have a comparable chemical composition and HAp properties. HAp has the potential to be used for many dental applications due to its high biocompatibility.

Hydroxyapatite (HAp) with the chemical formula $Ca_{10}(PO_4)_6(OH)_2$ is a mineral that occurs naturally in the inorganic components of human bones and teeth [17]. The constituent elements of HAp are mainly calcium and phosphorus, with the stoichiometric ratio of calcium to phosphorus is 1.667. In the synthesis of HAp, a sufficiently high source of calcium is required. Several materials can be used to make HAp, such as beef bone [18], eggshell [19], shells [20], limestone [21], and others. Of the several main ingredients, limestone is the largest source of calcium. Limestone is an inorganic mineral with calcium as the main constituent. The calcium carbonate ($CaCO_3$) content in limestone is around 95%. Calcium is obtained by purifying calcium carbonate. Therefore, limestone has many direct applications in various applications such as clinical, medical, biomaterial development and is suitable as a material in the manufacture of HAp. The structure of HAp consists of calcium, phosphorus, and hydroxyl ions $Ca_{10}(PO_4)_6(OH)_2$. It's often used as a substitute for bone mineral and dental tissue because of the chemical content that it has in common so that HAp can bind chemically with human bones and teeth [22]. About 65% of the mineral fraction in human bones consists of hydroxyapatite. HAp has been widely used to repair, fill, augment, and reconstruct damaged bone and tooth tissue as well as soft tissue [23].

Pure HAP cannot be used as an implant in load-bearing applications because of its highly brittle nature. Metal implants must also be coated to obtain biocompatibility properties with the local tissue environment. Therefore, coating metal surfaces with biocompatible materials such as HAp in dentistry is very promising. Many studies are carried out about the HAP coating of substantially load-bearing implants.A screw implant used in the maxillary anterior and posterior mandibula with a depth of more than 10 mm is recommended with HAp resurfacing. HAp coating is also recommended for the less dense cortical layer and spongiosis. HAp-coated cylindrical implants are recommended in the posterior maxilla or where the cortical layer is skinny with low density [24]. HAP coating on metal surfaces offers advantages in mechanical properties; it does not affect the load-bearing capacity due to the low density and the biocompatible texture. CP titanium and HAP-coated titanium have observed that the HAP-coated system develops five to eight times [25].

1.4. Hydroxiapatite (HAp) Based Limstone

Hydroxyapatite can be synthesized in various ways, including the wet method and dry method. There are three types of wet methods, namely the method of precipitation, hydrothermal, and hydrolysis. The damp method is commonly used for HAp synthesis because it is economical and straightforward. Also, the HAp crystals formed are easily regulated in composition and physical properties.

Another advantage, the byproduct of its synthesis is water, so the possibility of contamination during processing is very low [26]. Experiments on the manufacture of HAp from limestone in Lampung Province have been started in the last few years. The experiment with making HAp from the limestone of Mount Branti using a mechanochemical method done by smoothing limestone in a ball mill machine with various grinding time and sintering process at 600 °C for 2 hours for all samples.

The following procedure is mixing 5 grams of limestone powder after sintering with 5.34 grams of Na_2HPO_4, plus 10 ml of ethanol and stirring on a ball mill machine. In the last process, the sample is dried using an oven at 80 °C for 17 hours. Furthermore, the samples were tested for XRD, FTIR, and whiteness. The test results show that the highest degree of the white color is obtained from Mount Branti limestone, namely 85.7%, and the hardness obtained is 13.60 HV [27].

The effect of temperature and sintering time of limestone from Bandar Lampung on HAp characterization has also been studied. The highest temperature and the longest sintering time make the hardness of HAp increased. Xrd diffraction shows the same pattern as commercial HAp. However, the Ca / P ratio is still greater than 1.67 [28]. As an implant material, HAp has weak mechanical properties. Due to its low mechanical properties but bioresorbable properties, HAp is often used as a coating. For example, porous tantalum coating using HAp through the plasma spray method [29].

To obtain superior mechanical properties, HAp is usually combined with other materials to form a composite. The addition of 20% of silica increase the hardness of HAp 2 times compared to without the addition of silica [30]. HAp and polymer are combined to obtain a scaffolding product. The pectin extracted from the leaves of green grass jelly (Premna oblongifolia Merr) and hydroxyapatite (HA) as a scaffolding. The 3% Hap mixture with pectin was prepared.

The resulting scaffold had pore sizes ranging from 8.25 to 115 μm while the resistant to loads of 0.03 to 0.15 MPa. The porosity of the scaffold made is 15.33 to 40.97%, while the density is 0.69 to 1.02 g/cm^3 [31]. The use of HAp made from branti limestone in HAp - PLA composites also have been done. The production of HAp begins with an extraction process which is then continued with the hydroxyapatite synthesis stage at a temperature of 1,000 °C. The method of makes composites with the addition of PLA during the heating process. The results show the Ca / P ratio is 1.63, still below 1.67, which indicates that this process is still less than perfect [32].

2. The Potential Limestone and Basalt from Lampung Province

Limestone as the primary material for making HAp is often finding in Lampung Province. This limestone is mainly located on limestone hills which are mined by the community and industry. In the West Lampung area (Lemong District), there are hypothetical limestone resources of 20 million tonnes [33].

Tanggamus Regency is scattered in the Batu Mountain Block area with 45 million tonnes of limestone reserves. In Hilian Baji Block inferred 151 million tons, Cempaka Block 18 million tons [34], Pesawaran District, Tegineneng, and Teluk Pandan [35]. South Lampung district in the Gunung Branti area [32], Natar and others. Bandar Lampung district in the Bukit Sukamenanti area [36].

Complete data on basalt reserves in Indonesia are not yet available, only in several areas that have been researched. There are at least 12,036 million tons of basalt rock spread across Sumatra Island (1,310 million tons), Java (546 million tons), Kalimantan Island (4,138 million tons), Sulawesi (6,042 million tons) [37].

On the island of Sumatra, basalt which is a volcanic product, is often found in the Sipongi area (Mandailing Natal, North Sumatra), Silungkang (Sawahlunto, West Sumatra) [38]. According to the geological agency of the Ministry of Energy and Mineral Resources, basalt can be found in East Lampung, in the areas of district Sukadana, Mataram Baru, Bumi Agung, Marga Tiga, Jabung, and Labuhan Maringgai. Basalt resources in Lampung Province are estimated at more than 18 million tons. In the Sukadana district, basalt in the form of a lava flow, solid dark gray-black, the top contains many gas holes, locally has xenolith in the form of gas perforated basalt (vesicular), undergoes weathering, peels onions [39].

It is necessary to develop research from basalt and limestone as a material for glass-ceramics and Hydroxyapatite. The potential local utilization as a source of dental implant material will provide support for the independence of a nation in the welfare of its community.

3. Conclusions

Lampung Province has potential natural resources that can be used as dental implant materials. The bioceramic materials most commonly used as dental implant materials are glass-ceramics and hydroxyapatite (HAp). glass-ceramisc material. Basalt stone is a material that can be used as a glass-ceramics material. The basalt rock in Lampung Province is found in the Sukadana area and several other places. The potential of natural resources originating from basalt rock needs attention because its current use is only used as building stone. In contrast, the rock has tremendous benefits in the field of dental implants. Limestone has been mined by many companies and local communities in Lampung. Several companies use this limestone as an ingredient in cement and agricultural lime. The use of limestone as a HAp material is still very limited in research. And until now, the investigation has not reached maximum results.

Acknowledgement

This study is supported by BLU Universitas Lampung under the scheme of Penelitian Terapan 2021. YH is supported (in some part of this study) by the internal grant of LIPI Tanjung Bintang Lampung, Research Unit Mineral Processing.

References

[1] Meffert R M, Langer B, Fritz M E 1992 *Journal of periodontology.* **63** 859-870.
[2] Přikrylová J, Procházková J, Podzimek Š 2019 *BioMed research international.* 2519205
[3] Fitriani C Y, Wibawa A 2019 *Insisiva Dent. J.: Majalah Kedokteran Gigi Insisiva.* **8** 53-58.
[4] Jayaswal G P, Dange S P, Khalikar A N 2010 *The J. of Indian Prosthodontic Soc.* **10** 8-12.
[5] Hench L L 1991 *J. of the american cer. Soc.* **74** 1487-1510.
[6] Heness G, Ben-Nissan B 2004 *Materials forum* **27** 104 114
[7] Yamamuro T 1990 *Handbook of Bioactive Ceramics* (Florida: CRC Press Boca Ranton)
[8] Patel N, Piyush G A 2012 *Int. J. Emerging Tech. Adv. Eng.* **2** 2250-2459.
[9] Kokubo T, Soga W, Kato A 1991 *J. Ceramic. Soc.* 1991 **21** 500-518
[10] Kogan F, Nikitina O 1994 *Perspect.* **102** 205–206.
[11] Mcconnell E, Kamstrup O, Musselman R, Hesterberg T, Chevalier J, Miller W, Thevenas P 1994 *Toxicol* **6** 571–614.
[12] Matovic B, Boskovic S, Logar M 2003 *J. Srb.Chem.Soc* **68** 505–510.
[13] Yilmaz S, Bayrak G, Sen S, Sen 2006 *Materials* **27** 1092–1096.
[14] Sdvosb Https://Www.Reade.Com/Products/Trap-Rock-Basalt-Powder-Fiber
[15] Boshytska N V, Fedorenko Y O, Rekos A O, Kaplunenko N V 2017 *Powder Metall. Met. Ceram.* **55**, 574–579.
[16] Kaplunenko N V, Ulyanchych N V and Klipov V D 2016 *Powder Metallurgy and Metal Ceramics*

55 306-311.
[17] Kehoe S 2008 *Optimization Of Hydroxyapatite (Hap) For Orthopedic Application Via The Chemical Precipitation Technique* (Dublin City University).
[18] Warastuti Y, Budianto E, Darmawan 2014 *J. Sains Mater. Indonesia.* **16**, 83–90.
[19] Mozartha M, Praziandithe M, Sulistiawati 2015 *B-Dent* **2** 75–81.
[20] Khoirudin M 2015 Sintesis dan Karakterisasi Hidroksiapatit (HAp) dari Kulit Kerang Darah (Anadara Granosa) dengan Proses Hidrotermal *PhD Thesis*.
[21] Margareta M A H, Fuad A, Ilmiawati S A, Wonorahardjo S 2015 *J. Penelit. Fis. Dan Apl.* **5** 15-20.
[22] Arboleda A, Franco M, Caicedo J, Goyes C 2016 *Ing. Y Compet.* **78** 71–78.
[23] Sirait M, Sinulingga K, Siregar N, Siregar RSD 2020 *Journal Of Physics: Conference Series.* 012058.
[24] Ong J L, and Chan D C 2000 *Critical Reviews™ in Biomedical Engineering* **28** (5&6).
[25] Nasar A 2019 *Applications of Nanocomposite Materials in Dentistry* 145-160.
[26] Yenti S R 2016 Sintesis Hidroksiapatit dari Limbah Tulang Sapi menggunakan Metode Presipitasi dengan Variasi Rasio Ca/P dan Konsentrasi H3PO4 *Doctoral dissertation.*
[27] Priyanto A 2018 Pengaruh Sumber Batu Kapur Terhadap Kualitas Produk Hidroksiapatite Dengan Media Ethanol *Skripsi.*
[28] Sukmana I, Hendriyanto A, Savetlana S and Tarkono 2018 *AIP Conference Proceedings* **1977** 030039
[29] Safuan N, Sukmana I, Kadir M R A and Noviana D 2014 *Journal of Physics: Conference Series* **495** 012023
[30] Afriani F, Evi J and Tiandho Y 2020 *Journal of Engineering and Scientific Research* **2** 85-89.
[31] Habibie S, Tristiyanti Y, Gustiono D, Harahap M E, Chalid S Y and Effendi D 2019 J.of Eng. & Sci. R **1** 12-16
[32] Asmoro A D Pengaruh Ukuran Serbuk Hydroxiapatite Berbahan Batu Kapur Lampung Terhadap Kekerasan Komposit Hydroxiapatite/Polylatic Acid Sebagai Pengisi Tulang (Bone Filler) *Thesis.*
[33] Sukmawardany R S, Latif N A, Sutisna T, Endang R I and Logam S D M N 2016 *Iventarisasi dan Evaluasi Sumber Daya Mineral Di Daerah Kabupaten Lampung Barat dan Tanggamus - Provinsi Lampung* (ESDM Lampung).
[34] Zulfikar, Corry K, Martua R and Reza M D 2017 Eksplorasi Umum Batugamping di Kecamatan Pugung, Kabupaten Tanggamus, Provinsi Lampung (Bandung: Pusat Sumber Daya Mineral BatuBara dan Panas Bumi).
[35] bppesawaran 2018 http://potensi.pesawarankab.go.id.
[36] Naqiyya T 2019 Kajian Alih Fungsi Lahan Ruang Terbuka Hijau (RTH) Bukit Sukamenanti *Thesis*
[37] Isnugroho K, Hendronursito Y, Candra B D 2018 *Materials Science And Engineering* (Indonesia: IOP Conf. Series).
[38] Zulkarnain2001 *J. Teknology Minerals* **8** 37–52.
[39] Kusdarto S, Sukmawan S, Sutandi A S 2006 Inventarisasi Dan Evaluasi Bahan Galian Non Logam Di Kabupaten Lampung Tengah Dan Lampung Timur, Provinsi Lampung. (Lampung: Badan Geol. Kementeri. Energi Dan Sumber Daya Miner).

Effect of austempering temperatures on surface hardness of AISI 4140 steel

M Badaruddin *, B Bakti, B Prasetyo, Sugiyanto

Department of Mechanical Engineering, Faculty of Engineering, Universitas Lampung, Jalan Prof. S. Brojonegoro No. 1 Bandar Lampung 35145, Indonesia

* Corresponding author: mbruddin@eng.unila.ac.id

Abstract. AISI 4140 steel is widely used in various engine elements, such as connecting rods, crankshafts, axles, piston rods, and sprockets. The purpose of this study was to determine the hardness value and microstructure using temperature variations in the austempering process. The test results show that after the austempering process with temperature variations changed the hardness. The austempering process can increase the hardness value, but after the temperature is raised, the hardness value decreases. The greatest hardness value at 350°C was 264.4 BHN. After increasing the temperature to 400°C, the hardness decreased to 239.1 BHN. At a temperature of 450°C, the hardness was 226.7 BHN. Microstructure observations showed that the bainite is more uniform with increasing temperature. The test results show that the austempering temperature affects the hardness value and microstructure. When the austempering temperature is increased, the hardness value decreases.

Keywords: AISI 4140 steel, austempering, Brinell hardness, microstructure bainite

1. Introduction

AISI 4140 steel is widely applied in several engine components, such as connecting rods, crankshafts, axles, and gears [1]. Due to the application of AISI 4140 steel, it is required to be strong, tough, and able to withstand strong loads or moments [2]. One way to improve the mechanical properties of AISI 4140 steel is through the austempering process. According to research [3-6], the heat treatment process of austempering ductile iron can improve its mechanical properties. In this process, the resulting phase is bainite ferrite combined with high-carbon austenite [7,8].

A study was conducted by Long et al. [6] using 0.6CCrMn2Si steel. This steel is classified as high-carbon steel with 0.6% C and main alloying elements of 1.8% manganese, 1.24% chromium, and 2% silicon. The hardness of 0.6CCrMn2Si austempering steel tends to decrease more with the austempering temperature. Bilal et al. [9] studied medium-carbon steel with a chemical composition of 0.34 C, 1.52 Mn, 1.48 Si, 1.15 Cr, 0.71 Al, 0.93 Ni, and 0.40 Mo (wt.%). The temperature of Ms steel was 310°C. The research objective was to obtain steel with a bainite structure. The austempering temperatures used were 385°C, 350°C and 320°C. Jiang et al. [10] concluded that the bainite transformation at temperatures close to the Ms value has better mechanical properties. Based on several references, the effect of austempering temperature on hardness values and microstructure was the focus of the research.

In the present work, we have conducted austempering heat treatment on a low alloy steel, AISI 4140 steel, under different temperatures for austempering with constant holding time. The goal was to increase the surface hardness of the steel. The Brinell method was used to characterize the hardness property of austempered and non-austempered steel. Microstructural changes due to bainite structure formation in AISI 4140 steel after austempering heat-treated were characterized.

2. Experimental procedure

2.1 Sample preparation

We used a commercial, round, bar shape of AISI 4140 steel that was manufactured with hot forging in this experiment. The bar had a diameter of 16 mm. Raw steel was cut into a specimen with 20-mm length. A specimen surface area was ground using emery paper with 200 to1000 grade silica carbide until the specimen surface was mirror-like and smooth. Later, the specimen was ultrasonically cleaned by immersion into acetone solution for a few seconds.

2.2 Austempering process

Fifteen AISI 4140 specimens were heated in a box furnace at 850°C [1] under dry static air and atmospheric pressure (1 atm) for one hour of holding time. We used a Naberthem furnace with a maximum heating temperature of 1350°C. After the specimen of AISI 4140 steel had reached a uniform temperature of 850°C for one hour, the microstructure of the steel was in the austenite phase. All specimens were taken out from the furnace and directly immersed into a salt solution bath containing a composition of 50% $NaNO_3$ + 50 % KNO_3 (wt.%).

The austempering process was conducted by immersion into the salt solution bath with a holding time of a 1 hour at different temperatures: 350°C, 400°C, and 450°C [11]. After the austempering process, all specimens were taken out from the salt bath and cooled to room temperature. An ultrasonic cleaner was used for washing an austempered specimen by immersion into 500 mL of acetone in a glass beaker for five minutes. The cross-sectional area of each specimen of steel was polished using Autosol paste.

2.3 Sample characterization

RMU Hardness testing (made in Italia) was used to measure hardness specimen using Brinell method. A Brinell test was conducted using a 1-mm sphere-shaped hardened steel indenter (*D*) with 1.980 kN for loading indentation (*P*) for a constant time of 15 s. The diameter of indentation (*d*) formed on a specimen was measured using a profile projector Mitutoyo PJ-H30. Calibration was performed before measuring. Each specimen surface was measured for its hardness at five points in different areas. The hardness value was determined by Equation (1) [1], and the unit of the hardness value was BHN.

$$HB = \frac{2P}{\pi D\left[D-\sqrt{D^2-d^2}\right]} \quad (1)$$

The average hardness value for each specimen is displayed in Fig. 1. The microstructure of the austempered specimens and non-treated specimens were observed using an Olympus optical microscope. The preparation of the microstructure observation was similar to our previous research [2]. The microstructure of the steel was studied at the side of the cross-sectional area.

3. Results and discussion

3.1. Hardness test results

It can be seen in Fig. 1 that the temperature difference of the austempering heat treatment greatly affects the hardness value of the specimen. The hardness value for the raw steel is around 194.7 ± 7.13 BHN.

From the three temperature variations, it can be seen that the highest hardness value is 264.4 ± 6.07 BHN, which happens at 350°C. At a temperature of 400°C, the hardness decreased to 239.1 ± 8.33 BHN. Then, at a temperature of 450°C, the hardness value decreased to 226.7 ± 4.73 BHN.

This proves that in the austempering heat treatment, the higher the heating temperature, the smaller the hardness value is.

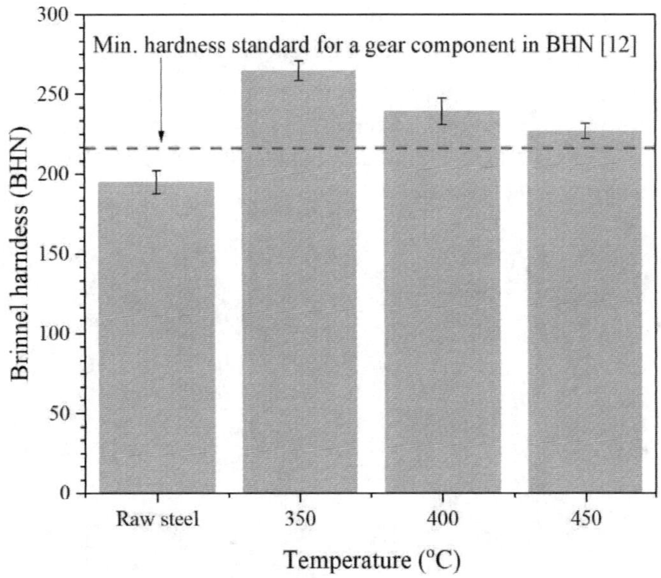

Figure 1. Variations of surface hardness of austempered AISI 4140 steel with different temperatures for austempering at one hour holding time

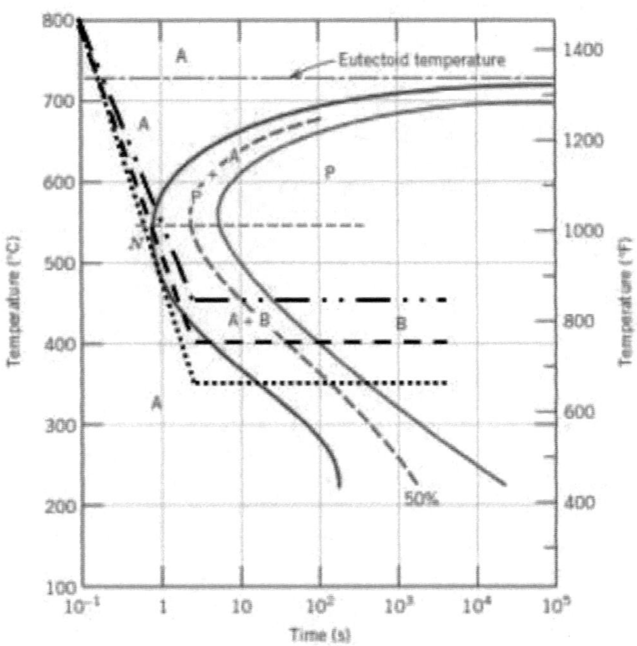

Figure 2. Transformation diagram of AISI 4140 steel austempering with temperature variations [1]

3.2. *Microstructural observation*

In a heat-treatment process, after the heating reaches a specified temperature and is given sufficient holding time, cooling is carried out at a certain rate. The austempering process is displayed in Fig. 2. The mechanical properties of the steel will depend on the microstructure that is determined by the rate of cooling during a heat-treatment. The microstructures of the steel before and after the austempering process are shown in Fig. 3.

The transformation of the austenite in cooling plays an important role in the properties of carbon steel, as demonstrated in phase diagram of Fe-Fe_3C [11]. When cooled slowly to the A3 temperature, austenite from hypoeutectoid steel begins to form a crystalline core of austenite. This transformation occurs due to a change from gamma iron (γ-austenite) to alpha (α-ferrite). Since ferrite can only dissolve very small amounts of carbon, the carbon content in the austenite will be even greater if there is much ferrite growth with decreasing temperature.

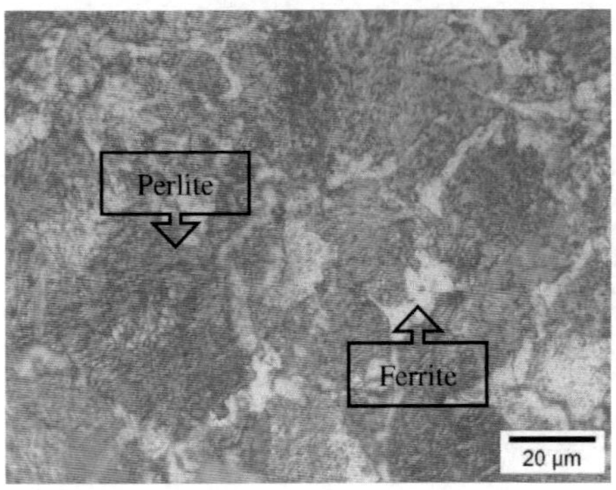

(a) Non-treated (raw steel)

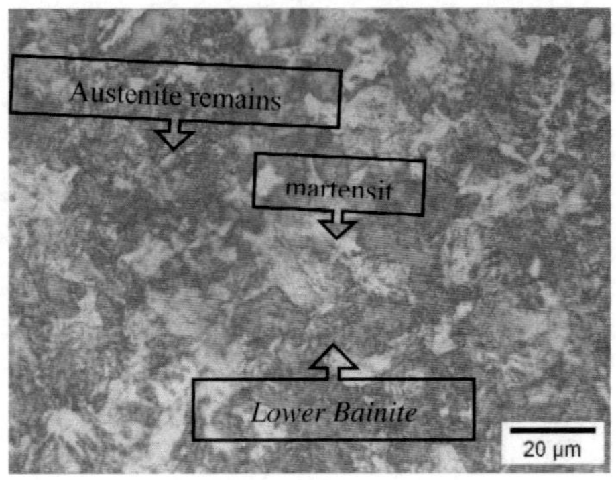

(b) 350°C

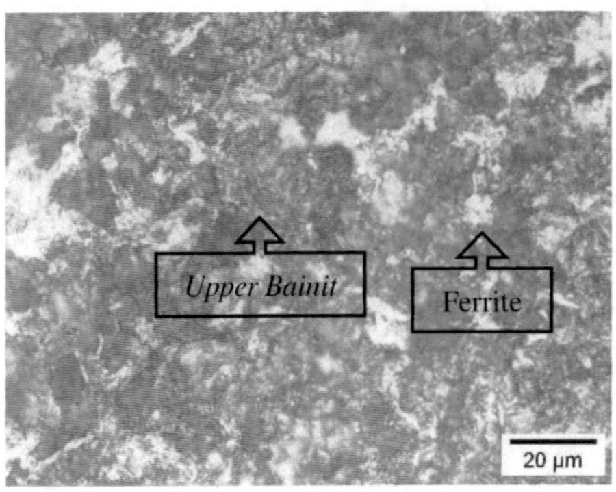

(c) 400°C

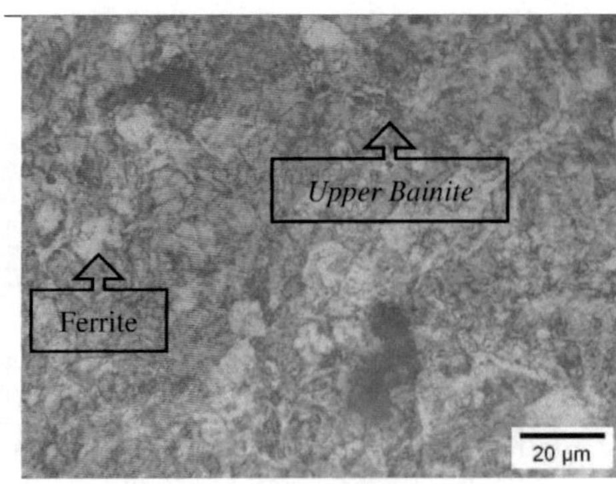

(d) 450°C

Figure 3. Microstructure of AISI 4140 steel subjected to austempering treatment

There is large carbon content at the critical temperature of A3, so when the temperature reaches the critical temperature of A1, the composition of the austenite is the same as the composition of the eutectoid composition, and at that time, the austenite deforms into pearlite. Pearlite is a special mixture of two phases and is formed when austenite with a eutectoid composition simultaneously transforms into ferrite and ferrous carbide [2,12]. The basic structure of pearlite is a lamellar structure composed of layers of ferrite and cementite [13].

The growth of pearlite begins with the growth of a cementite core at austenite grain boundaries. For cementite to grow, a large amount of carbon is required to be obtained from the surrounding austenite, so the austenite around cementite is poor in carbon and becomes ferrite. This atomic displacement takes place by diffusion, so it requires sufficient time.

In the tempering process, when the austenite is rapidly cooled, the transformed products of the austenite will change into new phases known as bainite and martensite. Bainite is formed when the austenite is subjected to a moderate cooling rate (slower than the rate required to form martensite but faster than the rate required to form pearlite) and the austemper temperature is kept constant. Bainite is a non-lamellar microstructure and is a mixed microstructure of ferrite and cementite phases [8]. This transformation of bainite is due partly to the diffusion process and partly to the non-diffusion process. The bainite structure obtained from the transformation at high temperatures has a feather-like shape, and at low temperatures, it has an accicular shape [14,15].

Martensite can occur when austenite is cooled very rapidly to a temperature below the temperature of bainite formation [1]. Martensite is a supersaturated solid solution of carbon in ferrite. It is very hard and brittle. With very fast cooling, there is not enough time for the carbon to diffuse out of the solid solution of the austenite, so the transformation occurs by shifting the atoms from face-centred cubic (FCC) structure to the body-centred tetragonal (BCT) structure.

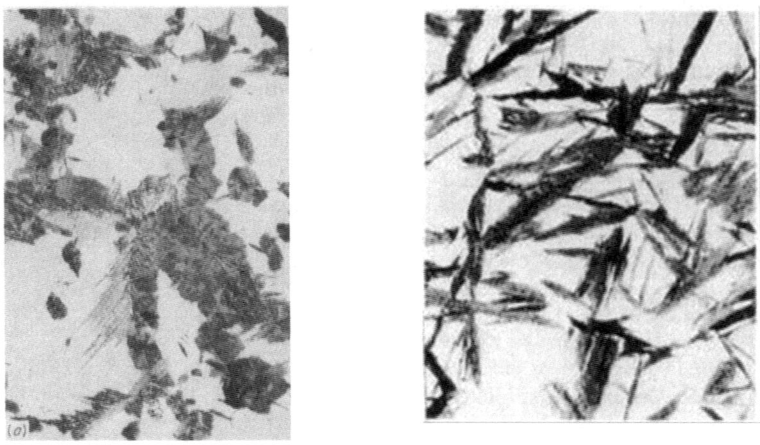

(a) (b)

Figure 4. Microstructures for (a) upper bainite-feathery like shape and (b) lower bainite-like needle-like shape [16]

In the AISI 4140 steel, the initial condition was tested by the Brinell hardness test, which obtained an average hardness value of 194.7 BHN. This hardness value is lower than that of the hardness value AISI 4140 steel for a gear component [12]. The results of the microstructure analysis show that the microstructure of AISI 4140 steel has a pearlite phase and a ferrite phase (Fig. 3a). The ferrite phase has a light colour and is soft. Meanwhile, perlite is dark in colour and hard compared to ferrite. The results of the hardness test showed that AISI 4140 steel that was subjected to the austempering process with a temperature variation of 350°C had the highest hardness value among other temperature variations (264.4 BHN). The austempering process can change the microstructure. This structural change is caused by a heating process with a high-enough temperature and then cooling it with a fast cooling rate.

The residual austenite formed white and uniform spots, ferrite structure was also visible, and the lower bainite microstructure was also formed (Fig 3b). The combination of martensite and the formation of bainite causes the hardness value of the specimen to increase. In this condition, the austenite does not have time to remove carbon from its crystals, which results in the structure becoming body cubic tetragonal (BCT).

According to Speer et al. [12], the classification of bainite is generally based on the temperature area of its formation. Upper bainite is formed isothermally in the temperature range of 400–550°C, while lower bainite is formed isothermally in the temperature range of 250-400°C [13,14]. Upper bainite microstructures in AISI 4140 steel formed at 400°C and at 450°C are presented in Fig. 3c and Fig. 3d, respectively. Upper bainite is a structure that is formed at relatively high temperatures and looks like feathers (Fig. 4a) [16]. Lower bainite is a structure formed at relatively low temperatures (close to Ms) and looks like needles, as seen in Fig. 4b [16].

4. Conclusion

The following conclusions about the effect of austempering on the hardness value of AISI 4140 steel were obtained that the greatest hardness value was 264.4 BHN, which was achieved at the austempering temperature of 350°C. Hardness and microstructure values were affected by the austempering temperature. When the austempering temperature was decreased, the hardness value increased. Also, observation of the microstructure showed that bainite is more uniform with increasing temperature.

Acknowledgements

The authors would like to thank the Ministry of Research and Technology-Board of National Indonesia Research (BRIN) for financially supporting the research project under contract no. 179/SP2H/AMD/LT/DRPM/2020 for the 2020 fiscal year.

References

[1] Callister Jr and Wlliam D Jr 2007 *Material Science and Engineering An Introduction-7th Edition* (New Jersey: John Wiley and Sons)
[2] Badaruddin M, Wardono H, Wang C J and Rivai A K 2019 Improvement of low-cycle fatigue resistance in AISI 4140 steel by annealing treatment, *Intl J Fat* ***125*** 406-417
[3] Chuaiphan W, Srijaroenpramong L and Pinpradub D 2013 The Effects of Heat Treatment on Microstructure and Mechanical Properties of AISI 4140 for Base Cutter Cane Harvester, *Adv Mater Res* **774** 1059-1067
[4] Cheng C, Kadoi K, Tokita S, Fujii H, Ushioda K and Inoue H 2019 Effects of carbon and chromium on microstructure evolution and mechanical properties of friction stir weldment in medium-carbon steel, *Mater Sci Eng A* ***762*** 138060
[5] Kappes M, Iannuzzi M, Rebak R B and Carranza R 2014 Sulfide stress cracking of nickel-containing low-alloy steels, *Corros Rev* ***32***(3-4) 101-128
[6] Long X Y, Kang J, Lv B and Zhang F C 2014 Carbide-free bainite in medium carbon steel, *Maters Des* ***64*** 237-245
[7] Gebril M A, Aldlemey M S and Kablan A F 2014 Effect of Austenization Temperatures and Times on Hardness, Microstructure and Corrosion Rate of High Carbon Steel In: Öchsner A, Altenbach H (eds) Design and Computation of Modern Engineering Materials, *Adv Struc Mater*, vol 54 Springer
[8] Cham M, Gebril M, Suliman and Khblan A F 2014 Effect of austenization temperatures and times on hardness, *Micros Corros Rate High Carbon Steel* **54** (2014) 421-428
[9] Bilal M M, Yaqoob K, Zahid M H, Ehsan ul haq, Tanveer W H, Wadood A and Ahmed B 2019 Effect of austempering conditions on the microstructure and mechanical properties of AISI 4340 and AISI 4140 steels, *J Mater Res Technol* **8**(6) 5194-5200

[10] Jiang B, Wu M, Zhang M, Zhao F, Zhao Z and Liu Y 2017 Microstructural characterization, strengthening and toughening mechanisms of a quenched and tempered steel: effect of heat treatment parameters, *Mater Sci Eng A* **707** 306-314
[11] Bhadeshia H K D H 1982 Thermodynamic analysis of isothermal transformation diagrams, *Met Sci* **16**(3) 159-166
[12] Speer J G, Edmonds D V, Rizzo F C and Matlock D K 2004 Partitioning of carbon from supersaturated plates of ferrite, with application to steel processing and fundamentals of the bainite transformation, *Curr Opin Solid State Mater Sci* **8**(3-4) 219-237
[13] Toji Y, Matsuda H and Raabe D 2016 Effect of Si on the acceleration of bainite transformation by pre-existing martensite, *Acta Mater* **116** 250-262
[14] Sanij M H K, Banadkouki S S G, Mashreghi A R and Moshrefifar M 2012 The effect of single and double quenching and tempering heat treatments on the microstructure and mechanical properties of AISI 4140 steel, *Mater Des* **42** 339-346
[15] Safi S M and Givi M K B 2004 A new step heat treatment for steel AISI 4340, *Met Sci Heat Treat* **56**(1) 78-80
[16] ASM Handbook 2004, *Metallography and Microstructures* **9** 136

Synthesis and characterization of hydroxyapatite bioceramics from shells of serai snail and mangrove crab in Tanjung Jabung beach: effect of milling process

R Utami[1,*], D Gustiono[2], M D Effendi[2], S Roseno[2], H D Fahyuan[1], and Nasri M Z[3]

[1] Departement of Physics Faculty of Sains and Technology University of Jambi Jl, Jambi-Ma. Bulian KM 15, Mendalo Darat, Jambi 36361

[2] Center of Technology for Materials, Building 224, Puspiptek, South Tangerang 15314

[3] Departement of Geophysics Faculty of Sains and Technology University of Jambi, Jl. Jambi-Ma. Bulian KM 15, Mendalo Darat, Jambi 36361

*Email: ikatami36@gmail.com, dwi.gustiono@bppt.go.id

Abstract. Hydroxyapatite (HA) is a calcium phosphate compound that is used as a substitute for damaged body tissues, such as bone and dental implants. HA has been synthesized from the serai snail shells and the mangrove crab of west Tanjung Jabung, Jambi province using the dry method. Calcination of the shell at 1000 °C for 12 hours produces CaO, then contact of CaO with air produces $Ca(OH)_2$ which is used as a source of calcium in HA synthesis. Milling time variations for 0 hours, 3 hours, and 6 hours. Sintering at 600 °C for 2 hours produces HA as characterized by using X-Ray Diffraction (XRD), Fourier Transform Infrared (FTIR), and particle size analysis using Particle Size Analyzer (PSA). HA which has the highest accuracy of lattice parameters is that which is milling for 3 hours. The diffraction pattern shows the formation of the HA phase accompanied by several other calcium phosphate phases at an angle of 2θ. The FTIR spectrum shows the presence of PO_4^{3-} and OH^- groups from HA. PSA analysis shows the longer the milling makes the smaller particle size.

Keywords: Serai snail shell, crab shell, hydroxyapatite, dry method, synthesis.

1. Introduction

Tanjung Jabung Regency is located in the mangrove ecosystem with an area of 5.009,82 km^2 or about ± 9.38% of the area of Jambi Province, Indonesia, that reaches about 53.435,72 km^2. The West Tanjung Jabung Regency located in a low-lying area with an altitude between 10-500 m above sea level. West Tanjung Jabung Regency especially the regency capital region, Kuala Tungkal, is a coastal area where the coastal waters of Tanjung Jabung Barat Regency reach 12 nautical miles from the outer coastline. In these vast waters living a wide variety of biological resources [5].

Many biological resources for the coastal area are natural resources from animals, especially fish. The types of fish resources that found in the coastal waters of Tanjung Jabung Barat Regency include pelagic fish, demersal fish, and non-pelagic fish. Serai snail and mangrove crabs are non-pelagic, they are large in number and also cause waste. The part that is consumed from the serai snail and the mangrove crab is the meat (the inside of the shell), while the serai snail shell and the mangrove crab are only disposed of by the community, and are said to be waste. Though the shell is composed of various minerals and the main mineral making up the shell is calcium. Based on Sasmita's research, the content calcium in the snail shells was 28.7%, magnesium was 2.14%, and ferrum was 0,033% [12]. According to Siahaya, et al, crab shells contained protein 15,60-23,90%, calcium carbonate 53,70-78,40%, and chitin 18,70-32,20% [14]. One way to utilize and maximize the use-value of serai snail shells and mangrove crabs is to synthesize HA material from the shells of mangrove snails and mangrove crabs in the Kuala Tungkal area by utilizing calcium in the animal's shell.

HA ($Ca_{10}(PO_4)_6(OH)_2$) has been synthesized since the early 1970 and is one of the most widely used biomaterials today [8]. The physical properties of HA have a hexagonal crystal structure and lattice parameters a=b= 9,418 Å and c = 6,884 Å so that the HA formed from the synthesis should have these physical properties. HA is proven to be biocompatible and can be tolerated by human oral tissue, has the osteoconductive ability and is proven to be able to stimulate osteoblast differentiation and bone formation [9]. HA prices on the market are very expensive because they are still imported from Japan and Korea. Therefore, it is necessary to innovate the development of HA material from local materials [1], such as by utilizing the shells of many snails and mangrove crabs that have not been utilized in the Kuala Tungkal area.

Much research has been carried out in synthesizing HA from various biomaterial, including chicken eggshell waste, shellfish, tutut shells, crab shells, golden snail shells, squid/cuttlefish bones, lamuru bone, and beef bones. Synthesized HA from cow bone waste using the wet method, the results of the characterization using XRF the most dominant HA content Ca and P, which was 88,88% and 8,78% [17]. Sintering temperatures, several studies state that the temperature of 900°C – 1200°C has the best HA results and the ball mill process makes the grain size smaller, but the processing time of the ball mill for 3 hours has not been able to make HA nano-sized [15].

Based on the description above, it is necessary to do research related to the synthesis of HA waste from serai snail shells and mangrove crabs' shells. HA synthesis method can be done in several ways including wet, dry, precipitation, mechanochemical, hydrothermal and sol-gel methods [21]. In this study, the method used is the dry method. The advantages of the dry method compared to other methods are that it can produce HA powder with fine grains and a high degree of crystallinity [3], the process is simpler and produces more HA and has economic value.

2. Experimental Method

2.1 Materials and instrument

The materials used in this study are samples of snail shells and mangrove crabs shells obtained from the coast of West Tanjung Jabung, NaOH 1%, acetone 96% , H_3PO_4, and distilled water.The instruments used in this study are digitas analytical balance, glassware, pans, furnace, planetary ball mill, X-Ray Fluorescence, X-Ray Diffraction (XRD), Fourier Transform Microscopy(FTIR), Particle Size Analyze (PSA).

2.2 Preparation of serai snail shells and mangrove crabs' shells

1 kg serai snail shells and mangrove crabs' shells are cleaned (washed) from dirt, boiled separately in a high-pressure pan, each shell is treated soaked in 1% NaOH solution for 24 hours and soaked in 96% acetone solution. The dried shell is calcined at 1000 °C for 12 hours and crushed using a mortar until smooth. After that, the levels of calcium (Ca) and other elements in serai snail shells and mangrove crabs' shells were tested using XRF. To identify compounds formed using XRD, and to identify functional groups using FTIR.

2.3 Synthesis of Hydroxyapatite

Snail and crab shells were given each treatment mixed with H_3PO_4 then milling process was carried out with variations for 0, 3 and 6 hours at a speed of 300 rpm. The molar ratio of $Ca(OH)_2$ and H_3PO_4 used is 0.3:0.18. Furthermore, the mixture is allowed to stand for 24 hours in a closed container before sintering at a temperature of 600 °C for 2 hours.

$$10Ca(OH)_2 + 6H_3PO_4 \longrightarrow Ca_{10}(PO_4)_6(OH)_2 + 18H_2O \quad (1)$$

2.4 Characterization of samples

The resulting products were characterized using FTIR, XRD, and PSA. FTIR analysis was applied to determine the functional groups that exist in the product. XRD analysis was undertaken to identify the crystal structure of the resulting products, and PSA analysis is used to determine the particle size distribution.

3. Result and Discussion

3.1 Shell Calcination Result

Based on XRF testing, the content of calcined snail and crab shellsis dominated by calcium, namely 99,3% in snail shell and 98,4% in crab shells and there is other element in small amounts as can be seen in Table 1.

Table 1. Shell content after calcination

No	Element	Amount (%)	
		Snail	Crab
1	Ca	99,3	98,4
2	P	0,0255	1,46
3	Cd	0,602	0,077

Based on the analysis of XRD, the compound $Ca(OH)_2$ was formed because the calcined samples originally predicted as calcium oxide (CaO) was allowed to contact with water vapor contained in the air. The diffraction pattern is presented in Figure 1a and FTIR spectra is presented in Figure 1b. Diffraction pattern can be seen in Figure 1a in accordance with the equation:

$$CaO + H_2O \longrightarrow Ca(OH)_2 \quad (2)$$

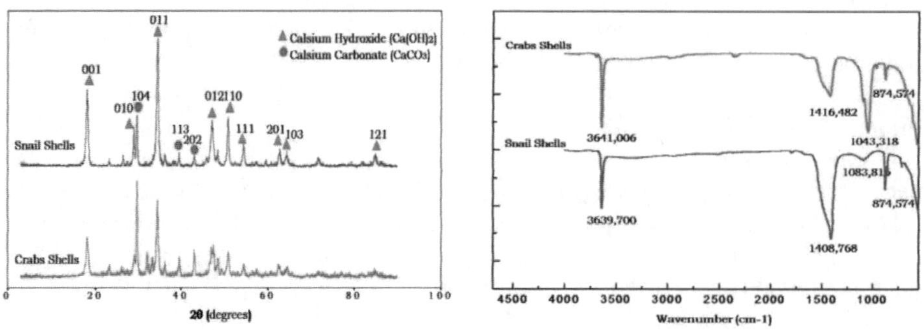

(a) (b)

Figure 1. (a) Diffraction pattern, (b) FTIR spectra of shell powder after calcinations

In the samples identified compounds $CaCO_3$ was allegedly due to having reached its equilibrium, so that a small amount of CaO react with CO_2 to form $CaCO_3$. Similar to the eggshell calcined at 900 °C for 8 hours had reached equilibrium to form the $CaCO_3$ phase [7]. Based on FTIR spectra, it was observed that the crab shell calcined for 12 hours had a peak in the area of 3641,006 cm^{-1} which was stronger than the snail shell. In the crab shell spectra, there were also observed CO groups of carbonates reinforced with a strong peak at wave number 1043,318 cm^{-1}. This could indicate more carbonate content in the calcium of the crab shell. The presence of carbonate in the FTIR results is thought to originate from carbon dioxide (CO_2) returning to the CaO when thermal equilibrium has been reached.

3.2 Result of XRD analysis

The diffraction pattern of the HA sample can be seen in Figure 2. Sample S0, S3 and S6 has a peak with high intensity at 2θ 31.8° hkl (211), 32.2° hkl (112), 25.9° hkl (002) indicating HA compound (COD 96-900-1234). Samples S3 and S6 look purer. This is caused by the sintering temperature needed by sample S6 being lower than other samples in forming HA compounds, because the particle size is smaller and homogeneous. Other studies mention that the smaller particles cause better particle homogeneity and high surface energy during the sintering process, this causes the sintering process to occur faster [19].

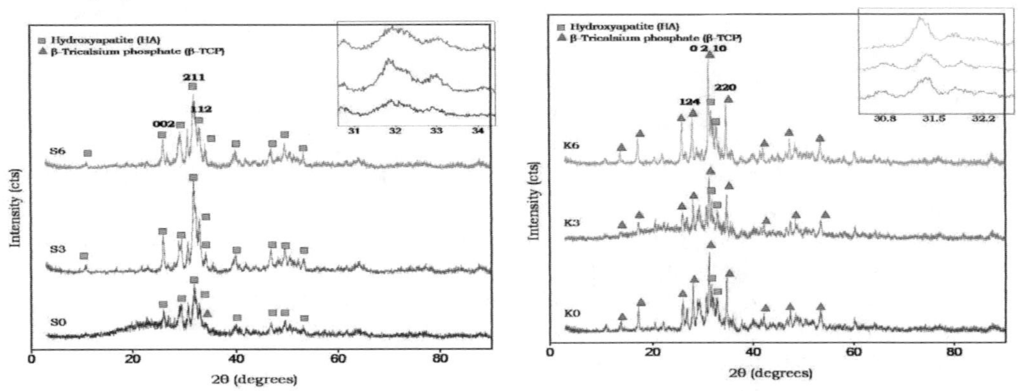

(a) (b)

Figure 2. The diffraction pattern HA synthesis product with a source of calcium (a) the shell of a snail on a variety of time milling 0 h (S0), 3 hours (S3), and 6 hours (S6), (b) the shells of crabs at the time variations milling 0 h (K0), 3 hours (K3) and 6 hours (K6)31.3 34.7 28.1

Sample K0, K3, K6 sample has a high-intensity peak at 2θ 31.3° hkl (0 2 10), 34.7° hkl (220), 28.1° hkl (124) indicating β-TCP (β-Tricalcium phosphate) compound (COD 96-901-2138). HA diffraction pattern with crab shell calcium source is dominated by β-TCP phase.

So, milling time variation does not shift the peaks that can be seen in the enlarged area 2θ 31-34 meaning that the variation of milling time does not change the phase formed, but decreases or increases the intensity of the peaks which will affect the crystallinity. Based on Nakano, et al, crystallinity also decreases with increasing length of time of milling [10]. Then other results showed that milling time affected pore size, porosity, compressive strength, and density of HA samples [11].

HA has a hexagonal structure with a = b = 9,432 Å and c = 6,881 Å [2]. The results of calculating lattice parameters and their accuracy can be seen in Table 2. The calculation results show that sampel S0, S3 and S6 have lattice parameters with accuracy value of lattice a ranging from 98,181 - 99,990% and accuracy of lattice c ranging from 98,185 - 99,176%, these values are still within the range of lattice parameters for calcium phosphate HA compounds. This is consistent with matching the maximum peaks of the sample which shows that the sample is dominated by apatite compounds which are present in the form of HA. The sample that has the highest accuracy value is the sample S3.

The lattice accuracy a of samples K0, K3, and K6 ranges from 89.220 - 90.436% while the lattice accuracy c is 0%. The lattice parameter of the three HA samples originating from the crab shell is far from the HA reference lattice parameter value, but the value is close to the lattice parameter value of β-TCP. Based on Shi, 2003. β-tricalcium phosphate crystals are rombohedral with a lattice parameter a = 10.439 Å and c = 37.375 Å. If the lattice parameters of the samples K0, K3, and K6 are compared with β-TCP, the accuracy of the values of lattice a is 99,413, 99,906, and 98,995 Å and the accuracy of lattice c are 99,614, 99,696, 99,328 Å which is still in the range of β-tricalsium phosphate (β-TCP) compounds.

The number of β-TCP phases and lattice parameter values that are similar to β-TCP in samples made with crab shells may be caused by the inaccurate Ca/P molar ratio for making HA which is 1,67. The lack of a molar ratio of 1,67 can lead to the formation of β-TCP because the molar ratio of Ca,/P to make β-TCP is 1.5.

Table 2. Lattice Parameters and Accuracy Value of HA

Sample	a = b		C	
	a (Å)	Accuracy (%)	c(Å)	Accuracy (%)
S0	9,380	99,447	6,848	99,613
S3	9,433	99,990	6,894	99,716
S6	9,603	98,181	6,999	98,185
K0	10,378	89,809	37,231	0
K3	10.448	89,220	37,489	0
K6	10,334	90,436	37,124	0

β-TCP compounds that are formed are not a problem because these compounds are also widely used as bone replacement material. The most widely used calcium phosphate is HA and tricalcium phosphate. Although both have the same chemical composition, these two materials have differences like resorbability (absorbability) in biological tissue [16]. HA and TCP are bioactive and osteointegrative when implanted into the body, but HA tends to be non-resorbable. The ideal bioceramics is not only bioactive but also bioresorbable, to provide space for new bone tissue growth. The combination of HA bioactive properties and TCP bioresorbable properties with certain compositions can better control the process of resorption and substitution of bone biomaterials [2].

Crystallinity is a term relating to the degree of regularity of the structure of a solid, the degree of crystallinity influences the physical properties of the solid. The more orderly the arrangement of atoms in the material, the higher the level of crystallinity. This is indicated by the higher intensity and narrower width of the peak half [13]. The degree of crystallinity of the sample can be quantitatively calculated by comparison of the top peak area to the total peak area as follows:

$$\text{Degree of crystalinity} = \frac{\text{area fraction of crystaline}}{\text{area fraction of crystaline+area fraction of amorphous}} \quad (3)$$

$$\text{Area fraction of crystalline/amorphous} = \text{FWHM x height} \quad (4)$$

Crystal size is calculated using the Debye Scherrer formula. Measurement analysis was carried out in the field hkl (002). Table 3 shows the calculated crystallinity and crystal size and the chart is shown in Figure 3. The size of the calculated crystal results is inversely proportional to FWHM, in accordance with the scherrer equation where if the FWHM value is small then the size of the sample crystal will be large.

Variations time milling does not give effect linear concerning for to crystallinity. HA made with snail shells has the highest crystallinity in S3 sample, that is, samples that are milling for 3 hours according to the diffraction pattern where the S3 sample has the highest intensity. In contrast to HA which is made with crab shells milling results for 3 hours (K3) makes the lowest degree of crystallinity seen in the diffraction pattern of the K3 sample has a low intensity. This is consistent with other studies that show the relationship between milling time, crystal size, and the degree of crystallinity [11].

Table 3. HA Crystallinity Degrees and Crystal Size

Sample	Crystallinity	Crystal Size	
		$\beta(°)$	D_{002} (nm)
S0	65,365	0,5862	15,140
S3	73,494	0,4519	19,658
S6	70,529	0,5682	15,158
K0	74,521	0,3984	22,298
K3	70,318	0,3892	22,827
K6	77,937	0,5636	15,767

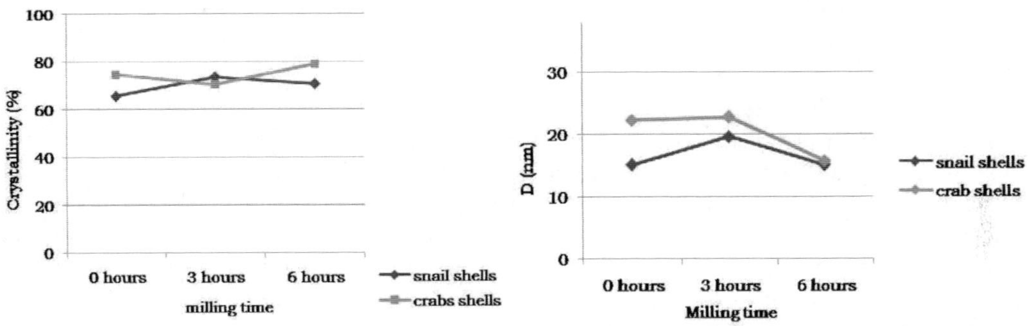

(a) (b)

Figure 3. Graph of differences in (a) crystallinity and (b) crystal size

The shape of the curve is not linear but very volatile. the size of the crystals in the sample increased at 3 hours of milling and then decreased at 6 hours of milling. Decreasing the size of the crystals at 6 hours of milling could have occurred because samples that have passed the peak point will rupture resulting in a decrease in crystallite size [18].

3.3 Result of FTIR analysis

The presence of carbonate phase in the sample was confirmed in the FTIR test results which showed carbonate absorption bands in the sample. FTIR spectra for all milling time variations can be seen in figure 4a and FTIR spectra of sample S3 is presented in Figure 4b. The FTIR spectra showed the peak absorption of the group of CO_3^{2-}, PO_4^{3-}, OH^- in the synthesis of HA samples. The functional groups that were characteristic of HA is PO_4^{3-}, OH^-, CO_3^{2-} and PO_4^{2-} which characterizes nonstoichiometric of HA [4]. The wavenumbers of each sample are presented in Table 4.

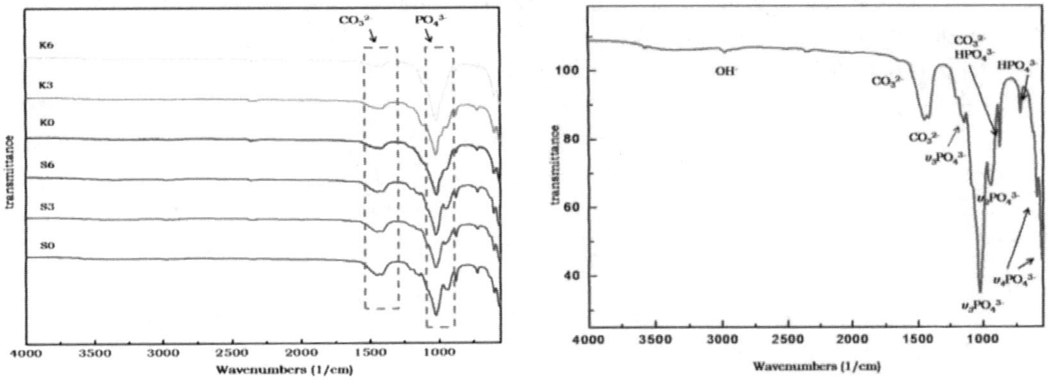

(a) (b)

Figure 4. FTIR Spectra for (a) all variations (b) sample S3

Table 4. Absorption Tape of FTIR Result

Group	Wavenumbers (cm^{-1})					
	S0	S3	S6	K0	K3	K6
$\upsilon_1PO_4^{3-}$	941,83	959,88	957,91	960,74	987,43	875,65
$\upsilon_3PO_4^{3-}$	1024,91	1024,04	1024,43	1021,33	1022,76	1019,86
$\upsilon_4PO_4^{3-}$	598,98; 558,93	599,18; 590,24	598,99; 559,32	598,58; 554,84	598,23; 552,81	599,19; 555,43
$\upsilon_2CO_3^{2-}$	1451,03; 1415,02	1451,49; 1414,79	1452,00; 1414,79	1420	1417,18	1422,28
$\upsilon_3CO_3^{2-}$,	876,00;	875,80;	876,06;	875,98;	875,55;	816,78;
HPO_4^{3-}	719,75	719,40	719,89	720,67	719,95	719,80
OH^-	3506,34	3514,68	3523,98	2967,07	2980,23	2990,12

The HA spectra obtained by each researcher did show different results but were still in close proximity. Absorption at wavenumbers around 3506 cm $^{-1}$ shows stretching vibrations OH^- The results of the analysis show that there are groups that should not be present in pure HA. The group is carbonate (CO_3^{2}) and its presence is known in the range of wave numbers 1700-1400 cm^{-1} and at 875 cm^{-1}. The presence of carbonate in HA synthesis results probably comes from sources of calcium snails and crabs that have been confirmed by XRD and FTIR indeed the presence of carbonate phases and groups in calcium sources and then the presence of carbonate groups may be derived from free carbon dioxide in the air. The presence of carbonate in HA synthesized results are suspected to originate from free carbon dioxide (CO_2) in the air because the mixing of the precursors is carried out in the open air. Carbonate that enters the crystal lattice will affect the value of the Ca/P ratio of HA and also the crystal fields. Carbonate ions into the HA crystal lattice will replace the hydroxyl ion (OH^-) or phosphate (PO_4^{3-}) and generate carbonate HA [6].

The PO_4^{3-} group is the highest intensity in all samples. In HA phosphate groups can be found in the four modes of vibration, namely: vibration stretching (v_1) at wave number of about 941 cm^{-1}, the vibrational bending (v_2) at wave number of about 363 cm^{-1} , vibration asymmetric stretching (v_3) at wave number of about 1020 cm^{-1} , and vibration antisymetric bending (v_4) at wave number of about 552-560 cm^{-1} this value indicates vibration P-O from the group PO_4^{3-}. PO_4^{3-} group seen in all samples can be observed in the range 1100-550 cm^{-1}. The varied milling time did not affect the FTIR test results. The sharpness of the peak FTIR spectra shows the same results, so also the wavenumbers do not show a striking difference.

The percent transmittance value is the amount of light transmitted (not absorbed) to the molecule. If all the light passes through the sample without the slightest absorption then the absorbance is zero, if all the light is absorbed then the percent transmittance is zero. In Table 5 we can see the percentage of transmittance of the $_3 PO_4^{3-}$ sample S0, S3, S6, K0, K3, and K6. The analysis was carried out on the group because it had the sharpest band. Based on table 6 it can be seen that there is no linear effect between the percentage of transmittance and the length of time milling. A small transmittance value indicates the amount of absorbance in the sample so that the peak in the FTIR spectrum is sharp.

Table 5. Value of %T CO_3^{2-} and $\upsilon_3PO_4^{3-}$ group Milling Time Variations

Sample	% T	
	CO_3^{2-}	$\upsilon_3PO_4^{3-}$
S0	86,060	34,958
S3	89,424	44,489
S6	89,572	35,413
K0	94,193	35,824
K3	93,768	35,542
K6	96,217	28,228

3.4 Result of PSA Analysis

Based on the particle size distribution curve (Figure 5), the sample S0 the highest frequency is at 1523.5 nm, sample S3 the left-hand curve with frequency the highest is at 691.5 nm, and as the sample S6 the curve shifts to the left and narrower with the highest frequency at 670.6 nm.

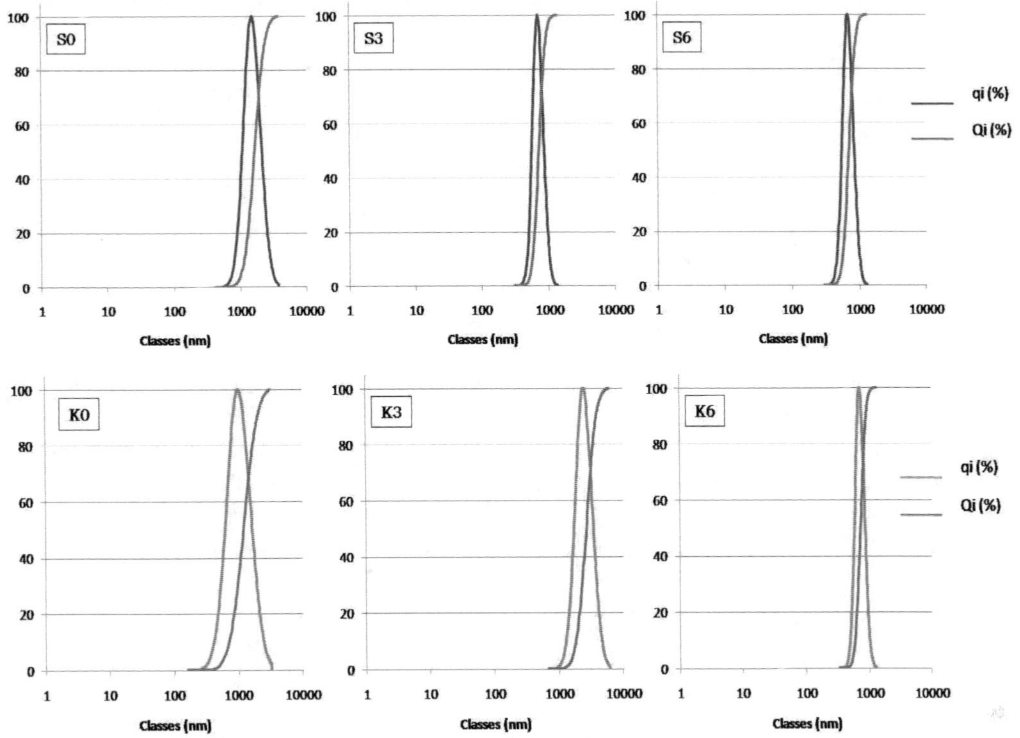

Figure 5. Particle size distribution curve

While the particle size distribution curve sample K0 is highest frequency at 1012,5 nm, sample K3 looks curve shifts to the right with the highest frequency there at 2406.5 nm, and sample K6 the curve seemed to shift to the left narrowing with the highest frequency at 705,3 nm. The narrowing of the curve indicates the increasingly homogeneous particle size in the sample. The average particle size results can be seen in Table 6.

Table 6. Average Particle Size of HA

Sample	$D_{average}$ (nm)
S0	1726,6
S3	727,3
S6	704,9
K0	1308,1
K3	2778,3
K6	895,5

The particle size of HA samples made with snail shells decreased with the increasing length of milling time. While the particle size of HA made with crab shells is non- linear, in the K3 the particle size increased to 2778,3 nm. The large particle size of the K3 sample may be due to the agglomeration of the particles due to imperfections during the process of dispersion of the sample to the liquid media. The average HA particles from each sample were not yet nano-sized, because the size was still above 100 nm.

4. Conclusion

Synthesis HA from the Serai snail shell and the Mangrove Crab Coast of Tanjung Jabung Barat, Jambi has been successfully carried out. Based on the results of the FTIR characterization, a functional group has formed which is a feature of HA. Lattice parameter calculation results show that the serai snail shell waste was successfully synthesized into HA using the dry method while the results of the synthesis of mangrove crab waste produced β-TCP. Based on testing using XRD, time measurement of the degree of crystallinity and crystal size of HA is also β-TCP. The degree of crystallinity of HA increases at 3 hours of milling, then decreases at 6 hours of milling, whereas β-TCP rises at 3 hours of milling, then increases at 6 hours of milling. Crystal size increases after 3 hours of milling and then increases at 6 hours of milling. Decreasing the size of the crystals at 6 hours of milling could have occurred because samples that have passed the peak point will break down in size. Then from tests conducted using PSA, it is known that the length of milling time affects the particle size, where the longer the milling makes the particle size smaller.

Acknowledgment

The author would like to thank the following parties: the Dean of the University of Jambi who has allowed them to undertake this research and The Director of the Center of Technology for Materials – BPPT who has granted them permission to use laboratory facilities and to conduct research within the Center of Technology for materials at the Agency for the Assessment and Application of Technology (BPPT). Futhermore, great appreciation directed to fellow senior researchers in biomaterial technology who have assisted in providing feedbacks and given fruitful discussions and technical guidance during the course of this research project and moreover to the laboratory staffs who have involved in this research project, ie. Mr. Lukmana, S.Si and Mrs. Henny Purwanti, S.Si.

References

[1] Anisah., Mutia, D., Nuning, A ., Dwi, G. 2018. *Pembuatan Graft Tulang Dengan Proses Ekstraksi Senyawa Hidroksiapatit dari Tulang Korteks Sapi*. Jurnal Fisika dan Aplikasinya. 3(1):31-36.

[2] Aoki, H. 1991. *Science and Medical Applications of HA*. The University of Michigan Tokyo (JP): Japanese Association of Apatite Science.

[3] Bhat, SV. 2002. *Biomaterials*. Springer: England.

[4] Cimdina dan Borodajenko. 2012. *Research of calcium phosphate using fourier transform infrared spectroscopy, infrared spectroscopy*. Material Science engineering and Technologgy. In Tech.

[5] Ermulan, U. 2011. *Rencana Pembangunan Jangka Menengah Tahun 2011-2016*. Pemkab Tanjab Barat: Kabupaten Tanjung Jabung Barat.

[6] Haris, Al., Ahmad, F., Silvia R.Y. 2016. *Sintesis Hidroksiapatit dari Limbah Tulang Sapi Menggunakan Metode dengan Variasi Rasio Ca/P dan Konsentrasi H_3PO_4*. Jurnal Online Mahasiswa FTEKNIK. 3(2): 1-10.

[7] Haryono., Christi, L.M., Rukiah., Yati, B. Y. 2018. *Kalsium Oksida Mikropartikel dari Cangkang Telur sebagai Katalis pada Sintesis Biodiesel dari Minyak Goreng Bekas*. Jurnal Material dan Energi Indonesia: 08(01).

[8] Hassan, A. dan Swaminathan, D. 2011. *An In Vitro Study to Evaluate the Genotoxicity of Value Added HA as a Bone Replacement Material*. Jurnal Sains Malaysia. 2(40) 163-171.

[9] Kattimani, VS., Chakravarthi, PS., Kanumuru, NR., Subbarao, V.V., Sidharthan, A, Kumar, TSS. 2014. *Egg shell derived HA as bone graft substitute in the healing of maxillary cystic bone defects: a preliminary report*. Journal of International Oral Health. 6(3):15-19.

[10] Nakano, T., Tokumura, A., Umakoshi, Y. et al. 2001. *Control of HA crystallinity by Mechanical Grinding Method*. Journal of Materials Science: Materials in Medicine 12, 703-706.

[11] Sa'adah, N.N., Rudyardjo, D.I., Siswanto. 2014. *Pengaruh Variasi Waktu Milling pada Sintesis Biokeramik Hidroksiapatit dengan Metode Solid State Reaction*. Skripsi. Departemen Fisika. Universitas Airlangga.

[12] Salma, K., Berzina-Cimdina, L., Norodajenko, N. 2010. *Calcium Phosphate Bioceramics Prepared from Wet Chemically Precipated Powders, Processing and Application of Ceramics*. 45-51.

[13] Sasmita, D. 2011. *Sintesis Hidroksiapatit dari Cangkang Keong Emas (Pomacea canaliculata Lamarck) Melalui Metode Hidrotermal*. Jurnal Saintek. 3(2): 129-135.

[14] Setianingsih, T., Sutarno., Masruroh. 2018. *Prinsip Dasar dan Aplikasi Metode Difraksi Sinar-X*. UB press: Malang.

[15] Siahaya, A.F., Tri, N., Emy, K. 2014. *Perbedaan Substitusi Tepung Kulit Udang, Cangkang Kepiting dan Kunyit dalam Pakan Komersial Terhadap Produksi dan Warna Kuning Telur Itik*. Jurnal Agroveteriner. 2(2): 139-146.

[16] Solechan dan Rubijanto, JP. 2015. *Pembuatan Material Sintesis Nano HA Untuk Aplikasi Scaffold Tulang Mandibula dari Tulang Cumi Sotong Menggunakan Metode Kalsinasi. Gardan*. 5(1): 1-12.

[17] Victoria, E.C., dan Gnanam, F.D. 2012. *Synthesis and Characterisation of Biphasic Calsium Phosphate*, Trends. Biomater. Artif. Organ, 16(1): 12-14.

[18] Wahdah, I., Wardhani, S., Darjito. 2014. *Sintesis Hidroksiapatit dari Tulang Sapi dengan Metode Basah-Pengendapan. Kimia*. Student Journal. 1(1):92-97.

[19] Wigayati, E. M., Raden I. P. 2015. *Analisis Pengaruh Mechanical Milling Menggunakan Planetary Ball Milling Terhadap Struktur Kristal dan Struktur Mikro Senyawa LiBOB*. Jurnal Sains Materi 16(3):126-132.

[20] Wu, S.C., Hsueh-Chuan, H., Shih-Kuang, H., Chien-Pei, T., Wen-Fu, H. 2017. *Preparation and Characterization of HA synthesized from oyster shell powders*. Journal Advanced Powder Technology. No 5 Model 5G.

[21] Yahya, M., Yelmida, A., Zultiniar. 2016. *Sintesis hidroksiapatit dari Precipitated Calcium Carbonate (PCC) Kulit Telur Ayam Melalui Proses Hidrotermal*. Jom FTEKNIK. 3(1).

The effect of preloaded on delamination of composite laminate subjected to low energy impact

A S A Sabli[1], M Morni[1], M Z H Ahmad[2], A H Noorhalidi[2], and Y Aminanda[1,*]

[1] Department of Mechanical Engineering, Faculty of Engineering, Universiti Teknologi Brunei, Jalan Tungku Higway Gadong BE1410, Brunei Darusalam

[2] Department of Mechanical Engineering, Faculty of Engineering, IIUM, Gombak Kuala Lumpur, Malaysia

* Email: yulfian.aminanda@utb.edu.bn

Abstract. The CFRP laminate composite is used widely especially in aerospace industry. The design concern of this type of structure is on its strength under impact loading even under low energy impact. Under low impact loading, delamination occurs and reduce the residual strength of laminate, therefore the necessity to predict this phenomena prior to its utilization. In practice the aircraft structure is subjected to pre-stress loading such as due to aerodynamics loading while in the same time loaded under impact loading due to, for example, birds impact. This paper deals with the investigation on the delamination trend of pre-stressed CFRP composite laminate subjected to low energy and low speed impact. A FE model is proposed and validated using published data. After validation confirmation, 2 types of pre-stressed; compression and tension are used to vary the pre-stress parameter. The pre-stress is modelled by prescribe displacement at both edge of the laminate with 5 different values of -0.01, -0.05, -0.1, -0.15, -0.2 mm for compression and 0.01, 0.05, 0.1, 0.3, 0.5 mm for tension. The FE shows that maximum force during impact increases with pre-stress in compression and decreases in tension. The contact time between impactor and laminate remains constant for pre-stress in tension and increases in compression. The impactor displacement during impact is lower for tension and higher for compression. Delamination area increases regardless whether in pre-stress tension or compression. The same trend of delamination shape evolution during impact is observed both for compression and tension; ellipse evolution in transverse or perpendicular direction to the pre-stress direction.

Keywords: FEA, CFRP laminate, impact, delamination, pre-stress.

1. Introduction

In 1961, the first ever composite material was patented, but it was several years before it was made available commercially. This composite improves the strength to weight ratio of structure, thus enable more applications in various industries such as aerospace, automotive, sporting goods and even consumer goods. Due to the development of the material science, various study has been done to learn exactly how the materials behave in different conditions especially during an impact. During an impact, metallic components can yield and dissipate energy through plasticity. As for laminate composites can only dissipate energy by a variety of damage modes, which usually degrades the strength of the structural

IOP Conf. Series: Materials Science and Engineering 1173 (2021) 012029 doi:10.1088/1757-899X/1173/1/012029

component. Low velocity impact such as tool falling down onto the composite laminate structures can affect the structure on the macroscopic level before its affect can be seen. The damage produces by the low velocity impact involved a complex failure mechanism: matrix crack, delamination and fiber break where matrix crack and delamination occur invisible on the surface of the structure. There were a quite significant number of researches and approaches have been done to study the effect of delamination but the prediction of the damage propagation is still up to date for investigation.

The traditional approaches to study the failure of the composite laminate is through experiments. The experiments should be tested from material, specimen and structure itself as recommended for certain application such as for aircraft structures through pyramid process testing for its certification. The number of tests is significant and costly. That is the reason, finite element analysis (FEA) has been developed to replace the testing. Furthermore, FEA can observe the effect on the macrostructure of the composite laminate plate due to the low velocity impact.

The damage caused by low velocity impact (LVI) in laminated composite plate is an important problem faced by designers using composite [1-10]. Not only the in-plane stresses, but the interlaminar normal and shear stresses also play a role in estimating the damage caused by the LVI. LVI is considered dangerous for a composite structure because the damage caused is difficult to be detected and tends to be created on the back face or within the laminate.

In the other hand, the published study relates to the laminate delamination due to impact loading alone [1-5,8-9]. In practice, the laminate structures used for all application especially aerospace ones are pre-stressed before impacted due to expected loading during its service. In this case, the matrix is over-stressed and local sub-critical cracking is produced. These micro-cracks are the main sources of stress concentration at the inter-ply regions and will lead to crack propagation, delamination and fracture of the structure [3,6-7].

Therefore, this paper proposes to observe the effect of pre-stressed on laminate composite subjected to low energy impact. The trend of peak contact force and delamination area in function of the pre-stress magnitude will be analysed using Finite Element simulation. This study on pre-stressed laminate is a first step of the investigation on laminate composite subjected to mechanical impact loading. Further study on hygro-thermal loading [11-12] producing ageing effect [13] on the laminate hence affect to the laminate strength under impact loading needs to be investigated.

2. Finite Element Model

The FE model is constructed based on the experiment specimen and set up proposed by L. Iannucci [10] and will be validated first. After validation confirmation, pre-stress conditions will be applied to analyse the peak force and delamination area trend during impact. The laminate properties used for the simulation is shown in Table I below which are taken form [10].

Table 1. Mechanical Properties of Composite Plate

Dimension (mm)	100 x 75	Poisson's Ratio	0.3
Laminate Thickness (mm)	1	Shear Modulus, G_{12} (GPa)	5.8
Density (kg/m3)	1600	Shear Modulus, G_{13} (GPa)	5.53
Young's Modulus, E_{11} (GPa)	140	Shear Modulus, G_{23} (GPa)	3.7
Young's Modulus, E_{22} (GPa)	9.5	Shear Strength, σ_{s13} (GPa)	0.085
Young's Modulus, E_{33} (GPa)	9.5	Shear Strength, σ_{s23} (GPa)	0.105
In-Plane Shear Strength, σ_{s12} (GPa)	0.105	Compressive Strength, σ_{c11} (GPa)	1.650
Tensile Strength, σ_{t11} (GPa)	2.000	Compressive Strength, σ_{c22} (GPa)	0.240
Tensile Strength, σ_{t22} (GPa)	0.070		

The solid spherical impactor is modelled using rigid body with the properties listed in the Table 2 below.

Table 2. Mechanical Properties of Composite Plate

Impact Energy (J)	0.67
Impactor Mass (kg)	0.412
Density (kg/m^3)	381438
Velocity (m/s)	1.8
Diameter (mm)	12.7
Poisson's Ratio	0.3

The laminate is modelled using Shell elements with dimension 100mm length, 75mm width and total thickness of 1mm. 2D modelling approach with shell elements is appropriate when the ratio of length and width dimension are very large compared to the thickness of the plate. As for the material, MAT 22 MAT_COMPOSITE_DAMAGE is used with Chang-Chang failure criteria.

The layup of the laminate of 8 plies [45/-45/90/0/0/90/-45/45] is used for the simulation and to be similar with L.Iannucci [10] work for the purpose of validation and further investigation. The delamination is modelled, at this stage of investigation, using tie-break contact, CONTACT_ AUTOMATIC_ SURFACE_ TO_SURFACE_TIEBREAK. Using this type of contact, it allows for the simulation of delamination based on the cohesive zone model (CZM). After defined the normal and shear failure stresses (NFLS, SFLS), damage is a linear function of the distance of two nodes initially in contact.

The delamination is modelled only at mid-plan of the laminate. Therefore, the 8 plies of laminate will be modelled by two plates of shell with 4 plies each; [+45/-45/90/0] at upper part and [0/90/-45/+45] at lower part. Globally the proposed finite element model is shown in Figure 1 below.

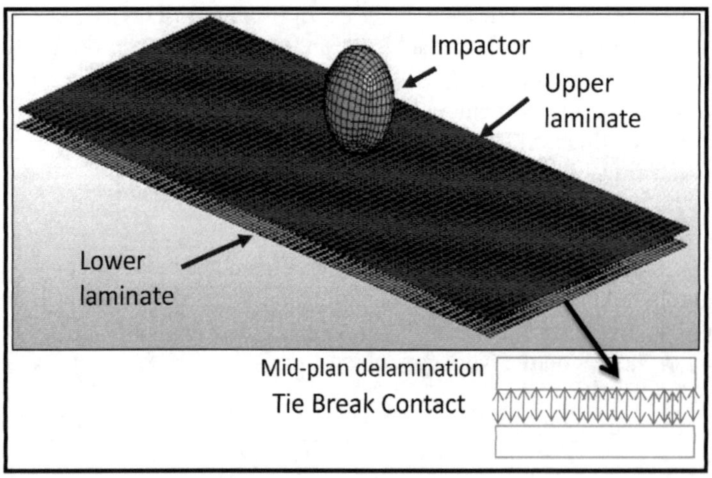

Figure 1. Finite Element model with delamination at mid-plan of the laminate

3. Validation

For validation, 2 cases will be computed; without delamination and with delamination element. The result of finite element will be compared to the experimental and simulation result of L. Iannucci works [10]. It should be noted that the boundary conditions used for the experiments are clamped at all edges of the laminate, as for the laminate properties and impactor specification, the simulation used the data shown at Table 1 and 2.

Figure 2 shows the comparison between Finite Element and experiment results. The Finite element results correlate the experimental one. Figure 2 shows as well that the delamination model using tie break elements indicates a good approach at the early stage of investigation on the delamination due to low energy impact loading. The tie-break element simulates well the effect of delamination by giving a different force history compared to without delamination (single laminate).

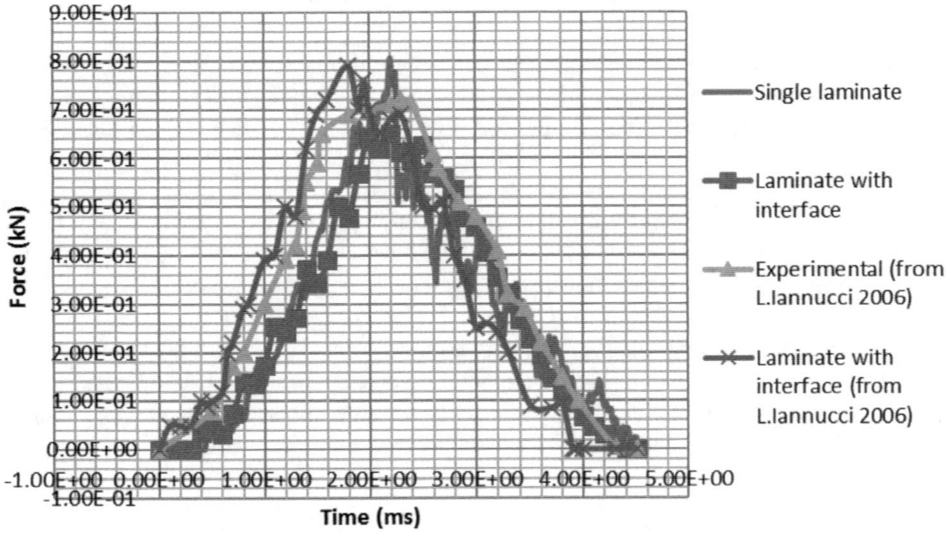

Figure 2. FEA-Experiments Comparison for validation

4. Pre-Stress Investigation

Once validated, the same laminate is investigated for the effect of pre-stress before subjected to low energy impact. The boundary conditions are changed to allow the pre-stress conditions to be applied; free displacement in longitudinal direction at 4 edges of laminates (Figure 3).

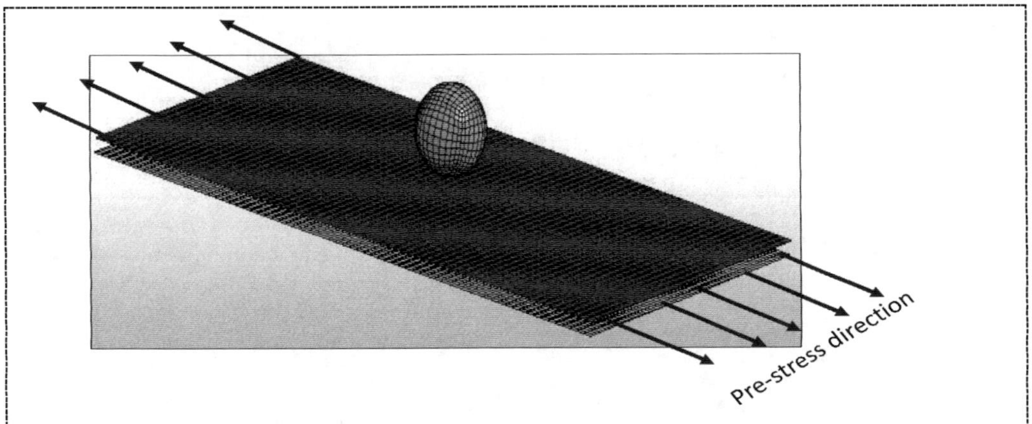

Figure 3. FE model with pre-stress

The pre-stress parameters used is in compression and tension. For stability of the computation, a prescribed displacement is employed instead of direct force. The magnitudes of prescribes displacement are shown in Table 3 where negative values represent compression and positive values for tension. The

FE computation integrating pre-stress is arranged in such that the pre-stress displacement occurs first and reaches the intended magnitude followed by the impact.

Table 3. Prescribed Displacement Simulating the Pre-Stress Conditions

Time of displacement (ms)	15
Displacement (tension) (mm)	0.01, 0.05, 0.1, 0.3, 0.5
Displacement (compression) (mm)	-0.01,-0.05,-0.1,-0.15,-0.2

In order to analyse the effect of pre-stress in compression and tension, the FE outputs such as; Maximum / peak force, contact time, impactor displacement and area of delamination is compared. From Figure 4, maximum force obtained by FE computation increases linearly with the magnitude of pre-stress in compression while in tension the force decreases insignificantly compared to compression. As for contact time of the impactor with the laminate during impact, it remains constant at 5.5ms (Figure 5) for pre-stress in tension case and increases linearly for compression case. It has been observed as well that local buckling occurs during the high value of pre-stress in compression. As for Impactor displacement (Figure 6), it varies insignificantly during tension and increases during compression.

Figure 4,5 and 6 shows that pre-stress in compression effects significantly the behaviour of laminate subjected to impact loading by producing higher maximum impact force, longer impact contact and therefore higher deflection of laminate (higher impactor displacement) to absorb the impact energy. As for pre-stress in tension, the effect is not as significantly as for compression even though the pre-stress gives an additional stiffness to the laminate before impact.

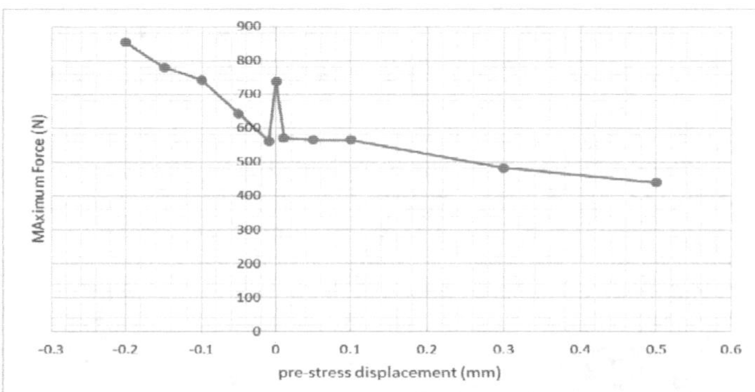

Figure 4. Maximum force in function of pre-stress displacement

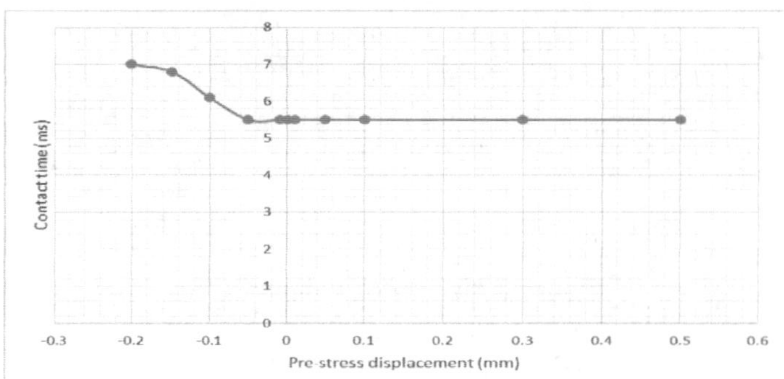

Figure 5. Contact time in function of pre-stress displacement

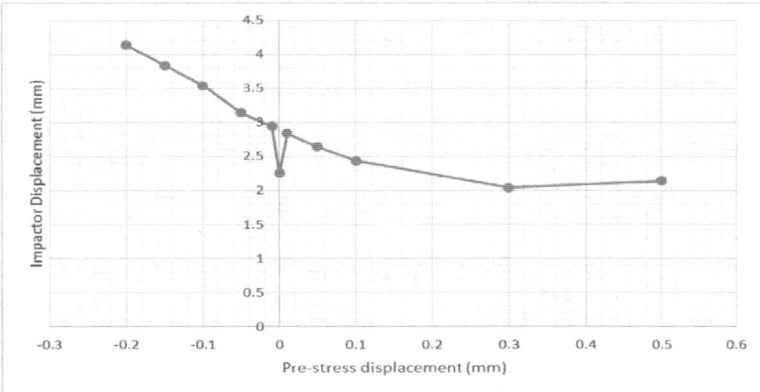

Figure 6. Impactor displacement in function of pre-stress displacement

Figure 7 shows that delamination area increases linearly with the magnitude of pre-stress displacement regardless of in tension or compression. The increment is significant for pre-stress in tension compared to compression. However, upon closer observation, under pre-stress in tension (Figure 8), the shape of delamination changes from circular (without pre-stress) to elliptical shape. The elliptical radius of delamination area in the direction of pre-stress increases with the magnitude of pre-stress. As for compression (Figure 9), it is observed that the evolution is not as regular compared to tension one. For both cases, in general the delamination shape becomes ellipse from circular without pre-stress after impact and under pre-stress loading.

From Figure 7-9, in can be concluded that pre-stress in compression has an effect more on the global behaviour of laminate during impact by having higher global deflection, contact time and maximum force during impact. As for pre-stress in tension, it produces the effects on the delamination area rather than the global behaviour of laminate during impact.

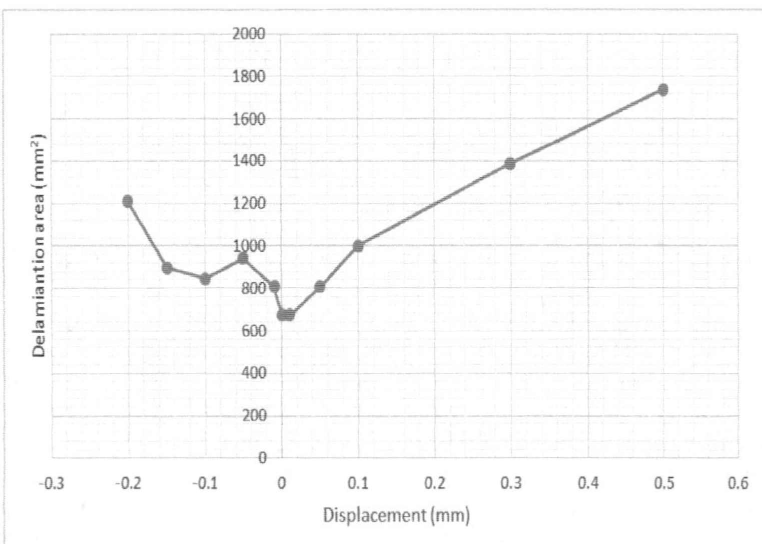

Figure 7. Delamination area in function of pre-stress displacement

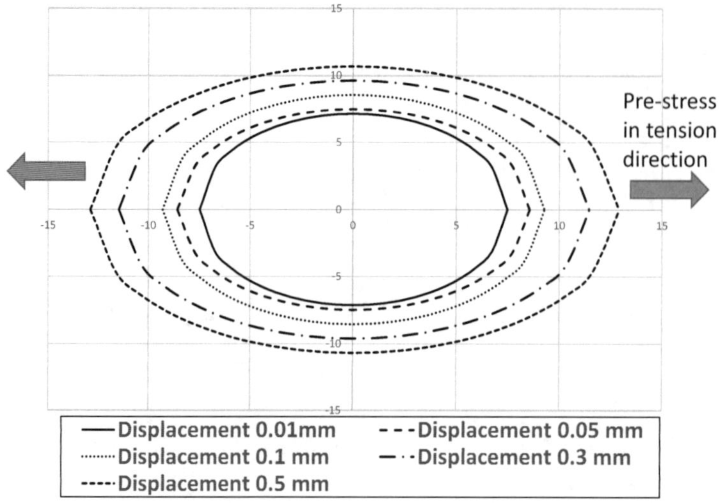

Figure 8. Delamination evolution of laminate pre-stressed in tension

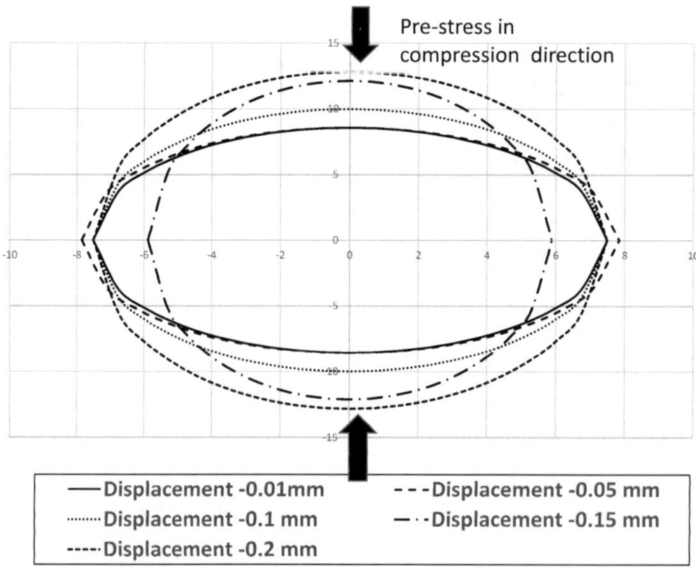

Figure 9. Delamination evolution of laminate pre-stressed in compression

5. Conclusion

This research work investigates the effect of pre-stressed in tension and compression on the delamination of laminate composite subjected to low energy impact. A finite elements model has been proposed and validated with published experiments data. The validated FE model was then used to compute the same laminate with different magnitude of pre-stress loading before impact. The pre-stress was model by prescribed displacement at the edges of the laminate. The Finite elements computation results shows that pre-stress in compression has an effect more on the global behaviour of laminate during impact by having higher deflection, contact time and maximum force during impact. As for pre-stress in tension, it produces the effects on the delamination area rather than the global behaviour of laminate during impact. The proposed FEA can be developed further for investigation on the prediction of delamination area at all interfaces of the proposed laminate.

References

[1] Balasubramani, S. Rajendra Boopathy, and R. Vasudevan, "Numerical analysis of low velocity impact on laminated composite plates," Procedia Eng., vol. 64, pp. 1089–1098, 2013
[2] R. Olsson, "Analytical model for delamination growth during small mass impact on plates," Int. J. Solids Struct., vol. 47, no. 21, pp. 2884–2892, 2010
[3] C. Bouvet, B. Castanié, M. Bizeul, and J. J. Barrau, "Low velocity impact modelling in laminate composite panels with discrete interface elements," Int. J. Solids Struct., vol. 46, no. 14–15, pp. 2809–2821, 2009.
[4] S. Heimbs, S. Heller, and P. Middendorf, "Simulation of Low Velocity Impact on Composite Plates with Compressive Preload," LS-DYNA Anwenderforum, pp. 11–24, 2008.
[5] K. H. Ji, "Direct Numerical Simulation of Composite Laminates Under Low Velocity Impact," 2006.
[6] H. H. T. D. P. B. F. Dogan, "Delamination of impacted composite structures by cohesive zone interface elements and tiebreak contact," Central European Journal of Engineering, no. ResearchGate, p. 4, 2012.
[7] F. F. A. L. C. M. G. P. M. Alfano, "Cohesive Zone Modelling of Mode I Fracture in Adhesive Bonded Joints," Vols. 348-349, 2007
[8] Aminanda, Y., Castanié, B., Barrau, J.-J., Thevenet, P, "Modelling of the static indentation of honeycomb core sandwichs with metallic skins", Mecanique et Industries,Vol6, Issue 5, Pages 487-498, 2005
[9] Castanié, B, Aminanda, Y, Barrau, J.-J, Thevenet, P , "Discrete modeling of the crushing of nomex honeycomb core and application to impact and post-impact behavior of sandwich structures", Solid Mechanics and its Applications, Vol 192, Pages 427-489, 2013
[10] L. Iannucci, "Dynamic delamination modelling using interface elements," Computers & Structures, vol. 84, p. 15, 2006
[11] Ali, J.S.M, Alsubari, Aminanda, "Hygrothermoelastic analysis of orthotropic cylindrical shells, Latin American Journal of Solids and Structures,Vol 13, Issue 3, Pages 573-589, 2016
[12] Alsubari, S, Mohamed Ali, J.S, Aminanda, Y, "Hygrothermoelastic analysis of anisotropic cylindrical shells", Composite Structures, Vol 131, Pages 151-159, 2015
[13] Mokhtar, H , Sicot, O, Rousseau, J, Aminanda, Y, Aivazzadeh, S, "The influence of ageing on the impact damage of carbon epoxy composites", Procedia Engineering, Vol 10, Pages 2615-2620, 2011

Determine the dynamic load acting on the chassis of multi-purpose forest fire fighting vehicle

L V Van* and N T Tung[2]

Department of Automotive Engineering, Vinh Long University of Technology Education Vietnam, 73 Nguyen Hue Stress, Ward 2, Vinh Long City, Vinh Long Province

* Email: vanlv@vlute.edu.vn

Abstract. Multi-purpose forest fire fighting vehicle is produced in Viet Nam include a combination of fire fighting equipment such as high-pressure water pump, create corridor fire insulation cutting machine, vacuum and high wind speed bowling machine, extinguish the fire sand blast apparatus that is mounted active three axles vehicle. The paper presents how to determine the dynamic load acting on the chassis by 3D modeling method. The results of this study are the input parameters for assessing the durability of the chassis of multi-purpose forest fire fighting vehicle.

Keywords: dynamic load, chassis, multi-purpose forest fire fighting vehicle, 3D vehicle model

1. Introduction

Multi-purpose forest fire fighting vehicle was designed on the basic URAL 4320, three active axle, used mainly in the military to increase mobility on the road surface. Shape of base vehicle Ural 4320 and multifunction forest fire fighting vehicle shown in Figure 1 [1,2].

a) Base vehicle Ural 4320

b) Multi-purpose forest fire fighting was built

Figure 1. Multi-purpose forest fire fighting vehicle.

The main equipment of multi-purpose forest fire fighting vehicle are shown in Figure 2 [3,4]

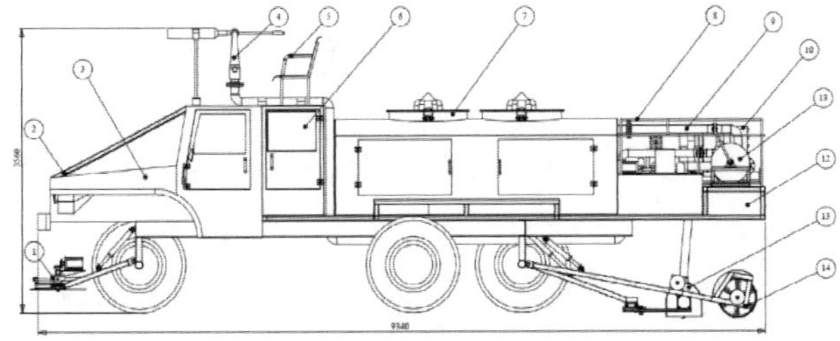

1) Tree cutting mechanism; 2) Tree cleaver; 3) Body; 4) Water sprayer; 5) Seat; 6) Cabin; 7) Water tank; 8) Back cover; 9) Pipe; 10) Table; 11) Plotting small nozzle; 12) Floorboard; 13) Dig system; 14) Lawn cutting mechanism.

Figure 2. General design multifunction forest fire fighting vehicle.

2. Mathematical modelling

2.1. Some initial assumptions

To building modelling, there are some assumptions:

- The wheels always contact the road surface;
- The modelling includes front axle, middle axle, rear axle are considered absolute solids;
- The resistance of tire is linear elastic;
- Mass of cab and container are distributed symmetry around the longitudinal and transverse planes;
- Speed of vehicle on the road is low, constant; therefore, inertial and air resistance low and considered zero [2].

2.2. Survey modeling

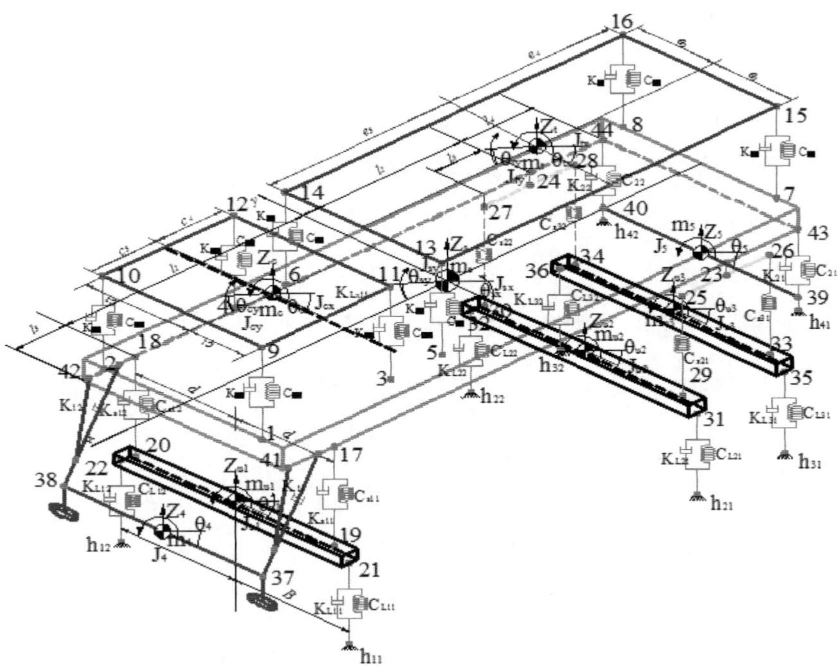

Figure 3. Full car suspension model

By applying Newton's second law of motion, we establish the differential oscillation system as follows [2]:

$$m_c\ddot{Z}_c + K_c\left[4\dot{Z}_c + 2\dot{\theta}_{cx}(c_1-c_2) + \dot{\theta}_{cy}(c_3-c_4) - 4\dot{Z}_s + 2\dot{\theta}_{sx}(c_2-c_1) - 2\dot{\theta}_{sy}(l_1+l_3) - 2\dot{\theta}_{sy}(l_1+l_3-c_3-c_4)\right]$$
$$C_c\left[4Z_c + 2\theta_{cx}(c_1-c_2) + \theta_{cy}(c_3-c_4) - 4Z_s + 2\theta_{sx}(c_2-c_1) - 2\theta_{sy}(l_1+l_3) - 2\theta_{sy}(l_1+l_3-c_3-c_4)\right] = 0 \quad (1)$$

$$J_{cx}\ddot{\theta}_{cx} + K_c c_1\left[2\dot{Z}_c + 2\dot{\theta}_{cx}c_1 + \dot{\theta}_{cy}(c_3-c_4) - 2\dot{Z}_s - 2\dot{\theta}_{sx}c_1 - \dot{\theta}_{sy}(2l_1+2l_3+c_3+c_4)\right]$$
$$-K_c c_2\left[2\dot{Z}_c - 2\dot{\theta}_{cx}c_2 + \dot{\theta}_{cy}(c_3-c_4) - 2\dot{Z}_s + 2\dot{\theta}_{sx}c_2 - \dot{\theta}_{sy}(2l_1+2l_3+c_3+c_4)\right]$$
$$+C_c c_1\left[2Z_c + 2\theta_{cx}c_1 + \theta_{cy}(c_3-c_4) - 2Z_s - 2\theta_{sx}c_1 - \theta_{sy}(2l_1+2l_3+c_3+c_4)\right]$$
$$-C_c c_2\left[2Z_c - 2\theta_{cx}c_2 + \theta_{cy}(c_3-c_4) - 2Z_s + 2\theta_{sx}c_2 - \theta_{sy}(2l_1+2l_3+c_3+c_4)\right] = 0 \quad (2)$$

$$J_{cy}\ddot{\theta}_{cy} + K_c c_3\left[2\dot{Z}_c + \dot{\theta}_{cx}(c_1-c_2) + 2\dot{\theta}_{cy}c_3 - 2\dot{Z}_s + \dot{\theta}_{sx}(c_2-c_1) - 2\dot{\theta}_{sy}(l_1+l_3)\right]$$
$$-K_c c_4\left[2\dot{Z}_c + \dot{\theta}_{cx}(c_1-c_2) - 2\dot{\theta}_{cy}c_4 - 2\dot{Z}_s + \dot{\theta}_{sx}(c_2-c_1) - 2\dot{\theta}_{sy}(l_1+l_3-c_3-c_4)\right]$$
$$C_c c_3\left[2Z_c + \theta_{cx}(c_1-c_2) + 2\theta_{cy}c_3 - 2Z_s + \theta_{sx}(c_2-c_1) - 2\theta_{sy}(l_1+l_3)\right]$$
$$-C_c c_4\left[2Z_c + \theta_{cx}(c_1-c_2) - 2\theta_{cy}c_4 - 2Z_s + \theta_{sx}(c_2-c_1) - 2\theta_{sy}(l_1+l_3-c_3-c_4)\right] = 0 \quad (3)$$

$$m_t\ddot{Z}_t + K_t\left[4\dot{Z}_t + 2\dot{\theta}_{tx}(e_1-e_2) + 2\dot{\theta}_{ty}(e_3-e_4) - 4\dot{Z}_s + 2\dot{\theta}_{sx}(e_2-e_1) - 2\dot{\theta}_{sy}(e_3+e_4-l_2-l_4) + 2\dot{\theta}_{sy}(l_2+l_4)\right]$$
$$+C_t\left[4Z_t + 2\theta_{tx}(e_1-e_2) + 2\theta_{ty}(e_3-e_4) - 4Z_s + 2\theta_{sx}(e_2\quad e_1) - 2\theta_{sy}(e_3+e_4-l_2-l_4) + 2\theta_{sy}(l_2+l_4)\right] = 0 \quad (4)$$

$$J_{tx}\ddot{\theta}_{tx} + K_t e_1\left[2\dot{Z}_t + 2\dot{\theta}_{tx}e_1 + \dot{\theta}_{ty}(e_3-e_4) - 2\dot{Z}_s - 2\dot{\theta}_{sx}e_1 + \dot{\theta}_{sy}(l_2+l_4) - \dot{\theta}_{sy}(e_3+e_4-l_2-l_4)\right]$$
$$-K_t e_2\left[2\dot{Z}_c - 2\dot{\theta}_{tx}e_2 + \dot{\theta}_{ty}(e_3-e_4) - 2\dot{Z}_s + 2\dot{\theta}_{sx}e_2 + \dot{\theta}_{sy}(l_2+l_4) - \dot{\theta}_{sy}(e_3+e_4-l_2-l_4)\right]$$
$$C_t e_1\left[2Z_t + 2\theta_{tx}e_1 + \theta_{ty}(e_3-e_4) - 2Z_s - 2\theta_{sx}e_1 + \theta_{sy}(l_2+l_4) - \theta_{sy}(e_3+e_4-l_2-l_4)\right]$$
$$-C_t e_2\left[2Z_c - 2\theta_{tx}e_2 + \theta_{ty}(e_3-e_4) - 2Z_s + 2\theta_{sx}e_2 + \theta_{sy}(l_2+l_4) - \theta_{sy}(e_3+e_4-l_2-l_4)\right] = 0 \quad (5)$$

$$J_{ty}\ddot{\theta}_{ty} + Ke_3\left[2\dot{Z}_t + \dot{\theta}_{tx}(e_1-e_2) + 2\dot{\theta}_{ty}e_3 - 2\dot{Z}_s - \dot{\theta}_{sx}(e_2-e_1) - 2\dot{\theta}_{sy}(e_3+e_4-l_2-l_4)\right]$$
$$-K_t e_4\left[2\dot{Z}_t + \dot{\theta}_{tx}(e_1-e_2) - 2\dot{\theta}_{ty}e_4 - 2\dot{Z}_s - \dot{\theta}_{sx}(e_2-e_1) + 2\dot{\theta}_{sy}(l_2+l_4)\right]$$
$$+Ce_3\left[2Z_t + \theta_{tx}(e_1-e_2) + 2\theta_{ty}e_3 - 2Z_s - \theta_{sx}(e_2-e_1) - 2\theta_{sy}(e_3+e_4-l_2-l_4)\right]$$
$$-C_t e_4\left[2Z_t + \theta_{tx}(e_1-e_2) - 2\theta_{ty}e_4 - 2Z_s - \theta_{sx}(e_2-e_1) + 2\theta_{sy}(l_2+l_4)\right] = 0 \quad (6)$$

$$m_s\ddot{Z}_s + K_c\left[4\dot{Z}_s + 2\dot{\theta}_{sx}(c_1-c_2) + 2\dot{\theta}_{sy}(c_3+c_4) - 4\dot{Z}_c + 2\dot{\theta}_{cx}(c_2-c_1) + 2\dot{\theta}_{cy}(c_4-c_3)\right]$$
$$+K_t\left[4\dot{Z}_s + 2\dot{\theta}_{sx}(e_1-e_2) + 2\dot{\theta}_{sy}(e_3+e_4) - 4\dot{Z}_t + 2\dot{\theta}_{tx}(e_2-e_1) + \dot{\theta}_{ty}(e_4-e_3)\right]$$
$$+K_{s1}\left[2\dot{Z}_s + 2\dot{\theta}_{sy}l_1 - 2\dot{Z}_{u1}\right] + K_1\left[2\dot{Z}_s + 2\dot{\theta}_{sy}(l_1+l_5) - 2\dot{Z}_4\right] + K_2\left[2\dot{Z}_s - 2\dot{\theta}_{sy}(l_2+l_4) - 2\dot{Z}_5\right]$$
$$+C_{c11}\left[Z_s + \theta_{sx}c_1 + \theta_{sy}(l_1+l_3) - Z_c - \theta_{cx}c_1 - \theta_{cy}c_3\right]$$
$$+C_c\left[4Z_s + 2\theta_{sx}(c_1-c_2) + 2\theta_{sy}(c_3+c_4) - 4Z_c + 2\theta_{cx}(c_2-c_1) + 2\theta_{cy}(c_4-c_3)\right]$$
$$+C_t\left[4Z_s + 2\theta_{sx}(e_1-e_2) + 2\theta_{sy}(e_3+e_4) - 4Z_t + 2\theta_{tx}(e_2-e_1) + \theta_{ty}(e_4-e_3)\right]$$
$$+C_{s1}\left[2Z_s + 2\theta_{sy}l_1 - 2Z_{u1}\right] + C_{s2}\left[2Z_s + 2\theta_{sy}(l_2-l_5) - 2Z_{u2}\right] + C_{s3}\left[2Z_s - 2\theta_{sy}(l_2+l_5) - 2Z_{u3}\right]$$
$$+C_1\left[2Z_s + 2\theta_{sy}(l_1+l_3) - 2Z_4\right] + C_2\left[2Z_s - 2\theta_{sy}(l_2+l_4) - 2Z_5\right] = 0 \quad (7)$$

$$J_{sx}\ddot{\theta}_{sx}+K_c c_1\left[2\dot{Z}_s+2\dot{\theta}_{sx}c_1+\dot{\theta}_{sy}(c_3+c_4)-2\dot{Z}_c-2\dot{\theta}_{cx}c_1-\dot{\theta}_{cy}(c_4-c_3)\right]$$
$$-K_c c_2\left[2\dot{Z}_s-2\dot{\theta}_{sx}c_2+\dot{\theta}_{sy}(c_3+c_4)-2\dot{Z}_c+2\dot{\theta}_{cx}c_2-\dot{\theta}_{cy}(c_4-c_3)\right]$$
$$+K_t e_1\left[2\dot{Z}_s+2\dot{\theta}_{sx}e_1+\dot{\theta}_{sy}(e_3+e_4-2l_2-2l_4)-2\dot{Z}_t-2\dot{\theta}_{tx}e_1-\dot{\theta}_{ty}(e_4-e_3)\right]$$
$$-K_t e_2\left[2\dot{Z}_s-2\dot{\theta}_{sx}e_2+\dot{\theta}_{sy}(e_3+e_4-2l_2-2l_4)-2\dot{Z}_t+2\dot{\theta}_{tx}e_2-\dot{\theta}_{ty}(e_4-e_3)\right]$$
$$+K_{s1}d\left[2\dot{\theta}_{sx}d-2\dot{\theta}_{u1}d\right]+K_1 d\left[2\dot{\theta}_{sx}d-2\dot{\theta}_4 d\right]+K_2 d\left[2\dot{\theta}_{sx}d-2\dot{\theta}_5 d\right] \quad (8)$$
$$C_c c_1\left[2Z_s+2\theta_{sx}c_1+\theta_{sy}(c_3+c_4)-2Z_c-2\theta_{cx}c_1-\theta_{cy}(c_4-c_3)\right]$$
$$-C_c c_2\left[2Z_s-2\theta_{sx}c_2+\theta_{sy}(c_3+c_4)-2Z_c+2\theta_{cx}c_2-\theta_{cy}(c_4-c_3)\right]$$
$$+C_t e_1\left[2Z_s+2\theta_{sx}e_1+\theta_{sy}(e_3+e_4-2l_2-2l_4)-2Z_t-2\theta_{tx}e_1-\theta_{ty}(e_4-e_3)\right]$$
$$-C_t e_2\left[2Z_s-2\theta_{sx}e_2+\theta_{sy}(e_3+e_4-2l_2-2l_4)-2Z_t+2\theta_{tx}e_2-\theta_{ty}(e_4-e_3)\right]$$
$$+C_{s1}d\left[2\theta_{sx}d-2\theta_{u1}d\right]+C_{s2}d\left[2\theta_{sx}d-2\theta_{u1}d\right]+C_{s3}d\left[2\theta_{sx}d-2\theta_{u1}d\right]$$
$$+C_1 d\left[2\theta_{sx}d-2\theta_4 d\right]+C_2 d\left[2\theta_{sx}d-2\theta_5 d\right]=0$$

$$J_{sy}\ddot{\theta}_{sy}+K_c(l_1+l_3)\left[2\dot{Z}_s+\dot{\theta}_{sx}(c_1-c_2)+2\dot{\theta}_{sy}(l_1+l_3)-2\dot{Z}_c+\dot{\theta}_{cx}(c_2-c_1)-2\dot{\theta}_{cy}c_3\right]$$
$$-K_c(l_1+l_3-c_3-c_4)\left[2\dot{Z}_s+\dot{\theta}_{sx}(c_1-c_2)-\dot{\theta}_{sy}(l_1+l_3-c_3-c_4)-2\dot{Z}_c-\dot{\theta}_{cx}(c_2-c_1)+2\dot{\theta}_{cy}c_4\right]$$
$$+K_t(e_3+e_4-l_2-l_4)\left[2\dot{Z}_s+\dot{\theta}_{sx}(e_1-e_2)+2\dot{\theta}_{sy}(e_3+e_4-l_2-l_4)-2\dot{Z}_t-\dot{\theta}_{tx}(e_2-e_1)-2\dot{\theta}_{ty}e_3\right]$$
$$-K_t(l_2+l_4)\left[2\dot{Z}_s+\dot{\theta}_{sx}(e_1-e_2)-2\dot{\theta}_{sy}(l_2+l_4)-2\dot{Z}_t-\dot{\theta}_{tx}(e_2-e_1)+2\dot{\theta}_{ty}e_4\right]$$
$$+K_{s1}l_1\left[2\dot{Z}_s+2\dot{\theta}_{sy}l_1-2\dot{Z}_{u1}\right]+K_1(l_1+l_3)\left[2\dot{Z}_s+2\dot{\theta}_{sy}(l_1+l_3)-2\dot{Z}_4\right]$$
$$-K_2(l_2+l_4)\left[2\dot{Z}_s-2\dot{\theta}_{sy}(l_2+l_4)-2\dot{Z}_5\right]$$
$$+C_c(l_1+l_3)\left[2Z_s+\theta_{sx}(c_1-c_2)+2\theta_{sy}(l_1+l_3)-2Z_c+\theta_{cx}(c_2-c_1)-2\theta_{cy}c_3\right]$$
$$-C_c(l_1+l_3-c_3-c_4)\left[2Z_s+\theta_{sx}(c_1-c_2)-\theta_{sy}(l_1+l_3-c_3-c_4)-2Z_c-\theta_{cx}(c_2-c_1)+2\theta_{cy}c_4\right]$$
$$+C_t(e_3+e_4-l_2-l_4)\left[2Z_s+\theta_{sx}(e_1-e_2)+2\theta_{sy}(e_3+e_4-l_2-l_4)-2Z_t-\theta_{tx}(e_2-e_1)-2\theta_{ty}e_3\right]$$
$$-C_t(l_2+l_4)\left[2Z_s+\theta_{sx}(e_1-e_2)-2\theta_{sy}(l_2+l_4)-2Z_t-\theta_{tx}(e_2-e_1)+2\theta_{ty}e_4\right]$$
$$+C_{s1}l_1\left[2Z_s+2\theta_{sy}l_1-2Z_{u1}\right]-C_{s2}(l_2-l_5)\left[2Z_s-2\theta_{sy}(l_2-l_5)-2Z_{u2}\right]$$
$$-C_{s3}(l_2+l_5)\left[2Z_s-2\theta_{sy}(l_2+l_5)-2Z_{u3}\right] \quad (9)$$
$$+C_{11}(l_1+l_3)\left[2Z_s+2\theta_{sy}(l_1+l_3)-2Z_4\right]-C_2\left[2Z_s-2\theta_{sy}(l_2+l_4)-2Z_5\right]=0$$

$$m_{u1}\ddot{Z}_{u1}+K_{s1}\left[2\dot{Z}_{u1}-2\dot{Z}_s-2\dot{\theta}_{sy}l_1\right]+K_{L1}\left(2\dot{Z}_{u1}-\dot{h}_{11}-\dot{h}_{12}\right)+C_{s1}\left[2Z_{u1}-2Z_s-2\theta_{sy}l_1\right]+C_{L1}\left(2Z_{u1}-h_{11}-h_{12}\right)=0 \quad (10)$$
$$J_{u1}\ddot{\theta}_{u1}+K_{s1}d\left[2\dot{\theta}_{u1}d-2\dot{\theta}_{sx}d\right]+K_{L1}B^2\left(2\dot{\theta}_{u1}-\dot{h}_{11}-\dot{h}_{12}\right)+C_{s1}d\left[+2\theta_{u1}d-2\theta_{sx}d\right]+C_{L1}B^2\left(2\theta_{u1}-h_{11}-h_{12}\right)=0 \quad (11)$$
$$m_{u2}\ddot{Z}_{u2}+K_{L2}\left(2\dot{Z}_{u2}-\dot{h}_{21}-\dot{h}_{22}\right)+C_{s2}\left[2Z_{u2}-2Z_s+2\theta_{sy}(l_2-l_5)\right]+C_{L2}\left(Z_{u2}-h_{21}-h_{22}\right)=0 \quad (12)$$
$$J_{u2}\ddot{\theta}_{u2}+K_{L2}B^2\left(2\dot{\theta}_{u2}-\dot{h}_{21}-\dot{h}_{22}\right)+C_{s2}d\left[+2\theta_{u2}d-2\theta_{sx}d\right]+C_{L2}B^2\left(2\theta_{u2}-h_{21}-h_{22}\right)=0 \quad (13)$$
$$m_{u3}\ddot{Z}_{u3}+K_{L3}\left(2\dot{Z}_{u3}-\dot{h}_{31}-\dot{h}_{32}\right)+C_{s3}\left[2Z_{u3}-2Z_s+2\theta_{sy}(l_2+l_5)\right]+C_{L3}\left(2Z_{u3}-h_{31}-h_{32}\right)=0 \quad (14)$$
$$J_{u3}\ddot{\theta}_{u3}+K_{L3}B^2\left(2\dot{\theta}_{u3}-\dot{h}_{31}-\dot{h}_{32}\right)+C_{s3}d\left[+2\theta_{u3}d-2\theta_{sx}d\right]+C_{L3}B^2\left(2\theta_{u3}-h_{31}-h_{32}\right)=0 \quad (15)$$
$$m_4\ddot{Z}_4+K_1\left[2\dot{Z}_4-2\dot{\theta}_{sy}(l_1+l_3)-2\dot{Z}_s\right]=0 \quad (16)$$
$$J_4\ddot{\theta}_4+K_1 d\left[2\dot{\theta}_4 d-2\dot{\theta}_{sx}d\right]=0 \quad (17)$$
$$m_5\ddot{Z}_5+K_2\left[2\dot{Z}_5+2\dot{\theta}_{sy}(l_2+l_4)-2\dot{Z}_s\right]+C_2\left[2Z_5+2\theta_{sy}(l_2+l_4)-2Z_s\right]=0 \quad (18)$$
$$J_5\ddot{\theta}_5+K_2 d\left[2\dot{\theta}_5 d-2\dot{\theta}_{sx}d\right]+C_2 d\left[2\theta_5 d-2\theta_{sx}d\right]=0 \quad (19)$$

The values of the geometry and dynamic parameters of multi-purpose forest fire fighting vehicle are given in Table 1 [1,3].

Table 1. Technical of multi-purpose forest fire fighting vehicle.

Symbol	Value	Symbol	Value	Symbol	Value
m_c	850 kg	m_5	1000 kg	K_{11}, K_{12}	2800 Ns/m
J_{cx}	200 kgm^2	J_5	1277 kgm^2	K_{21}, K_{22}	2800 Ns/m
J_{cy}	150 kgm^2	C_{c1}, C_{c2}, C_{c3}, C_{c4}	100000 N/m	c_1	0,7 m
m_t	8500 kg	K_{c11}, K_{c12}, K_{c21}, K_{c22}	750 Ns/m	c_2	0,6 m
J_{tx}	48700 kgm^2	C_{t1}, C_{t2}, C_{t3}, C_{t4}	500000 N/m	c_3	0,8 m
J_{ty}	13819 kgm^2	K_{t11}, K_{t12}, K_{t21}, K_{t22}	4000 N/m	c_4	0,7 m
m_s	14310 kg	C_{s11}, C_{s12}	401952 N/m	e_1	0,8 m
J_{sx}	64770 kgm^2	C_{s21}, C_{s22}	527042 N/m	e_2	0,8 m
J_{sy}	18852 kgm^2	C_{s31}, C_{s32}	527042 N/m	e_3	2,5 m
m_{u1}	1020 kg	K_{s11}, K_{s12}	3248 Ns/m	e_4	2,5 m
m_{u2}	1000 kg	C_{L11}, C_{L12}	569964 Ns/m	2d	1,5 m
m_{u3}	1000 kg	C_{L21}, C_{L22}	569964 Ns/m	2B	2,0 m
J_{u1}	1320 kgm^2	C_{L31}, C_{L32}	569964 Ns/m	l_1	2,538 m
J_{u2}	1227 kgm^2	K_{L11}, K_{L12}	6497 Ns/m	l_2	1,687 m
J_{u3}	1227 kgm^2	K_{L21}, K_{L22}	6497 Ns/m	l_3	1,847 m
m_4	800 kg	K_{L31}, K_{L32}	6497 Ns/m	l_4	1,9 m
J_4	1105 kgm^2	C_{21}, C_{22}	80000 N/m	l_5	0,55 m

3. Determination of dynamic load

In this survey, the author uses a simple sinusoidal shape is used because of its simple structure and easy to predict the results [2].

The height of the bump is determined by the formula:

$$h(x)=\begin{cases}\frac{1}{2}H\left(1-\cos\left(2\pi\frac{x}{L}\right)\right) & when \quad 0<x<L\\ 0 & when \quad x\le 0 \quad , \quad x\ge L\end{cases} \tag{20}$$

With a constant velocity of the car (v) is the height of the bump in time (h) calculated according to the formula:

$$h(t)=\begin{cases}\frac{1}{2}H\left(1-\cos\left(2\pi\frac{vt}{L}\right)\right) & when \quad 0<vt<L\\ 0 & when \quad vt\le 0 \quad , \quad vt\ge L\end{cases} \tag{21}$$

Inside: H - maximum bump height (m); L - bump length (m); v - movement velocity of the vehicle (m/s); t - Last time was bumpy (s).

IOP Conf. Series: Materials Science and Engineering 1173 (2021) 012030 doi:10.1088/1757-899X/1173/1/012030

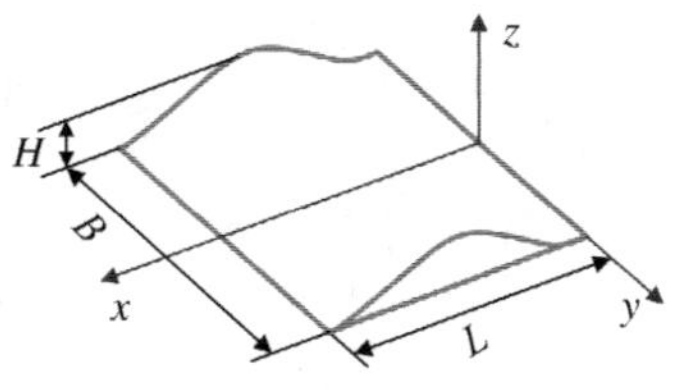

Figure 4. Description of the pavement according to the length.

Apply F_{z11}, F_{z12}, F_{z21}, F_{z22}, F_{z31}, F_{z32} is the dynamic load from the road surface acting alternately of the left front wheel, front right, left middle, right middle, rear left, right rear in verticality. Due to the high level of the forest road, sometimes the height of the road is over 0,35 m, and the movement velocity of the vehicle is about 25 km/h. Therefore, the survey when the wheels pass through the bump has a length of L = 0,5m and the height of the bump (H) varies from 0,1 to 0,4m, the vehicle velocity varies from 5 to 25 km/h, the vehicle is moving straight, not subject to horizontal forces.

3.1. Rough evenly two front wheels

Because two front wheels are evenly spaced, the vertical dynamic load acting the left and right wheels must be equal to the front and rear axles ($F_{z11} = F_{z12}$, $F_{z31} = F_{z32}$). On the car, the rear suspension design is balanced, to simply calculate whether vertical dynamic load acting on the middle wheel is equal to vertical dynamic load acting on the rear wheel in all moving conditions.

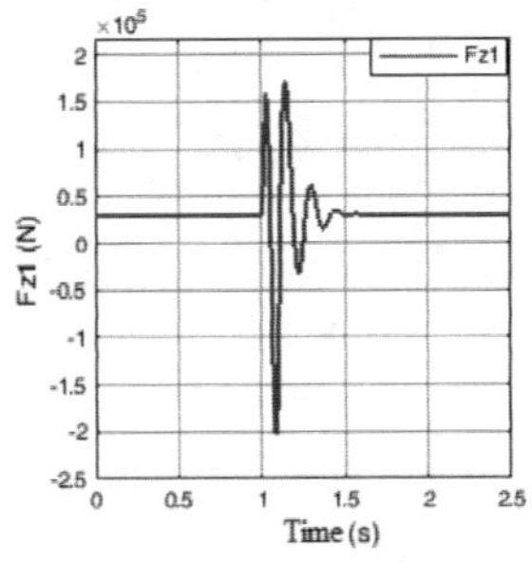

Figure 5. Dynamic load F_{z1} (v = 20 km/h, H = 0,4m).

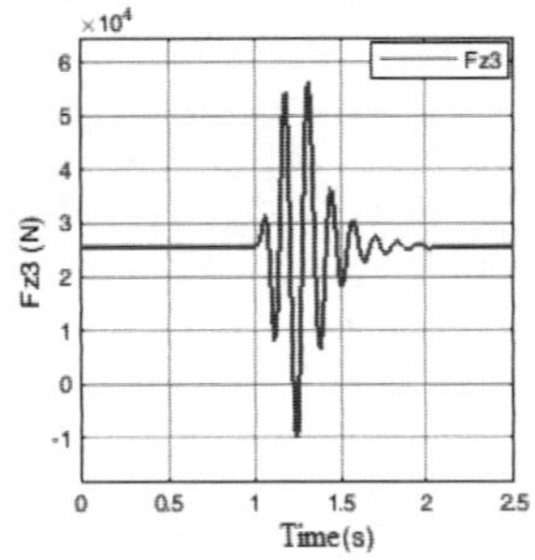

Figure 6. Dynamic load F_{z3} (v = 20 km/h, H = 0,4m).

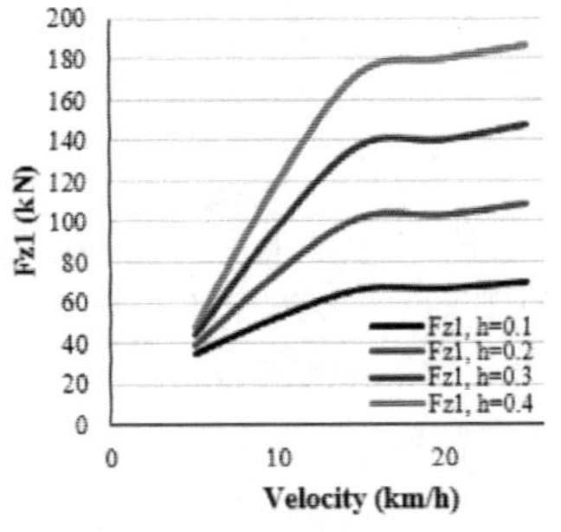

Figure 7. Dynamic load F_{z1} (v = 5 - 20 km/h, H = 0,1 - 0,4m).

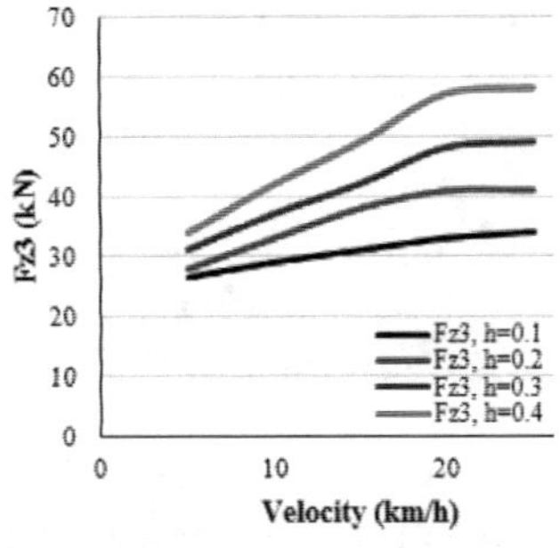

Figure 8. Dynamic load F_{z3} (v = 5 - 20 km/h, H = 0,1 - 0,4m).

The results in Figures 5 and 6 describle the vertical dynamic load at one side of the leaf spring at a speed v = 20 km/h when bumpy height of road surface 0,4 m in the front and rear axles, Figures 7 and 8 describe the change of maximum value of vertical dynamic load with different velocities, different heights on the front and rear axles. The results show that the law of physics: When the speed of movement of the vehicle increases, the vertical dynamic load will increase. With the speed v = 20 km/h, H = 0,4 m, the maximum value of vertical force F_{z1} = 170394N and F_{z3} = 56752N. Maximum vertical dynamic load is 5,62 times greater than in the case of static load (stationary vehicle) for front axle and 2,22 times greater for rear axle.

3.2. Single bump front wheel

In this case, the front right wheel must pass through the bump, The survey results will determine the dynamic load acting on the chassis through the axle and suspension. Because the two front wheels are not equal, the dynamic load on the left and right wheels is different ($F_{z11} < F_{z12}$).

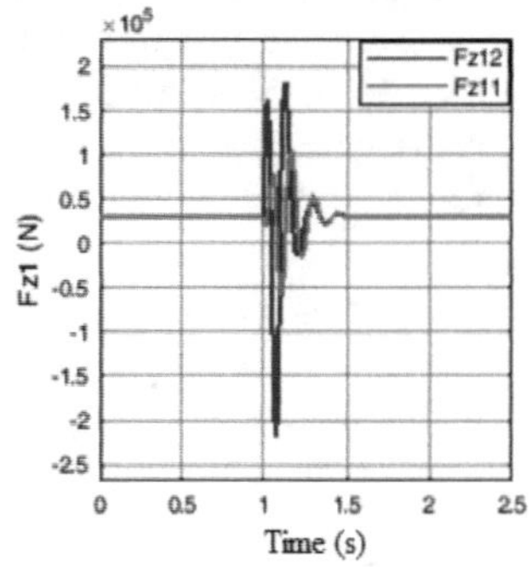

Figure 9. Dynamic load Fz1 (v = 20 km/h, H = 0,4m).

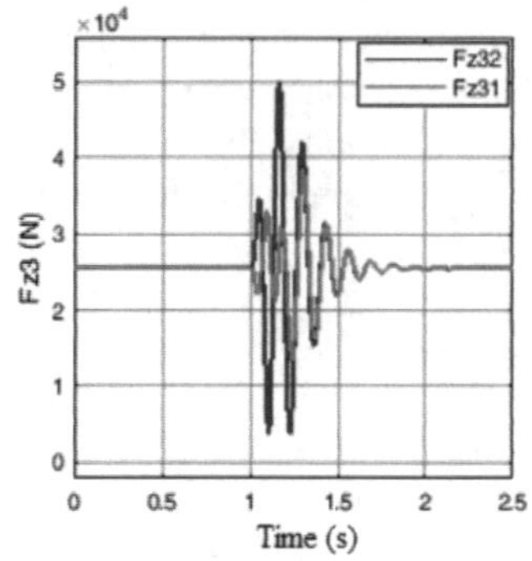

Figure 10. Dynamic load Fz3 (v = 20 km/h, H = 0,4m).

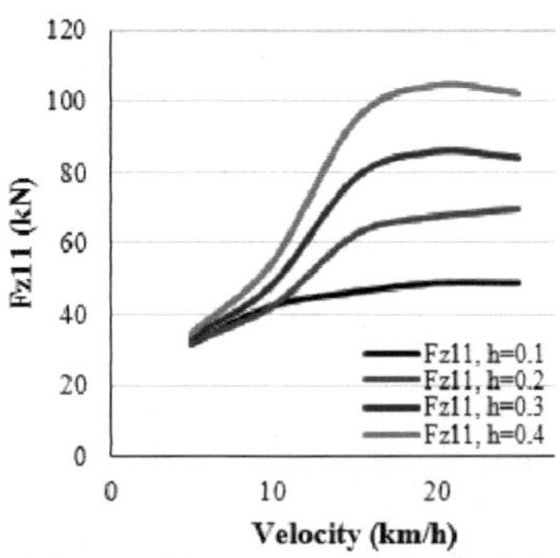

Figure 11. Dynamic load Fz1 (v = 5 - 20 km/h, H = 0,1 - 0,4m)

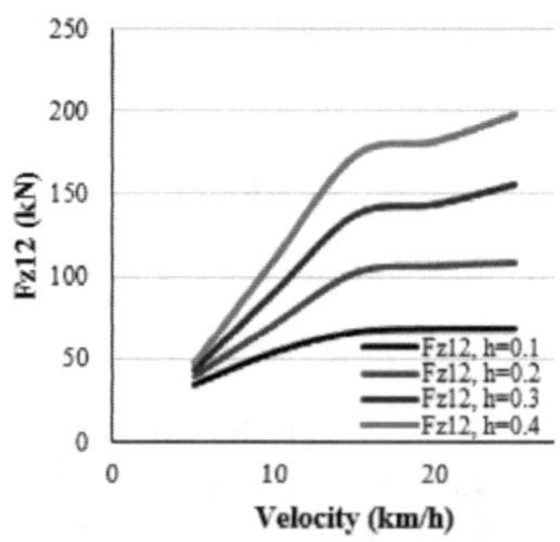

Figure 12. Dynamic load Fz3 (v = 5 - 20 km/h, H = 0,1 - 0,4m)

When the speed of movement of the vehicle increases, the vertical dynamic load will increase. With the speed v = 20 km/h, the maximum value of vertical force F_{z12} = 68150,6N. Maximum vertical dynamic load is 6,004 times greater than in the case of static load

3.3. Rough evenly two wheels cross (left front and right rear)

When moving at different speeds, rear wheel time contact with the surface roughness from the left front wheel contact road surface is different.

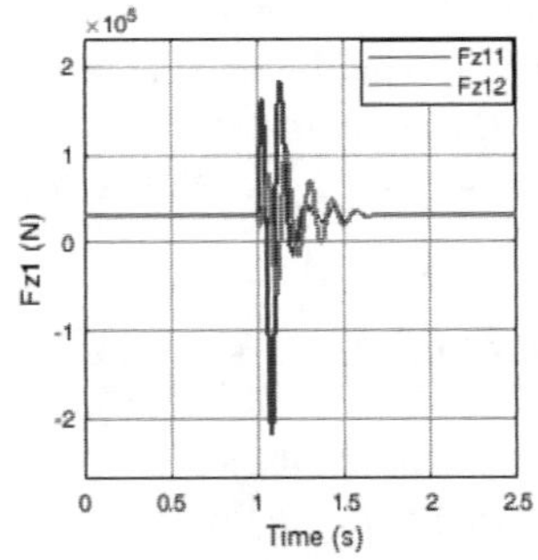

Figure 13. Dynamic load Fz1 (v = 20 km/h, H = 0,4m).

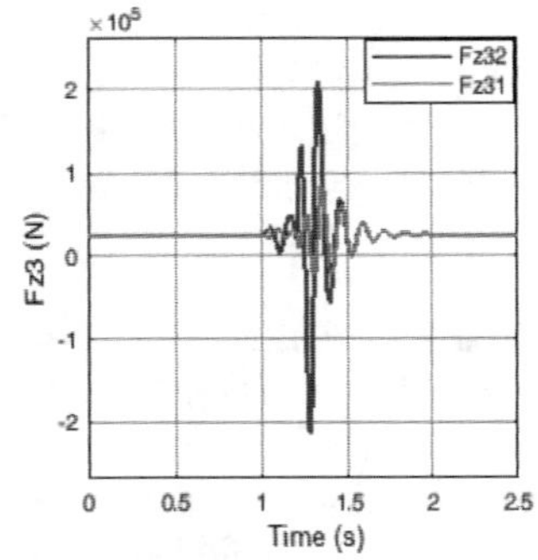

Figure 14. Dynamic load Fz3 (v = 20 km/h, H = 0,4m).

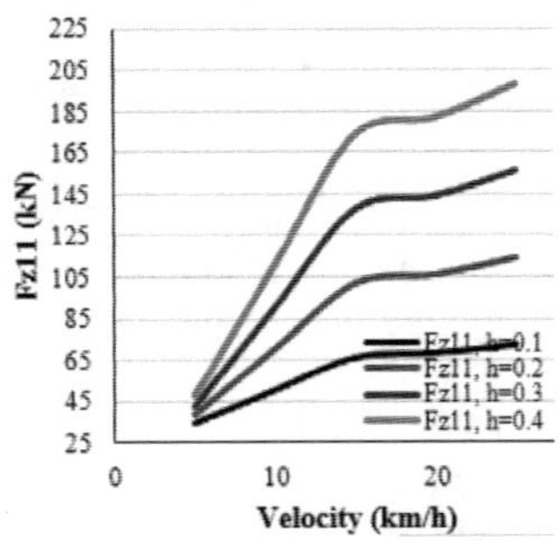

Figure 15. Dynamic load Fz1 (v = 5 – 20 km/h, H = 0,1 – 0,4m).

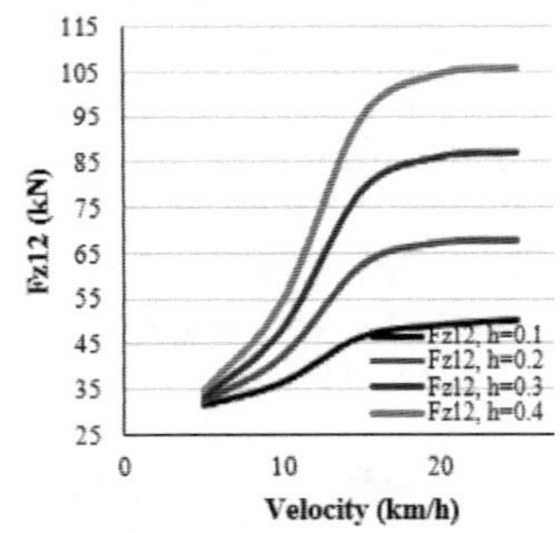

Figure 16. Dynamic load Fz3 (v = 5 - 20 km/h, H = 0,1 - 0,4m).

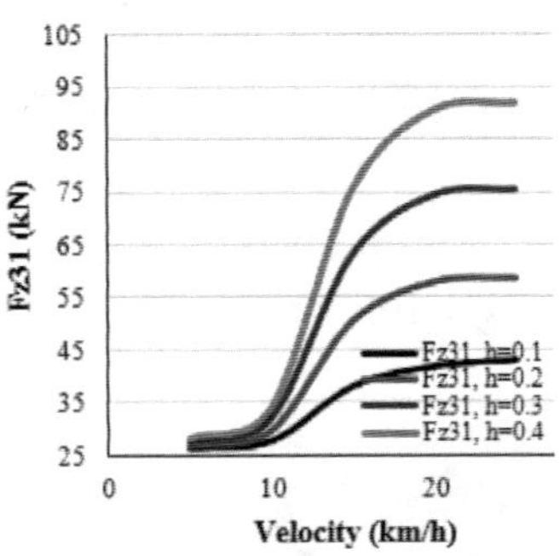

Figure 17. Dynamic load F_{z31} (v = 5 - 20 km/h, h = 0,1 - 0,4m).

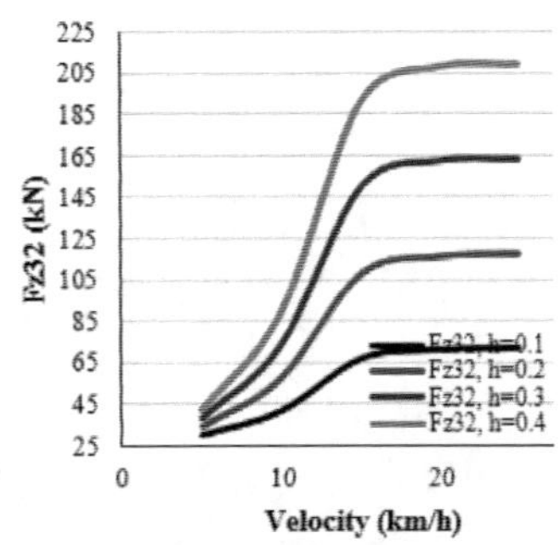

Figure 18. Dynamic load F_{z32} (v = 5 - 20 km/h, h = 0,1 - 0,4m).

When the speed of movement of the vehicle increases, the velocity v = 20 km/h, H = 0,4 m, the maximum value of vertical force F_{z11} = 181783N (Figure 14). Maximum vertical dynamic load is 6,004 times greater than in the case of static load. The maximum value of vertical force F_{z32} = 207840N (Figure 15), larger than 8,14 in the case of static load.

4. Conclusion

Multi-purpose forest fire fighting vehicle was designed on the basic URAL 4320, was installed specialized fire figtting to increase weight aa well as change the center of gravity of the vehicle. Because of the factor makes it become unstable while moving, at time reduce details when the vehicle has a large load, especially the chassis. The result of the study are the basic for stability of chassis to perfect the design of the multi-purpose forest fire fighting vehicle. However, this study only determines the dynamic load when the vehicle through sine-wave form. Therefore, further study is need to determine dynamic loads when vehicles are traveling on forest roads or on roads according to ISO 8608:1995.

References

[1] Luong V V, Nguyen Q Th, Tran T V 2018, *Proceeding of The 9th TSME International Conference on Mechanical Engineering*, Phuket, Thailand.
[2] Luong V V 2020. *Doctoral thesis Technology*, Vietnam National University of Forestry.
[3] Nguyen X H 2013, *Research dynamics of Multi-purpose forest fire fighting vehicle. Doctoral thesis Technology*, Vietnam National University of Forestry.
[4] Duong, T. V. (2010). *Research technology and design and manufacture specialized equipment for forest fire fighting*. National scientific and technological topics KC07.13/06-10. Vietnam National University of Forestry.
[5] Mitra, N. Benerjee, H. A. Khalane, M. A. Sonawane, D. R. Joshi, G.R. Bagul, *Journal of Mechanical and Civil Engineering (IOSR-JMCE).*
[6] Syabillah Sulaiman, Pakharuddin Mohd Samin, Hishamuddin Jamaluddin, Roslan Abd Rahman, Mohammad Safwan Burhaumudin 2012, "*International Conference on Automotive, Mechanical and Materials Engineering (ICAMME'2012) Penang (Malaysia).*
[7] M J Griffin, *Handbook of Human Vibration* (London Publication, ISBN-10: 0123030412. 1996).
[8] Galal Ali Hassaan, Nasser Abdul-Azim Mohammed 2015, *International Journal of Computer Techniques, Volume 2 Issue 1.*
[9] Michael Akpakpavi 2017, *Journal of Multidisciplinary Engineering Science Studies (JMESS).*
[10] D.J.M Sampson, G. McKevitt, D. Cebon 1999, *Proc. 16th IAVSD Symposium on the Dynamics of Vehicles on Roads and Tracks*, Pretoria, South Africa.
[11] P.M. Samin, H. Jamaluddin, R.A. Rahman, S.A.A. Bakar, K. Hudha 2007, *CADME07*, Putra Brasmana Hotel, Kuala Perlis, Malaysia

Degradation on mechanical properties of virgin and recycled polylactic acid ageing in aqueous environment

S Budin*, M H Koay, and N C Maideen

Faculty of Mechanical Engineering, Universiti Teknologi MARA, Cawangan Pulau Pinang, Jalan Permatang Pauh, 13500 Pulau Pinang

* Email: salinabudin@uitm.edu.my

Abstract. A growing demand in polylactic acid (PLA) has increased in recent years due to the increasing of non-degradable plastic wastes. The recycling of PLA waste is seen to be another permissible solution to conserve the resources. This work aims to study the degradation rate of virgin PLA (VPLA) and recycled PLA (RPLA) in terms of the mechanical properties after ageing in different aqueous environment. The samples were immersed in river water (RW) and seawater (SW) for the duration of 150 days. Tensile strength, transverse rupture strength (TRS), impact energy and hardness properties were investigated as a function of immersed time. The results revealed that the loss in the tensile strength accelerated in river water when compared to seawater. There was at least 50% of tensile strength sacrificed in recycled PLA. The TRS had dropped approximately half from its initial value after ageing in river water and about 30% after ageing in seawater. Besides, a huge decrease ranging around 70 - 80% from its initial impact energy was identified. However, the reduction in hardness was relatively low. The findings will help to elucidate the degradation process of PLA and the desired effect on the mechanical performance for further application.

Keywords: virgin polylactic acid, recycled polylactic acid, degradation, ageing, aqueous environment

1. Introduction

The issues towards the environment derived from petroleum-based plastic has brought serious attention around the world. It has been reported that 360 million tonnes of plastic were produced in 2018 and expected to grow to 2000 million tonnes by the year 2050 [1]. A large portion of these plastic will end-up as trash in the landfill and flow through the river and towards the ocean [2]. The demand for biodegradable plastic as single-used packaging material listed under Sustainability Development Goal (SDG) has sparked interests of many researchers to explore new alternative bioplastic material. Among those, polylactic acid (PLA) seen to be the most known renewable plastic with good biodegradability and biocompatibility.

PLA is a synthetic bio-based polyester derived from agriculture product. PLA exhibited excellent characteristics such as zero toxicity, appropriate compossibility, high mechanical strength and thermal stability and ease of fabrication [3]. However, it is said to be suffering on poor thermal stability, brittleness and high production cost, the ever-increasing diversion of starch feedstocks to PLA production is expected to pressure on land use on agriculture and waste management on worn part [4].

The degradation rate of PLA depends on the properties of the final products and disposal environment, but normally required 6 – 24 months to be fully degraded [5]. Recycling of PLA waste could probably help to support waste management challenges.

PLA with the presence of water is susceptible to hydrolytic degradation. Hydrolytic degradation of PLA primary will take place in two stages. At initial stage, random hydrolytic scission of ester bonds proceeds with the diffusion of water into amorphous regions which leads to the reduction of molecular weight. As the degradation continues, the crystallinity of the polymer increased. During second stage, hydrolytic attacks occurs from the edge towards the center of crystalline area with the degradation of the major portions of amorphous area [6]. The PLA degradation rate is strongly dependable on the particle size and shape, medium temperature, humidity and pH, moisture absorption, molecular weight of PLA, crystallinity and surface area [7].

Past studies that are related to hydrolytic degradation of PLA and other polyester polymer are reported as shown below. In a study done by Deroine et al. [8] have investigated the degradation mechanism and kinetics of PLA in distilled water and seawater. The researchers have remarked that the water uptake in each environment is according to Fickian behavior. However, the degradation rate in distilled water was higher when compared to seawater. In another work done by Iniguez-Franco et al., [9] have concluded that hydrolytic degradation of PLA in pure water and ethanol solution. The findings showed PLA experienced a faster hydrolytic degradation in 50% ethanol solution as compared in pure water. Higher percentage of crystallinity was also reported in similar samples. The alteration in molecular weight and crystallinity during hydrolytic degradation were significantly affected the mechanical properties of PLA.

Beltran et al. [10] have studied the effects of different mechanical recycling processes on the hydrolytic degradation of PLA immersed in phosphate buffered solution at the temperature of 37°C and 58°C. It is noted that the water absorption is higher on recycled PLA as compared to neat PLA. The rate of water absorption was also enhanced as the temperature of the solution increased. Consistent conclusions were drawn by Rapacz-Kmita et al. [11] which remarked on the accelerated hydrolytic degradation on PLA immersed in water at temperature 80°C. The hydrolytic degradation above T_g has increased the degree of crystallinity which consequently leads to an increase of rate in degradation of PLA. As the degradation rate increases, the mechanical properties will also reduce.

Sailema-Palate et al. [12] have investigated the degradation of poly(e-caprolactone) (PCL) in different extreme pH levels; pH 1 and pH 13. The results revealed that the water absorption in pH 1 is more dominant and increased hydrolytic rate which leads to intensive mass loss and porosity. Theoretically, as the amount of porosity increases, the mechanical properties will also deteriorate correspondingly. Similar finding was reported by Rajesh et al. [13] which observed a reduction of water absorption for PLA/Jute composite that has undergo alkaline treatment. As known, water absorption in biocomposites is considered as serious problem to be taken into consideration. The water absorption experience by the fibers inside the composite promote unusual swelling and dimensional instability that leads to the decreasing in mechanical strength due to degradation of interface of fibers and matrix.

In a study done by Rowe et al., [14] it was reported that the weight loss is inversely proportional to the initial solution pH which indicated that the degradation rate was lesser in alkaline environment as compared to acidic environment. However, a contradicting result have been debunked by Scaffaro et al. [15] which reported faster degradation kinetic on PLA/graphene nanoplatelets composite and PLA/lignocellulosic composite which are ageing in buffer solution with pH 10.

Although numerous studies on degradation of PLA had been published, very limited work on mechanical properties degradation of RPLA is being explored. With such versatile applications, huge usage of PLA has contributed to the high waste which mainly came from the worn parts. The recycling of PLA waste has been seen as a permissible solution to reduce the quantity of PLA waste in the landfills and those indiscriminately discarded into the environment. However, one of the most important conditioning factors when substituting a RPLA for a VPLA is whether the original characteristics of VPLA are still being preserved. This study aims to investigate the degradation in mechanical properties of VPLA and RPLA in aqueous environment; river water (RW) and seawater (SW). The degradation in mechanical properties is evaluated by tensile stress, transverse rupture strength (TRS), impact energy and hardness. Both samples; VPLA and RPLA are immersed in river water and seawater for a total duration of 150 days. The fracture morphology was also observed using scanning electron microscope.

2. Methodology

2.1. Material
The VPLA pallets were purchased from a local supplier in an injection moulding grade. The pallets were in cylindrical shape, with an average length and diameter of 3 mm. The RPLA were retrieved from the injection moulding waste. The waste was mostly collected from runners and crushed using a grinding machine. The recycled grounded material was in an angular shape with the average size less than 4 mm.

2.2. Sample preparation
Samples for tensile test, transverse rupture test (TRS), impact test and hardness were manufactured using the reciprocation injection moulding machine. The injection process was done continuously for 150 cycles. 50 samples were randomly selected for each type of mechanical testing. The process was repeated to prepare for the recycled PLA samples.

2.3. Ageing condition
Sample ageing was performed by immersing the samples in river water (RW) and seawater (SW). The river water and seawater were collected from nearby location with a coordinate of latitude 5.4°N and longitude 100.4° E. Both types of water were then poured in containers and placed in the laboratory. The temperature and pH of the baths are maintained and monitored weekly. The temperature and pH for river water are 27◦C ± 0.5 and 6 respectively. Meanwhile, the temperature and pH for seawater are being kept at 27◦C ± 0.5 and 7.5. The samples were aged continuously for 150 days.

2.4. Mechanical properties testing
The tensile strength was obtained using Universal Testing Machine based on ASTM D638 at room temperature. The gauge distance was 101.5 mm with a moving speed of 1 mm min^{-1}. A sampling rate was 20 pts s^{-1} and full-scale load range of 50 kN.

The TRS that was obtained using Universal Testing Machine based on ASTM D790 at room temperature. The span was set at 60 mm with a crosshead moving speed of 1 mm min^{-1}. A sampling rate was 20 pts s^{-1} and full-scale load range of 50 kN were recorded.The impact data was gained through Izod/Charpy Impact Tester according to the ASTM D256 set at room temperature. The hammer impact was fixed at 5.42 J. The mass of the pendulum is 0.906 kg.

The Rockwell hardness is determined using Rockwell Tester in R scale. The samples were placed on a steel anvil and brought into contact with indenting ball of a dimeter size 12.7 mm. A maximum of 60 kgf is loaded onto the sample for 15 seconds. After 15 seconds, the load is then removed and the indented surface is measured.

2.5. Fracture morphology
The fracture morphology of the VPLA and RPLA before and after ageing in river water and seawater environment were observed using scanning electron microscopy (SEM) with a constant acceleration voltage of 15 kV.

3. Results and Discussion

3.1. Tensile strength
The results of tensile test on VPLA and RPLA before and after ageing in river water and seawater are presented in Figure 1. The results depict an obvious drop in tensile strength after 150 days immersed in river water and seawater. For VPLA, the tensile strength has dropped 43.7% in river water whereas in

seawater the tensile strength has dropped approximately 31.9%. On the other hand, for RPLA, it can be observed that the tensile strength had drastically dropped in both aqueous environments, which is 75.8% in river water and 49.5% in seawater. As known, in aqueous environment, hydrolysis degradation will take place. The hydrolytic degradation process of polyester is based on a hydrolytic cleavage of ester bonds attached on the backbone of the polymer chains. When the water molecules attack the ester bonds in the polymer chains, the average length of degraded chains becomes shorter [12]. Consequently, reducing the ability of the polymer to resist the applied load and force is also decreased. Thus, reducing the tensile strength.

In other note, the findings indicated that the degradation in tensile stress is more accelerated in river water rather than in seawater. This could be due the difference in pH for both environments. As reported by Rowe et al. [14], the degradation rate is lesser in alkaline environment as compared to acidic environment. Since the seawater is more towards alkaline, the ability of water absorption is lower and thus, reducing the hydrolytic degradation of PLA.

The results also revealed that the tensile strength decreased on RPLA is higher when compared to VPLA. This behaviour can be influenced due to the thermal degradation of the recycling PLA during injection of moulding processing which generates shorter polymer chains. These shorter chains have enhanced mobility, faciliting the relaxation and swelling processes and thus increases the water absorption [10].

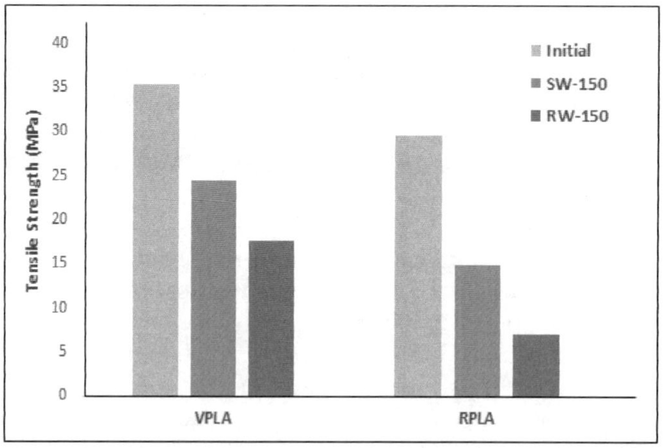

Figure 1. Comparison of tensile strength for VPLA and RPLA ageing in river water (RW) and seawater (SW).

3.2. Transverse rupture strength (TRS)

Figure 2 demonstrated the comparison of TRS for VPLA and RPLA after being immersed in river water and seawater for a duration of 150 days. As shown, the TRS value had decreased in both aqueous environments. However, the reduction in TRS in river water was more accelerated. Similar findings were observed for VPLA and RPLA. After being exposed to river water, the TRS for VPLA and RPLA has dropped to 43.7% and 45.4% respectively. On the other hand, for the immersion in seawater, the TRS for VPLA and RPLA had dropped 31.9% and 31.3%. Similarly, the results showed higher degradation rate in river water when compared to seawater. The findings are in agreement with previous work done by Deroine et al. [8] which reported higher degradation rate of PLA in distilled water as compared to seawater. The presence of high mineral salt in seawater could hinder the water absorption, consequently delaying the degradation process. When comparison between VPLA and RPLA are made, it is noticeable that decreasing in TRS in both environments were almost similar. Although RPLA was believed to have undergo excessive molecular chain scission from reprocessing and hydrolytic degradation which had led to the shorter molecular chain, but no significant differences were observed on TRS values.

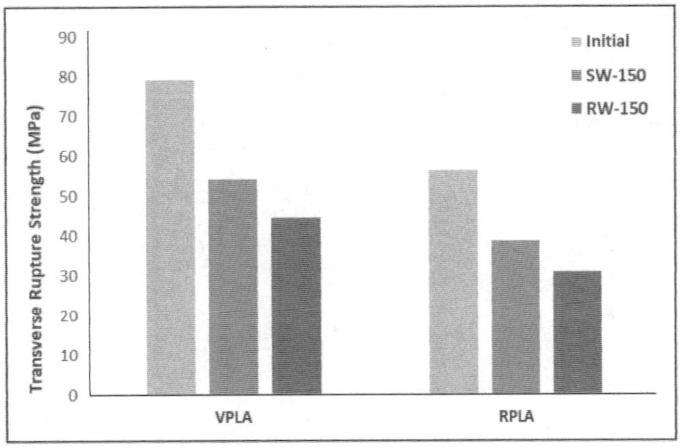

Figure 2. Comparison of TRS for VPLA and RPLA ageing in river water (RW) and seawater (SW).

3.3. Impact energy

Figure 3 is a portrayal of comparison in the change in terms of impact energy for VPLA and RPLA after being immersed in river water and seawater for a total duration of 150 days. As illustrated in Figure 3, it can be obviously seen that the impact energy had reduced drastically. Meanwhile, in river water, the impact energy had reduced to 78.9% for VPLA and 86.7% for RPLA. Similarly, in seawater, the impact energy for VPLA and RPLA had reduced to 80% and 71.9% respectively. When PLA was in contact with water, the water absorption would lead to structural change including swelling and increases the mobility of the polymer chains. As the creation of free volume due to swelling allows more water molecules to diffuse in and get absorbed in the PLA matrix. This accelerated the hydrolysis rate and deterioration of the molecular chains [9]. Consequently, resulting in the stiffness of the material [16]. The outcome had revealed that no significant differences in impact energy reduction after ageing for 150 days in river water and seawater.

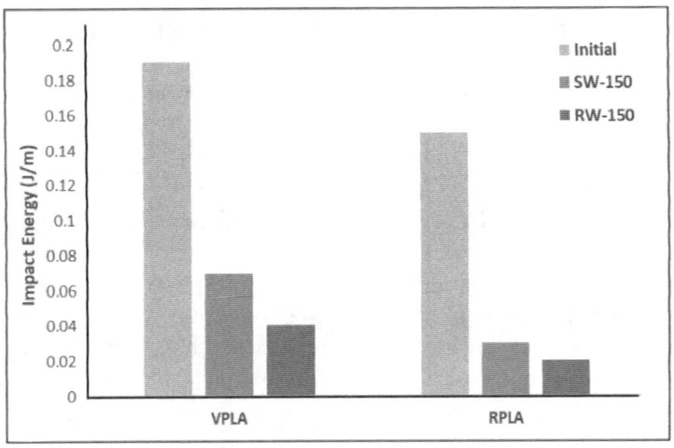

Figure 3. Comparison of Impact Energy for VPLA and RPLA ageing in river water (RW) and seawater (SW).

3.4. Hardness

The effect of ageing in aqueous environment on the hardness of VPLA and RPLA is revealed in Figure 4. As shown in the Figure, the mechanical properties degradation in term of hardness in both environment, river water and seawater were minimum for both VPLA and RPLA. After ageing in river water, the hardness drop was 7.5% for VPLA and 20% for RPLA. Meanwhile in seawater, the reduction in hardness properties for VPLA and RPLA were 7.6% and 12.8% respectively. The decreasing in hardness properties could be explain and correlated with tensile strength. The decreasing in tensile strength would reduce the ability of the polymer to withstand an applied load without failing and consequently reducing the ability to resist the deformation. However, the contradicting outcomes have been reported by Beltran et al. [10] which observed slight increment in hardness, approximately after ageing in aqueous solution with pH 7.4 at temperature 58°C. The researchers have reported the increment in of 5% for VPLA and 16.5% for RPLA.

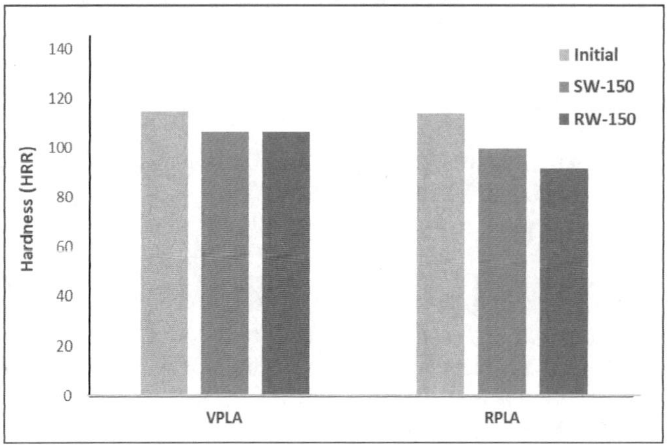

Figure 4. Comparison of hardness for VPLA and RPLA ageing in river water (RW) and seawater (SW).

3.5. Fracture morphology

The fracture morphology of VPLA and RPLA before and after ageing in river water and seawater were shown in Figure 5. At initial stage, the fractured area was rough and homogenous which showed signs of brittle fractures (Fig. 5a and 5b). After 150 days of being immersed in seawater, no significant differences were observed on the SEM images as illustrated in Figure 5c and Figure 5b. No microscopic abnormalities such as cracks, voids and pores had appeared on the area. On other note, no physical change and/or mass loss were noticed through visual observation on the physical samples which proved that no surface erosion had taken place and the mechanical properties degradation were believed due to the bulk degradation [6].

In contrast, after ageing for 150 days in the river water, the image of fracture area had revealed changes in fracture texture. The gullies were starting to be detected. The existence of the gullies became more obvious on RPLA as shown in Figure 5f. However, no physical change and/or mass loss were seen. This observation is correlated with drastic dropped in tensile strength which exhibited 75.8% of decrease. As previously reported, these gullies will act as stress concentration and core of crack growth which finally will lead to relatively low retention ability in mechanical properties [17].

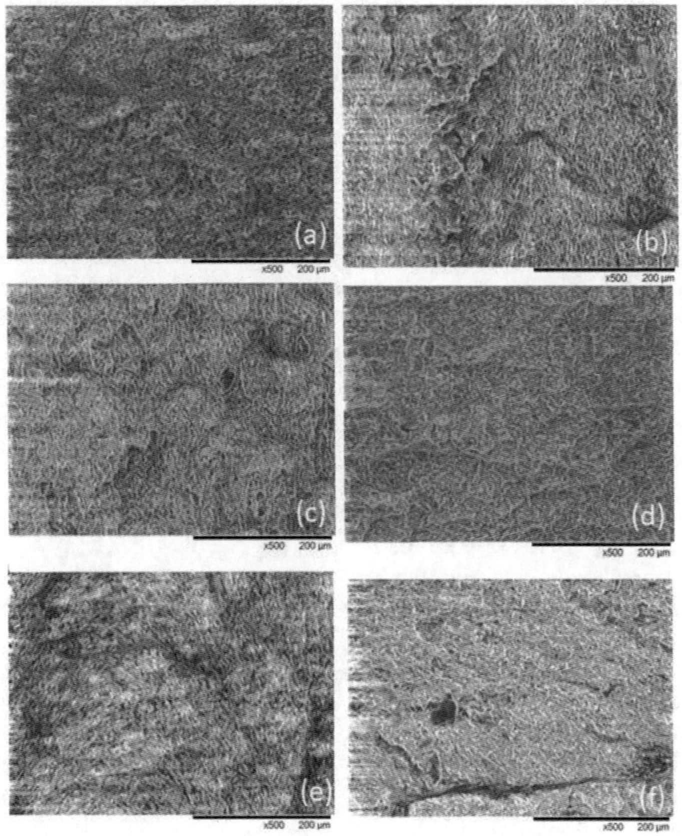

Figure 5. Fracture morphology. a) VPLA – 0 day b) RPLA – 0 day c) VPLA- 150 days (SW) d) RPLA – 150 days (SW) e) VPLA – 150 days (RW) f) RPLA – 150 days (RW).

4. Conclusion

This paper had focus on the mechanical properties degradation of virgin and recycled PLA set in two different aqueous environment; river water and seawater. The samples were continuously immersed in the environment for a total duration of 150 days. Based on the results obtained, it can be concluded that the monitoring of tensile strength has showed in both samples, VPLA and RPLA exhibited decreasing pattern in tensile strength after being immersed in river water and seawater. The tensile strength degradation is higher in river water compared to seawater. It is also noted that the RPLA underwent higher tensile strength degradation as compared to VPLA in both degradation environment. The analysis in TRS has pointed to similar findings. The TRS degradation is more accelerated in river water than in seawater. The RPLA was noted to experience a higher TRS degradation when compared with VPLA after going through 150 days ageing in both environment. The decreasing of TRS in river water for VPLA and RPLA are 43.7% and 45.4% respectively. Whereas for seawater ageing, the TRS for VPLA and RPLA had decrease to 31.9% and 31.3% correspondingly. Also, the evolution of the impact energy showed that the impact energy decreased as the samples underwent ageing in river water and seawater environment. After a duration of 150 days of ageing in river water and seawater environment, the impact energy for VPLA and RPLA showed decay more than three third of the initial impact energy. The impact energy degradation for VPLA and RPLA ageing under river water were 78.9% and 86.7% respectively. Wheares in seawater, the impact energy had reduced 80% for VPLA and 71.9% for RPLA. The study towards the hardness degradation after a duration of 150 days ageing in river water and seawater showed slight decreases for both VPLA and RPLA. After a duration of 150 days ageing in seawater, no alteration on the surfaces and fracture morphology were observed on the samples for both virgin and recycled PLA. However, gullies were starting to be seen on RPLA samples which were ageing in river water.

Acknowledgement
This research is supported by Faculty of Mechanical Engineering, Universiti Teknologi MARA, Cawangan Pulau Pinang.

References

[1] Coyle R, Hardiman G and O'Driscoll K 2020 *Case Studies in Chemical and Environmental Engineering* 100010
[2] Guillermo A and Noa S., Potential 2020 *Environmental Pollution* 115364
[3] Hamad K, Kaseem M, Ayyoob M, Joo J and Deri F 2018 *Progress in Polymer Science* **85** 83
[4] Karamanlioglu M, Preziosi R and Robson G D 2017 *Polymer Degradation and Stability* **137** 122
[5] Chen X and Yan N 2020 *Materials Today Sustainability* **7-8** 10003
[6] Elsawy M A, Kim K H, Park J W and Deep A 2017 *Renewable and Sustainable Reviews* **79** 1346
[7] Farah S, Anderson D G and Langer R 2016 *Advanced Drug Delivery Reviews* **107** 367
[8] Deroine M, Le Duigou A, Corre Y M, Le Gac P Y, Davies P, Cesay G and Bruzaud S 2014 *Polymer Degradation and Stability* **108** 319
[9] Iniguez-Franco F, Auras R, Burgess G, Holmes D, Fang X, Rubino M and Soto-Valdez H 2016 *Polymer* **99** 315
[10] Beltran F R, Lorenzo V, de la Orden M U and Martinez-Urreaga J 2016 *Polymer Degradation and Stability* **133** 339
[11] Rapacz-Kmita A, Stodolak-Zych E, Szaraniec B, Gajek M and Dudek P 2015 *Materials Letters* **145** 73
[12] Sailema-Palate G P, Vidaurre A, Campillo-Fernandez A J and Castilla-Cortazar I 2016 *Polymer Degradation and Stability* **130** 118
[13] Rajesh G, Ratna Prasad A V and Gupta A V S S K S *Material Today: Proceedings* **18** 1
[14] Rowe M D, Eyiler E and Walters K B 2016 *Polymer Testing* **52** 193
[15] Scaffaro R, Maio A, Gulino A F and Pitarresi G 2020 *Composites Science and Technology* **190** 108008
[16] Budin S, Maideen N C, Koay M H, Ibrahim D and Yusoff H *2019 Journal of Physics: Conference Series* **1349** 012002
[17] Lv s, Zhang Y, Gu J and Tan H 2017 *Colloids and Surfaces B: Biointerfaces* **159** 800

Densification and mechanical behaviour of zirconia containing flyash as dopant

R Singh[1,2*], S R Ya'akub[1], A D A Hamid[1], Z Ibrahim[1], and M Y Ali[1]

[1] Mechanical Engineering Programme Area, Faculty of Engineering, Universiti Teknologi Brunei, Tungku Highway, Gadong BE1410, Brunei Darussalam

[2] University of Malaya, 50603 Kuala Lumpur, Malaysia

* Email: ramesh.singh@utb.edu.bn

Abstract. The densification and mechanical behaviour of tetragonal zirconia doped with up to 2 wt% flyash as dopant were studied. In this work, the dopant was blended with commercial 3 mol% yttria-stabilised zirconia powders by attrition milling. Sintering was accomplished by firing the green bodies at various temperatures ranging from 1250 °C to 1500 °C under atmospheric conditions. Sintered bodies were evaluated for bulk density, phase retention, elastic modulus, Vickers hardness and fracture toughness. An improvement in the zirconia properties in terms of bulk density, Vickers hardness and matrix stiffness were evident at low sintering temperatures for samples containing up to 0.5 wt% flyash. In contrast, the fracture toughness and tetragonal phase stability of the zirconia was not influenced by the dopant and sintering temperature. The addition of 1 wt% and 2 wt% flyash, however were not effective in enhancing the properties of the sintered zirconia. This study also found that the addition of flyash as dopant was beneficial in suppressing the low-temperature degradation kinetics of tetragonal zirconia in steam environment.

Keywords: flyash-doped zirconia, sintering behaviour, mechanical properties, low-temperature degradation

1. Introduction

Zirconia-based ceramics, in particular when stabilized with 2.5 to 3 mol% yttria, results in the retention of a fully tetragonal structure that exhibits the highest mechanical strength and fracture toughness, suitable for many industrial applications [1-3]. This improvement in the mechanical behaviour can be attributed to a self-healing mechanism subsequently known as transformation toughening. According to this mechanism, the stress induced by a propagating crack will be absorbed by the tetragonal grains and transformed to the monoclinic symmetry with the accompanying volume expansion which occurs ahead of a propagating crack front [2-5]. The efficacy of the transformation toughening, however is dependent on several factors including yttria content and distribution, tetragonal grain size and grain boundary modification due to impurity present during powder manufacturing or added as a sintering additives to lower the densification temperature [6-10].

One of the limitations of tetragonal zirconia is the undesirable phase transformation resulting in formation of micro-cracks and property deterioration when the ceramic is exposed in steam environment, a phenomenon known as hydrothermal ageing or low-temperature degradation (LTD) [8, 11-13]. Many researchers have investigated this LTD phenomenon plaguing tetragonal zirconia and the general understanding is that the severity of LTD is predominantly governed by the tetragonal grain size after sintering [14, 15] and so there is a need to limit grain coarsening during powder sintering.

There are many methods that have been proposed to control grain growth but the most cost-effective and yet simple method has been to lower the densification temperature through the use of low-melting dopants or sintering additives. There were many dopants that have been explored by many researchers including that of MnO_2, CuO, TiO_2, SiO_2, Al_2O_3, GeO_2, CeO, etc. [16-18]. For example, Ramesh et al. [13, 14, 19, 20] have reported that the incorporation of low amounts of CuO and MnO_2 were effective in aiding low-temperature sintering, enhancing mechanical properties and suppressing LTD in tetragonal zirconia.

In the present work, the addition of up to 2 wt% flyash as dopant in a commercial 3 mol% yttria-stabilised zirconia was investigated.

2. Methods and Materials

A commercially available 3 mol% yttria-stabilised zirconia manufactured by Kyoritsu, Japan was used in the present work. The flyash dopant, varying up to 2 wt%, obtained from a local power station was refined by ball milling in ethanol for 2 h. The wet slurry was dried in an oven at 60 °C for 24 h and finally sieved to obtain fine powder. The chemical analysis by x-ray florescence performed on the as-refined flyash indicated that the powder comprised 60% silica, 15% alumina and 12% iron oxide as the major impurities. The balance 13% represents some minor impurities such as calcium and magnesium oxides.

In a typical doping, the flyash-doped zirconia (FZ) powders was prepared by blending both powders via attrition milling using ethanol as the mixing medium and zirconia beads (2 mm in diameter) as the milling medium. The rotation speed was kept constant at about 500 rpm. After the milling, the wet slurry was filtered, dried in an oven at 60 °C for 24 h and sieved to obtain ready to press FZ powder. Two different shape samples were prepared, i.e. rectangular bar for elastic modulus measurement and disc (20 mm diameter) for other tests, by uniaxial compaction at 10 kN to form green samples. Prior to sintering, the samples were subjected to cold isostatic pressing at 200 MPa. Pressureless sintering in air was accomplished using a box furnace. The samples were sintered over the range of 1250 °C to 1500 °C and a holding time of 2 h was used.

The phases present in the powders and sintered samples were determined by x-ray diffraction (XRD) performed under normal atmospheric conditions using Cu-Kα as the radiation source. The tetragonal and monoclinic phases were determined using the method proposed by Toraya et al. [21]. The sintered density of the samples was obtained by water immersion method based on Archimedes' principle and the relative density was calculated by taking the theoretical density of tetragonal zirconia as 6.1 g/cm^3. The Vickers indentation method was used to determine the hardness and fracture toughness of sintered bodies. An indentation load of 10 kg and a loading time of 10 s were employed for all samples. The Niihara et al. [22] equation was used to calculate the fracture toughness of the sample. The sonic resonance technique was used to determine the elastic modulus of the sintered bar samples in accordance to ASTM standard [23]. Finally, the LTD resistance of the FZ samples were evaluated by measuring the monoclinic phase present after exposure in superheated steam at 180°C/10 bar for periods ranging up to 50 h.

3. Results and discussion

The XRD analysis of the zirconia powders indicated that the monoclinic phase content did not changed significantly with the introduction of flyash as dopant. In all cases, the monoclinic phase content varied between 17% (undoped) to about 19% (FZ powders). Similarly, the XRD patterns of all sintered samples indicated the retention of a fully tetragonal structure regardless of dopant amounts and sintering temperatures.

The effect of sintering temperatures on the relative density of the zirconia are shown in Figure 1. In general, a similar density trend was observed for all samples i.e. the density increases rapidly with increasing sintering temperature up to 1350 °C for the FZ samples and 1400 °C for the undoped sample before remaining almost constant (for FZ up to 0.5 wt% and undoped samples) for sintering temperature up to 1500 °C. In contrast, the relative density of the 1 wt% and 2 wt% FZ started to declined when sintered above 1350 °C. Nevertheless, Figure 1 clearly shows the beneficial effect of flyash (up to 0.5 wt%) in promoting densification at low temperatures of 1250 °C, 1300 °C and 1350 °C. For instance, the relative density of the ceramic sintered at 1250 °C improved by about 5% with the additions up to 0.3 wt% flyash. The dopant somehow had negligible effect on the densification for samples sintered above 1350 °C. All the samples with the exception of the 1 wt% and 2 wt% FZ exhibited a relative density of > 99% when sintered > 1350 °C.

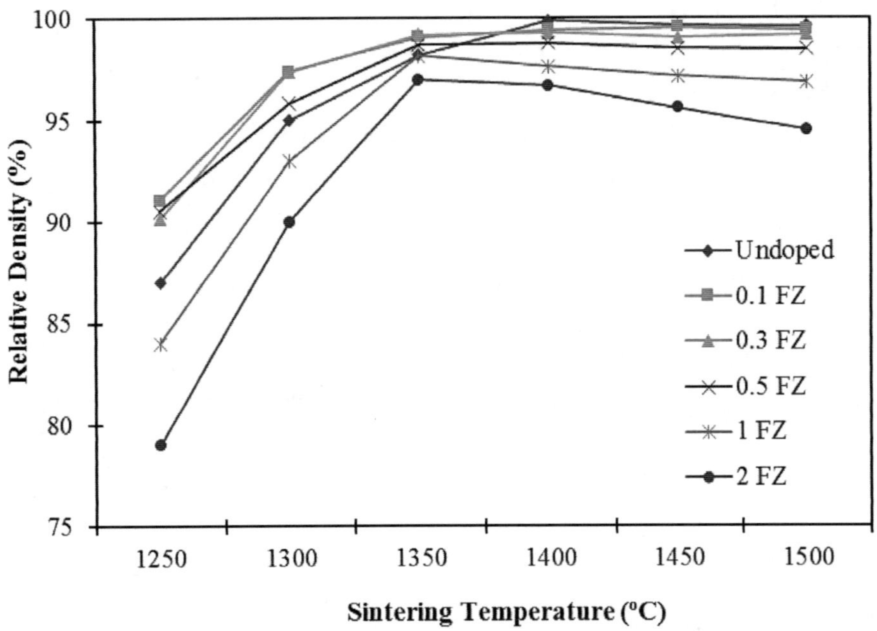

Figure 1. Effect of sintering temperature on the relative density of zirconias.

The variation in elastic modulus for the undoped and FZ samples with sintering temperature as shown in Figure 2 is in good agreement with the densification trend. Typically, all samples exhibited modulus values above 200 GPa when sintered at and above 1350 °C as depicted in Figure 2. The lower elastic modulus exhibited by the 1 wt% and 2 wt% FZ samples correlated well with the lower relative density of the samples.

In fact, the analysis between the relative density and the elastic modulus of the sintered samples revealed that a linear relationship exists between both properties as shown in Figure 3. It was found that the matrix stiffness of the sintered body increased linearly up to a maximum of about 200 GPa with increasing relative density up to about 98-99% of theoretical density, regardless of dopant amounts. A similar behaviour was also reported for hydroxyapatite bioceramic where the researchers noted that the stiffness of the sintered ceramics was influenced by the bulk density and not significantly affected by grain coarsening [24, 25].

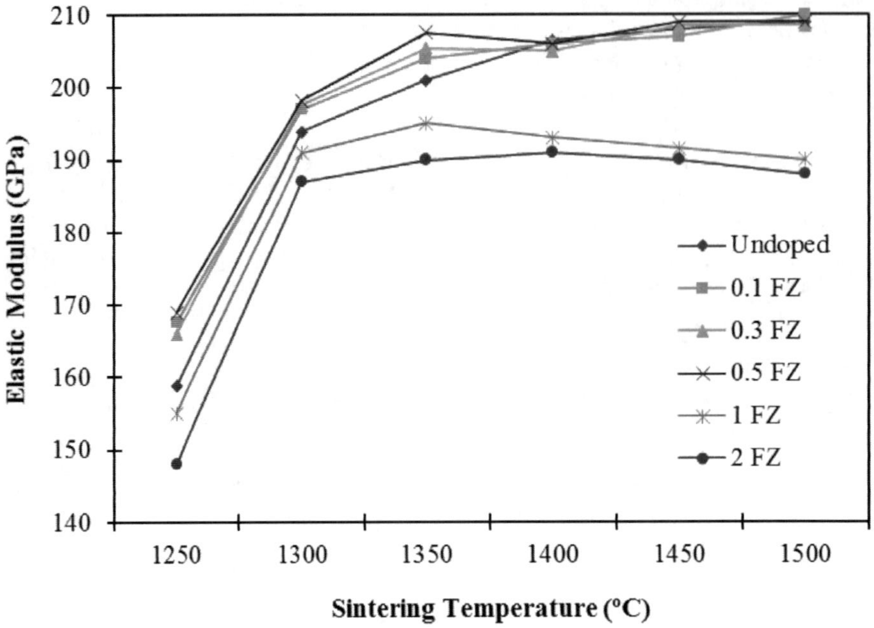

Figure 2. Effect of sintering temperature on the elastic modulus of zirconias.

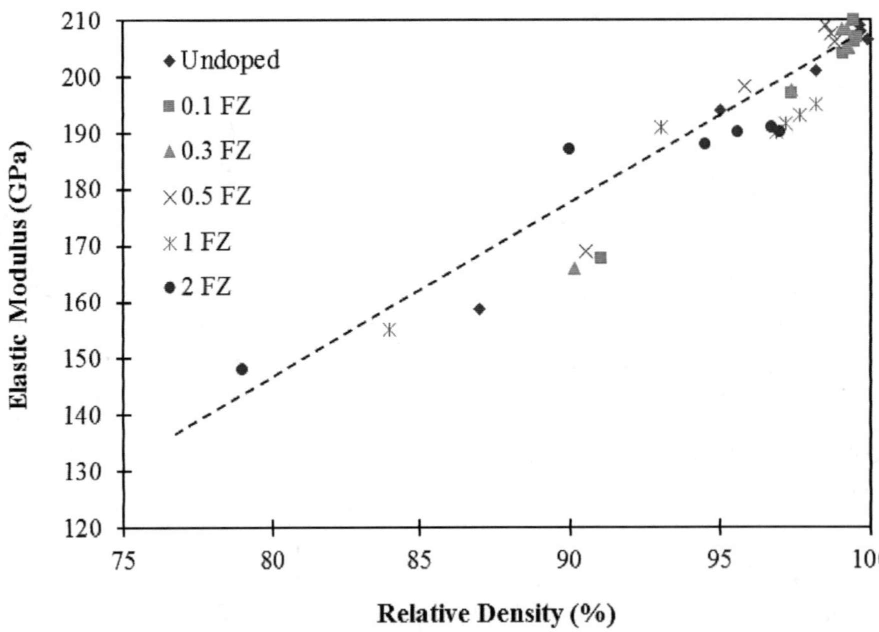

Figure 3. A linear relationship observed between relative density and elastic modulus of tetragonal zirconia.

The effect of dopant addition and sintering temperature on the Vickers hardness and fracture toughness of zirconia are shown in Figure 4. It has been found that regardless of dopant concentration, the sintering temperature has almost negligible effect on the fracture toughness of the tetragonal zirconia. However, a slightly lower toughness was noted at all sintering temperatures for the 2 wt% flyash-doped samples. On the contrary, the Vickers hardness was found to vary significantly with increasing temperatures. As can be noted from Figure 4, the hardness increases rapidly from about 9-9.8 GPa at 1250°C and reached a maximum of about 13.5 GPa at 1350 °C before decreasing slightly with increasing temperature to 1500 °C.

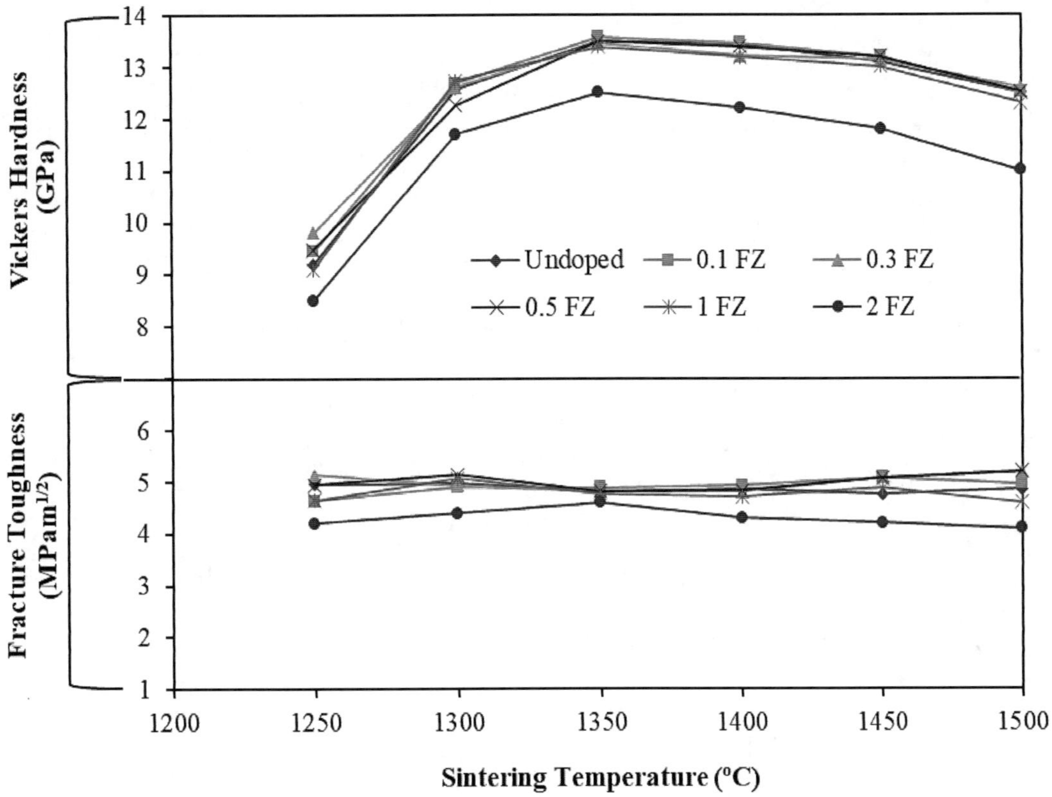

Figure 4. The effect of sintering temperature and flyash doping on the Vickers hardness and fracture toughness of zirconia.

The effect of superheated steam on the tetragonal phase stability of the zirconia as a function of flyash addition for samples sintered at 1350°C is shown in Figure 5. The beneficial effect of flyash in slowing the LTD kinetics in tetragonal zirconia has been revealed, particularly for the 0.5-2 wt% flyash doping. The results showed that the undoped zirconia undergo rapid phase transformation; attained about 65% and 90% monoclinic content after exposure in steam for 1 h and 3 h, respectively. On the other hand, the improved in LTD resistance for the 1 wt% and 2 wt% flyash-doped samples could be attributed to the lower matrix stiffness and lower relative density of the sintered body. It is unclear at this stage of this behaviour and more work is required to elucidate the actual role of the flyash in suppressing the LTD in zirconia.

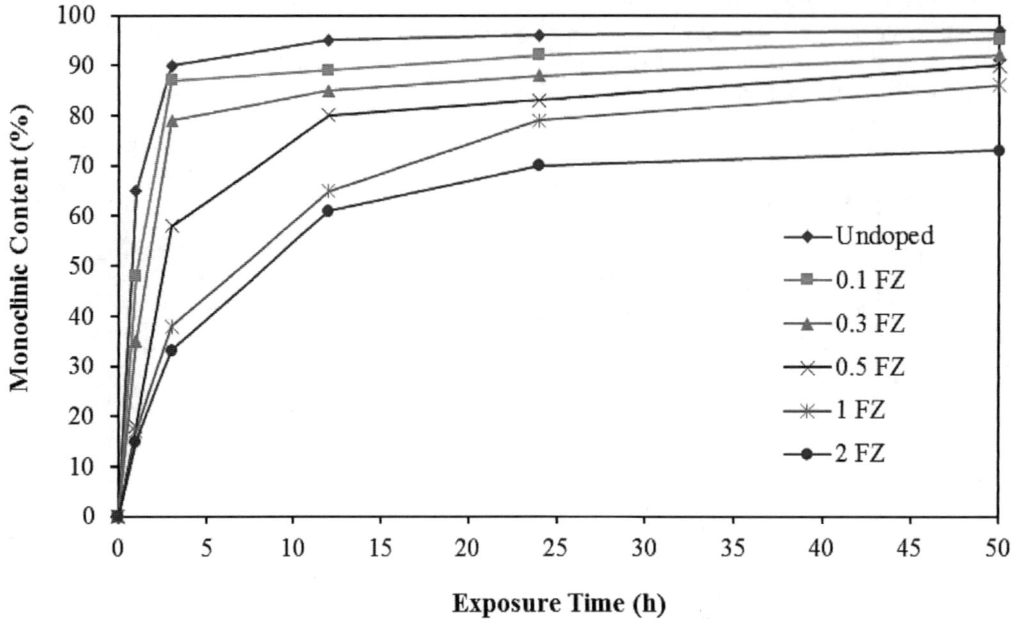

Figure 5. Monoclinic phase development with ageing time for zirconia exposed in steam environment for samples sintered at 1350 °C.

4. Conclusions

In the present work, the sintering and mechanical behaviour of flyash-doped yttria-stabilised zirconia was investigated. The results showed the addition of flyash up to 0.5 wt% was beneficial in aiding low-temperature sintering (1250-1350 °C), resulting in enhanced relative density, Vickers hardness and elastic modulus. The fracture toughness, however was not influenced by the sintering temperatures and dopant addition. The addition of higher content of flyash i.e. 1 wt% and 2 wt% was not effective in aiding sintering of zirconia although these samples exhibited improved LTD resistance when compared to other samples when exposed to superheated steam.

References

[1] Garvie R C, Hannink R H and Pascoe R T 1975 Ceramic steel? *Nature* **258** 703.

[2] Birkby I and Hodgson H 1991 Progress with zirconia ceramics *In 3rd European Symposium on Engineering Ceramics* Springer Dordrecht 167-198.

[3] Xue M, Liu S, Wang X and Jiang K 2020 High fracture toughness of 3Y-TZP ceramic over a wide sintering range *Mater Chem Phys* **244** 122693.

[4] Wang J, Rainforth M and Stevens R 1989 The grain size dependence of the mechanical properties in TZP ceramics *Brit Ceram Trans J* **88** 1-6.

[5] Theunissen G S A M, Bouma J S, Winnubst A J A and Burggraaf A J 1992 Mechanical properties of ultra-fine grained zirconia ceramics *J Mater Sci* **27** 4429-4438.

[6] Trunec M 2008 Effect of grain size on mechanical properties of 3Y-TZP ceramics *Ceram Silikaty* **52** 165-171.

[7] Ting C H, Ramesh S, Lwin N and Sutharsini U 2016 Sintering properties and low-temperature degradation behaviour of Y-TZP ceramics *J Ceram Proc Res* **17** 1265-1269.

[8] Sutharsini U, Thanihaichelvan M, Ting C H, Ramesh S, Tan C Y, Chandran H, Ahmed A D Sarhan and Urriés I 2017 Effect of two-step sintering on the hydrothermal ageing resistance of tetragonal zirconia polycrystals *Ceram Int* **43** 7594-7599
[9] Stemmer S, Vleugels J and Van der Biest O 1998 Grain boundary segregation in high-purity yttria-stabilized tetragonal zirconia polycrystals (Y-TZP) *J Eur Ceram Soc* **18** 1565-1570.
[10] Singh R, Gill C, Lawson S and Dransfield G P 1996 Sintering, microstructure and mechanical properties of commercial Y-TZPs *J Mater Sci* **31** 6055-6062.
[11] Wei C, Montagnac G, Reynard B, Le Roux N and Gremillard L 2020 Interplay between internal stresses and matrix stiffness influences hydrothermal ageing behaviour of zirconia-toughened-alumina *Acta Materialia* **185** 55-65.
[12] Kanellopoulos P and Gill C 2002 Hydrothermal ageing of yttria-stabilised zirconia, sintered at 1300° C–1325° C: the effects of copper oxide doping and sintering time variations. *J Mater Sci* **37** 5075-5082.
[13] Ramesh S, Gill C and Lawson S 1999 The effect of copper oxide on sintering, microstructure, mechanical properties and hydrothermal ageing of coated 2.5 Y-TZP ceramics *J Mater Sci* **34** 5457-5467.
[14] Ramesh S, Lee K S and Tan C Y 2018 A review on the hydrothermal ageing behaviour of Y-TZP *Ceram Int* **44** 20620-20634.
[15] Ran S, Vleugels J, Huang S, Vanmeensel K, Blank D H and Winnubst L 2010 Manipulating microstructure and mechanical properties of CuO doped 3Y-TZP nano-ceramics using spark-plasma sintering *J Eur Ceram Soc* **30** 899-904.
[16] Huang X J and Weppner W 1996 Characteristics of transition metal oxide doping of YSZ: structure and electrical properties *J Chem Soc, Faraday Trans* **92** 2173-2178.
[17] Zhang F, Chevalier J, Olagnon C, Batuk M, Hadermann J, Van Meerbeek B and Vleugels J 2017 Grain-boundary engineering for aging and slow-crack-growth resistant zirconia *J Dent Res* **96** 774-779.
[18] Mishra S, Kumar K, Patra A, Chowdhury A and Roy A 2020 Phase integrity of zinc oxide doped zirconia under low compacting pressure *J Alloys and Comp* 155927.
[19] Ramesh S, Meenaloshini S, Tan C Y, Kelvin Chew W J and Teng W D 2008 Effect of manganese oxide on the sintered properties and low temperature degradation of Y-TZP ceramics *Ceram Int* **34** 1603-1608.
[20] Ramesh S, Lee K S, Tan C Y, Wong Y H, Alengaram U J, Tarlochan F, Teng W D, Sutharsini, U and Sarhan A A 2018 Effect of microwave sintering on the properties of copper oxide doped Y-TZP ceramics *Ceram Int* **44** 19639-19645.
[21] Toraya H, Yoshimura M and Somiya S 1984 Calibration curve for quantitative analysis of the monoclinic-tetragonal ZrO_2 system by x-ray diffraction *J Am Ceram Soc* **67** C-119-C-121.
[22] Niihara K, Morena R and Hasselman D P H 1982 Evaluation of K_{Ic} of brittle solids by the indentation method with low crack-to-indent ratios *J Mater Sci Letts* **1** 13-16.
[23] ASTM E1876-97 1998 Standard test method for dynamic Young's modulus, shear modulus and Poisson's ratio by impulse excitation of vibration *Annual Book of ASTM Standards.*
[24] Ramesh S, Peralta C L, Tan C Y and Teng W D 2004 The effect of cold isostatic pressing on the sinterability of synthesized HA *J Biomed Eng: Appl, Basis & Comm* **16** 199-204.
[25] Adebisi A A, Maleque M A, Ali M Y and Bello K A 2016 Effect of variable particle size reinforcement on mechanical and wear properties of 6061Al–SiCp composite *Composite Interfaces* **23** 533-547.

Study of kinetic and thermal decomposition from municipal solid waste pellets using model-fitting methods

S R H Siregar, D Nursani, M I B Setyawan, and A Surjosatyo*

Department of Mechanical Engineering, University of Indonesia, Kampus UI Depok 16424, Indonesia

* Email: adisur@eng.ui.ac.id

Abstract. Reaction rate model could be determined from kinetic characteristics. Kinetic and thermal decomposition of municipal solid waste (MSW) pellets, consist of organic material, plastic, paper and leaves, with tapioca flour as binder (0%, 3% and 6%) were studied in thermogravimetric analyzer from temperature 50 °C to 800 °C at heating rate 10 °C/min in N_2 atmospheres. The kinetic parameter were determined by Model Fitting or Coats Redfern Methods with different kinetic reaction model. Results of this study have shown that MSW pellets in various binder has activation energy values 64.64 kJ/mol and 22.06 kJ/mol for 0%, 68.83 kJ/mol and 21.99 kJ/mol for 3%, and 69.29 kJ/mol and 22.62 kJ/mol for 6% at temperature ranges between 200 °C – 400 °C and 400 °C – 500 °C respectively in typical three dimensional diffusion (D3) mechanism kinetic reaction model.

Keywords: kinetic study, Model-Fitting methods, municipal solid waste, pellet, thermal decomposition

1. Introduction

Population growth and lifestyle changes are inseparable parts of life in the world. However, these developments often have a negative impact on the surrounding environment, one of which is the presence of large amounts of municipal solid waste (MSW) [1–3]. On average, organic waste (food scraps and garden waste) is the largest part of the world's MSW, followed by paper and cardboard, plastics, glass, metal, wood, rubber and leather, and other waste [4]. The composition of MSW depends on each country's current level of development; More developed the country often the lower amount of organic waste [5]. In Indonesia, the average waste generation reaches 0.24 - 0.33 kg / capita / day with a composition that is dominated by organic waste [6–8].

There are many technologies for treating municipal solid waste, such as landfill, recycling, biological treatment, and thermal treatment. The cost of biological treatments is cheaper than the latter, but it is difficult to control bacterial growth and it takes a long time to reduce MSW [5]. That is why thermal treatment is more commonly chosen as a treatment for MSW in developed country. MSW is a very variable and heterogeneous multi-component material [9], so its calorific values are inconsistent. The use of MSW as fuels in thermal treatment especially gasification process results the low energy efficiency, so that to increase the efficiency, uniformity and compaction is required. Therefore, MSW is pelletized to reach uniformity.

Kinetic and thermal decomposition of pelletized MSW are needed for modelling reactors, like gasification and incineration reactors. Reaction rate of pelletized MSW can determine on how devolatization begins and how much contents of devolatization process products are. Currently, MSW can be approached as lignocellulose fuel since organic waste is the highest share of MSW. Usually, researchers use Di Blasi's reaction rates to model a pyrolysis process of lignocellulose fuel [10]. However, the simulation results might be unreliable due to heterogeneity of MSW. That is why this research would find the kinetic of pelletized MSW.

This research would analyze kinetic characterization of pelletized MSW using thermogravimetric analysis (TGA). TGA is a powerful deformation tool commonly used in materials analysis [11]. There are researches that used TGA to determine thermal decomposition mechanism of organic wastes and MSW [12–16]. To perform kinetics analysis, there are many methods that can be used, but the method most often used is the Coats-Redfern method or call model fitting method. This method use an appropriate model of the integral function from various reaction mechanism [17,18]. Although many researches related to the kinetic characteristics of MSW have been conducted, the composition of MSW has unique variations in each region. In this study, which continues previous research [19], we examined the kinetics and thermal decomposition of MSW pellets using the composition of the waste entering the Bantargebang landfill from DKI Jakarta, Indonesia.

2. Material and methods

2.1. Material

The materials used are pellet from municipal solid waste with tapioca as binder. Composition of municipal solid waste simulated the composition of waste generation in Jakarta, which consist of 70% organic waste, 12% plastic, 10% paper and 8% leaves. Tapioca which one of favorite binder for mass production added 0%, 3%, and 12% in pellet production. Before used, tapioca flour diluted with water in a ratio of 1:4 and cooked.

2.2. Characterization of pellets

This study is a continuation of the previous paper[19]. In the previous paper, the physical characteristics of the MSW pellets have been described. Pellets have a diameter range of 7.8-8.2 mm with a length of 23.6-24.0 mm and a weight of 1.16-1.36 grams per unit of pellet. Other characteristics can be seen in **Table 1**.

Table 1. Characteristic of Pellet MSW [19]

Binder Adding	Density (kg/m^3)	Durability (%)	Moisture Content (%)	Hardness (kgF)	Caloric Value (kCal/kg)
0%	490.9	0.704	5.19	12.1	4360.84
3%	435.1	0.532	8.09	10.4	3783.02
6%	383.9	0.686	13.21	10.0	3290.71

2.3. Thermogravimetric Analyzer (TGA).

Pellets are tested in Center for Materials Processing and Failure Analysis (CMPFA) University of Indonesia using Simultaneous Thermal Analyzer (STA) 6000 - Perkin Elmer. Approximately 10-15 mg of sample was put into TGA (STA 6000) and heated from temperatures 50 ^{0}C to 800 ^{0}C at heating rate 10 ^{0}C/min in N_2 atmospheres.

2.4. Coats redfern methods

Kinetics is the study of the dependence of chemical reactions on time and temperature. Material kinetics can be studied by measuring the properties of a material when heated (non-isothermal) or under isothermal conditions [20]. In this process, there is a change in the sample weight which can be written by weight loss rate equation (1).

$$\alpha = \frac{W_0 - W_t}{W_0 - W_f} \tag{1}$$

Where α is the conversion or weight loss rate, W_0 is the initial weight of the sample (mg), W_t is the weight of the sample at a given time (mg) and W_f is the final mass of the sample (mg). In isothermal conditions, the solid fuel decomposition rate is a linear function of the reaction rate constant (k) with the conversion function ($f(\alpha)$) as expressed in Eq. (2). However, in non-isothermal conditions, the solid fuel decomposition rate will be affected by the heating rate ($\frac{dT}{dt}$ or β), so the equation may be define as in Eq (3).

$$\frac{d\alpha}{dt} = kf(\alpha) \tag{2}$$

$$\frac{d\alpha}{dt} = \frac{d\alpha}{dT}\frac{dT}{dt} = \beta\frac{d\alpha}{dT} \tag{3}$$

$$\frac{d\alpha}{dt} = \frac{d\alpha}{dT}\frac{dT}{dt} = \beta\frac{d\alpha}{dT} \tag{3}$$

The reaction rate constant is described in the Arrhenius equation as follows on Eq (4)

$$k = Ae^{\left(\frac{-E}{RT}\right)} \tag{4}$$

Where A is the pre-exponential factor (min^{-1}), E is the activation energy (kJ/mol), R is the unversal gas constant (8.314 $Jmol^{-1}K^{-1}$) and T is the absolute temperature (K). By substitution Arrhenius equation to solid fuel decomposition rate equation under non-isothermal conditions, gives on Eq (5).

$$\frac{d\alpha}{dT} = \frac{A}{\beta}e^{\left(\frac{-E}{RT}\right)}f(\alpha) \tag{5}$$

Equation (5) is known as the general equation of TGA curve [20]. Integration this equation, gives on Eq (6), where $g(\alpha)$ is the integrated reaction model and this is dependent on the reaction mechanism followed by the solid fuel.

$$g(\alpha) = \int_0^\alpha \frac{d\alpha}{f(\alpha)} = \frac{A}{\beta}\int_0^T e^{\left(\frac{-E}{RT}\right)}dT \tag{6}$$

Various reaction mechanism of $g(\alpha)$ are listed in **Table 2**. Coats-Redfern Methods uses asymptotic series for the resolution of Eq (6), gives Eq (7).

$$ln\left(\frac{g(\alpha)}{T^2}\right) = ln\left[\frac{AR}{\beta E}\left(1 - \frac{2RT}{E}\right)\right] - \frac{E}{R}\frac{1}{T} \tag{7}$$

Since asymptotic approximation used to solve integral equations, the number of $\frac{2RT}{E} \ll 1$, so this number is neglected [16,18] and gives the final equation for this method, shown in Eq (8).

$$ln\left(\frac{g(\alpha)}{T^2}\right) = ln\left(\frac{AR}{\beta E}\right) - \frac{E}{R}\frac{1}{T} \quad (8)$$

By linear fitting of $ln\left(\frac{g(\alpha)}{T^2}\right)$ versus $\frac{1}{T}$ for each reaction mechanism, the reaction order can be achieved. Energy activation (E) determined from the slope and pre-exponential factor (A) determined from the intercept of linear fitting equation. The highest coefficient of determination (R^2) among linear fittings represent a possible model of the reaction mechanism that occurs.

Table 2. Expressions of $g(\alpha)$ based on various reaction mechanisms [18,21]

Reaction Mechanism	Symbol	$g(\alpha)$
Nucleation reaction models		
Avrami-Eroféve (n=2)	A2	$[-ln(1-\alpha)]^{1/2}$
Avrami-Eroféve (n=3)	A3	$[-ln(1-\alpha)]^{1/3}$
Avrami-Eroféve (n=4)	A4	$[-ln(1-\alpha)]^{1/4}$
Power law	P2	$\alpha^{1/2}$
Power law	P3	$\alpha^{1/3}$
Power law	P4	$\alpha^{1/4}$
Phase boundary reaction models		
2D, shrinking cylinder	R2	$[1-(1-\alpha)^{1/2}]$
3D, contraction of sphere	R3	$[1-(1-\alpha)^{1/3}]$
Diffusion controlled models		
One-dimensional diffusion	D1	α^2
Two-dimensional diffusion, cylindrical symmetry	D2	$\alpha + [(1-\alpha) \times ln(1-\alpha)]$
Three-dimensional diffusion, spherical symmetry (Jander)	D3	$[1-(1-\alpha)^{1/3}]^2$
Three-dimensional diffusion, cylindrical symmetry	D4	$1 - \frac{2\alpha}{3} - (1-\alpha)^{2/3}$
Chemical reaction models		
First order	F1	$-ln(1-\alpha)$
Second order	F2	$[(1-\alpha)^{-1} - 1]$
Third order	F3	$[(1-\alpha)^{-2} - 1]/2$

3. Results and Discussion

3.1. Thermal decomposion

Figure 1 shows the results of thermo gravimetric (TG) analysis and derivative of thermo gravimetric (DTG) analysis for pellet of municipal solid waste in various binder composition. Based on **Figure 1**, the thermal behaviour between various binders composition have a same trends and range of peak temperature [16].

The TG curve shows three main stage of degradation. First stage in temperature less than 200 °C is evaporation of moisture content and highly volatile materials by 4% - 7% . The second stage in temperature ranges 200 °C - 400 °C is degradation of lignocellulose biomass by 33% - 46%. The third stage in temperature 400 °C - 500 °C is plastic degradation and hemicellulose degradation by 8% - 11%. The residue after 800 °C is ranging from 28% - 50%. The activation energy (E) is presented on Table 3.

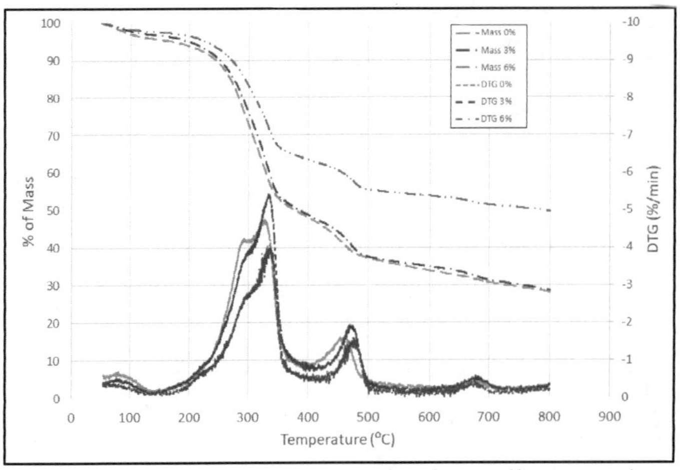

Figure 1. TG and DTG curves of various pellets sampel.

Table 3. Calculation of activation energy (E) and pre-exponential factor (A).

	0%			3%			6%		
	R^2	E	A	R^2	E	A	R^2	E	A
Peak 1 (Temperature 200 °C – 400 °C)									
A2	0.9343	10.67	7.26E-03	0.9516	11.71	9.57E-03	0.9549	11.86	1.01E-02
A3	0.8152	4.03	7.64E-04	0.8765	4.73	1.01E-03	0.8860	4.83	1.06E-03
A4	0.1893	0.71	7.12E-05	0.4563	1.24	1.36E-04	0.4974	1.31	1.47E-04
P2	0.8561	6.65	1.67E-03	0.9039	7.72	2.36E-03	0.9098	7.77	2.40E-03
P3	0.3351	1.35	1.32E-04	0.5807	2.07	2.30E-04	0.5995	2.10	2.35E-04
R2	0.9610	26.33	1.44E-01	0.9712	28.45	2.28E-01	0.9734	28.63	2.41E-01
R3	0.9637	27.70	1.42E-01	0.9728	29.80	2.22E-01	0.9748	30.02	2.37E-01
D1	0.9658	54.33	6.69E+01	0.9747	58.60	1.57E+02	0.9765	58.78	1.66E+02
D2	0.9701	59.11	1.20E+02	0.9777	63.35	2.76E+02	0.9793	63.65	3.02E+02
D3	0.9736	64.64	1.15E+02	0.9797	68.83	2.58E+02	0.9812	69.29	2.93E+02
D4	0.9715	60.94	4.34E+01	0.9785	65.17	9.90E+01	0.9802	65.52	1.09E+02
F1	0.9676	30.58	9.61E-01	0.9746	32.66	1.48E+00	0.9761	32.97	1.61E+00
F2	0.9696	40.47	1.46E+01	0.9715	42.41	2.11E+01	0.9714	43.05	2.49E+01
F3	0.9632	52.11	3.27E+02	0.9611	53.85	4.38E+02	0.9595	54.90	5.74E+02
Peak 2 (Temperature 400 °C – 500 °C)									
R2	0.7772	2.68	1.54E-04	0.5245	2.69	1.52E-04	0.4305	2.90	1.72E-04
R3	0.8959	5.07	3.28E-04	0.7184	5.04	3.18E-04	0.6312	5.36	3.61E-04
D1	0.9260	6.00	6.94E-04	0.8399	6.14	7.12E-04	0.7545	6.15	7.23E-04
D2	0.9650	12.28	3.06E-03	0.9020	12.37	3.03E-03	0.8496	12.56	3.24E-03
D3	0.9749	22.06	1.03E-02	0.9193	21.99	9.73E-03	0.8789	22.62	1.14E-02
D4	0.9703	15.44	1.71E-03	0.9105	15.48	1.66E-03	0.8639	15.81	1.83E-03
F1	0.9484	10.69	6.92E-03	0.8394	10.57	6.51E-03	0.7778	11.16	7.73E-03
F2	0.9684	34.18	2.90E+00	0.9008	33.57	2.44E+00	0.8608	35.51	3.70E+00
F3	0.9709	65.34	3.22E+03	0.9117	64.07	2.32E+03	0.8763	67.85	4.93E+03

3.2. Activation Energy

Activation energy is minimum amount of energy that is required to activate atoms or molecules to a condition in which they can undergo chemical transformation or physical transport. Table 3 shows the calculation of activation energy in various reaction mechanism for peak 1 or temperature range 200 °C - 400 °C and for peak 2 or temperature range 400 °C - 500 °C.

Based on Table 3, variations of binder composition do not affect the value of the activation energy. The activation energy value of pellets is in the range 64.64 kJ/mol - 69.29 kJ/mol for peak 1 and the range 21.99 kJ/mol - 22.62 kJ/mol for peak 2. Pre-exponential factor has a value of 1.15×10^{2} min^{-1} - 2.92×10^{2} min^{-1} and 9.73×10^{-3} min^{-1} - 1.03×10^{-2} min^{-1}, respectively.

3.3. Kinetic reaction model.

There are four major groups of reaction mechanisms calculated in this study, namely nucleation reaction models (A2, A3, A4, P2, P3, and P4), phase boundary reaction models (R2 and R3), diffusion controlled models (D1, D2, D3, and D4) and chemical reaction models (F1, F2, and F3). In this calculation the reaction mechanism which gives a negative activation energy value will be ignored [22]. The results show that the reaction mechanism that comes closest to the process that occurs is the Three-dimensional diffusion, spherical symmetry (Jander) mechanism (D3).

4. Conclusion

In this study, kinetics and thermal decomposition of municipal solid waste pellet were investigated using thernogravimetric analyzer. The results shown that thermal decomposition were divided in three stage, such as evaporation of moisture content, degradation of lignocellulose, and plastic degradation and hemicellulose degradation. There is no significant different due to addition of binder in thermal decomposition. In parallel results the activation energy values in various binder has 64.63 kJ/mol and 22.05 kJ/mol for 0 %, 68.83 kJ/mol and 21.99 kJ/mol for 3 %, and 69.29 kJ/mol and 22.62 kJ/mol for 6 % at temperature ranges between 200 °C – 400 °C and 400 °C – 500 °C respectively in typical three dimensional diffusion (D3) mechanism kinetic reaction model.

Acknowledgment

This work is supported by Hibah PITTA B 2019 funded by DRPM Universitas Indonesia No NKB-0692/UN2.R3.1/HKP.05.00/2019.

References

[1] Deng Y, Gong B, Chao Y, Dong T, Yang W, Hong M, Shi X, Wang G, Jin Y and Chen Z G 2018 Sustainable utilization of municipal solid waste incineration fly ash for ceramic bricks with eco-friendly biosafety *Mater. Today Sustain.*

[2] Salwa Khamis S, Purwanto H, Naili Rozhan A, Abd Rahman M and Mohd Salleh H 2019 Characterization of Municipal Solid Waste in Malaysia for Energy Recovery *IOP Conf. Ser. Earth Environ. Sci.* **264**

[3] Omari A M 2013 Characterization of municipal solid waste for energy recovery *J. Multidiscip. Eng. Sci. Technol.* **2** 230–7

[4] Sharma K D and Jain S 2020 Municipal solid waste generation, composition, and management: the global scenario *Soc. Responsib. J.*

[5] Sudibyo H, Majid A I, Pradana Y S, Budhijanto W, Deendarlianto and Budiman A 2017 Technological Evaluation of Municipal Solid Waste Management System in Indonesia *Energy Procedia*

[6] Saragih B R, Siregar S R H and Surjosatyo A 2018 Evaluation of Waste Potential in TPST Bantargebang Through Modified Triangular Method *E3S Web of Conferences* vol 67 (EDP Sciences) p 2040

[7] Sudibyo H, Pradana Y S, Budiman A and Budhijanto W 2017 Municipal solid waste management

in Indonesia-A study about selection of proper solid waste reduction method in DI Yogyakarta Province *Energy Procedia* **143** 494–9
[8] Dhokhikah Y, Trihadiningrum Y and Sunaryo S 2015 Community participation in household solid waste reduction in Surabaya, Indonesia *Resour. Conserv. Recycl.* **102** 153–62
[9] Ciuta S, Tsiamis D and Castaldi M J 2017 *Gasification of waste materials: technologies for generating energy, gas, and chemicals from municipal solid waste, biomass, nonrecycled plastics, sludges, and wet solid wastes* (Academic Press)
[10] Di Blasi C and Branca C 2001 Kinetics of primary product formation from wood pyrolysis *Ind. Eng. Chem. Res.* **40** 5547–56
[11] Chandrasekaran S R and Sharma B K 2019 13 - From Waste to Resources: How to Integrate Recycling Into the Production Cycle of Plastics *Plastics Design Library* ed S M B T-P to E Al-Salem (William Andrew Publishing) pp 345–64
[12] Carrier M, Loppinet-Serani A, Denux D, Lasnier J-M, Ham-Pichavant F, Cansell F and Aymonier C 2011 Thermogravimetric analysis as a new method to determine the lignocellulosic composition of biomass *Biomass and Bioenergy* **35** 298–307
[13] Varhegyi G, Jakab E, Till F and Szekely T 1989 Thermogravimetric-mass spectrometric characterization of the thermal decomposition of sunflower stem *Energy & Fuels* **3** 755–60
[14] Lai Z, Ma X, Tang Y and Lin H 2012 Thermogravimetric analysis of the thermal decomposition of MSW in N2, CO2 and CO2/N2 atmospheres *Fuel Process. Technol.* **102** 18–23
[15] Mutua N A, Ouma A B and Atieno O S 2017 Kinetic study of the thermal decomposition for mixed municipal solid waste using thermogravimetric analysis *Period. Eng. Nat. Sci.* **5** 355–63
[16] Omari A M, Kichonge B and Chaula Z A 2019 Kinetics Properties and Thermal Behavior of Pine Sawdust and Municipal Solid Waste *J. Energy Res. Rev.* **3** 1–9
[17] Cai J and Bi L 2008 Precision of the Coats and Redfern Method for the Determination of the Activation Energy without Neglecting the Low-Temperature End of the Temperature Integral *Energy & Fuels* **22** 2172–4
[18] Boumanchar I, Chhiti Y, M'hamdi Alaoui F E, Elkhouakhi M, Sahibed-dine A, Bentiss F, Jama C and Bensitel M 2019 Investigation of (co)-combustion kinetics of biomass, coal and municipal solid wastes *Waste Manag.* **97** 10–8
[19] Nursani D, Siregar S R H and Surjosatyo A 2020 Effect of Binder Adding to the Physical Properties of Municipal Solid Waste (MSW) Pellets *IOP Conf. Ser. Earth Environ. Sci.* **520**
[20] Çepelioğullar, Haykiri-Açma H and Yaman S 2016 Kinetic modelling of RDF pyrolysis: Model-fitting and model-free approaches *Waste Manag.* **48** 275–84
[21] Azam M, Jahromy S S, Raza W, Jordan C, Harasek M and Winter F 2019 Comparison of the combustion characteristics and kinetic study of coal, municipal solid waste, and refuse-derived fuel: Model-fitting methods *Energy Sci. Eng.* **7** 2646–57
[22] Mian I, Li X, Jian Y, Dacres O D, Zhong M, Liu J, Ma F and Rahman N 2019 Kinetic study of biomass pellet pyrolysis by using distributed activation energy model and Coats Redfern methods and their comparison *Bioresour. Technol.* **294** 122099

Growth inhibition of tropical fungi by silver nanoparticles incorporated polyurethane coating

K M Said[1,*], W M Z W Yunus[1,2,*], A Hidayah[3], K K Ong[4], A L N Shazwani[1], M S N N Shakira[5], and A Nurnadia[1]

[1] Centre for Tropicalization, National Defence University of Malaysia, Kem Sungai Besi, 57000, Kuala Lumpur, Malaysia

[2] Faculty of Defence, Science and Technology, National Defence University of Malaysia, Kem Sungai Besi, 57000, Kuala Lumpur, Malaysia

[3] Institute of Tropical Forestry and Forest Product (INTROP), University Putra Malaysia, 43400, UPM Serdang, Selangor, Malaysia

[4] Centre for Defence Foundation Studies, National Defence University of Malaysia, Kem Sungai Besi, 57000, Kuala Lumpur, Malaysia

[5] Faculty of Medicine and Defence Health, National Defence University of Malaysia, Kem Sungai Besi, 57000, Kuala Lumpur, Malaysia

* Email: najwa198@gmail.com, wanmdzin@upnm.edu.my

Abstract. Polyurethanes (PUs) have been widely used in the coating industry due to their high flexibility and good UV resistance. Extensive formulations of PU coatings have been reported previously in the literature. Scientists nowadays improvise products whereby nanomaterial additives such as silver nanoparticles are incorporated to obtain antibacterial and antifungal properties. In this study, spherical shaped silver nanoparticles with an average size of 46 nm were added into polyol (polypropylene glycol) which was cross-linked with toluene isocyanate to obtain PU nanocomposite coating films. Characterizations of the coating film were carried out using Fourier transform infrared (FTIR) spectroscopy to observe the urethane linkage that was formed during the cross-linking and Field Emission Scanning Electron Microscope with Energy Dispersive X-Ray (FESEM-EDX) Spectroscopy analysis to determine size, shape and distribution of the nanoparticles. The antifungal activity of the nanocomposite coating was studied against *Cladosporium cladosporiodes, Penicillium oxalicum* and *Aspergillus aculaetus*. The results showed that the AgNPs-incorporated PU coating film has the ability to inhibit the tropical fungi growth.

Keywords: digital image processing, fuzzy inference system, histogram equalization, mammography

1. Introduction

Polyurethanes (PUs) were discovered by Otto Bayer in the 1930s along with the introduction of basic ingredients for PUs synthesis; isocyanates, polyesters, polyethers and acrylics [1]. They are known for their wide range of applications including as rigid and flexible foams, thermoplastics and waterborne PU [2][3][4] due to their outstanding performances, such as highly flexible, good abrasion resistance and great adhesion to a surface. However, the drawback of these materials is that they can be degraded by fungi, bacteria and/or other microbes that can cause microbiologically-influenced-corrosion (MIC), hence limit their extensive industrial applications[5]. In the coating industry, PUs are favorable due to their chemical resistant properties which are contributed by the crosslinked structure. This behavior depends on polyol's molecular weight whereby a higher molecular weights leads to a more cross-linked structure of the coating thus exhibits excellent chemical stability[6][7]. For outdoor applications, the ability to resist chemical and microorganism attacks is necessary to prolong the coating effectiveness. Nowadays, a high-performance PU coating have been invented by incorporating of active ingredients into the PUs. Common additives that can inhibit microbial growth are metal cation (e.g. Ag^{+}, Au^{+} or Cu^{2+}) complexes, nanomaterials (e.g., metal nanoparticles or carbon nanotubes) and antibiotics[8][9][10]. The use of these additives has been demonstrated to boost the coating performances as well as improves the lifespan of the coating.

Silver (Ag) has been acknowledged a long time ago as one of the most effective germicides to prevent or reduce the growth of bacteria as well as fungi. To keep drinks fresh and unspoiled, the Greeks and Romans used containers made from silver. After years of researches, the nanomaterial experts discovered that silver works best in the form of nanoparticles (AgNPs) due to the high surface area to volume ratio. Furthermore, a small concentration of AgNPs, for example, less than 2 microgram/mL, does not show significant toxicity towards mammalian cells which in this case, monocytic cell [11].

AgNPs and PUs resin are commonly first synthesized individually, then mixed to form AgNPs-incorporated polyurethane composites (Atay et. al [12]). FTIR analysis which shows that the peak intensity of the organic bonds is independent on AgNPs concentration indicates that the presence of AgNPs does not affect the organic PUs structure. This suggests that the AgNPs are free to interact with the microorganisms. It has also been reported that Ag has the potential to reduce bio-aerosolization process of human pathogens in its surrounding by incorporating the nanoparticles into PU foams that were used as a part of air filtration device[13]. Lee et. al (2010) reported that at a high concentration of AgNPs, the probability of nanomaterial to be agglomerated is higher, hence the AgNPs were unable to penetrate effectively into the microorganism's cell membrane [14]. Other than that, a high concentration of AgNPs in polymeric coatings leads to a porous and rough coating surface [12].

The study on fungi growth inhibition by AgNPs that exist in the tropical climate has been carried out recently[15][16][17]. These nanomaterials possess an inherent ability to terminate the fungi cell through a reaction that damages protein and nucleic acid [18]. Therefore, AgNPs have a potential as an additive for coating materials which can be used to inhibit or prevent the fungi growth. This kind of concept is still limited as most works of literature were more involved with bacterial inhibition function. This report describes results of our work on synthesis of the AgNPs incorporated PU coating and its ability to inhibit three tropical fungi (*Cladosporium cladosporiodes, Penicillium oxalicum, Aspergillus aculaetus)* growth of on potato dextrose agar.

2. Materials and methods

2.1. Materials

Polypropylene glycol with an average molecular weight of 2000 g/mol was obtained from Sigma Aldrich; US. Toluene diisocyanate (TDI) (Toluene-2,4-diisocyanate, 80%, remainder 2,6-diisocyanate) was purchased from Alfa Aesar; US. Toluene, 99% purity was acquired from R&M; UK. Silicone oil was bought form ChemPur; Malaysia. Silver nanoparticles suspension was acquired form (Nanovex; Spain). Potato dextrose agar (PDA) was purchased from Oxoid; UK. Fungi (*Cladosporium cladosporiodes, Penicillium oxalicum, Aspergillus aculaetus*) samples were isolated from various corroded parts of an airplane used in tropical environment. All chemicals were used as received.

2.2. Methodology

2.2.1. Preparation of AgNPs incorporated polyurethane coating. Polypropylene glycol (4.0183 g) in toluene (2.0221 g) were first mixed in a beaker and heated in a silicone oil bath at 80.0 ℃ with constant stirring. PU prepolymer was then prepared by adding toluene diisocyanate dropwise into the mixture. After 2 hours of stirring, AgNPs suspension was added and mixed for 1 hour. Then, the prepolymer was spread onto a glass surface, and cured in a vacuum oven at 60 ℃ for 2 hours to obtain AgNPs incorporated PU coating film. The percentages of each material used are given in Table 1.

Table 1. Compositions used for PU coating preparation.

Materials	Mass percentage (%)
Polypropylene glycol	54%
Toluene	28%
Toluene diisocyanate	15%
AgNPs	3%

2.2.2. Antifungal activity. Three fungi (Cladosporium cladosporiodes, Penicillium oxalicum, Aspergillus aculaetus) were chosen for the assessment of antifungal activity of the AgNPs incorporated PU coating film. The antifungal test procedure was carried out by using PDA as the growth medium. The PDA was prepared by heating 39 g of PDA in 1000 mL of distilled water, poured into a 90 mm diameter petri dish just until the surface is covered, cooled it to a room temperature and stored at 4°C. In addition to AgNPs incorporated PU coating film, the PU film without the nanoparticles was also in the testing for a comparison. The agar plates were first divided into 4 areas that were used for control, Fungi 1 (Cladosporium cladosporiodes), Fungi 2 (Penicillium oxalicum) and Fungi 3 (Aspergillus aculaetus) (Figure 1). The PU film (1x1 cm^2) were placed in the middle of the plates. Then by using a sterilized cotton swab, the fungi were streaked onto the agar medium and films. Each plate was sealed with parafilm and kept in an incubator with an average temperature of 26°C for 8 days. All of the procedures were carried out in a sterilized laminar flow to prevent contamination.

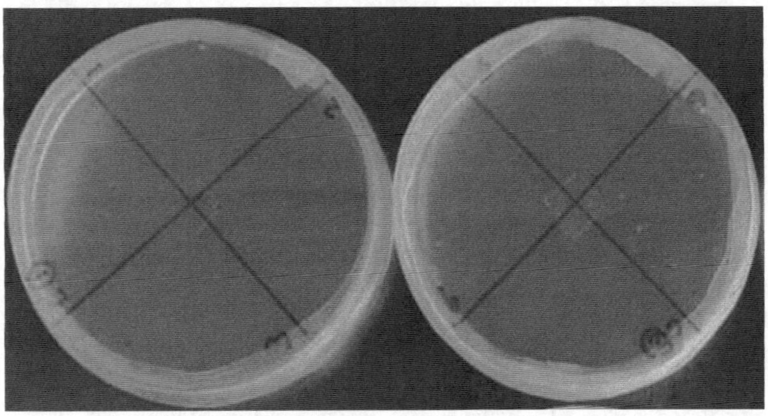

Figure 1. Preparation of antifungal activity.

2.3. Characterization of polyurethane coatings

2.3.1. Fourier transform infrared (FTIR) analysis. The interactions between polypropylene glycol and toluene diisocyanate were studied by using the FTIR spectrophotometer. The FTIR spectra were recorded by using Perkin Elmer Spectrum IR 10.6.1 and scanned from 4000 to 650 cm^{-1}. The cured sample that was peeled off from the glass surface was used for this FTIR study.

2.3.2. Field emission scanning electron eicroscope with energy dispersive X-ray spectroscopy (FESEM-EDX). The particle distribution of AgNPs in the PU coating films was examined using a field emission scanning electron microscope, FESEM-EDX (CARL ZEISS, Model: GeminiSEM 500) with gold-coated samples.

3. Results and discussion

3.1. Polyurethane synthesis

Figure 2 shows the chemical reaction equation for the PU synthesis from toluene diisocyanate (TDI) and polypropylene glycol (PPG) in the presence of toluene as solvent. The crosslinking process is through breaking of C=N bond in toluene diisocyanate and formation of urethane linkage (-RNHCOOR'-) in the final product.

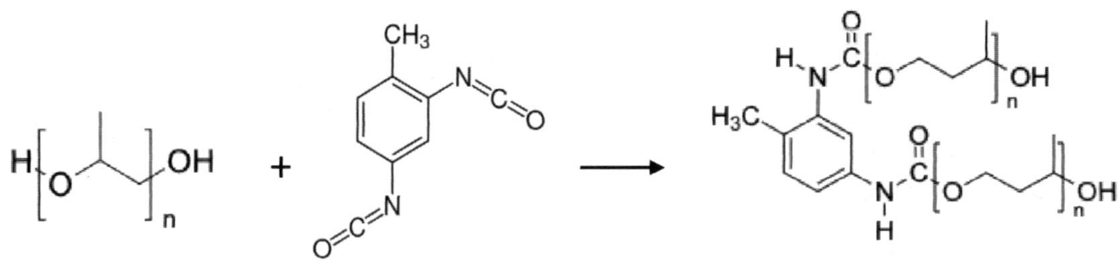

Figure 2. Synthesis of PU coating (n is ~2000).

In this polymer, polypropylene glycol and toluene diisocyanate represent as a soft segment and hard segment, respectively. In this study, polypropylene glycol with a molecular weight average of 2000 g/mol was used. According to the literature, polyols of an average of 1500-3000 g/mol molecular weights to get a smooth and flexible final product [19]. This is because low molecular weight of the polyol decreases the cross-linking density, hence increases the elasticity of the end product.

3.2. Fourier transform infrared spectroscopy (FTIR) analysis

The spectra of toluene diisocyanate, polypropylene glycol and cured PU coating, and their characteristics peaks are shown in Figure 3 and Table 2, respectively. The spectrum of toluene diisocyanate displays an intense and sharp peak at 2242 cm^{-1} which represents the isocyanate functional group of R-N=C=O. The polypropylene glycol spectrum shows stretching vibration of carbonyl group from secondary alcohol and alkene groups which are at 1095 cm^{-1} and 2858 cm^{-1}, respectively. Furthermore, the spectrum of the cured PU coating film displays the existence of a small and broad peak at 3300 cm^{-1} suggesting the presence of amine groups which were produced from the cross-linking of polypropylene glycol by toluene diisocyanate. The absence of the isocyanate peak at 2242 cm^{-1} also in line with the formation of urethane linkage. It also indicated that the diisocyanate was completely reacted in the PU formation. Carbonyl stretching of the urethane linkage was observed at 1725 cm^{-1}. All of the findings from the FTIR analysis were consistent and similar to that of the literature [7].

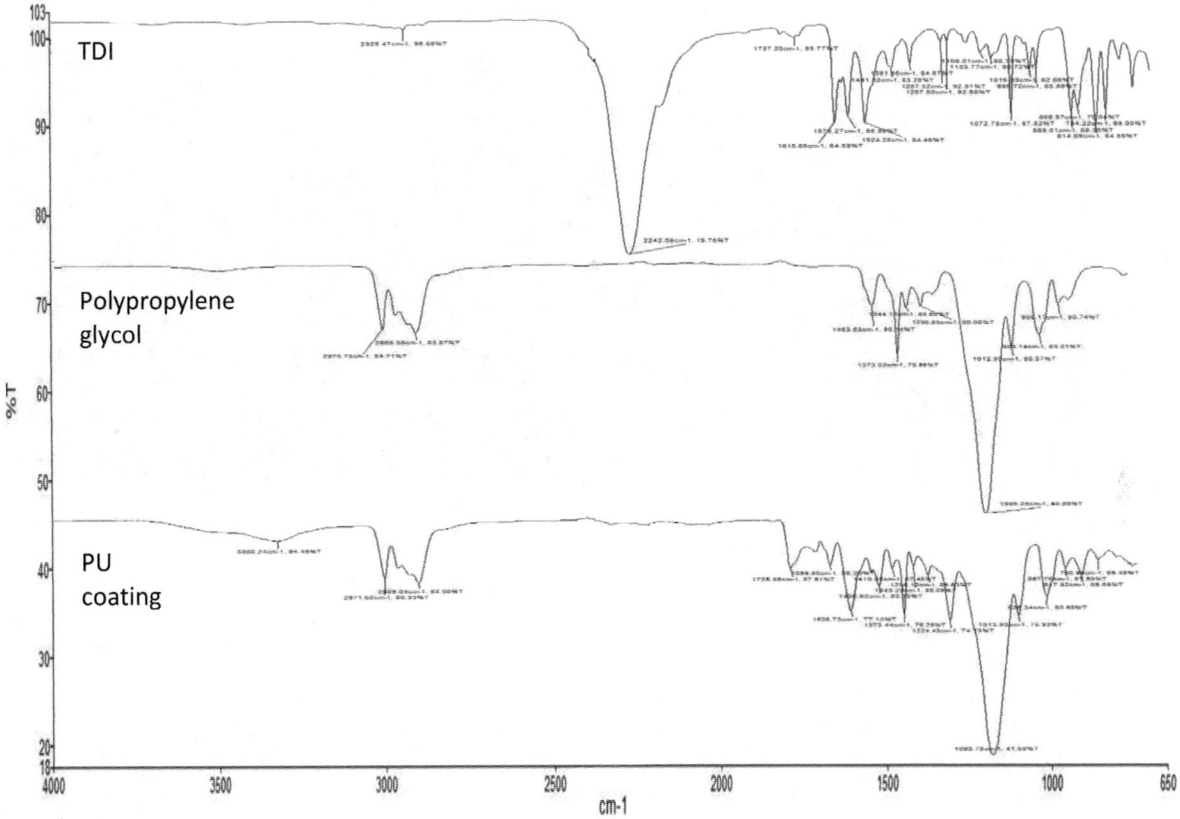

Figure 3. FTIR spectra of toluene diisocyanate, polypropylene glycol and cured PU coating.

Table 2. FTIR spectral assignments of toluene diisocyanate, polypropylene glycol and PU coating.

Functional groups	Wavenumber (cm^{-1})		
	TDI	PPG	PU coating
-N=C=O	2242	-	-
C=C (aromatic)	1576	-	1536
C-H str. Sym.	-	2868	2869
C-H str. Asym.	2925	2970	2971
C-H bending	1072	1453, 1373	1450, 1373
C-O-C str. (polyester)	-	1095	1093
-NH (urethane)-	-	-	3300
C=O (urethane)	-	-	1725

3.3. Field Emission Scanning Electron Microscope with Energy Dispersive X-Ray (FESEM-EDX) Spectroscopy

FESEM-EDX analysis on the AgNPs incorporated PU coating film was used to determine the Ag nanoparticle distribution and, their size and shape. The blue dots which represent the AgNPs show that the particles are well distributed with no obvious agglomeration (Figure 4). The well distributed particles are crucial as it will maximize the fungi growth inhibition function. The nanoparticles are in a spherical shape with the size range of between 58 – 74 nm (Figure 5).

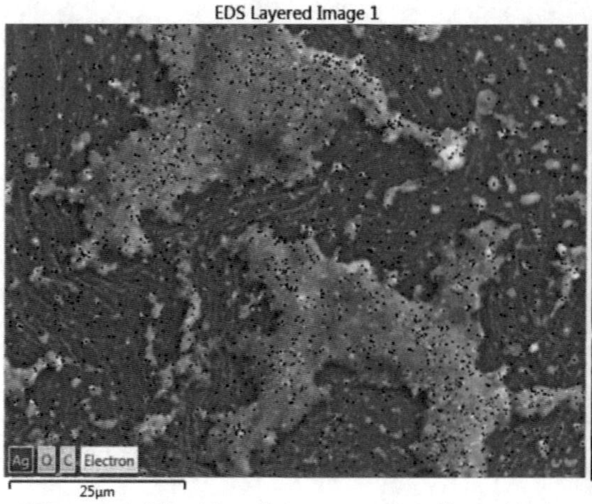

Figure 4. The blue dots represent silver (Ag).

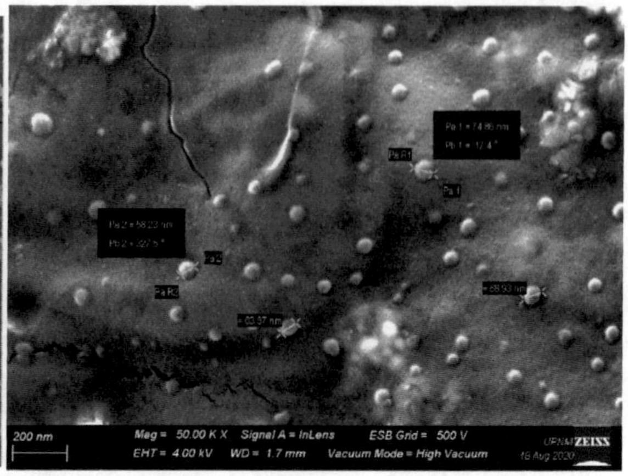

Figure 5. AgNPs size characterized by FESEM-EDX.

3.4. *Antifungal activity*

The antifungal activity was assessed using 3 fungi (*Cladosporium cladosporiodes, Penicillium oxalicum, Aspergillus aculaetus*) isolated from several corroded parts of an in service aircraft. As seen in Figure 6, the infected area for coating film without silver nanoparticles (Figure 6(a)) is wider than that of the silver nanoparticle incorporated film (Figure 6(b)). This indicates that the silver nanoparticle coating film inhibits the fungi growth. Silver nanoparticles has been known as one of the most effective germicides to inhibit the growth of bacteria and fungi. The presence of AgNPs damages to fungi's mycelia and leads to mycelial growth delay[20], destruction of hyphae and conidia distortion [21][18]. The silver nanoparticles disrupt the fungi's cell structure by damaging the cell membrane and stop the cell division process. This leads to cell death, hence inhibits the growth of fungi cells. Silver particles of nanosize is favorable because of their large surface area per volume ratio resulting in high number of cells' termination, thus, enhancing the antimicrobial activity[11][22].

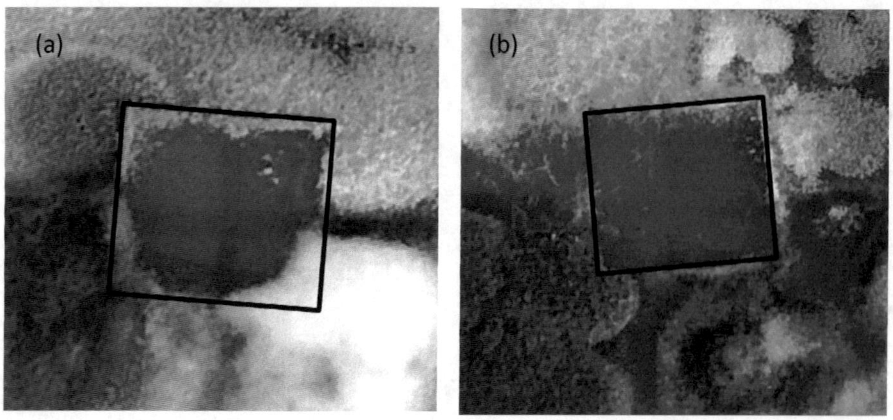

Figure 6. Fungi growth with (a) PU coating film without AgNPs and (b) AgNPs incorporated PU coating film.

4. Conclusion

This study has shown that AgNPs-incorporated PU coating film has the ability to inhibit the growth of fungi that are commonly found in tropical environment. The coating material can be prepared by entrapment of silver nanoparticles in a cross-linking reaction of polypropylene glycol by toluene diisocyanate. Initial observation indicates that the composite has the fungi growth inhibition behavior but further investigation is needed to determine the right formulation that is suitable to be used as an effective coating material.

Acknowledgments

The authors wish to acknowledge the financial support from the Fundamental Research Grant Scheme (Grant No. FRGS/1/2018/STG07/UPNM/01/2) and National Defence University of Malaysia (NDUM) for providing research facilities. This study would not have been possible without cooperation from the staff and students of the Centre for Topicalization, NDUM. Also, thanks to the Ministry of Higher Education, Malaysia for the grant funding.

References

[1] Bayer MaterialScience AG 2005 The Chemistry of Polyurethane Coatings 31

[2] Akindoyo J O, Beg M D H, Ghazali S, Islam M R, Jeyaratnam N and Yuvaraj A R 2016 Polyurethane types, synthesis and applications-a review *RSC Adv.* **6** 114453–82

[3] Kim H, Miura Y and MacOsko C W 2010 Graphene/polyurethane nanocomposites for improved gas barrier and electrical conductivity *Chem. Mater.* **22** 3441–50

[4] Zhang J, Zhang C and Madbouly S A 2015 In situ polymerization of bio-based thermosetting polyurethane/graphene oxide nanocomposites *J. Appl. Polym. Sci.* **132** 1–8

[5] Alasvand Zarasvand K and Rai V R 2014 Microorganisms: Induction and inhibition of corrosion in metals *Int. Biodeterior. Biodegrad.* **87** 66–74

[6] Paraskar P M, Prabhudesai M S and Kulkarni R D 2020 Synthesis and characterizations of air-cured polyurethane coatings from vegetable oils and itaconic acid *React. Funct. Polym.* **156** 104734

[7] Velayutham T S, Majid W H A, Ahmad A B, Kang G Y and Gan S N 2009 Synthesis and characterization of polyurethane coatings derived from polyols synthesized with glycerol, phthalic anhydride and oleic acid *Prog. Org. Coatings* **66** 367–71

[8] Vidales-Herrera J and López I 2020 *Nanomaterials in coatings: an industrial point of view* (INC)

[9] Mohammadi A, Hossein A, Beigi S and Amiri H 2020 Silver (I) complex with a Schi ff base ligand extended waterborne polyurethane : A developed strategy to obtain a highly stable antibacterial dispersion impregnated with in situ formed silver nanoparticles *Chem. Eng. J.* **381** 122776

[10] Paul D, Paul S, Roohpour N, Wilks M and Vadgama P 2013 Antimicrobial, Mechanical and Thermal Studies of Silver Particle-Loaded Polyurethane *J. Funct. Biomater.* **4** 358–75

[11] Martinez-Gutierrez F, Olive P L, Banuelos A, Orrantia E, Nino N, Sanchez E M, Ruiz F, Bach H and Av-Gay Y 2010 Synthesis, characterization, and evaluation of antimicrobial and cytotoxic effect of silver and titanium nanoparticles *Nanomedicine Nanotechnology, Biol. Med.* **6** 681–8

[12] Atay H Y, Yaşa I and Çelik E 2015 Antibacterial polymeric coatings with synthesized silver nanoparticles *Synth. React. Inorganic, Met. Nano-Metal Chem.* **45** 784–98

[13] Paladini F, Cooper I R and Pollini M 2014 Development of antibacterial and antifungal silver-coated polyurethane foams as air filtration units for the prevention of respiratory diseases *J. Appl. Microbiol.* **116** 710–7

[14] Lee S M, Lee B S, Byun T G and Song K C 2010 Preparation and antibacterial activity of silver-doped organic-inorganic hybrid coatings on glass substrates *Colloids Surfaces A Physicochem. Eng. Asp.* **355** 167–71

[15] Jaffri S B and Ahmad K S 2018 Augmented photocatalytic, antibacterial and antifungal activity

of prunosynthetic silver nanoparticles *Artif. Cells, Nanomedicine Biotechnol.* **46** 127–37

[16] Pulit J, Banach M, Szczygłowska R and Bryk M 2013 Nanosilver against fungi. Silver nanoparticles as an effective biocidal factor *Acta Biochim. Pol.* **60** 795–8

[17] Elgorban A M, El-Samawaty A E R M, Yassin M A, Sayed S R, Adil S F, Elhindi K M, Bakri M and Khan M 2016 Antifungal silver nanoparticles: Synthesis, characterization and biological evaluation *Biotechnol. Biotechnol. Equip.* **30** 56–62

[18] Sahar M O 2014 Antifungal Activity of Silver and Copper Nanoparticles on Two Plant Pathogens, Alternaria alternata and Botrytis cinerea **9** 34–42

[19] Badri K H 2012 Biobased Polyurethane from Palm Kernel Oil-Based Polyol *Polyurethane* 447–70

[20] Min J S, Kim K S, Kim S W, Jung J H, Lamsal K, Kim S Bin, Jung M and Lee Y S 2009 Effects of colloidal silver nanoparticles on sclerotium-forming phytopathogenic fungi *Plant Pathol. J.* **25** 376–80

[21] Lamsal K, Kim S W, Jung J H, Kim Y S, Kim K S and Lee Y S 2011 Inhibition effects of silver nanoparticles against powdery mildews on cucumber and pumpkin *Mycobiology* **39** 26–32

[22] Hwang E T, Lee J H, Chae Y J, Kim Y S, Kim B C, Sang B I and Gu M B 2008 Analysis of the toxic mode of action of silver nanoparticles using stress-specific bioluminescent bacteria *Small* **4** 746–50

Biodegradation of low-density polyethylene (LDPE) and starch – based plastic (SBP) by thermophiles *Bacillus subtilis* and *Candida tropicalis*

N Z Zahari*, S N Abdullah, P M Tuah, and F N Cleophas

Faculty of Science & Natural Resources, Universiti Malaysia Sabah, UMS Road, 88400 Kota Kinabalu, Sabah, Malaysia

* Email: zaidazahari@ums.edu.my

Abstract. This project was carried out to study the biodegradation of low-density polyethylene (LDPE) and starch-based plastics (SBP) by two types of thermophiles microorganisms namely *B.subtilis* and *C. tropicalis* in a lab scale method. A few specific objectives were set to identify the growth curve of both strains on LDPE and SBP films, changes in physical and chemical properties by weight loss, carbon dioxide (CO_2) evolved, topography changes of plastics surfaces and the efficiency on biodegradation rate of LDPE and SBP. The results showed that after 49 days incubation period, the optimum growth of *B. subtilis* for both LDPE and SBP is at week 5 of incubation with 8.9 x 10^8 CFU/mL and 9.1 x 10^8 CFU/mL respectively. While for *C. tropicalis* the highest growth was recorded at week 4 of incubation with 9.6 x 10^8 for both LDPE and SBP. The weight loss reduction percentage of LDPE and SBP by *C.tropicalis* was 3.2% and 22.3% respectively while for *B. subtilis* the results recorded were 4.6% and 12.9% respectively. SEM analysis revealed that there are topography changes for LDPE with bubbling on surface while cracks and holes formed on SBP surface. The strum test used to identify CO_2 evolved in SBP by *C. tropicalis* is 2.7 g/L which was 5-fold higher as compared to LDPE while in SBP by *B. subtilis* the results is 2.5 g/L which 5-fold higher compared to LDPE. Based on this study, it can be concluded that *C. tropicalis* have great potential in degrading SBP as compared to LDPE.

Keywords: low-density polyethylene (LDPE), starch-based plastic (SBP), thermophiles, *B. subtilis, C. tropicalis*, biodegradation

1. Introduction

Plastics or scientifically named as polymers had cover up the needs in human used nowadays. According to [1], the production of plastic has increased twenty fold and this number are expected to double in the next 20 years. Among this amount, it was reported that 12 million tones of plastics leak into the oceans. Polyethylene is the most commonly used plastics among people. Polyethylene can be categories to low-density polyethylene (LDPE), medium-density polythene (MDPE), high-density polyethylene (HDPE) and very low-density polyethylene (VLDPE) [2] Among this, LDPE is commonly used for making grocery bags, food wrapping material, power cable sheathing and laboratory containers, as it is excellently resistant to dilute and concentrated acids, ketones and vegetable oils. The resistant and characterization of LDPE make it harder for degradation and cause a major pollution. [3].

According to [1], there are 25 million tons of plastics accumulate in the coastal and terrestrial environment. Marine ecosystem are more adversely affected by the plastics waste that shares 60-80% of total marine waste. Biodegradable plastics has become the way for new considerations of waste management strategies since these materials are designed to degrade under environmental conditions or in municipal and industrial biological waste treatment facilities [4, 5]. In this study, the use of locally isolated Environmental-Relevant Microorganisms (ERM) is beneficial to treat the plastic pollution. This research focused on the inoculation of thermophiles microorganism's namely *B. subtilis* and *C. tropicalis* to degrade plastic samples for 49 days incubation period. The biodegradation efficiency was done by examined the biofilm cell growth, determination through physical and chemical analysis and spectrometric analysis of plastics surface changes by using scanning electron microscopy (SEM). From this research, the significance findings can be achieved through biodegradation rate by measuring the carbon dioxide evolved and weight loss measurement within 7 weeks period.

2. Materials and methods

2.1. Preparation of medium

There are four medium that being used for this study, which are nutrient broth (NB), medium salt mineral (MSM), Ramsay broth (RB) and nutrient agar (NA). These medium are used as nutrient supplied for microbial growth and further biodegradation studies.

2.2. Preparation of plastic films

Low-density polyethylene (LDPE) plastics bag is obtained from Giant Hypermarket 1 Borneo, Kota Kinabalu, Sabah. This plastic is manufactured by EH Packaging and Plastic Industry that located in Kota Kinabalu, Sabah, while starch-based plastics (SBP) bag is obtained from Maribumi Startech Sdn. Bhd. These plastics samples were cut into equal pieces 5cm × 3cm (l × b) and dried overnight at 60°C. For biodegradation studies, the plastics films were weighed before disinfest with 95% ethanol for 30 minutes. Ethanol was used as disinfectant to remove any impurities attached to its surface [6]. The initial weigh of plastic was recorded as day 0 to see the changes of plastics surfaces during biodegradation process.

2.3. Inoculation of thermophiles strains for biodegradation study

The strains used in this study were Thermophilic microoragnisms namely *Bacillus subtilis* [7] and *Candida tropicalis* [8]. The strains were obtained from beads at the Animal Physiology Lab, Faculty of Science & Natural Resources, Universiti Malaysia Sabah.

(a) (b) (c) (d)

Figure 1. The selected strains on basal medium (a) *B. subtilis* with LDPE (b) *B. subtilis* with SBP (c) *C. tropicalis* with LDPE and (d) *C. tropicalis* with SBP.

The experiment was performed by adding LDPE and SBP films in the two media, which are mineral salt medium (MSM) for *B. subtilis* and Ramsay broth (RB) for *C. tropicalis*, in accordance with the Standard test method BS EN ISO 846:1997 (with some modification). For biodegradation study, a total of 10% of bacterial colony *B. subtilis* were inoculated into each MSM flasks and *C. tropicalis* in RB flasks (Figure 1). Three strips of each type of plastic films are added to 2 different flasks that contain mineral salt medium (MSM) and Ramsay broth (RB) and then 2 flasks of LDPE and starch-based plastics (SBP) film are inoculate with thermophilic *B. subtilis* and *C. tropicalis* while the other 2 are leave and maintain as control. The experiment were conducted with *B. subtilis* were left in rotary shaker (180 rev min^{-1}) at 32°C for 49 days and *C. tropicalis* are left for 30°C at 200 rev min-1. Within this period, afew observations was made to study the biodegradation process of plastics by both strains.

2.4. Biofilm cell growth study

An aliquot for each of incubated flasks were taken at regular intervals for first 30 days of incubation for quantification of microbial cells by serial dilution technique. This is followed by plating on Nutrient Agar for viable colony count. For this study, biofilm layer formed on the plastics films is remove by immerse the plastics strips in 2mL of 0.9% NaCl for 2 hours and vortex for 5 to 10 minutes [9]. Aliquot from this suspension is taken and serially diluted and plated on Nutrient agar plates for colony counting. The colony forming units from biofilm cells is calculate according to equation below.

$$\log\left(\frac{CFU}{cm^2}\right) = \log\left(\frac{Average\ CFU}{Drop\ volume} \times dilution\ counted\right) \quad (1)$$

2.5. Carbon dioxide (CO_2) evolution

The degradation of plastics can be measured by determine the carbon dioxide evolve according to OECD Guidelines 1992 Method 301 B. After incubation, all the flasks including control flask is calculate gravimetrically as a result of degradation of polymeric chain that trap in absorption flask contains KOH (1.0 M). Barium chloride solution (0.1 M) was added into the flask containing KOH which resulted in barium chloride precipitation (using CO_2 released from the breakdown of polymer). CO_2 produced were calculated gravimetrically by measuring the amount of CO_2evolved by addition of $BaCl_2$. The changes of CO_2 evolved was recorded.

2.6. Determination of weight loss in LDPE and SBP films

The weight loss of LDPE and SBP films were calculated in form of percentage (%) biodegradation. For weight loss measurement, the plastic strips were sampling for every 7 days of interval. The residual of LDPE and SBP were washed in 70% of ethanol and rinsed through distilled water to remove any kind of impurity and organic matter films. Then the samples were dried overnight at 60°C to record the final reading. The weight loss was calculated by using the equation below.

$$Biodegradation\ (\%) = \frac{Initial\ weight - Final\ weight}{Initial\ weight} \times 100 \quad (2)$$

2.7. Spectrometric analysis of SEM

The changes of the surface topography for treated and untreated LDPE and starch- based plastic (SBP) films were examined by Scanning electron microscopy (SEM). The treated samples after 49 days of incubation were washed with 70% ethanol and distilled water for a few minutes to remove the cell. Then the samples were coated with gold for 40 second and were observed under high-resolution scanning electron microscope (EVO LS15; Carl Zeiss, Germany).

3. Results and discussion

3.1. Microbial cell growth on plastics

The results showed that the growth curves of *C. tropicalis* on both LDPE and SBP were slightly increase from week 1 to 7 along the biodegradation study. For *B. subtilis* the growth was increase from week 1 to 5 and slowly decrease along the incubation period. The highest growth curve of *B. subtilis* can be observed in SBP at week 5 with 9.1 x 10^8 CFU/ml. While for *C. tropicalis*, the highest growth rate can be shown at week 4 in SBP with 9.6 x 10^8 CFU/ml. The formation of biofilms by both plastics within 7 days had reduced the hydrophobicity of both polymers and yet it improved the degradation rate. This can be observed after a week of incubation whereas the microbial growth of both strains were slightly increased. This indicates that the biodegradation were rapidly progressed which means it has make an easier movement for microbial attack to degrade the plastic. In addition, the presence of plastics as sole source of carbon may favor the growing of both strains, especially for petroleum-based microbes. In this study, there were no other nutrients supplied to the cultures such as glucose [10] and the changes of microbial growth can be seen through the presence of LDPE and SBP.

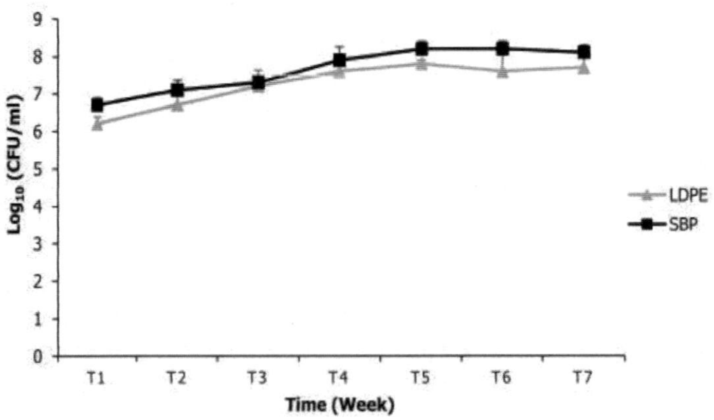

(a)

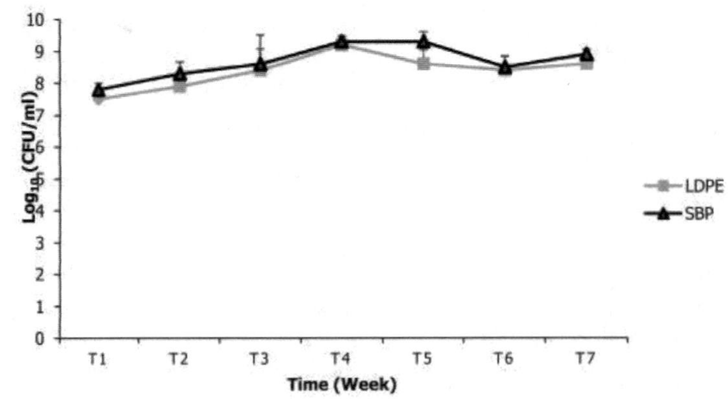

(b)

Figure 2. The microbial growth of (a) *B. subtilis* and (b) *C. tropicalis* on LDPE and SBP during incubation periods.

The graph also showed that both strains that supplied with SBP was rapidly developed as compared to LDPE. This is due to the composition of SBP which made by starch and contain high nutrient content as compared to LDPE. The results also showed that *C. tropicalis* was more efficient in degrading LDPE and SBP as compared to *B. subtilis*. According to [11], microorganisms effectively degrade the plastics when it attach to the surface. As SBP surface is hydrophilic, *C. tropicalis* was found to degrade this compound more effective as compared to *B. subtilis*.

3.2. Carbon dioxide (CO_2) evolution

Figure 3 showed the amount of CO_2 produced from the breakdown of LDPE and SBP after the biodegradation study. It was found that the CO_2 evolved by SBP is higher than LDPE for both cultured in *B. subtilis* and *C. tropicalis*. The data indicates that microbes utilized the oxidation products in the pre- oxidised films as carbon sole source. As SBP is made by starch, it becomes the sole of nutrients for microbial growth in turns increase the mineralization profile of plastics sample. The CO_2 evolved by C. tropicalis and B. subtilis in SBP was the highest with 2.69 g/L and 2.5 g/L respectively. The supplied nutrient by starch increase the favorable conditions for microbes. This finding showed that SBP degradation is more efficient than in previous study by [12]. He stated that the degradation of LDPE by *Aspergillus clavatus* produced only 2.32 g/L of CO_2. However, the degradation of LDPE in this study by both strains produced only 2.15 g/L and 2.19 g/L of CO_2 which is 1 – fold lower than the previous study.

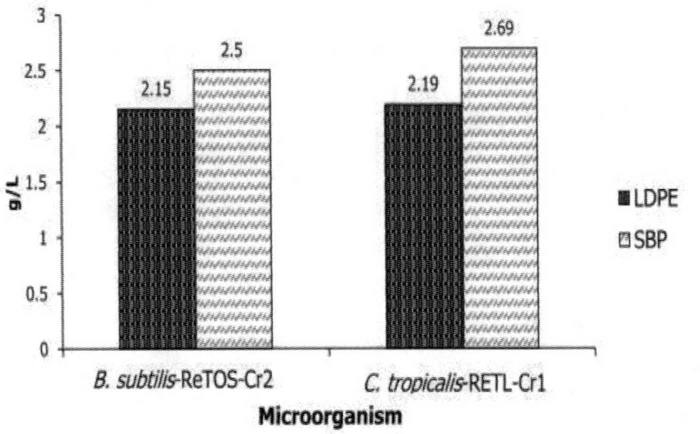

Figure 3. The amount of CO_2 produced after degradation of plastic at room temperature.

3.3. Biodegradation of plastic

The biodegradation of plastic was quantitatively assessed by measuring the percentage reduction of LDPE and SBP in shake flask experiment under control conditions. Figure 4 showed the percentage of weight loss in LDPE and SBP *by B. subtilis* and *C. tropicalis*. The data obtained indicates that the percentage of weight loss was increase as the incubation period. The highest percentage reduction was found in LDPE with 3.2% while SBP recorded 22.3% by *C. tropicalis*. For *B. subtilis*, the highest reduction was observed with 4.6% for LDPE and 12.9% for SBP respectively. This finding suggested that the presence of both strains *B. subtilis* and *C. tropicalis* as degrader had increase the process of plastic degradation. Microbes that grow utilizing the polymers lead to an increase in weight due to the adherence of microbes. The percentage weight reduction of LDPE and SBP films can be attributed to the breakdown of carbon backbone due to enzymatic degradation by both strains studies.

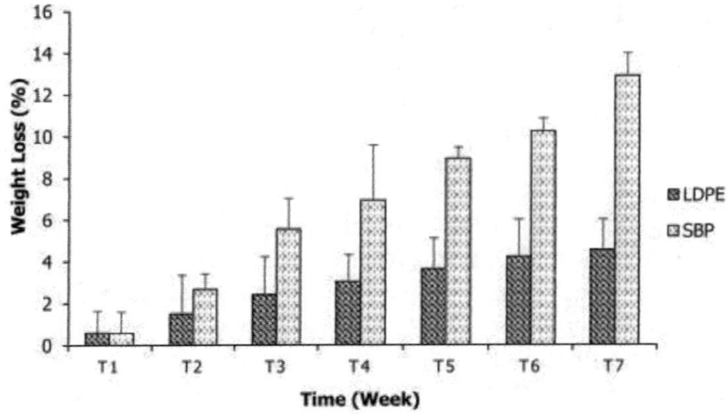

(a)

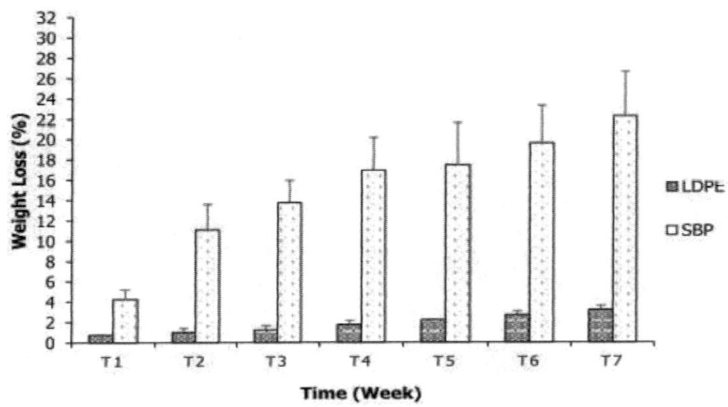

(b)

Figure 4. The percentage of weight loss (%) of LDPE and SBP within 7 weeks incubation periods by *B. subtilis* and (b) *C. tropicalis.*

3.4. Topography changes on polymeric surface

The extent of biodegradation was examined by observing the changes in surface morphology through SEM analysis. Figure 5 showed the physical changes of both polymeric surfaces before and after 49 days of incubation periods. Analysis on SEM showed that the topography of LDPE surfaces before inoculated with *C. tropicalis* was flatted and smooth which indicates that there was no reaction of biodegradation by *C. tropicalis*. However after being degraded, the surface was observed distinct in which a weakening surface of LDPE had been eroded with greasy surface. LDPE structure showed cracks and formed a micro bubbles in polymer matrix. This is due to scissoring of long chain polymer and evolution volatile products from LDPE surface [13]. On the other hand, the structure of SBP after biodegradation study was observed to be crumpled on the surface. It was found that there are a few holes formed on the surface with ragged edges and narrow gaps with smooth edges extending lengthwise along the film's surface. The destruction of surface was suggested due to the microbial attack of *C. tropicalis* that had weakened up the polymeric structure and penetrate through SBP matrix during the degradation process. According to [14], SBP films was easily disintegrated under mild pressure during the degradation process. The element starch in SBP film has increased the surface roughness compared to LDPE films. An increase in surface roughness could provide a plenty of anchoring places for microorganisms which is less possible in case of LDPE films [15]. The adherence of cells to SBP surface may due to biofilm formation [16], indicating strong adhering capabilities of *C. tropicalis* on SBP film until it caused the surface destroyed with a few holes.

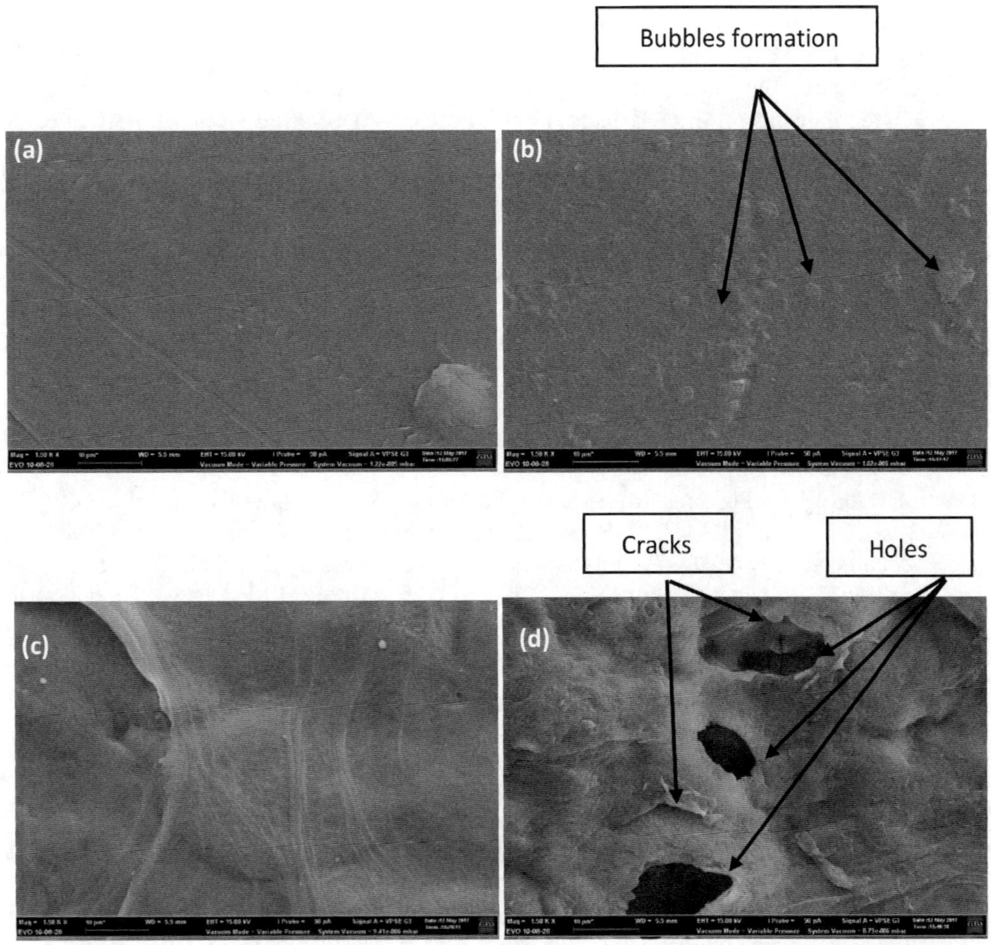

Figure 5. SEM images under 1.5 kV magnification of (a) LDPE surface before incubation (b) LDPE surfaces after incubated with *C. tropicalis* (c) SBP before incubation (d) SBP after incubated with *C. tropicalis*.

4. Conclusion

The present work indicates that both strains namely *B. subtilis* and *C. tropicalis* is capable in degrading low density polyethylene (LDPE) and starch- based plastic (SBP). Generally, *C. tropicalis* have a higher degrading effectiveness as compared to *B. subtilis*. The weight loss reduction (22.9%) in SBP by *C. tropicalis* proved that biodegradation of plastic had occur after 49 days of incubation. This showed that the relatively significant degradation was forced the strains to consume SBP as the only carbon source. Thus this condition had leads to development of biofilm and high growth rate on SBP surfaces.

Acknowledgment
We would like to express our gratitude to Faculty of Science & Natural Resources, Universiti Malaysia Sabah (UMS) for providing necessary facilities to complete this study successfully.

References

[1] World Economic Forum, 2016.The New Plastics Economy: Rethinking the future of plastics. Available online at: http://www3. weforum.org/docs/WEF_The_New_Plastics_Economy.pdf.

[2] Rivard, C., Moens, L., Roberts, K., Brigham, J. & Kelley, S. 1995. Starch esters as biodegradable plastics: Effects of ester group chain length and degree of substitution on anaerobic biodegradation. Enzyme and Microbial Tech. **17**: 848- 852.

[3] Usha R., Sangeetha T., and Palaniswamy M., (2011), Screening of Polyethylene Degrading Microorganisms from Garbage Soil, Libyan Agriculture Research Centre Journal International, 2(**4**), pp 200-204.

[4] Augusta, J., Muller, R. J., Widdecke, H. 1992. Biologisch abbaubare Kunstatoffe: Testverfahren und Beurteilungskriterion. *Chem Ing Tech*. **64:** 410-5.

[5] Witt, U., Muller, R. J., Deckwer, W. D. 1997. Biodegradation behavior and material properties of aliphatic/aromatic polysters of commercial importance. *J Environ Poly Degrad*. **15** :81-9.

[6] Rosenberg, M., Gutnick, D., Rosenberg, E., "Adherence of bacteria to hydrocarbons: a simple method for measuring cell-surface hydrophobicity", FEMS. Microbiol. Lett., 9(**1**): 29–33, (1980).

[7] Nurul Huda, Z. 2011. Pemencilan, penyaringan dan pencirian bakteria pengurai hidrokarbon dan kesan bioremediasi minyak diesel oleh konsortia bakteria yang diformulasikan. *Tesis Ijazah Sarjana Sains*. Universiti Malaysia Sabah.

[8] Nurul Huda, Z. & Piakong, M. T. 2007. Pertumbuhan microorganism terpencil RETL- Cr1, BAS- Cr1 dan RAS-Cr1 dalam medium mengandungi minyak mentah. *Proceedings of Seminar on Science and Technology 2007*. Universiti Malaysia Sabah. 150-155.

[9] Bhone, M. K., Ravi, C., Meena, K. S., Chu, S. L. & Kishore, R. S. 2011. . Biodegradation of Low- density polyethylene (LDPE) by *Pseudomonas* species. *Indian J Microbiol.* 52(**3**):411-419.

[10] Hadad, D., Geresh, S. & Sivan, A. 2005. Biodegradation of polyethylene by the thermophilic bacterium Brevibacillus borstelensis. *Journal of Applied Microbiology*. **98** :1093-1100.

[11] Arutchelvi, J., Sudhakar, M., Ambika, A., Dhobie, M., Bhaduri, S. & Uppara, P.V. 2008. Biodegradation of polyethylene and polypropylene. *Indian J Biotechnol.* **7** :9-22.

[12] Anudurga, G., Sharmila, K. & Jayanthi, A. 2016. Microbial degradation of low-density polyethylene (LDPE) by *Aspergillus clavatus* strain JASK1 isolated from landfill soil. *3 Biotech*. **6:** 52.

[13] Mumtaz, T., Khan, R. & Hassana, M. A. 2010. Study of environmental biodegradation of LDPE films in soil using optical and scanning electron microscopy. *Micron*. **41**: 430-438.

[14] Zahra, S., Abbas, S. S., Mahsa, M. T. & Mohsen, N. 2010. Biodegradation of Low- density polyethylene (LDPE) by fungi in solid waste medium. *Waste Management*. **30** : 396-401.

[15] Mehmood, C. T., Qazi, I. A., Hashmi, I. & Bhargava, S. 2016. Biodegradation of low density polyethylene (LDPE) modified with dye sensitized titania and starch blend using *Stenotrophomonas pavinii. International Biodeterioration & Biodegradation.* **113:** 276-285.

[16] Garima, M., Ashwani, M. & Ramasare, P. 2011. Colonization and degradation of thermally oxidized high-density polyethylene by Aspergillus niger (ITCC No 6052) isolated from plastic waste dumpsite. *Bioremediat. J.* **15**: 2.

Grafting the carboxyl groups onto DPNR by seeded emulsion polymerization to enhance the compatibility with silica filler

J Woowong, P Phinyocheep, and J Sakdapipanich*

Department of Chemistry and Center of Excellence for Innovation in Chemistry, Faculty of Science, Mahidol University, Bangkok, 10400, Thailand

* Email: jitladda.sak@mahidol.edu

Abstract. The modification of deproteinized natural rubber (DPNR) with maleic anhydride (MAH) was synthesized to enhance the better compatibility between silica filler and natural rubber (NR). The presence of carboxylic acid groups (–COOH) of MAH after hydrolysis in the grafted DPNR with MAH (DP-g-MAH) would directly interact with the hydroxyl groups (–OH) of silica filler, resulting from the formation of the ester bonds due to the esterification reaction. It is leading to form the chemical bonding, which is the significantly strong interaction between DPNR and silica filler. The structural characterization of DP-g-MAH was analyzed by Fourier transform infrared spectrometer (FTIR) and proton nuclear magnetic resonance (1H-NMR), respectively. The % grafting level of MAH approximately highs to 99.37 % with the low gel content approximate 14.45 (wt%). In addition, the DP-g-MAH latex was compounded with silica nanofiller 10 phr by latex compounding method. The morphology of silica filled DP-g-MAH (DP-g-MAH/Si 10) was characterized by field-emission scanning electron microscope (FE-SEM) in order to observe the silica dispersion into the DP-g-MAH rubber. The FE-SEM shows that silica filler could disperse in grafted rubber more than the DPNR. In addition, the crosslink density was investigated to confirm the chemical networks, which could occur from esterification between DP-g-MAH and silica filler.

Keywords: modification of DPNR, silica-rubber compatibility, chemical networks

1. Introduction

The chemical modification of NR has been interested in fabricating the satisfactory materials and approaching to various applications in many industrial fields. There are plenty of monomers for grafting NR such as methyl methacrylate (MMA), methyl methacrylate-styrene (MMA-ST), and styrene (ST). Interestingly, maleic anhydride (MA) becomes attractive due to it is one another of chemicals that can improve the polarity into the NR this chemical modification has been known as malenization reaction. However, a grafting efficiency of MA onto NR is inferior comparing with other monomers due to the homo-polymerization of MA difficultly occur onto NR backbone. Thus, there might be one unit of MA grafted on each active site of the polymer chains. Since MA is not reactive due to its chemical structure, which is symmetry and electron density at C=C make it stable.

Sealao et al. used styrene as a co-monomer to enrich the reactivity of MA and found that a small amount of styrene could achieve the higher % grafting MA at the beginning of reaction (~ 2 h) [1]. However, the additional reaction time would not affect the % grafting MA, although there were present of styrene or without styrene. The address of this result was that styrene and MA could react with each other, one-by-one molecule, to form a charge-transfer complex (CTC), which increased electrical asymmetry of the double bond of MA molecule [1] Moreover, there are many [2-5] showed that proteins on the rubber particles were one of the interferences of the chemical grafting onto NR backbone.

Thus, to increase the grafting efficiency, the DPNR latex was chosen for the grafting polymerization. Wongthong et al. found the optimal method to synthesize the MA grafting onto DPNR backbone via a differential microemulsion polymerization (DMP) technique with grafting condition DPNR: MA: BPO (85: 9: 6) at 60°C 8 h [2]. In addition, in 2014, they have used styrene as a co-monomer for increasing the reactivity of MA and observed that the percentage of grafting efficiency (G.E) was enhanced from 71.4 to 79.3% [3]. However, there is no further research about applying the Maleinized rubber in the practical fields. That would bring to the objective of this research. The introduction of some of the carboxyl groups into DPNR could achieve the esterification between rubber and silica filler, resulting in the increasing of compatibility by the occurrence of chemical networks, which would be studied.

2. Experimental section

2.1. Materials

Synthetic rubber (IR; *Bangkok Synthetics Co., Ltd supported cis-1,4 polyisoprene (98%)*. The 60% of dry rubber content was provided by Thai rubber latex group public Co., Ltd., Rayong, Thailand. Sodium dodecyl sulfate (SDS; 95%) was purchased from Ajax finechem pty. Ltd. Urea (99%) was purchased from ACL Labscan Ltd. (Thailand). Silica (HiSil®233) was purchased from OSC Silica Co., Ltd. Benzoyl peroxide (BPO; 75%) was purchased from Merck schuchardt OHG (Germany).

Maleic anhydride (MAH; 99%) was purchased from Acros Organics (Hungary). Sodium carbonate anhydrous (Na_2CO_3; 99.5%) was purchased from Loba Chemie Pvt. Ltd. (India). Sodium bicarbonate ($NaHCO^3$; 99.5%) was purchased from AnalaR® BDH laboratory supplies (England). Potassium hydroxide (KOH) was purchased from Ajax Finechem. Isopropanol (99.9%) was purchased from Beaker analyzed® A. C. S. reagent (USA). Formic acid, acetone and toluene analytical grade were purchased from QRëC® (New Zealand).

2.2. Preparation of deproteinized natural rubber graft maleic anhydride (DP-g-MAH)

The CNR, which consists of dried rubber content (DRC) 60% was diluted to 30% DRC and incubated with urea 0.1 wt% and SDS 1 wt% at room temperature for 60 min based on the work of Kawahara et al. [6]. The centrifugation 13,000 rpm 30 min (2 times) at 25 °C was required after finishing the reaction to obtain the purified DPNR latex.

First, fill 15% DRC of DPNR latex into glass reactor assembled with a reflux condenser, stirrer-hot plate, and temperature controller. The DPNR latex was stabilized with 1% (w/w) SDS, KOH and buffer solution ($NaHCO_3/Na_2CO_3$). Then nitrogen gas was purged into the glass reactor for 30 min in order to prevent an oxidative reaction. After that, the reaction was heated at 60 °C to activate the BPO. Next 6 phr of BPO was inserted into the reactor and left for 5 min to generate micro-radicals into the mixture. Then 9 phr of MAH solution was slowly dropped into the reactor, using dropping funnel and left the reaction occurred in 8 h. This suitable condition was obtained from the other published work [2]. The DP-g-MAH latex was purified by high-speed centrifugation at 10,000 rpm for 5 min and redispersed with 1% (w/w) SDS, which would be used in the further mixing with silica filler by latex compounding.

The % grafting level of MAH was defined utilizing the absorbance ratio of FTIR peaks between 1720 cm^{-1} and 836 cm^{-1} (A_{1720}/A_{836}). The wavenumber at 1720 cm has corresponded to C=O stretching of grafted MAH and 836 cm was referred to =C–H deformation of main rubber chain. The calculation was based on the following equation (1).

$$\%\,grafting\ level\ of\ MAH = \frac{amount\ of\ MAH\ in\ sample\ (phr)}{amount\ of\ MAH\ total\ feeded\ (phr)} \times 100 \quad (1)$$

2.3. Preparation of silica filled DPNR-g-MAH latex by latex compounding method

To prepare silica solution 10 phr, silica was weight and kept stirring with speed at 800 rpm in DI water at 80 °C for 1 h. Then the hot silica solution was stirred vigorously by homogenizer with speed at 14,000 rpm for one h. The silica solution 10 phr was mixed with DP-*g*-MAH under mechanical agitation with speed at 1,000 rpm at 80 °C for 1 h. Then the silica filled DP-*g*-MAH latex was sonicated at room temperature for 1 h. Next, the silica filled DP-g-MAH latex was solidified by formic acid/acetone coagulation and leached with DI water to remove the excess acid. Finally, the sample was dried in a vacuum oven at 40 °C.

2.4. Characterization

Determination of nitrogen content of rubber, referring to the number of proteins was determined utilizing nitrogen analyzer (LECO®: model FP528). The wholly dried rubber sample was weighed approximately 0.3000 g and subjected into the machine. Then the rubber combustion was started and obtained nitrogen gas from the nitrogen compound, which was calculated and converted to numerical values from built-in software.

The chemical function was investigated by FTIR analyzer (JASCO®: model FTIR 4100 spectrometer). The purified sample, which was reprecipitated with light petroleum ether/acetone, was dissolved in chloroform ($CHCl_3$) to obtain 1% (w/v) solution. Then the solution was cast on KBr disc and transmitted by IR sources with 32 scans at 25 °C.

The chemical structure was investigated by ^{1}H-NMR (Bruker®: model 400 Ultrashield) The purified sample, which was reprecipitated with light petroleum ether/acetone, was dissolved in chloroform ($CDCl_3$). The sample was run under the following condition; scan number 500 scans, pulse delay time 10 sec, temperature 25 °C)

The gel content of rubber was determined using the following condition; the exacted dried rubber was weighed 0.1000 g and soaked in toluene AR grade 30 ml at room temperature for 7 days. The insoluble gel was separated by high-speed centrifugation and coagulated by methanol. Then the obtained gel was dried in an air-flow oven for 24 h. The gel content was calculated by the following eq. (2) [7].

$$Gel\ content\ (wt\%) = \frac{weight\ of\ obtained\ gel}{weight\ of\ sample} \times 100 \qquad (2)$$

The crosslink density of rubber was determined using a swelling test. The exacted dried rubber was weighed 0.1000 g and soaked in toluene AR grade 10 ml. at 40 ± 3 °C for 48 h. The solvent would be renewable every 24 h. After that period, the swollen sample was collected and then removed the excess toluene by a cage before weighing in a capped weighing bottle. The crosslink density (ν) of each rubber film was estimated using the Flory-Rehner's equation (3) [8-9].

$$v = \frac{1}{2M_c} = (\ln(1 - V_r^0) + V_r^0 + \chi {V_r^0}^2)/2\rho_r V_0 \left({V_r^0}^{\frac{1}{3}} - \frac{V_r^0}{2} \right) \qquad (3)$$

where v is crosslink density per gram of rubber, M_c is the molecular weight between crosslinks, V_r^0 is the volume fraction of the vulcanized rubber, χ is Huggins interaction constant V_0 is the molar volume of the solvent, ρ_r is the density of rubber, ρ_s is the density of solvent.

The V_r^0 was obtained using equation (4) as follow:

$$V_r^0 = \frac{1}{\rho_r/\rho_s}\left(\frac{W_s - W_u}{W_u}\right) + 1 \qquad (4)$$

where W_u and W_s are weights of the unswollen rubber sample and weight of swollen rubber sample, respectively. The values of the constants used in above equation (*i.e.*, ρ_r = 0.930 g/cm^3, ρ_s = 0.886 g/cm^3, V_0 = 106.9 cm^3/mol, and χ = 0.39).

The silica dispersion on rubber was investigated by FE-SEM (Hitachi (SU-8010). The silica filled rubber and bare rubber were cut in a cross-section by cryogenic fracture surfaces and coated with Pt-Pd before investigated under FE-SEM [10].

3. Results and discussion

3.1. The confirmation of deproteinized natural rubber

The removal proteins of NR or DPNR was successfully prepared by Kawahara et al. method due to the urea, which is used as a protein denaturant, could change the conformation of proteins on NR particles. This physical attachment between urea and proteins on NR particles could be pulled the proteins out by the centrifugal force, which occur while washing by high-speed centrifugation. The proteins would be soluble in serum fraction, which completely separated from rubber fraction. Moreover, the confirmation of nitrogen content of DPNR is dramatically decreased to 0.044 (%wt), comparing with fresh natural rubber (FNR), which is 0.725 (%wt).

3.2. The elucidation of maleic anhydride grafting on natural rubber

Figure 1 shows the FTIR spectra of DPNR (b), which compares with DP-g-MAH (a) at reaction time 8 h. The characteristic peaks of DPNR main chain present at 1664 and 836 cm^{-1}, which are related to C=C stretching and =C–H deformation, respectively. Moreover, there are additional peaks present at 1720 and 3455 cm^{-1} that represented to C=O stretching grafted succinic acid groups and –OH groups of hydrolyzed MAH grafting on DPNR, respectively. Conforming to the presence of water in DPNR latex cooperated with temperature at 60 °C, grafted MAH could be hydrolysis by water, and the open-ring of MAH would occur following the chemical equation showing in Figure 2.

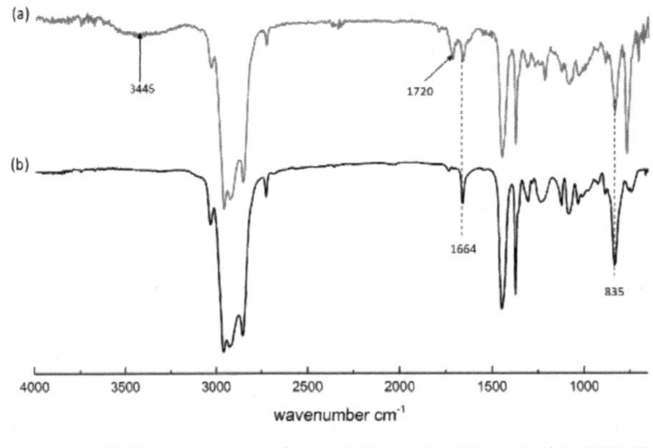

Figure 1. FTIR spectra of (a) DP-*g*-MAH and (b) DPNR.

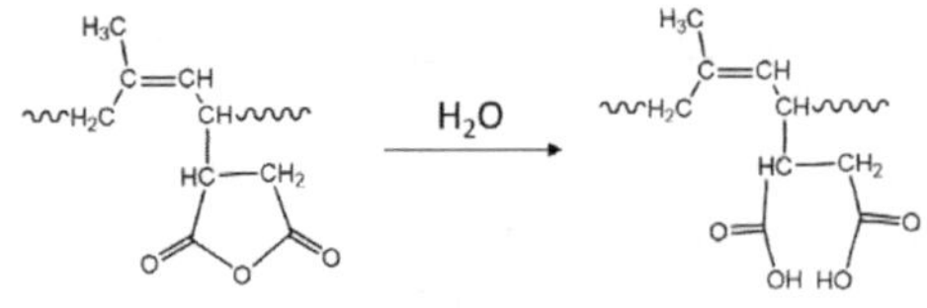

Figure 2. The chemical equation of DP-g-MAH, which was hydrolysis.

The % grafting level of MAH of DP-g-MAH is reported in Table 1, using the linear equation from the calibration curve ($y = 0.0382x$) that is obtained by plotting of the absorbance ratio between wavenumber 1720 and 836 cm^{-1} (A_{1720}/A_{836}) of IR blended MAH (IR/MAH) with the various concentrations of MAH (*i.e.*, 3, 5, and 9 phr). The absorbance peaks at 1720 and 836 cm^{-1}, which

IOP Conf. Series: Materials Science and Engineering 1173 (2021) 012036 doi:10.1088/1757-899X/1173/1/012036

corresponded to C=O stretching of blended MAH in the samples and –C=H deformation of IR main chain, respectively. In addition, the gel content of DP-*g*-MAH approximates 14.45 (wt%) as shown in Table 1, which is higher than DPNR 5.50 (wt%).

Table 1. The % grafting level of MAH and gel content of DP-*g*-MAH.

Grafted rubber	A_{1720}/A_{836}	grafting level of MAH (%)	Gel content (wt%)
DP-*g*-MAH	0.34165	99.37	14.54 ± 2

The chemical structure of DP-*g*-MAH is investigated by ^{1}H-NMR as demonstrated in Figure 3. It was found that the characteristic signals of DPNR presented at 1.68 (a), 2.04 (b, b*), and 5.13 (c) ppm, which related to methyl, methylene, and unsaturated methine proton, respectively. Moreover, the additional signals presented the doublet signal at 2.92 and 2.93 (d) ppm of the two methylene protons and the singlet signal at 3.49 (e) ppm of the methine proton of grafted MAH rings. In addition, some chemical shifts were reported about the resonances grafted MAH on DPNR that expected the two methylene protons of grafted MAH rings at the multiple positions along the DPNR chain, such as the triplet signals at 3.64 (f) and 3.71 (g) ppm. The possible reason for this consequence was suggested that the grafted succinic anhydride could subject to the allylic carbon atom. Resulting, the two methylene protons would be deshielded in the position of the C=C bond and presented at the higher chemical shift. At 7.26 (h) ppm, the chemical shift was the chloroform-d that utilized to dissolve the rubber sample.

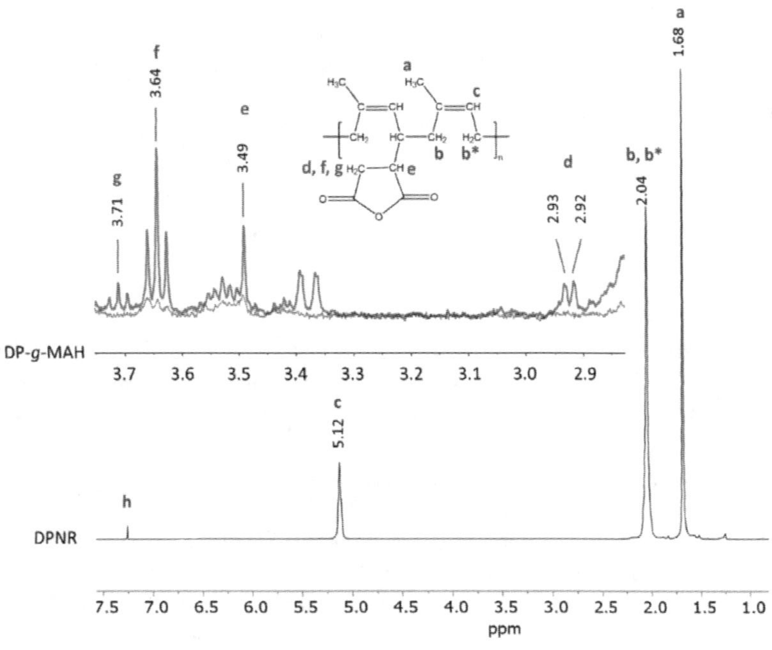

Figure 3. the comparison of ^{1}H-NMR spectra of DPNR and DP-*g*-MAH.

3.3. The effect of silica dispersion on DPNR and DP-g-MAH

The FE-SEM micrographs of DPNR/Si and DP-*g*-MAH/Si 10 phr are shown in Figure 3. It could be seen that the silica filler well dispersed on DP-*g*-MAH than DPNR. Since, the DP-*g*-MAH presents the higher polarity, which corresponds to be compatible with silica filler than DPNR.

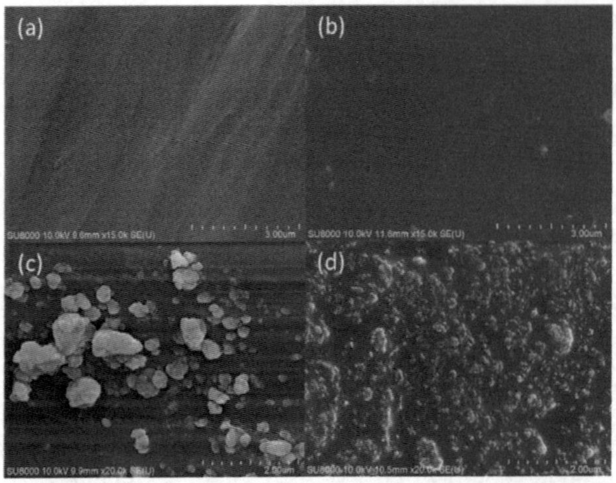

Figure 4. FE-SEM micrographs of plain DPNR (a), DP-*g*-MAH (b), DP/Si 10 (c), and DP-*g*-MAH/Si 10 (d).

3.4. The effect of chemical networks by esterification between silica filled DPNR, and DP-g-MAH

The crosslink density, used to affirm the chemical networks between silica filler and rubber, has been investigated. The result is presented in Table 2, which DP-*g*-MAH/Si 10 is slightly higher than DP/Si 10. Consequently, there would be the part of additional networks between silica filler and rubber, which are ester bonds of carbonyl groups ($-COO^-$) of grafted MAH and hydroxyl groups (–OH) of the silica surface.

Table 2. The crosslink density of silica filled DPNR and DP-*g*-MAH.

Silica filled rubber	Crosslink density (mol/g) (10^{-5})
DP/Si 10	2.85
DP-*g*-MAH/Si 10	3.28

4. Conclusion

The DP-*g*-MAH is successfully synthesized according to the FTIR and ^{1}H-NMR results and achieved the % grafting level of MAH approximates to 99.37% accompany with the low gel content around 14.45%. In addition, the silica filled grafted rubber could show the silica dispersion, which is better than the plain rubber. Resulting from the grafted rubber additionally consists of carboxyl groups (–COOH), which could interact with hydroxyl groups (–OH) on silica filler. Moreover, the crosslink density of grafted rubber is marginally increasing from 2.85 to 3.82 x 10^{-5} (mol/g). It is a consequence of the increment of chemical crosslink networks between rubber and filler. It should be noted that the value of crosslink density is obtained from the unvulcanized rubber or silica filled rubber without curing agents. If commonly used cross-linking agents have also crosslinked the rubber chains, the reinforcing efficiency would be stronger.

References

[1] Saelao J, Phinyocheep P. Influence of styrene on grafting efficiency of maleic anhydride onto natural rubber. Journal of Applied Polymer Science. 2005;95(1):28-38.

[2] Wongthong P, Nakason C, Pan Q, Rempel GL, Kiatkamjornwong S. Modification of deproteinized natural rubber via grafting polymerization with maleic anhydride. European Polymer Journal. 2013;49(12):4035-46.

[3] Wongthong P, Nakason C, Pan Q, Rempel GL, Kiatkamjornwong S. Styrene-assisted grafting of maleic anhydride onto deproteinized natural rubber. European Polymer Journal. 2014; 59:144-55.
[4] Phinyocheep P. 3 - Chemical modification of natural rubber (NR) for improved performance. In: Kohjiya S, Ikeda Y, editors. Chemistry, Manufacture and Applications of Natural Rubber: Woodhead Publishing; 2014. p. 68-118.
[5] Yusof NH, Kosugi K, Song TK, Kawahara S. Preparation and characterization of poly (stearyl methacrylate) grafted natural rubber in latex stage. Polymer. 2016; 88:43-51.
[6] Kawahara S, Klinklai W, Kuroda H, Isono Y. Removal of proteins from natural rubber with urea. Polymers for Advanced Technologies. 2004;15(4):181-4.
[7] Sriring M, Nimpaiboon A, Kumarn S, Higaki K, Higaki Y, Kojio K, et al. Film formation process of natural rubber latex particles: roles of the particle size and distribution of non-rubber species on film microstructure. Colloids and Surfaces A: Physicochemical and Engineering Aspects. 2020; 592:124571.
[8] Flory PJ. Statistical mechanics of swelling of network structures. The Journal of Chemical Physics. 1950;18(1):108-11.
[9] Sriring M, Nimpaiboon A, Kumarn S, Sirisinha C, Sakdapipanich J, Toki S. Viscoelastic and mechanical properties of large- and small-particle natural rubber before and after vulcanization. Polymer Testing. 2018; 70:127-34.
[10] Rattanasom N, Saowapark T, Deeprasertkul C. Reinforcement of natural rubber with silica/carbon black hybrid filler. Polymer Testing. 2007;26(3):369-77.

Prototype of automatic conveyor system using programmable logic controller for educational purpose

T W O Putri [1,*], and M I Mowaviq [1]

[1] Department of Electrical Engineering, Faculty of Electrical Power and Renewable Energy, Institut Teknologi PLN, Jakarta, Indonesia

* Email: triwahyu@itpln.ac.id

Abstract. The study of conveyor control and programming using PLC (Programmable Logic Controller) has been taught in many universities, especially in the field of Electrical Engineering. Unfortunately, the material obtained by students is only in the form of theory without direct application to the conveyor system in the industry because it is constrained by the company policy. Therefore, it is necessary to build a prototype of automatic conveyor system using PLC for educational purpose. This prototype is owned by the university so that students can directly implement the results of the PLC program design that they have learned in class. The prototype of the automatic conveyor system can increase students' interest and understanding of the industrial automation process so that they will be better prepared when they have implemented it directly to the industry. This research discusses the design and manufacture of a prototype of automatic conveyor system using PLC. The main controlling device used is the FX3U Mitsubishi PLC which is equipped with analogue inputs, analogue outputs, and high-speed counter output as PWM (Pulse Width Modulation) output. The conveyor used is a mini conveyor with a DC motor drive. Conveyor speed control is done by adjusting the rotational speed of the DC motor using direct PWM setting from PLC. To complete the automation system, infrared sensor was added as the input to detect the dummy load. Thus, full automation system of conveyor for educational purpose was done and it showed satisfying results.

Keywords: PLC, automation, prototype, conveyor, pulse width modulation

1. Introduction

Automation is intended so that machines in an industry move automatically without the need to be run by humans. Automation is a technology whose processes and procedures are completed without human involvement [1]. So that in general, an automation system can be defined as a process that can function automatically using technology related to electronic mechanical applications and computer-based systems without human intervention [2]. The use of humans in an automated production process is only to oversee the production process. Thus, the automation of machines in an industry provides an advantage for the company because it can reduce the number of workers which results in the company's low operating costs.

Programmable Logic Controller (PLC) is an automatic control device that is widely used in various sectors, especially the industrial sector. PLC according to Capiel (1982) is an electronic system that operates digitally and is designed for use in industrial environments, where this system uses programmable memory for internal storage of instructions that implement specific functions such as logic, sequence, timing, enumeration, and arithmetic operations to control machines or processes via digital and analog I / O modules [3]. Programmable Logic Controller (PLC) is an automatic control device that is widely used in various sectors, especially the industrial sector. PLC according to Capiel (1982) is an electronic system that operates digitally and is designed for use in industrial environments, where this system uses programmable memory for internal storage of instructions that implement specific functions such as logic, sequence, timing, enumeration, and arithmetic operations to control machines or processes via digital or analog I / O modules [3].

In general, the PLC functions are as follows [4]:

1. *Sequential Control*
   PLC processes the binary signal input into output which is used for technical processing purposes in sequence (sequential). PLC keeps all steps or steps in a sequential process taking place in the right order.
2. *Plant Monitoring*
   The PLC continuously monitors the status of a system (e.g.: temperature, pressure or altitude) and takes the necessary action in connection with the controlled process (e.g.: value has exceeded the limit) or displays these messages to the operator.

One of the most common industrial implementations of PLCs is its use for conveyor belt automation. In order to the conveyor can work automatically, it is necessary to design appropriate instruments and control systems using a PLC and its programming. The study of PLCs basically depends on the type and brand of PLC. Different types and brands, different PLC programs are used even though logically the program is still the same. Therefore, before the PLC can be implemented as a controlling device in the production process, it is necessary to do in-depth research and study regarding the PLC itself. This research is related to the syntax and block function of the PLC ladder diagram. Hardware research is also important. In hardware, research is carried out by studying the specifications and maps of the PLC input output pins.

The study of conveyor control and programming using PLC (Programmable Logic Controller) has been taught in many universities, especially in the field of Electrical Engineering. Unfortunately, the material obtained by students is only in the form of theory without direct application to the conveyor system in the industry because it is constrained by the company policy. Therefore, it is necessary to build a prototype of automatic conveyor system using PLC for educational purpose. This prototype is owned by the university so that students can directly implement the results of the PLC program design that they have learned in class. The prototype of the automatic conveyor system can increase students' interest and understanding of the industrial automation process so that they will be better prepared when they have implemented it directly to the industry.

This research discusses the design and manufacture of a prototype of automatic conveyor system using PLC. The main controlling device used is the FX3U Mitsubishi PLC which is equipped with analogue inputs, analogue outputs, and high-speed counter output as PWM (Pulse Width Modulation) output. The conveyor used is a mini conveyor with a DC motor drive. Conveyor speed control is done by adjusting the rotational speed of the DC motor using direct PWM setting from PLC. To complete the automation system, infrared sensor was added as the input to detect the dummy load.

2. Research Method

The research method in this study is divided into several stages, including: Phase I is a literature study related to conveyor mechanisms, infrared sensors, pulse width adjustment on DC motors, PLC programming, and PLC technical specifications. The expected output at this stage is to understand how to program the FX3U type PLC, how the infrared sensor works, adjust the pulse width of the DC motor, and PLC specifications. Phase II is the design stage of an automatic conveyor system including software design, hardware design, and instrumentation.

At this stage, it is hoped that the results of the software and hardware design and system instrumentation will be obtained. Stage III is the prototyping stage for the automatic conveyor system, including the manufacture of hardware, programs and instrumentation. At this stage it is expected to produce a conveyor system prototype. Finally, Phase IV is testing and correcting the prototypes that have been made. At this stage it also includes drawing conclusions and making scientific articles related to research.

3. Results and Discussion

In this section we discuss the result of each research's step start from I/O configuration, prototype designing and building, PLC ladder diagram, and the last step is testing section.

3.1. PLC I/O Configuration

The PLC input / output configurations for this conveyor prototype are:

- Power supply : 24 VDC
- Digital input : switch on / off (on pin X000); Infrared sensor (on pin X001)
- PWM output : DC motor (at pin Y000)
- Communication : RS-232 to USB

The pinout of I/O configuration based on the list above can be seen on Figure 1.

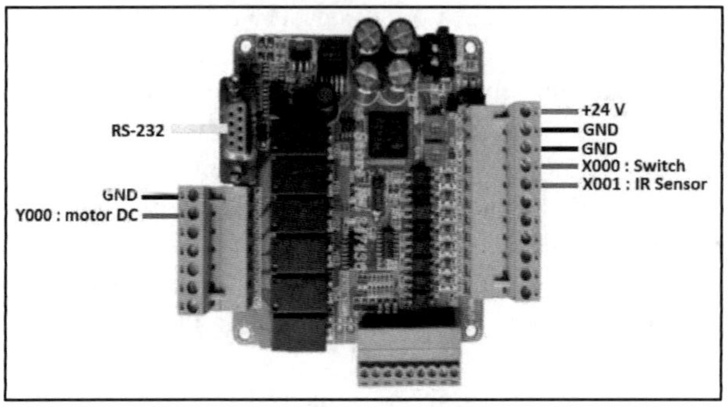

Figure 1. I/O Configuration for conveyor prototype.

3.2. Prototype designing and building

Prototyping is done in accordance with the research stages. Mini conveyors are made according to the desired dimensions. The DC motor used is a 24 Volt DC motor with a maximum rotation of 100 rpm and a torque of up to 16 kg. Infrared sensor was installed on the conveyor. Figure 2 shows the installed IR sensor. The manufacture of mini conveyors is carried out in a separate workshop. The next step is to install a tachometer as a motor rotation speed sensor / conveyor speed.

The installation of the speed sensor is not the part of the I / O configuration in Figure 1 and the speed sensor itself was not connected with PLC. The need for installing speed sensor is to test the characteristics of the DC motor used as a conveyor drive. After the speed sensor has been installed, the relation of motor voltage according to the motor speed (in RPM) could be get.

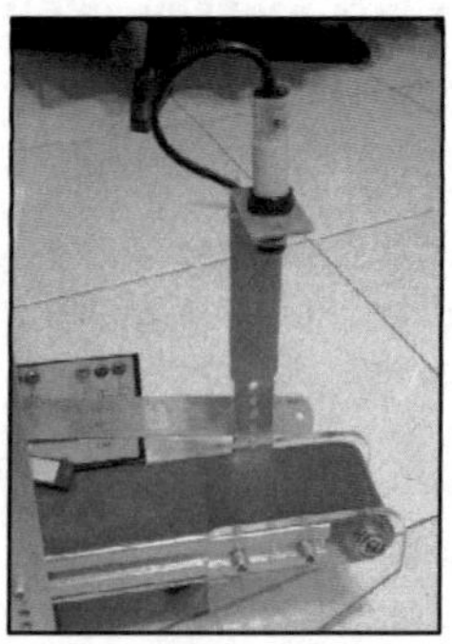

Figure 2. Installed IR sensos

The next step is the manufacture of a control module consisting of a PLC, switch, infrared sensor, and input / output terminals. The manufacture of the control module needs to be adjusted to the wiring diagram between the PLC and the conveyor. Figure 3 shows the conveyor prototype and control module that has been made.

Figure 3. I/O Configuration for conveyor prototype.

3.3. Ladder diagram

Ladder Diagram is a programming language for PLC in the form of a ladder diagram. Making ladder diagrams for a prototype automatic conveyor system with speed control based on the FX3U PLC was carried out using special software for the Mitsubishi FX3U type PLC, namely the GX-Developer software. Programming is done on a PC / Laptop that is connected to the FX3U PLC via an RS-232 serial cable.

The programming criteria are as follows:

1. If the switch is ON, the system will turn on, and vice versa.
2. If the infrared sensor detects a dummy load, the conveyor will run until the system is turned off (switch off)
3. The desired conveyor rotation speed is input on the ladder diagram in the "PWM" function. K3810 is the constant value of PWM where the duty cycle is 100%.
4. The output in the form of a high-speed counter on pin Y000 is connected to the DC motor as a conveyor drive
5. Internal memory in the form of M0 and M1 is used for the on / off system and sensor's latching.

Based on the criteria above, a ladder diagram can be developed using the GX-Developer software. The ladder diagram of the system can be seen in Figure 4.

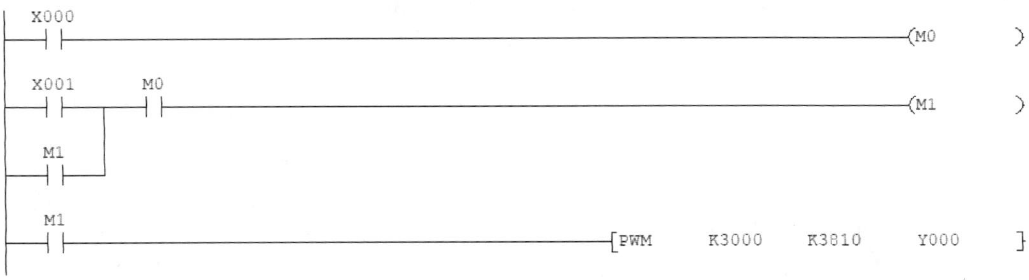

Figure 4. Ladder diagram

3.4. Overall system testing

Testing aims to determine whether the system that has been created is in accordance with the research objectives. System testing is divided into several stages. The first is testing to determine the relationship between the motor voltage and the conveyor rotation speed. The result of this stage is DC motor's characteristic. Figure 5 shows the process of testing the relationship between motor voltage and conveyor rotation speed using a speed sensor, laptop, and voltmeter.

Figure 5. Measuring motor voltage to get motor's characteristic

Based on the test results, the results are shown in Figure 6, where the motor voltage is directly proportional to the motor rotation. But there is a DC motor dead-zone, which is at voltages below 6 volts where voltages less than 6 volts will not be able to drive the DC motor. The dead-zone then being the limitation of desired speed input of the system.

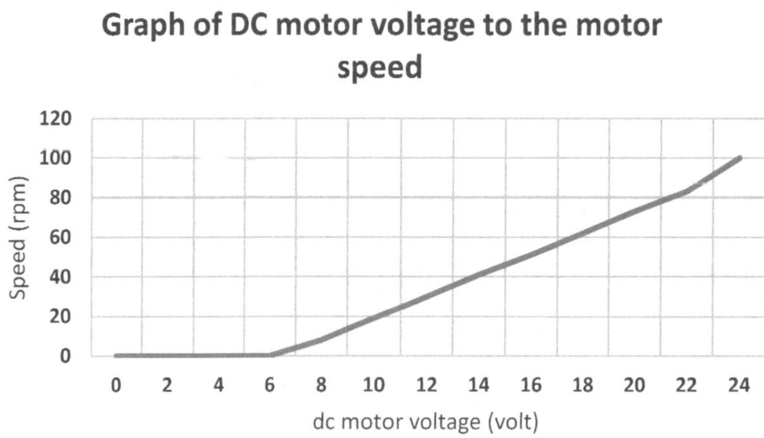

Figure 6. Graph of DC motor characteristic

The next testing is for the relationship between the PWM active pulse width and the motor voltage. The equation for motor rotation speed is as the eq. 1:

$$n = \frac{E_i - I_a R_a}{C\Phi} \quad (1)$$

where:

E_i = input voltage (V)

L_a = armature inductance (H)

R_a = armature resistance (Ω)

I_a = armature current (A)

Φ = flux

n = speed (rpm)

Based on the motor rotation speed equation, it can be seen that the motor rotation speed regulation can be done by changing the magnetic flux or changing the input voltage. Since the permanent magnet motor has a fixed magnetic flux value, adjustments can be made by changing the input voltage (E_i). The technique of changing the voltage is done by using PWM, namely by varying the width of the duty cycle (active cycle) of the motor input voltage. By setting 3810 ms as a constant 1 cycle, it can be calculated the PWM duty cycle which affects the PLC output voltage to drive a DC motor.

Based on the graph of the results of testing the relationship between the PWM active pulse width and the motor voltage in Figure 7, the results show that the relationship between the PWM active pulse width and the motor voltage is linear.

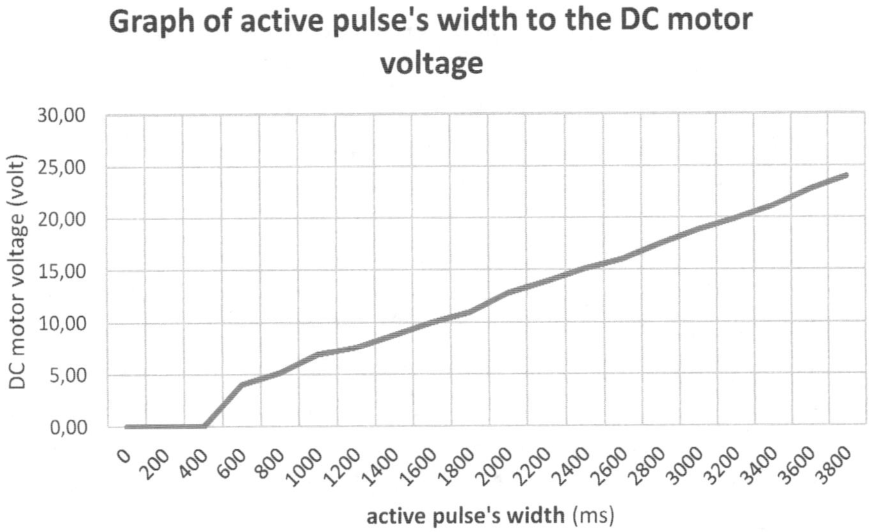

Figure 7. Graph of active pulse's width to the DC motor voltage

The last stage is testing the whole system using the ladder diagram that has been made. Selected some of the desired setpoint / conveyor speed with the setpoint limit must be more than 50rpm to avoid dead-zone and no external disturbance of the system. From the final testing, the results were satisfying, and the conveyor prototype worked as the criteria of the system, where the conveyor worked when the switch set to On and the infrared sensor detected the dummy load. The conveyor speed depended on the PWM setting. Then the conveyor would be stop working if the switch turned off. Thus, full automation system of conveyor for educational purpose was done and it showed satisfying results.

4. Conclusion

The design of an Automatic Conveyor System using a Programmable Logic Controller for educational purposes has been successfully made. The main controlling device used is the FX3U Mitsubishi PLC, which is equipped with an analog input, an analog output and a PWM (Pulse Width Modulation) output. The conveyor used is a mini conveyor with a DC motor drive and an infrared sensor. Conveyor speed control is done by adjusting the rotation speed of the DC motor using a PLC-based pulse width setting. The program is created with ladder diagrams using GX-Developer software. Conveyor speed regulation is done using the Pulse Width Modulation (PWM) method found on the PLC programming tool for FX3U PLCs, namely GX Developer. Based on the test data, the relationship between the conveyor motor voltage and the motor rotation speed is directly proportional. Likewise, the relationship between the PWM active pulse width and the motor voltage. System automation is obtained when the conveyor works automatically at the desired speed if the infrared sensor detects a dummy load. From the final testing, the results were satisfying, and the conveyor prototype worked as the criteria of the system, where the conveyor worked when the switch set to On and the infrared sensor detected the dummy load. The conveyor speed depended on the PWM setting. Then the conveyor would be stop working if the switch turned off. Thus, full automation system of conveyor for educational purpose was done and it showed satisfying results.

Acknowledgments
The author would like to thank the Ministry of Research and Technology-the National Research and Innovation Agency for providing the Beginner Lecturer Research Grants so that this research can be completed.

References

[1] Groover M P 2001 *Automation, Production System, and Computer Integrated Manufacturing* Surabaya: Guna Widya

[2] Kurniadi A N, Rachmat H, & Atmaja D S E 2015 *Perancangan Program Sistem Pengendali Untuk Otomatisasi Proses Pengepakan Teh Menggunakan Plc Omron Cp1E Di Pt. Pn Viii Unit Sinumbra*. eProceedings of Engineering, 2(1).

[3] Kusuma A 2013 *Pengertian PLC dan Jenis-jenis PLC.* (Online: http://kusuma-w-arya.blogspot.com/2013/05/pengertian-plc-dan-jenis-jenis-plc.html, accessed: 28 Februari 2018)

[4] Putri T W O 2014 *Pengendalian Suhu Pada Sistem Pasteurisasi Telur Cair Berbasis PLC (Programmable Logic Controller) Siemens Simatic S7-200 dan Hmi (Human Machine Interface) Simatic HMI Panel* Student Thesis Universitas Brawijaya Malang Indonesia

[5] Andriana Y & Triwiyatno A 2011 *Aplikasi Mikrokontroler AT89C51 untuk Pengendalian Kecepatan DC motor dengan menggunakan Gelombang Radio sebagai Media Transmisi* (Doctoral dissertation, Jurusan Teknik Elektro Fakultas Teknik).

iAMR: intelligent auto-dispenser mosquito repellent system

W A W A Bakar[1]*, M Man[2]*, L C Hwa[3], and L Y Wang[4]

[1] Faculty od Informatics and Computing, Universiti Sultan Zainal Abidin, Besut Campus, 22200 Besut, Terengganu, Malaysia

[2] Faculty of Ocean Engineering Technology and Informatics, Universiti Malaysia Terengganu, 21030 Kuala Nerus, Terengganu, Malaysia

[3] One Team Network Sdn. Bhd., 1-1B, Block 6, Jalan Pahat 15/H, Section 15, 40200 Shah Alam, Selangor, Malaysia

[4] DrMos Health Care Sdn. Bhd., 1st Floor, No 28, Jalan 4 Aman Square, Sungai Petani, 08000 Kedah, Malaysia

* Email: wanaezwani@unisza.edu.my, mustafaman@umt.edu.my

Abstract. Dengue fever has become an increasingly significant health threat worldwide. Currently, due to lack of specific treatment or vaccine against dengue, vector control, which is available in many forms, remains a key strategy for dengue fever prevention. Hence, we introduce an iAMR: Intelligent Auto-Dispenser Mosquito Repellent, an innovative system in line with "Industry 4.0" concept where it includes current trend of automation and data exchange. It also includes cyber-physical systems, the Internet of things, cloud computing and cognitive computing. The iAMR solves the issues of when to spray (by auto-spray using sensor) and long live batteries (with LORAWAN technology). When combined with the mini aerosol (X'MOS), it provides an even more effective mosquito control. This all-in-one system ensures a mosquito-free environment in your home. The iAMR device relies on the microcontroller and sensor in order to operate the intended spraying task. The microcontroller used is an IoT based device called ESP8266 which is a WiFi-embedded microcontroller that utilized standard everyday WiFi band frequency which is at 2.4 GHz. A DC geared motor sensor for dispenser device is used. The main contribution of iAMR is to monitor and track for the mosquito repellant (X'MOS) to be refilled. The implementation shows the iAMR system able to update the level of each X'MOS in iAMR devices which reducing the cost of manual human checking for X'Mos refill.

Keywords: dengue, auto-dispenser, mosquito repellent, motor sensor, ESP8266 microcontroller

1. Introduction

Dengue fever is a mosquito-borne tropical disease caused by the dengue virus from Aedes aegypti mosquito. The dengue symptoms are detected starting in three to fourteen days after infection. Dengue is severe and may affects infants, young children and adults, but seldom causes death. Because of its severity, the global efforts are always put in urgent matters. Many products on preventing dengue have been used in the market. In this paper, we introduce the auto-spray dispenser specially made for combating Aedes mosquito that constitutes to dengue disease.

This work is a collaboration between higher academic institutions and the industry that is directly involved in this vector-borne disease. The objective of this research is to provide a solution for fighting Aedes that is a main contributor of dangerous dengue disease by applying IOT technology. Research on Industry 4.0 or IR4.0 is trending for Z generation. Many emerging domains of fields especially medical, engineering, agriculture, marine and fisheries, automotive and computing and robotics require the demanding technologies to replace labor operational energy and cost.

The rest of the sections includes Section 2 for our related works on smart iot dispensing system, Section 3 illustrates our iAMR projects, Section 4 explains the experimentation being done, Section 5 describes on the result and discussion and Section 6 recapitulates the conclusion and future works.

2. Related Works

Agricultural Robot (Agribot) is developed by [1] that capable of doing activities of ploughing, seed dispensing, fertilizing and harvesting. They utilized the Arduino UNO microcontroller with the Wifi module whereas the other components include DC motor (control the wheels), stepper motor (control the seeds), relay board (control various activities) and a pump motor for an automatic flish and shut-off spraying and spreading the fertilizer. Effort by [2] proposed a wireless robot with various sensors for measuring different environmental parameters.

The robot is developed with Raspberry Pi 2 model B hardware for executing the whole process. The main feature is to execute tasks such as moisture sensing, scaring birds and animals, spraying pesticides, moving forward or backward and switching ON/OFF electric motor. The work by [3] aims at providing the Automated Medical Aid due to the high rate of death as the results of poor hospitality and medical services by the authority. The project used the components such as Raspberry Pi 3, temperature sensor (DS18B20), Load cell to measure weight, load cell amplifier (HX711), that is a 24 bit analog to digital converter (ADC), a pulse sensor to measure the heart rate, pressure sensor, servo motor for a vending machine with 360 degree rotation and SG9U SERVO that can rotate in 180 degree of rotation.

The four modes of communications define in [4] are wireless system, power line systems, hardwired systems and internet protocol systems. The authors developed the prototype of Smart Home Controlled System (SHCS) using Arduino IDE platform (APC220 wireless module and Ethernet W5100) and sensors like ultrasonic, Passive InfraRed (PIR) sensor, gas and temperature sensor. The experimental results show SHCS was successfully tested in the prototype as well as real environment. The results indicate very low error percentage and has improved the traditional home since the SHCS can be controlled through IOT or smartphone. The next section illustrates our iAMR in detail with the components that we use during project development.

An effort is initiated to present a holistic conceptual approach of an IOT system development and implementation to enhance bathroom safety [5]. The components that could be used as suggested are leak detection sensor to detect leakage at floor, digital light/LUX sensor to measure intensity of light radiation, voice detection sensor to detect acoustic and noise signals, pressure sensor to detect magnitude of contact pressure that able to detect person's status using toilet and bath tub, positional sensor (GYRO/Accelerometer/Magnometer) to detect tilt, pitch and inclination, motion sensor to detect movement or water flow sensor to detect water leakage. The paper demonstrates tremendous potential IOT especially for bathroom safety for elderly.

3. Project Description

In this section we discuss how the iAMR project is initiated with the collaboration of the product component from industry.

3.1. Project background

The iAMR project is a blended technology of auto spry dispenser, Aedes mosquito repellant and IOT equipment. The iAMR Dispenser is a product manufactured by DrMOS Healthcare Sdn. Bhd, the X'MOS spray is an Aedes Mosquito Repellent from On Team Network Sdn. Bhd. while the IOT equips are ESP8266 WiFi-embedded microcontroller and the motor sensor.

The components used are the iMOS dispenser with X'MOS spray, Node MCU ESP8266 Module as the Arduino microcontroller, DC Geared Motor for pressing the spray lid and the LoRaWAN [6] gateway for data transmission in cloud environment. These components are depicted in fig. 1 to fig. 4.

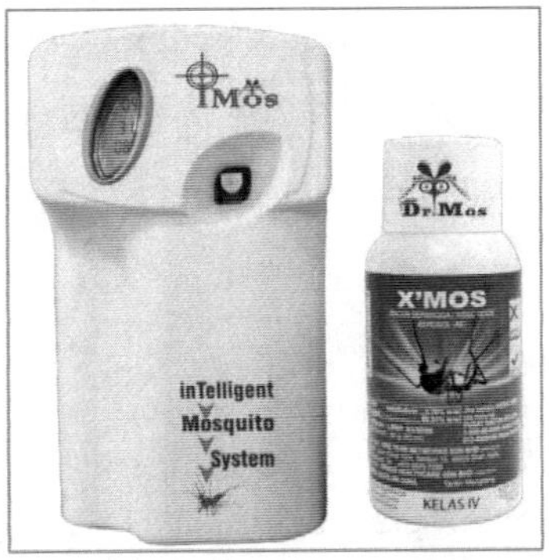

Figure 1. The iMOS dispenser and the X'MOS in iAMR project

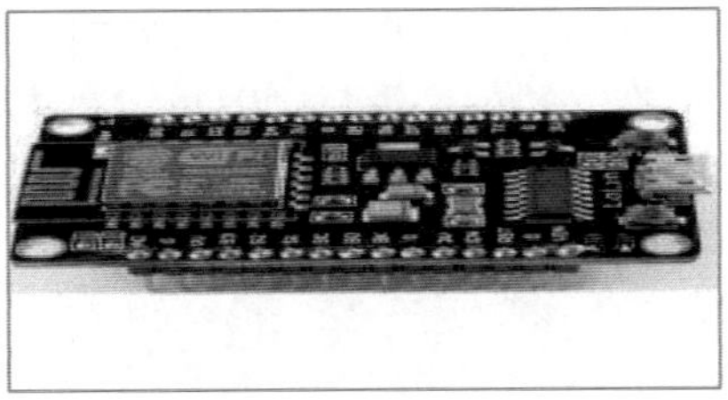

Figure 2. Node MCU ESP8266 Module

Figure 3. DC Geared Motor – 12V (10 to 1000 RPM)

Figure 4. LoRaWAN gateway

3.2. Setting Area for iAMR installation

After installing the iAMR, customers can enjoy the life without mosquitoes at home. Most wonderful of iAMR is, after a period of use, all the insects in the house, including cockroaches and ants, will stay away from home. For a 250ml of X'MOS, it could cover up to 300 ft. But it depends on the spraying frequency with the setting area of the location (room, living room or office area). The proposed setting time and area location of these combination is given in Table 1.

Table 1. Proposed setting for iAMR spray interval

Room Type (no of iMos to be installed)	**Start Time**	**End Time**	**Spray Interval**	**Xmos Exchange interval (day)**
Living Room (1)	A: 6.00 am B: 5.00 pm	A: 6.12 am B: 5.12 pm	1 min (12 sprays) 1 min (12 sprays)	33 days
Kitchen (1)	A: 6.00 am B: 5.00 pm	A: 6.12 am B: 5.12 pm	1 min (12 sprays) 1 min (12 sprays)	33 days
Bed Room (1)	A: 9.00 pm	A: 9.12 am	1 min (12 sprays)	66 days
Kindergarten/day care *entrance (1) *back door (1) *each class room (1)	A: 6.00 am B: 12.00 pm	A: 6.12 am B: 12.12 pm	1 min (12 sprays) 1 min (12 sprays)	33 days
Office *entrance (1) *back door (1) *each room (1)	A: 6.00 am B: 12.00 pm	A: 6.12 am B: 12.12 pm	1 min (12 sprays) 1 min (12 sprays)	33 days

Because of the manual operation, then a few precautions need to be undertaken to ensure the optimum effects (as referred in Figure 5 to Figure 7). The precautions include first, record the installation date, and pre-set X'MOS exchange date as reminder in dealer's smart phone calendar. Second, must check battery after 10 months. Third, must estimate the area of premises where 1 unit of iAMR for every 300 square feet. Fourth, to remind customer do not switch on the exhaust fan after the iAMR spraying time. And lastly, the iAMR should be set at the entrance (door) or near the window at about 6" (inches) from the ceiling.

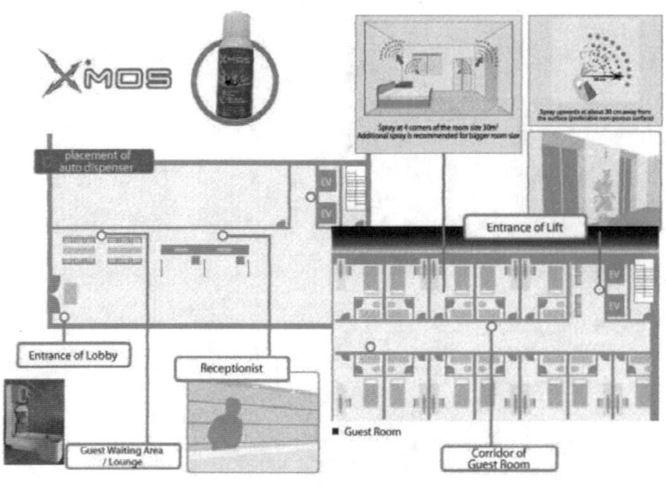

Figure 5. The illustration of iAMR installation in a building

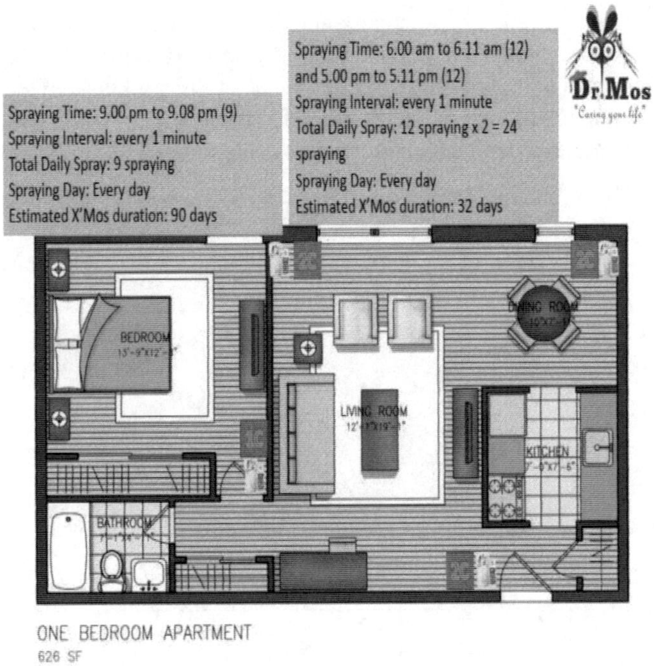

Figure 6. The illustration of iAMR installation in an apartment

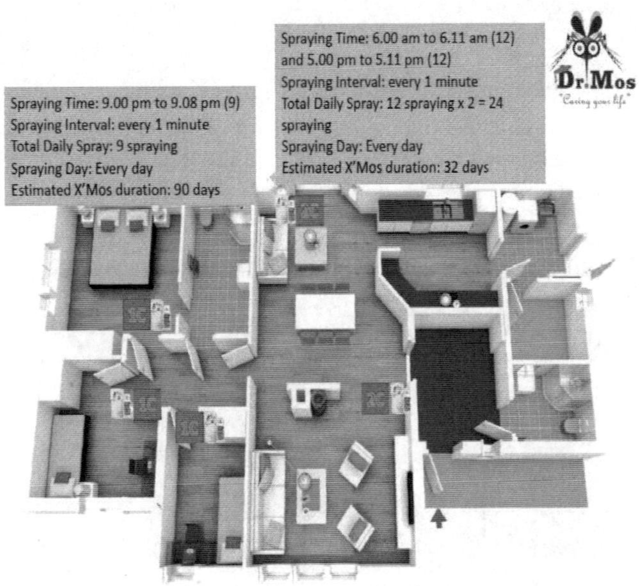

Figure 7. The illustration of iAMR installation in a bungalow

3.3. iAMR Network Architecture

The LoRaWAN® is a Low Power, Wide Area (LPWA) networking protocol designed to wirelessly connect battery operated 'things' to the internet in regional, national or global networks, and targets key Internet of Things (IOT) requirements such as bi-directional communication, end-to-end security, mobility and localization services (refer to Figure 8).

It is based on star-of-stars topology in which gateways relay messages between end-devices and a central network server. The gateways are connected to the network server via standard IP connections and act as a transparent bridge, simply converting RF packets to IP packets and vice versa. The wireless communication takes advantage of long-range characteristics of the LoRaWAN physical layer, allowing a single-hop link between the end-device and one or many gateways. The LoRaWAN® band rates range from 0.3 kbps to 50 kbps [6].

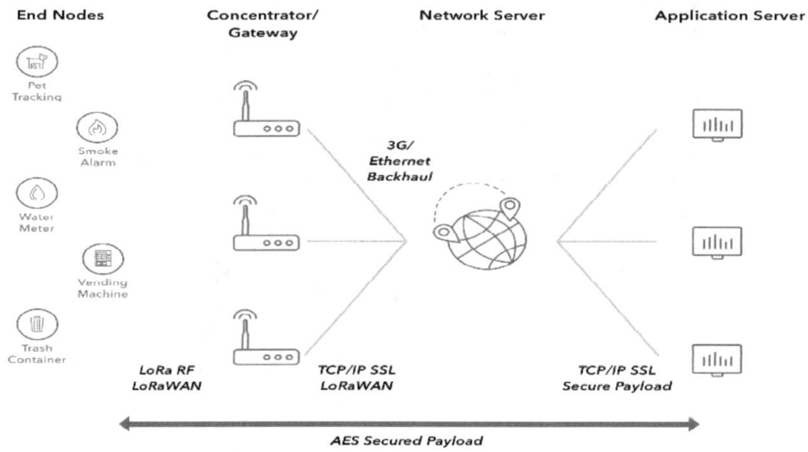

Figure 8. LoRaWAN network architecture deploys for iAMR

4. Results and Discussion

The iAMR system is developed with open-source programming language which is Php and MySQL Database Management System with Xampp control engine. The URL of the program is as follow https://mospryiot.musproject.com. The snapshot is given in Figure 9.

Figure 9. Main screen of iAMR system

When clicking on the dashboard, the status of X'MOS spray is forecasted using different color scheme i.e., Blue for high level of spray, Yellow that indicates the medium level of spray and Red in color for low level of liquid in X'MOS spray. This is shown in Figure 10.

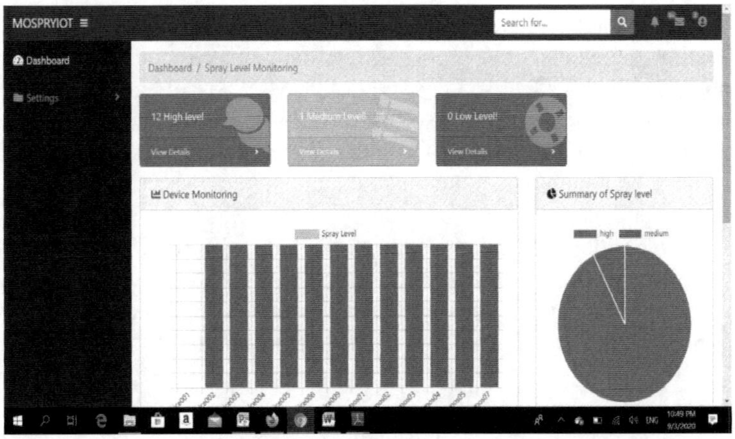

Figure 10. The level of liquid in X'MOS spray

Basically, when the level is getting LOW by red color, then we can click on the View Details to see the location by longitude and latitude as in Figure 11. The details consist of device id (iAMR device id), the water level (in cm), the location, the date and time and the View button.

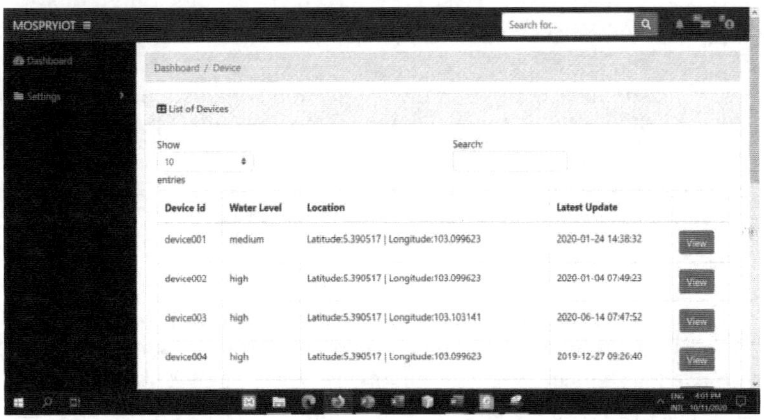

Figure 11. The detail location of the iAMR device

When we click on the View button it reaches us to this google maps page as in Figure 12.

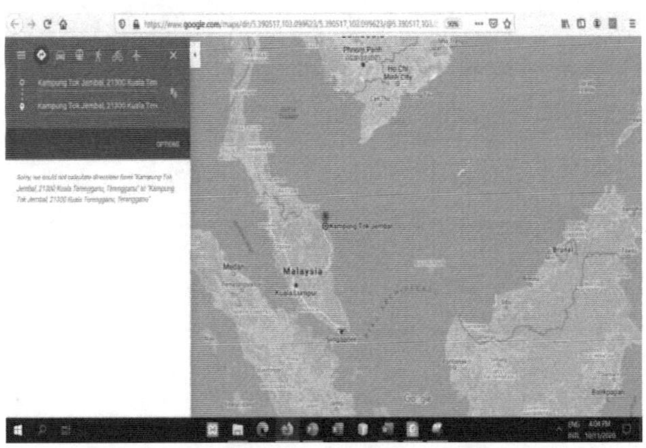

Figure 12. Google Maps showing the iAMR exact device location

Following this experimentation using LoRaWAN communication has achieved a success with a compact communication device can cover for long read range of communication data. This is attained by the iAMR dashboard where we can see the level of X'MOS spray and the battery are still on for a long period of time. On the other hand, the issue with the battery sustainability where the investigation is still continues on finding the right solution to sustain the long- lasting battery.

5. Conclusion

To resolve, we have two options i.e. first, we can use the battery in our iAMR or secondly, we can use direct current (DC) for the iAMR to continue its operation ability. Only when using DC, we have to investigate on the cost of the electricity bills charged from the usage. For future work, we would test on the different types of battery and try to come out with the cost analysis of electricity usage.

Acknowledgments

We wish to thank to our financial support under UniSZA grant with code UniSZA/2018/PKP/01 (An intelligent mosquito home system) and UMT Grant Vot: 53290. A sincere gratitude goes to AedesTechApps Mosquito Home System (AMHS) co-founder, One Team Network Sdn. Bhd. and Dr. MOS Healthcare Sdn. Bhd. and all faculty members in supporting our work.

References

[1] ArunaG G G and AnbuNivetha V 2017 Internet Of Things Based Innovative Agriculture Automation Using AGRIBOT *International Journal of Electronics and Communication Engineering* ISSN pp 2348-8549

[2] Lokesh K K et al 2017 Internet of Things application for implementation of smart agriculture system *2017 International Conference on I-SMAC (IoT in Social, Mobile, Analytics and Cloud)(I-SMAC)* IEEE pp 54-59

[3] Kumar R and Rajasekaran M P 2016 January An IoT based patient monitoring system using raspberry Pi *2016 International Conference on Computing Technologies and Intelligent Data Engineering (ICCTIDE'16)* IEEE pp 1-4

[4] Gunawan T S et al 2018 Performance evaluation of smart home system using internet of things *International Journal of Electrical and Computer Engineering 8* no 1 pp 400

[5] Koo D D et al 2016 An Internet-of-Things (IoT) system development and implementation for bathroom safety enhancement *Procedia Engineering* 145 pp 396-403

[6] LoRa Alliance 2021 LoRawan from https://lora-alliance.org/about-lorawan Accessed on 27 May 2021

Physical, mechanical and corrosion properties of Zn-15Al alloy coating deposited at different voltage using thermal spray process

L Yu- Song, T A A Bakar*, H Ghandvar, N A Fadil, and N Suhaili

School of Mechanical Engineering, Faculty of Engineering, Universiti Teknologi Malaysia (UTM), 81310 UTM Johor Bahru, Johor, Malaysia

*Email: tuty@utm.my

Abstract. Steel is the most widely used metal on earth, cheap and found everywhere. It has good properties such as machinability, strength, ductility, and compatibility with other materials. However, it has weakness such as high corrosion rate which need to improve to achieve satisfying properties. The coating is one of the enhanced methods to improve the quality of steel coating methods are dipping and thermal deposition processes. Both processes have their own advantages and disadvantages. However, the thermal deposition method can produce a coating with better characteristics and performance. Zinc and aluminium are common metals to be used in the thermal spray process; however, a few parameters need to be concerned in the thermal spraying process. They include spraying voltage, spraying distance, deposition rate and applied current. The basic concept is that it melts the material and burst it to the substrate with high velocity. The purpose of this research is to evaluate the physical and corrosion properties of thermal sprayed pure zinc, aluminum and Zn-15Al alloy coating on a mild steel plate. The experimental procedures include roughness, thickness, adhesion, SEM with EDX, and EIS tests. The results show that at 30V, the surface roughness becomes lower, zinc's decreases from 7.30μm to 6.55μm, Zn-15Al's is from 8.52μm to 7.93μm. For aluminium, its surface roughness increases at higher voltage, from 13.10μm to 18.25μm. For coating thickness, the value of zinc, aluminium and Zn-15Al of 20V and 30V spraying voltage decreased from 629μm to 601μm, 486μm to 452μm, 541μm to 536μm respectively. The reason for sprays the material at higher velocity to the substrate; these findings depicted smooth and dense coating. Based on SEM/EDX analysis on the cross-sectioned samples, there is no gap between the coating and substrate and demonstrates good coating adhesion. The Daimler-Benz adhesion test shows all coating samples demonstrate good adhesion as no failure at the indentation boundary. In the EIS corrosion test, the frequency range 0.1Hz to 100kHz in 3.5wt.% NaCl solution, the highest impedance shown by zinc, aluminium and Zn-15Al at 20V are 21, 2122 and 49ohms; at 30V, they possess 11, 1347 and 20ohms. It illustrates those coatings with lower spraying voltages demonstrate better corrosion resistance due to the higher coating thickness. Hence, the coating which owns the best properties in all aspects is pure aluminum coating.

Keywords: Wire Arc Thermal Spray, Steel Protection, Physical Properties, EIS.

1. Introduction

Steel is one of the principal metals used today in many industries due to its outstanding properties and low cost. With the discovery of new materials and steel processing technology, its properties can be changed and alter into different functions. Its advantages include high strength, easy to fabricate and produced, high machinability and durable product [1]. Metal coating techniques like thermal spray and dipping are important processes to enhance the properties of steel. Thermal Spraying demonstrated better quality than other coating processes because it provides dense and uniform coating, is portable, and requires less energy input to melt the material [2].

Corrosion, buckling issue and lack of function under high temperature is severe issue of steel product. So, the surface coating becomes essential. Zinc and aluminium are two common coating metals, but zinc does not provide good barrier protection, and aluminium does not provide cathodic protection. The hot dipping process is cheap; however, is possesses some limitation like heat distortion, only on-site process available, and the disability to coat on a particular part of the substrate [3].

The study aims to determine the characteristic of pure zinc, aluminium and Zn-15Al coating deposited by wire arc thermal spray of 20V and 30V. It includes coating thickness, surface roughness, coating adhesion, SEM/EDX observation, and corrosion test using Electrochemical Impedance Spectroscopy (EIS).

1.1 Protection of Steel by Coating

A protective coating like paint acts as barrier to steel substrate to protect against corrosion. It is also soft and vulnerable to the atmosphere, so a duplex coating technique combines it with a metal coating. Its corrosion resistance is known to be 1.5-2.3 times of both the methods combined. Besides thermal spray and hot dipping, the electro-galvanising process and coating in a vacuum are also good metal coating techniques [4].

1.2 Coating Materials

1.2.1 Polymer Coating

It has low strength in general, needs time to dry or cure after applied and the temperature resistance is about 280°C. Its advantage is cheap and easy to apply. Some common types of polymers are nylon, alkyds and acrylics, polyurethane [5].

1.2.2 Ceramic Coating

They are good insulator with great abrasive resistance and hardness. Ceramic can be coated using plasma spray, detonation gun, oxygen acetylene powder, rod and CVD [6].

1.2.3 Metallic Coating

Metalizing means applying a metal coating on the substrate. It enhances the physical and corrosion resistant of the component, its friction characteristic can also be improved [7]. Some common types of coating materials are zinc (good cathodic protection), aluminium (good barrier protection), chromium, nickel and NiCoTef.

1.3 Methods of Coating

1.3.1 Dipping

Dip coating consists of 3 stages: immersion, dwell time (film formation) and withdrawal. The rate of withdrawal, and viscosity determine the quality of coating.

1.3.2 Thermal Spray

The material is melted and the atomization jet will accelerate it to the surface in high speed. Nearly all materials are available for this method. There is minimal thermal distortion of the component because no high energy input is needed. Some different characteristics of thermal spray processes are shown in Figure 1.

Type of System	Flame or Plasma Exit Temperature (^{0}C)	Heat Transfer or Substrate (^{0}C)	Particle Impact Velocity (m/s)	Oxide Content (%)	Porosity (%)	Adhesion (Bond Strength)
Detonation Gun	3000	20-150 MAX	800-1000	0.1	0.1 – 1.0	Extremely High
HVOF	2500-3100	500- 700	500-800	0.2	1.0 - 10	Very High
Plasma Spraying	5500-8300	700-1000	200-600	0.1-1	1.0 - 10	Very High
Wire Arc	4000-6000	500-800	240	0.5-3	10-20	High
Flame Spraying	2500-3000	500-700	30-180	4-6	10	Low

Figure 1. The comparison of different thermal spray process [8]

1.4 Corrosion Test (Electrochemical Impedance Spectroscopy, EIS)
The concept EIS is by applying low AC signal to an electrode in the electrolyte, then the electrode (sample) response is analysed by comparing the phase shift and measuring their amplitudes. The dynamic of moving charges of the interface can be determined, which resembles corrosion mechanism. Then, the properties of metal can be represented by electronic circuit with impedance under alternating current (AC), the results can be expressed in Ohm's law.

From the Nyquist plot, the larger ~~the~~ diameter of the semicircle, the higher ~~the~~ resistance and hence, the lower the corrosion rate. The impedance at low frequency response was related to the charge transfer process, considering that the response at high frequencies was attributed to the presence of a surface film [9]. Environment with high resistivity such as concrete can be measured. It is non-destructive testing, can monitor the evolution of the passive or active state over time. It is accurate and quick to get the results [10].

2. Materials and Methods
In this experiment, wire arc of 2mm diameter of pure zinc, aluminium and Zn-15Al were deposited on a mild steel plate, spraying voltage of 20V and 30V were tested, on 6 specimens altogether.

2.1 Mechanical Testing
Surface Roughness Test was carried out using Mitutoyo-SJ310 profilometer and Daimler-Benz Adhesion Test was taken by Rockwell-C indenter with 100kg load.

2.2 Morphology, Microstructure and Chemical Phase Analysis
Sample preparation was carried out by cross-sectioning the sample using CNC, hot mounting, grinding using sandpaper of 300 to 1200 roughness, polishing and etching. Then, the coating thickness was measured under optical microscope. Scanning Electron Microscope (SEM) with Energy Dispersive X-Ray (EDX) analysis was used to determine the microstructure and surface adhesion between the coating and substrate to confirm the element content of the materials.

2.3 Corrosion Test
Electrochemical Impedance Spectroscopy (EIS) method in VersaSTAT was used by determining its coating impedance when different frequency of current is applied. To simulate marine condition, 3.5wt.%

NaCl solution was used as electrolyte as shown in Figure 2. Start and End Frequency applied were 100kHz and 0.1Hz respectively with 10s measurement delay. The result was plotted in Nyquist plot and Bode Diagram.

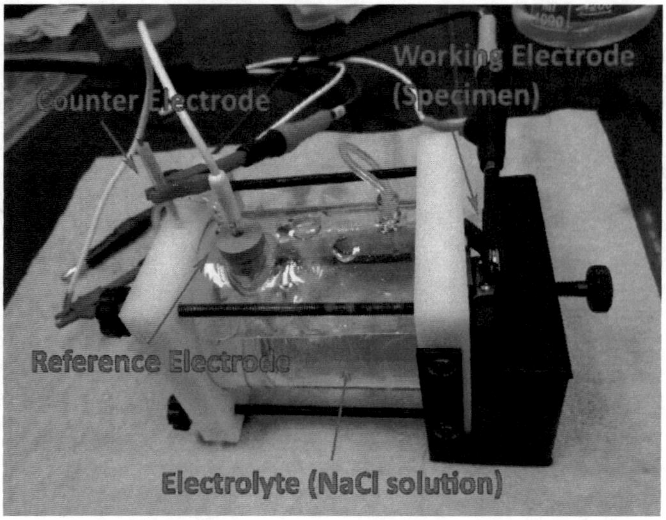

Figure 2: EIS setup for corrosion test

3. Results and Discussion

3.1 Surface Roughness

It shows a trend of decreasing surface roughness as the voltage of thermal spraying increases, except pure aluminium coating as 30V is not the optimum voltage to melt the material, as shown in Table 1. High voltage means high energy uses to melt and pump the coating material at a higher velocity; hence, the size of the droplet of the melted material would be smaller [11].

Table 1. Surface Roughness of different materials and spraying voltage.

Type of coating	Spraying Voltage	Surface Roughness(µm), Average Roughness (Ra)			Average
Pure Zinc	20V	7.708	7.212	6.969	7.296
	30V	6.610	6.186	6.838	6.545
Pure Aluminium	20V	12.844	13.499	12.960	13.101
	30V	17.349	17.8751	9.514	18.246
Zn-15Al	20V	8.103	9.543	7.906	8.517
	30V	6.312	7.988	9.490	7.930

3.2 Daimler-Benz Adhesion Test

Figure 3 presents the adhesion quality of pure zinc, in which Zn-15Al coating is HF3, while aluminium is HF2. Referring to the Daimler-Benz Adhesion chart if it is under HF5 and HF6 it means that the coating quality and adhesion are poor. Using a thermal spray, they all have good bonding with the substrate with minimal cracks and delamination. For aluminium, it possesses the best adhesion quality without delamination. Spraying voltage does not seem to have any effect on them.

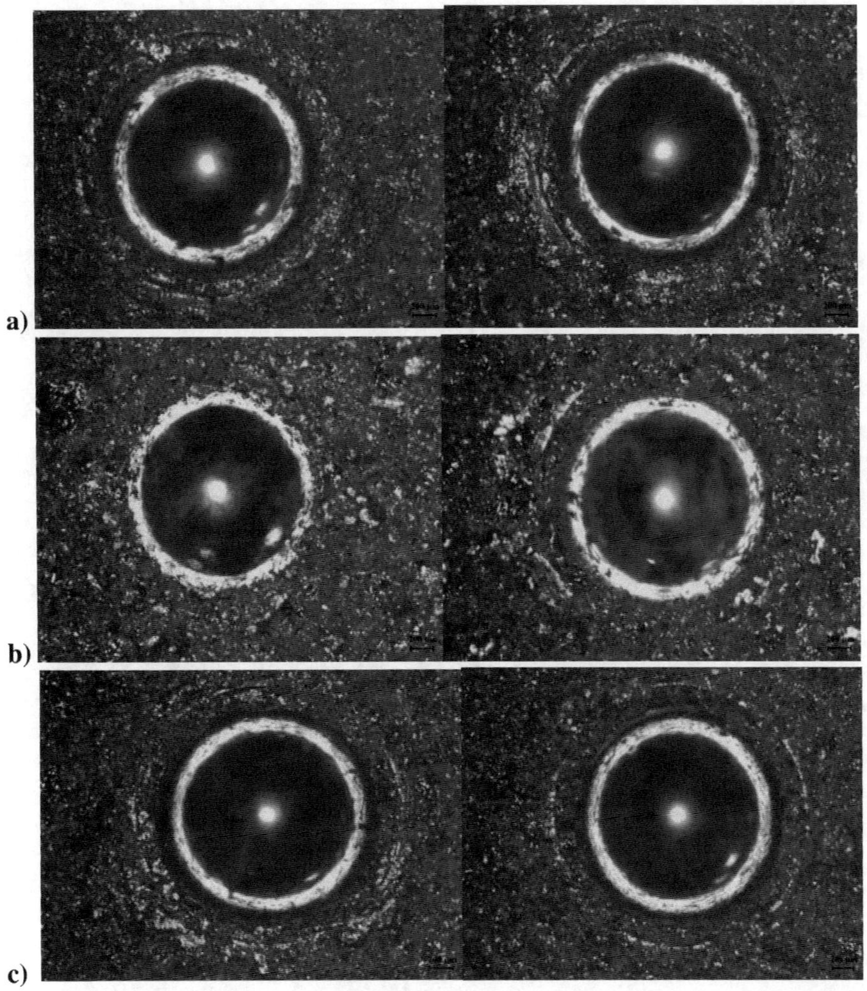

Figure 3. Rockwell C indentation of 100kg on the specimen's surface. (a) pure zinc. (b) pure aluminium and (c) Zn-15Al.

3.3 Coating Thickness

A trend of coating thickness is shown in Table 2 when the higher the spraying voltage, the lower coating thickness. In fact, the high voltage provides higher energy to melt and spray the materials in smaller particles so there is a lower mass of them deposited on the surface. Also, the lightest coating material which is aluminium deposited on the surface, followed by Zn-15Al and zinc [12].

Table 2: Coating thickness of different materials and spraying voltage

Type of coating	Spraying Voltage	Coating Thickness (µm)			Average
Pure Zinc	20V	645	634	609	629
	30V	592	603	607	601
Pure Aluminium	20V	460	468	529	486
	30V	402	500	455	452
Zn-15Al	20V	535	556	532	541
	30V	537	527	543	536

3.4 Coating Surface Morphology and Composition

3.4.1 Scanning Electron Microscope (SEM)

The coating morphology in Figure 4 shows dense and good bonding with the steel substrate. There are no porosity and cracks at the interface between the coating layer and substrate of pure zinc and aluminium. The Zn-Al alloy shows irregular and not uniform patterns because the content element is more complicated. There are differences in atomic size and melting point.

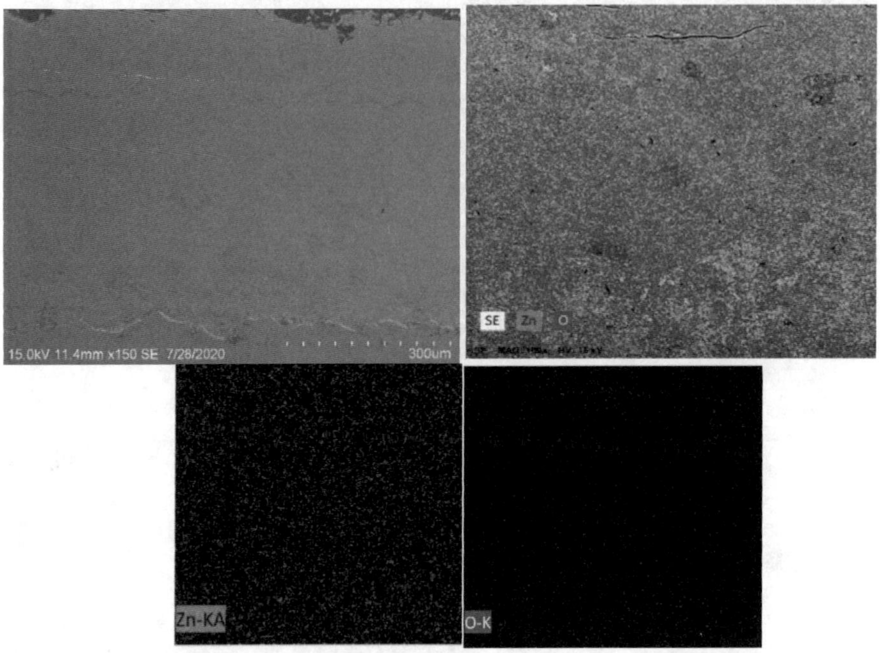

(a)

(b)

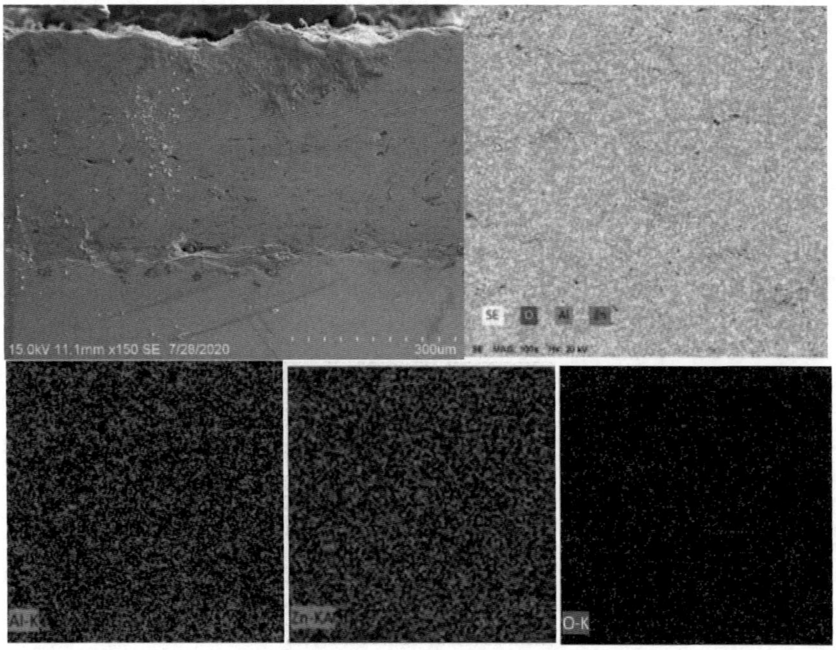

(c)

Figure 4. SEM microstructures and element mapping on the cross section. (a) pure zinc. (b) pure aluminium. (c) Zn-15Al.

3.4.2 Chemical Composition Analysis by EDX

Pure Zinc. The chemical composition of zinc is shown in Figure 5. The presence of the silicon is due to silicon carbide from SiC paper that contaminated during grinding process. The EDX result confirmed that the coating is pure zinc.

Element	At. No.	Line s.	Netto	Mass [%]	Mass Norm. [%]	Atom [%]	abs. error [%] (1 sigma)	abs. error [%] (2 sigma)	abs. error [%] (3 sigma)	rel (
Zinc	30	K-Serie	16073	2.96	84.20	69.59	0.13	0.25	0.38	
Silicon	14	K-Serie	2048	0.56	15.80	30.41	0.05	0.11	0.16	
Aluminium	13	K-Serie	0	0.00	0.00	0.00	0.00	0.00	0.00	
			Sum	**3.52**	**100.00**	**100.00**				

Figure 5. Elemental composition of zinc coating.

Pure Aluminium. The carbon could be from the steel substrate or impurity of aluminium and the oxygen content is the oxides of aluminium. Hence, the result in Figure 6 shows it is pure aluminium coating.

Element	At. No.	Line s.	Netto	Mass [%]	Mass Norm. [%]	Atom [%]	abs. error [%] (1 sigma)	abs. error [%] (2 sigma)	abs. error [%] (3 sigma)	r
Aluminium	13	K-Serie	413361	18.90	87.10	76.17	0.90	1.80	2.70	
Carbon	6	K-Serie	2059	2.13	9.82	19.29	0.43	0.85	1.28	
Oxygen	8	K-Serie	2713	0.67	3.08	4.54	0.14	0.29	0.43	
			Sum	**21.70**	**100.00**	**100.00**				

Figure 6. Element content of aluminium coating.

Zn-15Al. The chemical composition of Zn-15Al is shown in Figure 7. It consists of all the elements of impurities and bulk materials, the iron is expected from the substrate.

Element	At. No.	Line s.	Netto	Mass [%]	Mass Norm. [%]	Atom [%]	abs. error [%] (1 sigma)	abs. error [%] (2 sigma)	abs. error [%] (3 sigma)	re (
Aluminium	13	K-Serie	45672	5.13	50.14	43.40	0.27	0.53	0.80	
Carbon	6	K-Serie	3588	1.68	16.38	31.84	0.30	0.61	0.91	
Zinc	30	K-Serie	8952	1.56	15.28	5.46	0.08	0.16	0.24	
Oxygen	8	K-Serie	5235	0.83	8.14	11.88	0.15	0.31	0.46	
Silicon	14	K-Serie	5409	0.80	7.80	6.48	0.06	0.12	0.18	
Iron	26	K-Serie	278	0.23	2.26	0.95	0.04	0.08	0.13	
			Sum	**10.24**	**100.00**	**100.00**				

Figure 7. Element content of Zn-15Al coating.

3.5 Corrosion Test by Electrochemical Impedance Spectroscopy, EIS

3.5.1 Nyquist Plot

According to Figure 8, the first curve represents the coating and the second one is the substrate. The spraying voltage of 20V exhibits a bigger semicircle which means it has better corrosion-resistant properties, theoretically higher voltage produces a denser and more compact coating that protects the substrate from corrosion. The result is the opposite due to its more increased thickness, which avoids contact and dissolution with NaCl solution.

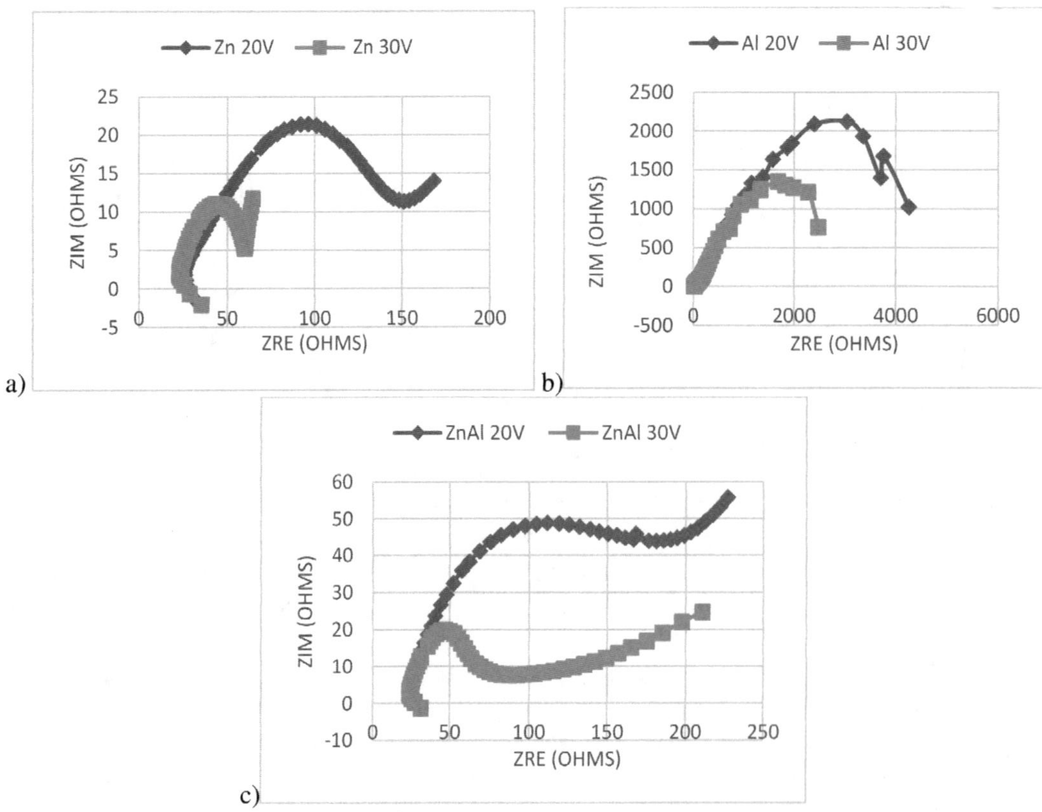

Figure 8. Nyquist plot from EIS of different material: (a) pure zinc. (b) pure aluminium and (c) Zn-15Al.

3.5.2 Bode Diagram

The impedance is inversely proportional to the frequency applied, as shown in Figure 9. High frequency with high penetrating power can detect coating properties without affecting by the deposition of corrosion products on the surface. Generally, the impedance of coatings with 20V is only slightly higher than 30V, except for zinc at low frequency, which is more significant [13].

In conclusion, the effect of coating thickness is more significant than the coating density. The corrosion resistance of each material arranges in ascending order in such a way Zn < ZnAl < Al. The formation of dense aluminium oxide protects the substrate from the contact with the electrolyte for corrosion process. At the same time, the zinc acts as a sacrificial anode that is more prone to corrosion than the steel substrate.

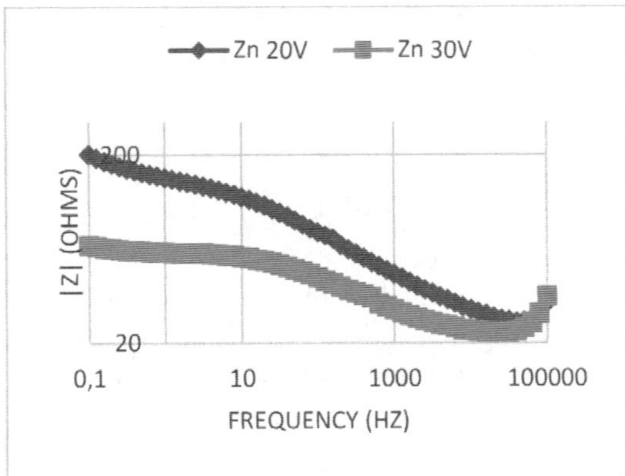

(a)

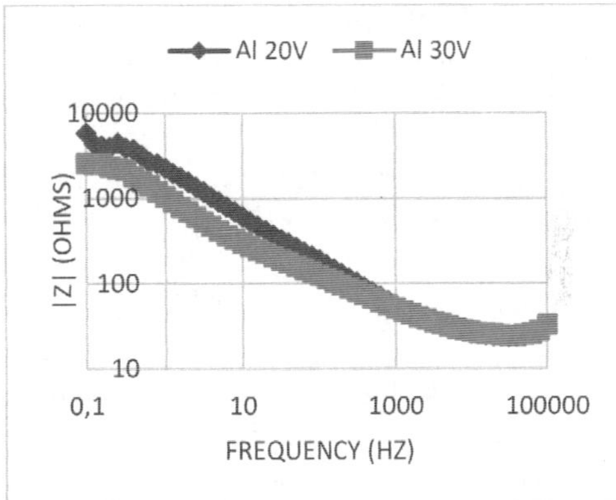

(b)

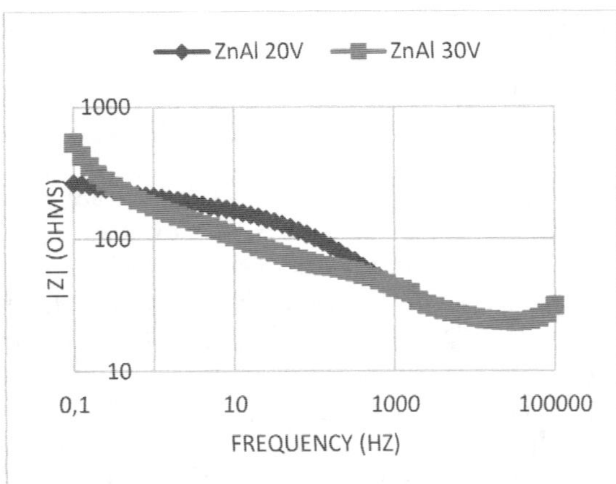

(c)

Figure 9. Bode diagram of impedance as function of frequency of each specimens (a) pure zinc, (b) pure aluminium and (c) Zn-15Al.

4. Conclusions

For surface roughness, the result shows a trend of decreasing roughness as the voltage increases for pure zinc and Zn-15Al coating. The 30V spraying voltage provides more constant and denser surface properties. For pure aluminium, it is the opposite. The reason is that 30V is not the optimum voltage to melt and spray the aluminium to the substrate. In Daimler-Benz Adhesion Test, all coatings on the samples are accepted as good adhesion bonding to the substrate. The pure aluminium coating shows the best quality without significant defects such as delamination and crack. About coating thickness, the coating becomes thinner and denser when the spraying voltage increases from 20V to 30V. Higher voltage melts and sprays the materials at higher velocity to the substrate, making more a compact coating. For microstructure and chemical composition, the interface between coating and substrate is observed. The coating quality is good without porosity. There are porosity and microcrack in the Zn-15Al layer due to the content of more complex elements. In EDX, the coating materials' content consist of a small number of impurities, except Zn-15Al. In the corrosion test (Electrochemical Impedance Spectroscopy, EIS), the aluminium coating has the highest corrosion resistance, followed by Zn-15Al and zinc coating by analysing the Nyquist plot. Generally, lower spraying voltage provides better corrosion resistance because the more increased thickness provides more barrier protection for the substrate.

Acknowledgement

The authors gratefully acknowledge the financial support for this research provided by Universiti Teknologi Malaysia under the Research Grant Scheme (TDR) vote number 06G22.

References

[1] Northern Weldarc. 2018 *Advantages and Disadvantages of Steel Structure*. Retrieved from: http://northern-weldarc.com/advantages-disadvantages-structural-steel-structures/
[2] Muhamad Hafiz, et al. 2014 *Critical process and performance parameters of thermal arc spray coating*. International Journal of Materials Engineering Innovation.
[3] Yan Li. 2001 *Corrosion behaviour of hot dip zinc and zinc–aluminium coatings on steel in seawater*. Chinese Academy of Sciences.
[4] American Galvanizers Association. (n.d). *Duplex Systems for Corrosion Protection*. Retrieved from: https://galvanizeit.org/corrosion/corrosion-protection/duplex-systems.
[5] Apex Instruments. 2016 *Dip Coating Technology*. Retrieved from: http://www.apexicindia.com/technologies/dip-coating-technology.
[6] THOMAS. (n.d). *Ceramic Coatings for Metal Parts*. Retrieved from: https://www.thomasnet.com/articles/chemicals/ceramic-coating-metals/.
[7] Machine Design. 2002 *Metal Coating*. Retrieved from: https://www.machinedesign.com/basics-design/metal-coatings.
[8] Han-Seung Lee, et al 2019 *Corrosion mechanism and kinetics of Al-Zn coating deposited by arc thermal spraying process in saline solution at prolong exposure periods*. Scientific Report.
[9] D.V. Ribeiro, C.A.C. Souza & J.C.C. Abrantes. 2015 *Use of Electrochemical Impedance Spectroscopy (EIS) to monitoring the corrosion of reinforced concrete*.
[10] V. Encinas-Sánchez, et al. 2018 *Electrochemical impedance spectroscopy (EIS): An efficient technique for monitoring corrosion processes in molten salt environments*. Spain: Complutense University of Madrid. Handbook of Thermal Spray Technology. (2004). Introduction to Thermal Spray Processing. ASM International.
[11] Fred F. Farshad, et al. 2001 *A Comparison of Surface Roughness of Pipes as Measured by Two Profilometers and Atomic Force Microscopy*. USA: University of Louisiana.
[12] Maria Oksa, et al. 2011 *Optimization and Characterization of High Velocity Oxy-fuel Sprayed Coatings*. Finland: VTT Technical Research Centre of Finland.
[13] Tung-Yuan Yung, et al. 2018 *Thermal Spray Coatings of Al, ZnAl and Inconel 625 Alloys on SS304L for Anti-Saline Corrosion*. Taiwan: Ming Chi University of Technology.

Reduced-order modelling of unsteady vortex lattice method

P Bundith[1], P Pimanprom[1], W Janthornsirijan[1], C Punkun[1], S Sleesongsom[1,*], and S Bureerat[2]

[1]Department of Aeronautical Engineering, International Academy of Aviation Industry, King Mongkut's Institute of Technology Ladkrabang, Bangkok 10520, Thailand
[2] Sustainable and Infrastructure Development Center, Department of Mechanical Engineering, Faculty of Engineering, KhonKaen University, KhonKaen City 40002, Thailand

* Email: suwin.se@kmitl.ac.th

Abstract. The purpose of this research is to present a new idea of reduced-order modelling of unsteady flows without static correction, which is based on fluid eigenmodes. The new reduced order model is formulated based on a new form of wake vortices without static correction. This technique is expected to perform a reduced-order model similar but faster than the previous techniques. The new method is used to analyse unsteady flows over a three-dimensional wing and comparing with the previous techniques. The performances of the present method are demonstrated with numerical examples. The results show the accuracy of the new reduced-order model of unsteady flow with higher computational efficiency. In conclusion, the new method can be considered as an alternative technique to perform the reduced-order models of unsteady flow.

Keywords: unsteady vortex lattice, reduced-order, flow, three-dimensional wing

1. Introduction

The purpose of this research is to present the new reduced-order modeling (ROM) based on Eigenanalysis of an unsteady vortex lattice method (UVLM)about plate like wing. In the past, this technique was proposed based on an idea, which needed a static correction technique to suppress the quasi-steady term making it obtain acceptable results [1-2]. Later, the technique has been improved to the reduced-order modeling of unsteady flows without static correction requirements [3]. The efficiency of this technique as a result from the quasi-steady or zero eigenvalue is removed by rearranging the UVLM equations to a new form of wake vortex strength.

The difference between these two methods is that one is a conventional method with static correction and another is without static correction. The performance of the latter has been proved by many researches [4-8]. It is found that inefficiency of the technique without static correction as it needs matrix inversion of the new equation form [3]. It is thus interesting to study how to discard the inversion of the matrix from the UVLM equation, which expect to increase the performance of the computational efficiency. In the past, it had two models of the vortex in the wake. The first model considers the unsteady wake vortices shedding in the wake with their strengths being proportional to the time rate of change of circulation about the body wing [1-8].

Another one states that the vortex strength of the wake is shed from the trailing edge remains unchanged. This implies that the vortex strength at a position downstream is equal to the vortex strength forward at the previous time step [9-12]. The formulations may cause an arrangement of the reduced-order modeling (ROM) without static correction technique.

2. Unsteady vortex lattice method

In this study, unsteady aerodynamic analysis of an aircraft wing can be modeled with vortex ring method [13]. The unsteady aerodynamic is based on potential flow, which is called UVLM. In general, the aircraft wing with airfoil can be modeled as a lifting surface. A simple lifting surface without camber is represented with a plate like wing as shown in figure 1.

The model of UVLM consists of vortex rings on the body and wake, while their vortex strengths are denoted by Γ_i and $\Gamma_{w,i}$, respectively. All panels have chordwise length Δx and spanwise width Δy. Aircraft fly with velocity V and the time increment is Δtwhich means $\Delta x = \mathrm{V}\Delta t$. According to the boundary condition that there is no flow normal to the lifting surface and the velocity induced by those vortex rings, the UVLM equation is of the matrix form.

$$A\Gamma^{n+1} + B\Gamma^{n} = w^{n+1} \tag{1}$$

where w is the known downwash and A and B are known sparse matrices.

The introduction has revealed two models of vortex wake that lead to an idea of the ROM without static correction as follows.

2.1. First model

The model of tailing edge wake can be considered as unsteady vortex which is shed in the wake. Its strength is proportional to the time rate of change of circulation about the wing. The strength of the first vortex point in the wake at the n + 1times level is given by

$$\Gamma_{W,i+1}^{n+1} = -\sum_{j=1}^{M}(\Gamma_j^{n+1} + \Gamma_j^{n}) \tag{2}$$

where M is the number of vortex elements on the lifting surface.

Once the vortex strength has been shed into the wake, it is described numerically by

$$\Gamma_{w,i}^{n+1} = \Gamma_{w,i-1}^{n} \tag{3}$$

Finally, because the vortex sheet is infinitely long, but the computational model has a finite length wake, special treatment is required at the last vortex element. Otherwise, the starting vortex would disappear abruptly when it reaches the end of the computational wake producing a discontinuous change in the induced wash at the airfoil. To alleviate this difficulty, the convective vortices, upon reaching the end of the wake, are allowed to dissipate smoothly using the following relationship:

$$\Gamma_i^{n+1} = \Gamma_{i-1}^{n} + \alpha\Gamma_i^{n+1} \quad ,i = N \tag{4}$$

where α is a relaxation factor. For the plate like wing problem, the relaxation factor from [1] is used as $\alpha = 0.5$.

From the induced vortex on body and wake equation (2-4), the UVLM equation can be expressed as in equation (1).

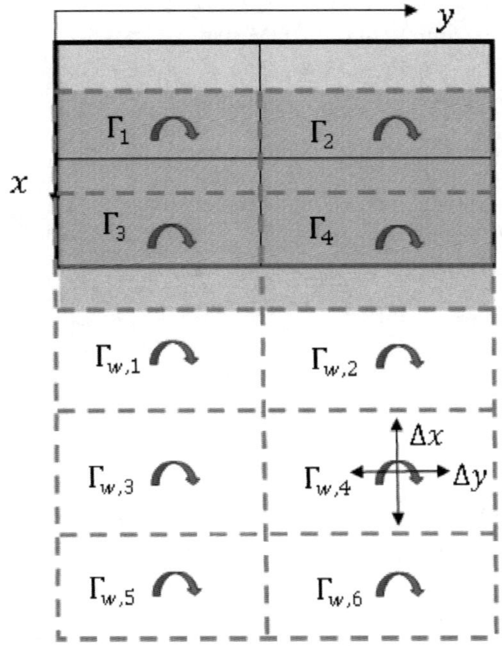

Figure 1. Three – dimensional vortex lattice model of the wing and the wake element

2.2. Second model

The second one states that the vortex strength of the wake shed from the trailing edge remains unchanged. This implies that the vortex strength at a position downstream is equal to the vortex strength forward at the previous time step. This means that the strengths of the vortex rings at the trailing edge at time step n are equal to their corresponding wake vortex ring in the first row at time step n+1 and similarly for the strength of the wake vortex rings in following row.

$$\Gamma_{W,i+1}^{n+1} = \Gamma_j^n \tag{5}$$

3. ROM

3.1. ROM without static correction technique

From the equation (1) with the 1st wake model, by using the technique in [3], ROM without static correction can be obtained as follows. Due to matrices B_{11} and B_{12} are 0, the UVLM equation can be formulated as

$$[A_{22} - A_{21}A_{11}^{-1}A_{12}]\Gamma_w^{n+1} + [B_{22} - B_{21}A_{11}^{-1}A_{12}]\Gamma_w^n = -A_{21}A_{11}^{-1}W_b^{n+1} - B_{21}A_{11}^{-1}W_b^n \tag{6}$$

$$A_{new} = A_{22} - A_{21}A_{11}^{-1}A_{12} \tag{7}$$

$$B_{new} = B_{22} - B_{21}A_{11}^{-1}A_{12} \tag{8}$$

$$W_{new} = -A_{21}A_{11}^{-1}W_b^{n+1} - B_{21}A_{11}^{-1}W_b^n \tag{9}$$

$$A_{new}\Gamma_w^{n+1} + B_{new}\Gamma_w^n = W_{new}^{n+1} \tag{10}$$

Equation (10) is based purely on Γ_w, in which its corresponding eigenvalues will never be zero. The new eigensystem can lead to accurate ROM without using a static correction technique. The drawback of this equation may cause inefficient computation due to the need of inverse of matrices solved in equation (7-9).

3.2. New ROM without static correction technique

To increase efficiency in computation of ROM without static correction technique, the new form of UVLM is proposed. From the equation (1) with the 2nd wake model, the new UVLM for ROM can be performed by multiplying equation (6) with A_{11}leading to

$$[A_{11}A_{22} - A_{21}A_{12}]\Gamma_w^{n+1} + [A_{11}B_{22} - B_{21}A_{12}]\Gamma_w^n = -A_{21}W_b^{n+1} - B_{21}W_b^n. \quad (11)$$

Due to matrix A_{21}being zero, the new UVLM can be formulated as follows:

$$[A_{11}A_{22}]\Gamma_w^{n+1} + [A_{11}B_{22} - B_{21}A_{12}]\Gamma_w^n = -B_{21}W_b^n \quad (12)$$

where

$$A_{new} = A_{11}A_{22}$$
$$B_{new} = A_{11}B_{22} - B_{21}A_{12}$$
$$W_{new} = -B_{21}W_b^n.$$

This method can be calculated faster than the previous technique because there is no matrix inversion required but the wing elements and the wake vortex elements must be equal so that the matrix A_{11} and B_{22} can be operated.

4. Results and Discussion

4.1. Eigenanalysis

To study the performance of the proposed technique, the eigenanalysis of the conventional method, the method based on Γ_w, and the present method are compared. The eigenvalues of the plate like wing for the conventional UVLM with the 1st model of wake are shown in figure. 2.

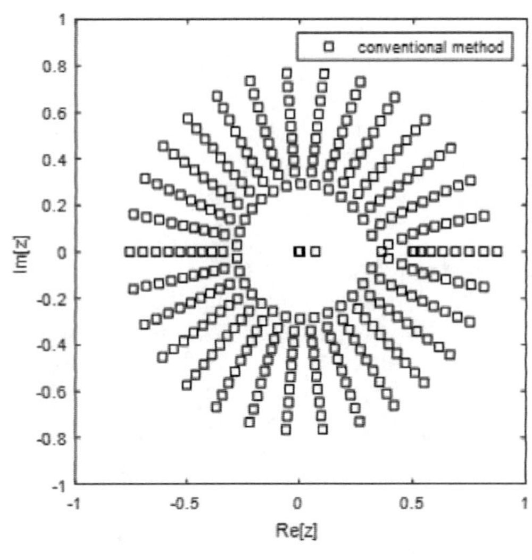

Figure 2. Eigenvalues of the plate like wing for the conventional UVLM with 1st model of wake.

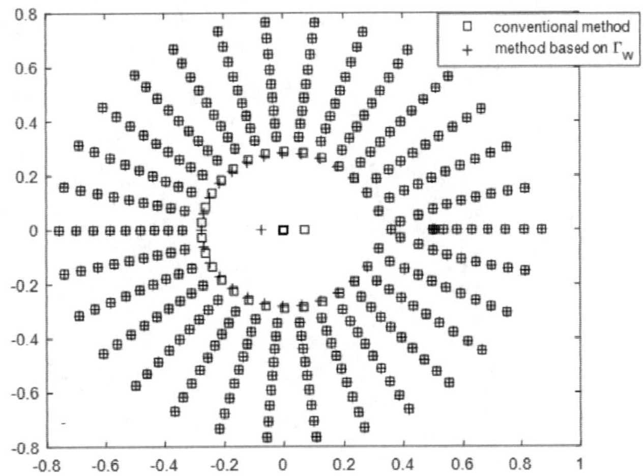

Figure 3. Eigenvalues of vortex lattice model of plate like wing -o conventional method + method based on Γ_w.

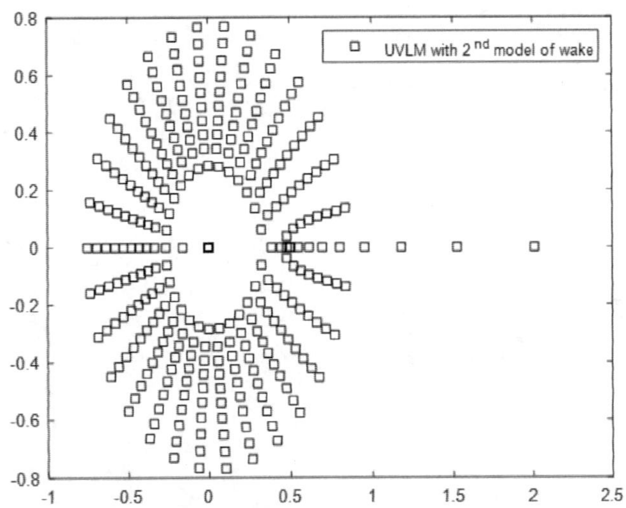

Figure 4. Eigenvalues of the plate like wing for the conventional UVLM with 2nd model of wake.

The eigenvalues of plate like wing using equation (11) and 1st model of wake are shown in figure. 3.Note that the nonzero eigenvalues of the conventional method are removed. The eigenvalues of the plate like wing for the conventional UVLM with the 2nd model of wake are shown in figure4, while the conventional UVLM with 1st model and 2nd model of wake are shown in figure 5. The comparative results show only a few eigenvalues are different between UVLM with 1st and 2ndwake models. The comparison of the UVLM with 2nd model of wake and new ROM is shown in figure 6. The results show the new ROM can remove zero eigenvalues without using static correction technique like the work by

[3], but the new equation requires no matrix inversion, which is expected to reduce the computational time and increase performance in computation. The result of the computational time will be present in the next section.

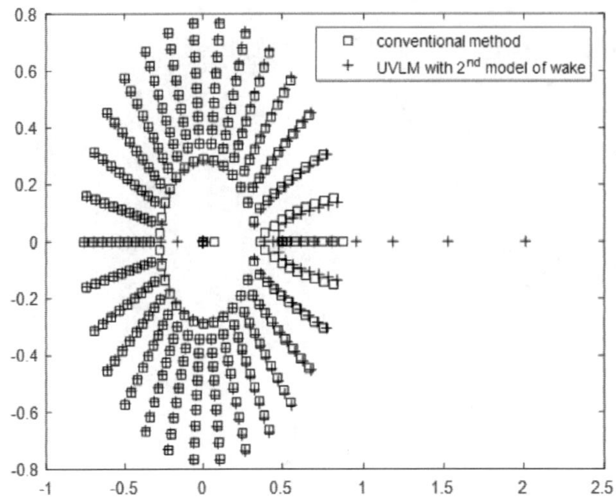

Figure 5. Eigenvalues of the plate like wing for the conventional UVLM with 1st model of wake (conventional method) and + 2nd model of wake.

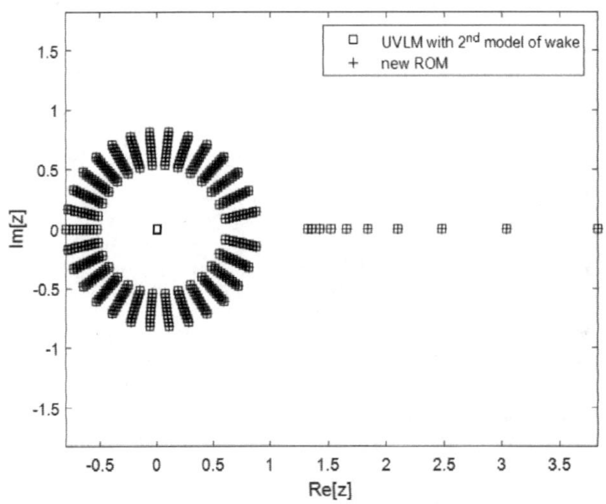

Figure 6. Eigenvalues of the plate like wing for the conventional UVLM with 2nd model of wake and the new ROM.

4.2. *Computational Time Comparison between Conventional and New-ROM*

The comparison of the computational time difference between the method based on Γ_W and New-ROM is shown in table 1. The results lead to the conclusion that the new technique can sufficiently reduce computation time and help to increase the performance in computation in cases of the design of aircraft wings with using an optimization technique especially for the method based on evolutionary algorithms (EAs). For example, aerodynamic design of aircraft member using EAs with 100 populations and run for 500 times, it needs 50,000 times of activating UVLM analyses. It means, when performing with the new ROM, computational time can be reduced for several hours.

Table 1. The Computational Time between Conventional and New-ROM Method.

Method	N-Chord	N-Span	N-Wake	Time (s)
Method based on Γ_w	35	10	35	3.40554
New-ROM	35	10	35	3.343399
Method based on Γ_w	30	30	30	26.547974
New-ROM	30	30	30	25.744878
Method based on Γ_w	45	20	45	26.632752
New-ROM	45	20	45	25.827317
Method based on Γ_w	50	25	50	59.779007
New-ROM	50	25	50	58.986076

5. Conclusions

This research presents a new reduced-order model for unsteady flow without static correction technique. This method can construct the eigensystem in the new form without matrix inversion, which can remove zero eigenvalues from the eigensystem. Furthermore, the new method can reduce computation time when compared with the conventional ROM without static correction. In conclusion, the new ROM can be considered as an alternative technique to perform the reduced-order models of unsteady flow. Future work, the present method is extended to design of a high-lift mechanism, which is combination of aerodynamics design and mechanism motion generation synthesis.

Acknowledgments

The authors are grateful for financial support provided by King Mongkut's Institute of Technology Ladkrabang and the National Research Council Thailand.

References

[1] Hall KC 1994 *AIAA J.* **32** 2426.
[2] Florea R, Hall K C and Cizmas P G 1998 *AIAA J.* **36** 1039.
[3] Behbahani-nejad M, Haddadpour H and Esfahanian V 2005 *J. Aircr.* ***42*** 882.
[4] Esfahanian V and Behbahani-Nejad M 2002 *J. Fluids. Eng.* **124** 988.
[5] Shahverdi H, Nobari A S, Behbahani-Nejad M and Haddadpour H 2007 *J. Fluid. Struct.* **23** 143.
[6] Shahverdi H, Nobari A S, Behbahani-Nejad M and Haddadpour H 2009 *J. Fluid. Struct.* **25** 1243.
[7] Dardel D and Bakhtiari-Nejad F 2010 *Aero. Sci. Tech.* **14** 95.
[8] Sleesongsom S and Bureerat S 2013 *J. Eng.* **1** 1.
[9] Lee HT, Kroo LM and Bieniawski S 2002 Flutter Suppression for High Aspect Ratio Flexible Wings Using Micro flaps (AIAA. 2002-1717)
[10] Dimitriadis G, Giannelis NF, and Vio GA 2018 *J. Fluid. Struct.* **76** 216.
[11] Pareteau M, and Laurendeau É 2018 *AIAA J.* **56** 2242.
[12] Winyangkul S, Sleesongsom S, and Bureerat S 2021 *Appl. Sci.***11** 6463.
[13] Katz J and Plotkin A 2001 *Low-Speed Aerodynamics 2nd ed.* (Cambridge: Cambridge Univ. Press).

Effect of 3D-printing parameters on the tensile strength of acrylonitrile butadiene styrene (ABS) polymer

S R Ya'akub[1*], N Ibrahim[1], and R Singh[1,2]

[1] Mechanical Engineering Programme Area, Faculty of Engineering, Universiti Teknologi Brunei, Tungku Highway, Gadong BE1410, Brunei Darussalam

[2] University of Malaya, 50603 Kuala Lumpur, Malaysia

* Email: serirahayu.yaakub@utb.edu.bn

Abstract. In this research, the tensile strength behaviour of 3D-printed ABS polymer was studied. In particular, the following printing parameters were investigated i.e. angle orientation of 0 °, 30 ° and 45 °, layer thickness of 0.2 mm, 0.3 mm and 0.4 mm, and printing speeds of 60 mm/s, 40 mm/s and 20 mm/s. The samples were subjected to tensile loading in order to determine the stress-strain curve, UTS and the elastic modulus. The results indicated that the speed has no major impact on the strength of the 3D-printed samples. On the other hand, the 0.4 mm layer thickness gave the highest stiffness, whereas the angle orientation of 45 ° gave the highest tensile strength. However, the combination of 3D parameters that resulted in the overall best results was determined as 0.4 mm layer thickness with an angle orientation of 30 °.

Keywords: tensile test, ABS polymer, 3D printing

1. Introduction

Over the years, the manufacturing industries have revolutionised their process capabilities to be more efficient and cost-effective including the use of IoT and 3D-printing [1-3]. In particular, 3D-printing which is also known as additive manufacturing, is a rapid-prototyping technique used to produce geometrical shape items by simply depositing thin layers of materials over and over again until the intended geometry is formed. This near-net shaping process which is widely used to produce any polymer product including very thin sections, requires data obtained from computer aided design (CAD) based 3D model or a computed tomography scanner [4, 5].

The main advantages of 3D-printing technologies are that it can produce complex shaped product precisely, and within shorter span of time if compared to the traditional manufacturing processes which requires considerable amount of time and effort to produce a good and decent quality product [6]. Another advantage of 3D printed product is that it produces less waste material, so they will cause less harm to the environment and the ecology [7].

A variety of different materials have been used for 3D-printing including that of ceramics, gypsum, metals and concrete, however the most recommended materials are polymers such as polyamide, acrylonitrile butadiene styrene (ABS), polylactic acid (PLA) and polyvinyl alcohol (PVA) [3, 8]. Among these materials, PLA and ABS have been widely used for 3D-printing to produce various products because of their excellent thermal, mechanical and biodegradable properties [9, 10]. However, it has been reported that the printing parameters can have an influence on the properties of the printed polymers [11]. In this study, the effect of selected 3D printing parameters on the tensile strength behaviour of ABS have been studied.

2. Methods and Materials

In this work, the design of the sample was done using the Autodesk Inventor Professional 2016 CAD software. The initial dimensions selected for the sample were based on previous work [12] and to suit the 3D printing machine model 'UP Plus 2'. The final dimension of the dog bone shape ABS sample used in this study is as shown in Figure 1.

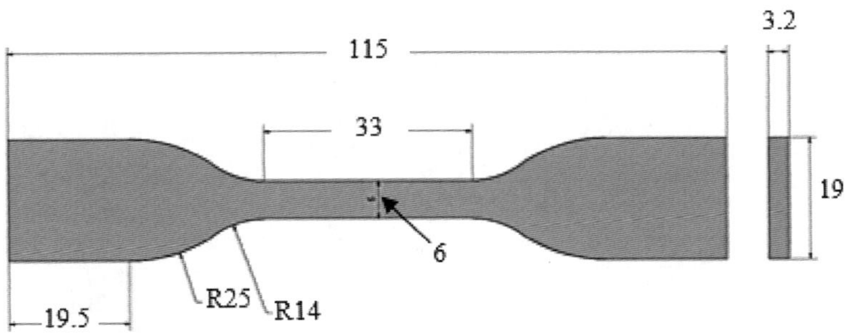

Figure 1. Schematic diagram of the sample used in the present work. All dimensions shown are in mm.

In a typical sample preparation procedure, the sample was sketched using a CAD software and subsequently exported as a stereolithography file (stl) into the available 3D printer software. The filament used to print the sample was a commercial acrylonitrile butadiene styrene (ABS) polymer. Samples were prepared by varying the printing parameters i.e. speed (20 mm/s, 40 mm/s and 60 mm/s), layer thickness (0.2, 0.3 and 0.4 mm) and angles of printing or the raster orientation (0°, 45° and 90°). A typical 3D printed ABS sample is as shown in Figure 2. A total of 54 samples were manufactured and the final dimensions of the printed samples were measured prior to tensile test.

The tensile test was conducted at room temperature based on ASTM D695 using a universal testing machine model Instron 5565. The strain of the samples was measured from a gauge length of 33 mm using an extensometer. The sample was loaded at a rate of 0.5 mm/min. and data were collected continuously by the system software to calculate the ultimate tensile strength, percentage elongation and modulus of elasticity.

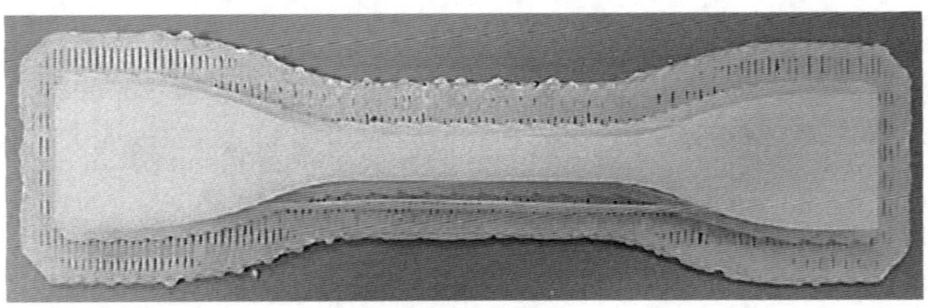

Figure 2. A typical 3D printed ABS tensile test sample used in the present work.

IOP Conf. Series: Materials Science and Engineering 1173 (2021) 012041 doi:10.1088/1757-899X/1173/1/012041

3. Results and discussion

The stress-strain diagrams for the samples fabricated with different thickness as a function of raster orientation and printing speed are as shown in Figures 3, 4 and 5.

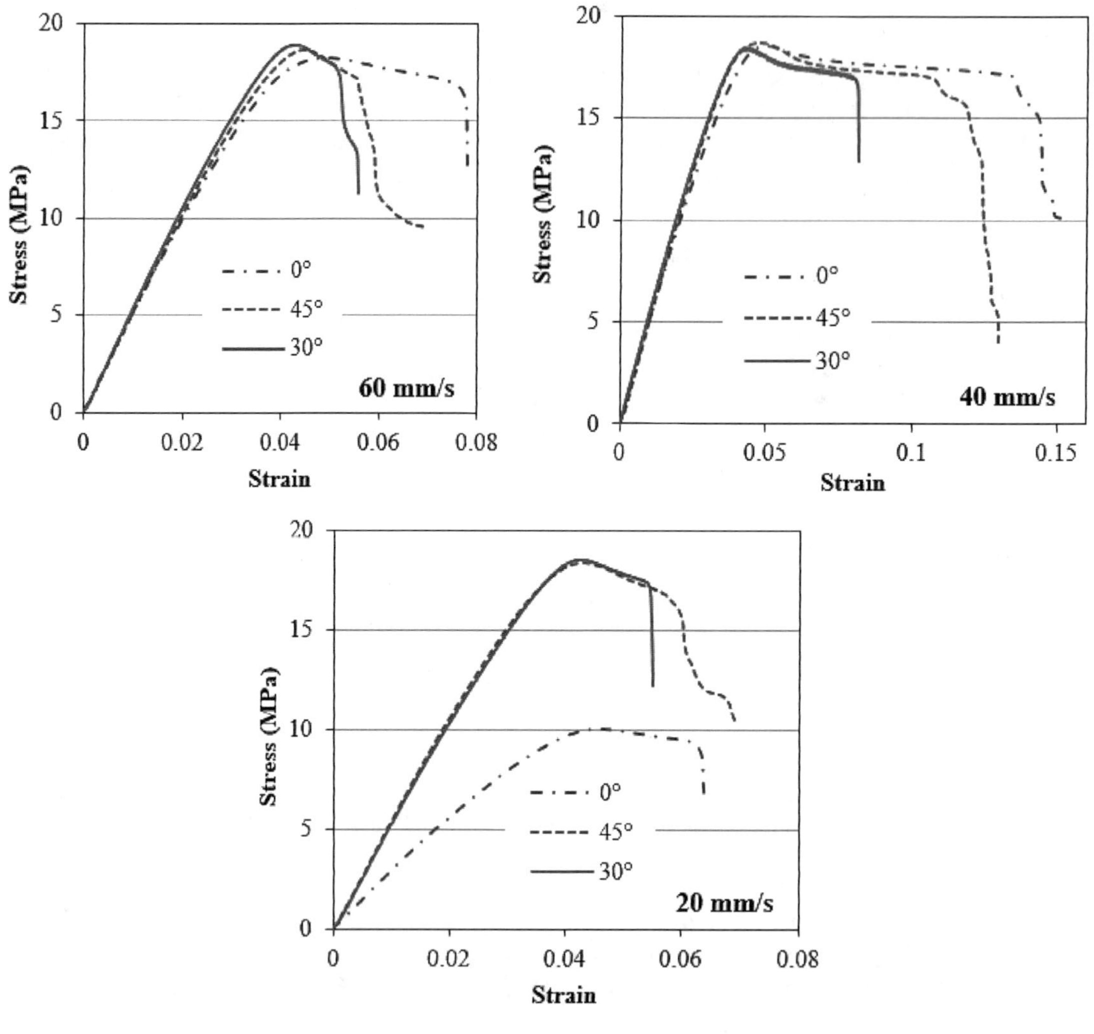

Figure 3. Stress-strain curves for samples thickness of 0.2 mm as a function of raster orientation and printing speed.

Figure 3 shows that for sample of thickness 0.2 mm, the strain of the sample reduced with the angle of orientation regardless of printing speed. It was also found that the UTS of the sample with orientation of 0 ° was not affected at printing speed of 60 mm/s and 40 mm/s (i.e. the value remained at about 18 MPa) but decreased to about 10 MPa when the printing speed reduced to 20 mm/s. Similarly, the stiffness of this sample also decreased as evident from the slope of the linear section of the curve. In contrast, the UTS of the samples having orientation of 30 ° and 45 ° remained almost unchanged at above 18 MPa regardless of printing speed. The results also showed that sample printed with a raster orientation of 30 ° exhibited the lowest strain value.

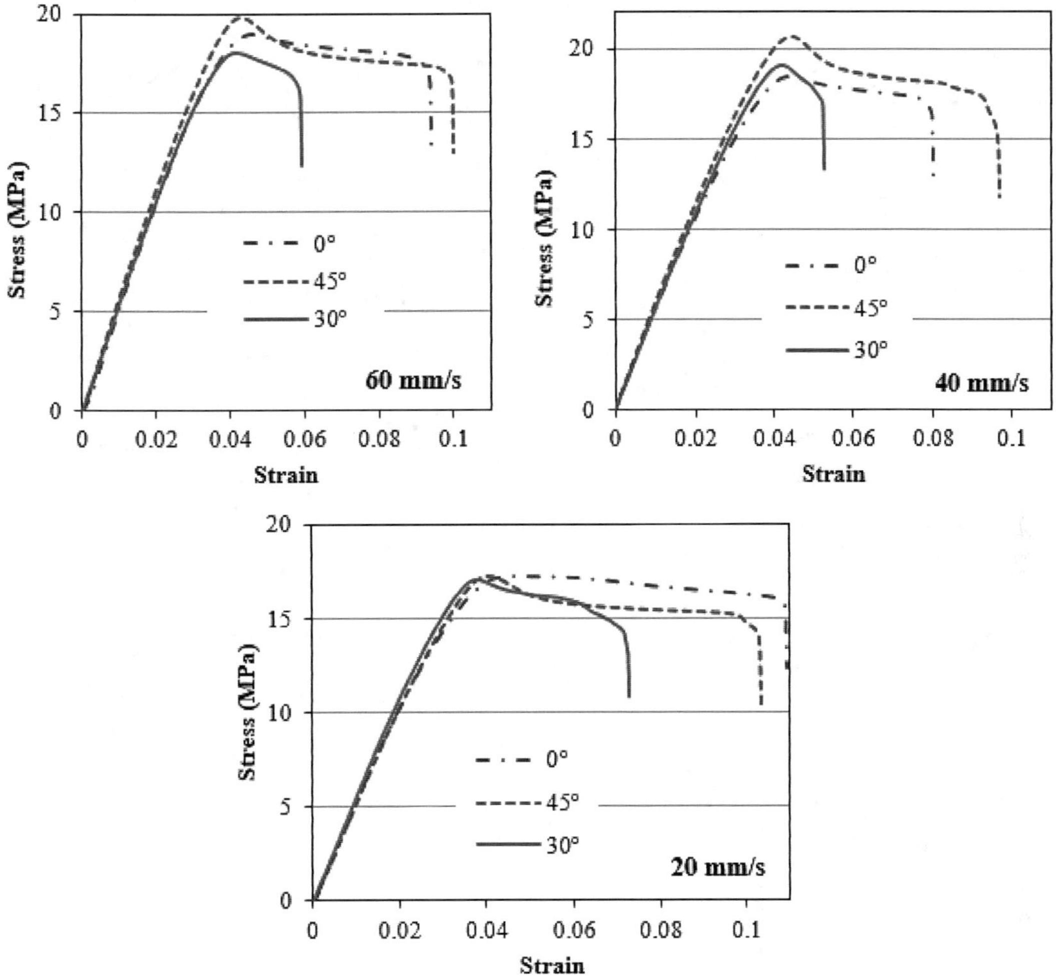

Figure 4. Stress-strain curves for samples thickness of 0.3 mm as a function of raster orientation and printing speed.

Comparison of the stress-strain graphs shown in Figure 4 for samples having 0.3 mm thickness indicated that regardless of the printing orientation and printing speed, the UTS did not change very much (18 – 19 MPa) between the samples. On the other hand, sample having orientation of 30 ° exhibited the lowest strain before fracture for all printing speed as observed for the 0.2 mm thickness samples. It was also revealed that the sample printed at an orientation of 45 ° exhibited the highest strain value for speeds of 60 mm/s and 40 mm/s, while sample of orientation 0 ° had the highest value when printed at 20 mm/s.

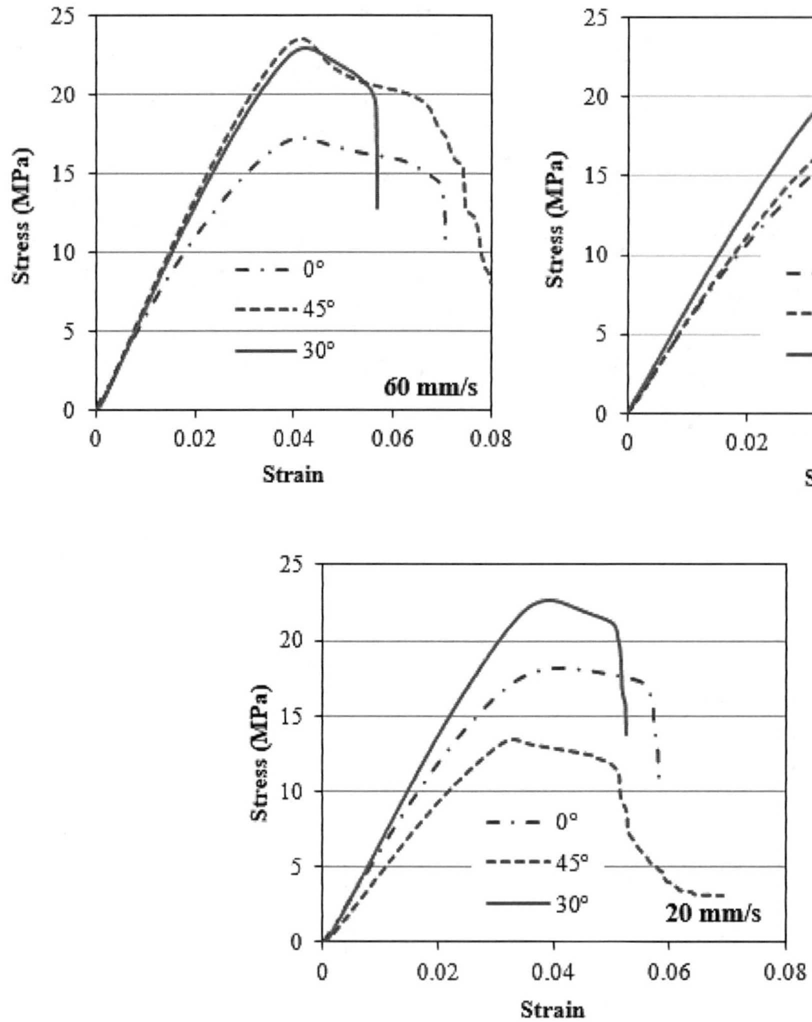

Figure 5. Stress-strain curves for samples thickness of 0.4 mm as a function of raster orientation and printing speed.

For 0.4 mm layer thickness, the stress-strain graphs shown in Figure 5, exhibited a wider range of UTS data for different speeds and orientation angles. A different behaviour could be seen in these graphs if compared to that observed for the lower thickness samples shown in Figures 3 and 4. Nevertheless, the 30 ° printing orientation samples exhibited the highest UTS (above 20 MPa) when compared to the lower thickness samples. For different printing speeds, there were minor differences that could be observed in Figure 4. Therefore, it is assumed that the speed of printing does not have any effect on the strength of the samples. As mentioned in the previous section, speed consumes the most time to complete printing of one specimen.

Typical view of the fracture morphology of the samples as shown in Figure 6 indicated that a brittle type fracture was inevitable for all samples regardless of sample thickness, printing orientation and speed.

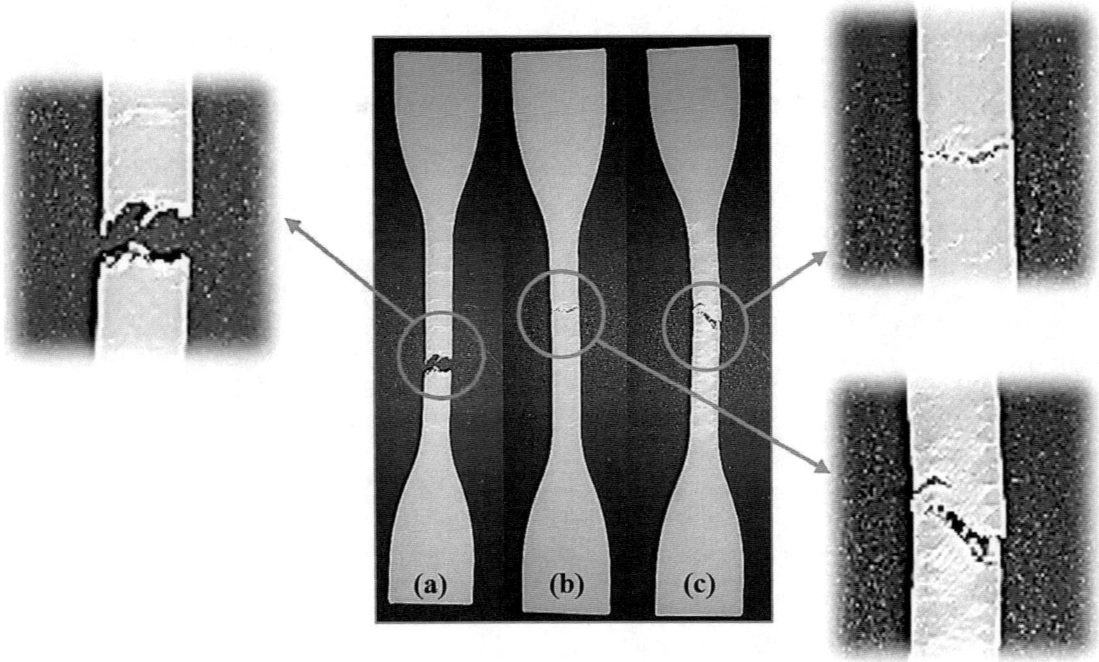

Figure 6. Typical brittle fracture of the 3D printed samples of thickness 0.4 mm and printing speed of 60 mm/s. (a) 0 ° orientation, (b) 30 ° orientation and (c) 45 ° orientation.

4. Conclusions

In the present work, the effects of 3D printing parameters on the tensile behaviour of ABS were investigated. The results showed that the ÚTS of the printed ABS was affected by the layer thickness and the printing or raster orientation. Within the boundary set for this study, the findings indicated that the layer thickness of 0.4 mm and at a raster orientation of 30 ° resulted in the overall best tensile properties. It was also found that the speed of printing did not have a major influence on the tensile behaviour of the ABS polymer.

References

[1] Bourell D L 2016 Perspectives on additive manufacturing *Annu Rev Mater Res* **46** 1-18.

[2] Wang X, Jiang M, Zhou Z, Gou J and Hui D 2017 3D printing of polymer matrix composites: a review and prospective *Compos B Eng* **110** 442-458.

[3] Manu K, Rahul M A V, De Schutter G and Tittelboom K V 2021 Extrusion-based concrete 3D printing from a material perspective: a state-of-the-art review *Cement and Concrete Compos* **115** 103855.

[4] Chen Z, Yang M, Ji M, Kuang X, Jerry Qi H and Wang T 2021 Recyclable thermosetting polymers for digital light processing 3D printing *Mater Des* **197** 109189.

[5] Yeong W, Chua C, Leong K, Chandrasekaran M and Lee M 2006 Indirect fabrication of collagen scaffold based on inkjet printing technique *Rapid Prototyping J* **12** 229-237.

[6] Chao X, Qinghua W, L'Espérance G, Louis L and Therriault D 2018 Environment-friendly and reusable ink for 3D printing of metallic structures *Mater Des* **160** 262-269.

[7] Alicia A T, Elaine L F and van der Ploeg M J C 2021 Regulatory developments and their impacts to the nano-industry: a case study for nano-additives in 3D printing *Ecotoxicology and Environmental Safety* **207** 111458.
[8] Ilbey K and Liwei L 2020 3D printing technologies: techniques, materials, and post-processing *Current Opinion in Chem Eng* **28** 134-143.
[9] Alexander K N, Narayan R J and Shafiee A 2019 3D printing in the biomedical field *Encyclopedia of Biomed Eng* 275-280.
[10] Harshitha V and Seeram S R 2019 Design and analysis of ISO standard bolt and nut in FDM 3D printer using PLA and ABS materials *Mater. Today: Proc.* **19** 583-588.
[11] Antonio J G, Jan L-L, Alex K, Kleinfehn A, Ciro A R, Luiz H C, Matthew L B, Joaquim C and David D 2019 Optimization of photocrosslinkable resin components and 3D printing process parameters *Acta Biomaterialia* **97** 154-161.
[12] Melenka G, Cheung B, Schofiels J, Dawson M and Carey J 2016 Evaluation and prediction of the tensile properties of continuous fiber-reinforced 3D printed structures, *Compos Struc* **153** 866-875.

The application of Tukey's mean difference (Tukey's HSD) for identification of local genetic sources of corn in selecting prospective parental lines

S Hikam[1,*], P B Timotiwu[2], D Sudrajat[3], and S Alvianti[2]

[1] Biosystem Technics Study Program. Institute Teknologi Sumatera. Jalan Terusan Ryacudu. Jati Agung. Lampung Selatan. 35365. Indonesia.

[2] Department of Agronomy and Horticulture. College of Agriculture. University of Lampung. Jalan Sumantri Brojonegoro. Bandar Lampung. 35144. Indonesia.

[3] Department of Estate Crops. State Polytechnics of Lampung. Jalan Raya Soekarno-Hatta Bandar Lampung. 35145. Indonesia

*Email: saiful.hikam@tbs.itera.ac.id

Abstract. Extensive use of corn hybrids in Indonesia started in 1980 became expensive to farmers. In fact, the hybrids became obsolete and were replaced with the new ones. The obsolete hybrids, however, expressed superior traits toward detrimental hindrances *in situ*. The traits were identified, the obsolete hybrids were worthy to select for being local genetic sources (LGS) in a corn breeding program. The research was done to identify the LGS collected from Padang, Palembang, Jogjakarta, West Kalimantan and Lampung. In 2018 the LGS were planted in two rows and open-pollinated among themselves. The seeds were harvested separately to make female lines. A synthetic line was made by mixing 100 seeds of each female line. In 2019 the six lines: five female and one synthetic, and Pioneer36 (P36) F1-hybrid as control were planted in a randomized complete block design with 3 replicates. The data were collected for traits: plant height, leaf number plant^{-1}, time to tasseling, time to silking, ear length, ear diameter, 100 seed weight and seed yield m^{-2}. The data were variance-analysed and Tukey's HSD was applied to separate and group the lines. A dendrogram was constructed to decide lines suitable to recombine. The result indicated that the Padang, Palembang, Lampung and Synthetic lines could be used as the male parents to recombine with the Jogjakarta and West Kalimantan lines. The Tukey's HSD and dendrogram statistics could be useful to identify line superior and pairing appropriate lines as parents in a breeding program.

Keywords: dendrogram, local genetic source, statistical corn breeding, Tukey's HSD

1. Introduction

Developing better new varieties of corn had the second importance in plant breeding program in Indonesia after rice [1]. The dependency of animal feeds on corn as raw material increased from year to year following the Government program of increasing animal protein intake in diet to fight stunting among children. As the facts that Indonesia had been introducing hybrid corn varieties since 1975, the corn breeding program had successfully introduced no less than 200 hybrid varieties, exotic and own hybrids as well [2]. The speed of the introduction was rather high, in an average of five hybrids every year. This depleted the parental lines required for corn breeding also rather sooner and impacted on the price of hybrid-corn seeds. The price of the hybrid seeds nowadays was 28 – 30 times higher than that of 20 years ago.

The introduction of the hybrid corn varieties in 1975 was to fight *Peronosclerospora* downy-mildew disease [2]. The disease was soil-borne type hard to defeat and fast to disperse. Later the *Fusarium* and *Colletotrichum* also soil-borne diseases infected the corn hybrids. Therefore the hybrids released and marketed should have the resistance against the diseases. And the resistance so complete might not be found easily in one or several parents maintained by the breeders. Following the dispersion of the hybrids, the hybrids released earlier became obsolete. They were of superior recombination only to be put aside due to the release of the newer hybrids. The obsolete hybrids existed, however, planted and managed by the local farmers. In time the hybrids became the gene pool 1 *in situ* having traits well adapted and considered superior locally. The local genetic source (LGS) of corn varieties expressed superiority on resistant to drought, low pH, lower fertilizer requirement, and pest infestation, while maintained acceptable seed yield [3].

In time that the lines beneficial for recombination became less and less due to their extensive use in hybridization, the genetic relationship among the hybrids so developed became closer [4]. This resulted in their superiority against the diseases and unfavorable planting conditions reduced and resulted in using more insecticides and fertilizers. In this concept the use of the LGSs as parent lines in a breeding program was considerable. The study intended to evaluate the potential of the LGSs of corn collected from five Indonesia's provinces as parent lines in the breeding program.

The problems in using the LGSs were each of the LGS might be superior in its localization and would expressed wide differences in preferred characters when planted *ex situ* [5]. The differences mostly due to adaptation ability on a different localization. To overcome the adaptation limitation the LGSs were planted in the first localization in Lampung province before were tested for their performances in the second localization [6]. Data were collected on vegetative and reproductive traits to decide the best LGS or several LGSs for recombination in later programs.

Tukey's mean difference (Tukey's HSD) might be the simplest statistics to differentiate the trait's performance means among the LGSs, however the Tukey's HSD differentiated only for each performance [7]. This led to the problem of deciding which LGSs were prospective based on all tested trait's performances. The study analysis offered a solution to the problem by arranging the trait's performance done with Tukey's HSD into ranks of the LGSs. The performance data would also be used to construct a dendrogram which would explain supposed recombination among LGSs in developing prospective corn hybrids [8].

2. Materials and methods

The study was done in rainy season, March – June 2018 in Gaya Baru, East Lampung Regency and March – June 2019 in the State Polytechnics of Lampung's research field. The soil in both places was of Red-Yellow Podsolic ultisol having low pH and fertility. Five LGSs were collected from Padang, Palembang, Jogjakarta, West Kalimantan and Lampung provinces. The sixth LGS, a synthetic line was made as a mixture of 100 seeds of each LGSs. In 2018 the LGS were planted in two rows and let open-pollinate among themselves. The seeds were harvested separately to make superior-female lines. In 2019 the six lines: five female and one synthetic lines, and Pioneer36 (P36) F1-hybrid as control were planted in a randomized complete block design with 3 replicates. The data were collected for traits: plant height (cm), leaf number plant^{-1}, time to tasseling (days after planting; dap), time to silking (dap), ear length (cm), ear diameter (cm), 100 seed weight (g), seed number ear^{-1} and seed yield m^{-2} (kg). The data were analyzed with ANOVA and Tukey's HSD was applied to separate the means so that to construct the rank of the LDSs, respectively. A dendrogram was constructed based on the distances among lines to decide lines suitable to recombine.

3. Results and discussion

3.1. ANOVA's mean squares analyses for plant traits

Table 1 indicated that there were mean differences for LGS lines meant that that would be averaged value which would ideal for recombination. The CV value ≤ 10 % would indicate that the averaged means fit the ideal expectation whereas ≥ 15 % indicated that the LGSs needed to be reselected to decrease the CV for the trait. The data indicated that the average value for each traits was within the range of new released national corn hybrids, only the standard deviations for 100 seed weight, seed weight ear^{-1} and seed yield m^{-2} were high ≥ 15 %. Furthermore, the data indicated that the traits: plant height, leaf number, days to tasseling, days to silking would not need further improvement because their average values were already within the national range. However, the ear diameter trait would need to increase to 5.0 – 5.5 cm through a recombination with new hybrid varieties [9].

Table 1. Mean square analyses for plant traits.

Source of variation	df	Plant height	Leaf number	Days to tasseling	Days to silking	Ear diameter	Ear length	Seed row ear^{-1}	100 seed weight	Seed number ear^{-1}	Seed yield m^{-2}
Replicate	2	832.73	1.92	2.89	2.65	0.01	0.59	0.08	2.63	1162.91	0.002
LGS lines	6	719.15*	7.33**	40.62**	39.42**	0.52**	1.86**	4.59**	72.24*	8624.83**	0.13**
Error	12	214.89	0.55	5.80	8.05	0.07	0.10	0.75	16.48	1292.19	0.02
Total	20	8558.99	54.45	319.16	338.44	4.02	13.50	36.74	636.40	69581.11	0.96
CV %		8.33	4.91	4.52	5.31	6.41	1.86	6.26	13.69	11.81	16.58
Average value		176.06 cm	15.11	53.32 dap	53.41 dap	4.14 cm	17.02 cm	13.84	29.65 g	304.36	0.78 kg
Standard deviation %		11.47	10.65	7.31	7.52	10.56	4.71	9.56	18.57	18.91	27.49

* and ** indicated there were mean differences at $P \leq 0.05$ and $P \leq 0.01$, respectively

3.2. Mean difference analyses and grouping of the LGS

The Tukey's HSD was used to separate the variable means and resulted in that the traits separated into two classes, except seed yield m^{-2} which separated into three classes, in which West Kalimantan LGS was the lowest (horizontal data in Table 2).

Table 2. Mean analysis for plant traits using Tukey's $HSD_{0.05}$ and rank of the LGSs

Trait	LGS					Synthetic	P36	$HSD_{0.05}$
	Padang	Palembang	Jogjakarta	West Kalimantan	Lampung			
Plant height (cm)	177.6 ab	166.93 ab	187.47 a	145.27 b	181.43 ab	184.43 ab	189.3 a	41.89
Leaf number	16.2 a	14.87 a	15.67 a	11.8 b	16.47 a	15.67 a	15.13 a	2.12
Days to tasseling (dap)	57 a	54.47 a	52.27 ab	45.67 b	54.6 a	53.73 a	55.53 a	6.88
Days to silking (dap)	56.8 a	54.93 a	52.47 ab	45.8 b	54.4 a	53.8 ab	55.67 a	8.11
Ear diameter (cm)	4.08 a	4.21 a	4.47 a	3.28 b	4.57 a	4.17 a	4.22 a	0.76
Ear length (cm)	17.13ab	17.00ab	16.33bc	15.67c	17.53a	17.57a	17.93a	0.41
Seed row ear^{-1}	13.73 b	13.6 b	13.6 b	12.13 b	16.31 a	13.63 b	13.87 ab	2.48
100 seed weight (g)	30.79 a	30.71 a	31.13 a	18.86 b	33.84 a	30.17 ab	32.02 a	11.60
Seed number ear^{-1}	298.26 ab	399.22 a	248.14 b	247.74 b	292.11 b	345.31 ab	299.72 ab	102.72
Seed yield m^{-2} (kg)	0.8 ab	1.05 a	0.66 bc	0.4 c	0.85 ab	0.9 ab	0.81 ab	0.37
∑a of the LGS	9	9	6	0	9	9	10	
Rank of the LGS	2	2	3	4	2	2	1	

LGS traits (horizontal data) followed by the same letter were not different at Tukey's $HSD_{0.05}$

Table 2 (vertical data) indicated the superiority of a certain LGS over the others. Following the reasoning in studying the data vertically in which the means having "a" would be considered the best, Table 2 would present the rank of the LGSs as well based on how many "a"s a certain LGS owned [10].

Thus Table 2 indicated that the hybrid P36 would be in rank 1; the Padang, Palembang, Lampung and Synthetic LGSs in rank 2; Jogjakarta LGS in rank 3 and West Kalimantan in rank 4. That the hybrid P36 was in the rank 1, or the best, would not be surprising because it was a new national-hybrid and had been tested and certified to be suitable on a low-pH soil (Red-Yellow Podzolic ultisol) planting in Lampung. Furthermore, the data also indicated that the Padang, Palembang, Lampung and Synthetic LGSs were in the same rank 2 which indicated that the LGSs expressed a good adaptability on the Lampung soil.

3.3. Dendrogram Analysis
A dendrogram was made based on the correlation values among the lines. Table 2 showed that the lines were separated by the Tukey's HSD into 4 ranks and the dendrogram resulted in five cluster: (1) the Lampung line/P36; (2) Padang/West Kalimantan; (3) Synthetic; (4) Jogjakarta; and (5) Palembang (Figure 1). The farthest distance of the lines marked the Lampung line/P36 to Palembang was 0.29 % may indicate that the LGS had one or more common ancestors in the hybridization of each. The Lampung line (ranked 2 in Table 2) which in fact an LGS was closely related with P36 (ranked 1) one of the popular hybrids planted by the Lampung corn-farmers. The natural cross-hybridization may have improved The Lampung line inasmuch to improve its performance.

The dendrogram analysis gave different result from the mean analyses based on the Tukey's HSD. The statistical actuality may give different perspective to a plant-breeder to choose which line would better recombine to another line.

In the Table 2 the LGS's: Padang, Palembang, Lampung and Synthetic might not recombine well since they were ranked the same. They would be better used as the male parents to recombine with the Jogjakarta and West Kalimantan lines. For themselves, they would be better off when recombine with the P36 hybrid or other new hybrids.

On the other hand, the dendrogram informed that the Lampung line would be appropriate to hybridize as male parent with the West Kalimantan and Jogjakarta lines; whereas the lines: Lampung, Padang, Palembang and Synthetic could be hybridized as the male or female parents to each other. The P36 as the newer line could be used as male parent to hybridize with the LGS lines.

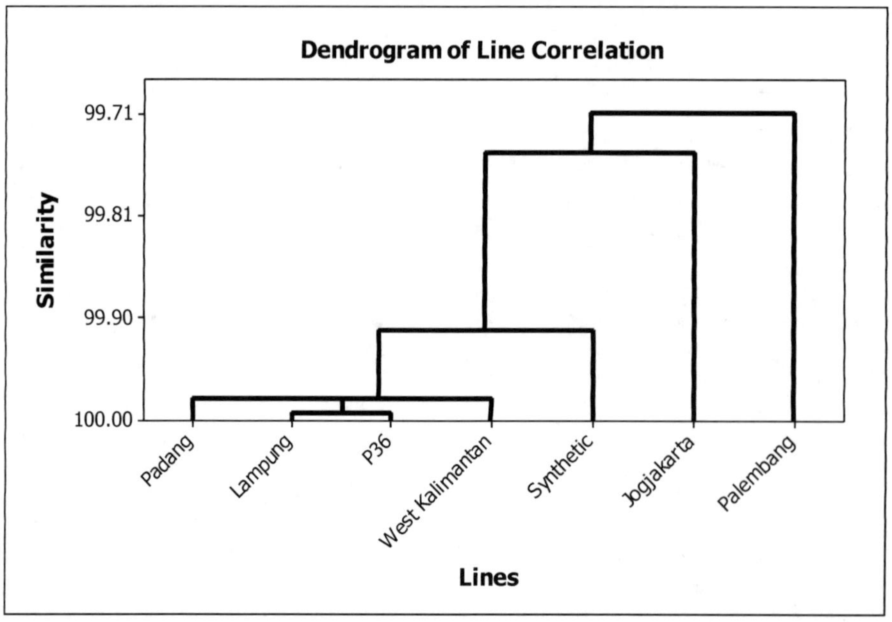

Figure 1. Dendrogram constructed based on the correlations for all traits among lines.

4. Conclusions

The study concluded that the Tukey's HSD was useful in determining superior line (or lines) as identified by traits of each; how many traits a line would have which were bests as compared to those of other lines. The dendrogram statistics were useful to determine which lines were suitable to become hybrid parents in a recombination. The study indicated that the Tukey's HSD could rank the LSD lines according to the performance of each line measured of all traits. The rank of the lines from the best down to the poor: the hybrid P36 was ranked 1; the Padang, Palembang, Lampung and Synthetic lines were ranked 2; Jogjakarta was ranked 3 and West Kalimantan was ranked 4. The dendrogram differenciated the lines into the Lampung lines/P36 hybrid, Padang/West Kalimantan, Synthetic, Jogjakarta and Palembang; and concluded that the Padang, Palembang, Lampung and Synthetic lines could be used as the male parents to recombine with the Jogjakarta and West Kalimantan lines.

Acknowledgements

The authors wished to express their appreciation to Dr. Djaenuddin Kartahadimadja of the State Polytechnics of Lampung, Bandar Lampung, Indonesia for his permission of using the Polytechnics Research Field and helps provided during the study.

References

[1] Swastika D K S. 2002. Corn Self-Sufficiency in Indonesia: The Past 30 Years and Future Prospects. Center for Agro-Socio-Economic Research and Development. Jurnal Litbang Pertanian, 21(3):75-83.

[2] Balitsereal Maros. 2019. Perkembangan Jagung Hibrida Badan Litbang Pertanian. Indonesian Agency for Agricultural Research and Development. Maros. South Sulawesi.

[3] Wijayanto T. 2007. Karakterisasi Sifat-Sifat Agronomi Beberapa Nomor Koleksi Sumber Daya Genetik Jagung Sulawesi. Jurnal Penelitian dan Informasi Pertanian, 11(2):75-83.

[4] Syafii M, Ruswandi D. 2019. Analisis Estimasi Jarak Genetik dan Hubungan Kekerabatan Genotipe Jagung Unpad Toleran Naungan Berdasarkan Marka SSR. Prosiding Seminar Nasional Agroteknologi. Fakultas Sains dan Teknologi. UIN Sunan Gunung Jati Bandung. DOI: 10.15575/psa.v1i0.128.

[5] Azeez M A, Adubi A O, Durodola F A. 2018. Landraces and Crop Genetic Improvement. Rediscovery of Landraces as a Resource for the Future. Oscar Grillo IntechOpen. DOI: 10.5772/intechopen.75944.

[6] Hallauer A R, Carena M J, Filho J B M. 2010. *Quantitative Genetics in Maize Breeding.* Handbook of Plant Breeding. Springer-Verlag Publisher. New York.

[7] Hutauruk J N, Kuswanto K, Sugiharto A N. 2017. Uji Daya Hasil Pendahuluan 9 Galur Jagung (*Zea mays* L.). Jurnal Produksi Tanaman. Universitas Brawijaya. Vol 5:12(2017)

[8] Andarini Y N, Sutoro. 2018. Pengelompokan Plasma Nutfah Jagung Lokal Berdasarkan Karakter Kuantitatif Tanaman (*Local Corn Germplasm Grouping Based on Plant's Quantitative Characters*). Informatika Pertanian, Vol. 27(1):15 – 24

[9] Yasmanidar. 2019. Proses Pelepasan Varietas Unggul. Cybex. http://cybex.pertanian.go.id /mobile/ artikel/59123/

[10] Hikam S, Timotiwu P B. 2016. Roles of Calcium and Magnesium as Selection Factors in Sweet Corn Quality Improvement on Acidic Red-Yellow Podsolic Soil. AGRIVITA. Journal of Agricultural Science. 2016. 38(2): 163-173.

Microbial chitosan for the fabrication of piezoelectric thin film

A T Hazmi[1], F B Ahmad[1,2,*], M H Maziati Akmal[3], and A A M Ralib[4]

[1] Department of Biotechnology Engineering, International Islamic University Malaysia, 53100 Kuala Lumpur, Malaysia

[2] Center for Energy Science, Department of Mechanical Engineering, University of Malaya, 50603 Kuala Lumpur, Malaysia

[3] Department of Science in Engineering, International Islamic University Malaysia, 53100 Kuala Lumpur, Malaysia

[4] Department of Electrical and Computer Engineering, International Islamic University Malaysia, 53100 Kuala Lumpur, Malaysia

* Email: farahahmad@iium.edu.my

Abstract. Chitin has proven to have a good mechanical and electrical properties to be used in making piezoelectric thin films. However, due to the restriction in solubilizing chitosan in many solvents, there is increasing interest in exploring the used of chitosan in producing thin films. Chitosan, compared to chitin, can be easily solubilized in certain dilute acids. Chitosan that has been extracted from fungal biomass can be used for the fabrication of biomaterial thin films. There are different ways that can be used to fabricate a thin film such as electrospinning, spin-coating, solvent casting and also the hot press technique.

Keywords: Chitosan; chitin; electrospinning; fungi; piezoelectric; thin films

1. Introduction

Chitin is the second most abundant natural polysaccharide that can be extracted from the shells of crab, shrimp, and prawn [1]. It is a linear 1,4-linked polymer that is made up of N-acetyl-D-glucosamine residues [2]. As chitin is a semi-crystalline polymer with high hydrogen bond, the solubility of the chitin in most organic solvents is not feasible [3].

There are two allomorphs of the chitin which are α-chitin and β-chitin [4]. The most abundant one is the α-chitin where it can be found in the cell wall of yeast and fungi, krill, lobster, crab and shrimp whereas the β-chitin is the rare type, naturally occurring in squid pens, vestimetiferan worms and pogonohoran [5]. Both allomorphs of chitin are insoluble in most solvents making it one of the challenges for it [5]. Only few solvents that can solubilize chitin are hexafluoro-isopropanol, hexafluoroacetone, chloroalcohols and dimethylacetamide, that are typically toxic and dangerous to nature [6].

Chitosan can be produced from chitin through process called deacetylation. Compared to chitin, chitosan can be solubilized in dilute acids. The solubility of this material depends on the types (mineral or organic) and concentration of the acids used [3]. One of the main sources of chitin is from the crustacean shells where annually it has been produced around 1.2 mega tons [7]. As there are limitations in the supply of this material due to season and limitation in the fishing industry itself, fungal biomass-derived chitin and chitosan has gained more attention [8]. This study reviews various methods for extraction of chitosan from fungal biomass and fabrication of thin film from chitosan.

2. Chitin isolation and chitosan extraction from microbial biomass

Crustacean-based chitin extraction involves three processes involving demineralization, deproteinization and deacetylation [9]. In demineralization step, the inorganic calcium carbonate is removed while the protein is removed during deproteinization. The demineralization and deproteinization processes can be performed either using chemical treatments or enzymatic treatments [5].

The use of fungi as the source of chitin and chitosan is promising due to the sustainability of fungal cultivation that is not limited by the supply from fishing industry [10]. Fungal biomass is acquired from the harvesting of fungal cell at the end of cultivation. Fungi can be cultivated in large scale bioreactor, in which large scale of production of fungal biomass does not requires large area of available land [11]. The pathway of extraction of chitin from fungi biomass is similar to that from crustacean. However, the chitin extracted from fungal biomass does not require demineralization process [8].

The first step in converting chitin to chitosan is the deproteinization as the chitin has a high protein content [10]. This step is said to be difficult to disrupt the bond between chitin and protein. The most preferred reagent used by many studies is NaOH [5].

The next important step is the removal of *B*-glucan from fungal biomass. In fungal cell wall, the major polysaccharide existed is the *B*-glucan which consists of (1→3) backbone with (1→6) branches [12]. Most of the glucan int the biomass can be degraded with the help of harsh acid treatment [4].

During the process of deacetylation of chitin, chitosan is produced by removing the acetamide group in chitin and replacing with amino group [13]. Figure 1 shows the chemical structure for deacetylation of chitin where chitosan will be produced at the end of the reaction.

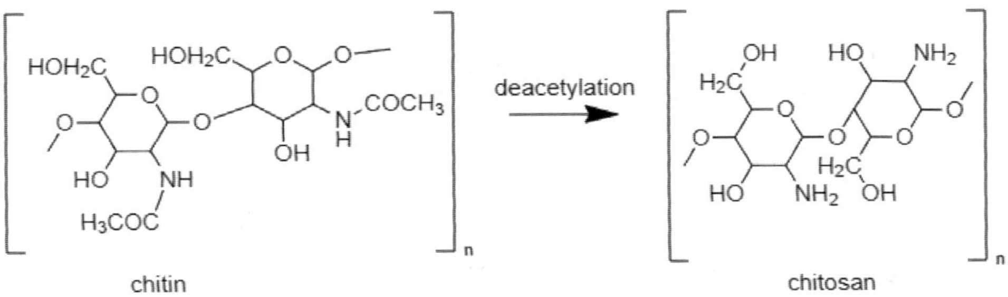

Figure 1. The chemical structure of chitin and chitosan in deacetylation process.

Table 1 shows several methods of extraction of chitosan from fungal biomass and chitosan yield obtained through respective methods. Although most of the methods for the extraction were using the alkaline treatment, method developed by Zamani et al. involved the application of dilute sulfuric acid treatment. In the extraction method from that study, chitosan was being recovered by precipitation at lower temperature after treating fungal biomass with strong acid [14]. There are studies that used enzymatic reaction in extracting chitosan from fungal biomass, where the α-amylase was used [15]. The enzyme used in the study help to breakdown the chitosan-glucan complex and produce a better quality of chitosan [15]. However, the enzymatic extraction method was time-consuming and costly.

Table 1. Methods of chitin isolation and chitosan extraction, chitosan yield and the degree of deacetylation (DDA).

Fungal strain	Method of extraction		Chitosan yield (%)	DDA (%)	Ref.
	Pre-treatment	Extraction			
Rhizomucor pusillus	-	Alkaline treatment using NaOH 0.5 M at 120 °C for 20 min. AIM was treated with 72 mM sulfuric acid at 22 °C for 10 min and proceed with same treatment for another 45 min.	28.7	-	[14]
Gongronella butleri	Biomass was dried at 45 °C	Treated with 10 mL of 1 M NaOH per g of mycelia at 40 °C for 16 h. AIM was washed and dried before subjected to 200 mL of 0.35 M of acetic acid at 40 °C for 16 h.	1.9	13	[15]
		Treated with 10 ml of 11 M NaOH containing 0.05 g of sodium borohydride per g of mycelia (40 °C, 13 h). AIM washed until neutral pH and dried before subjected to 200 mL of 0.35 M acetic acid at 95 °C for 5 h. pH of AIM was adjusted to 4.5 using 1 M NaOH and 4% (*v/v*) *α*-amylase and incubated in water bath shaker at 65 °C, 200 rpm for 3 h.	5.9	13	
Penicillium chrysogenum	Dry biomass was pulverized	Alkaline treatment using 1 M NaOH for 1:30 (*w/v*) and autoclaved at 121 °C for 15 min.	12.2	82.4	[16]
Aspergillus niger	Dry biomass was milled with blender	Alkaline treatment using 1 M NaOH for 1:40 (*w/v*) at 90 °C for 3 h. AIM was treated with 10% acetic acid at 1:40 (*w/v*) at room temperature for 6 h on 200 rpm shaker.	7.57	83.64	[17]
Mucor genevensis (chitin-chitosan thin film)	Biomass blended for 5 min with water	Alkaline treatment using 1 M NaOH at 65 °C for 3 h, cooled and neutralised	-	Partially deacetylated	[18]

3. Method to fabricate chitosan thin film

One of the biodegradable natural polymers is chitosan. The degree of deacetylation of the chitin should reach about 50% to ensure that the chitins were successfully converted into chitosan [1]. Chitosan thin film has various application including for piezoelectric application [1]. The properties of chitosan are biocompatibility, biodegradability, non-toxic and can be easily formed as film, making it suitable to be used in piezoelectric application. Thin film fabricated using pure chitosan can produce higher piezoelectricity compared to thin film from composite materials [6]. However, there are issues concerning the process of fabrication of chitosan thin film as it only dissolves in polar acidic solutions [19]. There are several methods in fabrication the chitosan thin film such as electrospinning, spin coating and also solvent casting. To date, all studies on chitosan thin film for piezoelectric application were focusing on the use of solvent casting [1,20].

3.1. Electrospinning

Electrospinning is one of the process used to synthesis polymer fibres that utilizes high voltage to develop electrostatic force. As thin films are widely explored, electrospinning has been one of the attractions in film fabrication as it can synthesize material customized for diameters ranging between nanometers and micrometers. The basic setup of this method, as shown in figure 2, consists of injection pump, power supply, fixture that support the spinneret and plate [21]. Chitosan has difficulty to solubilize in aqueous and organic solution as it has high crystallinity properties. Due to this, chitosan is mixed with other polymer to help with the spinnability of chitosan. It has been reported that medium and high molecular weight of chitosan will produce a uniform surface using electrospinning [3]. The drawback of this method is the formation of beads along with the fibers [20].

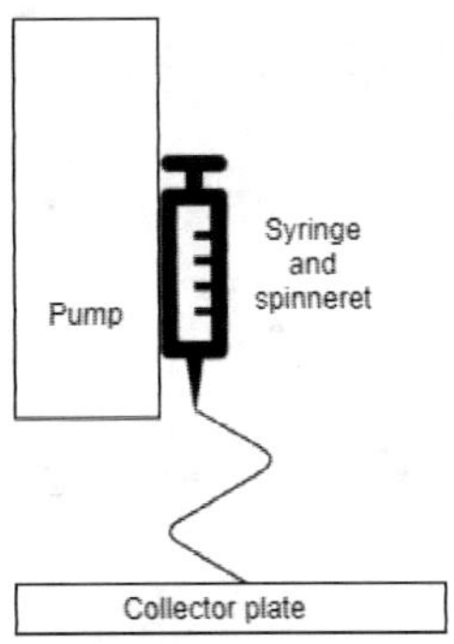

Figure 2. The basic setup of electrospinning.

3.2. Spin-coating

Spin-coating is a method used to make a thin film that is uniform and the thickness can be in nanoscale [19]. There are five stages of spin coating which are solution deposition, acceleration, viscosity-controlled thinning, evaporation-controlled thinning and lastly drying [22]. As the solution is deposited on a substrate, the disc will be rotated until it reaches the desired speed. The common speed used in spin coating are between 2000 to 6000 rpm [22]. The centrifugal forces and adhesion forces from the solution-substrate will create strong sheering that leads to radial flow and ejection of the solution from the substrate, resulting in a uniform and thin chitosan film [19]. With this method, the thickness of the film created can be controlled easily by changing the spin speed on the substrate. Figure 3 shows the setup of spin-coating method. However, as the size of the substrate increases, the thinning of the film becomes hard as it is harder to control the high-speed spinning of the disc [23].

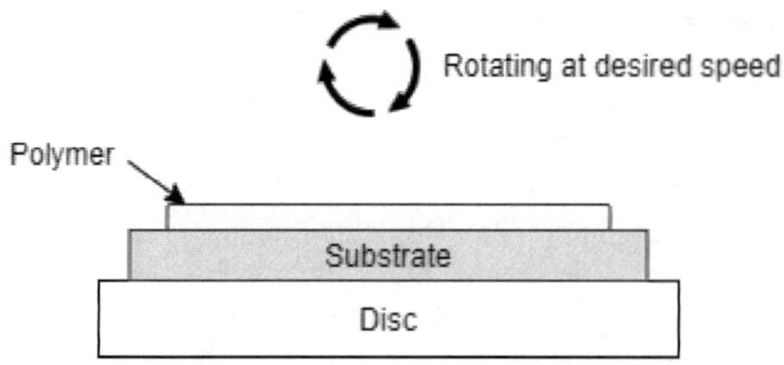

Figure 3. The setup of spin-coating method in fabricating thin film.

3.3. Solvent casting

The most basic method for fabricating thin film is the solvent casting method. The polymer needs to be soluble in water or volatile solvent before doing the casting. It is also important that the mixture need to have minimum solid content and viscosity and the material used must be able to form homogeneous film and be separated from the casting surface [24]. The drawback from this method is that it cannot produce very thin films from polymers without stretching the film during drying and the films produced are more hydrophilic compared to electrospun films [24]. Figure 4 shows the solvent casting process of thin film fabrication.

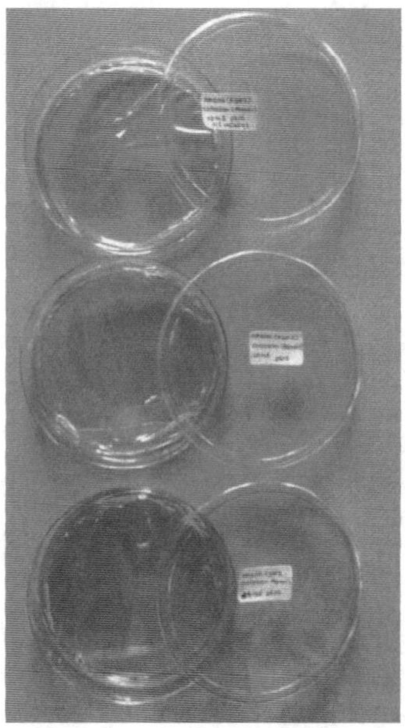

Figure 4. The solvent casting for thin film.

3.4. Hot press technique

This technique is not a common choice as it is hard to get a low thickness film. However, a study by Hashizume et al. reported on fabrication of chitosan thin film using hot press. The main point of this technique is that the mixture of chitosan needs to be freeze dry prior to hot press [25]. During the process, polyethylene terephthalate (PET) sheets and polytetrafluoroethylene (PTFE) sheets were used as separator between the sample and hot press stage in order to avoid contamination of the thin film. In the study by Hashizume et al., the temperature and pressure used for of the hot press technique were set at 120 °C and 20 MPa respectively [25].

4. Conclusions

The production of chitosan from microbial biomass is attractive due to the sustainable supply chain. There are many ways to successfully extract chitosan from microbial biomass. The fabrication of chitosan into thin film is crucial for piezoelectric application. The use of chitosan in piezoelectric application is significant as chitosan is a biomaterial that is biocompatible which enable its application as biomedical piezoelectric devices. All of the methods of fabricating the thin films mentioned before has proved to successfully producing a good quality of chitosan thin film.

Acknowledgement

The authors would like to acknowledge financial assistance from Ministry of Education under Fundamental Research Grant Scheme (FRGS) (FRGS/1/2019/TK02/UIAM/02/3).

References

[1] Hänninen A, Sarlin E, Lyyra I, Salpavaara T and Kellomäki M (2018). Nanocellulose and chitosan-based films as low cost, green piezoelectric materials. *Carbohydrate Polymers*, **202**(February), 418–424

[2] Jayakumar R, Prabaharan M, Sudheesh Kumar P T, Nair S V and Tamura H (2011). Biomaterials based on chitin and chitosan in wound dressing applications. *Biotechnology Advances*, **29**(3), 322–337.

[3] Salas C, Thompson Z and N. Bhattarai (2017). Electrospun chitosan fibers. In Electrospun Nanofibers. *Elsevier Ltd.*

[4] Nawawi W M F B W, Jones M, Murphy R J, Lee K, Kontturi E and Bismarck A (2019). Nanomaterials derived from fungal sources - is it the new hype ?

[5] Younes L and Rinaudo M (2015). Chitin and chitosan preparation from marine sources. Structure, properties and applications. *Marine Drugs*, **13**(3), 1133–1174.

[6] Akmal M H M and Ahmad F B (2020) Bionanomaterial Thin Film for Piezoelectric Applications. In: A. Jameel, A. Yaser (eds) Advances in Nanotechnology and Its Applications. *Springer*, Singapore.

[7] Ospina Álvarez S P, Ramírez Cadavid D A, Escobar Sierra D M, Ossa Orozco C P, Rojas Vahos D F, Zapata Ocampo P and Atehortúa L (2014). Comparison of extraction methods of chitin from ganoderma lucidum mushroom obtained in submerged culture. *BioMed Research International*

[8] Akila R M (2014). Fermentative production of fungal Chitosan, a versatile biopolymer (perspectives and its applications). **5**(4), 157–170.

[9] Kumar S and Rath P K (2014). Extraction and Characterization of Chitin and Chitosan from (Labeo rohit) Fish Scales. *Procedia Materials Science*, **6**(Icmpc), 482–489.

[10] Ahmad F B, Akmal M H M, Amran A and Hasni M H (2020). Characterization of chitosan from extracted fungal biomass for piezoelectric application Characterization of chitosan from extracted fungal biomass for piezoelectric application.

[11] Ahmad F B, Zhang Z, Doherty W O S and Hara I M O (2015). A multi-criteria analysis approach for ranking and selection of microorganisms for the production of oils for biodiesel production. *Bioresource Technology*.

[12] Fesel P H and Zuccaro A (2016). b -glucan : Crucial component of the fungal cell wall and elusive MAMP in plants. *Fungal Genetics and Biology*, **90**, 53–60.

[13] Chen G, Zhao L, Qi Y, & Cui Y L (2014). Chitosan and its derivatives applied in harvesting microalgae for biodiesel production: An outlook. *Journal of Nanomaterials*, 2014(1).

[14] Zamani A, Edebo L, Niklasson C and Taherzadeh M J (2010). Temperature shifts for extraction and purification of Zygomycetes chitosan with dilute sulfuric acid. International Journal of Molecular Sciences, **11**(8), 2976–2987.

[15] New N and Stevens W F (2002). Production of fungal chitosan by solid substrate fermentation followed by enzymatic extraction. *Biotechnology Letters*, **24**(2), 131–134.

[16] Osarumwense J O, Akpaja E O J and Ekhosuehi J A (2018). Preparation and characterisation of chitosan from Penicillium chrysogenum Thom for decolourisation of Congo red dye in aqueous solution. *Journal of Applied Sciences and Environmental Management*, **21**(7), 1365.

[17] Abdel-Gawad K M, Hifney A F, Fawzy M A and Gomaa M (2017). Technology optimization of chitosan production from Aspergillus niger biomass and its functional activities. *Food Hydrocolloids*, **63**, 593–601

[18] Jones M, Weiland K, Kujundzic M, Mautner A, Bismarck A and John S (2019). Chitin-chitosan Thin Films from Microbiologically Upcycled Agricultural By-products. July, 11–14.

[19] Niemczyk A and Kaczorowski P (2015). Spin-coated chitosan on copolyester substrates. November.

[20] Praveen E, Murugan S and Jayakumar K (2017). RSC Advances property of a bio-polymer – chitosan and its. *RSC Advances*, **7**, 35490–35495.

[21] Veerabhadraiah A, Ramakrishna S and Angadi G (2017). Development of polyvinyl acetate thin films by electrospinning for sensor applications. *Applied Nanoscience*, **7**(7), 355–363.

[22] Mcilwee H A (2008). Chitosan Thin Films as Metal Ion Sensors and Structurally Colored Coatings. Chitosan Thin Films as Metal Ions Sensors and Structurally Colored Coatings, May, 1–83.

[23] Yilbas B S, Al-Sharafi A and Ali H (2019). Surfaces for Self-Cleaning.

[24] Deng L, Kang X, Liu Y, Feng F and Zhang H (2018). Characterization of gelatin/zein films fabricated by electrospinning vs solvent casting. In *Food Hydrocolloids. Elsevier Ltd*, **74**

[25] Hashizume M, Kobayashi H and Ohashi M (2011). Preparation of free-standing films of natural polysaccharides using hot press technique and their surface functionalization with biomimetic apatite. *Colloids and Surfaces B: Biointerfaces*, **88**(1), 534–538

Solar Photovoltaic Power Generation for Distillation Process

M I Mowaviq[1,*], A Junaidi[1], T W O Putri[1], and D Rosanda[1]

[1] Department of Electrical Engineering, Institut Teknologi PLN, Jalan Duri Kosambi, Jakarta Barat, Indonesia

* Email: mowaviq@itpln.ac.id

Abstract. Indonesia is an archipelago which is one of the countries with the longest beaches in the world. One of the great potentials of the sea can be used to produce freshwater in coastal areas. One of the processes of converting seawater into freshwater vapor is the distillation process. The freshwater vapor condenses and turns into water. Distillation is carried out by evaporating seawater so that the freshwater is separated from the salt. In this research, the distillation process is assisted by a solar power plant with photovoltaic panels. The hardware design consists of a solar panel, solar charge controller, battery, inverter, pump, electric heater, temperature sensor, and level sensor. The solar panel uses a total of 1200Wp with a voltage of 12 V. The battery uses a 12V 100Ah battery and a 1000W electric heater. Meanwhile, the pump power is 48W. From the test results, it was found that in one hour, 1.3 liters of freshwater were obtained with a total capacity of 10 liters of water. The use of solar photovoltaic in a distillation system would be better if combined with direct solar energy heating methods.

Keywords: Distillation system, solar photovoltaic, electric heater, seawater

1. Introduction

The need for clean water is a very basic need in human life. The use of clean water is very important for household, public and industrial consumption. Fresh water needs can be met in several ways. In urban areas, the provision of clean water is managed by the Drinking Water Company or carried out independently by the community by taking groundwater. However, in coastal areas that are far from springs, there are still many people who do not get clean water facilities. Meanwhile, in coastal communities, sea water is so abundant.

Coastal areas have abundant water potential, it is seawater that cannot be consumed directly. So, there needs to be another solution for provision clean water that can be consumed. Indonesia as an archipelago has a sea area of 5.8 million km2 or 70.8% of the total area of Indonesia. Besides having the potential for abundant sea water, it also has the potential for high sunlight intensity. These two potentials can be utilized to produce clean and consumable fresh water.

The method that can be used to get fresh water from seawater is the distillation method. The water distillation is one of the feasible solutions to overcome the freshwater crisis[1]. Distillation is the settling of sea water by evaporating the water and condensing the vapor so that it separates between salt minerals and produces fresh water. In Indonesia, with an average of 12 hours of sunlight a day, it can be used as an energy source for distillation of sea water. In general, the use of solar energy is done conventionally

IOP Conf. Series: Materials Science and Engineering 1173 (2021) 012044 doi:10.1088/1757-899X/1173/1/012044

to evaporate sea water. The distillation process using solar still work in six hours[2] or during the peak solar hours[3]. Meanwhile, the need for fresh water continues for a full day.

This study aims to answer the problem of the need for fresh water in coastal areas by utilizing solar energy to carry out the seawater distillation process. As well as the use of solar power plants to increase the yield and efficiency of the distillation process. Desalination plants powered by renewables can in principle be used to address some of the challenges arising from depleting water reserves around the world [4]. Solar energy is a relatively free renewable, clean, green, and environmentally friendly energy resource produced from the sun, using different technologies like solar thermal and photovoltaic (PV) modules to generate heat and electricity, respectively[5]–[8]. Solar energy refers to the radiant energy emitted, transmitted or received by the sun in the form of electromagnetic energy[9]. This research focus on solar power generation using photovoltaic for distillation system.

2. Design and Method

2.1. Distillation Systems

Distillation is one of the oldest and most common methods for the purification of liquid. Distillation columns are made up of several components, each of which is used either to transfer heat energy or enhance material transfer. A typical distillation unit contains several major components:

a) A vertical shell where the separation of liquid components is carried out.

b) Column internals such as trays/plates and/or pickings, which are used to enhance component separations.

c) A reboiler to provide the necessary vaporization for the distillation process.

d) A condenser to cool and condense the vapour leaving the top of the column.

e) A reflux drum to hold the condensed vapour from the top of the column, so that liquid (reflux) can be recycled back to the column [10].

In this research, sea water distillation design can be shown in Fig.1.

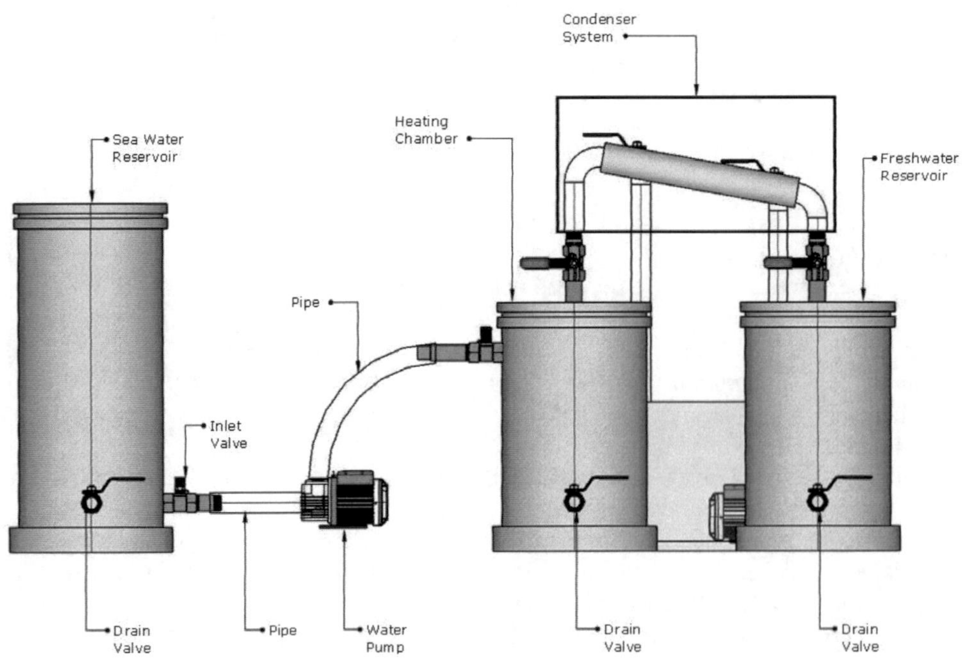

Figure 1. Front View of Distillation System Design

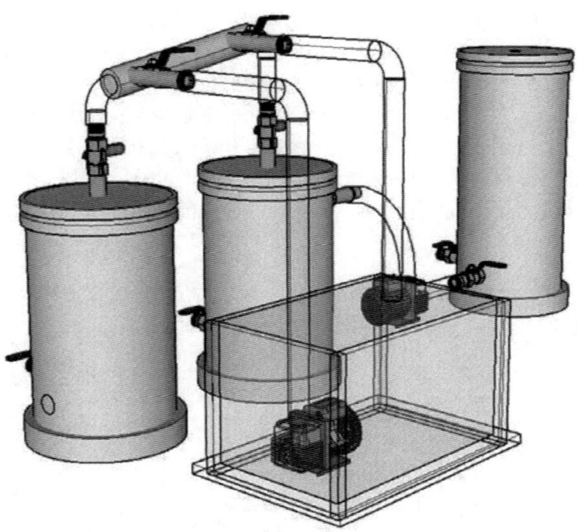

Figure 2. Back View of Distillation System Design

The process of distillation process of sea water in this research is below:
1) Water pump suck up sea water of the reservoir to heating chamber until full.
2) After heating chamber is full, the pump stop, the heater is on, and the valve is closed.
3) The electric heater work until 90°C and the condensation pump is on.
4) The water vapor enter the condenser and change to water dew.
5) The water dew flow to collection drum.

To find out how much energy is needed can be calculated by the formula:

$$Q = m \,.\, c \,.\, \Delta T + m \,.\, U \tag{1}$$

$$Q = W \tag{2}$$

$$W = P \,.\, t \tag{3}$$

From those formulas, we can calculate amount of energy, how long the system work, and how the design of photovoltaic solar generation. The calculation can be shown in Table 1.

Table 1. The Value of Quantities in Distillation Process per Litre

Symbol	Quantity	Value	Unit
c	heat capacity of water	3,850	J/kg.°C
T_1	First temperature	25	°C
T_2	Last Temperature	120	°C
U	Laten heat of vaporization	2,260,000	J/kg
P_h	Power of heater	1,000	W
P_p	Power of pump	48	W
t	Time	0.83	h
W_h	Work of heater	0.83	kWh
W_{pc}	Work of condenser pump	0.028	kWh

2.2. Solar Power Generation

Solar Power Generation based on working principle of photoelectric. When the sun shines on the semiconductor p-n junction, a new hole-electron pair is formed. Under the action of the p-n junction electric field, electrons flow from the p region to the n region, and the holes flow from the n region to the p region[8]. A conducting wire connects the cathode, the load, and the anode in series to form a circuit. Hence, an electric current is generated to supply the external load. Solar panel or solar module is the combination of multiple solar cells connected together and a solar array is the combination of multiple solar panels or solar modules in concert [5]. The diagram of solar power generation can be shown in Figure 3.

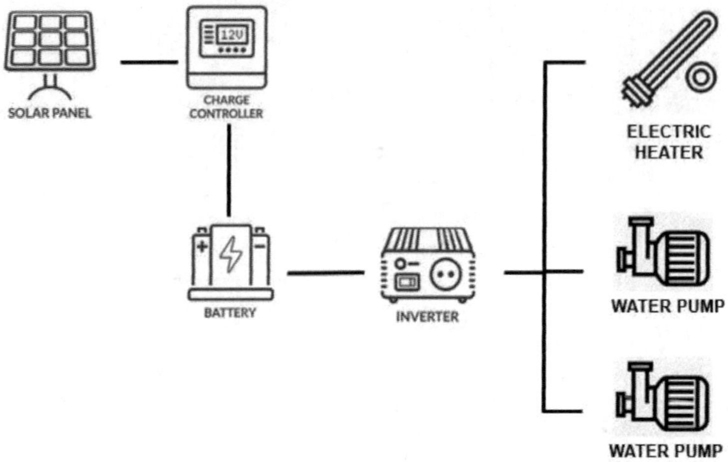

Figure 3. Diagram of Solar Power Generation

3. Result and Discussion

The main function of inverter is to change the variable dc voltage input of the transformed power supply into non-interference ac sine wave output, which can be supplied to the corresponding equipment or fed back to the power grid. Inverter in addition to the conversion of ac and dc voltage, but also can perform such as: disconnect the circuit, to avoid the current surge damage to the circuit. In addition, functions such as tracking the maximum power point (MPPT) and storing data and controlling the charge and discharge of batteries can help improve the efficiency of power generation [9].

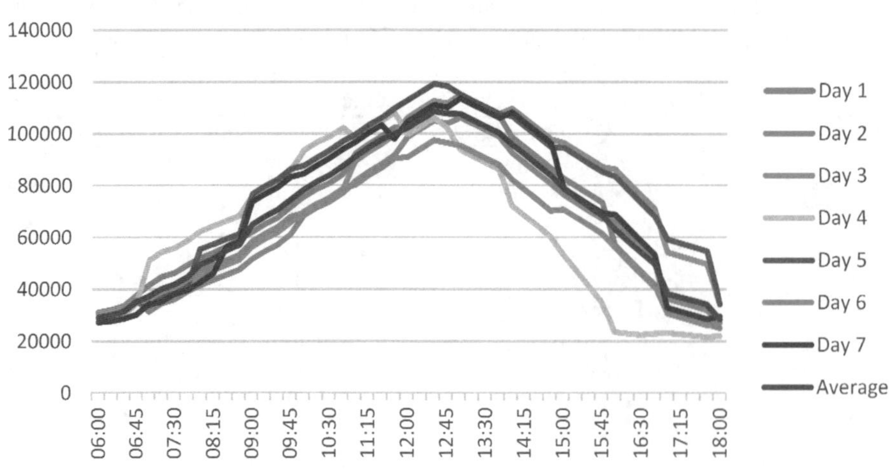

Figure 4. Graphic of Light Intensity Measurement

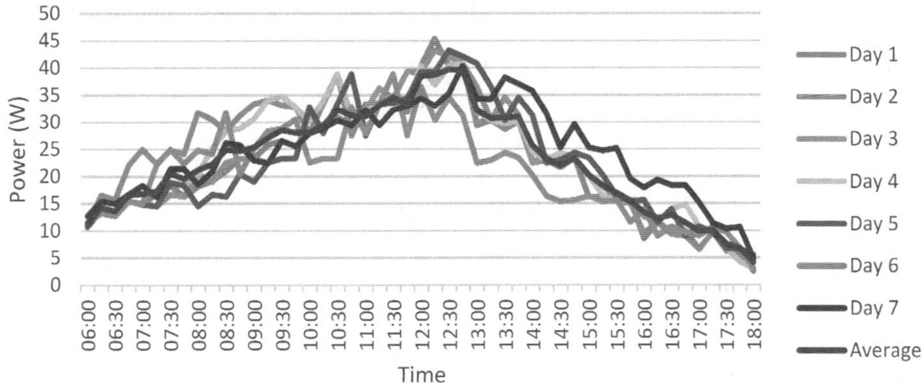

Figure 5. Graphic of Power Output Measurement

Solar panel testing is done by measuring the light intensity and power produced. The test was carried out for 7 days. The solar panels used are 300Wp solar panels. From the results of the tests carried out, it can be obtained a suitable solar power plant design. The results of the sunlight intensity test can be seen in Figure 4. Meanwhile, the resulting output power can be seen in Figure 5. From the data obtained, the average sunlight intensity is 69141 lumens. Meanwhile, the average energy produced by solar panels 300Wp for a day is 280 Wh. So, to meet the needs of the distillation system there are at least 4 of 300Wp solar panels. Meanwhile, in the distillation test, in one hour, 1.3 liters of fresh water can be obtained.

4. Conclusion

The energy required for the distillation process is quite large, so that if only photovoltaic solar panels are used, it will absorb a lot of costs. Therefore, the use of photovoltaic solar panels would be better if combined with direct solar heating. In addition, there is not too much fresh water obtained from the distillation process. So, it is advisable to use a heater with a larger capacity to speed up the evaporation.

References

[1] M. F. Remeli, B. Singh, N. Amirah, M. S. Meon, and W. N. Fadilla, "Solar Distillation Thermoelectric Power Generation," *IOP Conf. Ser. Earth Environ. Sci.*, vol. 268, no. 1, 2019.

[2] P. A. Mehta, A. Vyas, N. Bodar, and D. Lathiya, "Design of Solar Distillation System," *Int. J. Adv. Sci. Technol.*, vol. 29, pp. 67–74, 2017.

[3] S. Abhishek, A. S. Kumar, E. Anjana, M. Rahul, and S. Jisma, "Water purification using solar thermal and solar PV," in *2018 International Conference on Emerging Trends and Innovations In Engineering And Technological Research, ICETIETR 2018*, 2018, pp. 1–4.

[4] V. Rallabandi, N. Alawhali, O. Akeyo, and D. M. Ionel, "Simulation Studies for a Multi-MW Hybrid Wind-Solar PV System for Desalination Plants," *7th Int. IEEE Conf. Renew. Energy Res. Appl. ICRERA 2018*, vol. 5, pp. 1413–1416, 2018.

[5] R. Liu, S. Bimenyimana, G. N. Osarumwense Asemota, M. Xu, J. De Dieu Niyonteze, and L. Li, "Assessment of power generation performance characteristics using different solar photovoltaic technologies," *J. Phys. Conf. Ser.*, vol. 1311, no. 1, 2019.

[6] Z. Ma, Y. Zhang, and H. Li, "Energy efficiency analysis of inland ship photovoltaic system based on PVsyst," *IOP Conf. Ser. Earth Environ. Sci.*, vol. 242, no. 2, pp. 6–11, 2019.

[7] D. Van Tai, "Solar photovoltaic power output forecasting using machine learning technique," *J. Phys. Conf. Ser.*, vol. 1327, no. 1, 2019.

[8] R. Li, "China's photovoltaic power generation technology and application," *IOP Conf. Ser. Earth Environ. Sci.*, vol. 300, no. 4, 2019.

[9] Y. Wang, "Solar Photovoltaic Power Generation Technology Research," *IOP Conf. Ser. Mater. Sci. Eng.*, vol. 677, no. 3, 2019.
[10] K. Patil, P. Choudhary, and T. Bhatia, "Distillation Operations : Methods , Operational and Design Issues," *Natl. Conf. Adv. Heat Mass Transf. FAMT*, no. June 2014, p. 10, 2009.

Aerodynamic drag reduction on the application of suction flow control on vehicle model with varied upstream velocity

R Tarakka[1*], N Salam[1], Jalaluddin[1], W Rauf[1,2] and M Ihsan[3]

[1] Department of Mechanical Engineering, Faculty of Engineering, Hasanuddin University, Jalan Poros Malino, KM.6, Gowa, 92172, Indonesia
[2] Department of Mechanical Engineering, Faculty of Engineering, Gorontalo University, Jalan Jenderal Sudirman, Gorontalo 96211, Indonesia
[3] Sekolah Tinggi Teknik Baramuli, Pinrang, 91215, Indonesia

*Email: rustan_tarakka@yahoo.com

Abstract. The innovation in the automotive industry must prioritize the issue of energy security and the environment. One aspect of the innovations is regarding the development of more aerodynamic vehicle design. The study of vehicle aerodynamics has been considered as very important, due to its direct relation to the safety and efficiency factors of energy utilization. Applying active control on the onset region of flow separation is one example of the efforts. This study aims to analyse the effect of the application of active control by suction to the aerodynamic drag of the vehicle model through qualitatively observations of flow dynamics and quantitatively pressure fields. The test model is a modified Ahmed body with 35° slant angles and varied flow orientations. This study has been conducted in both numerical computation and experimental testing at a suction velocity of 1.0 m/s and upstream velocity of 11.1 m/s, 13.9 m/s, and 16.7 m/s, respectively. The results have found out that the attachment of active control is proven capable able to delay flow separations, to increase the pressure coefficients on the back wall, and to reduce aerodynamic drag by 10.8487% for computational methods and 10.9748% for experimental methods.

Keywords: flow control, pressure coefficient, aerodynamic drag

1. Introduction

Innovations in the automotive industry must prioritize the issue of the energy crisis and environmental resilience. One aspect of the innovations is focused on developing more aerodynamic vehicle designs. Vehicle aerodynamics is a study related to the pattern of airflow through the surface to affect vehicle performance when driving. This study has been conducted as an effort to improve the safety and efficiency of fuel consumptions [1]. Efficiency on fuel consumptions of 700 to 1000 litres every year can be achieved just by reducing aerodynamic drag by 2-3% [2].

Aerodynamic drag analysis has been supported by visualizing flow characteristics and pressure fields on the rear wall of the vehicle. Airflow patterns describe the separation process occurring at the top edge of the back. The faster the process of flow separation occurs; the aerodynamic drag of the vehicle will be even greater due to the high intensity of wake formation, which is the main cause of the rearward traction phenomenon.

The pressure field on the back wall also determines the intensity of the aerodynamic drag. The flow separation will generate a reduction in pressure distribution to cause considerable differences of pressure between the front and rear walls of the vehicle, which is a major contributor to aerodynamic drag. One effort to overcoming this problem is to delay separation. The delay of the flow separation process is intended to reduce the intensity of wake formation in the back wall so that the minimization of negative pressure and the reduction of drag can be achieved. Methods that can be carried out as an effort to delay the process of flow separation include the application of active control. The use of active control is considered effective because it involves the addition of external energy in the form of the application of suction techniques in the initial area of the flow separation process. Part of the fluid flow that reaches the separation area will undergo suction resulting in less intense wake formation and subsequently minimizing the phenomenon of rearward traction.

Mousa et al investigated of the impact of applying rectangular suction active control placed on the upper side of the back of the vehicle model through a numerical simulation approach (Computational Fluid Dynamics). The test model used is the result of modelling a sport utility vehicle (SUV) with a 1:10 dimension ratio to real vehicles. The results of the study revealed the application of suction active control could reduce aerodynamic drag by 9% [3]. Another study was carried out by Wessen & Thiele who placed active control on the upper end of the Ahmed body model with a 25° angle of attack. The study applied a numerical computational approach with the Large Eddy turbulence model. The results showed a reduction in flow separation, an increase in the distribution of pressure on the rear wall of vehicle models, and obtained an aerodynamic drag reduction of 9.4% [4].

2. Materials and Methods

This study is devoted to assessing the magnitude of the reduction in aerodynamic drag of vehicles that are modelled referring to the Ahmed model by altering the direction of flow orientation and reducing the dimensions of the test model. Vehicle modelling to obtain relevant features related to the fluid dynamics in real vehicle has been widely studied [5, 6, 7, 8, 9, 10, and 11]. The test model has a dimension ratio of 0.17 (1:6) to the actual Ahmed model with a slope of the front geometry angle (θ) of 35°. The dimensions of the test model are written length (l) =174mm, height (h) =48mm, and width (w) =64.83mm. The suction active control applied is in the form of five apertures with a diameter of each circle (Ø) 7mm and the distance between apertures (s) of 10.81mm.

The apertures are designated as suction 1 (Sc_1), suction 2 (Sc_2), suction 3 (Sc_3), suction 4 (Sc_4), and suction 5 (Sc_5). The suction velocity (Usc) applied was 1.0 m/s at an upstream velocity (U_0) of 11.1 m/s, 13.9 m/s, and 16.7 m/s. The test model is shown in figure 1.

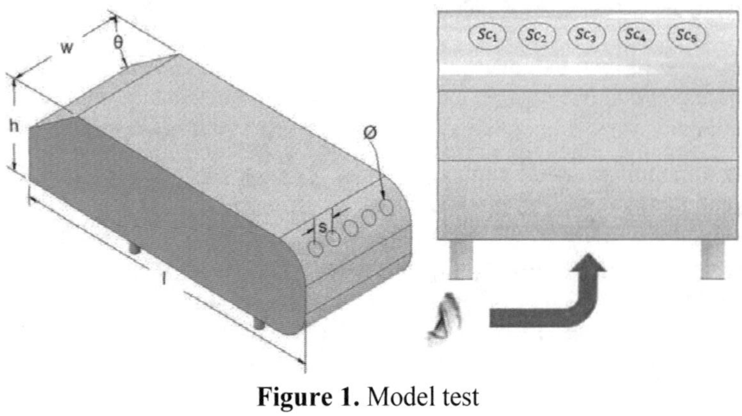

Figure 1. Model test

IOP Conf. Series: Materials Science and Engineering 1173 (2021) 012045 doi:10.1088/1757-899X/1173/1/012045

To support the study of the aerodynamic drag of the vehicle model, fluid flow patterns generated on the rear parts of the model and the pressure field are displayed through the display of the minimum pressure coefficient table (C_p) on the rear wall of the vehicle model through a computational approach. Pressure data retrieval is focused on the rear wall of the vehicle model. The data were taken as many as five lines in the location in the direction of the vertical axis.

Respective locations are defined as z/w=-1/2, z/w=-1/4, z/w=0, z/w=1/4, and z/w=1/2, where z/w is the ratio of the grid width (z) to the model width (w). For horizontal axis direction, 5-lines pressure field data are obtained (y/h=0.83, y/h=0.67, y/h=0.50, y/h=0.33, and y/h=0.17) on models without active control, whereas 4-lines pressure field data are obtained (y/h=0.67, y/h=0.50, y/h=0.33, and y/h=0.17) on models with applied active control. y/h is the ratio of grid height (y) to model height (h). Figure 2 shows the details of the location of pressure field data retrieval. The value of the pressure field will be displayed in a dimensionless number, the pressure coefficient (C_p) through the application of equation 1 [12].

$$C_p = \frac{P - P_0}{\frac{1}{2}\rho V^2} \quad (1)$$

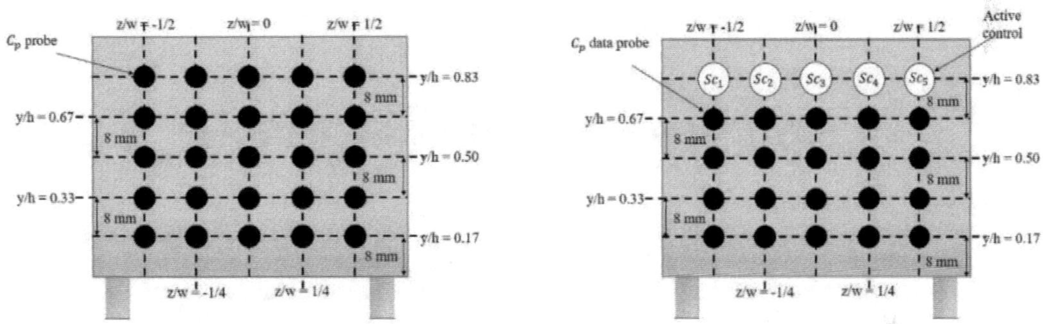

(a) Without control (b) With suction

Figure 2. Location for pressure field retrieval

The vehicle model was designed using the Autodesk inventor software as shown in figure 1. Then the model was assigned into the computational domain and the meshed was generated using Gambit software. The computational domain and mesh display are shown in figure 3.

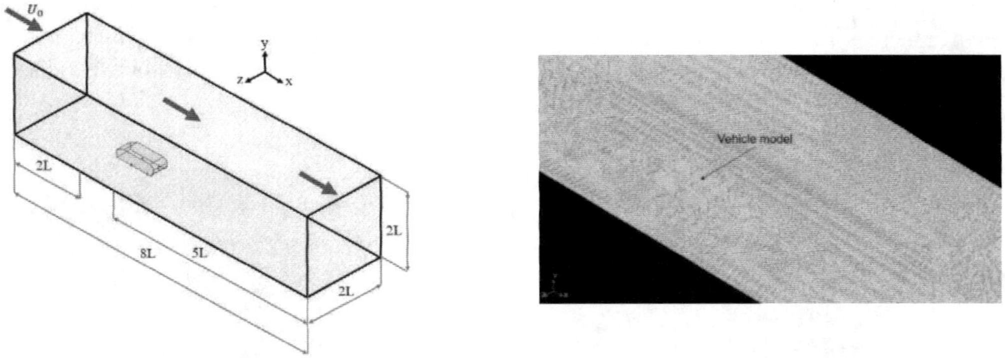

(a) Computational domain (b) Mesh display

Figure 3. Computational domain and mesh display

The numerical computational results for aerodynamic drag data are validated through the experimental laboratory testing process using the subsonic wind tunnel facility as shown in figure 4. The dimensions of the physical model and numerical model are identical. The measurement of the drag force utilizes a load cell device that is connected to the vehicle model via aluminium support rods. The force received by the vehicle model will be read on the computer display as a drag force (F_d) with newton unit (N). Retrieval of data is determined for 120 seconds to obtain a total of 120 drag data for each upstream velocity on the model with and without suction active control. From the 120 data, the average value is then taken as a comparison of the results obtained through the previous numerical computational approach. The average value of the drag force obtained was then converted into a dimensionless number of drag coefficients (C_d) through the application of equation 2 [12].

$$C_d = \frac{F_d}{\frac{1}{2}\rho v^2} \quad (2)$$

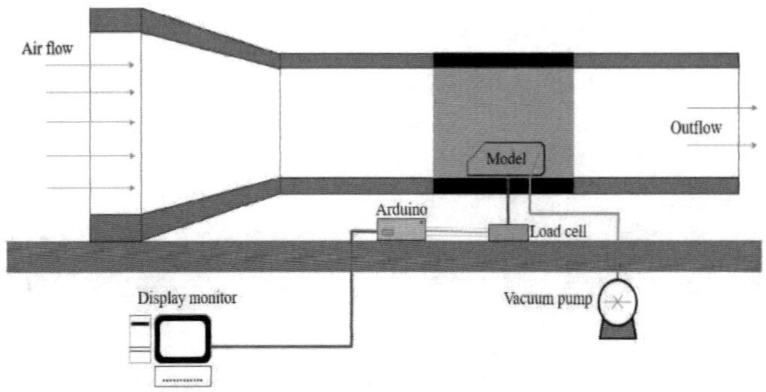

Figure 4. Experimental setup

3. Results and discussion

3.1. Flow field

Comparison of flow models of the model without control and with the application of active control at upstream velocity (U_0) of 11.1 m/s, 13.9 m/s, and 16.7 m/s are shown in figures 5, 6, and 7 respectively. For models without control at upstream velocity, U_0=11.1 m/s, 13.9 m/s, and 16.7 m/s as shown in figure 5(a), 6(a), and 7(a) shows large wake formation due to the process of flow separation that occurs right on the upper wall of the rear part of the vehicle model.

Flow that initially moves regularly experiences separation when it reaches the rear end of the model. As a result, there is a slowing down of flow on the middle side resulting in backflow. This slowing flow triggers the formation of a longitudinal vortex due to significant speed differences between the centre and side parts of the vehicle model.

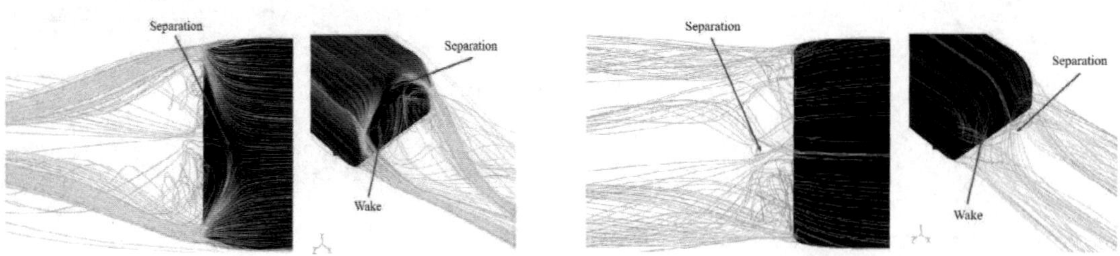

(a) Without control (b) With suction

Figure 5. Flow field at upstream velocity, U_0 = 11.1 m/s.

In figure 5(b), 6(b), and 7(b) for models with suction active control application, the wake formation is apparently smaller than that without control, due to the delay in the process of flow separation. The separation process tends to move away from the back wall resulting in minimization of backflow. This also causes a decrease in the intensity of longitudinal vortex formation. These results are consistent with the findings of other researcher [13], revealing that the application of suction active control could minimize wake formation. Furthermore, the flow field at upstream velocity of U_0=13.9 m/s and U_0=16.7 m/s is presented respectively on Fig. 6 and Fig. 7.

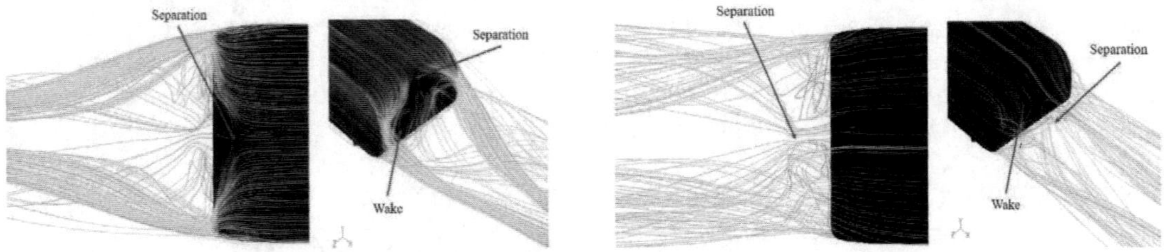

(a) Without control (b) With suction

Figure 6. Flow field at upstream velocity, U_0=13.9 m/s.

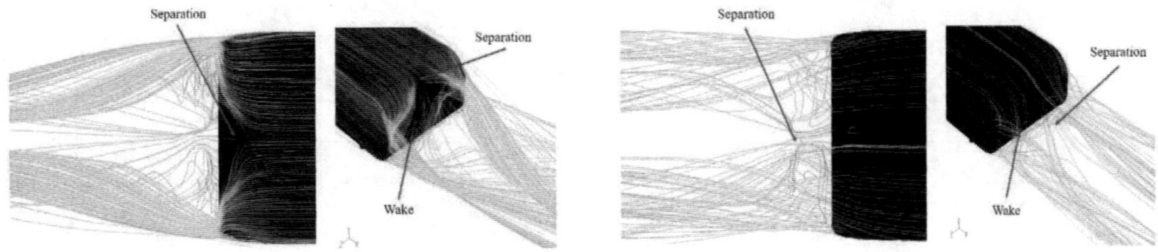

(a) Without control (b) With suction

Figure 7. Flow field at upstream velocity, U_0=16.7 m/s.

3.2. Pressure field

In table 1, the lowest average minimum pressure coefficient for upstream velocity, U_0=11.1 m/s is obtained in the model without control at C_p=-0.4295. This minimum pressure coefficient is obtained on the upper rear part (y/h=0.83) because in this position the beginning of the flow separation occurs as shown in figure 5(a).

Table 1. Minimum pressure coefficient on upstream velocity, U_0=11.1 m/s

Position	Minimum pressure coefficient, C_p	
	Without control	With suction
z/w=-1/2	-0.4516	-0.2939
z/w=-1/4	-0.4148	-0.2717
z/w=0	-0.4148	-0.2717
z/w=1/4	-0.4148	-0.2939
z/w=1/2	-0.4516	-0.2939
Rate	-0.4295	-0.2850
Increase in C_p (%)	-	33.6422

Meanwhile, the model with the application of suction control shows 33.6422% increase in the average minimum pressure coefficient of 33.6422%, where the average minimum pressure coefficient value of -0.2850. This is caused by the separation delay where the separation process tends to retreat from the rear wall of the vehicle model. The lowest pressure coefficient value is obtained at position y/h=0.17 because of the backflow that appears from the bottom side of the model.

At the upstream velocity, U_0=13.9 m/s as shown in table 2 the lowest average minimum pressure coefficient is also obtained on the model without control of at C_p=-0.4099. The lowest pressure coefficient value is obtained at y/h=0.83 because this position is the beginning of the separation process as shown in figure 6(a). Meanwhile, the model with the application of suction active control shows an increase in the average minimum pressure coefficient of 17.0041%, where the average minimum pressure coefficient is recorder at C_p=-0.3402. The minimum pressure coefficient is obtained at position y/h=0.17 due to backflow arising from the bottom part of the model.

Table 2. Minimum pressure coefficient on upstream velocity, U_0 = 13.9 m/s

Position	Minimum pressure coefficient, C_p	
	Without control	With suction
z/w=-1/2	-0.4411	-0.3598
z/w=-1/4	-0.4112	-0.3272
z/w=0	-0.3747	-0.3272
z/w=1/4	-0.3747	-0.3272
z/w=1/2	-0.4477	-0.3598
Rate	-0.4099	-0.3402
Increase in C_p (%)	-	17.0041

Table 3 shows that the lowest average minimum pressure coefficient of -0.3769 for upstream velocity, U_0=16.7 m/s obtained in the model without control. The minimum pressure coefficient is also found on the direction axis y/h=0.83, because in that position the beginning of the flow separation occurs as shown in figure 7(a). For models with the application of suction active control an increase in the average minimum pressure coefficient of 24.2239% was achieved. This is because of the delay in the process of flow separation as shown in figure 7(b).

The minimum pressure coefficient is achieved in the direction of the axis y/h=0.17 because it is affected by the backflow that appears from the bottom part of the model. The results obtained in this study, are consistent with previous studies, which revealed that embedding active control on the upper rear part of the vehicle model could increase the basic pressure [14].

Table 3. Minimum pressure coefficient on upstream velocity, U_0=16.7 m/s

Position	Minimum pressure coefficient, C_p	
	Without control	With suction
z/w=-1/2	-0.4132	-0.3022
z/w=-1/4	-0.3769	-0.2841
z/w=0	-0.3405	-0.2788
z/w=1/4	-0.3405	-0.2788
z/w=1/2	-0.4132	-0.2841
Rate	-0.3769	-0.2856
Increase in C_p (%)	-	24.2239

3.3. Aerodynamic drag

3.3.1. Computational method

Both the model without control and the model with suction active control show a decrease in the drag coefficient along with the increase in upstream velocity as shown in figure 8. The drag coefficient for the model without control is greater than the model with the application of suction active control at each upstream velocity. This corresponds to the flow field shown in figures 5, 6, and 7 and the average minimum pressure coefficient shown in tables 1, 2, and 3, where the model without control is the model with the largest wake as well as the model with lowest minimum pressure coefficient at each upstream velocity.

Drag coefficients for models without control at each of the upstream velocity (U_0) 11.1 m/s, 13.9 m/s, and 16.7 m/s of 1.9894, 1.8733, and 1.8486. Meanwhile, for models with a suction active control implementation, drag coefficients are sequentially achieved at 1.7736, 1.7083, and 1.6493 as presented on Fig. 8.

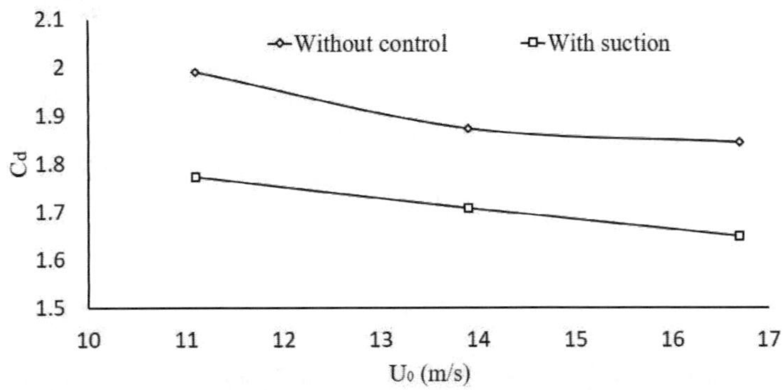

Figure 8. Relationship of drag coefficient and upstream velocity for computational methods

3.3.2. Experimental method

Similar to the computational approach, the experimental approach also shows a decrease in the drag coefficient along with an increase in upstream velocity. The highest drag coefficient is obtained in the model without control for each level of upstream velocity. The drag coefficient values of the model without control for each upstream velocity (U_0) of 11.1 m/s, 13.9 m/s, and 16.7 m/s are obtained at 1.8818, 1.7642, and 1.7628. Drag coefficients for the model with the application of suction active control at each upstream velocity are recorded respectively 1.6753, 1.6353, and 1.5748, refer to Fig. 9.

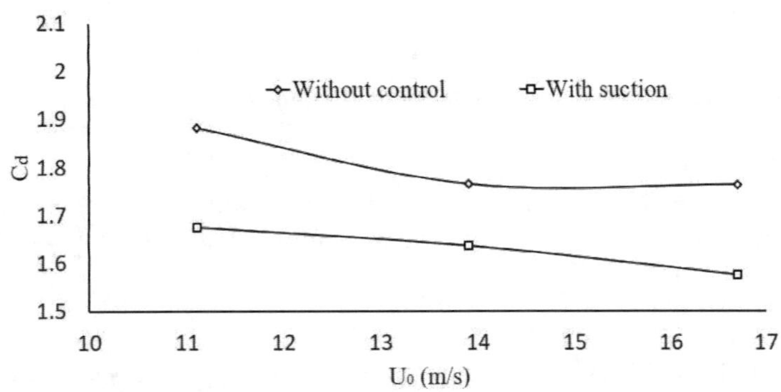

Figure 9. Relationship of drag coefficient to upstream velocity with experimental methods

3.3.3. *Reduction ratio*

The presence of suction active control in both the computational and experimental approaches at all upstream velocity produced reductions of the drag coefficient as shown in table 4. Reductions for the computational approach at upstream velocity (U_0) 11.1 m/s, 13.9 m/s, and 16.7 m/s are 10.8487%, 8.8079% and 10.5877%, respectively. From experimental approaches, the reductions of drag coefficients were recorded at 10.9748%, 7.3064%, and 10.6648%.

Table 4. Reduction ratio

U_0 (m/s)	Reduction (%)	
	Computational	Experimental
11.1	10.8487	10.9748
13.9	8.8079	7.3064
16.7	10.5877	10.6648

From these results, it was also found out that the highest drag coefficient reduction occurred at an upstream velocity of 11.1 m/s on both computational and experimental results. These results are consistent with previous research, which found out that the active control implementation could reduce the aerodynamic drag of the vehicle model [15-17]. Reduction of drag will improve overall vehicle performance [18].

4. Conclusion

The application of suction active control with a velocity of 1.0 m/s on the back of the vehicle model has a significant impact on the flow field, pressure field, and aerodynamic drag. It was found out that there was a reduction in wake formation due to a delay in the process of flow separation, an increase in the highest-pressure coefficient of 33.6422% at the upstream velocity U_0=11.1 m/s. The largest drag coefficient reductions of 10.8487% and 10.9748% were obtained at the same velocity for both the computational and experimental approaches, respectively.

Acknowledgments

This research was funded through the 2020 University Basic Research (PDU) Scheme, managed by the Hasanuddin University Research and Community Service Institute. We extend our gratitude to the Head and Staff of the Fluid Mechanics Laboratory of the Hasanuddin University, Faculty of Engineering.

References

[1] Watkins S and Vino G 2008 *The effect of vehicle spacing on the aerodynamics of a representative car shape* Journal of Wind Engineering and Industrial Aerodynamics. **96** 1232-1239. DOI:10.1016/j.jweia.2007.06.042
[2] Kumar R, Varshan N and Kannan 2019 *Aerodynamic design optimization of an automobile car using computational fluid dynamics approach* Australian Journal of Mechanical Engineering. DOI:10.1080/14484846.2019.1654963
[3] Mousa A A, Yadav R and Fischer J *Aerodynamic drag reduction for a generic sport utility vehicle using rear suction* Journal of Engineering Research and Applications. **4** 101-107.
[4] Wessen E and Thiele F 2009 *Road vehicle drag reduction by combined steady blowing and suction* 39th AIAA Fluid Dynamics Conference. DOI:10.2514/6.2009-4174
[5] Hassan S M R, Islam T, Ali M and Islam MQ 2014 *Numerical study on aerodynamic drag*

reduction of racing cars Procedia Engineering **90** 308-313 DOI:10.1016/j.proeng.2014.11.854
[6] Kulkarni S, Gandole G, Gaikwad P and Yadav J N 2018 *A review of aerodynamic study for reduction of drag force in cars in order to enhance the performance* International Journal on Emerging Trends in Technology. **5.**
[7] Meile W, Brenn G, Reppenhagen A, Lechner B and Fuchs A 2010 *Experiments and numerical simulation on the aerodynamics of the Ahmed body* CFD Letters **3** 32-39.
[8] Nallusamy S P and Mahali P K 2016 *Numerical and experimental investigations of drag force on scaled car model* Thermal Science. **20** 1153-58. DOI: 10.2298/tsci16s4153p
[9] Tarakka R, Jalaluddin, Mire B and Umar M N 2015 *Effect of turbulence model in computational analysis of active flow control on aerodynamic drag of bluff body van model* International Journal of Applied Engineering Research. **10** 207-219.
[10] Fourrie G, Keirsbulck L, Labraga L and Gillieron P 2011 *Bluff-body drag reduction using a deflector* Experiments in Fluids **50(2)** 385-395 DOI:10.1007/s00348-010-0937-6
[11] Harinaldi, Budiarso, Tarakka R, and Simanungkalit S P 2011 *Computational analysis of active flow control to reduce aerodynamics drag on a van model* IJMME-IJENS. **11** 24-30.
[12] Cengel Y A and Cimbala J M 2014 *Fluid mechanics: fundamental and applications* (Mc Graw Hill)
[13] Harinaldi, Budiarso, Warjito, Kosasih E A and Tarakka R 2012 *Modification of flow structure over a van model by suction flow control to reduce aerodynamics drag* MAKARA. **16(1)** 15-21. DOI:10.7454/mst.v16i1.1021
[14] Krajnovic S and Fernandes J 2011 *Numerical simulation of the flow around a simplified vehicle model with active flow control* International Journal of Heat and Fluid Flow. **32(1)** 192-200. DOI:10.1016/j.ijheatfluidflow.2010.06.007
[15] Lehugeur B and Gillieron P 2006 *Active control of vortex breakdown phenomenon in the wake of a simplified car geometry* ASME 2006 2nd Joint U.S.-European Fluids Engineering Summer Meeting. DOI:10.1115/fedsm2006-98349
[16] Tarakka R, Salam N, Jalaluddin, Ihsan M 2019 *Effect of blowing flow control and front geometry towards the reduction of aerodynamic drag on vehicle models* FME Transactions. **47(3)** 552-559. DOI:10.5937/fmet1903552T
[17] Tarakka R, Salam N, Jalaluddin, Ihsan M 2018 *Active Flow Control by Suction on Vehicle Models with Variations on Front Geometry* International Review of Mechanical Engineering. 12(2) 128-134. DOI:10.15866/ireme.v12i2.13876
[18] Hyams D G, Sreenivas K, Pankajakshan R, Nicholas S D, and Briley W R 2011 *Computational simulation of model and full scale class 8 truck with drag reduction devices* Computers & Fluids 41(1) 27-40. DOI:10.1016/j.compfluid.2010.09.015

Flow drags across three minibus car models arranged in tandem in four configurations

N Salam[1], R Tarakka[1]*, Jalaluddin[1], M Ihsan[2], and M A Jimran[1]

[1] Department of Mechanical Engineering, Hasanuddin University, Makassar, Indonesia
[2] Sekolah Tinggi Teknik Baramuli, Pinrang, Indonesia

*Email: rustan_tarakka@yahoo.com

Abstract. The purpose of this study was to investigate the drag coefficient of flow on three tandem minibuses in four configurations and to determine the optimum drag coefficient across the three cars. Flow drags measurements have been conducted on a wind tunnel by force balance measurement. The minibus car models with a 1:40 ratio to the originals were constructed from iron material with a thickness of 1 mm. The three minibus cars are arranged in four configuration models where car 1 and car 2 are arranged in series while car 3 position changeable according to the configuration. Each configuration model was treated with 6 changes in the distance of the car to 3, while the distance between car 1 and 2 was constant, with similar 7 levels of treatment of velocities ranging from 8 m/s to 20 m/s. Results show a similar drag coefficient (C_d) pattern in all configuration models, where smaller C_d related to greater velocities and smaller distances of the three cars. Furthermore, at a similar Re=49,608, the smallest Cd was obtained in model III (C_d=0.78), and followed by model II (C_d=0.80), model IV (C_d=0.81) and model I (C_d=0.84).

Keywords: flow drag coefficient, three tandem minibus cars, four configuration models

1. Introduction

The flow across three minibuses arranged in tandem is a form of fluid flow applicable in transportation and infrastructure engineering, particularly on expressways. Aerodynamic load is one of the major factors that must be addressed in a design, with special characteristics on tandem objects. Due to coupled interferences of flow around grouped or tandem objects, various interesting and unexpected phenomena can be observed.

When fluids flow through objects, for example, a minibus, energy loss will be developed in the influence of drag forces by either boundary layers or flow separations. In boundary layer terms, drag is a direct viscous effect of tangential stress and is therefore termed as viscous or frictional drag. The latter category is due to the effect of pressure by normal forces and is therefore termed shape or pressure drag, although it can also indirectly be due to viscosity. In avoiding energy losses, a cross-sectional shape of the object should be precisely designed. Interaction of the object with other objects can also be engineered, allowing the air to flow through the object without creating separation and resulting uniform flow beyond the objects. This research is about reducing the drag force in cylinders either single or arranged in tandem with various methods.

A research on the reduction of drag on circular cylinders in airflow has incorporated the installation of disturbance rods embedded on the upstream area before the cylinder, and found out that the flow pattern alterations depend on the diameter, distance, and Reynolds number. Reynolds numbers based on the cylinder diameter were ranging from $1.5 \cdot 10^4$ to $6.2 \cdot 10^4$. The total drag reduction was 63% less when compared to the one of a single cylinder [1].

Research on the efficacy of installing a minute control rod upstream of the cylinder with a focus on the drag and flow pattern has also been conducted, in Reynolds number of approximately Re=20,000, based on the diameter of the main cylinder. The highest reduction of the total drag coefficient of the total arrangement consisting main cylinder and control rod has been investigated to approximately 25%. Furthermore, the research was carried out by varying the values of L/D and d/D, resulting in total reduction of drag coefficient. From the research, it was also found that the ideal ratio of the diameter of the disturbance rod was at d/D=0.233. Installation of the minute control rod were ideal in the ratios of distance to cylinder diameter L/Ds=2.0-2.08 [2].

Research on the aerodynamic characteristics of the flow due to the interaction of square cylinders installed in tandem at a laminar flow or low Reynolds number, the eddy flow is affected by the Reynolds number, while the force actions are different between the up-stream cylinder compared to the down-stream cylinder, producing differences drag coefficient value [3].

Research on the interaction of alternately arranged circular and square cylinders installed upstream in the wind tunnel. The distances between the two cylinders were varied with the ratio S/d from 0 to 10. It was found out that the pressure drop characteristics were influenced by the ratio of the diameters and distances of the two types cylinders (S/d). Furthermore, the optimum results were obtained at S/d=1.0-1.5, indicating the lowest pressure reduction for overall treatments of changes in positions, diameters, and Reynolds numbers [4].

A research on the use of a circular cylinder as a disturbance cylinder before 2 square cylinders in a tandem arrangement was conducted. From this study, the optimum ratio of the distances between a disturbance cylinder to the diameter of tandem cylinders (L/D) was of 0.43 and the ratio of the disturbance cylinder's diameter to the tandem cylinder's diameter (d/D) was of 0.14, where a reduction in the drag coefficient (C_d) was obtained by 21.596% [5].

A research on flow characteristics around an array of four circular cylinders near the wall in the laminar boundary layer was carried out by numerical simulation. The pressure distribution across the walls and on each cylinder surface was observed at L/D=2.7 (L was the distance to the centre of each cylinder; D was the diameter of the cylinder), G/D=0.3 (G was the gap distance between the cylinder surface and the wall), with a variation of x/D=5, 10, and 15 (x is the distance from the leading edge from the wall to the centre of the upstream cylinder) with a Reynolds number of $1.743 \cdot 10^4$ based on the cylinder diameter. The numerical simulation method employed the 2D Unsteady Reynolds Averaged Navier Stokes (U-RANS) approach and the k-ω SST viscous model utilizing FLUENT 6.3.26 software to observe the flow phenomenon around the cylinder. The results of the study indicated that there was an effect of changes in boundary layer thickness to the pressure distribution in each cylinder [6].

Three square cylinders were arranged in series and parallel configurations. Each configuration was given 2 distance treatment models, namely, Model I, where the distance between cylinder 1 and cylinder 2 (M) changed while the distance between cylinder 2 and cylinder 3 (N) was set fixed at 6 cm, and Model II, where M and N changed with the same distance. then treated with the same 7 levels of speed ranging from 8 m/s to 20 m/s. Research took place at the Reynolds numbers (Re) from 9,395 to 62,634, or laminar flow for external flow. The experimental results showed that the characteristic of the drag coefficient (C_d) in the series configuration increases when the distance of three-square cylinders in a tandem arrangement is enlarged, while in the parallel configuration the value of C_d decreases when the distance is enlarged. The value of Cd obtained in each of the series and parallel configuration models is smaller than the Cd value of a single square cylinder [7]. Salam et al. have also investigated flow separation on three tandem square cylinders in serial and parallel configuration [8]. Also, other researchers have defined the value of the drag coefficient of drag for minivan cars to be $C_d = 0.4$, and for sedan (passenger car) to be $C_d=0.3$ [9,10].

2. Materials and Methods

This research method is experimental in the form of a wind tunnel, to measure the drag force of fluid flow. Measurement of the flow drag force is done by using forces balance measurement. The test objects were 3-piece minibus car models, constituting a car model with a length of 121 mm, a width of 45 mm, a height of 43 mm and a hydraulic diameter (D) 44 mm and 1:40 ratio of the model to the prototype. The car models were manufactured from iron material with a thickness of 1 mm. The three minibuses were arranged in four configuration models where the position of car 1 and car 2 was arranged in series at a constant distance L, while the position of car 3 changes in the Y direction or moving sideward at the distance of M, and in the X direction or moving forward at a distance of N depending on the configuration models, namely the configuration model I (car 3 parallel to car 2 or at N/D=0), configuration model II (car 3 parallel in between car 1 and car 2 or at N/D=2.75), configuration model III (car 3 parallel to car 1 or at N/D=5.5), and configuration model IV (car 3 parallel and ahead to car 1 or at N/D=8.25).

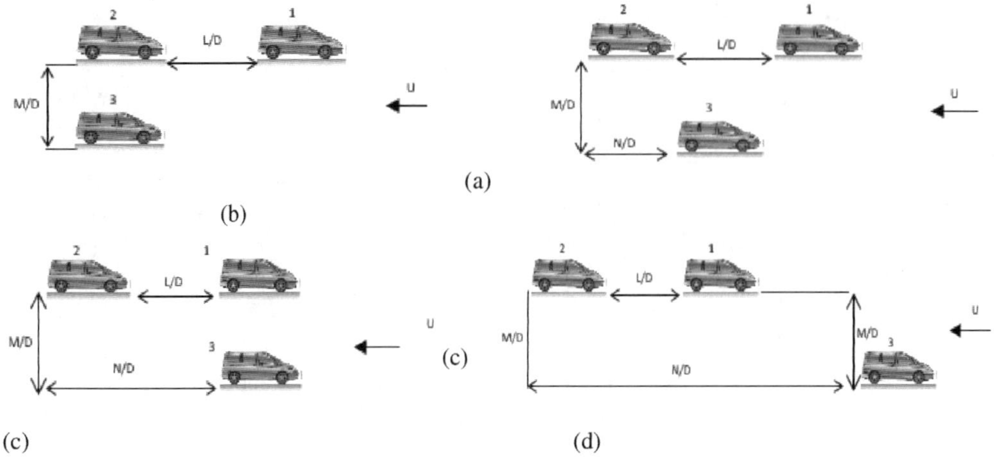

Figure 1 Tandem positions of 3-minibus in four configuration models (a) configuration model I, (b) configuration model II, (c) configuration model III and (d) configuration model IV

Furthermore, each configuration model was given 6 variations in M distance and 4 variations in N distance from car 1 with car 2 arranged in series, then given the same 7 levels of speed treatment ranging which were 8 m/s, 10 m/s, 12 m/s, 14 m/s, 16 m/s, 18 m/s and 20 m/s. Figure 1 shows the position of the three minibuses arranged in tandem in four configuration models, with a distance ratio M/D of 6 levels, (namely, 0.57; 1.13; 1.70; 2, 72; 2.84 and 3.41) and the ratios of the distance between N/D of 4 levels, (namely, 0; 2.75; 5.50 and 8.25) while the ratio of distance L to D (L/D) was constant of 2.75.

The equipment was international standard testing equipment in the Laboratory of Fluid Mechanics, Department of Mechanical Engineering, Faculty of Engineering, Hasanuddin University. The wind tunnel used in this study is a low-speed wind tunnel made by Plint & Partners LTD Engineers [10], where the velocity of airflow through the test section (300 mm x 300 mm) is set maximum at 22 m/s. This equipment is equipped with a forces balance measurement equipment and airflow velocity measurement, as shown in Figure 2.

Remarks:
1. *Prototype Screen;*
2. *Diffuser;*
3. *Test section;*
4. *Divergence section;*
5. *Fan;*
6. *Electric motor;*
7. *Flow velocity measurement;*
8. *Electric voltage regulator;*
9. *Force's balance measurement*

Figure 2 Wind tunnel used in the research with its elements [10]

To determine the value of the drag coefficient and describe the characteristics of fluid flow across a minibus, the Reynolds formula was used as shown in Eq. 1 [9],

$$R_e = \frac{U \cdot D}{\upsilon} \quad (1)$$

The variables and parameters in Eq. 1 are the airflow velocity before the test models (U), the hydraulic diameter of the minibus car (D), and the air kinematic viscosity (υ).
In determining the drag coefficient (C_d), the Eq. 2 was used [9],

$$C_d = \frac{2 \cdot F_d}{\rho_{air} \cdot U^2 \cdot A} \quad (2)$$

The variables in Eq. 2 are the experimental drag force (F_d), the frontal surface area of the minibus car (A), and the density of the air (ρ_{air}). The values of kinematic viscosity and density of air are defined based on the pressure and room temperature in the laboratory.

Based on the hydraulic diameter of the minibus model and the treatment of airflow velocities, ranging from 8 m/s to 20 m/s, or from 28.8 km/h to 72.2 km/h, the experiments were in a laminar flow area or at $Re<10^5$ for external flow. Moreover, the flow velocity was maintained in a constant condition at each speed level, and the speed treatments for each configuration model were 7 levels of U velocity, namely (8, 10, 12, 14, 16, 18, and 20) m/s.

3. Results and discussion

The experimental results of airflow across three minibuses arranged in tandem in four configuration models show that the characteristics of the drag coefficient (C_d) tend to be the same, that is, if the airflow velocity U or the Reynolds number (Re) increases, the C_d value decreases. The following describes the experimental results for the four configuration models, in the form of a graph of the relationship between C_d and airflow velocity U or the Re number at the same M/D. The results for four configuration models are subsequently compiled in a graph of the relationship of C_d and M/D at the same velocity.

The experimental results of the configuration Model I are depicted in Figure 3 in the form of a graph of the relationship between C_d and Re at the same M/D.

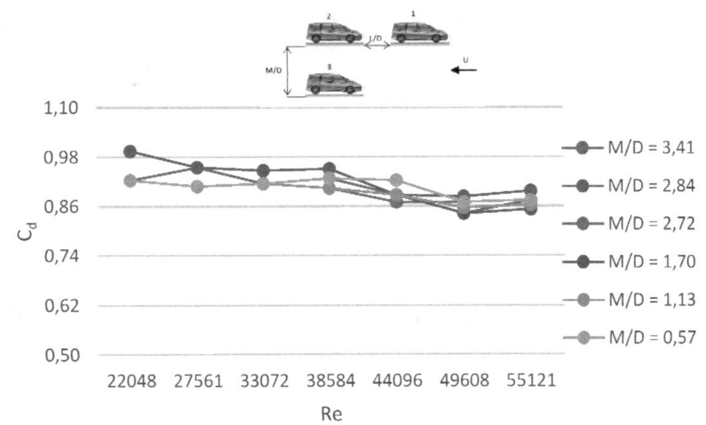

Figure 3 Relationship of C_d and Re for configuration model I for 6 levels of M/D and for N/D=0

Based on Figure 3 above, it is shown that the alterations in M/D affects the value of Cd at all levels of Reynolds numbers. These results show that the largest drag coefficient of C_d=0.99 was at M/D = 2.84 and at Re=22,048, while the smallest resistance coefficient value of Cd=0.84 was at M/D=1.70 and at Re=49,608. These results indicate that when car 3 is closer to car 2, the flow separation was slowed down causing the boundary layer to shrink, and the vortex of the airflow was dampened after crossing between car 2 and car 3. However, when the position of car 3 is very close to car 2 or at M/D=0.57, the value of C_d increases again. The difference in values between the largest and the smallest C_d shows the effect of changing the distance between car 3 and car 2, and this proves that, in the right position, cars arranged in tandem in configuration model I can reduce aerodynamic drag. The right position of the car, of course, will result in more efficient fuel consumption.

Figure 4 shows the experimental results of configuration model II, depicting the relationship between C_d and Re at the same M/D.

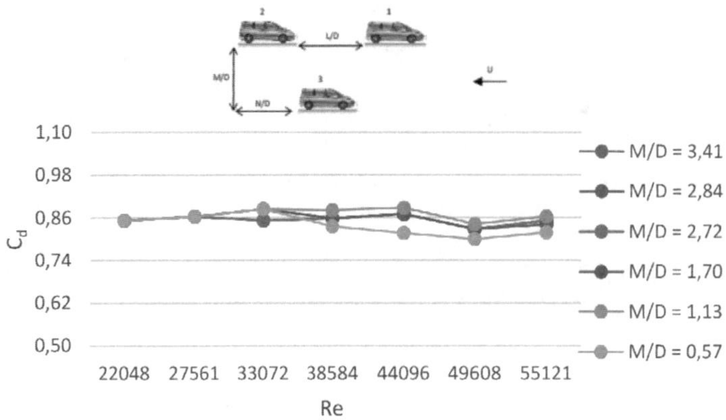

Figure 4 Relationship of C_d and Re for configuration model II for 6 levels of M/D and for N/D=2.75

Figure 4 also describes that varying M/D could affects the values of Cd at all levels of the Reynolds numbers. These results indicate that the largest flow drag coefficient value was C_d=0.89, occurring at M/D=1.13 and Re=44,096, while the smallest drag coefficient value was C_d=0.80 at M/D=0.57 and Re=49,608. These results indicate that when car 3 is closer to car 2, the flow separation was delaying causing the boundary layer to shrink, and the vortex of airflow was dampened after crossing between car 2 and car 3. The difference of the largest and the smallest Cd shows the impact if varying the distances between car 3 and car 2, and this also proves that, in proper position, cars arranged in tandem configuration model II can reduce aerodynamic drag which also contribute to the reduction of fuel consumptions.

The same results of configuration model III are shown in Figure 5, which is a graph of the relationship between C_d and Re at the same M/D.

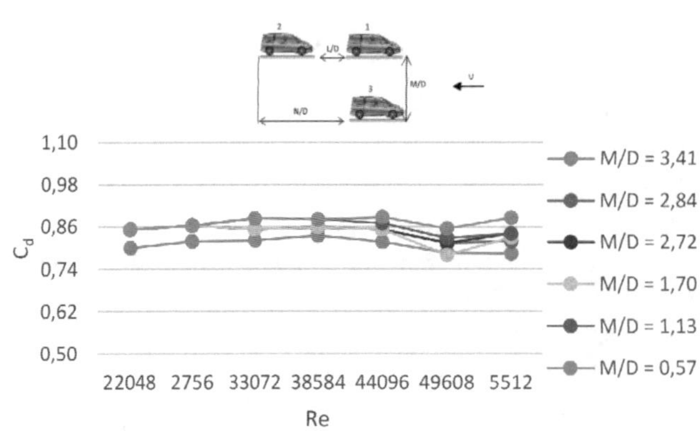

Figure 5 Relationship of C_d and Re for configuration model III for 6 levels of M/D and for N/D=5.5

It is also shown in Figure 5 that the changes in M/D have affected the magnitude of C_d at all levels of the Reynolds numbers. These results also show that the largest flow drag coefficient value was at

Cd=0.89 and at M/D=0.57 and Re=44,096, while the smallest drag coefficient Cd of 0.78 occurred at M/D=1.70 and Re=49,608. These results also prove that when car 3 is closer to car 1, the flow separation was delaying, resulting on shrunken boundary layer and dampened airflow vortex after the flow passing through the path between car 1 and car 3. The difference value of the largest and the smallest C shows the effect of altering the side-by-side distances of car 3 to car 1, and this also proves that the car arranged in tandem configuration model III is effective to reduce drag in airflow and of course, will result in savings car fuel consumption.

The experimental results of the configuration model IV is shown in Figure 6 below, as a graph of the relationship between C_d and Re at the same M/D.

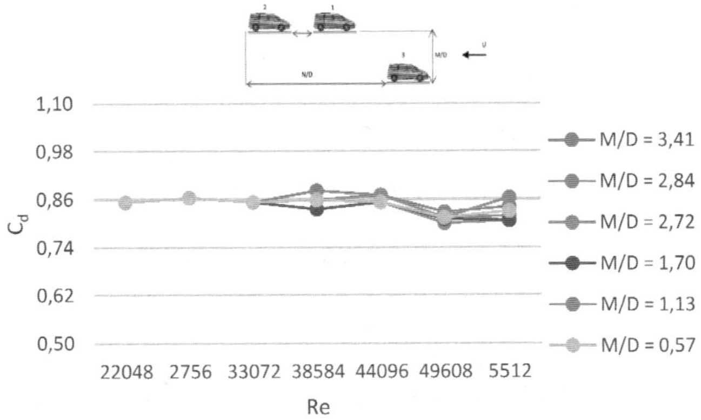

Figure 6. Relationship of C_d and Re for configuration model IV for 6 levels of M/D and for N/D=8.25

Figure 6 shows that the change in M/D will affect the drag coefficient Cd at all levels of the Reynolds numbers. It gives information that the largest flow drag coefficient was Cd=0.88 at M/D=3.41 and Re=38,584, while the smallest resistance coefficient value is Cd=0.80 at M/D=1.70 and Re=49,608. Similar to previous configurations, the results indicate that when car 3 is closer to car 1, the flow separation was again slowed down, resulting the boundary layer to shrink and the vortex of the airflow to dampen after flowing through car 1 and car 3.

The difference in value between the largest and the smallest Cd shows the effect of changes in the distance between car 3 and car 1, and this also proves that the car arranged in tandem model configuration IV can reduce aerodynamic drag, if set the right position. The expected results is also the reduction fuel consumption.

The compilation results of the four configuration models are shown in Figure 7 below in the form of a graph of the relationship between c_d and M/D at the same Re=49,608.

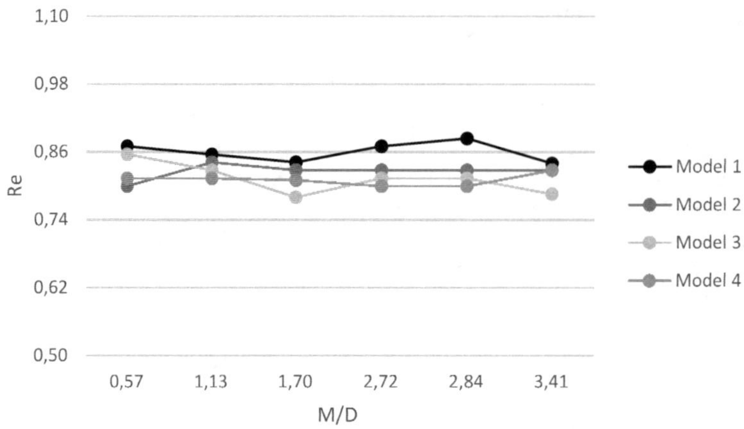

Figure 7 Relationship of C_d and M/D for the four configuration models and for Re=49,608

Based on Figure 7 above, it is indicated that at the same M/D and Re, the value of C_d will be different in different configuration models. Configuration model I shows the value of Cd which was always greater for all levels of M/D. The configuration model II, configuration model III, and configuration model IV gave smaller C_d from the former to the latter. However, at M/D=1.70 the smallest Cd value was on configuration model III. The same results occurred for M/D=3.41, where the smallest Cd value was also occurring in configuration model III.

These results indicate that when car 3 is closer to car 1 and car 2 or at M/D=1.7 the configuration model III is the most optimum one in reducing drag coefficient. This is because the flow separation was slowed down and resulted the boundary layer to shrink. The vortex of airflow was also dampened after crossing between car 1 and car 2 and car 3. Whereas when car 3 was in the farthest position from car I and car 2 or at M/D=3.41, the configuration model III was the best. This is also because the flow separation was slowed down, producing shrunken boundary layer and dampened the airflow vortex after crossing between car 1 and car 3.

The difference in the value between the largest and the smallest Cd shows the effect of changing the distance of car 3 next to car 1, and this proves that the car is arranged in tandem model configuration IV, which can reduce airflow resistance when in the right position. The right position of the car, of course, will result in savings in car fuel consumption.

The characteristic values of Cd towards the M/Ds for the four configuration models are as shown in Figure 7 above. The results are then compiled with the CFD flow path line simulation results at M/D=1.7 and Re=49,608, as shown in Figure 8.

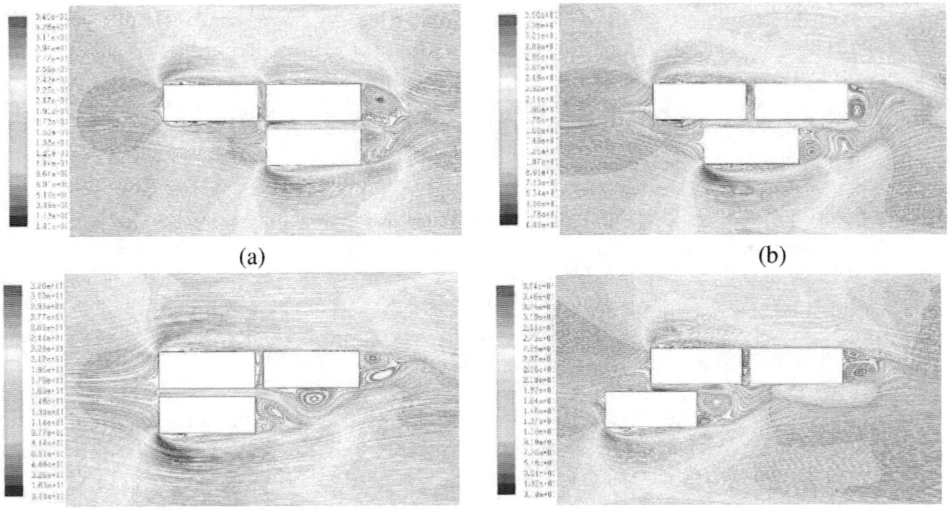

Figure 8. Simulated CFD pathlines of flows at M/D=1.7 and Re=49608 for (a) configuration model I, (b) configuration model II, (c) configuration model III, and (d) configuration model IV

Based on Fig. 7 and Fig. 8 above, the best configuration model was configuration model III. This can be seen from the smallest flow vortex that occurs after passing through car 2 and car 3 in configuration model III. The largest flow vortex is in configuration model I. The results of this compilation show conformity to the experimental results. This is because the flow vortex after passing through car 3 in the configuration model III loss its effect on the flow vortex after passing through car 2, creating dampened or more stable flow vortex.

4. Conclusions

It can be concluded that the characteristics of the flow drag coefficient across three minibus cars arranged in tandem in four configuration models have been examined and shows the smallest C_d value at M/D=1.70 and Re=49,608 or at airflow velocity U=18 m/s (equal to the car speed of 64.8 km/h). An exception is for configuration model II at M/D=0.57. Also, the smallest drag coefficients of flow across three minibuses arranged in tandem for respective configuration model I, II, III, and IV respectively are: C_d=0.84 at M/D=1.70 and Re=49,608; C_d=0.80 at M/D=0.57 and Re=49,608; C_d=0.78 at M/D=1.70 and Re=49,608; and C_d=0.81 at M/D=1.70 and Re=49,608. On the condition of the minibus moves independently, the C_d value for each car was 0.4 resulting total C_d value of 1.2. Whereas if the three minibuses move in tandem with configuration model I, II, III, and IV, the total average smallest value was C_d=0.8075, resulting the reduction of C_d by 31.71%.

Acknowledgement

This research was funded by Hasanuddin University Basic Research scheme, 2020 Fiscal Year based on the Unhas Chancellor's Decree Number 2649 / UN4.1 / KEP / 2020, May 22, 2020, managed by the Institute of Research and Community Service, Hasanuddin University.

References

[1] Tsutsui, T. and Igarashi, T. 2002. Drag reduction of a circular cylinder in an air-stream. *Journal of Wind Engineering and Industrial Aerodynamics*, **90**, pp.527-541.DOI: 10.1016/s0167-6105(01)00199-4

[2] Lee, S., S. Lee, and C. Park 2004, Reducing the drag on a circular cylinder by upstream installation of a small control rod, *Fluid Dynamics Research*, **34**. Pp.233-250. DOI: 10.1016/j.fluiddyn.2004.01.001
[3] Etminan A., Moosavi, M. and Ghaedsharafi, N. 2011. Characteristics of aerodynamics forces acting on two square cylinders in the streamwise direction and its wake patterns, *Advances in Control, Chemical Engineering, Civil Engineering and Mechanical Engineering*. pp. 209-217. DOI: 10.1016/0167-6105(92)90522-c
[4] Daloglu, A. 2008. Pressure drop in a channel with cylinder in tandem arrangement, *International Communication in Heat and Mass Transfer*, **35**, pp.76-83. DOI: 10.1016/j.icheatmasstransfer.2007.05.011
[5] Salam, N., Tarakka, R., Jalaluddin, and Bachmid, R. 2017, The effect of the addition of inlet disturbance body (IDB) to flow drag through the square cylinders arranged in tandem, *International Review of Mechanical Engineering (I.RE.M.E.)*, **11**(3), pp. 181-190. DOI:10.15866/ireme.v11i3.11338
[6] Wailanduw A. G., Adiwibowo, P. H. and Achmad, H.B. 2018. Numerical simulation of cross flow around four circular cylinders in an in-line square configuration near a plane wall at laminar boundary layer, *Prosiding SNTTM XVII*, pp.162-167.
[7] Salam, N, Tarakka R., Jalaluddin, Sukardin, M.S., and Machfud, A. 2019. Karakteristik koefisien tahanan aliran melintasi tiga silinder persegi tersusun tandem konfigurasi seri dan paralel. *Prosiding Seminar Nasional Teknik Mesin Politeknik Negeri Jakarta* (2019), ISSN 2085-2762. pp.1244-1251.
[8] Salam, N., Tarakka, R., Jalaluddin, Ihsan, M. 2020. Flow separation across three square cylinders arranged in serial and parallel tandem configuration. *International Journal on Engineering Applications*, **8**(3), pp. 96-106. DOI: 10.15866/irea.v8i3.18302
[9] Cengel, Y.A., and Cimbala, J.M. 2006. *Fluid Mechanics Fundamentals and Applications*, Published by The McGraw-Hill Companies, Inc. New York.
[10] Plint & Partner LTD Engineer. 1982. *Manual Educational Wind Tunnel*, England.

IOP Conf. Series: Materials Science and Engineering 1173 (2021) 012047 doi:10.1088/1757-899X/1173/1/012047

Oils and water absorption behavior of Biduri (*Calotropis gigantea*) fibers

A Sukmawati[1,*], and W Septiani[1]

[1] Balai Besar Tekstil, Jalan Jend. A Yani No. 390 Bandung 40272, Indonesia

*Corresponding e-mail: anne.sukmawati@gmail.com

Abstract. This article reports on the oil and water absorption behavior of Biduri (*Calotropis gigantea*) fibres at room temperature. The material measured is a 100% raw material for Biduri (Calotropis gigantea) fibre, from 0.3 g to 1.5 g, and nonwoven structures made from 50/50, 60/40, 70/30, 80/20, and 90/10 mixture of Biduri and polyester. FTIR characterized the Biduri fibre, and the viscosity of used oil measured using Brookfield viscometer. The higher oils absorption at weight 0.3 g, 45.53 g/g, and water absorption at weight 0.3 g, 1.47 (g/g). The higher oil absorption at nonwoven structure showed at 90/10 composition, 42,86 g/g. It showed Biduri fibre is a natural absorbent that exhibits excellent hydrophobic and oleophilic properties.

Keywords: *Calotropis gigantean, water absorption, oleophilic, fibers*

1. Introduction

The most significant cause of oil pollution in the oceans by humans is waste that comes from the exploration and processing of oil production and oil storage tanks on ships. Due to the ship's engines and tanks cleaning, oily waste illegally spilled from the vessel. Tank cleaning must be submitted to a licensed waste processing service. However, there are still many parties who are not responsible for processing waste illegally. Spilled oil waste is moved by wind and currents and often stays in motion for days and sometimes weeks. The oil on the surface of the water usually spreads out to form a very thin layer. The size of the affected area will increase over time, and it may not be easy to find some or all of the oil. Oil waste is a hazardous and toxic material (B3).

If left unchecked and continuously, the sea waters can be polluted, and environmental damage occurs. Waters contaminated by waste oil must be addressed immediately not to produce more severe marine pollution. One way to handle waste oil in the waters is to use porous materials, i.e., absorbent materials, which are readily available, simple to operate, and cheap [1]. Sorption is a physical method of oils removal in which the process of their absorption take place.

Generally, oil-absorbing materials are made from synthetic materials such as polypropylene fibers, polyester fibers, and polyurethane foam. Still, not a few researchers have developed natural materials that are more environmentally friendly and inexpensive [2]. Polypropylene offers a great absorbency capacity relative to cost. It is lightweight and clean to handle. Materials containing cellulose have high absorption and can be used as absorbents, such as absorbing oil or oil spills. Absorbents from cellulosic materials have similar capabilities to synthetic polymer (polypropylene) absorbents. Cellulose materials that have been studied include corn stalk [3], durian skin [4], sawdust [5], kapok fiber [6-7], coir [8], and others.

Indonesia is a country with high biodiversity. Animals and plants can live well and have various types because the sun continues to shine throughout the year. Biodiversity is very beneficial for human life as a source of food, medicine, and industry. However, many plants have not been used optimally, including the Biduri (*Calotropis gigantea*). This plant is a wild shrub with a height of 3-4 meters and is found in dry areas such as grasslands, low mountain slopes, and sandy beaches[9]. Biduri is similar to kapok in that both fibres are found around the plant seeds rather than extracted from the stem or leaves.

Biduri fibres are lustrous and made of a mix of lignin and cellulose. Each fibre is about 2.5 cm long, has a wide lumen (central cavity), and thin walls covered with waterproof wax. The Biduri seeds' fiber is the staple fiber, lustrous, has a cavity in the middle like the cotton fiber. The presence of hollow causes the fibrous to be lightweight and have good insulation properties[10]. Biduri fibers have a waxy coating on the fibers' surface, which has great potential to absorb oil [10].

Other research shows that Biduri fibers have excellent hydrophobic-oleophilic properties and the highest absorption capacities between 22.6 and 47.6 g / g [11]. Also, in the other study showed that soaking in an ethanol solution containing PFOTES for 1 second resulted in a high absorption capacity of 47-87 times its weight [12]. Tu reported that Biduri fiber (*Calotropis gigantea*) was processed into carbon fiber using pyrolysis resulting in oil absorption of up to 130 g / g [13]. The potential of Biduri fiber is enormous, so further research is still needed. The research aimed to remove waste oil from seawater using Biduri fibres with various variations and different treatments.

2. Materials and Methods

2.1. Materials

The study used Biduri fiber was collected from Bandung, West Java. Polyester fiber and polypropylene pad are purchased from a supplier of synthetic fibers. Polypropylene fiber are sample from PT. Hilon Indonesia. Hydrofob FC, demineralized water, NaCl, and NaOH were purchased from a local supplier. Waste oil was collected from local workshops.

2.2. Seawater preparation

Artificial sea water was used to imitate the condition of oil spills on seawater with a concentration of 3.5 wt. % was prepared. Seventy grams of NaCl are dissolved in a beaker containing 2000 ml of demineralized water. NaCl solution and demineralized water were filtered to remove suspended matter.

2.3. Treatment of the sorbent

Mercerization of the absorbent is carried out using a NaOH solution. The alkaline mercerization treatment is one of the most frequently used organic absorbent modification methods in the literature due to its simple and effective manner. The fiber was added to the NaOH solution (2%) in a glass flask and shaken at room temperature for 1 hour. After the treatment is complete, the obtained fibers are dried and then oven-dried at 105°C for 3 hours. Biduri fibers were added with Hydrob FC water, and oil proofing agent soaked for 1 hour and dried in an oven at 105°C for 3 hours.

2.4. Methods

Density is the mass of a unit volume of oil, usually expressed as grams per milliliter (g/ml). The density of an oil sample in g/ml, is measured using pycnometer. The Biduri fibers are cleaned and separated from the seeds, then weighed and put into plastic rollers of known dimensions. Prepare a tube-shaped plastic container, enter the artificial seawater then pour the used oil until it reaches a layer of 10 mm. Put the roller filled with fiber into the oil in the container with a specific contact time. Remove and drain on wire mesh for 5 minutes, then weigh.

The Biduri nonwoven is obtained from a mixture of Biduri fibers with polyester fibers and Biduri fibers with polypropylene. Nonwoven is made on a hot-press machine with a temperature of 130-140°C for 5 minutes. Nonwoven is cut to a specific size, then weighed and put in a container filled with oil and seawater. Let it sink and remove it after reaching the desired contact time. Drain the Biduri nonwoven for 5 minutes and weigh it.

Absorption capacity calculated using Eq. (1) below:

$$q = \frac{(m_a - m_i)}{m_i} \quad (1)$$

where q is the absorption capacity (g/g), m_a is the weight of sample fibers after the absorption and m_i is the initial weight of sample fibres before to absorption.

3. Results and Discussion

The density of most oils ranges from 0.7 to 0.99 g/ml. In this research, density waste oil is 0.877 g/ml, oils typically float on water. The density of sea water is 1.03 g/ml, even heavier oils will usually float on it.

3.1. Properties of fibre

Biduri fiber was characterized using attenuated total reflectance Fourier Transform Infrared (FTIR) summarized in table 1.

Table 1. Fourier Transform Infrared (FTIR) summarized

Spectra	Description
3384, 2898, 1377, 1319, 1163, 1052	Cellulose
1734, 1427, 1247	Hemicellulose and lignin

The main characteristic of an oil-absorbing sorbent is the high oil absorption capacity. Besides, the oil must be firmly bound by the sorbent, but it can still be recovered quickly, and the sorbent can be reused. For oil spills in water, the oil sorbent must be hydrophobic and have good buoyancy. Hydrophobic is defined as a low ability to absorb water. The lower the ability of a sorbent to absorb water, the higher its hydrophobicity. In the case of an oil spill in water, there will be competition between water molecules and oil molecules to bind to the sorbent. Hydrophobic sorbents tend to bind to oil molecules and repel water molecules, while the good buoyancy properties make it easier to separate sorbents already bound with oil from water. The structure of Biduri fibers is composed of cellulose.

The absorption of water by the lignocellulosic material depends on the number of free -OH groups present in the fiber. The reduction in water absorption capacity is thought to occur due to a decrease in the free -OH group on the lignocellulose, thereby reducing hydrogen bonds with water molecules. In this process, it is estimated that the compound that plays a role in cellulose because lignin is naturally hydrophobic.

3.2. Effect of sorbent weight

The effect of absorbent weight on oil absorption was carried out in a weight range of 0.3 to 1.5 g after 10 minutes of contact time with 10 mm oil film thickness using waste oil and hanging for 5 minutes, as shown in Fig. 1. Based on Figure 1, it can be seen that the addition of sorbent weight from 0.3 to 1.5 g decreased the absorption capacity of Biduri fiber from 45.53 g/g to 18.22 g/g and decreased the absorption efficiency from 97.89% to 94.75%. The addition of fiber weight at a constant volume will increase the fibers' density so that the space between the fibers gets smaller, and oil does not enter the fibers.

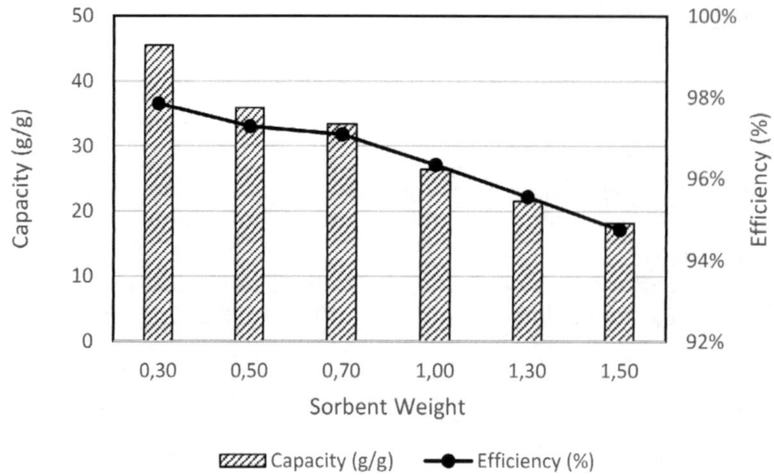

Figure 1. Oil absorption capacity and efficiency at different sorbent weight

Figure 2 shows the water absorption capacity of Biduri fibers at a sorbent weight of 0.3 to 1.5 g. Based on Figure 2, it can be seen that the addition of sorbent weight from 0.3 to 1.5 g decreased the water absorption capacity of the Biduri fiber from 1.47 g/ g to 0.39 g /g and decreased the absorption efficiency from 59.47% to 27.87%. The addition of fiber weight at a constant volume will increase the fibers' density so that the space between the fibers is getting smaller and water does not enter the fibers. The absorption capacity of Biduri fibers to oil is higher than water. The same result as previous research that Biduri fiber can be used as an oil absorbent material.

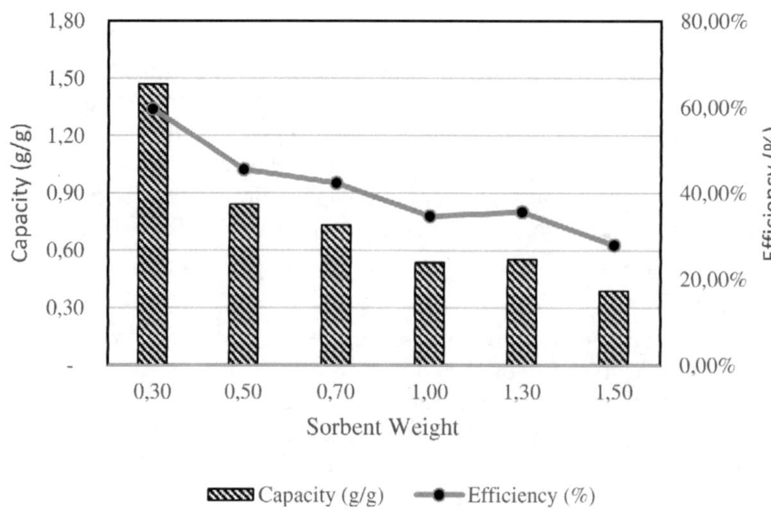

Figure 2. Water absorption capacity and efficiency at different sorbent weight

3.3. Effect of Pre-treatment

The effect of the comparison of treatment results on oil absorption is presented in Fig. 3. Absorption of Biduri fibers without treatment resulted in the absorption of 24.91 g /g, while the absorption power of Biduri fibers from 2% NaOH mercerization decreased to 17.75 g /g. The decrease can occur because lignin has been reduced in the Biduri fiber. After all, it dissolves with NaOH. Meanwhile, the literature states that lignin is a supporter of oil absorbents.

In the pre-treatment, using the Hydrob FC waterproof agent resulted in lower absorption. The waterproof agent envelops the fiber and enters the cavity in the middle of the fiber so that oil cannot stick to the fiber's surface. However, this needs to be proven by morphological tests using SEM. In this study, the morphological test has not been carried out.

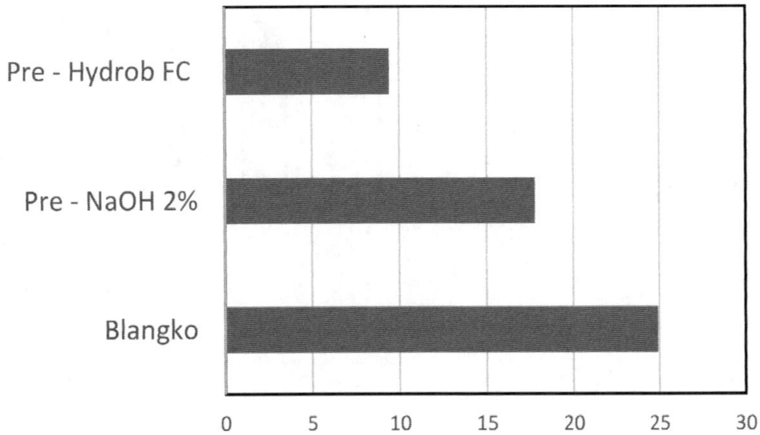

Figure 3. Oil absorption capacity at different pre-treatment

3.4. Effect of contact time

The contact time effect was made with five different time variations: 5, 10, 30, 60, and 90 minutes. The absorption power of Biduri can be seen in Fig. 4. At the contact time of 5 minutes, the absorption capacity was 17.17 g /g. The absorption capacity increased at the contact time of 10 minutes and reached a peak of absorption power at the contact time of 30 minutes, namely 24.91 g /g. After 60 minutes and 90 minutes, there was no increase in absorption and even more stable because the fiber has started to saturate not to absorb oil anymore.

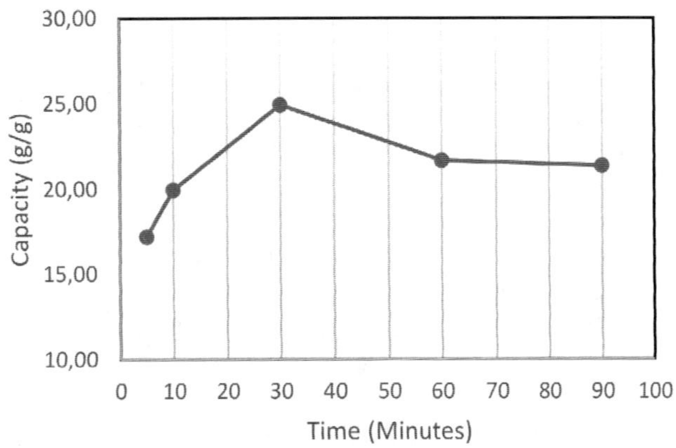

Figure 4. Oil absorption capacity at different contact times

3.5. Biduri nonwoven

For efficiently handling, Biduri fibers are made into a nonwoven as in commercial products. The Biduri nonwoven consists of Biduri fibers and polyester fibers, with the composition of Biduri fibers being 50, 60, 70, 80, and 90%. The absorption capacity and efficiency of various Biduri fibers and polyester fibers nonwoven are shown in Figure 5.

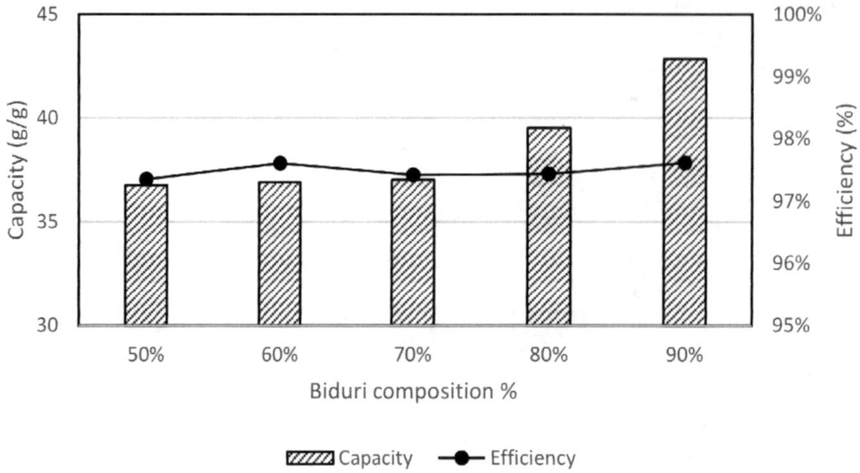

Figure 5. Oil absorption capacity and efficiency at different composition

The absorption capacity of the Biduri nonwoven to oil at the composition of 50% was 36.76 g / g. The composition's addition to 60% resulted in 36.90 g / g absorption, and 70% composition resulted in 37.03 g / g absorption. The increase in absorption capacity from 50% to 70% has only a small difference. At the 80% composition, the absorption capacity increased to 39.52 g / g, and at the 90% composition, the oil absorption rate was 42.86 g / g, which was the highest oil absorption in this study. At 80% composition, the absorption rate increased to 39.52 g / g, and at 90% composition, the oil absorption rate was 42.86 g / g. The highest oil absorption occurred in the 90% composition of the Biduri nonwoven because the arrangement of cellulose fibers, Biduri, was more significant than polyester fibers. The Biduri fiber, which has a cavity in the middle, is a place that can be used as a trap.

In this research, Biduri and polypropylene nonwoven were also made with a composition ratio of 50:50. The absorption capacity of Biduri-polypropylene nonwoven was 29.27 g / g, and absorption efficiency was 96.70%. Compared to the absorption capacity of 50% Biduri-polyester nonwoven, the absorption capacity of Biduri-polypropylene nonwoven is lower so that the variations in other compositions do not follow suit. It is assumed that the overall absorption capacity will be lower than that of Biduri-polyester nonwoven.

3.6. Comparison with commercial products.

The commercial product commonly used is polypropylene fiber compared to a Biduri membrane with a composition of 70%, as in Table 2.

Table 2. Comparison Biduri nonwoven and Polypropylene Pad (Commercial Product)

Product	Gramasi (g/m^2)	Capacity (g/g)
Biduri Nonwoven	179.34	18,79
Polypropylene Pad	178.39	12,40

Biduri-polyester nonwoven is manufactured with every effort to approach the gramation of the polypropylene absorbent material. The absorption capacity of the Biduri-polyester nonwoven was 18.70 g/g and higher than that of polypropylene pad, which only reached 12.4 g / g. This is very exciting and wonderful because the Biduri fiber can outperform commercial materials.

4. Conclusion

This study concludes that Biduri fibers have a better absorption performance than commercial polypropylene materials. Additional treatment NaOH and Hydrofob agent does not help the absorption of oil. The best composition for the manufacture of Biduri membranes with the best absorption is 90% Biduri fibers and 10% polyester fibers. This results show a wider applications of Biduri fibers.

References

[1] Gołub A and Piekutin J, 2018 Use of porous materials to remove oil contaminants from water *Sci. Total Environ.* **627** p. 723–732.
[2] Cao S Dong T Xu G and Wang F mei, 2017 Oil Spill Cleanup by Hydrophobic Natural Fibers *J. Nat. Fibers* **14**, 5 p. 727–735.
[3] Febri, Melinda ; Putra A, 2015 Pemanfaatan Limbah Tongkol Jagung Untuk Menanggulangi Tumpahan Oli Di Air Laut **4**, 3 p. 312–315.
[4] Nururrahmah and Rosnita, 2015 Uji efektivitas limbah kulit durian sebagai adsorben tumpahan minyak pelumas *Pros. Semin. Nas.* **02**, 01 p. 843–849.
[5] Nurliyana Che Mohamed Hussein S Hidayati Othman N Dollah A Nazihah Che Abdul Rahim A Shuhadah Japperi N and Syamimi Mohd Asymawi Ramakrishnan N, 2019 Study of Acid Treated Mixed Sawdust as Natural Oil Sorbent for Oil Spill *Mater. Today Proc.* **19** p. 1382–1389.
[6] Quek C S Ngadi N and Ahmad Zaini M A, 2020 The oil-absorbing properties of kapok fibre – a commentary *J. Taibah Univ. Sci.* **14**, 1 p. 507–512.
[7] Lugito G Kustiana A Martuani R and Wenten I G, 2020 Kapok fibre as potential oil-absorbing material: Modification mechanism and performance evaluation *IOP Conf. Ser. Mater. Sci. Eng.* **823**, 1.
[8] Teli M D Valia S P and Mifta J, 2017 Application of functionalized coir fibre as eco-friendly oil sorbent *J. Text. Inst.* **108**, 7 p. 1106–1111.
[9] Faradilla M and Maysarah H, 2019 Potensi Biduri (Calotropis gigantea (L.) WT Aiton) sebagai Tanaman Obat (Prospect of Biduri (Calotropis gigantea (L.) WT Aiton) as Medicinal Plant) *Ilmu Kefarmasian Indones.* **17**, 2 p. 246–250.
[10] Hassanzadeh S and Hasani H, 2016 A review on milkweed fiber properties as a high-potential raw material in textile applications *J. Ind. Text.* **46**, 6 p. 1412–1436.
[11] Zheng Y Zhu Y Wang A and Hu H, 2016 Potential of Calotropis gigantea fiber as an absorbent for removal of oil from water *Ind. Crops Prod.* **83** p. 387–390.
[12] Zheng Y Cao E Zhu Y Wang A and Hu H, 2016 Perfluorosilane treated Calotropis gigantea fiber: Instant hydrophobic-oleophilic surface with efficient oil-absorbing performance *Chem. Eng. J.* **295** p. 477–483.
[13] Tu L Duan W Xiao W Fu C Wang A and Zheng Y, 2018 Calotropis gigantea fiber derived carbon fiber enables fast and efficient absorption of oils and organic solvents *Sep. Purif. Technol.* **192**, August 2017 p. 30–35.

Indonesian glider GL-1 spoiler preliminary design and computational fluid dynamics analysis using EASA CS-22 performance requirements

T Mulyanto*), and M F Zulkarnain

Aircraft Design, Operation, and Maintenance Research Group, Faculty of Mechanical and Aerospace Engineering, Institut Teknologi Bandung, Jl. Ganesha 10 Bandung 40132, Indonesia

* Email: taufiq.mulyanto@ae.itb.ac.id

Abstract. GL-1 is an Indonesian glider which is currently being developed using EASA CS-22 as its basis requirements. Spoiler is an important control mechanism in glider as it is used to control the glide slope of a glider. The design of spoiler includes configuration and dimension such as span, height, and position in wing. One of the requirements for spoiler design in EASA CS-22 is approach and dive performance. Compliance to the performance requirements needs aerodynamics analysis to find lift loss and drag gain due to spoiler extension. Several variable of spoiler dimension, such as the length, the height, and the gap, is varied to obtain a spoiler design which comply to the requirements. To simplify design and aerodynamic analysis process, a method of obtaining 3D aerodynamic characteristics from 2D airfoil with spoiler simulation is developed. 3D CFD simulations is used as a validation and to develop correcting factors. The spoiler dimension which complies to requirements are: 1.4-meter span, 90mm height, and location at 24 to 43% wing span.

Keyword: sailplane, performance, CFD simulation, spoiler design, EASA CS-22

1. Introduction

GL-1 is a single seat, mid-performance, Indonesian National Glider which has been developed since 2014, from design by other previous study [1], and other study using EASA CS-22 [2] as requirement basis. The glider is currently still in development with a half scale prototype has been flown in 2017 [3]. A full scale prototype is currently being built. Studies concerning GL-1 includes: aerodynamic analysis [4-8]; structural analysis [9-11]; and internal configuration design [12].

Spoilers has a function to control glide slope because it is not only increase drag, but also decrease lift [13-14] for cloud flying as well as for landing aid [15]. The design of spoiler must comply to CS-22, including: performance, pilot control forces, and inadvertent extension [2] [13], as shown in Table 1. Initial study on spoilers has been conducted by other [1], while preliminary design of spoiler control mechanism has been conducted further by other researchers [12]. Only CS 22 inadvertent extension has been discussed in those studies [12]. The effect of spoiler on glider's aerodynamic and performance has not yet been analyzed. Fig.1 illustrates the GL-1 view drawing.

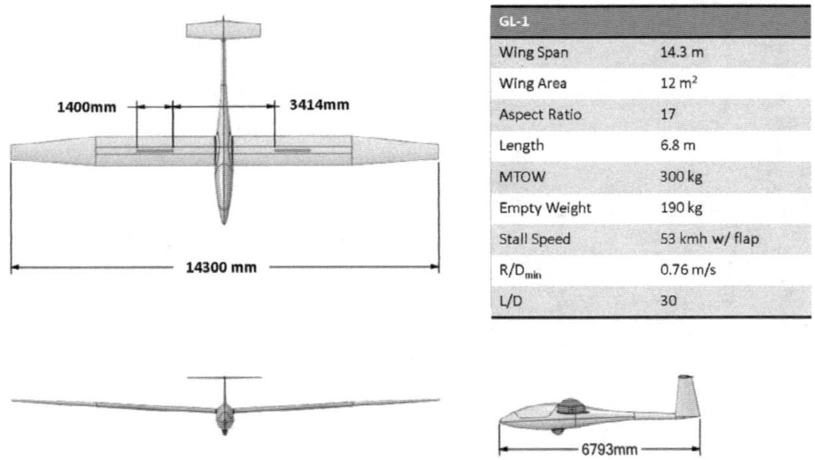

GL-1	
Wing Span	14.3 m
Wing Area	12 m²
Aspect Ratio	17
Length	6.8 m
MTOW	300 kg
Empty Weight	190 kg
Stall Speed	53 kmh w/ flap
R/D_{min}	0.76 m/s
L/D	30

Figure 1. GL-1 Three View Drawing

This paper aims to discuss the process of designing GL-1 spoiler which comply to CS 22.73 and CS 22.75 with concern to their performance requirements. Also, CS 22.697(a)(2) requirements about pilot load was not analyzed as it needs combination of aerodynamic and mechanical analysis, which will need more in-depth studies.

Table 1. Regulation for Spoiler Design

CS 22.75 Descent approach	Glide slope not flatter than 1:7 at 1.3VS0
CS 22.73 Descent, High speed	With airbrake extended, sailplane will not exceed VNE either in a dive of 30° or at sink rate of more than 30 m/s.
CS 22.143 (c)	[list of limit pilot fore for various control surface, including airbrake]
CS 22.697(a)(2) wing flaps and airbrake controls	Must be possible to retract airbrakes up to VT [min 125km/h], but not less than 1.8VS1, with a hand force not exceeding 20 daN
CS 22.697(b)	Each wing-flap and airbrake must be designed to prevent inadvertent extension or movement.

The glider aerodynamics characteristics in several different configurations are needed to analyzed performance requirements, i.e. in clean configuration, flap extended, airbrake extended, and combination of flap and airbrake configuration. Clean aerodynamics coefficients will use previous works [5] [6] [4] using CFD simulations. The results of the studies are a C_L, C_D, and C_M at various angle of attacks. Flap aerodynamics coefficient estimation has also been conducted by Pratama [1] using Datcom. The spoiler aerodynamic coefficient will be analyzed in this paper by obtaining lift loss (dC_L) and drag gain (dC_D) due to spoiler extension. This is obtained by simulating aircraft wing with clean configuration and aircraft wing with spoiler extended and then subtracting the resulting C_L and C_D.

Combination of clean configuration, flap deflections, and spoiler extensions will be used to calculate approach and dive performance. The performance then will be compared to CS-22.73 and CS 22.75 requirements to check for compliance.

2. Methodology

The design of spoiler includes the location (Y_{sp}), span (b_{sp}), and height (h_{sp}), as shown in Figure 2. Initial location and span of the spoiler are obtained by studying other sailplane from reference [16]. Preliminary span and height sizing are then calculated by using CFD simulations and then calculating descent requirements from CS-22.

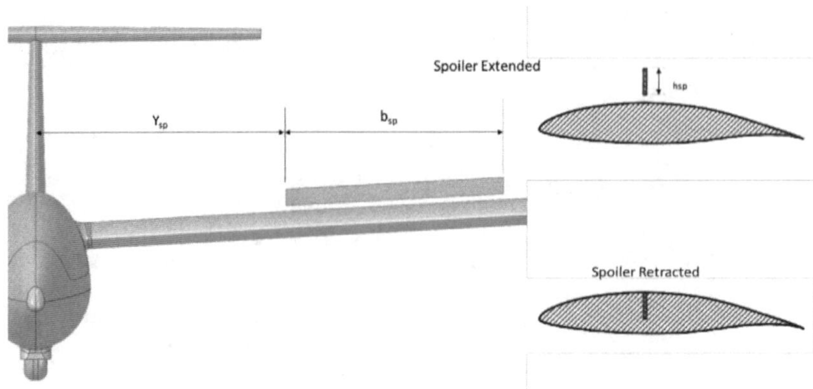

Figure 2. Spoiler dimensions

As discussed in previous section, the requirements checked for spoiler design is CS 22.73 and CS 22.75. The descent requirements are then reformulated to requirements for Lift to Drag ratio (L/D), glide slope, and sink speed at certain airspeed, as shown in Table 2.

Table 2. Performance Requirements

No.	Simulation Case	Ref.	Airspeed	Requirements		
				L/D	glide slope	sink speed
1	Approach-Spoiler	CS 22.75	19.5 m/s	< 7	>1:7 rad (8.2 degree)	
2	Dive-Spoiler	CS 22.73	48.1 m/s		>30 degree	OR > 30 m/s

Both 2D and 3D CFD simulations is used to calculate aerodynamic characteristics. ANSYS WORKBENCH is used for domain building, mesh building, and numerical simulation. D-domain and structured mesh is used for 2D simulation, while half wing domain model and unstructured mesh are implemented for 3D simulations. Numerical simulation uses ANSYS FLUENT as solver for CFD Analysis use RANS as governing equation and SST as turbulence model, as previous studies [17] as well as other studies glider [18] prove these combinations as most fit for modelling and also has been used to predict spoiler performance [13].

The convergence criterion is set to be 10-4. As seen in Table 2, approach and dive performance happened at different flight conditions. Thus, a careful combination of angle of attack and speed flight cases is chosen to be able to cover both conditions of spoiler requirements, while limiting total number of case to save time. The cases are shown in Table 3.

Table 3. Simulation Case for Spoiler

Case	Velocity	Angle of Attack
I	19.5 m/s	6, 7, 8
II	25 m/s	0, 1, 2
III	50 m/s	-4, -3, -2, -1

Simulation of clean configuration and wing with spoiler configuration is done to obtain dCL and dCD. 2D simulations is used to obtain two dimensional dCl and dCd. Spoiler height is varied in this simulations. The 2D characteristics will then multiplied by %lift and %drag for a certain %span to obtain 3D characteristics for various span. The %lift and %drag, as illustrated in figure 3, are the percentage of lift and drag, compared to total, for a certain span of spoiler, and is assumed be constant for various angle of attack. 3D simulation is then used as a validation for this method. Several correction factors are then can be obtained to correct transformation from 2D to 3D characteristics.

The 3D characteristics of various configuration is then obtained by superposition of clean, flap, and spoiler 3D characteristics. This method is used for preliminary design as there are various configuration and dimension will be analyzed, a quicker result can be obtained compared to only 3D simulations used.

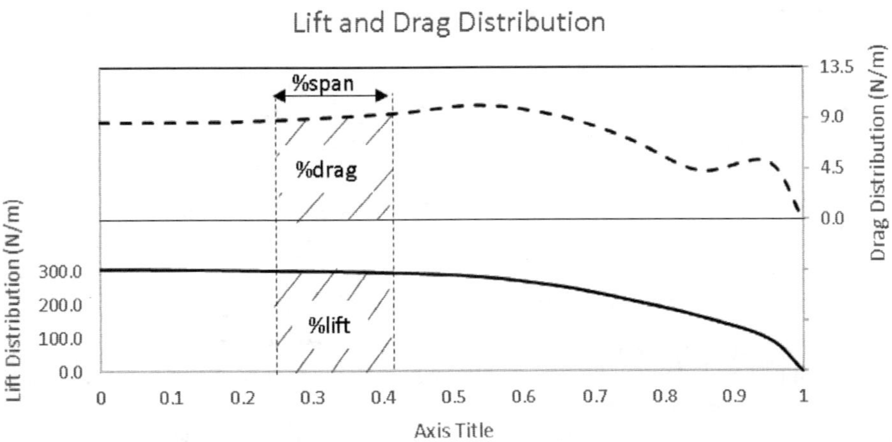

Figure 3. Drag and lift distribution along wing span.

Descent performance is calculated using equations in presented by other [16]. The typical low glide angle assumptions is not used, as the descent performance are calculated at relatively high glide angle, as seen in Table 2.

3. Results and Analysis

Study of various gliders spoiler dimension has been conducted and shown in Table 4. It is shown that spoiler span is between 1.0 to 1.4 meter, with average of 1.2 meter. While spoiler inner location (Y_{sp}) is located in 15 to 33% span, with average of 24% span. The variation of spoiler span will be used for calculation. While the spoiler inner location of 24% span is chosen.

Table 4. Comparative study results

Aircraft	h_{sp}	b_{sp}	Spoiler inner location	Spoiler outer location
Various gliders with L/D between 30-50, and AR between 17-25	n/a	1.0 to1.4 meter	15% to 33%	33% to 52%
Value Tested	70 to and 90 mm	1.0 to 1.4 meter	24% (1.4m)	41%

There are three results of CFD simulations with spoiler configuration, which corresponds to the spoiler itself and its effects to the wing. The spoiler will produce extra drag (dC_D) while producing negligible dC_L effects. The wing produces two things, loss of lift dC_L and induced drag dC_D due to spoiler affecting pressure distribution and detached flow behind the spoiler.

It is found during 3D simulations, that the 3D characteristics obtained from 2D characteristics resulted in lower dC_D and dC_L compared to 3D simulations, as shown in Figure 4. This is may be due to the initial assumption that the spoiler will only effect along the span of which the spoiler located. 3D CFD simulations shown that the effect of spoiler extends out of the spoiler span, as shown in Figure 5. Due to this, a correction factor, shown in Figure 6, is made so that the 2D to 3D method results will be as close to 3D simulations, as shown in Figure 4.

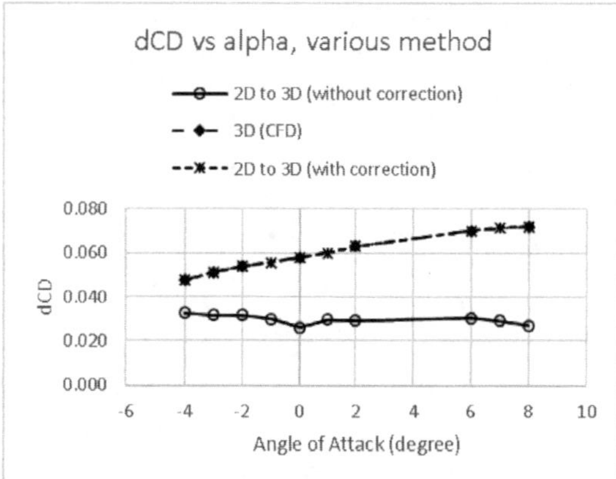

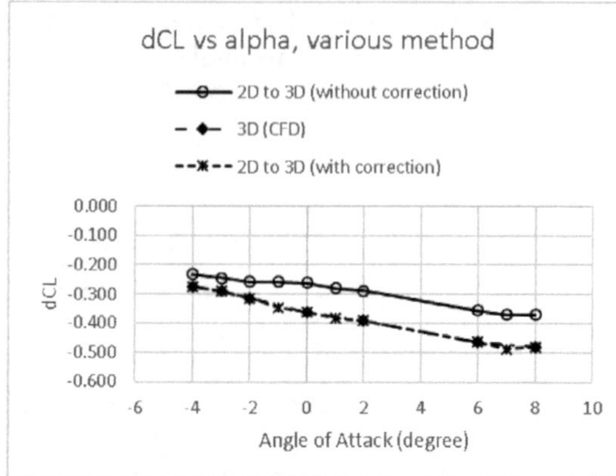

Figure 4. dC_L and dC_D with various method.

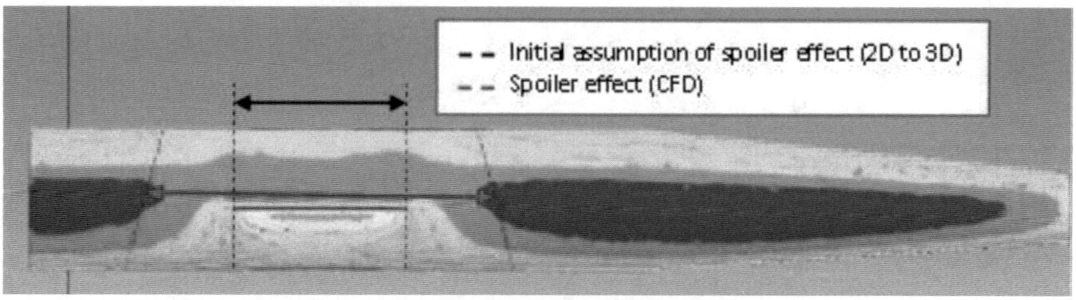

Figure 5. Pressure Distribution Upper Wing with Spoiler Extended

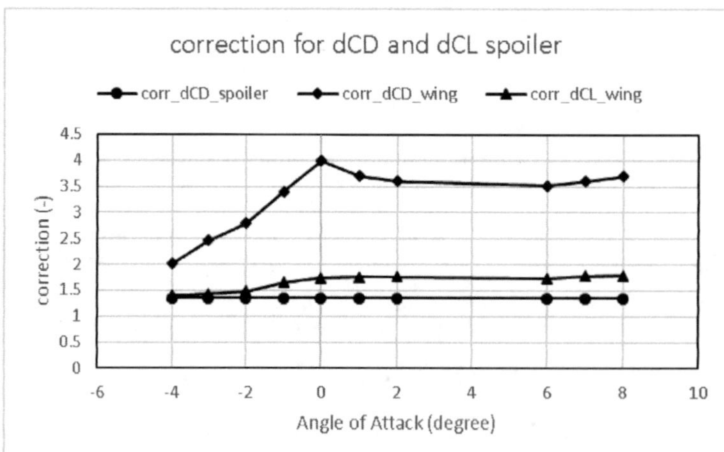

Figure 6. Correction factor for 2D to 3D method.

The spoiler height was varied 70, 80, and 90 mm. As seen in Table 5, as the height increase, dC_D has very little effect. However, the effect is apparent on the loss lift dC_L. Spoiler height of 90 mm is chosen for GL-1, as it gaves higher spoiler performance, while still small for wing limited space.

Table 5. dC_L and dC_D, various spoiler height

h_{sp}	Spoiler Pos	b_{sp}	AoA	dCD	dCL
70 mm	Fully extended	1.2 meter	8 deg	0.102	-0.409
80 mm	Fully extended	1.2 meter	8 deg	0.103	-0.449
90 mm	Fully extended	1.2 meter	8 deg	0.102	-0.491

The next variable to analyze is spoiler span, which will be compared according to descent requirements. The high-speed descent performance requirement are shown in Table 6, while approach-descent performance are shown in Table 7. Out of the two descent requirements, the high-speed descent is shown to be more stringent, as only spoiler with 1.4 m span has glide slope more than 30 degrees. All variation of spoiler span, complies to approach descent requirements, with glide slope of more than 7. However, it obtained with spoiler fully extended and flap down at 30 degrees, and the glide slope has value relatively larger than the requirements.

Table 6. High-speed descent, various spoiler span.

b_{sp}	Spoiler Config	Flap Config	h_{sp}	Airspeed m/s	L/D	glide slope degree	sink rate m/s
1.1 m	Fully extended	Retracted	90 mm	48.1	2.2	24	19
1.2 m				48.1	2.0	26.5	21
1.3 m				48.1	1.8	28.5	23
1.4 m				48.1	1.65	31	25

Table 7. Approach descent, various span.

b_{sp}	Spoiler Config	Flap Config	h_{sp}	Airspeed m/s	L/D	glide slope degree	sink rate m/s
1.1 m	Fully extended	30 degrees	90 mm	19.5	5.5	10.1	3.5
1.2 m				19.5	5.1	11.1	3.7
1.3 m				19.5	4.8	11.8	4.0
1.4 m				19.5	4.2	12.7	4.4

A variation of spoiler and flap configuration are analyzed in Table 8. It is shown that for spoiler span 1.4 m to comply approach-descent requirements, it need to have at least 20% spoiler extension and 10 degrees' flap.

Table 8. Approach descent for 1.4m spoiler span, various spoiler and flap configuration.

b_{sp}	Spoiler Config	Flap Config	h_{sp}	Airspeed m/s	L/D	glide slope degree	sink rate m/s
1.4 m	8% extend	Retracted	90 mm	19.5	10.7	5.4	1.83
1.4 m	20% extend	10 degrees	90 mm	19.5	6.7	9.5	2.9
1.4 m	30% extend	20 degrees	90 mm	19.5	5.4	10.4	3.5
1.4 m	50% extend	30 degrees	90 mm	19.5	4.4	12.8	4.3
1.4 m	full extend	30 degrees	90 mm	19.5	4.2	12.7	4.4

4. Conclusion

Preliminary design of GL-1 spoiler has been conducted, with two performance requirements from CS-22: approach-descent (CS 22.75) and high speed-descent (CS 22.73). The spoiler aerodynamics characteristics (dC_L and dC_D) are obtained using CFD simulations, by subtracting wing-spoiler configuration with clean configuration. Spoiler 3D aerodynamics characteristics are obtained using 2D characteristics. Correction factor has been developed by validating the results to a 3D CFD simulations. From this study, the spoiler is located from 24% chord to 43%. The spoiler has 90 mm height and 1400 mm span. The approach-descent is complied with spoiler extended at least 20% and flap configuration at least 10 degrees down.

References

[1] Hendi Pratama 2015 *Conceptual Design of Indonesia National Glider GL-01* Institut Teknologi Bandung Master Thesis
[2] EASA 2008 *Certification Specifications for Sailplanes and Powered Sailplanes CS-22*
[3] Zulkarnain, M.F, Rahman, M.F., Nurhakim, M.L.I, Arifianto, O., Mulyanto, T. 2017 Flight Test of GL-1 Glider Half Scale Prototype *5th Int. Seminar of Aerospace Science and Technology*
[4] Amalia, E., Moelyadi, M.A, Julistina, R 2019 *Dokumen Teknis Aerodinamika Glider GL-1* Institut Teknologi Bandung
[5] Amalia, E., et al 2018 Aerodynamics Characteristics of Glider GL-1 Based on Computational Fluid Dynamics. *ISAST Proceedings*
[6] Amalia, E., Moelyadi, M. A., Julistina, R. 2018 Aerodynamics Performance of Glider GL-1 Based on Computational Fluid Dynamics in Optimum Flight Conditions. *ICASA Proceedings*
[7] Vega Handojo 2013 *Sailplane Performance Estimation: Student Project Report*, Institut Teknologi Bandung
[8] Azizah, N. A. D., Nurhakim, M. L. I., Arifianto, O., Mulyanto, T. 2016 Glider Uncentered Drag Polar Estimation Based on Flight Manual Data. *Advances in Aerospace Science and Technology in Indonesian Vol.I 2016*
[9] Darsono, R 2017 *Desain dan Analisis Struktur Sandwich Fuselage Pesawat Glider GL-1 Berbasis Metode Elemen Hingga dan Analitik* Institut Teknologi Bandung Bachelor Thesis
[10] Nisa, A.K.. et al 2018 Flutter Analysis of High Aspect Ratio Wing of Sailplane Aircraft. *International Journal of Engineering and Technology, Vol. 7.*
[11] Rahmah, A. 2018 *Pengujian Three Point Bending pada Komposit Sandwich Kayu Balsa dan Fiberglass untuk Rancang Bangun Pesawat Glider* Institut Teknologi Bandung UndergraduateFinal Assignment.
[12] Mulyanto, T., Zulkarnain, M.F. 2019 Preliminary Design of GL-1 Sailplane Cockpit Arrangement and Control Mechanism Based on EASA CS-22 Requirements. *IOP Conference Series: Materials Science and Engineering, Vol. 645, p. 012005*
[13] Greiner, M., Bangga, G. 2018 Dilemma resolved: Airbrakes tamed. *Technical Soaring, Vol. 42, pp. 23-30*
[14] FAA 2013 *Glider Flying Handbook*. Oklahoma City : United States Department of Transportation
[15] Gedeon, J. 2007 A Few Words on Airbrakes. *TECHNICAL SOARING, vol. 31, no. 4, pp. 110-113*
[16] Thomas, Fred 1999 *Fundamentals of Sailplane Design* Maryland : College Park Press
[17] Moelyadi, M. A., et al. 2020 Computational fluid dynamics based winglet design and analysis of GL-1 glider. *AIP Conference Proceedings* .
[18] T, Hansen 2014 Modeling the Performance of the Standard Cirrus Glider using Navier-Stokes CFD. *Technnical Soaring vol 38 no 1.*
[19] Souckova, N. and Matejka, M. 2010 Experimental Investigation and Numerical Analysis of Flow Past Airfoils with Spoilers and High Lift Devices. *Technical Soaring Vol. 34 no. 4 pp. 110-117*

Preliminary design of bionic flapping wing vehicle

M A Moelyadi[1], E Amalia[1*], A D Tanoto[1], L M Septiani[1], M Rafie[2], A S Perdana[1], and R Agung[1]

[1] Faculty of Mechanical and Aerospace Engineering, Bandung Institute of Technology, Jalan Ganesha No. 10 Bandung 40132, Indonesia

[2] School of Electrical Engineering and Informatics, Bandung Institute of Technology, Jalan Ganesha No. 10 Bandung 40132, Indonesia

*Corresponding E-mail: ema@ftmd.itb.ac.id

Abstract. This paper deals with the preliminary design of bionic flapping wing vehicle. The design is driven by its requirement and objective namely the maximum weight should be less than 0.5 kg with 1.5 m wingspan. In the conceptual design, the wing planform, wing structures and system of flapping mechanism will be considered to find the initial configuration and then continued the sizing of the flapping vehicle. The analysis of designed flapping vehicle is increasingly complex due to firstly the generation of time-dependent aerodynamic forces and moments from unsteady flow around the flapping wing, secondly, flexible wing structure which generates higher thrust may cause a structure failure and thirdly, a flapping mechanism system generating differential flapping motion for the vehicle maneuver. The aerodynamic analysis for given flapping motion model is carried out using Computational Fluid Dynamics method based the solution of unsteady Reynold Averaged Navier-Stokes equations and the structure analysis is conducted with the input of aerodynamic loads using Finite Element method to find critical stresses that may cause a structure failure. As a solution of differential flapping two controlled servos are used. The architecture of electrical system is made to analysis of the distribution of data signal and power. The flight maneuver is achieved by changes in flapping frequency and sweeping angle. The selection of system components is performed by considering weight constraint. The flight maneuver is achieved by changes in flapping frequency and sweeping angle. The wing platform has elliptical shape with airfoil maximum camber of 6% chord at a quarter chord. The outer body and wing are made of foam laminated by glass fibre with the foam density of 0.045 g/cm^3. The horizontal tail is made of mylar film with density of 1.38 g/cm^3 and its area of 168.3 cm^2. As simulated results, the amount of total lift is 4.18 Newton generated at 6 Hz flapping frequency.

Keywords: bionic flapping, aerodynamics analysis, flapping mechanism, electrical system

1. Introduction

The development of bionic flapping wing vehicle has been inspired by bird since the beginning of its development [1]. Before developing a bionic flapping wing vehicle, researchers were studying about many aspects of bird-like flapping wings. For example, in [1] it is studied by using CFD about the thrust production of flapping wing for a bird-like wing with wing span 1.1 m and chord 0.6 m. The results are the thrust produced by the flapping wing for frequency of 1 Hz, 2 Hz, and 4 Hz. Reference [2] studied about the aerodynamics performance of bird flight with retracted wing to increase thrust produced by bird wing at various frequencies.

While, other researcher is reporting study of unsteady aerodynamics simulation of bird wing for various frequencies. Here RANS solver was used for simulation with quasi-steady and unsteady cases [3], and also other studied about endurance efficiency for various motion model of flapping bird wing which are pure flapping, pure flapping-retraction, and symmetrical pure flapping-pitching [4]. In those reports, CFD simulations were carried out to calculate endurance efficiency and the higher endurance efficiency is for the flapping-pitching motion which is about ten times the other motion models. For theoretical study about the drag of flapping wing, reference [5] gives a good approach of drag prediction.

After having good understanding about how birds fly with their wings, then the principle is to be applied to an unmanned aerial vehicle (UAV). Further study gives a comprehensive review about some bionic UAV that used flapping wing [6], where it has been chosen for the product from FESTO "SmartBird" as our model in developing bionic flapping wing vehicle in this study. In developing our bionic flapping wing vehicle, previous studies reported by others have been used as a base for knowledge-based development of bionic flapping wing vehicle reported in this study [1-4].

2. Design Process

Preliminary design of our bionic flapping wing vehicle was started by using FESTO's "SmartBird" as a model. By referring to reference [6], we make a list of design requirement and objective (DRO) as presented in table 1. Although "SmartBird" is used as the main model, DRO is opened for some modification that might be needed in design process.

Table 1. Design Requirement and Objective

Span Length : 30 – 150 cm
Maximum Weight : 0.4 kg
Fight duration : min 5 s

Based on DRO in table 1, design process was conducted with steps as follows: 1. Conceptual Design and Sizing; 2. Analyses of Aerodynamics, Structure, and System of Flapping Mechanism 3. Design Results. First we design planform, flapping mechanism, and electronic system. Then, we were conducting CFD simulation based on our accumulated knowledge. From reference [1], a general planform shape could be determined. After the planform is determined, then CFD simulation with FSI (Fluid Structure Interaction) approach is carried out for calculating lift and thrust produced by the planform. This process is referring to reference [1] to [4], reference [7] and [8].

After we make sure that the lift and thrust is sufficient, then we design the flapping wing mechanism based on reference [9], [10], and [12] to [16]. We considered some alternatives on this flapping wing mechanism and finally choose to use two-servo mechanism. After that, electronic system was designed. Finally, a structural analysis was conducted by using Finite Element Method to check whether the structure fails or not.

2.1 Aerodynamic Analysis of designed bionic wing configuration

2.1.1 Designed bionic wing configuration

The design of Flapping wings consider two aspects as given in Figure 1, firstly time-dependent aerodynamics loads subjecting on the surface that is influenced by some motion parameters such as air speed, flapping frequency, flapping dihedral angle, pitching angle and airfoil shape, and secondly structure deformation of the wings due to elasticity of the wing structure. When it is assumed as a decouple problem then we may reduce some important parameters such as a rigid wing [17] and simple oscillating plate.

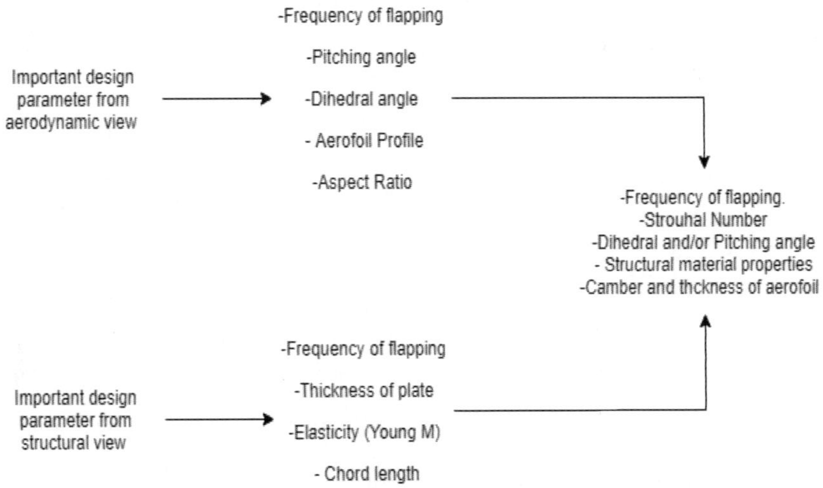

Figure 1. Important Aspects of Design of the Flapping Wing Vehicle

Those parameters provide effects lift and thrust over time. From the previous work we know that flapping frequency heavily affect the thrust. Too high or near a resonated fluid-structure frequency produce complex structural respond, therefore will destroy thrust generation, but too low it produces no thrust at all. Strouhal number consisting of parameters of span length, frequency, and freestream velocity affects all variable roughly the same. For one geometrical dimension there must be a range of Strouhal number in which flapping wing will behave well (producing high amount of average lift with some amount of average thrust left).

Dihedral angle mostly affects maximum value of lift and the pitching angle assist to increase thrust when down stroking and to reduce drag when up stroking. Both angles could have different values when down stroking and up stroking. On the other hand, the camber and thickness mainly affect lift generation which behaves accordingly to airfoil theory. The material properties determine the structure response whether it assist thrust production or not. More rigid the material is, more lift produces at the expense of thrust.

The bionic flapping wing configuration is initially obtained based on early literature [7] namely a designed wing like a dove bird wing, which has a 20cm semispan and a 15cm root chord wing (see figure 2). In order to provides adequate aerodynamic lift to carry the target weight, the wing is modified by enlarging its span and area. The elliptic wing shape is used to minimize aerodynamic drag, so that the required power for driving the wing is lower and the battery weight and servo size can be reduced. The modification is done by taking the same dimensionless Strouhal number to maintain aerodynamic performance of the flapping wing without much change [9]. An iterative approximation to determine the size of the wing planform is as shown in flowchart of Figure 3.

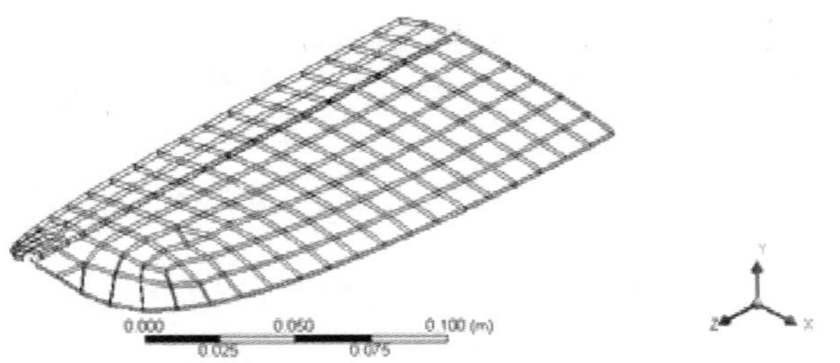

Figure 2. Designed wing

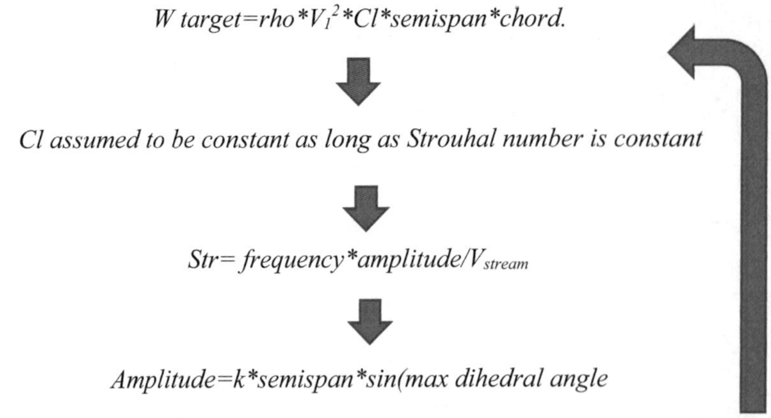

Figure 3. Iterative Approximation for obtaining flapping wing Planform

The curvatures of the leading and trailing edges are made using Ellipse equation where the semi-major axis is the wing semispan and the semi-minor axis is starting from mean aerodynamic chord (MAC) to edge (either trailing or leading). The distribution of wing section along the spanwise is carried out with a sinus function distribution which is denser to the direction of the wing tip. The line of aerodynamic center of 25% of the chords is kept a straight for allocating the wing spar and the leading edge and trailing edge lines have the distances from the aerodynamic center with the ratio of 1:3 as shown in the figure 4.

This is to make the maximum chamber be always at 25% of the chords throughout the span. Because this wing will take a form as a thin shell wing, so this planform will take a chambered plate as a baseline instead of an airfoil. The important parameters that can be adjusted to obtain the geometry of the flapping wing are semispan, root chord, and maximum chamber. Table 2 shows the resulted geometry planform for the UAV Flapping wing and the flapping motion condition is given in Table 3.

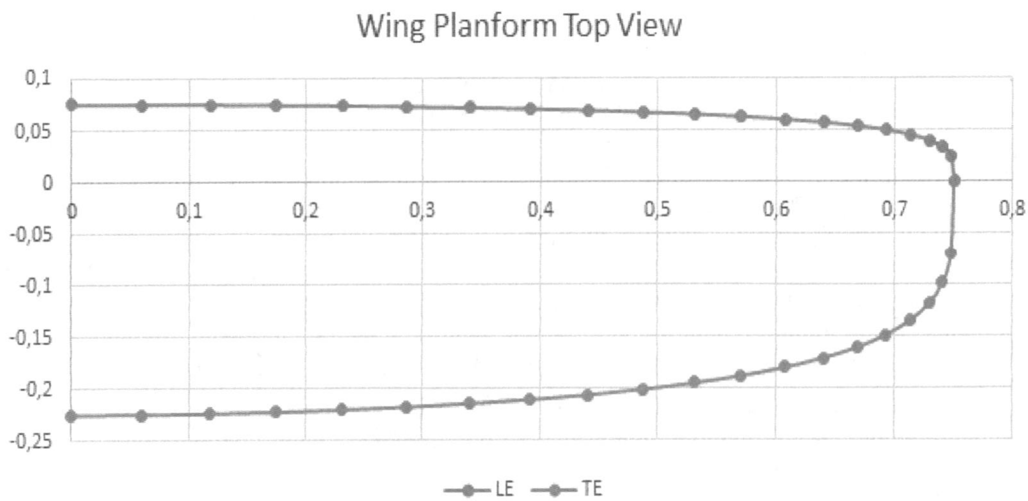

Figure 4. Planform of the Flapping Wing

Table 2. Geometry of Planform of the Flapping Wing

Semispan: 75 cm
Root Chord: 30cm
Thickness: 2 mm
Max chamber: 6%

Table 3. Flying Conditions of the Flapping Wing for Simulation

V_{body}: 10 mph
Frequency: 4.5.6 Hz
W_{target}: about 200 grams.
Total max dihedral angle: 60 degree

This assumption has a drawback because even though the Strouhal number is kept constant (which is an important parameter in periodic unsteady flow), there is a very large change in the aspect ratio value, from 2.66 to 5 and there is a minor change in the Reynolds number.

2.1.2 Aerodynamic Analysis of Flapping wing

2.1.2.1 Governing equation of RANS, Turbulence Model and FSI Coupling

Fluid and structure interaction is a quite complicated involving both structural dynamics and fluid dynamics. They provide an influence each other. The changes in structure affect the surrounding fluid and the fluid flow producing a pressure on the body surface trasfering to the structure.The equation for motion of the continuum body of the structure is as shown in equation (1).

$$\tilde{\nabla}^T \boldsymbol{\sigma}_s + \mathbf{b}_s = \rho_s \frac{\partial^2 \mathbf{u}_s}{\partial t^2} \tag{1}$$

where σ_s is the diagonal value of the stress tensor in the structural material, bs is the body force (in this case there is only gravitational force), and us is the location of the structural elements themselves.

For this case the fluid is assumed to be incompressible flow so that there is no coupling between the mass and momentum conservation equations with the energy equation as in equation (2). This is done by performing a simulation on CFX using the Air at 25ºC option. So that the Navier stokes equation will be coupled with the remaining structure.

$$\nabla \cdot \vec{V} = 0$$
$$\rho \frac{\partial \vec{V}}{\partial t} + \rho \left(\vec{V} \cdot \nabla\right) \vec{V} = -\nabla p + \mu \nabla^2 \vec{V} + \rho \vec{g} \tag{2}$$

Equation above also supplemented with general fluid boundary conditions such as no slip conditions on the walls, inlet conditions and surroundings. Initial conditions must also be determined in advance. Then the equation that combines the two systems is to equalize the fluid velocity with the structure on the wall (specifically no slip condition) and iteration between displacement and force.

The structure will accept the wall stress tensor and process it into displacement and the displacement will become a new boundary condition for the fluid to be analysed to produce a wall stress tensor (Cauchy stress tensor). In analytic equations the above must be done simultaneously. In CFD / CSD this will be done until it reaches convergence in each timestep as shown in Figures 5 and 6.

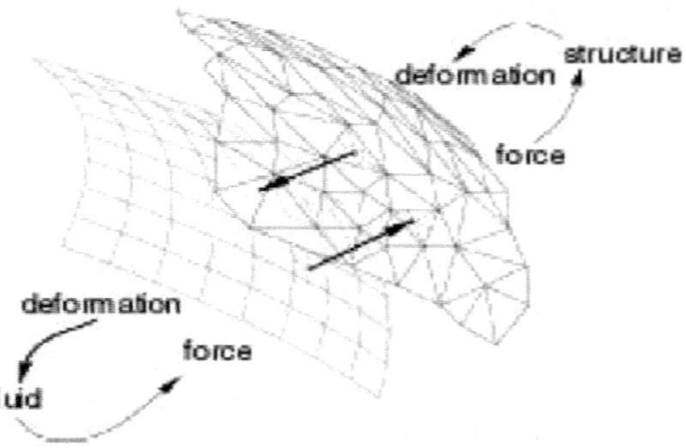

Figure 5. External CFD-CSD Coupled Simulation

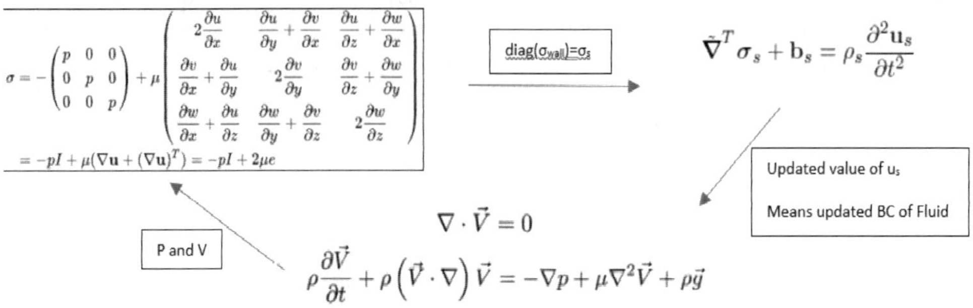

Figure 6. Equations that used in CFD/CSD Coupling System

2.1.2.2 Simulation of Flapping wing

The simulations are carried out with Ansys CFX and Ansys Transient Structural through the coupling system between fluid and structure analysis as shown in Figure 5 and the computational domain is as shown in Figure 7. The simulations are done in a half-symmetrical model to save computational resources. For the structure model, a polyfoam material with density is 50g / cm3 is used obtained from references [9]. Elasticity is assumed to be an isotropic material with a modulus Young value of 3250 MPa and a Poisson ratio of 0.32.

The Movement are assigned at 25% chords assuming the line stretches across 25% of the chords (i.e. none other than the spars) remains rigid. The ratio of the maximum flap up and flap down is 2: 3. This value is taken from statistical data which have the max flap-down is greater than the flap-up and the ratio is quite significant but not too extreme.

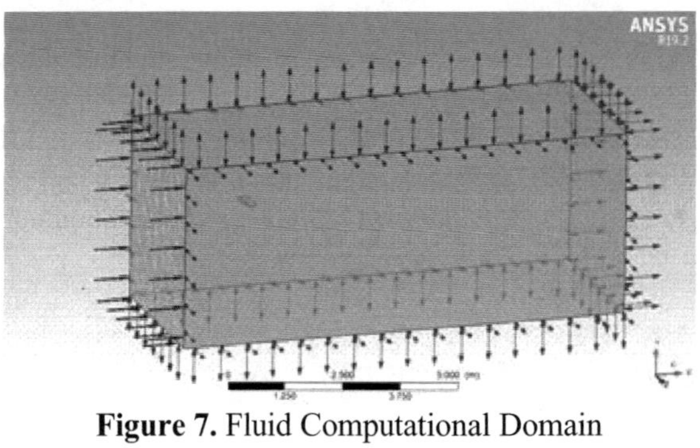

Figure 7. Fluid Computational Domain

2.2 Structure of Bionic Flapping Wing

2.2.1 Structure Design of Bionic Flapping Wing

The design of flapping wing structures considers critical loads that may be occurred in several components including wings, supporting part and front fuselage. For the flapping wings, the cyclic aerodynamic loads subjects to the wing surface. The loads are then transferred to the wing spar which is connected to the flapping mechanism devices such as joiners and servos. The spar should be designed to have a shape and size for maintaining bending moment and torsion. Figures 8.and 9 show the structure of wing and the structure of spar and joiner, respectively.

Figure 8. Structure of wing and fuselage

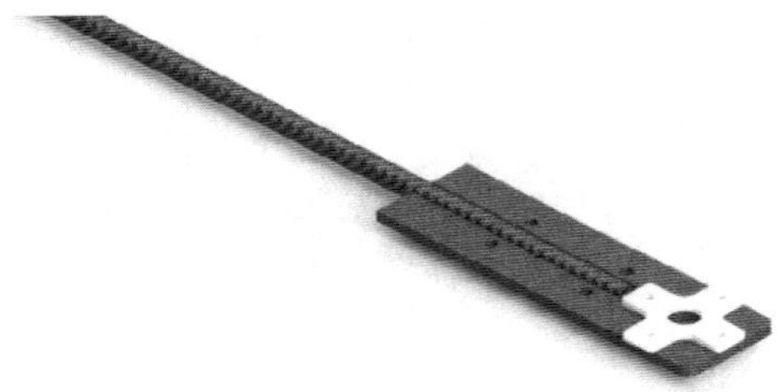

Figure 9. Flapping Wing's Servo-Spar Joint

Next, the fuselage is designed to have a streamline shape with the rounded nose shape and to have a minimum weight. The fuselage nose may be subjected to the stagnation air load that may be critical causing the deformation of surface nose as shown in Figure 10 Inside the fuselage there is a main frame to support all electronic components including servo and the joint of the tail. This main frame may be also critical due transferring loads from the servos and the tail joint. In order to reduce its weight, the structure of the main frame is made some holes as depicted in figure 11.

Figure 10. Flapping Wing's Body

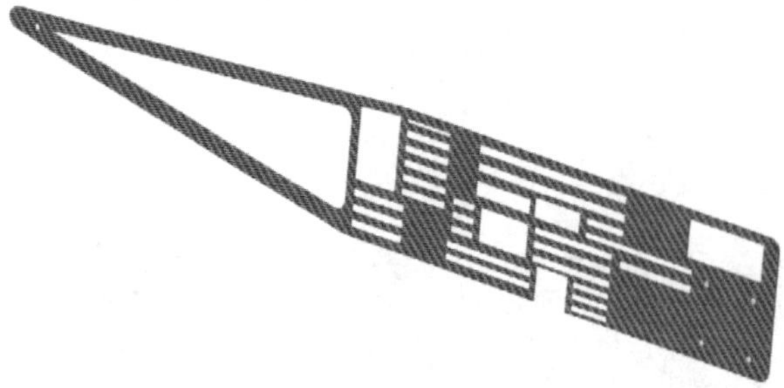

Figure 11. Main Frame of supporting component

2.2.2 Structural Analyses with Finite Element

The critical load on flapping wing structure occurs on wing spar. This is due to the aerodynamic load transferred to only one spar and connected to servo. For structure simulation, the model is developed and analyzed using finite element ABAQUS software with the first step is how to distribute loads in the spar. The lift distributing along spar is applied as uniform pressure on the bottom wing skin. This is to check if the designed spar is fail due to the bending moment of the lift distribution from wing skin.

The main materials used for flapping wing vehicle are carbon fiber and polyfoam. the structure materials are used as follows: carbon rod for the spar, carbon sheet for the servo-spar joint, polyurethane for wing, and polycarbonate for servo horn. The mechanical properties are shown in Table 4. two different size of diameter of carbon rod are compared namely, 4 mm and 6 mm.

Table 4. Mechanical properties of carbon rod, carbon sheet, polyurethane, and polycarbonate

Material	Parameter	Unit	Value
Carbon/Epoxy Composite Rod	Longitudinal modulus	GPa	140
	Transverse modulus	GPa	10
	Poisson's ratio	-	0.3
	Shear modulus	GPa	5
Carbon/Epoxy Composite Sheet	Longitudinal modulus	GPa	70
	Transverse modulus	GPa	70
	Poisson's ratio	-	0.1
	Shear modulus	GPa	5
Polyurethane	Modulus	GPa	3.45
	Poisson's ratio	-	0.3
Polycarbonate	Modulus	GPa	2.48
	Poisson's ratio	-	0.38

2.3 Design of System of Flapping Wing Mechanism

2.3.1 Differential Flapping Motion

Differential flapping motion is preferred than BLDC (Brushless DC Motor) for driving flapping mechanism, since its compactness and simplicity in term of mechanical parts. Its simplicity is achieved by two servos configuration. Reducing number of parts to conduct in-flight maneuver will reduce overall AUW (All Up Weight) of the vehicle. Two servos are being used to conduct differential flapping motion.

Differential motion is being used to control frequency and center angular position of each servo independently. The control of servo movement will be programmed via an MCU (Microcontroller Unit). There are several proven examples of previous differential flapping wing vehicle, including: Robo Raven of University of Maryland and University of Southern California [6], and eMotionButterflies of Festo. Designed differential flapping motion is pictured as shown in Figure 12.

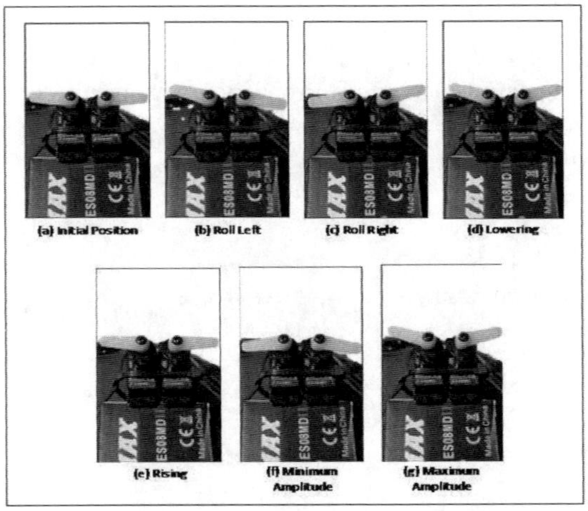

Figure 12. Stages of Differential Flapping Wing Motion

2.3.2 Design of Wing Mechanism

The servo is selected as a part of flapping mechanism to receive trenfered load and driven the movement of the wing directly without any mechanical gear addition. However, it still need an additional part to link the servo's gear with the wing's spar. The critical thing that will be faced is the joint should be design properly to the complex form of the servo's gear. To overcome this problem, the selection of 3D printed material and great design approximation should be conducted carefully.

2.3.3 Electric Components

This flapping wing vehicle is intentioned to have two main features, surveillance and autonomous flight. A FPV (First-Person View) camera is used for surveillance and GNSS (Global Navigation Satellite System) is used to determine geospatial position of vehicle. Figure 13 shows a diagram of power and data lines for integrating power and signal process of all electronic components of the bionic flapping wing. The requirement for electronic component selection is component weight with the maximum limit weight is 0.1 kg. The list of all used electronic components and each component weight data are given in Table 6

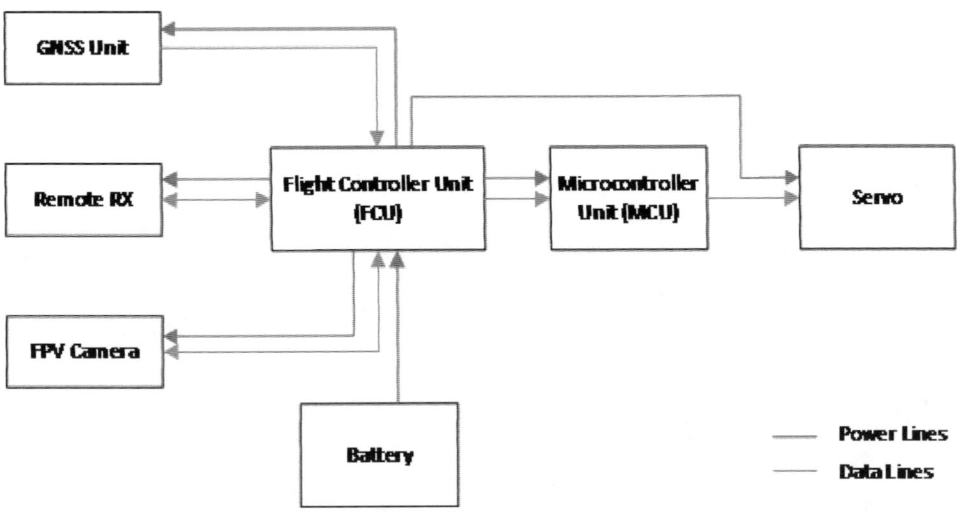

Figure 13. Power and Data Lines

2.3.3.1 Differential Flapping Control

Differential flapping control act simultaneously as propulsion sub-system and surface control. It is done by creating certain output signal mixing for flapping motion of two-independent servos. The generation of motion from differential flapping control covers from: thrust, lift, roll, yaw, rising-lowering, to gliding. Designed output signal mixing is shown in Figure 14.

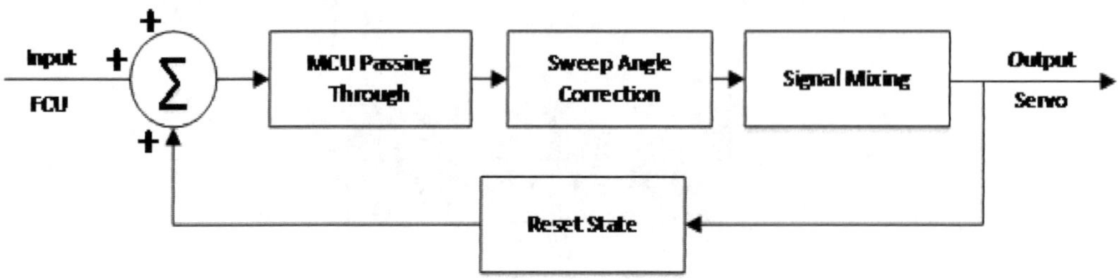

Figure 14. Differential Flapping Control

The output signal mixing is produced by outcoming signal post-processing from Flight Controller Unit (FCU) into Microcontroller Unit (MCU). The MCU modifies outcoming PWM signal from FCU into sinusoidal PWM signal to drive servo as flapping mechanism. A sweep angle correction is utilized to correct current servo position when exceeding determined minimum or maximum position. The exceeding position is as a result of lagging sequence of flapping command relative to signal mixing process. The reset state option is functioned for two purposes, first to reset previous mixing state, second to switch gliding mode when throttle input is below minimum threshold value.

3. Results and Discussions

3.1 The Bionic Flapping Wing Vehicle

The final design had made some change, specifically on the wing camber design due to achieve desired lift and thrust force. The final design of tail mechanism is quite simple, without any requirement to perform yawing then it was fixed on a particular angle of attack due to create some forces to make it has great longitudinal stability performance. Figure 15 shows the final design of the model.

Figure 15. The Model of Flapping Wing Vehicle's CAD

3.2 CFD/CSD Simulation Results

Figure 16 shows the drag/thrust versus time graph and Figure 17 shows the lift versus time graph from the simulation as explained in the above section.

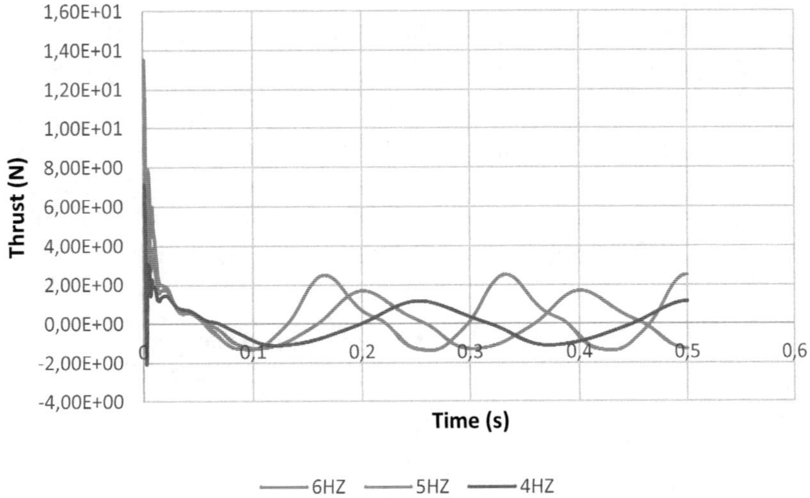

Figure 16. Drag/Thrust versus Time of the Flapping Wing Vehicle

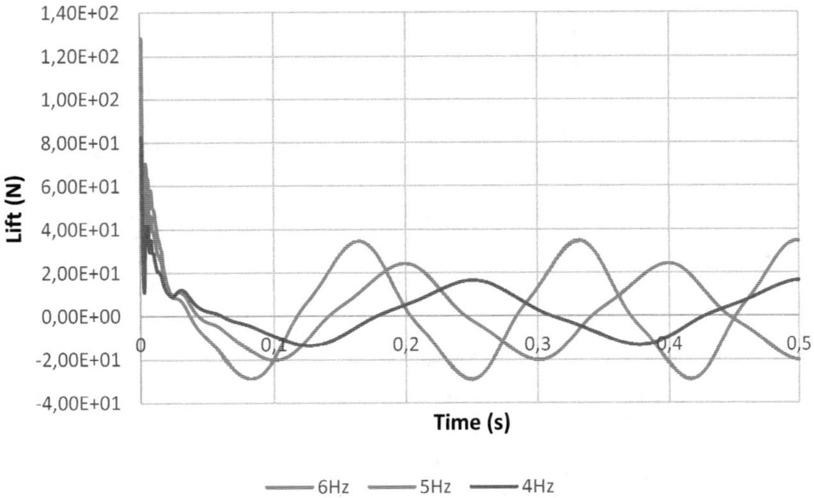

Figure 17. Lift Versus Time of the Flapping Wing Vehicle

There is some numerical initialization error, so we are ignoring them. And then we do average per cycle, and because each simulation has different frequency and should be average based its own period. Table 5 shows the summary of simulation results for the flapping wing vehicle.

Table 5. Summary of CFD Simulation Results for the Flapping Wing Vehicle

	Average drag/thrust(N)	Average Lift (N)	Average motor torque (Nm)
4 hz	4.55E-02 (drag)	1.24E+00	5.28E-01
5hz	3.87E-02 (thrust)	1.41E+00	5.96E-01
6hz	2.78E-01 (thrust)	2.09E+00	8.88E-01

3.3 Electric Components

Table 6 shows the electric components and weight that is used in the bionic flapping wing vehicle as explained in paragraph 2.3.3. Electric components is chosen to fulfill system requirement based on preceding simulation and further calculation.

The fundamental parts of flapping wing systems are propulsion-control servo and servo driver. KST MS320 as servo, capable to produce 5.50 kg.cm of torque. KST MS320 is able to produce flapping frequency 6.25 Hz. Teensy 3.2 as servo driver, is preferred because availability of interrupt pins across all digital pins. Interrupt pins are used to read/write PWM. The remaining components are chosen with performance and weight consideration. In total, it will be 94.6 gr of electric components weight. Figure 18 shows main frame with component assembled.

Table 6. Electric Components of Flapping Wing Systems

Type	Name	Gross Weight (gr)
Propulsion-control servo	 KST MS320	41.6
Servo driver	 Teensy 3.2	3
Flight controller unit	 Matek F722 SE F7	10
Receiver	 Frsky XM+	1.6
FPV camera All-In-One (AIO)	 Caddx Firefly	4.1
GNSS unit	 Matek SAM M8Q	7
Battery	 Emax Tinyhawk S 300mAh	15.3
Cable and connector	Various	±12
Estimated Total		94.6

(a) (b)

Figure 18. Main frame with component assembled (a) Right side (b) Left side

3.4 Finite Element Result

After conducting the numerical simulation of this wing structure, the result shows some element which have high stress value. The element with the high stress value will be more tended to experience failure due to the given load. By comparing the 2 different diameters (4mm and 6mm), the Tsai-Wu Failure Criterion conducted to check either the structure will fail or not. Table 7 will show the Tsai-Wu Failure Criterion of the finite element analysis

Table 7. Tsai-Wu Failure Criterion Calculation

Diameter 6 mm			Diameter = 4 mm		
σ_1	432.716	MPa	σ_1	1180	MPa
σ_2	9.184	MPa	σ_2	13.644	MPa
τ_{12}	33.884	MPa	τ_{12}	78.88	MPa
X_T	1900	MPa	X_T	1900	MPa
X_C	1300	MPa	X_C	1300	MPa
Y_T	50	MPa	Y_T	50	MPa
Y_C	250	MPa	Y_C	250	MPa
S	70	MPa	S	70	MPa
F_1	0.001296	MPa^{-1}	F_1	0.001296	MPa^{-1}
F_2	0.024	MPa^{-1}	F_2	0.024	MPa^{-1}
F_{11}	-4E-07	MPa^{-2}	F_{11}	-4E-07	MPa^{-2}
F_{22}	-0.00008	MPa^{-2}	F_{22}	-0.00008	MPa^{-2}
F_{66}	0.000204	MPa^{-2}	F_{66}	0.000204	MPa^{-2}
Tsai-Wu	0.934385		Tsai-Wu	2.553909	

The Tsai-Wu Failure Criterion is satisfied with the composite material that was used in this spar and joint of wing in this vehicle and it is also satisfied with the bending phenomenon which is become the boundary condition of the Finite Element Analysis. The Tsai-Wu value should be less or same as 1 to ensure the structure will not experience any failure. Hence, the Table 7 shows that the structure which has larger diameter will be safe due to the inertia of the structure is larger and could stand with the given load. The definition of sigma 1, sigma 2, and tou 12 are the maximum stress in x direction, the maximum stress in y direction, and the maximum shear stress in xy plane. All of this stress value is obtained by Finite Element Analysis which is already described in previous section.

4. Conclusion

A model of bionic flapping wing vehicle has been designed with span length of 1.5 m and with flapping mechanism using 2-servo. It produced totally 4.18 N generated at 6Hz flapping frequency that has been fulfilled the requirement of the flapping wing. The structure of wing spar has been analyzed using finite element and the failure criteria of Tsai-Wu showing it remain safe for the given loads.

References

[1] M. A. Moelyadi, D. A. Priyanto and G. Sachs,Simulation of Flexible Flapping Wing Bird in Producing Thrust, Journal of Unmanned System Technology, Vol. 2, No.2, September, 2014, pp. 62-69.
[2] M. A. Moelyadi, F. Rhamadiansyah and G. Sachs, Simulation of Aerodynamics of bird wing retraction, 9th International Conference on Intelligent Unmanned Systems 2013 (ICIUS 2013) Jaipur, India, 25-27 September 2013.
[3] M. A. Moelyadi, H. A. Putra and G. Sachs, Unsteady aerodynamics of Flapping Wing of a Bird, ITB Journal of Engineering Science, April 2013.

[4] R. Julistina and M. A. Moelyadi, CFD Based Prediction of Endurance Efficiency for Various Motion Models of Flapping Bird Flight, International Conference on 14th Intelligent Unmanned Systems 2018 (ICIUS 2018), Jeju, South Korea, 20-23 August. 2018
[5] Sachs G., Aerodynamics Cost of Flapping, Journal of Bionic Engineering, 12 (2015), 61-69.
[6] Chen C., Zhang T. (2019) A Review of Design and Fabrication of the Bionic Flapping Wing Micro Air Vehicles. Micromachines.
[7] Namhun Lee et al. (2018) "Effect of flexibility on flapping wing characteristics in hover and forward flight" in Elsevier.
[8] Alex E.H., et al. 2017. Characterizing and Modeling The Enhancement of Lift and Payload Capacity Resulting from Thrust Augmentation in a Propeller-Assisted Flapping Wing Air Vehicle. International Journal of Micro Air Vehicles.
[9] Mohd Firdaus Bin Abas (2016) "Flapping wing micro-aerial-vehicle: Kinematics, membranes, and flapping mechanisms of ornithopter and insect flight" in CJA at CSAA.
[10] Sutthiphone S., Woei-Leong C. 2015. Ornithopter Type Flapping Wings for Autonomous Micro Air Vehicles. Aerospace 2015.
[11] Chen, C., & Zhang, T. (2019). A Review of Design and Fabrication of the Bionic Flapping Wing Micro Air Vehicles. Micromachines, 4.
[12] Jackowski, Z. J. (2009). Design and Construction of an AutonomousOrnithopter. Boston: MASSACHUSETTS INSTITUTE OF TECHNOLOGY.
[13] Kim, D. K., & Han, J. H. (2006). Smart Flapping Wing Using Macro-Fiber Composite Actuators. In Y. Matsuzaki (Ed.), SPIE, 6166-6174, pp. 1-9. San Diego, California. doi:http://dx.doi.org/10.1117/12.658117
[14] Send, W., & al, e. (2012). Artificial hinged-wing bird with active torsion and partially linear kinematics. International Congress of the Aeronautical Sciences, 5.
[15] Srigrarom, S., & Chan, W.-L. (2015). Ornithopter Type Flapping Wings for Autonomous Micro Air Vehicles. aerospace.
[16] Wang, P. L., & McCarthy, J. M. (2018, April). Design of a Flapping Wing Mechanism to Coordinate Both Wing Swing and Wing Pitch. Journal of Mechanisms and Robotics, 10, 1-6.
[17] Tantono, A.D.,(2019), Preliminary Wing Mechanism Design and CFD Analysis of Flapping Wing
[18] Mini Unmanned Aerial Vehicle, Bachelor Final Project, Bandung Institute of Technology.

Maneuverability analysis on flying vehicle with thrust vectoring system

V H Al Aviev[1,*], **R A Sasongko**[1,*]

[1] Department of Aerospace Engineering, Institut Teknologi Bandung, Jl. Ganesha No. 10, Bandung 40132, Indonesia

* Email : valeriaviev@gmail.com ; sasongko@ae.itb.ac.id

Abstract. Fighter aircraft often require unconventional maneuverability that civil aircraft cannot perform. The conventional flight control devices are not sufficient for this maneuver. Thus, to be able to perform unconventional maneuvers, additional control devices may be required, such as the thrust vectoring system (TVS). This paper will discuss the effect of TVS implementation in improving maneuverability. In this work, a dynamic model of F-4 Phantom is used as the basis for the study. The dynamic model of the aircraft with TVS is constructed analytically and numerically. The nonlinear model obtained from the modeling process then linearized at a particular flight condition and some TVS deflection values. Further, the linear system and control approach is employed for evaluating and designing the linear controllers for stabilization and maneuvering tasks. The closed-loop system, which is implemented in the MATLAB/Simulink environment, then numerically simulated and observed for analyzing the effect of implementing different TVS deflection on the required control effort in high angle of attack maneuvers. The simulation results show that the TVS can affect the performance of the closed-loop system in stabilization tasks. While at some deflection values, TVS can also help to reduce the required control effort to tracking attitude reference command when executing the high angle of attack maneuvers.

Keywords: Thrust vectoring system, high angle of attack maneuver, flight control system, flight dynamics, aircraft modeling

1. Introduction

In carrying out its mission, fighter aircraft often require unconventional maneuverability that civilian aircraft cannot perform. With unconventional maneuvers, unconventional control devices are needed. The thrust vectoring system (TVS) is an example of unconventional control devices, which direct the thrust of the jet engine in the desired direction to control the aircraft's attitude.

In this research, the use of TVS is expected to increase the maneuverability of the fighter aircraft. This research will present and analyze how using TVS can help the aircraft to achieve a high angle of attack maneuver. The scopes of the research are the aircraft model used is the F-4 Phantom aircraft, the flight control system is developed using the linear approach, flights are simulated during high angle of attack maneuver, the dynamics of the control devices, including the TVS, are neglected or simplified, the control devices' responses are instantaneous, TVS' are assumed fixed in certain setting angles, not as a control input, although the setting angles are variated, and motion is considered only in longitudinal mode.

2. Modeling of the Aircraft

The aircraft is modeled for four different thrust vectoring system deflections, δ_{TVS} = -5°, 0°, 5°, and 10°. The F-4 Phantom aircraft model is used, as presented on Fig. 1 and Table 1.

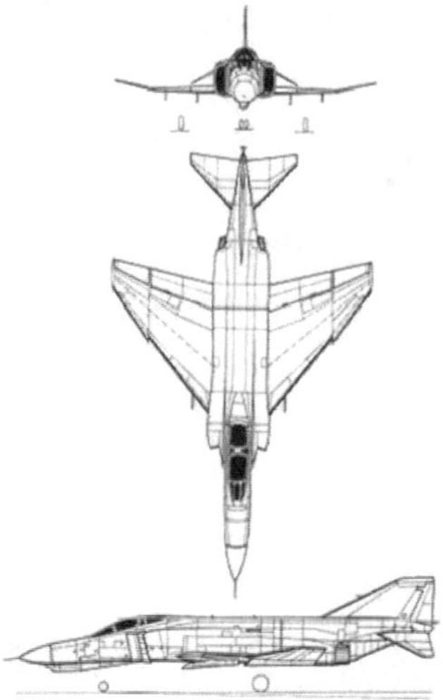

Figure 1. F-4 Phantom Aircraft [1]

Table 1. F-4 Phantom Specifications [1-2]

Parameter	Value	Unit
Mass (m)	17656	kg
Inertia X; Y; Z (I_X; I_Y; I_Z)	33855; 165667; 189543	kg.m^2
Inertia ZX (I_{XZ})	1593	kg.m^2
Length	19.2	m
Height	5	m
Mean Aerodynamic Chord ($\bar{c}$)	4.3	m
Center of Gravity Reference (CG_{ref})	0.298 $\bar{c}$	-
Engine to CG (X_B-axis) Distance	6	m
Engine to CG (Z_B-axis) Distance	0.0005	m
Power Plant: 2 x General Electric J79- GE-17A after-burning turbojet	79.38	kN

The basic equations of motion on the body reference frame, in the time domain, considering only longitudinal motions are presented in equations 1 to 3 [1,3-4]:

$$-W\sin\theta + \bar{q}SC_X + T = m(\dot{u} + qw - rv) \quad (1)$$

$$W\cos\theta\cos\varphi + \bar{q}SC_Z = m(\dot{w} + pv - qu) \quad (2)$$

$$\bar{q}SC_m = I_y\dot{q} + (I_x - I_z)rp + J_{xz}(p^2 - r^2) \quad (3)$$

The thrust component is separated from the aerodynamic coefficient. This component will have the effect of adding TVS.

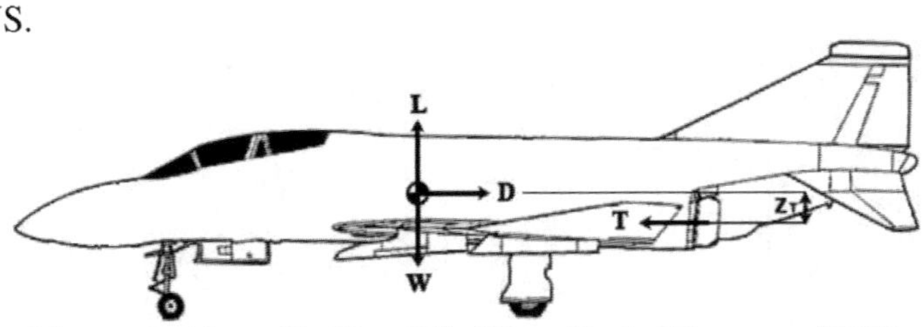

Figure 2. Aircraft's Simplified Free-Body Diagrams (FBD)

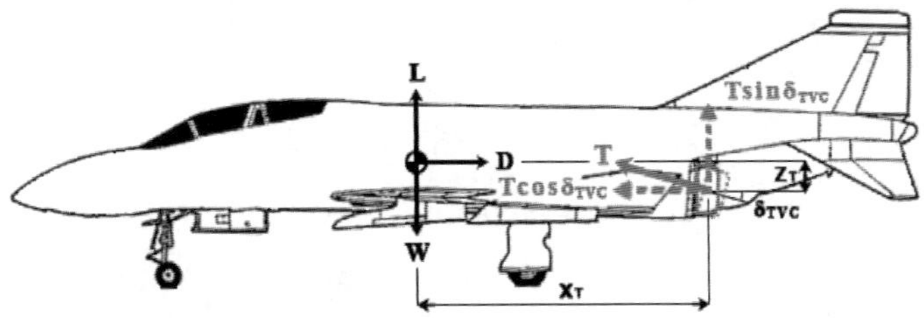

Figure 3. Aircraft's Simplified Free-Body Diagrams (FBD) with TVS

By reviewing Figures 2 and 3, Equations 1, 2, and 3 will be modified due to TVS addition.

$$-W\sin\theta + \bar{q}SC_X + T\cos\delta_{TVC} = m(\dot{u} + qw - rv) \tag{4}$$

$$W\cos\theta\cos\varphi + \bar{q}SC_Z - T\sin\delta_{TVC} = m(\dot{w} + pv - qu) \tag{5}$$

$$\bar{q}SC_m + T\sin\delta_{TVC}.x_T + T\cos\delta_{TVC}.z_T = I_y\dot{q} + (I_x - I_z)rp + J_{xz}(p^2 - r^2) \tag{6}$$

A positive deflection of δ_{TVS} means TVS is deflected downward. This deflection will produce thrust force in Z-axis upward. It will help the tail to move upward and the aircraft to pitch down. The non-dimensional aerodynamic force and moment coefficients already covered flight conditions on the high angle of attack maneuvers.

The $15^o < \alpha \leq 30^o$ coefficients are used in the model. These coefficients represent the trimmed condition at $\alpha = 30^o$. Small changes in α values are covered in the same coefficients. They are computed as follows [1].

$$C_X = 0.141 - 0.0154\alpha + 2.96\,x\,10^{-4}\alpha^2 + 1.82\,x\,10^{-3}\delta_e - 7.3\,x\,10^{-5}\delta_e\alpha + \left(\frac{180q\bar{c}}{\pi 2V}\right)(-0.0602 + 2.04x\,10^{-3}\alpha) \tag{7}$$

$$C_Z = -0.608 - 0.022\alpha - 6.77\,x\,10^{-3}\delta_e + 9.7\,x\,10^{-5}\delta_e\alpha + \left(\frac{180q\bar{c}}{\pi 2V}\right)(1.136 - 0.1418\alpha + 3.1\,x\,10^{-3}\,\alpha^2) \tag{8}$$

$$C_m = 0.0549 - 6.08x10^{-3}\alpha - 8.14x10^{-3}\delta_e + 1.1x10^{-4}\delta_e\alpha + \left(\frac{180q\bar{c}}{\pi 2V}\right)(-0.0951 + 1.4x10^{-3}\alpha) + (x_{c.g.,ref} - x_{c.g.})C_Z \tag{9}$$

The numerical modeling is done with help of MATLAB and Simulink device. In constructing the equations of motion, the 6DOF Quaternion block is used compared to Euler to avoid singularity due to a 90 degree of pitch angle.

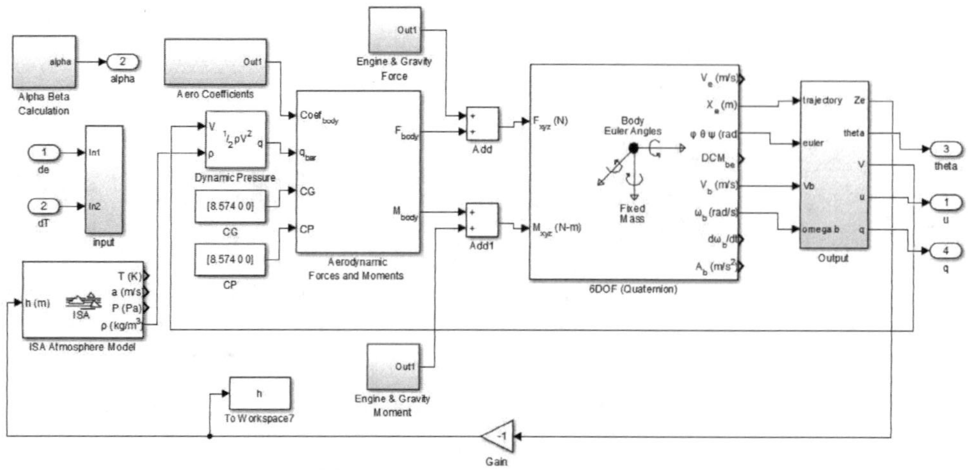

Figure 4. Numerical Model [5]

The parameters used in the Quaternion block are 8000 m altitude with initial velocities u = 55 m/s and w = 31.7543 m/s so that α becomes 30°. Initial Euler angles and body rotation rates are set as zero. The aerodynamic and the control surfaces components are included in the Aerodynamic Forces and Moments block. The gravity and thrust components are independent of the aerodynamic coefficients, also not included in the calculations in the 6DOF Quaternion block, so these components are modeled in separate subsystems.

3. Controller Design

To design a controller, the nonlinear aircraft models are trimmed and linearized with the Linear Analysis Tool by Simulink Control Design. After the trimming process is completed, an operating point has been created. Trim values of inputs and outputs are gained and system characteristics are checked.

The systems with δ_{TVS} = -5° and 0° are considered as stable systems. While the systems with δ_{TVS} = 5° and 10° are otherwise. They need stabilization with the Linear Quadratic Regulator (LQR) control method. Apart from stability issues, the damping ratio of the system have to meet Military Specification: Flying Qualities of Piloted Airplanes by U.S. Military (MIL-F-8785C). [6] The high angle of attack maneuver is classified as a Class IV airplane with Category A. To achieve Level 1 of the short-period response, the damping ratio shall be within the limits of 0.35 to 1.3.

Before the LQR control method is performed, the controllability and observability of the system need to be checked. The systems are considered as a fully controllable and observable system, so the LQR control can be performed. Tables 2 and 3 below show the poles and damping ratio, both for open-loop and closed-loop system (results of the LQR method).

Table 2. LQR Results for δ_{TVS} = -5° and 0°.

δ_{TVS} = -5°		δ_{TVS} = 0°	
Poles (O.L; C.L)	ζ (O.L; C.L)	Poles (O.L; C.L)	ζ (O.L; C.L)
-4.57e-02; -6.51e-02	1; 1	-5.9e-02; -7.95e-02	1; 1
-6.16e-03+1.37e-01i; -9.77e-01	4.49e-02; 1	-3.58e-03+1.2e-01i; -1.33+1.11e+001	2.99e-02; 7.68e-01
-6.16e-03-1.37e-01i; -3.21	4.49e-02; 1	-3.58e-03-1.2e-01i; -1.33-1.11e+001	2.99e-02; 7.68e-01
-1.07; -1.98e+01	1; 1	-1.06; -2.01e+01	1; 1

IOP Conf. Series: Materials Science and Engineering 1173 (2021) 012050 doi:10.1088/1757-899X/1173/1/012050

Table 3. LQR Results for δ_{TVS} = 5° and 10°

$\delta_{TVS} = 5°$		$\delta_{TVS} = 10°$	
Poles (O.L; C.L)	ζ (O.L; C.L)	Poles (O.L; C.L)	ζ (O.L; C.L)
-8.44e-02; -1.06e-01	1; 1	1.82e-02+7.96e-02i; -1.52e-01	-2.22e-01; 1
4.23e-03+9.96e-02i; -1.43+8.72e-01i	-4.25e-02; 8.54e-01	1.82e-02-7.96e-02i; -6.14e-01	-2.22e-01; 1
4.23e-03-9.96e-02i; -1.43-8.72e-01i	-4.25e-02; 8.54e-01	-1.24e-01; -3.98	1; 1
-1.05; -2e+01	1; 1	-1.04; -1.95e+01	1; 1

Equations 10 to 13 show the gain results (K) from the LQR method.

$$K_{\delta TVS\,-5} = \begin{bmatrix} -0.9241 & -0.1246 & -49.4509 & -17.1285 \\ 2.1053 & -0.5551 & 12.0508 & -7.1696 \end{bmatrix} \quad (10)$$

$$K_{\delta TVS\,0} = \begin{bmatrix} 0.0219 & -0.0039 & -59.5057 & -31.9083 \\ 2.2428 & -0.6196 & 37.9755 & -3.2288 \end{bmatrix} \quad (11)$$

$$K_{\delta TVS\,5} = \begin{bmatrix} 2.0782 & -0.3928 & -2.4523 & -39.0860 \\ 0.9072 & -0.4915 & 79.8256 & 22.3168 \end{bmatrix} \quad (12)$$

$$K_{\delta TVS\,10} = \begin{bmatrix} 2.1280 & -0.5901 & 49.9387 & -20.7811 \\ -0.8449 & -0.1263 & 59.5931 & 28.8254 \end{bmatrix} \quad (13)$$

4. Simulation and Analysis

The aircraft model is built with the tracking concept for α and γ through V. Figure 5 shows its Simulink model. The reference values for both α and γ are set. α is related to the elevator deflection input, while γ affects the V that is related to the throttle deflection input.

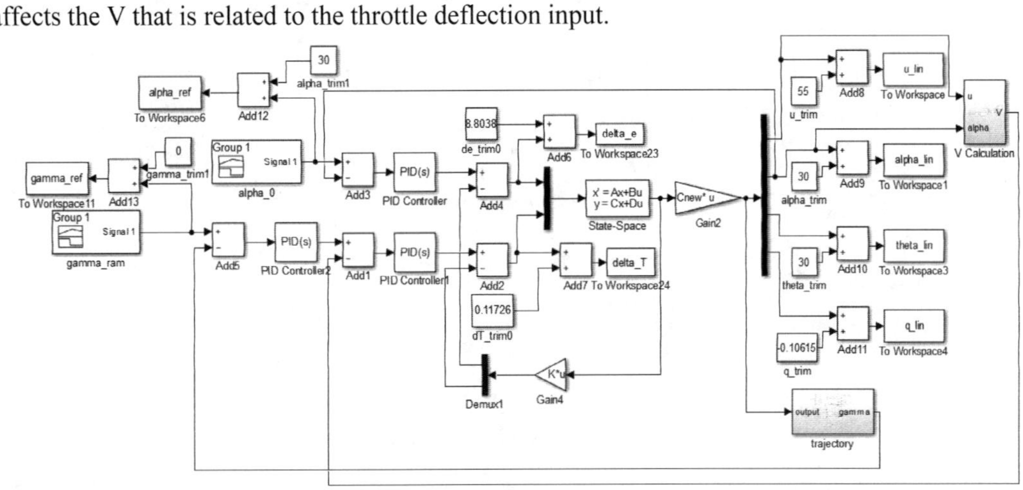

Figure 5. Tracking Model [6]

The state variables are being feed-backed by the gain K from the LQR control method. α, V, and γ, are being feed-backed with PID control. α_{ref} is set as the signal in Figure 6 below, while γ_{ref} is set to a constant value of 0°.

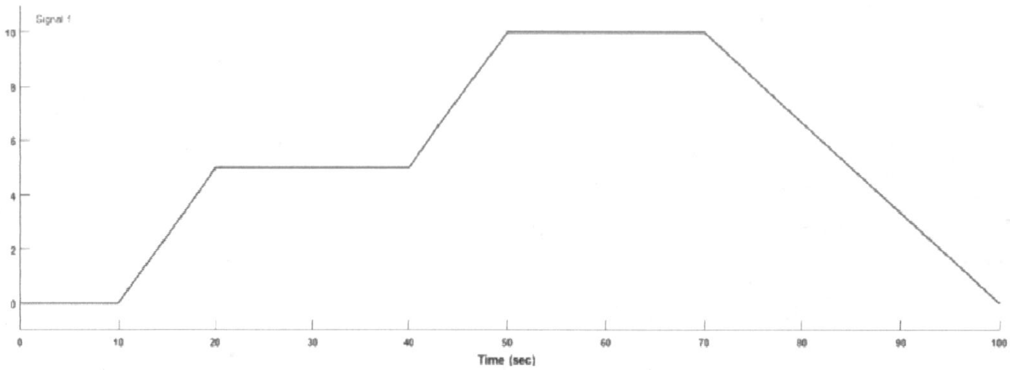

Figure 6. Reference Signal

The PID Controllers are used for α and γ tracking. PID blocks in Simulink are used and automatically tuned for the tracking model. Figure 7 shows the outputs for tracking variables, Figure 8 shows other outputs from the model, while Figure 9 shows the control variables responses.

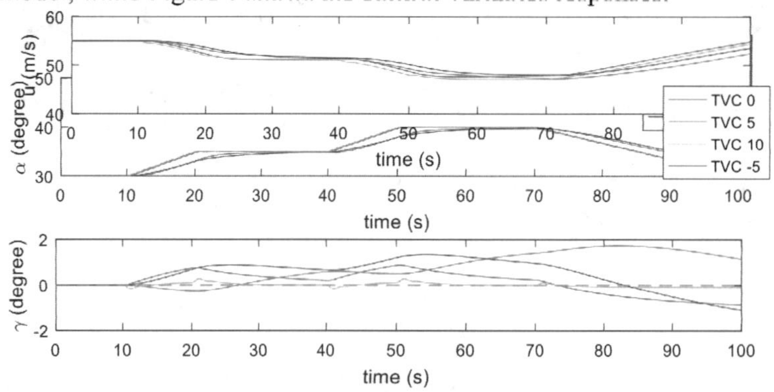

Figure 7. Tracking Outputs

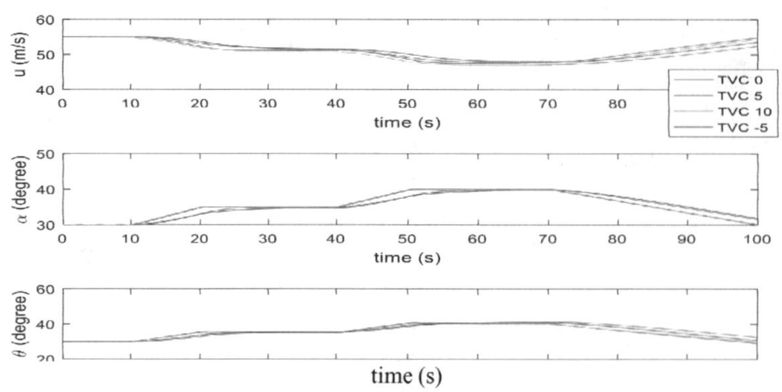

Figure 8. Other Outputs

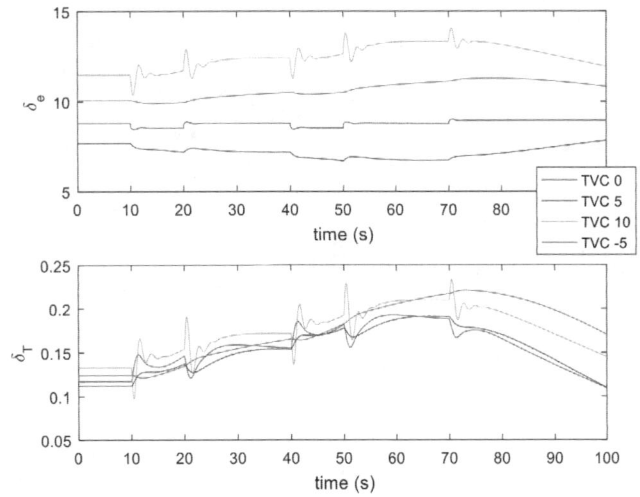

Figure 9. Control Variables.

The α responses follow the reference signal. While the γ responses disturbed because it is difficult to keep the γ to 0° while the α keeps changing. The system needs a slower time for α changes because there is a time delay to adjust the γ. It also can be caused by limited control for thrust. For α, there is an elevator as a control input, that can be deflected upward and downward. For V, there is the thrust that is effective only for increasing the airspeed, but not decreasing it. Other control inputs for V, such as airbrake or spoiler is needed so the γ tracking can perform better.

In this case, as the δ_{TVS} increases, the responses of u and θ quite similar. The responses of q increase. The responses of X_e similar, the aircraft keeps moving forward. The responses of h end in the lower values, but the response of $\delta_{TVS} = 10°$ does not follow the path. It may be caused by the limits of aircraft performance. For the control variables, as the δ_{TVS} increases, the efforts of δ_e are increase. It means that the model with δ_{TVS} = -5° has the lowest effort for δ_e, as it helps the aircraft pitching up. While the responses of δ_T quite similar to greater value for greater δ_{TVS}.

5. Conclusion

The dynamics of fighter aircraft using thrust vectoring system (TVS) as one of its control devices are successfully modeled. The flight control systems, involving the use of TVS, are also successfully designed. The closed-loop behaviors of the aircraft under the TVS implementation are simulated and analyzed. To analyze the effect of using TVS on fighter aircraft, the aircraft is modeled in four variations of δ_{TVS}, which are -5°, 0°, 5°, and 10°. From Tables 2 and 3, it can be seen that the stability controllers are well performed and result in stable responses. From Figures 7 to 9, it can be seen that negative δ_{TVS} helps the aircraft to pitch up, and vice versa. Greater δ_{TVS} led to lower δ_e. It can help the works of δ_e by minimizing elevator efforts. It means that TVS can help the aircraft to do the high angle of attack maneuver. For the better results of the research, the TVS should be dynamically modeled in the aircraft dynamic model so that the results can be more realistic. TVS also can be used as the control input to the aircraft. The deflection can change by time in a certain flight condition.

References

[1] Garza F R and Morelli E A 2003 *A Collection of Nonlinear Aircraft Simulations* (Virginia: National Aeronautics and Space Administration) p 2-6, 39-45

[2] Green W and Swanborough G 2001 *The Great Book of Fighters* (St. Paul: MBI Publishing)

[3] Muhammad H and Jenie Y I 2019 *AE 3220 Flight Dynamics: Lecture Notes* (Bandung: Institut Teknologi Bandung) chapter 3

[4] Mulder J A, Staveren W H J J V, Vaart J C V D, Weerdt E D, Veld A C I and Mooij E 2011 *Flight Dynamics: Lecture Notes* (Delft: TU Delft) chapter 2-3

[5] U.S. Military 1980 Military Specification Flying Qualities of Piloted Airplanes (U.S. Military) p 1-7, 11-16

[6] Mathworks 2020 *Help Center* (Mathworks [Online]. Available: http://www.mathworks.com/help/). [Accessed June 2020].

IOP Conf. Series: Materials Science and Engineering 1173 (2021) 012051 doi:10.1088/1757-899X/1173/1/012051

Dynamic response analysis of aircraft with actuation system failure: open-loop and closed-loop cases

M F D Syamsul, and R A Sasongko*

Department of Aerospace Engineering, Bandung Institute of Technology, Jalan Ganesha
No. 10, Bandung 40132, Indonesia

* Email: sasongko@ae.itb.ac.id

Abstract. On flight control implementation, the control surface actuator system plays one of the key roles that will significantly affect the closed-loop performance of the aircraft. The actuators must provide the required effort for moving the control surface so that the desired control action can be reached. Any failure on the actuator system will limit or degrade the operation of the control surface, so that it will affect the performance of the system in controlling the aircraft, or even worse, may bring the aircraft into an unsafe condition. In this project, the effect of some control surface problems due to actuator failures are modeled, simulated, and analyzed. To reconstruct the situation, the control device failure model is integrated into the flight dynamic model of an aircraft. The numerical model of the equation then is used for simulating the aircraft response under some failure conditions. It can be seen from the simulation results that control surface failure can significantly affect the trim condition of the aircraft, since the failures can change the balance of forces and moments that work on the aircraft. The simulation results also show that for some control surface failures may initiate an asymmetrical response from the aircraft dynamics. On closed-loop simulations, where some type of flight control systems for stabilization and maneuvering tasks are implemented, control surface failures may affect the performance of the control system. The results of the study suggest that any possibilities of control surface failure must be anticipated, by using a robust control scheme or other reconfigurable schemes that can cope with the change in the dynamic characteristics due to the failures.

Keywords: actuation system failure, dynamic response, stability, control, flight control system

1. Introduction

Failure in the actuator will change the dynamic characteristics of the aircraft itself. So it is very necessary to have an input that can re-adjust these dynamic characteristics so that it does not allow LOC (Loss Of Control) to occur. LOCs can endanger passengers on aircraft due to uncontrollable aircraft.

To analyze further, it is necessary to look at the effect of this actuator failure if it is implemented in a flight control system either to see its stabilization or to see the response to achieve the desired variable value. As the initial hypothesis where in case of failure there will be the inaccuracy of the aircraft output issued by the flight control system when compared with the given input. This is due to an error when reading the response of the deflection of the control surface failure. Therefore, there is a deflection of the other control surface to replace the performance of the control surface failure.

For modeling, it is necessary to make an aircraft flight control system design that can match the failure of the actuator. Therefore, to develop the existing AFCS system, response data from the actual condition of the aircraft due to actuator failure are needed which will later be used in the aircraft control system model. Based on the explanation above, this research will focus on designing an aircraft control system in normal conditions and various failure cases. Then, it will be used to analyze the dynamics and control of the aircraft with the hope that in the future the results of this research can help pilots in controlling the aircraft.

The model to be used in this study is the F4 phantom by modeling all control fields independently with a stick-fixed assumption so the actuator and control surface as one integrated system, then forming a non-linear model which is then linearized so that a linear model can be used to perform failure analysis with various desired cases.

2. Modeling of Aircraft

In this research, the F4 phantom is used as an airplane model and the specification contains the properties of the aircraft can be seen in the reference [1, 2] . In addition to the data, the F4 phantom also has several limitations of its control surface deflection which are presented in Table 1 below.

Table 1. Control Surface Deflection Limit [2]

Control surface deflection
$-21^o \leq \delta_e \leq 7^o$
$-30^o \leq \delta_r \leq 30^o$
$-15^o \leq \delta_a \leq 15^o$
$0 \leq \delta_{T_L} \leq 1$

The non-linear mathematical equation of motion described the aircraft motion in the body-fixed reference frame, it consists of 12 equations [3]. In each equation, there are the non-dimensional aerodynamics force and moment coefficients for the F4 model [2] relative to the mean aerodynamics center and the wingspan. In this research, the aerodynamic coefficient applies to fly with a maximum angle of attack of 15^0 [2]. In this research, there will be a failure model also. Failure that occurs in actuators will induce other unwanted movements. The effects of these failures are summarized in table 2 below [1]

Table 2. Actuator Failure Effect [1]

	Control Surface						
Aircraft motion	Left Elevator	Right Elevator	Rudder	Left Aileron	Right Aileron	Left Engine	Right Engine
Surge	V	V		V	V	V	V
Sway			V	V	V		
Heave	V	V		V	V		
Roll	V	V	V	V	V		
Pitch	V	V		V	V	V	V
Yaw	V	V	V	V	V	V	V

Failure cases and their effects are then modeled as follows

$$\delta_e = \frac{1}{2}(\delta_{e_L} + \delta_{e_R}) \tag{1}$$

$$\delta_a = \frac{1}{2}(\delta_{a_L} + \delta_{a_R}) \tag{2}$$

$$T = (T_L + T_R) \tag{3}$$

Due to the above failure modeling, the control surface deflection limit is also affected, shown in Table 3.

Table 3. Control Surface Deflection Limit due to Actuator Failure.

Control surface deflection
$-21^o \leq \delta_{e_L} \leq 7^o$
$-21^o \leq \delta_{e_R} \leq 7^o$
$-30^o \leq \delta_r \leq 30^o$
$-15.5^o \leq \delta_{a_L} \leq 15.5^o$
$-15.5^o \leq \delta_{a_R} \leq 15.5^o$
$0 \leq \delta_{T_L} \leq 1$
$0 \leq \delta_{T_R} \leq 1$

Due to the adjustment of the control surface resulting in adjustments to the equation of motion of the aircraft and its aerodynamic coefficient. Some of them shown as follows

$$\dot{u} = -qw + vr + g\,sin\,\phi + \frac{F_{AX} + T_L + T_R}{m} \tag{4}$$

$$\dot{r} = \frac{\left[qrI_{ZX}(I_Y - I_Z - I_X) + pq\left({I_{ZX}}^2 + I_X^2 - I_XI_Y\right) + \bar{q}Sb(C_lI_{ZX} + C_nI_X)\right] + T_L(Y_{\delta_e}) - T_R(Y_{\delta_e})}{I_ZI_X - {I_{ZX}}^2} \tag{5}$$

$$\begin{aligned} C_X = {} & -0.0434 + 2.39\,x\,10^{-3}\alpha + 2.53\,x\,10^{-5}\beta - 1.07\,x\,10^{-6}\alpha\beta^2 \\ & + 4.75\,x\,10^{-4}\delta_{e_L} + 4.75\,x\,10^{-4}\delta_{e_R} - 4.25\,x\,10^{-7}\delta_{e_L}\beta^2 \\ & - 4.25\,x\,10^{-7}\delta_{e_R}\beta^2 \\ & + \left(\frac{180q\bar{c}}{\pi 2V_t}\right)(8.73\,x\,10^{-3} + 0.001\alpha - 1.75\,x\,10^{-4}\,\alpha^2) \\ & + 1.695\,x\,10^{-5}\,\delta_{a_L} + 1.695\,x\,10^{-5}\delta_{a_R} \end{aligned} \tag{6}$$

$$\begin{aligned} C_Y = {} & -0.012\beta + 1.55\,x\,10^{-3}\delta_r - 8\,x\,10^{-6}\delta_r\alpha \\ & + \left(\frac{180b}{\pi 2V_t}\right)\big(2.25\,x\,10^{-3}p + 0.0117r - 3.67\,x\,10^{-4}r\alpha \\ & + 0.875\,x\,10^{-4}\,r\delta_{e_L} + 0.875\,x\,10^{-4}\,r\delta_{e_R}\big) \end{aligned} \tag{7}$$

$$\begin{aligned} C_Z = {} & -0.131 - 0.0538\alpha - 2.38\,x\,10^{-3}\delta_{e_L} - 2.38\,x\,10^{-3}\delta_{e_R} \\ & - 1.65\,x\,10^{-5}\delta_{e_L}\alpha - 1.65\,x\,10^{-5}\delta_{e_R}\alpha - 3.75\,x\,10^{-5}{\delta_{a_L}}^2 \\ & + 7.5\,x\,10^{-5}\delta_{a_L}\delta_{a_R} - 3.75\,x\,10^{-5}{\delta_{a_R}}^2 \\ & + \left(\frac{180q\bar{c}}{\pi 2V_t}\right)(-0.111 + 5.17\,x\,10^{-3}\alpha - 1.1\,x\,10^{-3}\,\alpha^2) \\ & + 0.735\,x\,10^{-3}\delta_{a_L} + 0.735\,x\,10^{-3}\,\delta_{a_R} \end{aligned} \tag{8}$$

After the analytical model, the author builds a numerical model of the aircraft with failure conditions that are already included. The numerical model of the aircraft is designed using Simulink-Matlab software as shown in Figure 1.

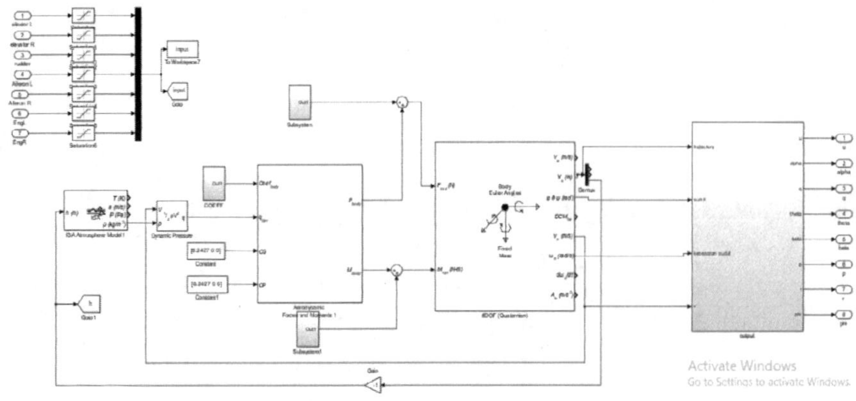

Figure 1. F4 Phantom Numerical Model

3. Controller design

3.1 Trimming and Linearization

In the trimming process [4], it is set to maintain the velocity value on the XB-Axis and the altitude of the aircraft, which means we maintain the aircraft in a cruise condition. Besides that, the deflection limit of the control plane and the angle of attack value is set according to the reference and the cases. Trimming will be carried out in several cases presented in Table 4.

Table 4. Simulation Cases

Failure Case	Failure Description
Case 1	Normal Case, all the actuator/control surface are healthy and can be used normally
Case 2	Elevator Failure means the right elevator of the airplane cannot be deflected
Case 3	Aileron Failure which means the right aileron of the airplane cannot be deflected
Case 4	Elevator-Aileron Failure which means the left elevator and the right aileron cannot be deflected
Case 5	Elevator- Rudder Failure which means the Left Elevator and the Rudder cannot be deflected
Case 6	Aileron-Rudder Failure which means the left aileron and the rudder cannot be deflected

The failure condition referred to above is a jamming failure condition where the control surface is stuck even though it has been given input so that before doing the trimming process the control surface deflection limit is adjusted to a value of 0 on the control surface/actuator that is a failure. trimming is done with the value of operation/holding conditions at an altitude of 1000 m above sea level with a speed of 215 m / s at XB-axis and the maximum angle of attack is 15°. After that, the linearization process is carried out and produces a state-space model, there are 6 state-space models for each case.

In a state-space or linear model that produce by the linearization process. There are some variables such as A, B, C, D,x, and y, A is State matrix with size 8x8, B is input matrix with size 8x7, C is output matrix with size 8x8, and D is zero matrix. While x is a state vector with size 8x1 consist of u, w, θ, q, v, p, r, and φ. Y is output vector with size 8x1 consist of u, α, θ, q, β, p, r, and φ. And the last one u is the input vector with size 7x1 consist of control surface deflection that symbolized as in table 3. After that, an open-loop analysis is carried out by providing input impulses to see the stability of the aircraft. When giving input still sees the deflection limit of the linear model control surface in table 3 with an adjustment for each case to value 0 as mention before. The location of poles for each of the linear models, all of them located in the stable area. so we just need to improve the stability of the system by design a stability augmentation system.

3.2. Closed-Loop Model

To perform closed-loop analysis, gain control is required. Determination of control gain using the Linear Quadratic Regulator (LQR) method [5]. LQR can be determined by solve optimation problem as follow

$$J = \int_0^\infty \bar{x}^T (Q + K^T R K)\bar{x}\, dt \tag{9}$$

where x is the state variable and u input variable, the input signal that fulfills the optimization problem (9)) can be defined as follows

$$u(t) == -Kx(t) = -R^{-1}B^T P x(t) \tag{10}$$

where P is a definite positive matrix solution of the Riccati equation defined as follow

$$\mathrm{A}^{\mathrm{T}}P + PA + Q - PBR^{-1}B^T P = 0 \tag{11}$$

There are 7 gain controls obtained from each state space. The gain control obtained is used for closed-loop which will then be carried out the simulation process.

4. Results and Discussions

4.1 Closed-Loop Stability Analysis and Failure Cases

The closed-loop model is made using a linear model and the gain control has been get, the gain control will process the feedback signal [5] in the form of the open-loop system state variable value. the closed-loop model will follow Figure 2 below

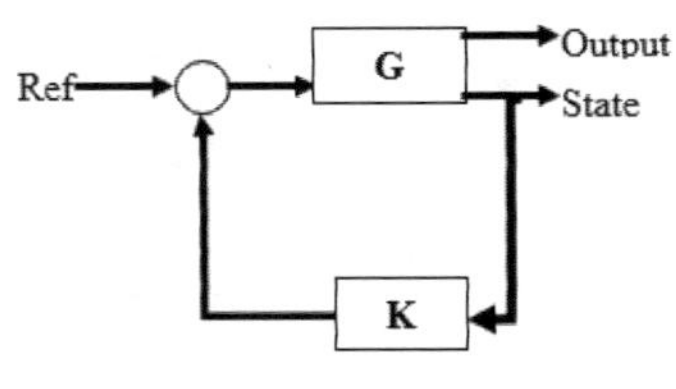

Figure 2. Closed-Loop Model

By using the model in Figure 2, several simulations were carried out with a linear model (state space) and a different controller except for G_Norm (State Space Model in normal cases) the controller only from K_Norm (The gain is obtained by calculating using the state variable on G_Norm), but for the other, the closed-loop model will follow G_Failure(State Space Model in each failure cases, i.e elevator failure = G_Elev)+K_Norm, and G_Failure+K_Failure(The gain is obtained by calculating using the state variable on G_Failure, i.e. K_Elev from G_Elev).

There will be 15 simulations consist of 5 simulations of G_Norm+K_Norm with input for each control surface deflection and 10 simulations for each failure case (for every G_Elev simulation only get elevator deflection, for every G_Ail simulation will get aileron deflection), the input given is doublet signal. The result for each simulation with the same input will be compared, for example, G_Norm+K_Norm (elevator input) will be compared with G_Elev+K_Norm. after that G_Elev+K_Norm will be compared with G_Elev+K_Elev, the comparison will follow for case 2 until 6 in table 4.

IOP Conf. Series: Materials Science and Engineering 1173 (2021) 012051 doi:10.1088/1757-899X/1173/1/012051

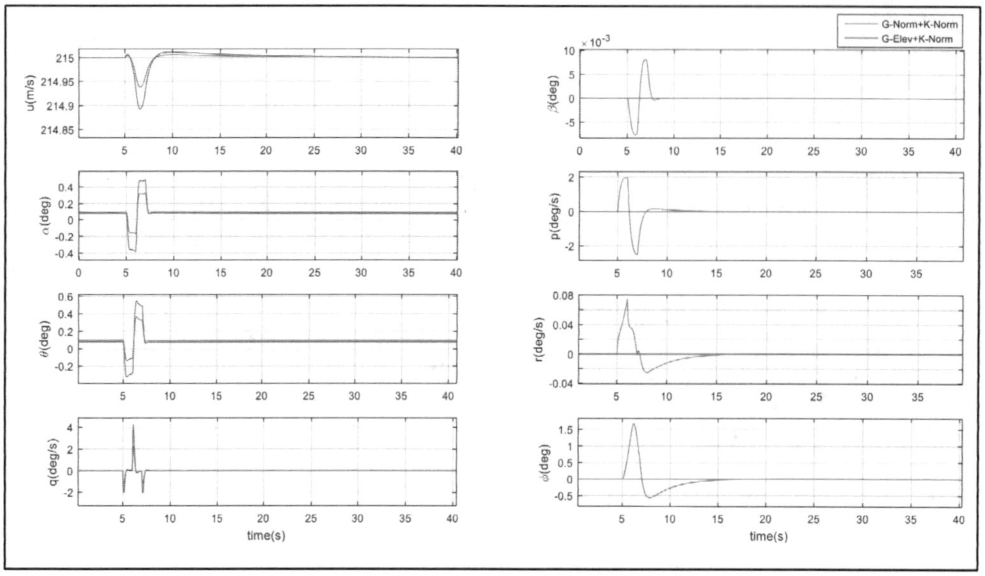

Figure 3. Output Comparison of simulation G_Norm+K_Norm with G_Elev+K_Norm

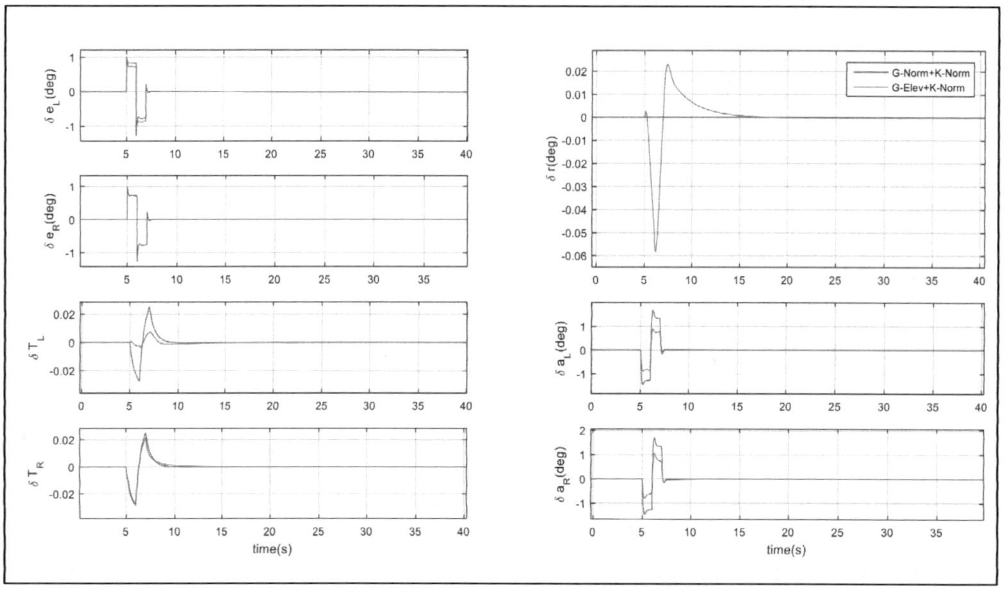

Figure 4. Actuator Respond Comparison of simulation G_Norm+K_Norm with G_Elev+K_Norm

From the simulation results, it can be seen that with normal gain control when an elevator failure occurs, the actuator cannot produce the same energy so that the control surface in this case the right elevator does not give any response due to the change in deflection. The output in the flight motion plane that corresponds to the control surface failure, in this case, the longitudinal flight motion plane being reduced from the normal response, so that the control surface that responds due to the feedback, in this case, the thrust produces a different response from the normal.

Then due to the failure, it causes an asymmetric condition so that the coupling effect occurs which makes the response to other flight motion planes, in this case, a lateral-directional flight motion plane.

Also because of the coupling condition, to keep the aircraft in a trim condition, the control surface that not compatible with the longitudinal flight motion plane, in this case, the rudder will be deflected.

This condition also applies to compare simulations G_Norm+K_Norm with G_Ail+K_Norm or G_EA+K_Norm or G_ER+K_Norm or G_AR+K_Norm. while when compared between simulations for each failure linear model with different controllers for example G_Elev+K_Norm with G_Elev+K_Elev, it is found that using a different controller in the linear model the same failure will not affect the output obtained.

4.2 Flight-Path Angle Hold and speed hold

To create a flight-path angle hold model and speed hold. The closed-loop model in figure 2 will be an inner closed loop and there will be 2 outer loops, so there will be 2 feedbacks, different from figure 2 this time the feedback comes from the output. The feedback value is the aircraft velocity obtained by transforming the variable u (Velocity at XB-axis) by using equation (12), and the second is the flight path angle value obtained through the transformation of the variable values for pitch angle (θ) and angle of attack (α) by using equation (13).

Then the feedback variable is compared with the reference value. After that, the result of the comparison of the two values will be transformed using the PID controller into an input signal to the actuator so that there are 2 PID controllers for speed hold and flight-path angle hold. The flight path angle hold and speed hold model is shown in figure 5.

$$V = \frac{u}{\cos\alpha} \tag{12}$$

$$\gamma = \theta - \alpha \tag{13}$$

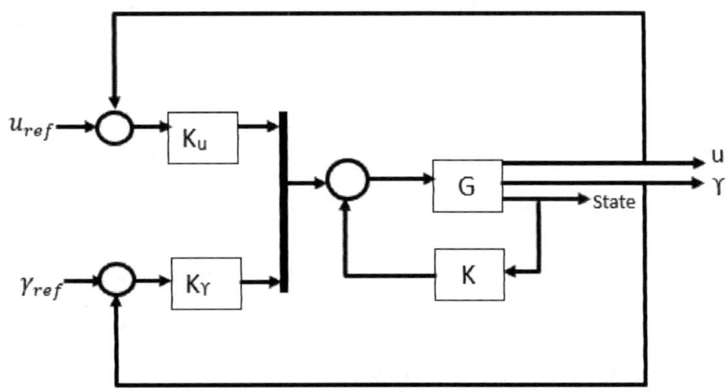

Figure 5. Closed-Loop Model for flight path angle hold and speed hold

As mentioned before there will be 2 PID controllers, but because the PID for speed hold and flight-path hold is in the same closed-loop model design so that it is assumed to be 1 unit so that the names become 1 unit, and are used simultaneously. naming for the PID controller follows each case. After that, the simulation is carried out as in table 4, each linear model is paired with its PID for example, G_elev + K_Norm + PID_elev will be simulated, then G_elev + K_elev + PID_Elev and so on.except for G_Norm+K_Norm will only use PID_Norm

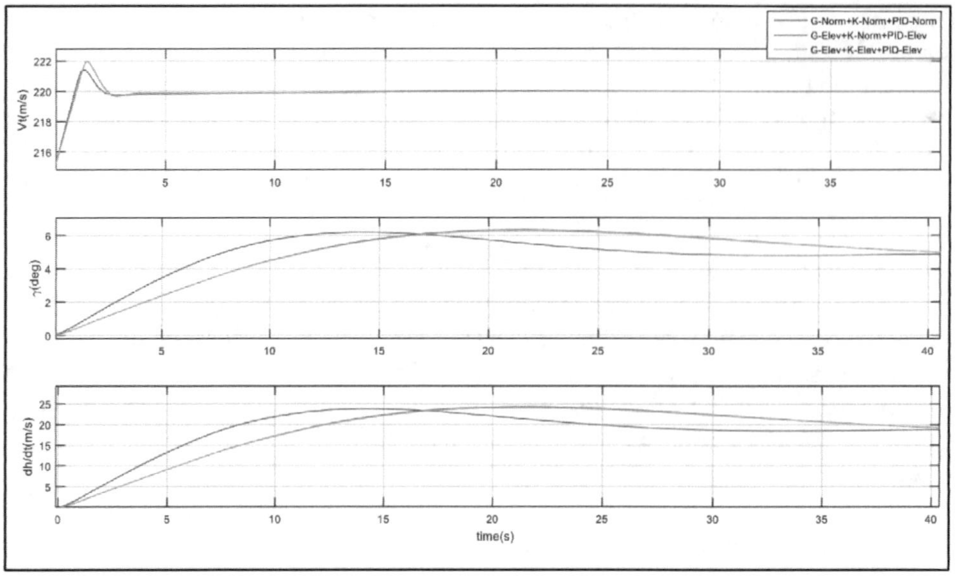

Figure 6. Tracking Output Comparation of Simulation G_Norm+K_Norm+PID_Norm with G_Elev+K_Norm+PID_Elev and G_Elev+K_Elev+PID_Elev

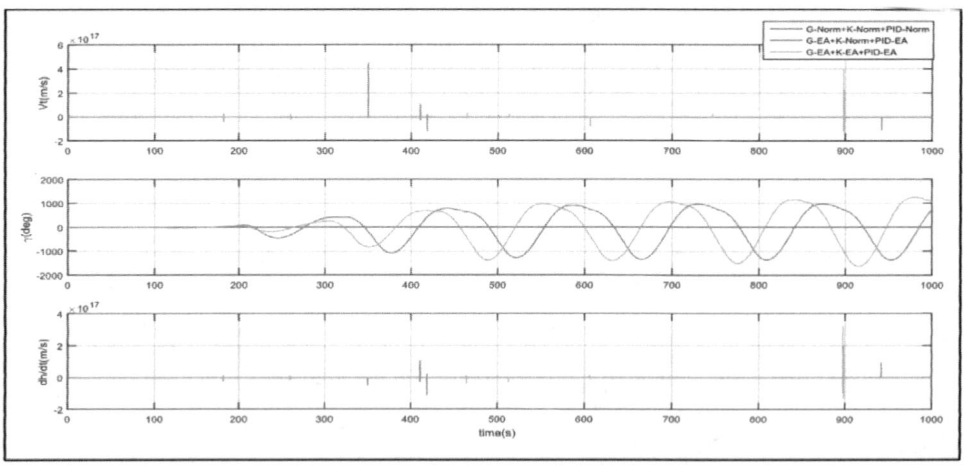

Figure 7. Tracking Output Comparation of Simulation G_Norm+K_Norm+PID_Norm with G_EA+K_Norm+PID_EA and G_EA+K_EA+PID_EA

Based on the result obtained, it can be seen that the normal model has the best performance so that a faster response is obtained to reach the desired reference value, according to the analysis in the previous section when there is a dynamic response failure there will be a decrease in performance and due to the failure of the control response, it produces an asymmetric effect that makes the response appear in the lateral-directional flight motion plane while the flight control system is designed for dynamics in a longitudinal flight motion plane.

According to the results obtained, it can be seen in the elevator-Aileron failure model the flight control system cannot be applied which has been designed using either normal gain or failure gain for the inner loop. This condition because of the PID controller that uses in the flight-path angle hold and speed hold mode. The PID controller is SISO while the model is MIMO, although the PID controller for each flight-path angle hold and speed hold but each PID does not work as a unit so it does not care about other PID conditions which result in the desired response being difficult to obtain because each PID "works" to achieve the desired value.

5. Conclusions

Actuator failure that results in jamming conditions or the control surface is unable to deflect can be modeled using a normal controller or a controller from a model with failure condition included. In the failure condition, the aircraft can reach its trim condition even though there are limitations, therefore the linearization process can be calculated and the linear model can be obtained. By using the linear model and the controller that the aircraft has obtained, it can still be stabilized even though it creates unwanted dynamics (coupling effect) due to actuator failure. Besides, the aircraft can still be controlled to obtain the desired value, but there are 2 conditions when the failure of the elevator-aileron and aileron-rudder of the aircraft cannot be controlled. For future work, dynamic modeling of the actuator can be carried out (stick-free condition) so that the results are more accurate, besides that, analysis can also be carried out on other types of failure such as changes in the dynamic rate of control surface deflection, after that in the future work, automatic control modeling can be carried out which detects damage so that it can be anticipated to produce dynamic as desired and for the flight-path angle hold and speed hold can use another type of controller.

References

[1] A. A. Arif, A. R. Sasongko and O. Arifianto, "Undergraduate Thesis: Development Of Simulation Platform Anf Stability Analysis Of Aircraft With System Failures," Faculty of Mechanical And Aerospace Engineering, Department of Aerospace Engineering, Institut Teknologi Bandung, Bandung, Indonesia, 2018.

[2] F. R. Garza and E. A. Morelli, A Collection of Nonlinear Aircraft Simulations, Virginia: National Aeronautics and Space Administration, 2003.

[3] J. A. Mulder, W. H. J. J. v. Staveren and J. C. V. der Vaart, Flight Dynamics: Lecture Notes, TU Delft, 2011.

[4] "MathWorks: Trimming and Linearizing Simulink Models," [Online]. Available: https://www.mathworks.com/help/slcontrol/gs/trimming-and-linearizing-simulink-models.html. [Accessed 13 August 2020].

[5] K. Ogata, Modern Control Engineering - Fifth Edition., New Jersey: Prentice-Hall, 2010.

IOP Conf. Series: Materials Science and Engineering 1173 (2021) 012052 doi:10.1088/1757-899X/1173/1/012052

Distributed robust adaptive output-feedback synchronization

M R Rosa[*], **and E Susanto**

School of Electrical Engineering Telkom University Bandung, Indonesia

* E-mail: mridhorosa@telkomuniversity.ac.id

Abstract. This paper addresses the output synchronization of heterogeneous uncertain linear systems in the presence of input disturbance. The synchronization problem is solved by introducing the distributed model reference adaptive control. The distinguishing features of the proposed approach are the capability to synchronize the entire network output by using only the control input and the output of neighbouring systems. It also provides synchronization in the presence of bounded control input disturbance. The control input disturbance is handled by extending the proposed method with the sigma-modification. By using the distributed matching condition assumptions, the coupling gains between the agents can be updated adaptively. The distributed robust adaptive controller is first analysed by using the distributed Lyapunov function. Its effectiveness is validated in the simulation of the cart inverted pendulum systems stabilization problem.

Keywords: linear system, synchronization, control input, robust adaptive, Lyapunov system

1. Introduction

In the presence of model dynamics uncertainties, a simple controller cannot stabilize the system. Here, adaptive control can be the solution to control the system with large uncertainties. The stabilization problem arises in many applications such as the inverted pendulum system. The stabilization problem of inverted pendulum systems such as wheeled inverted pendulum [1], double inverted pendulum [2], flying inverted pendulum [3] has been discussed extensively. These stabilization problems represent a potential solution for many applications such as Automated Guided Vehicle (AGV) with heavy payload [4], lower limb exoskeleton for paraplegics [5], stabilization of offshore platform [6], stabilization of unmanned aerial vehicle [7] and a Segway [8].

The systems with adaptive control may lead to an unstable system in the presence of disturbances. A lot of effort has been proposed to handle the disturbance problem, which leads to a research work called robust adaptive control [9]. The stabilization problem becomes more interesting if the systems are distributed [10]. This distributed stabilization problem requires synchronization to achieve its objective. The main objective of the synchronization is to reduce the states' or the output's error of the entire agent to zero by the time goes to infinite. The synchronization can be divided into the centralized scheme and distributed scheme. In a centralized scheme, there is a central node that receives all the information of each agent then uses it to control the agents [11–13]. In the distributed scheme, each agent only using the local information, i.e. neighbour information's control input or states/output [14]. The distributed synchronization of linear heterogeneous uncertain agents without any sliding mode has been proposed in [15].

In this work, the distributed adaptive reference model control is proposed to converge each uncertain heterogeneous agent's dynamics to the reference dynamics. This approach only requires the control input and the output of the neighboring agents. Note that the proposed approach does not require any global information and we are assuming a Directed Acyclic Graph (DAG). The bounded input disturbance can be handled by introducing the σ-modification to the proposed control law. The proposed distributed output synchronization with σ-modification shows the synchronization error converges asymptotically to zero. Here, the Lyapunov-based approach is used to analyze the stability of the proposed method.

The simulation of inverted pendulum systems demonstrated the effectiveness of the proposed approach. This manuscript is organized as follows: problem formulation and the system's dynamics, proposed method for the synchronization of the uncertain heterogeneous agent under bounded disturbance, presents the numerical simulation of the test case: inverted pendulum systems, then concluded and propose directions for further research.

2. Problem formulations

In order to support the proposed method, we propose the communication graph shown in Figure 1 that describes the communication flow in a network. Here, Agent 0, the reference model, does not receive any signal from other agents. Agent 1, the leader, synchronizes its output to the reference model output. Simultaneously, agent 2 and agent 3, synchronizes its output to the predecessor agents, agent 1 and then agent 2.

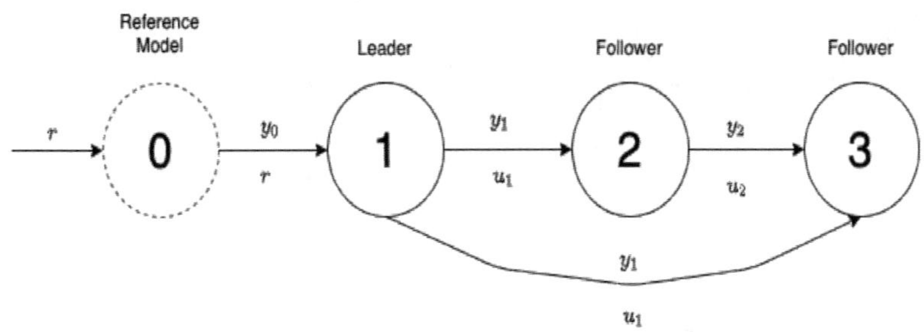

Figure 1. DAG of multi-agent system

From the fig. 1, agent 2 and agent 3 synchronize its output to the reference model indirectly by using local information, the output, and the neighboring agent's input. Note that the communication graph does not contain any cyclic communication, Directed Acyclic Graph (DAG). In cyclic networks, the distributed model reference adaptive control with parameter projection can achieve the synchronization [16]. Then, let us define the agent's dynamics in the network. The dynamics of the reference model satisfying the following dynamics

$$y_0 = G_0(s) = k_0 \frac{N_0}{D_0} r \quad (1)$$

where N_0 and D_0 known monic polynomials, $y_0 \in \mathrm{R}$ is the output of the reference model, $r \in \mathrm{R}$ is the reference input, and k_0 is the high-frequency gain. Then, let us define the dynamics of a leader, and the followers denoted with the subscript 1, 2, and 3 satisfying the following dynamics

$$y_1 = G_1(s) = k_1 \frac{N_1}{D_1} u_1, \quad y_2 = G_2(s) = k_2 \frac{N_2}{D_2} u_2 \quad , \\ y_3 = G_3(s) = k_3 \frac{N_3}{D_3} u_3 \tag{2}$$

where $u_1,u_2,u_3 \in$ R are the control inputs and $y_1,y_2,y_3 \in$ R are the outputs, $N_1,N_2,N_3,D_1,D_2,$ and D_3 are unknown monic polynomials, and k_1,k_2,k_3 are the high frequency gains. In this work, the N_1,N_2,N_3 and D_1,D_2,D_3 may have different dynamics. To achieve of our main objective, we propose the assumptions for the agents and the reference model as follows:

Reference model Assumptions:
(i) The Agent 0, the reference model, transfer function, G_0, is monic Hurwitz polynomials.
(ii) The relative degree of the reference model, G_0, is the same as the agents' relative degree, G_i.
Agents Assumptions:
(i) The Agent does not have any positive zero.
(ii) The degree of the denumerator D_i is known.
(iii) The sign of the high frequency gains k_i, $i \in 1,2,3$ is known.

3. Distributed Robust adaptive output-feedback synchronization

3.1. Leader - reference synchronization

This section will discuss the preliminary result of the distributed adaptive controller in line with [17]. It is known that we can synchronize the leader's output to the reference's output by using a well-known model reference adaptive control in Chapter 5.3 in [18] which leads to the controller

$$u_1 = f_1' \frac{\beta}{\Lambda(s)} u_1 + g_1' \frac{\beta}{\Lambda(s)} y_1 + h_1 y_1 + l_1 r \tag{3}$$

where f_1', g_1', h_1, and l_1 are the estimates for the ideal gain f_1^*, $g_1^{*\prime}$, h_1^*, and l_1^*, respectively, $\Lambda(s)$ is a Hurwitz monic polynomial which is defined as

$$\Lambda(s) = s^{n-1} + \lambda_{n-2} s^{n-2} + \ldots + \lambda_1 s + \lambda_0 \tag{4}$$

and β is defined as

$$\beta(s) = \begin{bmatrix} s^{n-2} & s^{n-3} & \ldots & s & 1 \end{bmatrix} \quad \text{for } n \geq 2 \\ \beta(s) = 0 \quad \text{for } n = 1. \tag{5}$$

The adaptive law for the leader is defined as follows:

$$\dot{\theta}_1 = -\Gamma_1 e_{10} \omega_1 sgn(\frac{k_0}{k_1}) \tag{6}$$

where

$$\theta_1' = \begin{bmatrix} f_1' & g_1' & h_1 & l_1 \end{bmatrix} \\ \omega_1 = \begin{bmatrix} \omega_{u1} & \omega_{y1} & y_1 & r \end{bmatrix}' \tag{7}$$

$\Gamma_1 = \Gamma_1' > 0$ is a positive diagonal matrix, $e_{10} = y_1 - y_0$. Then, let us defined $\dot{\omega}_{u1}$, $\dot{\omega}_{y1}$, F, and d as follows

$$\dot{\omega}_{u1} = F\omega_{u1} + Ru_1 \quad F = \begin{bmatrix} -\lambda_{n-2} & \cdots & -\lambda_0 \\ I_{n-2} & & 0 \end{bmatrix} \\ R = \begin{bmatrix} 1 & 0 & \ldots & 0 \end{bmatrix}' \\ \dot{\omega}_{y1} = F\omega_{y1} + Ry_1 \tag{8}$$

By using the control law u_1, the convergence of $e_1 \to 0$ for $t \to \infty$ can be achieved. The Lyapunov approach is derived to prove the asymptotic convergence of synchronization error analytically. First, let us rewrite the state–space of the Agent 1 (the leader) in the closed-loop form by adding and subtracting it by the ideal control law:

$$\dot{\bar{x}}_1 = \bar{A}_1\bar{x}_1 + \bar{B}_1 l_1^* r + \bar{B}_1(u_1 - \theta_1^{*\prime}\omega_1) \qquad y_1 = \bar{C}_1 \bar{x}_1 \tag{9}$$

where $\bar{x}_1 = \begin{bmatrix} x_1' & \omega_{u1}' & \omega_{y1}' \end{bmatrix}^T$. Then $\bar{A}_1$, $\bar{B}_1$, and $\bar{C}_1$ are defined as

$$\bar{A}_1 = \begin{bmatrix} A_1 + B_1 h_1^* c_1' & B_1 f_1^{*\prime} & B_1 g_1^{*\prime} \\ R h_1^* c_1' & F + R f_1^{*\prime} & R g_1^{*\prime} \\ R c_1' & 0 & F \end{bmatrix}$$
$$\bar{B}_1 = \begin{bmatrix} B_1 \\ R \\ 0 \end{bmatrix}$$
$$\bar{C}_1' = \begin{bmatrix} c_1' & 0 & 0 \end{bmatrix}. \tag{10}$$

The transfer function of Agent 1 can be matched to the transfer function of Agent 0 which is defined as follows $\bar{C}_1(sI - \bar{A}_1)^{-1}\bar{B}_1 c_1^* = C_0(sI - A_0)^{-1}B_0$. Thus, the state–space of Agent 1 can be written as follows

$$\dot{\bar{x}}_1 = A_0\bar{x}_1 + B_0 r + B_0\gamma_1^*(u_1 - \theta_1^{*\prime}\omega_1) \qquad y_1 = C_0\bar{x}_1 \tag{11}$$

where $\gamma_1^* = \frac{1}{l^*} = \frac{k_0}{k_1}$. Then, we define the state error and the output error between the reference and the leader $\tilde{x}_{10} = \bar{x}_1 - x_0$ and $e_{10} = y_1 - y_0$, respectively. The error equation is defined as

$$\dot{\tilde{x}}_{10} = A_0\tilde{x}_{10} + B_0\gamma_1^*\tilde{\theta}_1'\omega_1 \qquad e_{10} = C_0\tilde{x}_{10} \tag{12}$$

where the difference between the estimate and ideal parameters is defined as $\tilde{\theta}_1 = \theta_1 - \theta_1^*$.

The following Lyapunov function is used to proof the asymptotic convergence of the synchronization error between the leader's (Agent 1) output and the model reference's (Agent 0) output analytically

$$V_1(\tilde{\theta}_1, \tilde{x}_{10}) = \frac{\tilde{\theta}_1'\Gamma_1^{-1}\tilde{\theta}_1}{2}|\gamma_1^*| + \frac{\tilde{x}_{10}'P\tilde{x}_{10}}{2} \tag{13}$$

where $P = P^0 > 0$ such that

$$PA_0 + A_0'P = -qq' - wH \qquad PB_0 = C_0 \tag{14}$$

where $H = H^0 > 0$, and $w > 0$. Then, the time derivative of V_1 can be defined as follows

$$\dot{V}_1 = -\frac{\tilde{x}_{10}'qq'\tilde{x}_{10}}{2} - \frac{w}{2}\tilde{x}_{10}'H\tilde{x}_{10} + PB_0\tilde{x}_{10}\gamma_1^*\tilde{\theta}_1'\omega_1$$
$$+ \tilde{\theta}_1'\Gamma_1^{-1}\dot{\tilde{\theta}}_1|\gamma_1^*|. \tag{15}$$

Since $PB_0\tilde{x}_{10} = C_0\tilde{x}_{10} = e_{10}$ and $\gamma_1^* = |\gamma_1^*|sgn(\gamma_1^*)$, we can define the $\dot{\tilde{\theta}}_1$ such that

$$\dot{\tilde{\theta}}_1 = \dot{\theta}_1 = -\Gamma_1 e_{10}\omega_1 sgn(\gamma_1^*) \tag{16}$$

thus we have

$$\dot{V}_1 = -\frac{\tilde{x}'_{10}qq'\tilde{x}_{10}}{2} - \frac{w}{2}\tilde{x}'_{10}H\tilde{x}_{10}. \tag{17}$$

From (17), we obtain that V_1 has a finite limit, so $\tilde{x}_{10}, \tilde{\theta}_1 \in L_\infty$. We have $\tilde{x}_{10} = \bar{x}_1 - \bar{x}_0 \in L_\infty$ and it is known that $\bar{x}_0 \in L_\infty$, therefore we have $\bar{x}_1 \in L_\infty$. This implies $x_1, y_1, \omega_1, \omega_2 \in L_\infty$. Then, we have $u_1 \in L_\infty$. It can be seen that all signals are bounded and leads to $\dot{V}_1 \to 0$ as $t \to \infty$. Thus we have $e_{10} \to 0$ for $t \to \infty$, which concludes the proof.

3.2. Follower-leader robust synchronization

This section's primary focus is to discover the control law u_2 of the Agent 2 (the follower) that synchronizes its dynamics to the Agent 1 (the leader) under input disturbance. First, let us propose the control law u_2 to match the follower dynamics to the leader dynamics in the absence of input disturbance

$$\begin{aligned} u_2 = {} & f'_{21}\frac{\beta}{\Lambda(s)}u_1 + g'_{21}\frac{\beta}{\Lambda(s)}y_1 + h_{21}y_1 + l_{21}u_1 \\ & + f'_2\frac{\beta}{\Lambda(s)}(u_2 - u_1) + g'_2\frac{\beta}{\Lambda(s)}(y_2 - y_1) \\ & + h_2(y_2 - y_1) \end{aligned} \tag{18}$$

where the parameter of the control law f'_{21}, f'_2, g'_{21}, g'_2, h_{21}, h_1, and l_{21} are the estimates for f'_{21}, f'_2, g'_{21}, g'_2, h^*_{21}, h^*_1, and l^*_{21}, respectively. Then let us proposed the following adaptive law

$$\dot{\theta}_2 = -\Gamma_2 e_{21}\omega_2 sgn(\frac{k_1}{k_2}) \tag{19}$$

where

$$\begin{aligned} \theta'_2 &= \begin{bmatrix} f'_{21} & g'_{21} & h_{21} & l_{21} & f'_2 & g'_2 & h_2 \end{bmatrix} \\ \omega_2 &= \begin{bmatrix} \omega'_{u_1} & \omega'_{y_1} & y_1 & u_1 & \omega'_{u_{21}} & \omega'_{y_{21}} & y_2 - y_1 \end{bmatrix}' \end{aligned} \tag{20}$$

$\Gamma_2 = \Gamma'_2 > 0$ is a positive diagonal matrix, and $e_{21} = y_2 - y_1$. Then, let us defined $\dot{\omega}_{u21}$ and $\dot{\omega}_{y21}$ as follows

$$\begin{aligned} \dot{\omega}_{u21} &= F\omega_{u21} + R(u_2 - u_1) \\ \dot{\omega}_{y21} &= F\omega_{y21} + R(y_2 - y_1) \end{aligned} \tag{21}$$

By using the control law, u_2, we can synchronize the follower's output to the leader's output without input disturbance. In the presence of input disturbance, some bounded error will occur in the control law (18). This input disturbance leads to the modification of follower control law u_2 as follows

$$\begin{aligned} u_2 = {} & f'_{21}\frac{\beta}{\Lambda(s)}u_1 + g'_{21}\frac{\beta}{\Lambda(s)}y_1 + h_{21}y_1 + l_{21}u_1 \\ & + f'_2\frac{\beta}{\Lambda(s)}(u_2 - u_1) + g'_2\frac{\beta}{\Lambda(s)}(y_2 - y_1) \\ & + h_2(y_2 - y_1) + d_{21} \end{aligned} \tag{22}$$

where d_{21} is the bounded control input error which satisfies $|d_{21}| < \bar{d}$ for some unknown $\bar{d}$. So that we proposed the adaptive law with leakage or σ-modification as follows

$$\dot{\theta}_2 = -\Gamma_2 e_{21}\omega_2 sgn(\frac{k_1}{k_2}) - \sigma\Gamma\theta_2 sgn(\frac{k_1}{k_2}) \tag{23}$$

where $\sigma > 0$ is a scalar to be designed by the user that allows robustness concerning input disturbance. In order to prove the asymptotic convergence of the synchronization error analytically in the presence of input disturbance, the Lyapunov approach is derived. First, we define the follower, Agent 2, dynamics by adding and subtracting it by the ideal control law

$$\dot{\bar{x}}_2 = \bar{A}_2\bar{x}_2 + \bar{B}_2 l^*_{21}\bar{u}_2 + \bar{B}_2(u_2 - \theta^{*\prime}_2\omega_2) \quad y_2 = \bar{C}_2\bar{x}_2 \tag{24}$$

where $\bar{x}_2 = \begin{bmatrix} x'_2 & \omega'_{u_1} & \omega'_{y_1} & \omega'_{u_{21}} & \omega'_{y_{21}} \end{bmatrix}'$ and $\bar{u}_2 = \begin{bmatrix} u_1 & y_1 \end{bmatrix}'$.

$$\bar{A}_2 = \begin{bmatrix} A_2 + B_2 h^*_2 c'_2 & B_2 f^{*\prime}_{21} & B_2 g^{*\prime}_{21} & B_2 f^{*\prime}_2 & B_2 g^{*\prime}_2 \\ 0 & F & 0 & 0 & 0 \\ Rc'_2 & 0 & F & 0 & 0 \\ Rh^*_2 c'_2 & Rf^*_{21} & Rg^*_{21} & F + Rf^*_2 & Rg^*_2 \\ Rc'_2 & 0 & 0 & 0 & F \end{bmatrix},$$

$$\bar{B}_2 = \begin{bmatrix} B_2 & \frac{B_2(h^*_{21} - h^*_2)}{l^*_{21}} \\ \frac{R}{l^*_{21}} & 0 \\ 0 & \frac{R}{l^*_{21}} \\ R(1 - \frac{1}{l^*_{21}}) & \frac{R(h^*_{21} - h^*_2)}{l^*_{21}} \\ 0 & -\frac{R}{l^*_{21}} \end{bmatrix} \quad \bar{C}_2 = \begin{bmatrix} c^T_2 & 0 & 0 & 0 & 0 \end{bmatrix}. \tag{25}$$

From Equation (25), the transfer function of Agent 2 can be matched to the transfer function of Agent 1 which is defined as follows $\bar{C}_2(sI - \bar{A}_2)^{-1}\bar{B}_2 c^*_{21} = \bar{C}_1(sI - \bar{A}_1)^{-1}\bar{B}_1 c^*_1$. Thus, the state–space of Agent 2 can be written as follows:

$$\dot{\bar{x}}_2 = A_0\bar{x}_2 + B_0 r + B_0\gamma^*_2(u_2 - \theta^{*\prime}_2\omega_2) \qquad y_2 = C_0\bar{x}_2 \tag{26}$$

where $\gamma^*_2 = \frac{1}{c^*_{21}} = \frac{k_1}{k_2}$.

Then, we define the state error and the output error between the follower and the leader $\tilde{x}_{21} = \bar{x}_2 - \bar{x}_1$, and $e_{21} = y_2 - y_1$, respectively. The error equation is defined as

$$\dot{\tilde{x}}_{21} = A_0\tilde{x}_{21} + B_0\gamma^*_2\tilde{\theta}^{*\prime}_2\omega_2 \qquad e_{21} = C_0\tilde{x}_{21} \tag{27}$$

where $\tilde{\theta}^*_2 = \theta_2 - \theta^*_2$. The following Lyapunov function is used to proof the asymptotic convergence of the synchronization error between the follower's (Agent 2) output and the leader's (Agent 1) output analytically

$$V_2(\tilde{\theta}_2, \tilde{x}_{21}) = \frac{\tilde{x}'_{21}P\tilde{x}_{21}}{2} + \frac{\tilde{\theta}'_2\Gamma_2^{-1}\tilde{\theta}_2}{2}|\gamma^*_2| \tag{28}$$

where $\Gamma_2 = \Gamma'_2 > 0$ and $P = P^0 > 0$ such that (14) holds. Then, the time derivative of V_2 can be defined as follows

$$\dot{V}_2 = -\frac{\tilde{x}'_{21}qq'\tilde{x}_{21}}{2} - \frac{v}{2}\tilde{x}'_{21}L\tilde{x}_{21} + PB_0\tilde{x}_{21}\gamma^*_2\tilde{\theta}'_2\omega_2 \\ + \gamma^*_2\tilde{\theta}'_2\tilde{\theta}_2\sigma + \tilde{\theta}'_2\Gamma_2^{-1}\dot{\tilde{\theta}}_2|\gamma^*_2|. \tag{29}$$

Since $PB_0\tilde{x}_{21} = C'_0\tilde{x}_{21} = e_{21}$ and $\gamma^*_2 = |\gamma^*_2|sgn(\gamma^*_2)$, we can define $\dot{\tilde{\theta}}_2$ such that

$$\dot{\tilde{\theta}}_2 = \dot{\theta}_2 = -\Gamma_2 e_{21}\omega_2 sgn(\gamma^*_2) - \sigma\Gamma\theta_2 sgn(\gamma^*_2) \tag{30}$$

thus we have:

IOP Conf. Series: Materials Science and Engineering 1173 (2021) 012052 doi:10.1088/1757-899X/1173/1/012052

$$\dot{V}_2 = -\frac{\tilde{x}'_{21} q q' \tilde{x}_{21}}{2} - \frac{v}{2}\tilde{x}'_{21} L \tilde{x}_{21}. \quad (31)$$

From (31), we obtain that V_2 has a finite limit, so $\tilde{x}_{21}, \tilde{\theta}_2 \in L_\infty$. We have $\tilde{x}_{21} = \bar{x}_2 - \bar{x}_1 \in L_\infty$ and previously known that $\bar{x}_1 \in L_\infty$, we have $\bar{x}_2 \in L_\infty$. This implies $x_2, y_2, \omega_{u1}, \omega_{y1}, \omega_{u21}, \omega_{y21} \in L_\infty$. Then, we have $u_2 \in L_\infty$. It can be seen that all signals are bounded and leads to $\dot{V}_{21} \rightarrow 0$ as $t \rightarrow \infty$. Thus, we have $e_{21} \rightarrow 0$ for $t \rightarrow \infty$ which conclude the proof.

4. Numerical Simulation

This section will consider the synchronization of cart inverted pendulum as a test case for the proposed distributed robust output synchronization control algorithm. Figure 1 shows the communication graph of cart inverted pendulum.

The cart inverted pendulum dynamics are divided into rotational, horizontal, and vertical motions, which are defined as follows:

- Rotational motion of pendulum

$$I\ddot{\theta} = f_y l sin\theta - f_x l cos\theta \quad (32)$$

- Horizontal motion of pendulum

$$m\frac{d^2}{dt^2}(x + lsin\theta) = f_x \quad (33)$$

- Vertical motion of pendulum

$$m\frac{d^2}{dt^2}(lcos\theta) = f_y - mg \quad (34)$$

- Horizontal motion of the cart

$$M\frac{d^2}{dt^2} = F - f_x - b\dot{x} \quad (35)$$

where I is moment of inertia of pendulum, m is the mass of pendulum, M is the mass of cart, l is the length of pendulum, g is gravitational force, θ is the pendulum angle, x is the cart position coordinate, F is the external force applied to the cart, f_x and f_y are the force applied to the pendulum, and b is force caused by friction.

The cart inverted pendulum is shown in Figure 2 We linearized the cart inverted pendulum dynamics by assuming the angle of the pendulum is slightly change $sin\theta = \theta$ and $cos\theta = 1$.

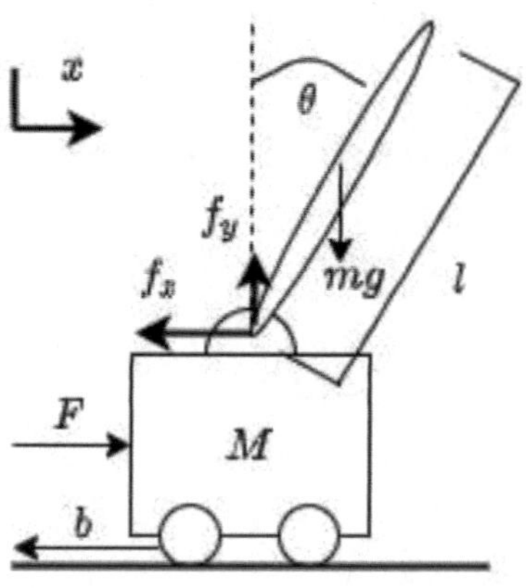

Figure 2. Cart Inverted Pendulum

Therefore, we have the cart inverted pendulum dynamics defined as

$$Ml\ddot{\theta} = (M+m)g\theta + bx^{\cdot} - F$$
$$Mx^{\cdot\cdot} = F - mg\theta - bx.^{\cdot} \tag{36}$$

The representation of cart inverted pendulum dynamics in state-space form can be written as follows

$$\begin{bmatrix}\dot{x}\\ \ddot{x}\\ \dot{\theta}\\ \ddot{\theta}\end{bmatrix} = \begin{bmatrix}0 & 1 & 0 & 0\\ 0 & -\frac{b}{M} & -\frac{mg}{M} & 0\\ 0 & 0 & 0 & 1\\ 0 & \frac{b}{Ml} & \frac{(M+m)g)}{Ml} & 0\end{bmatrix}\begin{bmatrix}x\\ \dot{x}\\ \theta\\ \dot{\theta}\end{bmatrix} + \begin{bmatrix}0\\ \frac{1}{M}\\ 0\\ -\frac{1}{Ml}\end{bmatrix}F$$
$$y = \begin{bmatrix}0 & 0 & 1 & 0\end{bmatrix}. \tag{37}$$

Table 1. Cart inverted pendulum initial conditions and parameters

	Initial cond. x,	mass (m)	length (l)	mass (M)	(b) Pendulum	Cart Friction
Agent 1 (Leader)	[1,0,1,0]'	0.002			50	0.02 0.1
Agent 2 (Follower 1)	[-1,0,-1,0]'	0.003			75	0.0030.1
Agent 3 (Follower 2)	[0.5,0,0.5,0]'	0.003			70	0.0030.1

Table 1 shows the cart's parameters inverted pendulum, which are used only for the sake of simulations. In practice, the parameters are unknown and heterogeneous among the cart inverted pendulum. The reference model dynamics which satisfied the assumption 1 and the assumption 2 is defined in state-space representation as follows

$$\begin{bmatrix}\dot{x}\\ \ddot{x}\\ \dot{\theta}\\ \ddot{\theta}\end{bmatrix} = \begin{bmatrix}0 & 1 & 0 & 0\\ 0 & 0 & 1 & 0\\ 0 & 0 & 0 & 1\\ -1 & -4 & -6 & -4\end{bmatrix}\begin{bmatrix}x\\ \dot{x}\\ \theta\\ \dot{\theta}\end{bmatrix} + \begin{bmatrix}0\\ 1\\ 0\\ -1\end{bmatrix}F$$
$$y = \begin{bmatrix}0 & 0 & 1 & 0\end{bmatrix}. \tag{38}$$

Figure 3 shows that the synchronization of cart inverted pendulums without input disturbance can be achieved by using control law (18) and the adaptive law (19). Figure 4 shows the synchronization of cart inverted pendulums with input disturbance, $\bar{d} = 200$. It can be seen that the synchronization cannot be achieved if we used the control law (18) and the adaptive law (19).

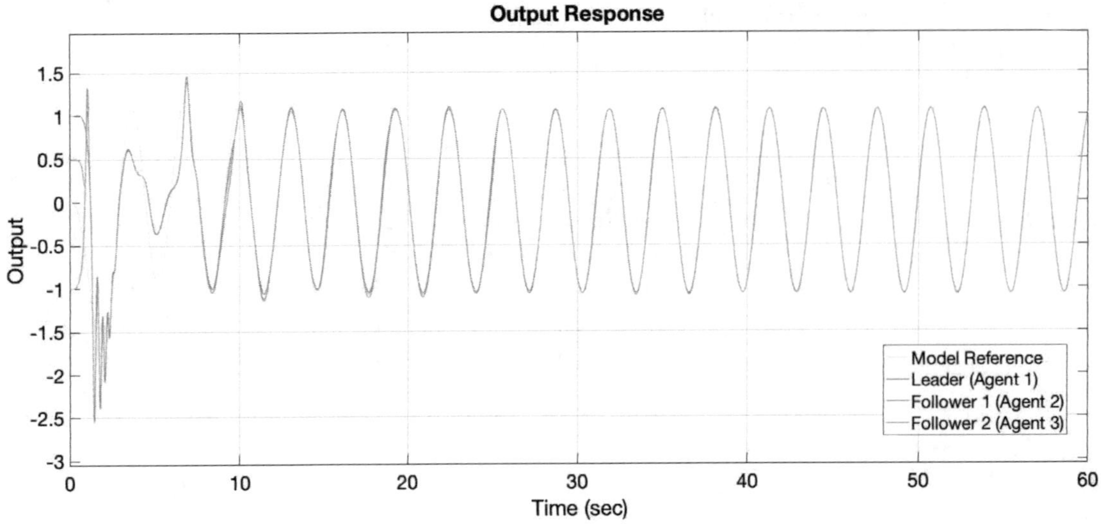

Figure 3. Output Synchronization without input disturbance

In the presence of input disturbance, the proposed control law (22) and the adaptive law (23) guarantee the synchronization and the stability of the entire network. Figure 5 shows the synchronization in the presence of input disturbance with $\sigma = 1$. It can be seen that the output synchronization of all agent to the reference model output can be achieved for sinusoidal reference input [19-21].

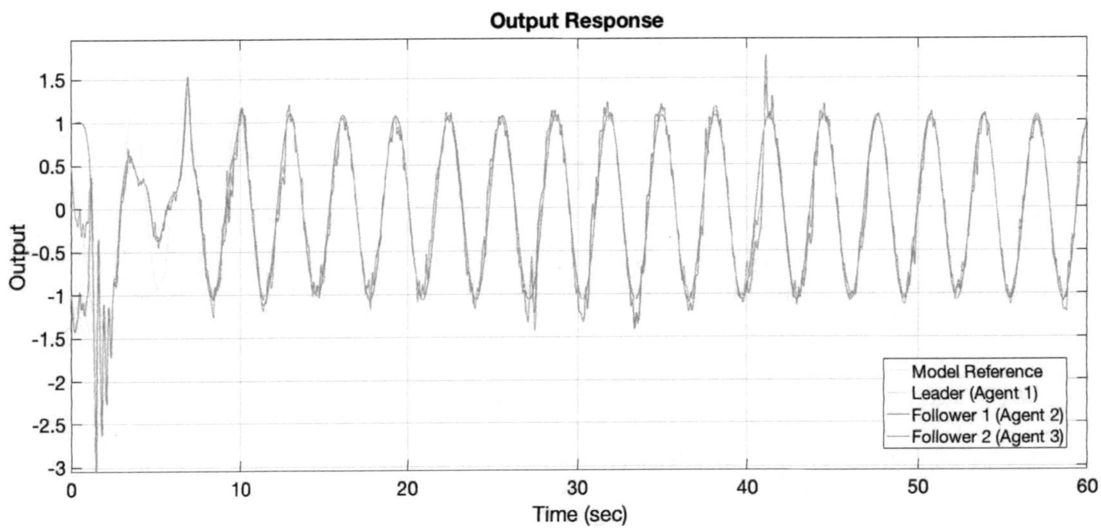

Figure 4. Output Synchronization with input disturbance, $\bar{d} = 200$

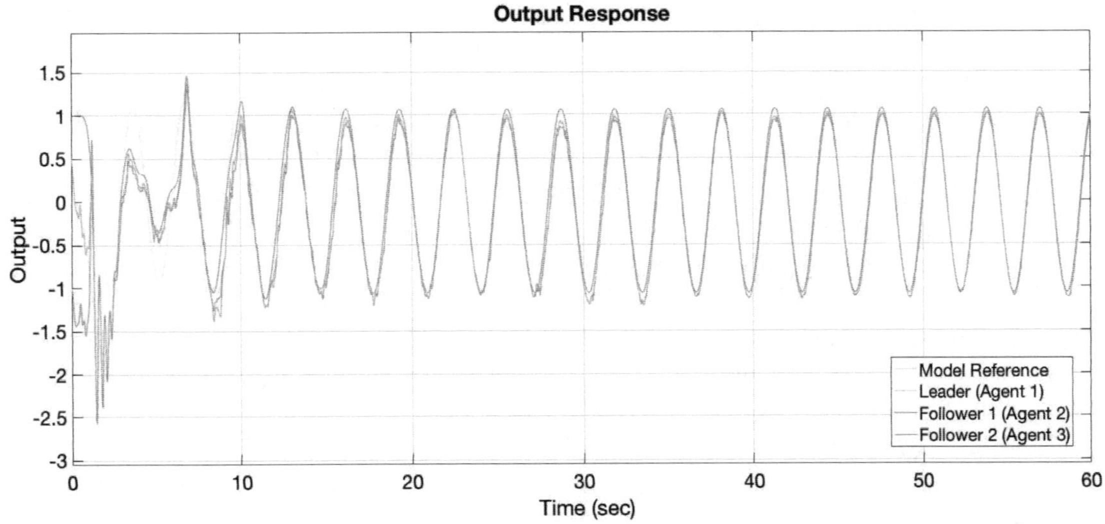

Figure 5. Output Synchronization with input disturbance, $\bar{d} = 200$ and $\sigma = 1$

5. Conclusion

This work has shown the potential solution of synchronizing the uncertain heterogeneous agents in the presence of input disturbance. The σ-modification is distributed among the entire network to handle input disturbance. The synchronization is achieved without any global information and applicable only for the DAG network. The Lyapunov-based approach was derived analytically to show that error converges to zero in the presence of input disturbance. Future work will include the distributed σ-modification for uncertain Euler-Lagrange system that consider the input constraint and the state constraint.

Acknowledgments

The authors express their gratitude to the Ministry of Research and Technology of the Republic of Indonesia (KEMENRISTEK/BRIN) for providing the research grant No.25/E1/KPT/2020. This work is registered under PPM Telkom University with contract No. 037/PNLT2/PPM/2020.

References

[1] Kim Y, Kim S and Kwak Y 2005 *Journal of Intelligent and Robotic Systems* **44** 25–46
[2] Wei Zhong and Rock H 2001 *Proceedings of the 2001 IEEE International Conference on Control Applications (CCA'01) (Cat. No.01CH37204)* pp 896–901
[3] Hehn M and D'Andrea R 2011 *2011 IEEE International Conference on Robotics and Automation* pp 763–770
[4] Silvirianti, Krisna A S R, Rusdinar A, Yuwono S and Nugraha R 2017 *2017 2nd International Conference on Frontiers of Sensors Technologies (ICFST)* pp 426–430
[5] Wang L, Wang S, van Asseldonk E H F and van der Kooij H 2013 *2013 IEEE/RSJ International Conference on Intelligent Robots and Systems* pp 965–970
[6] Zhang B L, Hu Y H and Tang G Y 2012 *Nonlinear Dynamics* **70**
[7] Qasim M, Susanto E and Wibowo A S 2017 *2017 5th International Conference on Instrumentation, Control, and Automation (ICA)* pp 109–114

[8] Nguyen H, Morrell J, Mullens K, Burmeister A, Miles S, Farrington N, Thomas K and Gage D 2004

[9] *Proceedings of SPIE - The International Society for Optical Engineering* **5609**

[10] Lavretsky E and Wise K A 2013 *Robust Adaptive Control* (London: Springer London) pp 317–353

[11] Ritz R and D'Andrea R 2013 *2013 IEEE/RSJ International Conference on Intelligent Robots and Systems* pp 3465–3471

[12] Seyboth G S, Ren W and Allgo¨wer F 2016 *Automatica* **68** 132 – 139 ISSN 0005-1098 [12] Das A and Lewis F L 2010 *Automatica* **46** 2014 – 2021 ISSN 0005-1098

[13] Olfati-Saber R, Fax J A and Murray R M 2007 *Proceedings of the IEEE* **95** 215–233

[14] Baldi S, Rosa M R and Frasca P 2018 *7th IFAC Workshop on Distributed Estimation and Control in Networked Systems (NecSys18)*

[15] Baldi S and Frasca P 2018 *Journal of the Franklin Institute* ISSN 0016-0032

[16] Baldi S, Rosa M R, Frasca P and Kosmatopoulos E B 2018 *7th IFAC Workshop on Distributed Estimation and Control in Networked Systems (NecSys18)*

[17] Rosa M R 2018 *Machines* **6** ISSN 2075-1702

[18] Ioannou P and Fidan B 2006 *Adaptive Control Tutorial (Advances in Design and Control)* (Philadelphia, PA, USA: Society for Industrial and Applied Mathematics) ISBN 0898716152

[19] Rosa M R, Baldi S, Wang X, Lv M and Yu W 2019 *European Journal of Control* **48** 52 – 65 ISSN 0947-3580 advanced Control Theory and Applications for Next-Generation Engineered Systems

[20] Rosa M R 2020 *Aerospace* **7** 127

[21] Romdlony M Z and Jayawardhana B 2016 *Automatica* **66** 39 – 47 ISSN 0005-1098

Stochastic modeling of aircraft flight parameters in terminal control area based on automatic dependent surveillance-broadcast (ADS-B) data

R Medianto[1,*], J Jusuf[1], R Oktafianto[1], R Atiqah[1], J Sembiring[1], H M Pasaribu[1,*], Y I Jenie[2], and H Muhammad[2]

[1]Aircraft Design, Operation, and Maintenance Research Group, Faculty of Mechanical and Aerospace Engineering, Bandung Insitute of Technology, Bandung, Indonesia

[2]Flight Physics Research Group, Faculty of Mechanical and Aerospace Engineering, Bandung Insitute of Technology, Bandung, Indonesia

* E-mail: rullymedianto@gmail.com, hmpasaribu@ae.itb.ac.id

Abstract. A critical step in developing good traffic management (ATM) system simulation model is to build a kinematic model of an aircraft movement that can represent the actual conditions. This study aims to obtain flight parameters for aircraft operating in the terminal control area of Soekarno-Hatta International Airport using Automatic Dependent Surveillance-Broadcast (ADS-B) data, which is openly available on FlightRadar24. Flight parameters were measured for several flight phases, such as take-off, initial climb, climb, cruise, descend, and final approach. In addition to these flight phases, flight parameters at waypoints on the arrival and departure were also measured. Furthermore, all flight parameters are modeled stochastically through the probability distribution function (PDF) approach. The best model for each parameter is obtained using the Maximum Likelihood method (MLE) with some relevant Kosmolgrove Smirnov Test criteria. The use of ADS-B data along with the stochastic model presented in this paper provides comprehensive information on flight behavior, which can be further utilized for an aircraft simulation model of an Air Traffic Management system.

Keywords: ADS-B, flight parameters, terminal control area, simulation model

1. Introduction

Aircraft Traffic Management (ATM) gives safe, economic, and efficient air traffic and airspace during all phases of aircraft operations through integrated air traffic services, airspace management, and air traffic flow management. The integrated management and collaboration with all parties and involving airborne and ground-based functions through facilities and seamless services[1]. One effective method for studying and analyzing complex systems such as ATM systems is a simulation model that provided visualization to support the validation process [2]. The flight parameters model is required to develop an aircraft kinematics model as part of the ATM simulation model. Estimating the aircraft flight parameter using stochastic modeling based on ADS-B data is the advanced method used in this research.

Open access to flight data from ADS-B has provided researchers with more insights into air traffic management. Specific flight data such as position, altitude, velocity, and vertical speed of common aircraft types are broadcast and continuously received around the world by these ground receiver networks. Nowadays, as recommended by the International Civil Aviation Organization (ICAO) [3], almost all commercial aircraft are equipped with ADS-B that makes it easier to collect ADS-B data. The biggest challenge is that ADS-B data are in large quantities. Machine learning and data mining methods have provided efficient ways to process these data into organized segments based on its flight phases.

Aircraft flight parameters from the flight phases observed include take-off, initial climb, climb, cruise, descend, and final approach. The significance of the performance indicators differs from one flight phase to another. In this research, five aircraft types are used to analyze flight parameters at specific flight phases and waypoints. The paper aims to obtain the stochastic flight parameters model for aircraft operating in the terminal control area of Soekarno-Hatta International Airport. The step of obtaining the flight parameters at the specified flight phases and waypoints will be explained.

2. Literature Review

The ATM System's simulation model helps analyze operational, tactical, and strategic levels to improve the ATM system performance. The ability to represent social and technical factor provide a simulation model of an ATM system a comprehensive and complete perspective of the system [4][5]. The ATM simulation model needs a good model of aircraft kinematics to represent the valid aircrafts traffic and movement in the control area. The kinematic flight model allows the simple generation of stochastic representations for crucial parameters. It is easily incorporated into various simulations with these representations changed to reflect empirical data describing the performance of various aircraft [6].

The proper aircraft kinematics model needs a stochastic flight parameter model representing any aircraft type that operates on the ATM control area. Commonly, flight parameter can be gathered from aircraft performance data provided by aircraft manufactures and flight performance model that developed by research/commercial institution such as Eurocontrol Experimental Centre that develops Base of Aircraft Data (BADA) [7] and Lissys Ltd. that develops PIANO-X [8]. These flight parameter models need additional cost and have license limitations.

Collecting and extracting data from ADS-B gives an alternative method to obtain a valid stochastic flight parameter model. The advantage of this method is that ADS-B data can be accessed easily at a lower cost. ADS-B data can be gathered freely using off-the-shelf receivers. There are also ADS-B data providers that provide ADS-B data worldwide for a lower cost and can be downloaded through the internet, such as Flightradar24, FlightAware, and ADS-B Exchange [9]. In order to process large amounts of ADS-B data, machine learning methods can be used, such as identifying flight phases [10]. The stochastic modeling applied in this method is used to handle dynamic and large-scale problems whose main characteristics are quite challenging to quantify or identify [11]. Stochastic modeling has been used in present aviation-related research, such as the queuing model of air traffic [12], airport operation [13], and air traffic management [14].

Sun et al. utilize large amounts of ADS-B data to construct accurate aircraft flight parameter models. A comprehensive set of methods is used for extracting complete operational and limitation models solely based on the open data approach. The method uses machine learning and data mining to efficiently process loose data into organized segments based upon trajectories and respective flight phases. The Maximum Likelihood Estimation (MLE) method is used to estimate the parameter's best value based on the observation. The results are compared with the existing models such as BADA and the Eurocontrol Aircraft Performance Database [15] to validate the results [16]. Wind speed, wind direction, and temperature may be extracted from massive ADS-B and Mode S data using the Meteo-Particle model [17] or from local METeorological Aerodrome Reports (METARs) data.

3. Materials and Methods

3.1 Data collection

The ADS-B data are downloaded from the FlightRadar24 database in Comma Separated Values (CSV) format. Data are classified based on arrival/departure from the specific airport and aircraft types. In this research, the data collection and processing based on fights departed and arrived at Jakarta Soekarno-Hatta International Airport (WIII) [18]. Data from FlightRadar24 contains information of timestamp, the date-time format in universal time coordinate, flight callsign, aircraft position in latitude-longitude, aircraft altitude in meter, aircraft speed in meter per second, and aircraft heading in degree

Data preparation used wind data model compiled from METARs data that contains weather reports coming from airports or permanent weather observation stations. METARs data used in this research is taken from Aviation Meteorological Information System in Meteorological, Climatology, and Geophysical Agency (BMKG) website (aviation.bmkg.go.id) on time range from May 2019–May 2020 and recorded on WIII weather station [19]. METARs data is collected automatically using Python scripts and Selenium module as a proxy browser to connect to the website. Retrieved data is a two-line METARs standard that will not be entirely processed, and only wind speed and direction data will be used to make the atmospheric model in an attempt to retrieve true airspeed.

Some data preparation and assumption was made in order to anticipate some limitations on retrieved METARs data. Limitations of METARs data and corresponding data preparation, including:

- Low data frequency, METARs data is recorded twice an hour, so the assumption on wind data is adjusted to the first recorded METARs data prior to recorded flight.
- The occurrence of variable wind direction and gust, some of the recordings, including natural weather phenomena in gust (G) and variable wind direction (VRB). Data preparation used for this limitation is to make the data value the same as prior recorded wind direction and speed. The variable wind is omitted as null and later imputed using K-Nearest Neighbors (KNN).
- Data units, standard METARs are using knots and degrees as wind speed and direction unit. For the simplifying parameter calculations, it will be converted to SI units.

3.2 Data preparation

The purpose of data preparation is to obtain some parameters that not available in the original ADS-B data. This parameter is required for further analysis, such as probability distribution function (PDF) estimation. Input data required for data preparation are ADS-B data available in flightradar24.com and METARs database for the weather analysis. METARs database is used for calculating TAS and Mach numbers. Both input data will be prepared for flight metadata, time point database for flight phase analysis, and flight parameter database for PDF estimation.

The original ADS-B data contains only altitude (in ft), ground speed (in knots), and direction (in degree). To get the precision parameter flight data, especially for flight phase analysis, interpolating is necessary. This action is required due to the non-uniform of each timestamp interval. The interpolation procedure will follow the linear interpolation based on the interval for each row. The calculating parameters and interpolating procedures will be done parallel. Some parameters could be obtained based on the original ADS-B data, such as vertical rate (rate of climb) and flight path angle.

PDF estimation requires a lot of data, such as 3000 data. Hence, classification is a must for better distribution. Hence, it is classified into two categories, such as depart from WIII and arrive at WIII. Besides the departure and arrival category, each data will also be classified by aircraft type—this required metadata for ease of data management. The metadata will consist of naming convention, original name file, category departure or arrival, aircraft type, departure and arrival airport, and departure and arrival runway. The original ADS-B data does not inform the departure and arrival airport and its runway. K-Nearest Neighbors (KNN) algorithm is necessary to detect airports and runways in use.

KNN is a non-parametric classification algorithm that did not make presumptions of the dataset and classify data based on the number of closest neighbors dataset given[20]. For a given dataset that consists of two categories with two specific parameters. In this research, the dataset comes from an airport or runway location. To detect which airport and runway from each ADS-B data, it is required the latitude and longitude of take-off and landing position. For runway detection, it is required additional data such as aircraft heading and runway heading.

This research analyzes flight parameters for each aircraft type. However, each flight parameter for take-off, climb, cruise, and others are not identical. It is required to detect the flight phase for ease of analyzing segmentation. The result for each flight phase will be the index number. The index number is shown in the first column (0,1,2,3,...). In this research, six flight phases must be analyzed: initial climb, climb, cruise, descent, approach, and final approach. Each flight phase will be detected based on its criteria, as shown in Table 1. Figure 1 shows the result of phase-detection for a specific flight. The red dot indicates the start point of the flight phase.

Table 1. The criterion for Flight Phase

Flight Phase	Altitude	Vertical Rate
Initial climb	±35 ft	positive
Climb	±200 ft	positive
Cruise	-	0
Descent	-	less than -2 m/sec
Approach	±100 ft	negative
Final approach	±35 ft	negative

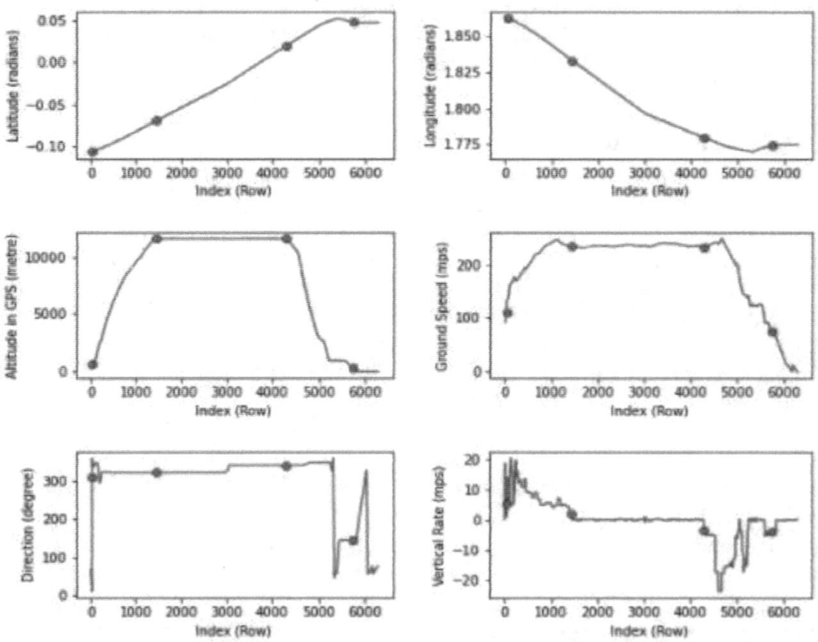

Figure 1. Six graphics that show flight phase detection results with index (row) on X-axis and latitude, longitude, altitude, ground speed, direction, and vertical rate on Y-axis.

The preparation of data gives ease of use for flight parameter modeling. There are three probability distribution functions (PDF) that are used to estimate the flight performance parameter value by applying the maximum likelihood estimation (MLE) method. Each flight parameter will be fitted to three distribution, i.e., Normal Distribution, Beta Distribution, and Gamma Distribution. Kolmogorov-Smirnov Test (K-S Test) is intended to find the best distribution. The D score results from K-S Test will be the largest distance of two functions, as shown in Equation 1. The smaller D score will give a better fit for the statistical data[21].

$$\max_{x} |F_n(x) - F(x)| \tag{1}$$

The two functions $F(x)$ are the cumulative distribution function (CDF) of a statistical model. $F_n(x)$ is the empirical distribution function (EDF). Figure 2 shows the example of three distributions that fit with the statistical model. The best distribution is shown as Beta Distribution, which has the smallest D value from its figure.

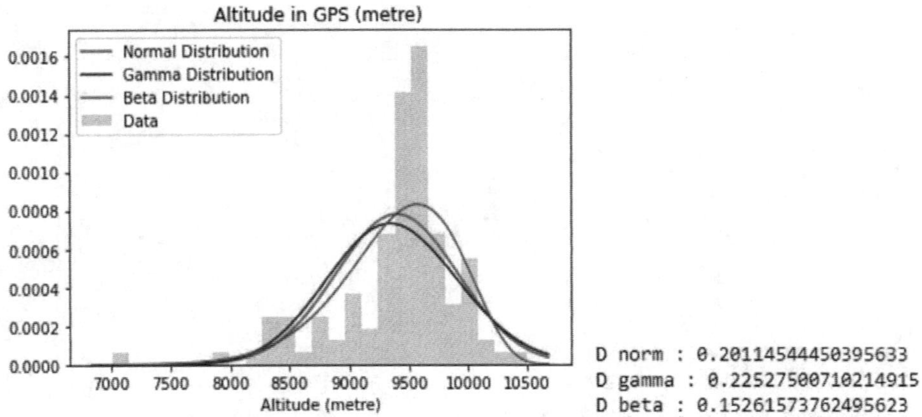

Figure 2. Distribution of Altitude in GPS (meter)

Estimated flight performance parameters for each flight phase are:

- Cruise: altitude, ground speed, and Mach number;
- Descend: descend range, CAS, Mach number, and time to descend;
- Final approach: ground speed and vertical speed;
- Initial climb: ground speed and vertical speed;
- Climb: time to climb, climb range, CAS, and Mach number;

This research also analyzes flight parameters for specific waypoints. This process will filter which flights that went through the intended waypoint. First of all, the waypoints near Soekarno-Hatta International Airport need to be classified by runway and category (Departure or Arrival). This classifying process is listed in the Soekarno Hatta International Airport Terminal Chart published by the Indonesia Directorate General of Civil Aviation (DGCA)[22]. The flight parameter that will be analyzed is the closest position to the waypoint, and the closest position is less than 1 NM to its waypoint using the KNN algorithm.

For example, flight_27698 is intended to land at Runway 25R, as shown in Table 2. This flight went through waypoint BUNIK, RAMAL, PRIOK, ARKAP. This flight trajectory is shown in Figure 3. The closest position to its waypoint will be row numbers 5091, 5301, 6135, and 6263. Flight parameters for these row numbers are then extracted to get PDF estimation.

Table 2. Waypoint detection on arrival Runway 25R

	File Name	Waypoint Through	Index Number	Aircraft Type	Runway
0	flight_27698	[BUNIK, RAMAL, PRIOK, ARKAP]	[5091, 5301, 6135, 6263]	A20N	25R
1	flight_27702	[BUNIK, RAMBU, PRIOK, ARKAP]	[5349, 6097, 6330, 6454]	A20N	25R
2	flight_27716	[BUNIK, ARKAP]	[5473, 6392]	A20N	25R
3	flight_27727	[BUNIK, RAMAL, NOKTA, RAMBU, PRIOK, ARKAP]	[5284, 5448, 5625, 6414, 6640, 6758]	A20N	25R
4	flight_27729	[BUNIK, PRIOK, ARKAP]	[4744, 5611, 5737]	A20N	25R
...	...	...	...	...	...
7737	flight_48767	[PRIOK, ARKAP]	[25252, 25359]	B789	25R
7738	flight_48771	[BUNIK, RAMAL, NOKTA]	[24412, 24627, 24857]	B789	25R
7739	flight_48783	[RAMBU, PRIOK, ARKAP]	[26027, 26288, 26424]	B789	25R
7740	flight_48784	[BUNIK, ARKAP]	[27690, 28595]	B789	25R
7741	flight_48785	[RAMBU, PRIOK, ARKAP]	[26304, 26516, 26639]	B789	25R

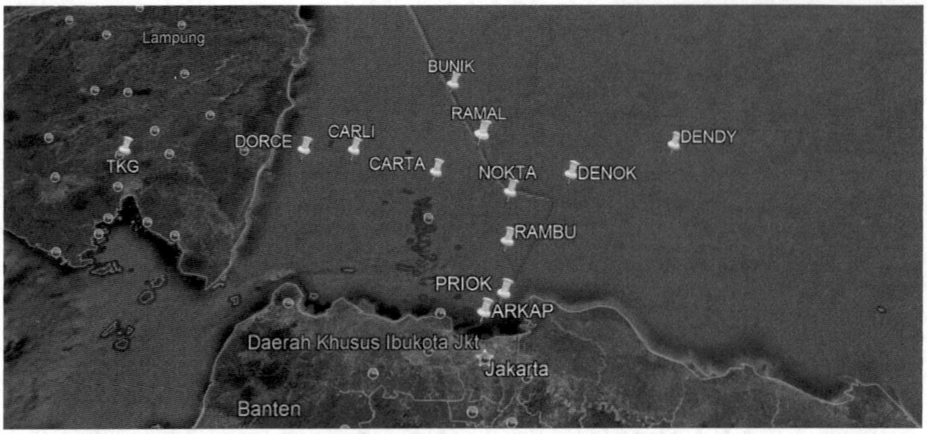

Figure 3. Flight_27698 trajectory with its waypoints while arriving at Runway 25R

4. Result and Discussion

Estimated flight performance parameters are categorized and classified into arrival/departure, aircraft type, and flight phase. Each parameter is plotted for analyzing and validating the process. Figure 4 shows a sample of the estimated flight parameters of the Airbus A320. It can be shown there are three types of the distribution function, Normal distribution function (red color), Gamma distribution function (blue color), and Beta distribution function (green color). Table 3 shows the mean and standard deviation of flight parameters from five types of aircraft.

Figure 4. Histogram of some flight performance parameters of Airbus A320 with estimated probability distribution function

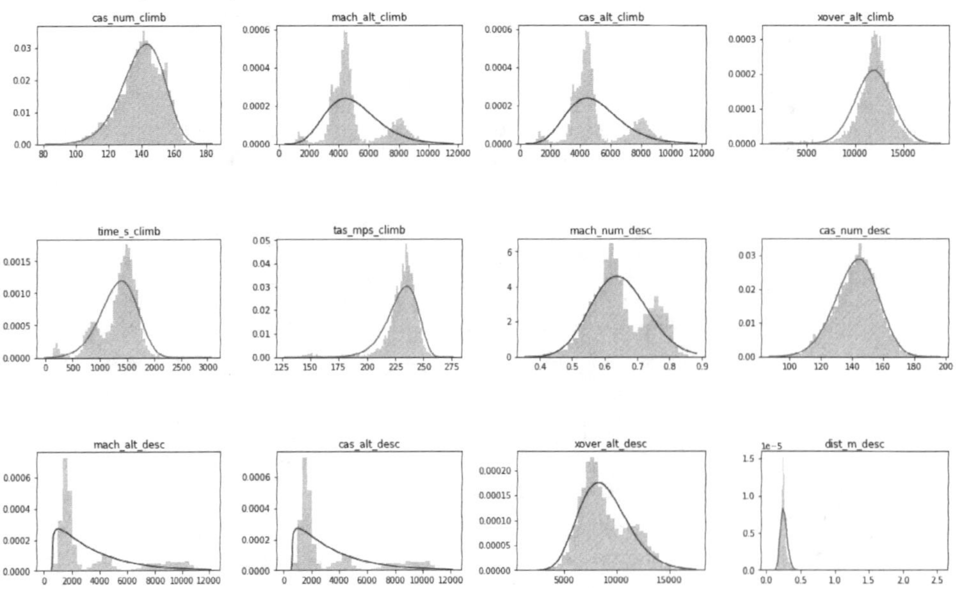

Table 3. The estimated flight parameters for five aircraft type

Parameter	B737-NG		B787		B777		A330		A320	
	Parameter	PDF	Parameter	PDF	Parameter	PDF	Parameter	PDF	Parameter	PDF
Ground Speed at Initial Climb (m/s)	K:12.4 α:15.03	Gamma	α:2.86 β:1.89	Beta	α:0.54 β:1.62	Beta	α:0.81 β:2.18	Beta	α:0.66 β:1.50	Beta
Vertical Speed at Initial Climb (m/s)	α:909.8 β:11	Beta	α:190.51 β:8.25	Beta	μ:3.8 σ:4.3	Norm	α:6.94 β:82130639	Beta	α:9.66 β:10.68	Beta
Distance to Climb (km)	K:24661 α:13.7	Gamma	α:214216265 β:4.54	Beta	α:7.65 β:5701467	Beta	K:30751 α:17.11	Gamma	α:737.2 β:324.3	Beta
Mach Number at Climb	α:2994132 β:5.1	Beta	α:994905 β:2.71	Beta	α:5884.57 β:16.87	Beta	α:8194613 β:10.13	Beta	α:307786640 β:4.73	Beta
Calibrated Airspeed at Climb (m/s)	α:4859871.8 β:30.5	Beta	α:57.31 β:8.7	Beta	α:166.89 β:38.33	Beta	μ:141.6 σ:13.13	Normal	α:349.3 β:15.4	Beta
Time to Climb (s)	K:101.8 α:16.05	Gamma	α:35200.6 β:4.8	Beta	α:12.78 β:8142880	Beta	α:16.42 β:9843043	Beta	α:4046323 β:25.8	Beta
Average Cruise altitude (m)	α:2314968 β:4.1	Beta	μ:11849 σ:721.5	Normal	μ:11313 σ:752.71	Normal	α:62.99 β:4.06	Beta	μ:10386 σ:1087	Norm
Average Ground Speed at Cruise (m/s)	α:4842283.7 β:20.03	Beta	α:8431641.8 β:10.15	Beta	α:5563655.0 β:27.19	Beta	α:1613 β:34.5	Beta	α:16.51 β:7.17	Beta
Mean Mach Number at Cruise	α:374242 β:10.75	Beta	α:3198144.8 β:7.8	Beta	α:5144939.7 β:24.4	Beta	α:405.6 β:22.77	Beta	μ:0.78 σ:0.087	Normal
Mach Number at Descend	K:0 α:832	Gamma	α:10.18 β:8.95	Beta	K:0 α:1268.87	Gamma	α:1822.08 β15.09:	Beta	K:0.01 α:66.2	Gamma
Calibrated Airspeed at Descend (m/s)	α:58044.1 β:195.6	Beta	K:0.28 α:4497.3	Gamma	μ:146.85 σ:13.63	Norm	α:697559.45 β:608.16	Beta	α:221.1 β:44.5	Beta
Distance to Descend (km)	α:2.07 β:388810.64	Beta	K:1386948 α:0.53	Gamma	α:0.66 β:182.17	Beta	α:1 β:38.08	Beta	α:14.2 β:180091012582810	Beta
Time to Descend	α:2.95 β:203494858	Beta	K:4494 α:0.59	Gamma	α:0.92 β:11053.9	Beta	α:1.18 β:66.29	Beta	α:4.3 β:248661902835.1	Beta
Ground Speed at Final Approach	α:232588.18 β:216283.54	Beta	α:6.34 β:2.62	Beta	α:24.43 β:1.61	Beta	K:50.95 α:2.4	Gamma	α:1.88 β:1.21	Beta
Vertical Speed at Final Approach	K:0.18 α:35.52	Gamma	α:467.21 β:0.61	Beta	μ:-1.07 σ:2.45	Norm	α:23718076 β:45.72	Beta	α:1568.7 β:62.4	Beta

An analysis of estimated parameters on all five aircraft types shows a good fit for overall aircraft climb and cruise parameters. Estimation on the cruise Mach number differs below ~3% across all aircraft types, while estimation on cruise altitude differs ~15% compared to a database value. Estimation on climb parameter also shows adequate measures on the rate of climb, constant Mach number, and constant CAS number estimation. The estimation of descending flight phase parameters is inaccurate compared to a database value. Generally, estimated descend parameters resulting in less descend rate, lower constant Mach number, longer descend time, and longer distance compared to the value in the database. Parameter estimation also shows the struggle in the initial climb and final approach phase, which results in high error across all aircraft types. Similar to Airbus A320 results, parameter estimation from other aircraft types in distance and time on the climb and descend phase yields high standard deviation value.

Based on the waypoint detection and aircraft types, Table 4 below indicates 16 waypoints for arrival on Runway 25R/L as a sample of estimated flight parameters on a specific waypoint using Airbus A320. These estimated parameters are compared with the reference parameter from WIII Terminal Chart issued by Indonesia DGCA[22].

IOP Conf. Series: Materials Science and Engineering 1173 (2021) 012053 doi:10.1088/1757-899X/1173/1/012053

Table 4. The PDF parameters, mean and standard deviation of estimated flight parameters on specific waypoint for arrival Runway 25L/R compare to reference parameters from WIII Terminal Chart

Waypoint	Altitude (meter)		Calibrated Airspeed (m/s)		Track Angle (degree)		Reference Parameters from WIII Terminal Chart		
	Parameters	PDF	Parameters	PDF	Parameters	PDF	Altitude (meter)	Speed Limit (m/s)	Track Angle (degree)
ARKAP	α:2.5 μ:678.7 K:44.9 σ:71.1	Gamma	α:7.635e+7 μ:94.7 β:105.4 σ:10.8	Beta	α:512.5 μ:233.7 K:1687.0 σ:9.5	Beta	609.6	-	247
BUNIK	α:39.7 μ:6949.0 K:1098.1 σ:619.4	Beta	α:3.126 μ:145.8 β:2.918 σ:13.4	Beta	α:44.6 μ:160.8 K:5.398e+6 σ:8.6	Beta	6705.6 (-)	138.9	-
CARTA	μ:4277 σ:406.1	Normal	α:0.207 μ:130.9 β:0.568 σ:8.1	Beta	α:0.356 μ:155.9 β:0.397 σ:17.9	Beta	-	-	107
DAPIK	μ:1262.4 σ:151.2	Normal	μ:105.1 σ:12.7	Normal	μ:254.8 σ:7.4	Normal	1219.2	97.7	248
DENOK	μ:3532.6 σ:150.0	Normal	μ:124.3 σ:3.5	Normal	μ:193.7 σ:25.9	Normal	4267.2	128.6	248
GAPRI	α:8.792e+5 μ:6908.6 β:9.812e+5 σ:536.8	Beta	α:9.373 μ:138.2 β:9.400 σ:11.0	Beta	α:2.924e+6 μ:278.6 β:54.7 σ:16.0	Beta	7010.4	136.3	214
GETES	μ:7941.7 σ:505.6	Normal	μ:141.6 σ:9.0	Normal	μ:306.6 σ:12.7	Normal	7924.8	136.3	314
IMU	α:447.0 μ:8812.8 K:30.0 σ:634.0	Gamma	α:2046.1 μ:142.4 β:151.1 σ:8.6	Beta	α:2765.0 μ:299.3 K:0.2 σ:13.4	Gamma	8534.4	136.3	270
KIDET	α:1421.5 μ:8734.1 K:15.9 σ:599.4	Gamma	α:6463.5 μ:218.8 β:289.1 σ:11.4	Beta	μ:301.2 σ:12.4	Normal	-	-	-
KURUS	α:6.668e+6 μ:9418.4 β:8.9 σ:476.3	Beta	α:27.8 μ:144.9 β:4.1 σ:8.0	Beta	α:7.7 μ:273.3 β :2.69e+11 σ:10.9	Beta	-	-	-
NOKTA	α:3.2 μ:3636.6 K:367.1 σ:659.5	Gamma	α:9.142 μ:125.3 K:3.135 σ:9.5	Gamma	α:9.3 μ: 170.2 β:9.643e+6 σ: 11.7	Beta	3352.8 (-)	123.5	180
PRIOK	α:1.5 μ:1078.3 β:2.064e+8 σ:398.0	Beta	α:100.313 μ:106.4 K:1.613 σ:11.6	Gamma	α:3.8 μ:201.5 K:8.7 σ:17.1	Gamma	914.4	97.7	180
RAMAL	α:14.3 μ:5343.4 β:5.449e+6 σ:763.0	Beta	α:10.652 μ:136.2 β.51745 σ:11.9	Beta	α:6.169e+6 μ:158.4 β:130.6 σ:8.1	Beta	5181.6 (-)	138.9	155
RAMBU	α:1195.1 μ:2104.6 β:107.5 σ:277.3	Beta	α:4921 μ:118.9 K:0.132 σ:9.2	Gamma	α:29.9 μ:178.6 K:5.496e+6 σ:10.0	Beta	2133.6	-	180
TEGAR	α:2.009e+5 μ:218.9 β:4.387e+4 σ:11.6	Beta	α:917.3 μ:115.0 K:0.362 σ:11.0	Gamma	μ:259.4 σ:6.3	Normal	2133.6	108.0	270
WETES	μ:5062.1 σ:536.7	Normal	A:39.560 μ:133.8 K:1.796 σ:11.3	Gamma	μ:269.3 σ:7.8	Normal	5181.6	128.6	270

From Table 4, it can be seen that the estimated parameters for altitude are close enough to the reference altitude. For airspeed, calibrated airspeed from estimated parameters has a value that closes enough to the reference speed limit note in indicated airspeed. The estimated parameter has a significant difference in the track angle. This difference may occur because aircraft movement did not exactly fly over the waypoint, so the track angle did not have a track angle as specified in the terminal chart. Some waypoints did not have enough data because many aircraft fly directly (with the Air Traffic Controller's approval) without passing these waypoints.

5. Conclusion

This research utilizes ADS-B data that are obtained from FlightRadar24 to estimate valid stochastic flight parameters. Flight parameters are extracted for specific flight phases and waypoints. The flight data consist of information on the aircraft that arrive and depart from Soekarno Hatta International Airport. Each data file has been equipped with metadata to facilitate file classification and search. The ADS-B and METARs data processing step is explained to get flight performance parameters for a specific flight phase. It is challenging to work on raw data because every bit of information provided may not be useful. Data preparation will manage data into a homogenized form to be easily matched to those significant figures. Managing data will be resulting in well-structured data and reduces time-consuming with a low risk of losing essential data. Data preparation filters a large amount of raw data collected from FlightRadar24 into detailed data that will be useful in the future analysis of the flight trajectory. Data management and data preparation are essential to make analysis easier. The validation process confirms that most of the estimated flight parameters at any flight phase close to value when to compare to EuroControl Aircraft Performance Database. The estimation of the vertical rate parameter needs accuracy by developing advanced methods. Estimated flight parameters for specific waypoints show that altitude and airspeed have a slight difference with the terminal chart's reference values.

Acknowledgments
We gratefully thank to Program Penelitian, Pengabdian kepada Masyarakat dan Inovasi ITB (P3MI-ITB) program of the Institut Teknologi Bandung, year of 2020 for financial support.

References

[1] ICAO 2016 *Doc 4444, Procedures for Air Navigation Services — Air Traffic Management* (Montréal, Quebec, Canada)
[2] Medianto R, Pasaribu H M and Muhammad H 2019 Development of Hybrid Simulation Model of Air Traffic Management in the Terminal Control Area *IOP Conf. Ser. Mater. Sci. Eng.* **645**
[3] Zhang J, Liu W and Zhu Y 2011 Study of ADS-B data evaluation *Chinese J. Aeronaut.* **24** 461–6
[4] Stroeve S H and Everdij M H C 2017 Agent-Based Modelling and Mental Simulation for Resilience Engineering in Air Transport *Saf. Sci.* **93** 29–49
[5] Gurtner G, Bongiorno C, Ducci M and Miccichè S 2017 An empirically grounded agent based simulator for the air traffic management in the SESAR scenario *J. Air Transp. Manag.* **59** 26–3
[6] McGovern S M, Cohen S B, Truong M and Fairley G 2007 Kinematics-based model for stochastic simulation of aircraft operating in the national airspace system *AIAA/IEEE Digit. Avion. Syst. Conf. - Proc.* 1–11
[7] Csanda J 2018 Using BADA 4 Performance Model for Aircraft Simulation
[8] Lissys Ltd 2008 Piano-X Aircraft Emissions and Performance User's Guide 1–10
[9] Opitz F, Dästner K, Roseneckh-köhler B V H and Schmid E Data Analytics and Machine Learning based on Trajectories 1–12
[10] Sun J 2016 Large-Scale Flight Phase identification from ADS-B Data Using Machine Learning Methods PhD student, ATM Control and Simulation, Aerospace Engineering Large-Scale Flight Phase identification from ADS-B Data Using Machine Learning Methods
[11] Shone R, Glazebrook K and Zografos K G 2020 Applications of stochastic modeling in air traffic management: Methods, challenges and opportunities for solving air traffic problems under uncertainty *Eur. J. Oper. Res.*
[12] Simaiakis I and Balakrishnan H 2016 A queuing model of the airport departure process *Transp. Sci.* **50** 94–109
[13] Jacquillat A, Odoni A R and Webster M D 2017 Dynamic control of runway configurations and of arrival and departure service rates at jfk airport under stochastic queue conditions *Transp. Sci.* **51** 155–76
[14] Hentzen D, Kamgarpour M, Soler M and González-Arribas D 2018 On maximizing safety in stochastic aircraft trajectory planning with uncertain thunderstorm development *Aerosp. Sci. Technol.* **79** 543–53
[15] EuroControl EuroControl Aircraft Performance Database *Eurocontrol*
[16] Sun J, Ellerbroek J and Hoekstra J 2017 Modeling aircraft performance parameters with open ADS-B data *12th USA/Europe Air Traffic Management R and D Seminar* (Seattle, Washington, USA)
[17] Vuˆ, Huy, Junzi Sun, Joost Ellerbroek J H 2018 *ADS-B and Mode S Data for Aviation Meteorology and Aircraft Performance Modelling* (Delft University of Technology,)
[18] FlightRadar24 FlightRadar24 Data/History Flights
[19] BMKG 2020 METAR/SPECI & Trend Forecast *Pus. Meteorol. Penerbangan, Badan MEeteorologi, Klimatologi, dan Geofis.*
[20] Taunk K, De S, Verma S and Swetapadma A 2019 A brief review of nearest neighbor algorithm for learning and classification *2019 Int. Conf. Intell. Comput. Control Syst. ICCS 2019* 1255–60
[21] Sun J, Ellerbroek J and Hoekstra J M 2019 WRAP: An open-source kinematic aircraft performance model *Transp. Res. Part C Emerg. Technol.* **98** 118–38
[22] Directorate General of Civil Aviation 2019 AIRAC AIP Supplement Nr : 14/19 - The Establishment And Revision Of Instrument Flight Procedures at Soekarno Hatta International Airport – Jakarta

Dynamic obstacle avoidance system for the unmanned aerial vehicle (UAV)

S S Rawikara*, and R A Sasongko

Flight Physics Research Group, Faculty of Mechanical and Aerospace Engineering, Institut Teknologi Bandung, Jalan Ganesha 10, Bandung, Indonesia.

* E-mail: sahisnu@ae.itb.ac.id

Abstract. This paper presents a method for collision avoidance for moving obstacle based on ellipsoid geometry. Ellipsoid algorithm itself has been implemented in small unmanned aerial vehicle (UAV) for static obstacle avoidance. In extending the algorithm to dynamic object, two adjustments have been made, first is to calculate safe heading and selection of head or tail contact point. Using these two adjustments, the algorithm succesfully generates safe waypoint for dynamic obstacle avoidance and select which direction to use, whether it is to the head or the tail of obstacle moving direction.

Keywords: collision avoidance, ellipsoid zone, flight guidance

1. Introduction

In recent years, UAV's has been deployed not only for military purpose but also for civilian use, ranging from agricultural mapping, asset surveillance and inspection, and even logistic delivery for remote areas. The relatively low operating cost and simplicity of UAV's deployment are main advantage of UAV's over manned aircraft, especially for operation in remote areas and dangerous missions [1]. Nowadays, there are efforts to integrate UAV's into civil airspace, which would be beneficial especially for air taxi operation and urban drone logistics delivery. However, to fully integrate UAV's into civil airspace, UAV's will need to comply to the policies of aviation authorities [2]. Even though UAV's now able to autonomously operate from take-off to landing, conflict situation for both betweeen UAV's and manned aircraft still presents challenging problem [3]. There have been research in developing conflict resolution algorithm for UAV using several approaches and paradigms.

Generally, there are two different aprroaches in collision avoidance algorithm [4]. The first category is the global planner, which requires knowledge about the aircraft operation environment. The advantage of this approach is that the trajectory generated by the algorithm can always yield to optimal path, even though this approach have difficulties in dealing with sudden changes in the environement of UAV operation. Another approach is the local trajectory planner, which the algorithm runs continuously to deal with unknown obstacle during the flight. In [5] discuss the evolution of UAV obstacle avoidance philosophy, with several methods being evaluated, namely graph search, rapidly exploring random tree, minimum effort guidance, and potential field method. One of the main conclusion in this research is that even though an optimal path can be generated from the collision avoidance algorithm, the computational time required to generate such path must be taken into account, since fixed-wing UAV doesn't have the luxury to stop in mid air. Proportional navigation which is popular for missile termination guidance is also adapted to generate avoidance guidance in [6]. In this research, Porportional Navigation Collision Avoidance Guidance (PNCAG) is generating guidance velocity vector based on object position and velocity, with predetermined object size. Another approach for real time collision avoidance is presented in [7]. In this paper collision avoidance trajectory is generated using B-splines, based on optimal control problem. While the generated trajectory can guarantee optimal solution, the optimizer could be trapped in local minimal, hence increasing computational time required to generate the trajectory. There is also an approach using reachable set as elaborated in [8]. This research focused on defining control constraints set that are reachable by the UAV, so that the maneuver can always be executed by aircraft. However, the computation requirement for solving reachable set is relatively expensive problem and requires significant computation time.

In this paper, a reactive collision avoidance method is developed using ellipsoid geometry to determine the object size and compute safe waypoint to avoid collision. First, the methodology of the obstacle avoidance for static object is elaborated in the methodology section, also the extension from static to dynamic obstacle avoidance and the adjustments that have been made. Afterward, the method that has been explained is tested in simulation. Finally, results of the simulation for dynamic obstacle avoidance algorithm will also presented and analyzed.

2. Methodology

In this section, the collision avoidance method will be discussed with short introduction to static object avoidance method. Its extension from static to dynamic object will be presented with the adjustment of contact point selection from static to dynamic object avoidance. From the following dynamic object avoidance method, selection of contact point will also be discussed.

2.1. Static Obstacle Avoidance

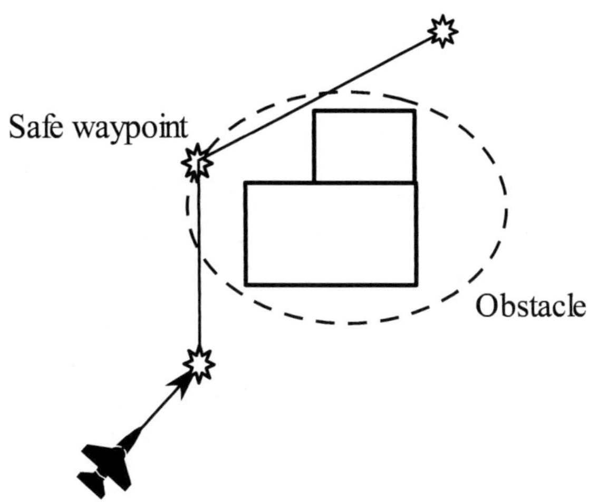

Figure 1. Static Obstacle Avoidance

Suppose a UAV is going from point A to point B (Figure 1). Along the way, an obstacle is detected, so the UAV need to compute a new waypoint to avoid the obstacle. This waypoint can be calculated given that the size and the orientation of the obstacle is known. As in [9], the obstacle is defined as ellipsoid, and using the velocity vector of the UAV a new waypoint that lie in boundary of the ellipsoid can be generated.

A 2-D ellipsoid of the obstacle can be defined as:

$$\bar{x}^{\mathrm{T}}\mathrm{A}\bar{\mathrm{x}} + \mathrm{B}^{\mathrm{T}}\bar{\mathrm{x}} = c \tag{1}$$

where $\bar{\mathrm{x}} = [\mathrm{x}_1 \ x_2]^T, A \in \mathrm{R}^{2x1}$ is a symmetric positive definite matix, $B \in \mathrm{R}^{2x1}$ is the vector of linear term, and $c \in \mathrm{R}$ is a scalar. The eigenvalues of A matrix representing length of major and minor axis of ellipsoid. The ellipse can be rotated to represent obstacle geometry using matrix V which defined as:

$$\mathrm{V} = \begin{bmatrix} cos\psi_E & -sin\psi_E \\ sin\psi_E & cos\psi_E \end{bmatrix} \tag{2}$$

where ψ_{E} is the obstacle heading. To find contact point, following formula is used [10]:

$$\overline{x}_{cp} = -\left[\sqrt{(V^T A^{-1} V c^{-1})} A\right]^{-1} V + \overline{x}_{obs} \quad (3)$$

where $\overline{x}_{obs}$ is the obstacle position relative to the UAV and $\overline{x}_{cp}$ is the contact point for safe waypoint guidance. Since the vector used in contact point calculation is based normal vector of UAV heading, there will be two solutions of contact points, which corresponds to +90- and -90-degree rotation of velocity vector. In static obstacle avoidance, contact point can be selected which have minimum heading change.

2.2. Extension from Static to Dynamic
In general, calculation for contact point between static and dynamic is the same, with moving obstacle size is becoming the parameter for ellipse and heading of the obstacle is heading of the ellipse. From static obstacle avoidance that calculation of contact point will always yield to two contact point solutions. While for static obstacle avoidance the selection of contact point can be simply selected based on difference between current heading and contact point heading reference, in moving obstacle avoidance this is not the case. Since the obstacle is moving, selection of contact point can lead dangerous waypoint crossing the ellipsoid zone. In the next section, proposed methods of selecting contact point will be elaborated.

2.3. Contact Point Selection

2.3.1. Unsafe Heading Requirement
For dynamic obstacle avoidance, since the ellipse is moving, contact points calculated from ellipse algorithm can lead to dangerous position. As shown in Figure 2, because of the velocity vector from the UAV, both contact points calculated from avoidance algorithm are not guiding UAV to safe position. Instead of following calculated contact point, UAV should directly go to final waypoint as the obstacle is already avoided.

To mitigate this behaviour, an unsafe heading calculation for determining which heading reference to follow is proposed. First, limit of obstacle is calculated based on relative vector between UAV and the obstacle. From relative vector of UAV and obstacle, contact points is determined using same formula for finding contact points in avoidance scheme. Relative vector between obstacle and UAV can be defined as:

$$\overline{v}_c = \begin{bmatrix} x_{obs} - x_{acf} \\ y_{obs} - y_{acf} \end{bmatrix} \quad (4)$$

where $\overline{v}_c \in R^{2x1}$ is the relative range. To find unsafe contact point, the following formula same as static avoidance is used:

$$\overline{g} = \begin{bmatrix} \cos(\pi/2) & -\sin(\pi/2) \\ \sin(\pi/2) & \cos(\pi/2) \end{bmatrix} \begin{bmatrix} x_c \\ y_c \end{bmatrix} \quad (5)$$

$$\overline{x}_{sv} = -\left[\sqrt{g^T M^{-1} g L^{-1}} M\right]^{-1} + \overline{x}_{obs} \quad (6)$$

$$\psi_{unsafe} = \tan^{-1}\left(\frac{x_{sv} - x_{acf}}{y_{sv} - y_{acf}}\right) \quad (7)$$

In dynamic avoidance, $\bar{x}_{sv} \in R^{2x1}$ is the contact point. After that, the heading for computed contact point from avoidance scheme is compared between both unsafe heading reference. If the heading for avoidance is inside unsafe heading reference, then the heading from avoidance become the heading reference for the guidance system. The following criteria is used to determine if the heading reference is inside unsafe heading:

$$min(\psi_{unsafe}) \leq \psi_{waypoint} \wedge \psi_{waypoint} \leq max(\psi_{unsafe}) \quad (8)$$

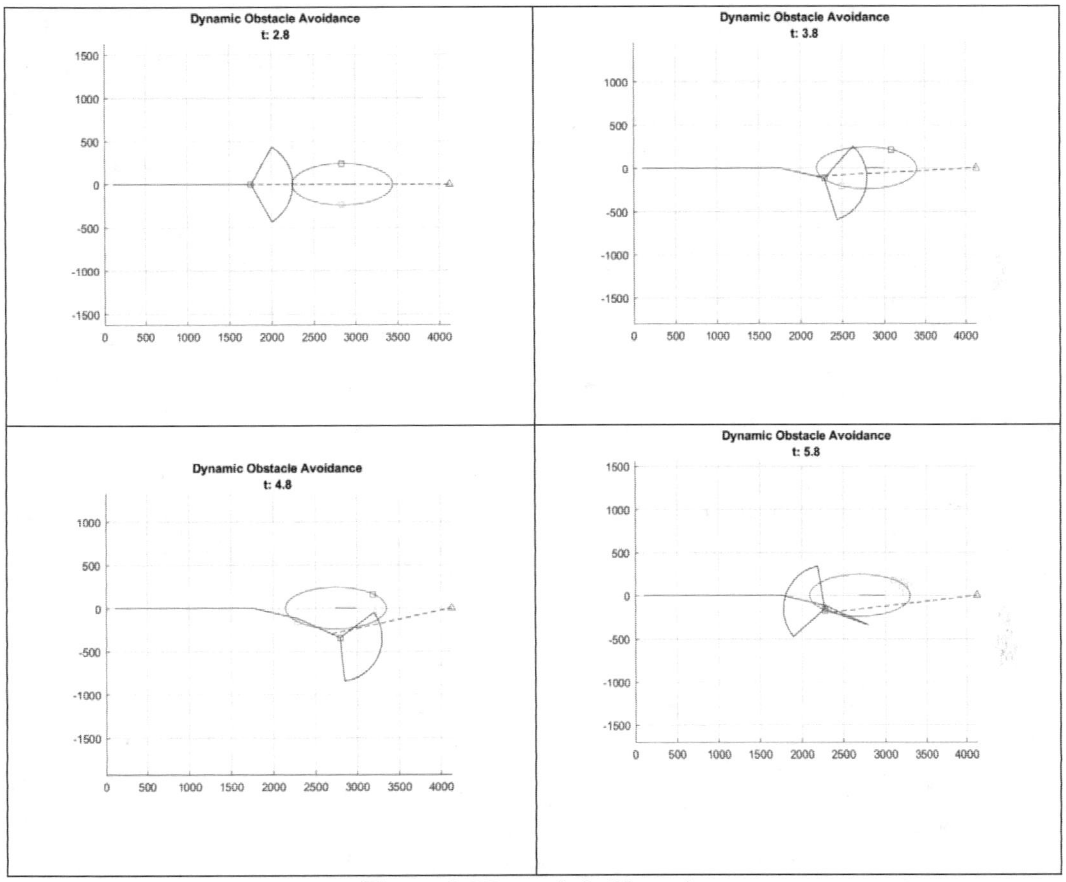

Figure 2. Wrong Safe Waypoint

2.3.2. Head-or-tail Selection

In dynamic obstacle, selection of contact points will determine whether UAV wil go same or opposite direction with the obstacle. Example of importance in contact point selection can be seen in Figure 2, where the contact point selected by the algorithm is into the head of the obstacle direction, even though it is still safe to use that contact point for waypoint, other contact point which is in tail position of the obstacle is relatively safer.

First, the preference for head or tail selection of contact point is defined. Then, relative heading between UAV and obstacle is calculated using:

$$\psi_{rel} = w_{360}(\psi_{obs} - \psi_{acf}) \quad (9)$$

where w_{360} is a function to wrap from 0 to 360 degrees. After relative heading is found, then the orientation of the obstacle needs to be checked whether it is in quadrant 1 and 2 or in quadrant 3 and 4, using the following:

$$c_{factor} = 1 - 2 \cdot (180 < \psi_{rel} \wedge \psi_{rel} < 360) \qquad (10)$$

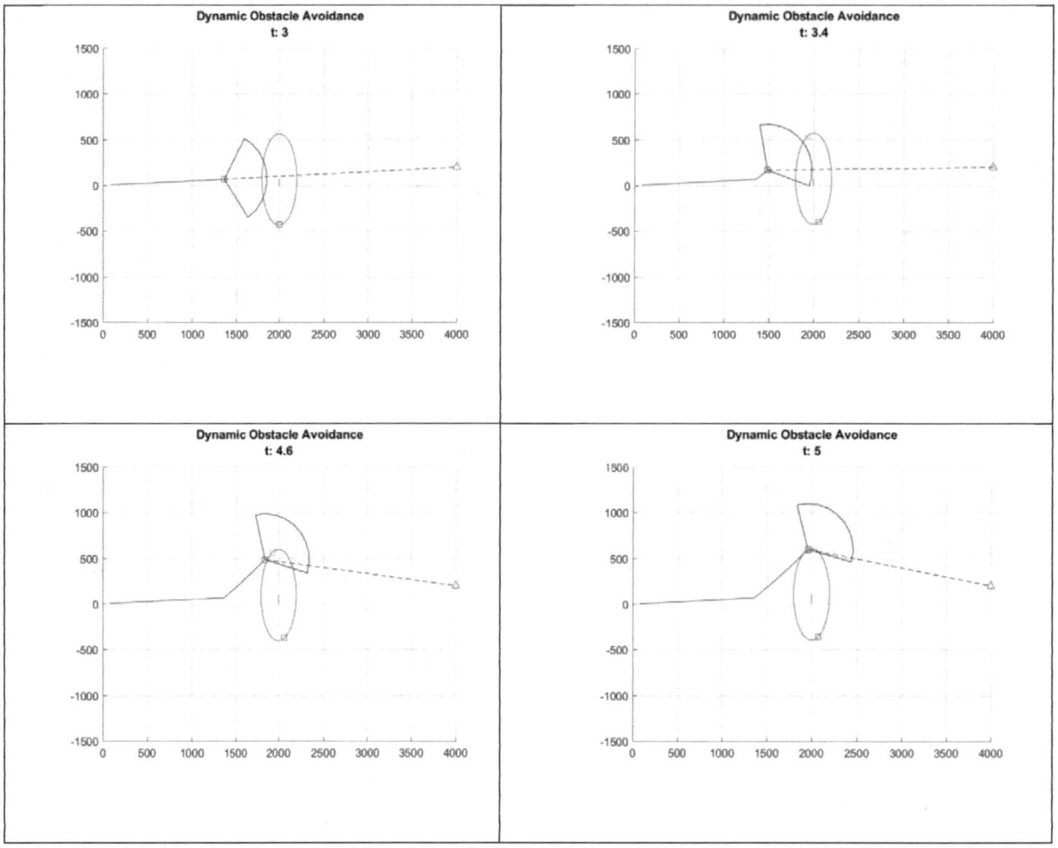

Figure 3. Head Case Waypoint

Value check of relative heading will give true (1) or false (0), so eq. will result in 1 if it is true or -1 if it is false. From the value check of relative heading, to determine normal vector that will be used in contact point calculation, the following transformation is used:

$$g = \begin{cases} \begin{bmatrix} \cos(c_{factor} * \pi/2) & -\sin(c_{factor} * \pi/2) \\ \sin(c_{factor} * \pi/2) & \cos(c_{factor} * \pi/2) \end{bmatrix}; \textit{if tail} \\ \begin{bmatrix} \cos(-c_{factor} * \pi/2) & -\sin(-c_{factor} * \pi/2) \\ \sin(-c_{factor} * \pi/2) & \cos(-c_{factor} * \pi/2) \end{bmatrix}; \textit{if head} \end{cases} \qquad (11)$$

In general, the addition of unsafe heading check and head-or-tail selection as flowchart is shown in Figure 4.

3. Results and Discussions

In this section the collision avoidance algorithm is demonstrated with several cases. This simulation is only consider the kinematics of the UAV in 2D X-Y plane, because the focus is in testing the algorithm in various cases. In this simulation, initial speed and heading of the UAV is fixed, while the speed and heading of the obstacle is calculated so that collision will occur at 20 second.

Three cases will be considered, the first case is perpendicular case to show that without unsafe heading check, waypoint from obstacle avoidance calculation will lead to oscillation. Second is head-to-tail case, where the obstacle is in pursuit by the UAV, and last is the head-to-head case.

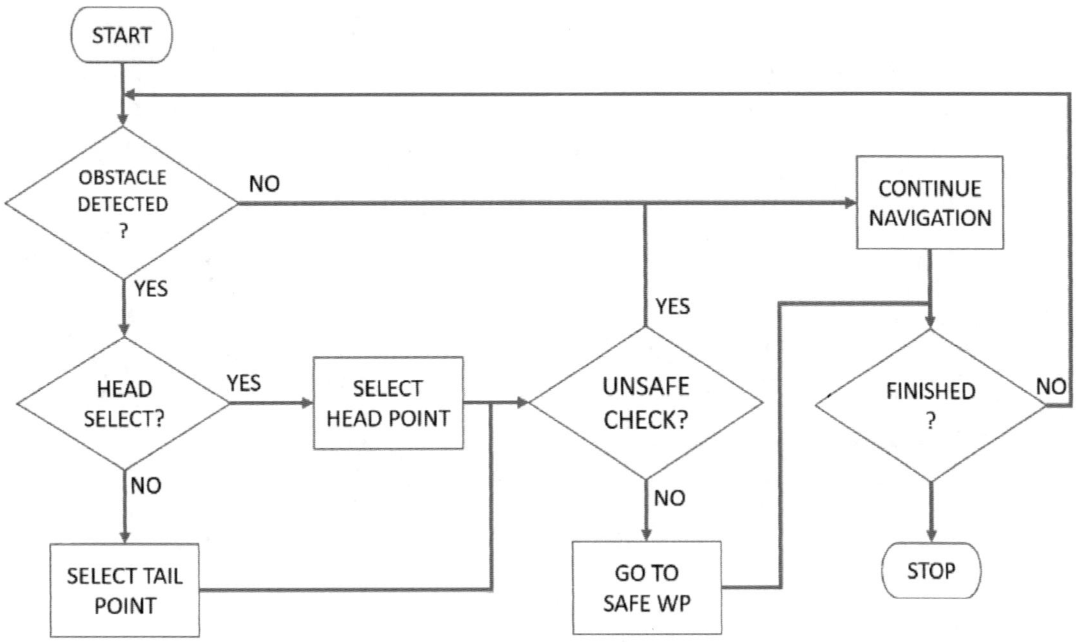

Figure 4. Dynamic Obstacle Avoidance Flowchart

3.1. Perpendicular Case

In this case, the obstcle is moving perpendicular to the UAV. In the Figure 5, unsafe heading check is turned off. The waypoint generated by static obstacle avoidance algorithm will lead to oscillation. This is caused by the change of sign while calculating normal vector from ellipse contact point. In Figure 6, using unsafe heading check as described in section 2.3.1, UAV can return to its original waypoint.

This has shown that even the generated waypoint from ellipse algorithm can give a safe guidance for the UAV, since the obstacle is moving, there is a chance that the UAV is trapped in oscillation because of switching in normal vector of UAV velocity. Unsafe heading check ensure that the contact point given by ellipsoid algorithm is the actual safe waypoint from obstacle.

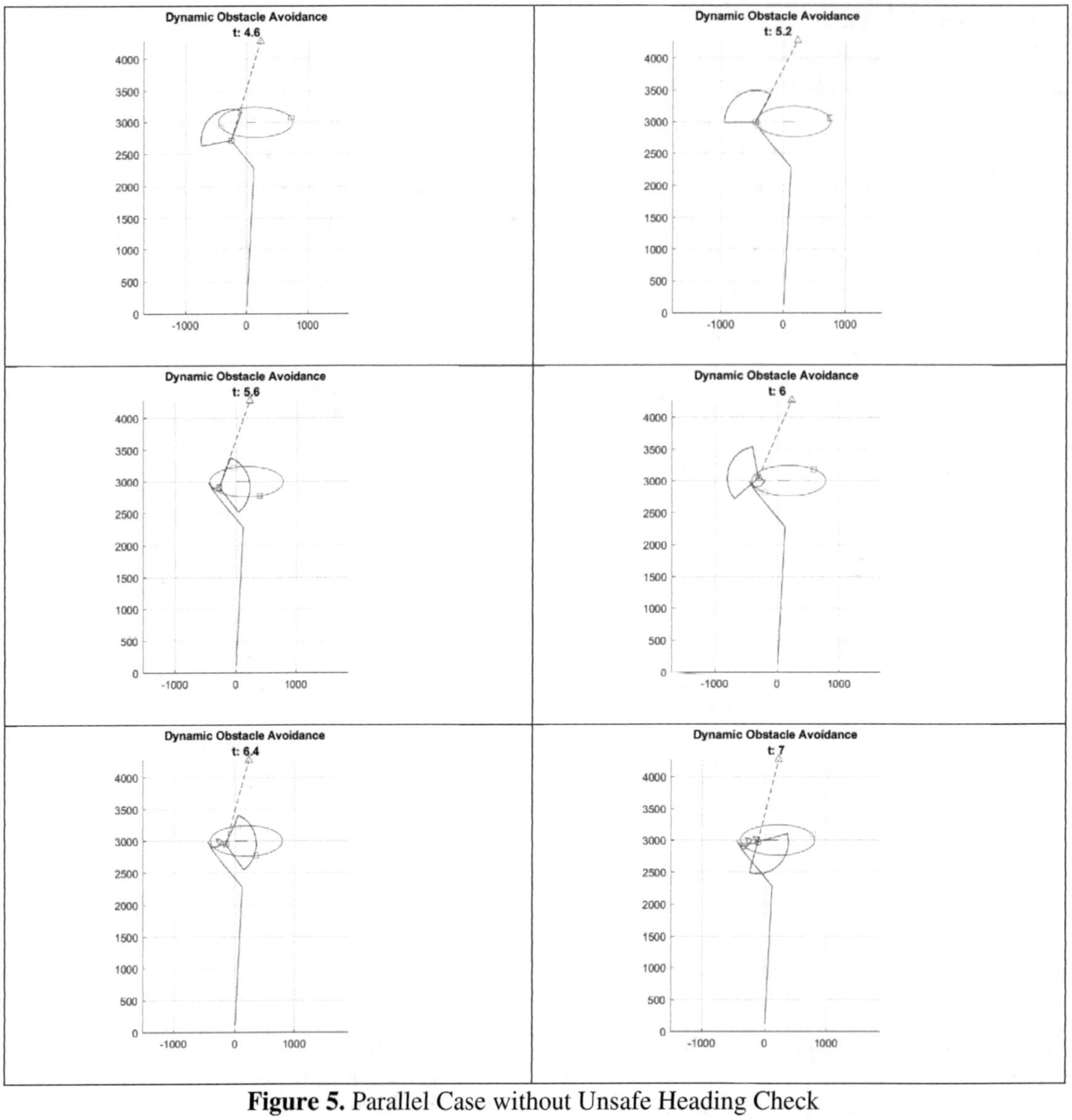

Figure 5. Parallel Case without Unsafe Heading Check

3.2. Head-or-Tail Selection

In this simulation, the selection of head or tail contact point is shown in Figure 7 and Figure 8. Simulation parameter of both the UAV and the obstacle are the same. Initial condition in both head and tail simulation are also the same. Figure 7 shows that by selecting head as the contact point preference, UAV will try to avoid the obstacle by going to the same direction with the obstacle.

This will lead to relatively smaller heading rate, but also by going to the same direction with the obstacle, UAV will need to regulate its velocity since at some point UAV will overtake the obstacle to avoid the obstacle.

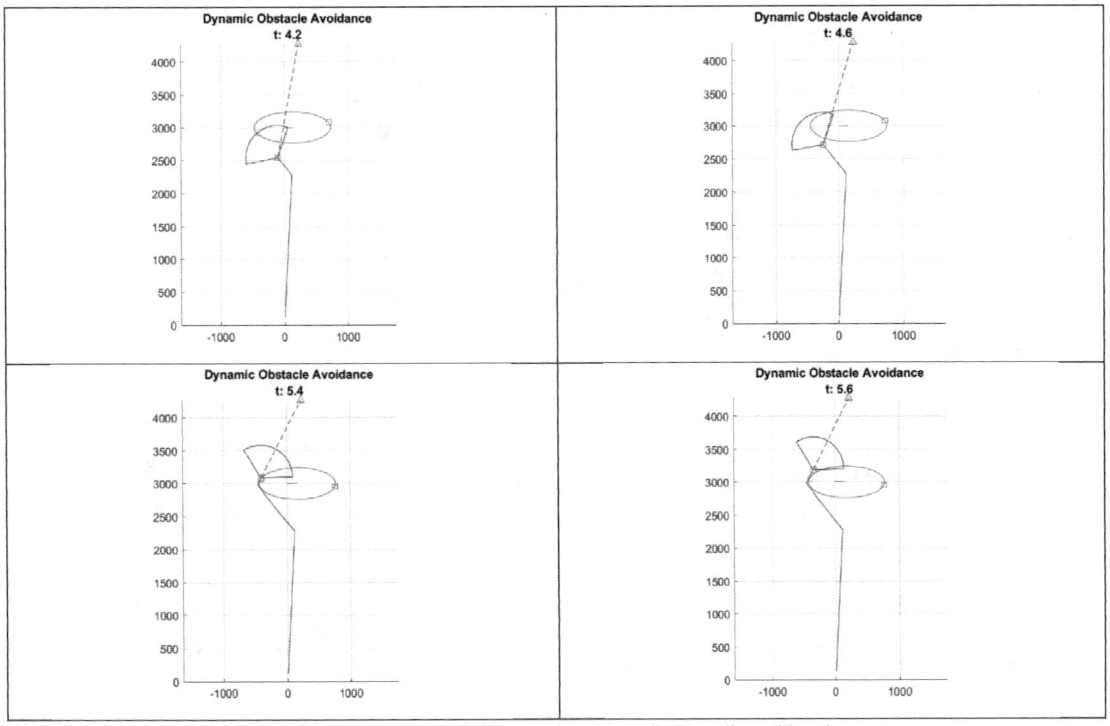

Figure 6. Parallel Case with Unsafe Heading Check

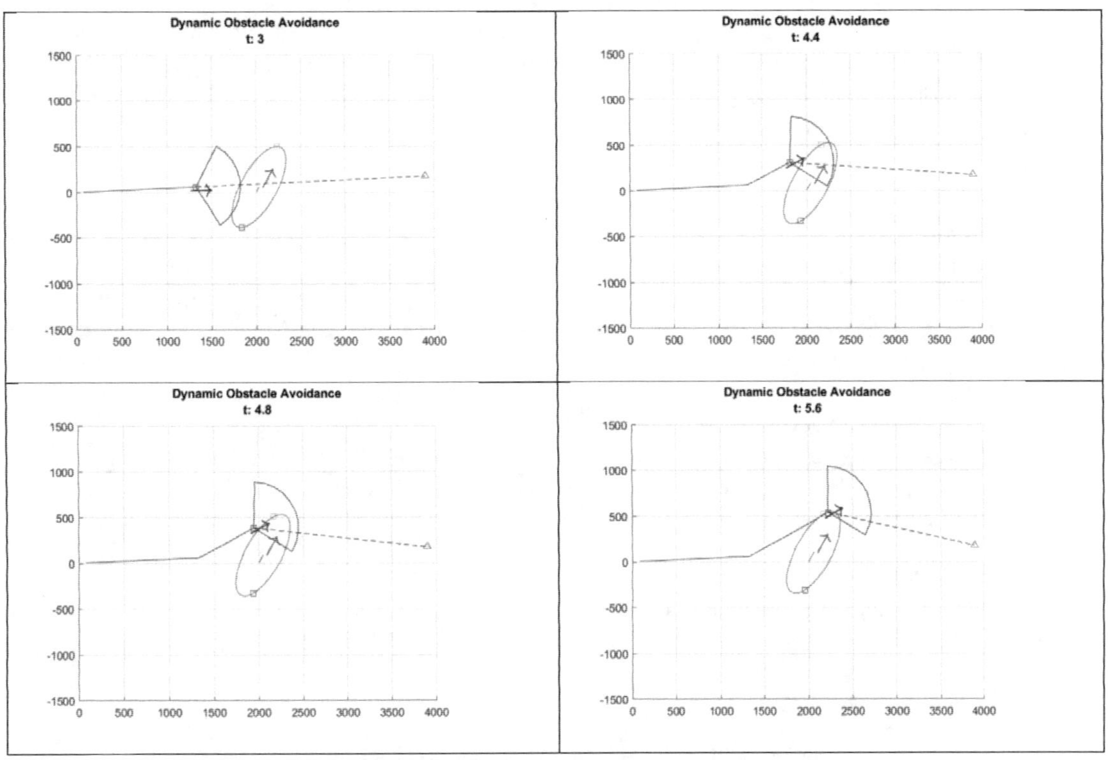

Figure 7. Head Contact Point Selection

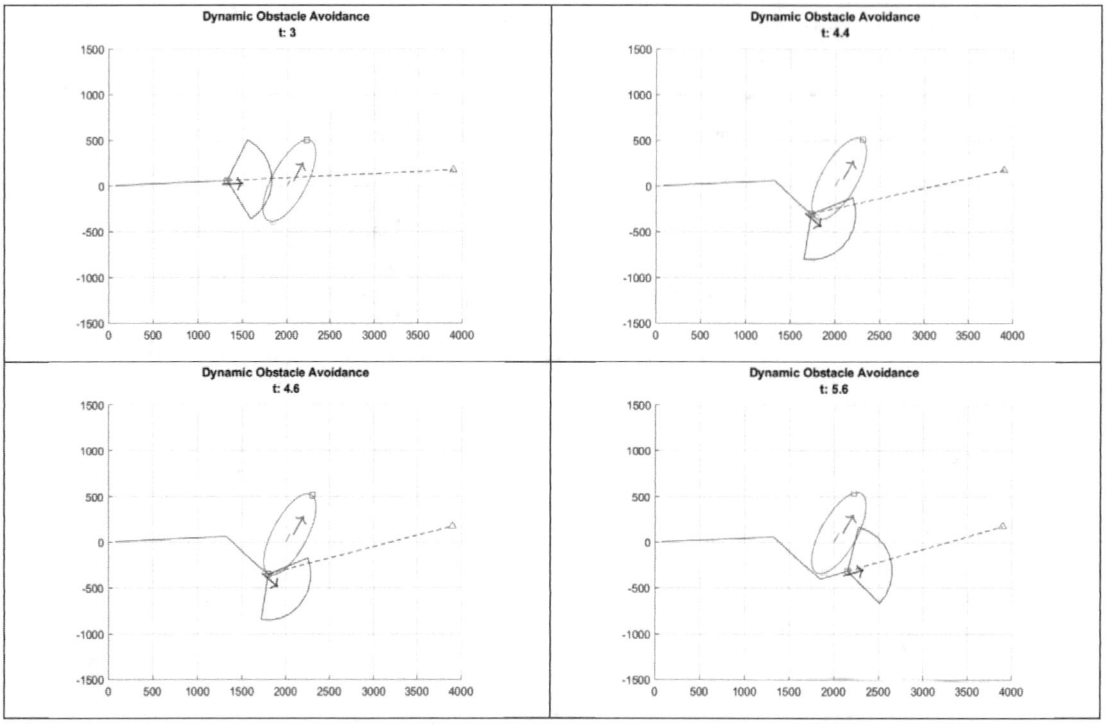

Figure 8. Tail Contact Point Selection

In Figure 8, the UAV is going to opposite direction of the obstacle by selecting tail as the contact point preference. In this case, since the direction is in opposite direction of the obstacle, the UAV is avoiding the obstacle at much faster rate than head case. But depending on the distance of the obstacle and the UAV, larger heading rate might be needed so that the UAV will be able to reach calculated safe waypoint at given time.

4. Conclusions

Collision avoidance system for dynamic obstacle has been developed and simulated. Although the static obstacle avoidance can be readily extended to dynamic avoidance problem, there are adjustments that needs to be done. There are two adjustment that proposed, first is to use the vector between the UAV and obstacle to calculate unsafe heading check so that oscillation is avoided and using head and tail criteria in selection of safe waypoint alternative to accomodate direction of obstacle movement. Next, the effect of tail and head selection for the avoidance will be explored further so that the algorithm can automatically determine in which direction will give the optimal route for the avoidance. Also, implementation of avoidance algorithm to aircraft dynamic model as guidance will be investigated.

Acknowledgements

We gratefully acknowledge the support from Flight Physics Ressearch Group Institut Teknologi Bandung for the P3MI funding.

References

[1] S. M. Adams and C. J. Friedland, "A SURVEY OF UNMANNED AERIAL VEHICLE (UAV) USAGE FOR IMAGERY COLLECTION IN DISASTER RESEARCH AND MANAGEMENT," p. 9.

[2] J. S. McCarley and C. D. Wickens, "Human Factors Implications of UAVs in the National Airspace," p. 63.

[3] P. Riseborough, “Automatic Take-Off and Landing Control for Small UAV’s,” p. 9.

[4] H. Alturbeh, “COLLISION AVOIDANCE SYSTEMS FOR UAS OPERATING IN CIVIL AIRSPACE,” Cranfield University, 2014.

[5] A. Mujumdar and R. Padhi, “Evolving Philosophies on Autonomous Obstacle/Collision Avoidance of Unmanned Aerial Vehicles,” *Journal of Aerospace Computing, Information, and Communication*, vol. 8, no. 2, pp. 17–41, Feb. 2011, doi: 10.2514/1.49985.

[6] S.-C. Han and H. Bang, “Proportional Navigation-Based Optimal Collision Avoidance for UAVs,” *Journal of Control, Automation and Systems Engineering*, vol. 10, no. 11, pp. 1065–1070, Nov. 2004, doi: 10.5302/J.ICROS.2004.10.11.1065.

[7] H. Alturbeh and J. F. Whidborne, “Real-time obstacle collision avoidance for fixed wing aircraft using B-splines,” in *2014 UKACC International Conference on Control (CONTROL)*, Loughborough, UK, Jul. 2014, pp. 115–120, doi: 10.1109/CONTROL.2014.6915125.

[8] Y. Zhou and J. S. Baras, “Reachable set approach to collision avoidance for UAVs,” in *2015 54th IEEE Conference on Decision and Control (CDC)*, Osaka, Dec. 2015, pp. 5947–5952, doi: 10.1109/CDC.2015.7403154.

[9] R. A. Sasongko and S. S. Rawikara, “Collision Avoidance System for Small Unmanned Aerial Vehicle (UAV),” p. 6.

[10] D. P. Bertsekas, *Constrained optimization and Lagrange multiplier methods*. Belmont, Mass: Athena Scientific, 1996.

High altitude long endurance of unmanned aerial vehicle (UAV) ITB solar panel conceptual design

M A Moelyadi[1, *], M A Sulthoni[2], M F Zulkarnain[1], M F Akbar[2], and B K Assakandari[2]

[1] Flight Mechanics Research Group, Faculty of Mechanical and Aerospace Engineering, Bandung Institute of Technology, Jalan Ganesha No. 10 Bandung 40132, Indonesia

[2] School of Electrical Engineering and Informatics, Bandung Institute of Technology, Jalan Ganesha No. 10 Bandung 40132, Indonesia

*Corresponding E-mail: moelyadi@ae.itb.ac.id

Abstract. One of the factors for the long endurance of high altitude long endurance (HALE) of an unmanned aerial vehicle (UAV) that belong to Institut Teknologi Bandung, Indonesia or called as HALE UAV ITB is the design of battery and solar panel. This paper discusses the method to calculate battery and solar panel requirements for HALE UAV ITB to fly for more than 24 hours at 20.000 feet. Several designs, configuration, and type of solar panels are analysed to fulfil HALE UAV ITB requirements. The variable compared is solar panel area, weight, and types. As a result, the use of Monocrystalline Silicone solar cell was found to be much more efficient than Polycrystalline Silicon, CIGS Thin Film, and Amorphous Silicon Thin Film with power generated per area of 224 Watt/m2 and power per weight of 436 Watt/kg. The minimum area required for a Monocrystalline Silicone type solar cell to generate power is 5.94 m2 with a total cell weight of 5.31 kg.

Keywords: unmanned aerial vehicle, high altitude long endurance, battery, solar panel

1. Introduction

Development of a High Altitude and Long Endurance (HALE) UAV may align to current issues such as regional border, disaster, and foreign areas. However, the development of a HALE UAV has its own obstacle, notably in obtaining high altitude flight and long flight duration. To obtain these capabilities, several aspects need to be considered: aerodynamic efficiency, motor-prop efficiency, battery and solar panel, lightweight airframes, efficient control, and optimum flight mission [1].

One of the most successful HALE UAV is Zephyr UAV [2] which set a record with Zephyr-s at 2018 flying for 26 days with recorded highest altitude at 74,000 feet [3]. Zephyr UAV uses carbon construction, flying at a very high altitude during the day while descending to a lower altitude during night to conserve energy. Other development including: Atlantik Solar, which record 3 days endurance flying at constant altitude of 2,000 feet [4]; Solar Impulse, manned HALE which has a record of 5 days of flight with the highest altitude of 28,000 feet [5]; and Korean Aerospace Research Institute with EAV-3 which has reached 18,000 km with using own design propeller at 2018 [6].

Institut Teknologi Bandung is currently developing a HALE UAV. It has MTOW of 50 kg and a wing loading of 2.1 kg/m2. The aircraft has twin wing and triple boom configuration. Propulsion is provided by two electric motors. Battery is used as energy storage and the solar cell is used to charge the battery, to provide endurance of more than 24 hours. HALE ITB has a mission to climb and cruise at 20,000 feet, while at night descend to 5,000 feet to reserve energy. The HALE ITB UAV is shown in figure 1.

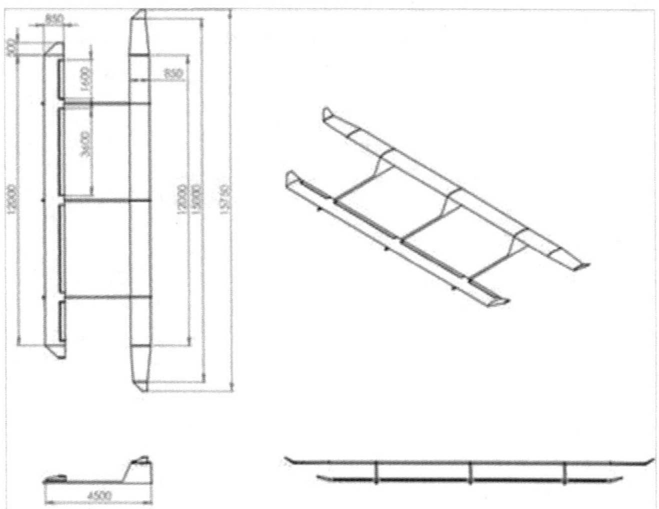

Figure 1. Hale UAV ITB

This study aims to discover the design, configuration, and type of solar panels which fulfil HALE UAV ITB requirements by comparing the solar panel area, weight, and types.

2. Materials and Methods

The method to calculate total solar panel and battery are shown in figure 2. It is an iterative method by calculating the required power and energy for given MTOW, performance, mission profile, and motor data. Power and energy required can then be used to calculate solar cell weight and area. The resulting weight and area are then checked so that the MTOW does not exceed initial guess, and solar panel area does not exceed available space on the wing. If the MTOW exceeds the initial guess, then the larger MTOW will be the new guess and then the iteration will restart.

Required power and energy will be calculated using the concept of total drag experienced equal to total thrust required. Total drag is obtained by assuming thrust to weight ratio for each phase of mission profile. It is assumed 0.03 for cruise and 0.22 for climb.

Total thrust is provided by the motor and propeller. Thrust provided are a function of airspeed, altitude, propeller efficiency, airspeed, and motor rpm. Motor requirements are obtained from its website [7]. However, as the specifications provided are only for static and sea level conditions, the motor-propeller is then replaced with NACA propeller with similar specifications from references [8], which has specifications for varying speed and rpm. It is found that T-motor 30" and 32" is similar to NACA Prop with 14 degre pitch. The corrected thrust from motor and propeller combination, at various airspeed, rpm, and altitude then can be obtained. This corrected thrust will then correspond to a certain throttle and electric power required.

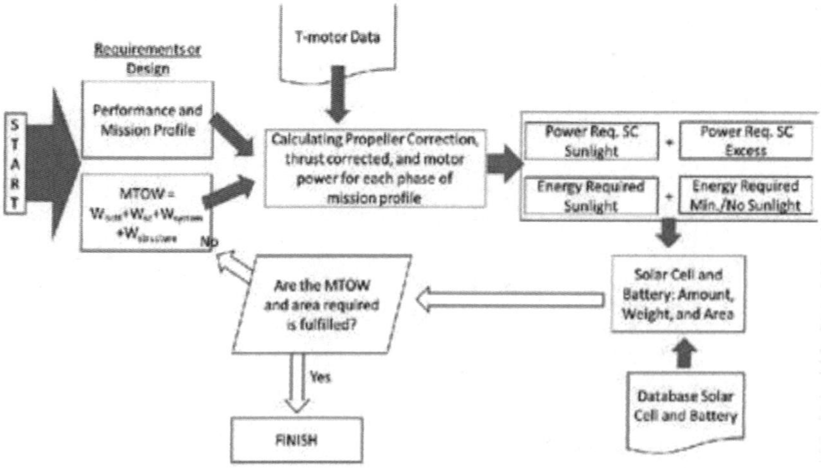

Figure 2. Methodology for calculating solar cell

The electric power required will be calculated for every phase in the mission profile. The mission profile is shown in figure 3. There are day cruise at 20,000 feet for 8 hours and night cruise at 5,000 feet for 10 hours. For climb, will only be calculated at sea level and for 2 hours. Electric power required times the duration of each phase will results in energy required. The total energy required for limited sunlight operation is used to calculate battery, while the total energy required is used to calculate solar cell.

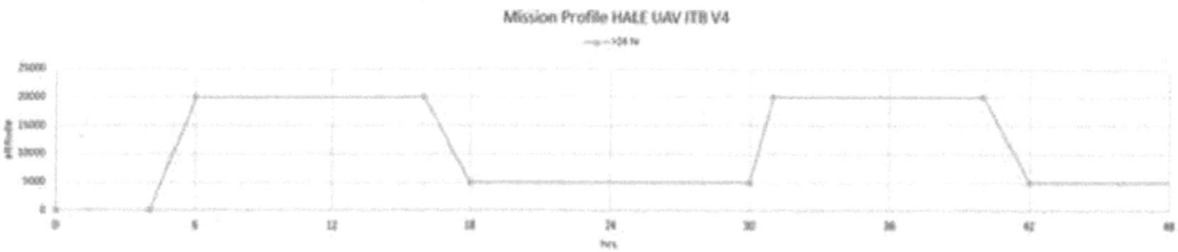

Figure 3. Mission profile of HALE UAV ITB

In this section, some data are taken from several types of solar panels. solar panel data is obtained from measurements and datasheets. Some parameters to consider are the length, width, the weight of each cell, and efficiency. One additional parameter namely purchasing power is taken into account as cost - effective also an important factor in the production. These solar panels are obtained from several marketplaces.

Table 1. Solar Panel Data [12-17

Name	Type	Width (mm)	Length (mm)	Cell Weight (g)	Efficiency (%)
Sunpower Maxeon C60 Cell	Monocrystalline Silicon	125	125	8	22.4
Xinpuguang PV Cell	Monocrystalline Silicon	125	125	8	19.6
Sunyima Solar	Monocrystalline Silicon	125	125	5	17.28
Xiamen Esunpower Solar	Polycrystalline Silicon	156	156	12	18.4
Sunpower Flexible	Monocrystalline Silicon	540	1050	1750	17.6367
SRFLEX04-N Solar	Amorphous Silicon	665	1710	1700	14

SN-CIGS90W	CIGS Panel	546	960	1000	17.2

From the analysis of mission profile and other factors, the power requirements for 24 hours flight obtained to be around 6984 Wh. This power requirement must be met by charging through solar panels. The amount of energy received at ground level from the Sun at its peak depends on the distance to the Sun and the time in. assuming that the value of the Earth-Sun distance is one astronomical unit, the direct sunlight on the Earth's surface when the Sun is at its peak will be around 1000 W / m2.

Through analyzing data from the previous table and assuming that 60% of the sunlight from 12-hours(day) illumination will be effectively converted into electricity, this table will show the ability of each solar panel to receive power from the sun. 60% effective sunlight is assumed by using modified irradiance model from Reference [9].

Table 2. Solar Panel Efficiency

Name	Efficiency (%)	Efficiency (W/m^2)	12-hours Efficiency (Wh/m^2)
Sunpower Maxeon C60	22.4	224	1175.36
Xinpuguang PV	19.6	196	1028.44
Sunyima Solar Cell	17.28	172.8	906.7
Xiamen Esunpower Solar Cell	18.4	184	965.47
Sunpower Flexible Solar Panel	17.6367	176.367	925.42
SRFLEX04-N Solar Panel	14	140	734.6
SN-CIGS90W	17.2	172	902.51

The battery sizing is done by stating some requirements needed for the battery. The propulsion subsystem uses two T-Motor MN1005 as the main thruster. This propulsion subsystem is planned to use 12S (around 48V) Lithium-Ion batteries as its power source. The first requirement is the battery capacity must fulfill total power needed when there is not enough sunlight. The second requirement is to have a combination of batteries that have voltage around 12S or 48V. This combination can be achieved by making twelve 1S batteries in series, six 2S batteries in series, four 3S batteries in series, three 4S batteries in series, two battery 6S in series, or one 12S battery. The 6S in series configurations are preferred because of simplicity of the wiring. The third requirement is the dimension of the battery.

The battery should be small enough to fit inside the main wing and rear wing. The fourth requirement is C rating. C rating of a battery is the maximum output current that the battery can wish to stand [10]. The battery combination must be capable of delivering up to 80A of current, which is the total maximum draw current from both motors. To meet the first requirement, the battery's capacity must be at least the same or more than total energy needed when there is not enough sunlight, which is 4200 Wh. Lithium-Ion batteries, with 243 Wh/kg, achieve the battery capacity at 17.2 kg.

3. Results and Discussion

3.1. Battery

The Navi Power 6S 7000mAh 2.5C Lithium-Ion battery is chosen as the battery to power the propulsion subsystem because it fulfills all the requirements. It weighs 640 grams and has power density 243 Wh/kg. Its size which is 144 x 42 x 62 mm makes the battery fit into both main wing and rear wing [11]. The number of batteries needed for perpetual flight is 26 batteries. The number of batteries is determined based on the total energy needs of the ITB HALE UAV to fly at night. During the day, the solar panel subsystem operates the vehicle's motor & subsystems and charges the battery for the night time.

The total energy is obtained from the calculation between the mission profile and the capabilities of the existing motor. The power needed for HALE UAV ITB to climb is 922 W, for high altitude cruise, it takes 141 W of power, while low altitude cruise requires a power of 88 W. Total energy needed during daytime is 2781 W while at night is 4203 W. So that the total energy required for 24 hours of flying is 6984 W. That way, the battery capacity is 109% compared to the needs.

3.2. Solar Panel Sizing

The weight of the solar panel is an important parameter as the solar panels will be installed on the UAV. Based on the area of the wing that will be used, the number of cells needed to reach the entire wing area could be obtained. By conducting small-scale experiments with a series of solar panels composed of only a few cells and using laminated plastic as the panel's cover and base, we can get the total laminate weight and the weight of lamination for each area per square meter.

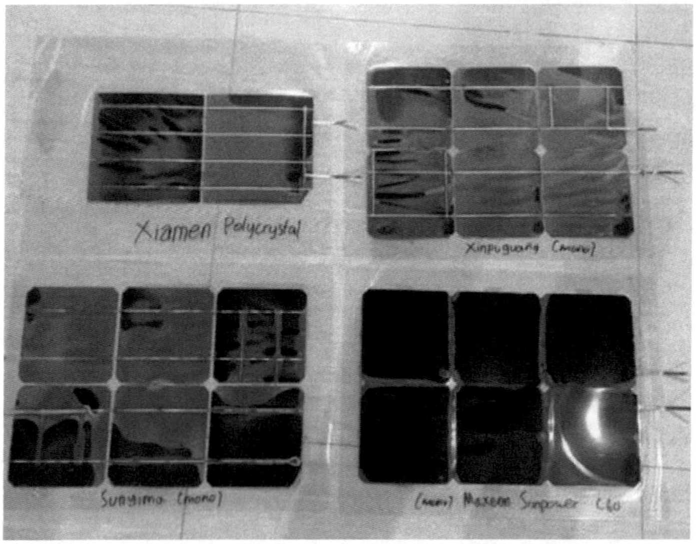

Figure 4. Small-scale experiments result using laminated plastic as solar cell's cover and base

Table 3. Solar Panel Sizing

Name	Area(m^2)	Cell per Panel	Total Cell Weight (kg)	Lamination Weight per Area (g/m^2)	Total Lamination Weight (kg)	Total Panel Weight (kg)
Sunpower Maxeon C60	5.94	381	3.05	277.33	2.26	5.31
Xinpuguang PV	6.79	435	3.48	258	2.95	6.43
Sunyima Solar Cell	7.70	493	2.465	154.8	3.8	6.26
Xiamen Esunpower Solar Cell	7.23	298	3.576	771.36	3.16	5.73
Sunpower Flexible Solar Panel	7.55	14	24.5	-	-	24.5
SRFLEX04-N Solar Panel	9.51	9	15.3	-	-	15.3
SN-CIGS90W	7.74	15	15	-	-	15

IOP Conf. Series: Materials Science and Engineering 1173 (2021) 012055 doi:10.1088/1757-899X/1173/1/012055

From the table, it is shown that the Sunpower Maxeon C60 solar cell has the greatest efficiency (22.4%) compared to other types of solar cells / solar panels followed by Xinpuguang PV (19.6%). Sunpower Maxeon C60 also has the lowest total weight. Xiamen Esunpower Solar Cell has the lowest cell needed per panel with 298 cells followed by Sunpower Maxeon with 381 cells per panel. Xiamen Esunpower also has lowest lamination weight per area making it possibly the lightest solar cell. Sunpower Maxeon C60 has the lowest area needed with the highest power efficiency. All types of solar cells / solar panels need less area than HALE UAV ITB wing total area

3.3. Weight Sizing

By calculating structural weight, battery weight, and solar panel weight, we can determine which solar panel that meets the HALE UAV ITB MTOW (50 kg). By subtracting structural weight, which is predicted to be 22.5 kg, subsystems weight, which is 3 kg, and battery weight, which is 17.3 kg, the maximum weight for solar panels can be obtained. This maximum weight for the solar panels is 7.2 kg. This means Sunpower Maxeon C60, Xinpuguang PV, Sunyima Solar Cell, and Xiamen Esunpower Solar Cell achieved the weight requirements. The best solar cell that has the lowest total panel weight which is 5.31 kg and lowest area which is 5.94 m2 is the Sunpower Maxeon C60. These solar cells have low lamination weight by using laminated plastic.

4. Conclusion

In conclusion, there are four solar panels that meet HALE UAV ITB requirements based on our data. These are Sunpower Maxeon C60, Xinpuguang PV, Sunyima Solar Cell, and Xiamen Esunpower Solar Cell. All of these are monocrystalline and polycrystalline solar cells. These solar cells need an area less than the total wing area of HALE UAV ITB and weigh below maximum weight for solar panel requirement. This is due to high efficiency and low lamination weight by using laminated plastic in covering those solar cells. Sunpower Maxeon C60, which is monocrystalline solar cells achieve the lowest area with 5.94 m2 and lowest total weight with 5.31 kg.

References

[1] Moelyadi, M. A., Rosid, N. H., Hakim, M. L., Rumansyah, K. K., Manurung, F. N., & Darmansya, R. 2018 Preliminary Design and Analysis of Hybrid Solar Powered High Altitude Long Endurance UAV. International Conference on Intelligent Unmanned Systems.
[2] Jane's 2017. High-flying bird: Zephyr remains in the vanguard of solar-powered flight.
[3] Thisdell, D. 2018. Airbus sets flight endurance record with Zephyr UAV. Retrieved 8 26, 2020, from www.flightglobal.com/airbus-sets-flight-endurance-record-with-zephyr-uav/129186.article
[4] Oettershagen, P. et al 2015. A solar-powered hand-launchable UAV for low-altitude multi-day continuous flight. IEEE International Conference on Robotics and Automation.
[5] Solar Impulse Foundation Our Adventure Our Adventure.aroundtheworld. solarimpulse.com/adventure.
[6] Park, D., et al 2018. Design and Performance Evaluation of Propeller for Solar-Powered High-Altitude Long-Endurance Unmanned Aerial Vehicle. International Journal of Aerospace Engineering.
[7] T-Motor. T-Motor Store. store-en.tmotor.com
[8] McCormick, B. W. (1979). The analysis of propellers including interaction effects. NASA.
[9] Duffie, J. A. (2013). Solar Engineering of Thermal Processes. John Wiley & Sons.
[10] Huai Chuangfeng, Liu Pingan,, Jia Xueyan, "Measurement and analysis for lithium battery of high-rate discharge performance," Procedia Engineering 15 (2011) 2619 – 2623.
[11] Navi Power battery store, navi-power.com/product/navi-power-7000mah-22-2v-2-5-6c-6s2p
[12] Sunpower C60 Product Datasheet, www.taipo-tech.com/wp-content/uploads/2017/06 /Sunpower-solar-cell.pdf
[13] Xinpuguang product store, www.aliexpress.com/item/32957112785.html

[14] Sunyima product store, www.aliexpress.com/item/4000277722639.html
[15] Xiamen product store, www.aliexpress.com/item/32626393408.html
[16] Sunpower flexible panel store, www.enfsolar.com/pv/panel-datasheet/crystalline/33938
[17] SRFLEX04-N solar panel product store www.alibaba.com/product-detail/90w-250w-275W-High-Quality-rollable_60634545182.html
[18] SN-CIGS90W product store www.alibaba.com/product-detail/CIGS-Rollable-Solar-Panel-Semi-Flexible_62435314206.html

Take off simulation and analysis of aircraft with twin floats

H A Jakaria, and T Indriyanto*

Department of Aerospace Engineering, Faculty of Mechanical and Aerospace Engineering, Institut Teknologi Bandung, Jalan Ganesha No. 10 Bandung 40132, Indonesia

*Corresponding e-mail: t.indriyanto@ae.itb.ac.id

Abstract. Amphibious aircraft is a multi-function aircraft which can be operated on land and water. This type of aircraft helps transportation sector due to its ability to reach remote areas. Indonesia as the largest archipelago in the world has problems in connecting remote areas. Land transportations and conventional aircraft have difficulties to reach remote areas due to lack of facility. Hence, small aircraft capable of operating on land or water with take off and landing distance less than 1000 m are required. This study conducted take off simulation and analysis for DHC-6 Twin Otter Series 400 equipped with twin float. It tries to propose a simpler method to initiate the analysis of take of performance for float planes with limited data constraints. Several programs were used to generate required data, such as DatCom and SolidWorks. Float plane take-off simulation used in this research utilized Gudmundsson's method. Results showed that the method is suitable and correctly produce take off distance and time similar to those of the real aircraft. Aircraft weight and thrust affect take-off distance and required time, as well as hydrodynamic forces produced. Reduction of weight 10% reduces the distance approximately 17%, while 10% thrust reduction increases take off distance about 15%. In the future this research will be expanded to include other float shapes and sizes, as well as flying boats.

Keywords: take off, twin floats, simulation, DHC-6 Twin Otter

1. Introduction

Indonesian Aerospace Industries (PT. DI) is currently developing a 19 passenger turboprop aircraft. As requirements increase, this aircraft is then developed to fulfil the need of float planes in eastern Indonesian area. The development was performed by simply replacing landing gears with two floats to enable it to take off and land on the water.

Just like any other aircraft, comprehensive analysis is required to assess the performance of this new flying vehicle. Especially for float plane, take off performance analysis is more challenging because of the involvement of water characteristics in the process. Water tunnel testing is also required, in addition to thorough wind tunnel testing. These tests will take a lot of time and money. This paper tries to propose a simpler method to initiate the analysis of take of performance with limited data constraints such as those usually encountered in the beginning of such research.

Gudmundsson's method [1] is recently used in more and more research related to amphibian aircraft design and analysis, such as in [2]. This method is used as the baseline simulation algorithm in this research. Aircraft modelled in this study is the de Haviland Canada DHC-6 Twin Otter which is very similar to PT. DI's aircraft in terms of size and configurations. Simulation were performed for three values of aircraft weight and two throttle settings. Results were then analyzed to check for consistency and accuracy of the method compared to available trusted data.

2. Materials and Methods

This chapter describes the float components of float plane, float plane in take-off maneuver, aerodynamic forces, water forces and aircraft data for simulation.

2.1 Floatplane in Take off

Typical floatplane is given in Figure 1. Floatplanes are conventional landplanes that have been fitted with separate floats or pontoons in place of the wheels. The de Havilland Canada DHC-6 Twin Otter is an example of twin-float plane used widely around the world.

Figure 1. DHC-6 Twin Otter float plane [www.airplane-pictures.net]

Float components are shown in Figure 2. A keel along the bottom center guides the float in a straight line through the water and resists sideways motion. The keel under the front portion is intended to bear the weight of the floatplane when it is on dry land. A step in the middle of the float reduces water drag during take off and high-speed taxi. The step is located slightly behind the center of gravity, approximately at the point where the main wheels are located on a landplane with tricycle gear. The sharp break along the underside of the float concentrates structural stress into this area, and the disruption in airflow produces considerable drag in flight. Without a step, the flow of water aft along the float would tend to remain attached all the way to the rear of the float, creating unnecessary drag. Skeg located behind the step makes it more difficult to tip the floatplane backward when it is on land. Chine guides the water out and away from float, reduces spray and increases hydrodynamic lift. Two sister keelsons add strength to the structure and also function as additional keels. The front of float has a rubber bumper to cushion minor impacts with docks, etc. Spray rails along inboard forward portions of chines reduce the amount of spray hitting the propeller [3].

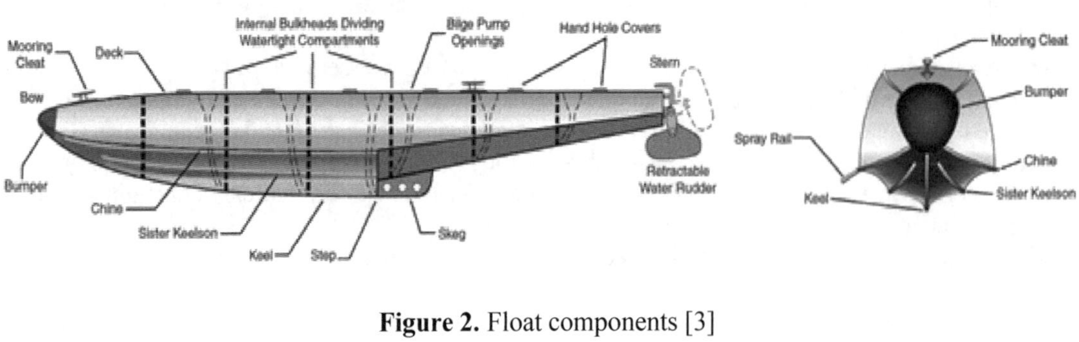

Figure 2. Float components [3]

For light propeller-driven airplanes take off can be defined as the maneuver by which the airplane is accelerated from rest to climb out speed V_C over a 50 ft obstacle [4]. Floatplane take off maneuver can be divided into four distinct phases, i.e. displacement, plowing, planing and lift off phases [3].

Figure 3. Displacement phase.

Figure 4. Plowing phase.

Figure 5. Planing phase.

When taxying at low RPM, floatplane will be in displacement condition similar to being at rest on the water (Figure 3). In this phase the floats support the weight of the airplane through buoyancy where they displace a weight of water equal to the weight of the airplane.

When the power is increased, the floatplane will be in nose-up or plowing position (Figure 4). In this phase pilot holds elevator control in full aft position. Aerodynamic and hydrodynamic lift begins to support a certain amount of the weight, while the rest is still supported by buoyancy. Hydrodynamic lift increases with speed, but hydrodynamic drag increases more quickly.

Accelerating the aircraft faster makes it passes through the hump speed where the water drag is maximum. To minimize water drag, pilot relaxes elevator back pressure, thus allowing the seaplane to assume a pitch attitude that brings the aft parts of the floats out of the water. The step makes this possible. The floats are now riding the water in a level position on the step (Figure 5). Higher speed will increase aerodynamic lift, while hydrodynamic lift will decrease as the float displacement lowers. When running on the step, a small part of the front float supports the seaplane.

Lift off phase transfers support from floats to wings by applying back elevator pressure. The seaplane will lift off the water and become airborne. To avoid porpoising, proper pitch angles need to be maintained. Too much back elevator pressure will force the aft part of the floats to dive deeper into the water and retard the take off. On the contrary, insufficient back elevator pressure will cause the front part of the floats to remain in the water and result in excessive water drag.

2.2 Aerodynamic forces

Aerodynamic forces acting on aircraft are lift and drag. These forces will depend on the air density, airspeed, as well as the shape and configuration of wing, fuselage, etc. Equations for lift and drag forces are as follows [4]:

$$L = \frac{1}{2}\,\rho\,V^2\,S\,C_L \tag{1}$$

$$D = \frac{1}{2}\,\rho\,V^2\,S\,C_D \tag{2}$$

where ρ is air density, S is aircraft wing area, V is aircraft airspeed, C_L is lift coefficient and C_D is drag coefficient. Values of aircraft's C_L and C_D as function of angle of attack, airspeed and aircraft configurations can be predicted using software such as DatCom or AAA.

2.3 Water forces

Two water forces acting on floats are also lift and drag. Water lift is generated from buoyancy (hydrostatic) and movement of float above the water (hydrodynamic force). Buoyancy or displacement Δ is experienced by any object immersed in a fluid and equals to the weight of the displaced fluid [1]:

$$\Delta = \rho_w\,g\,\nabla \tag{3}$$

where ρ_w is the water density, g is the gravitational acceleration and ∇ is the volume of displaced fluid. The ∇ value as function of the depth of float immersion or draft can be calculated using SolidWorks.

Hydrodynamic lift can be formulated similar to equation (1). However, the development of water lift coefficient will be very complicated since it involves various factors such as the float shape, angle of attack, air-water interaction, etc. Prediction of this lift coefficient should use methods such as computational fluid dynamics or water tunnel models. Furthermore, since the water density is approximately one thousand times the air density, water lift will quickly build up as the speed increases. This water lift can overcome aircraft weight, while actually the aerodynamic lift is not yet ready to support the weight. This condition will lead to porpoising phenomenon in which the airplane is thrown several times to the air by the water. Actual take off of floatplanes is such that the transition from full water support to full air support can be performed smoothly. This can be achieved when the take off maneuvers are performed such that the lift of water can be maintained as small as possible, so that only the aerodynamic lift and buoyancy support the weight of aircraft. Transition from floating to flying occurs only by an exchange between the two forces. This approach will be used in this simulation.

In [1] it is assumed that the water drag can be represented by water resistance and Froude resistance. Water resistance depends on the design and shape of the float, water density and speed. Typical example of water resistance from NACA tank test is given in Figure 6 for two types of float. Water resistance will increase from zero at rest to maximum at hump speed. It will then decrease to zero again at lift-off speed when aircraft move from plowing to planing attitude.

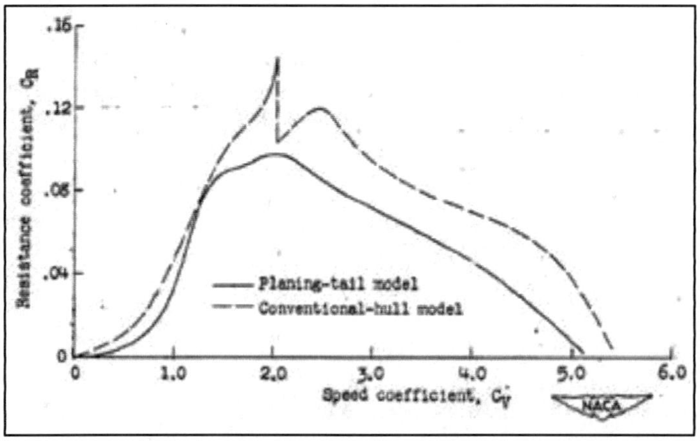

Figure 6. Typical values of resistance coefficient as function of speed coefficient [5]

Based on such experiments, water resistance R_w can be calculated as follows [1]:

$$R_W = C_R \frac{\Delta}{C_{\Delta 0}} \quad (4)$$

where C_R is water resistance coefficient and $C_{\Delta 0}$ is displacement coefficient at zero speed. The speed coefficient C_V is defined as:

$$C_V = V\,(g\,B)^{-\frac{1}{2}} \quad (5)$$

Froude or hydroplanning resistance depends on immersed area and type of float, as well as water speed. As the name implies, this resistance is taken into account when hydroplanning occurs and the float is getting on the step, i.e. from hump speed to lift-off speed. This force can be formulated as [1]:

$$R_F = f\,S_{wet}\,V^n \quad (6)$$

where f is coefficient of frictional resistance, S_{wet} is wetted area of the float and n is a constant which depend on surface quality. The f coefficient varies depending on surface quality, but values between 0.012 – 0.010 can be applied for smooth surfaces such as floats. The n value of 2 is accurate enough for most applications.

2.4 Data for Aircraft Simulation Model

General configuration for DHC-6 Twin Otter amphibian is given in Figure 7. It is a high-wing, twin-turboprop aircraft with twin float and retractable landing gears. The floats do not have water rudder.

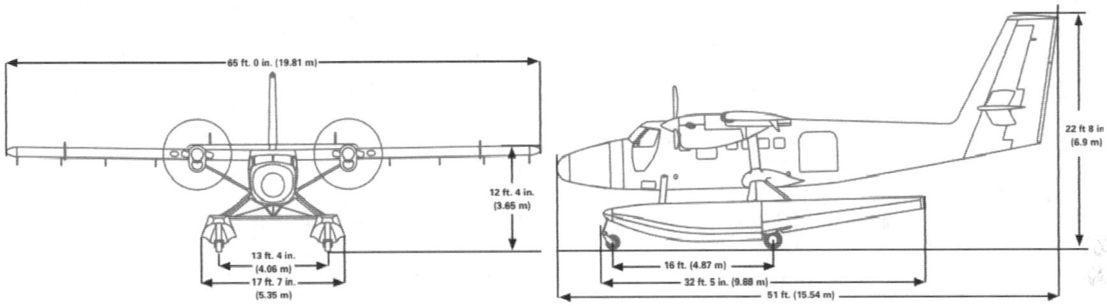

Figure 7. Front- and side-view of DHC-6 Twin Otter floatplane [6].

Table 1 shows some key data for the aircraft. The first 4 data are used in simulation, while the last 3 are used to compare the results.

Table 1. General data for DHC-6 Twin Otter floatplane [6].

Data	Value	Unit
engine power	750	shp
engine efficiency	0.85	-
specific fuel consumption	0.06	kg/s
wing area	39	m^2
take off distance to 50ft	599	m
V_{LOF}	35.8	m/s
V_2	41.15	m/s

2.5 Aircraft Modeling in DatCom

Figure 8 shows the modeling of Twin Otter and its float in DatCom.

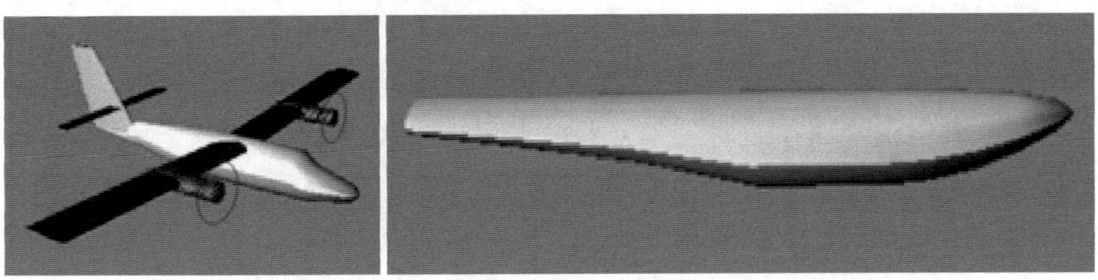

Figure 8. Modeling Twin Otter and float in DatCom [7].

Typical result from DatCom is shown in Table 2. DatCom is capable of estimating values of aerodynamic lift and drag coefficients as function of aircraft angle of attack, airspeed, altitude and aircraft configurations. Such values are used to compose aerodynamic database to calculate aerodynamic lift and drag in simulation.

Table 2. Example of aerodynamic coefficients [7].

Flap deflections [°]	C_L	C_D
0	0.317	0.0270
10	0.446	0.0290
15	0.497	0.0300
20	0.540	0.0320
25	0.579	0.0350
30	0.614	0.0390
37	0.647	0.0440

2.6 Float Modeling

Based on [1] the resistance coefficient of the float can be expressed as:

$$C_R = 0.0011\, C_V^3 - 0.0221\, C_V^2 + 0.1062\, C_V - 0.0149 \quad (7)$$

However, in this case care must be taken, so that C_R values are always positive. Depending on the type of float, water tunnel result such as [5] can also be employed.

SolidWorks software is then used to model the floats as represented in Figure 9.

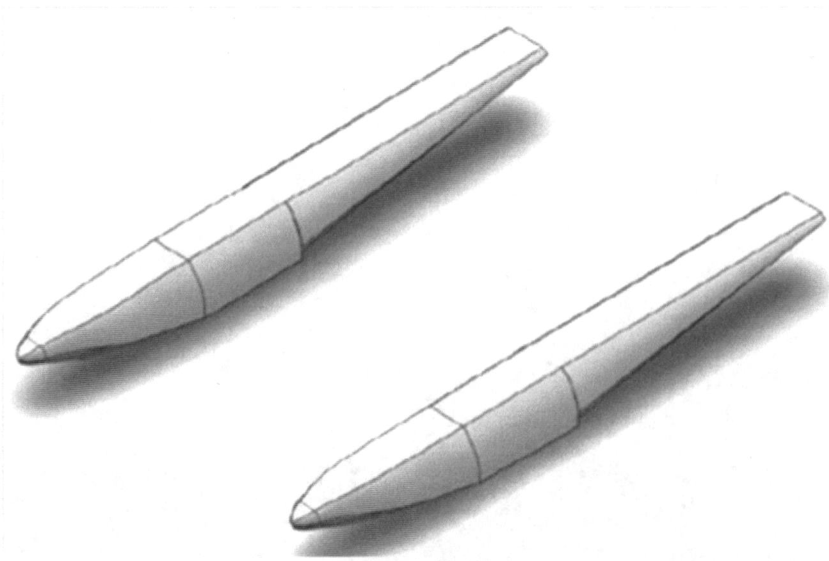

Figure 9. Modeling twin floats in SolidWorks [7].

Based on SolidWorks result, the draft d in [m] and wetted area S_{wet} in [m^2] of the float are then formulated as functions of displacement Δ:

$$d = -1x10^{-9}\, \Delta^2 + 6x10^{-5}\, \Delta \quad (8)$$

and

$$S_{wet} = -6x10^{-9}\, \Delta^2 + 8x10^{-4}\, \Delta \quad (9)$$

2.7 Engine Modeling
Using equation in [8] the maximum thrust obtained from the two engines can be estimated to be 13.8 kN. Variation of engine thrust with airspeed can also be modeled using approach described in [9]:

$$T = T_s \left\{1 - 0.96(M_0 - 0.1)^{1/4}\right\} \tag{10}$$

where T_s is the maximum static thrust. Note that equation (10) is only valid for Mach number M_0 higher than 0.1.

2.8 Simulation
The take-off simulation algorithm as described in [1] can be summarized in Figure 10. Unique characteristics for float planes in take off is shown in the presence of buoyancy, water and Froude resistance calculation. Take-off simulation was started with the aircraft at maximum take-off weight where pilot apply full thrust for take-off until the obstacle height of 50 ft was achieved.

The aircraft weight is then varied to 90% MTOW and 80% MTOW. The last value of weight represents flight condition with 2 crews and some amount of fuel. Engine thrust is also varied at 100% and 80% of its maximum value.

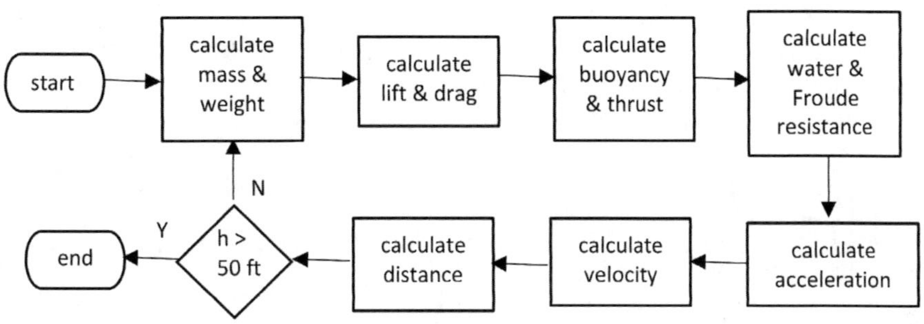

Figure 10. Flowchart for float plane take-off simulation.

3. Result and Discussions

3.1 Effect of Weight on Take-off Performance
Twin Otter's take-off performance for various values of weight is shown in Table 3. From this table it is clear that Gudmundsson method is consistent in estimating the take off distance and time. The heavier the aircraft, the longer the distance and time required to achieve obstacle height, as well as other variables. Heavier weight requires higher lift-off speed and V_2. Heavier aircraft will also result in higher water and Froude resistance, since the floats will be immersed deeper in the water. The resulted values of take off distance, lift-off speed and V_2 shown in this table are quite close to those in Table 1.

Table 3. Take-off performance with weight variation, 100% thrust, standard sea-level.

Variables	Unit	MTOW	0.9 MTOW	0.8 MTOW
Distance	m	603	494	399
Time	s	24.5	20.8	17.5
V_{LOF}	m/s	36.3	35.3	33.8
V_2	m/s	42.1	41.0	40.3
R_w max	N	2767	2649	2507
R_F max	N	347	287	232

3.2 Effect of Thrust on Take-off Performance

Take-off performance of Twin Otter for two values of thrust is given in Table 4. The table shows that the method gives consistent simulation results. The higher thrust applied, the shorter take off distance and time required to clear the obstacle. Other variables are also consistent with previous results. Lift-off speed almost unaffected by thrust, since it is very much dependent on the aerodynamic lift. V_2 will be lower for lower thrust, since the aircraft flies slower. Slower speeds also give higher water resistance since more amount of water will be plowed in front of the floats.

Table 4. Take-off performance with thrust variation, MTOW, standard sea-level.

Variables	Unit	100% thrust	80% thrust
Distance	m	603	789
Time	s	24.5	32.5
V_{LOF}	m/s	36.3	36.8
V_2	m/s	42.1	40.9
R_w max	N	2767	2824
R_F max	N	347	347

3.3. Take-off Distance of Other Float Planes

Comparison of take-off distances for 5 float planes is shown in Table 5 assorted according to their weight. Smaller float planes tend to need longer distances since they have smaller and less powerful engine with lower power loading.

Table 5. Take-off distance for various aircraft with floats.

Aircraft type	Weight [kg]	Power [hp]	Power loading [kg/hp]	Take off distance [m]
DHC-6 Twin Otter [6]	5670	620	9.1	599
DHC-6 Twin Otter simulation	5670	620	9.1	603
Cessna Grand Caravan [10]	4110	867	4.7	610
DHC-2 Turbo Beaver [11]	2721	680	4.0	635
Airshark [12]	1598	300	5.3	610
Cessna 172 [13]	1007	160	6.3	658

4. Conclusions

Gudmundsson method is suitable to be used as simple scheme to analyze take off performance of float planes. It can be applied for take off analysis of other float planes in early design phase. However, to obtain accurate results the simulation still need to be supplied with correct data particularly for float used. Weight of aircraft affects the take-off distance and time, as well as water resistance produced. Reduction of weight by 10% will result in decrease of distance approximately 17% and time about 15%. Thrust also affects take-off distance and time. Thrust reduction of 10% gives approximately 15% longer distance and about 16% more time to reach obstacle height. Smaller float planes tend to need longer distances since they have smaller and less powerful engine with lower power loading. In the future this research will be expanded to include other float designs, configurations and sizes, as well as flying boats.

References

[1] Gudmundsson S 2014 *General Aviation Aircraft Design: Applied Methods and Procedures* (Amsterdam: Elsevier)

[2] Stanislav K 2017 *KR-1 Multi-Mission Amphibian* (Daytona Beach: Embry-Riddle Aeronautical University)

[3] Anon 2004 *FAA-H-8083-23 Seaplane, Skiplane, and Float/Ski Equipped Helicopter Operations Handbook* (U.S. DoT, FAA, FSS)

[4] Ruijgrok G J J 2004 *Elements of Airplane Performance* (Delft: Delft University Press)
[5] Suydam H B 1952 *Hydrodynamic Characteristics of A Low-Drag, Planing-Tail Flying-Boat Hull* (Langley Field: Langley Aeronautical Laboratory)
[6] Viking Air 2018 *Twin Otter Series 400 Brochure* (Wichita: Viking Aerospace)
[7] Jakaria H A 2020 *Simulasi dan Analisis Fase Take-off untuk Pesawat Udara dengan Twin Float* (Bandung: ITB)
[8] Walsh P P and Fletcher P 2004 *Gas Turbine Performance 2nd ed* (Oxford: Blackwell Science)
[9] Mattingly J D, Heiser W H and Pratt D T 2002 *Aircraft Engine Design 2nd ed AIAA Education Series* (Reston: AIAA)
[10] Cessna 2019 *Cessna Grand Caravan EX* (Wichita: Textron Aviation)
[11] Viking Air 2019 *Viking Turbo Beaver Brochure* (Sydney: Viking Air)
[12] Shark Aviation 2019 *The Shark Seaplane* (https://www.flytheshark.com/the-shark-seaplane/ accessed August 4th 2020)
[13] Cessna 1981 *Pilot Operating Handbook Floatplane Model 172P* (Wichita: Cessna Aircraft Company)

Missile threat assessment system using Naive Bayes classifier approach

W Dewangga, and R A Sasongko*

Department of Aerospace Engineering, Bandung Institute of Technology, Jalan Ganesha No. 10 Bandung 40132, Indonesia

*Corresponding e-mail: sasongko@ae.itb.ac.id

Abstract. In the modern era, missiles have been used to serve both defence and offense purposes. Numerous research projects have been conducted to improve various aspects of missile systems. One area of main interest regarding the development of missiles in general is automation, particularly surrounding a feature known as guidance system. This feature had been proven to relieve a missile user of several workloads, particularly during aim and launch phase. One key component of a guidance system is known as threat assessment system. Such system is assigned to assess the threat situation that a missile's engaged in. In this research paper, the developed guidance system uses an approach that is based on the naive Bayes classifier concept. The idea is to establish a system that first describes relationship between a missile and its' target during mission in the form of a set of combat parameters. These combat parameters are then treated as evidence that is used by the Bayesian algorithm to update an assigned prior belief, yielding an updated posterior belief. The belief comes in a form of probability distribution with each probability representing a predefined threat situation class.

Keywords: missile, threat assessment, combat parameters, naive Bayes, probability distribution

1. Introduction

This paper introduces a missile's threat assessment system that is developed based on naive Bayes classifier approach. The developed threat assessment system translates inertial geometric properties of the missile and its' target to relative geometry between the missile and the target. Quantitative result of this translation procedure is defined as combat parameters and the physical situation depicted by these combat parameters is defined as combat geometry [1]. The previously obtained combat parameters are then treated as a bag of evidence and each parameter is assumed to be independent of the rest [2]. Each combat parameter is then set to become input for a corresponding likelihood function. Such likelihood function is usually obtained through the process of training a machine learning program [3].

In this paper, the utilized likelihood functions are assumed to be already obtained in order to preserve simplicity. After the system obtained each combat parameter's likelihood value, these values are processed by the naive Bayes classifier. This process updates an assigned prior belief using the collected evidence, yielding a posterior belief [4]. These beliefs are presented in the form of probability distribution and each probability in the distribution represents a threat situation outcome which acts as a class in a Bayesian classification approach. In this paper, the simulated scenarios involve two missiles and a target.

However, the demonstrated threat assessment program is limited to interaction between the involved missiles with both missiles being featured with identical threat assessment system to preserve simplicity. As additional information, for comparison, another probabilistic approach using similar concept that is known as *Probability of Kill* (*PK*) was used by [5]. The developed *PK* threat assessment program was developed for one-on-one air combat scenarios involving two fighter aircrafts and takes guns or armaments into account.

2. Materials and Methods

2.1 Combat Parameters

Statistical computation performed using any variation of Bayesian approach requires data, acting as set of evidence, as the inputs. In this case, these features can be obtained by defining a set of combat parameters. The established formulas of combat parameters are set to perform translation from independent geometry of two separate agents to expression of one agent's state relative to the other. From here on, the involved agents will be labeled as B (blue missile), R (red missile), and T (target). Since combat parameters implies the interaction between two agents, B's state with respect to R is chosen to be used in the following definition for convenience and demonstrative purposes. The set of combat parameters being used consists of the following terms:

a. *Line of sight (σ)*, which is defined as the angle between B's velocity vector and R's range vector from B's perspective.
b. *Aspect angle (ζ)*, which is defined as the angle between R's velocity vector and R's range vector from B's perspective.
c. *Heading crossing angle (ε)*, which is defined as the angle between B's velocity vector and R's velocity vector.
d. *Range (r)*, which is defined as the distance between B and R. *Range* in vector form is simply R's position relative to B's.
e. *Closure ($\dot{r}$)*, which is defined as the decrement of distance between B and R. Decreasing distance yields positive closure value while increasing distance yields negative closure value.

Mathematical formulation of all five combat parameters above in order are as follows:

$$\sigma_B = acos\left\{\frac{dot(\overrightarrow{r_B},\overrightarrow{V_B})}{\|r_B\|\|V_B\|}\right\} \tag{1}$$

$$\zeta_B = acos\left\{\frac{dot(\overrightarrow{r_B},\overrightarrow{V_R})}{\|r_B\|\|V_R\|}\right\} \tag{2}$$

$$\varepsilon_B = acos\left\{\frac{dot(\overrightarrow{V_B},\overrightarrow{V_R})}{\|V_B\|\|V_R\|}\right\} \tag{3}$$

$$r_B = \|\overrightarrow{pos_R} - \overrightarrow{pos_B}\| \tag{4}$$

$$\dot{r}_B = \frac{-\left(\left\|\overrightarrow{pos_R} - \overrightarrow{pos_B} + \left(\overrightarrow{V_R} - \overrightarrow{V_B}\right)\Delta t\right\| - r_B\right)}{(\|V_B\| + \|V_R\|)\Delta t} \tag{5}$$

and those five combat parameters can be calculated with the assistance of the following two auxiliary parameters, position vector and range vector, which in order are formulated as follows:

$$\overrightarrow{pos_B} = \{x_B, y_B, z_B\}^T \; ; \; \overrightarrow{pos_R} = \{x_R, y_R, z_R\}^T \tag{6}$$

$$\overrightarrow{r_B} = \overrightarrow{pos_R} - \overrightarrow{pos_B} \tag{7}$$

Combat parameters obtained by solving equations (1) through (5) are next treated as a bag of evidence. According to naive Bayes principal, the order of presentation of the evidence is not important. This implies that the order in which the combat parameters are presented by equations (1) through (5) is arbitrarily chosen. All five combat parameters are also independent from one another primarily to comply with one of naive Bayes algorithm's requirements, which is independence between one component of evidence and the other. Secondary advantage of establishing the combat parameters in such manner is isolability, resulting in enhanced convenience during analysis of each parameter's influence to the threat assessment outcomes. In addition, the following figures are visualization of the previously defined and formulated combat parameters. Note that these figures assume missile B's perspective against missile R.

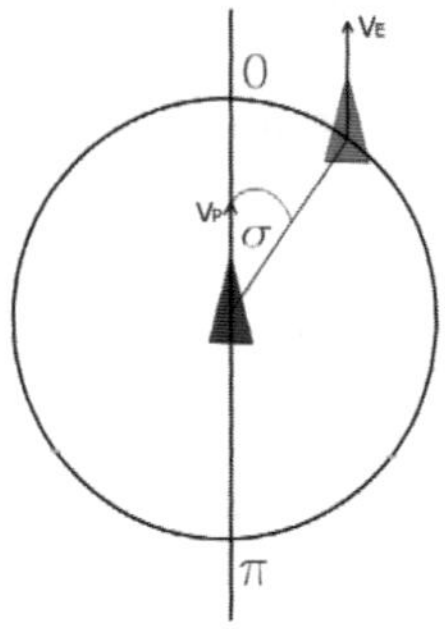

Figure 1. Line of sight

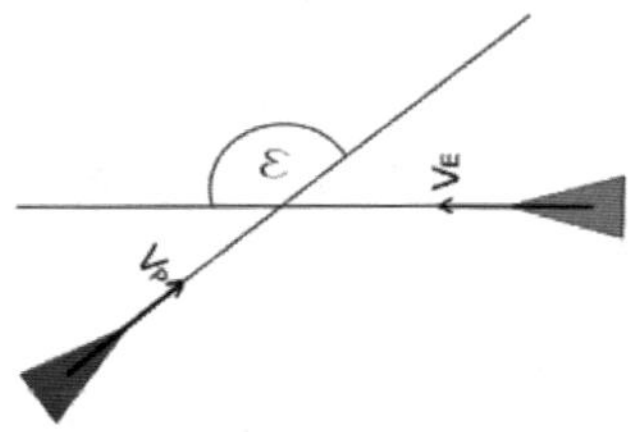

Figure 2. Heading crossing angle

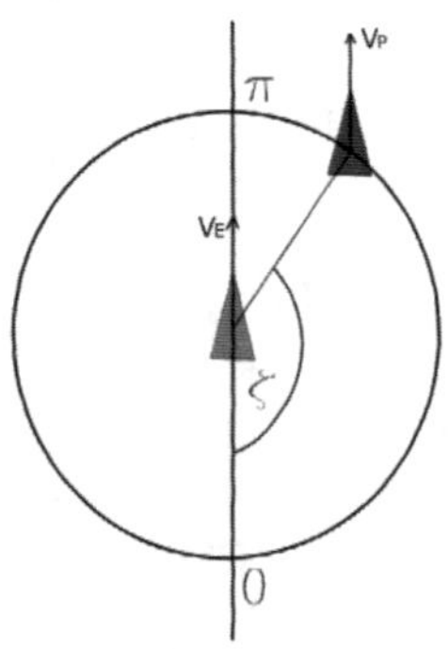

Figure 3. Aspect angle

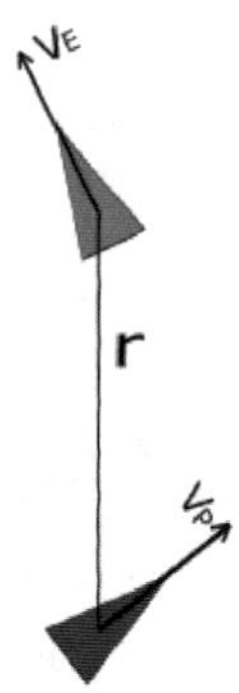

Figure 4. Range

Figure 5. Closure

2.2 Threat Situation Outcomes

In this paper, there are 6 classes that are defined as possible threat situation outcomes. These 6 possible threat assessment outcomes are expansions of the idea of using 4 classes as demonstrated by [6]. Definition of those 6 possible outcomes are as follows:

a. *Miss (M)* situation. When engaged in this situation, it is likely for σ, ε, and r to have high values while ζ value is likely to be low. The value of $\dot{r}$ is likely negative.
b. *Chase (C)* situation. When engaged in this situation, it is likely for σ, ε, ζ, and r to have small values. The value of $\dot{r}$ is likely positive.

c. *Undershoot (U)* situation. When engaged in this situation, it is likely for σ, ε, ζ, and r to have small values. The value of $\dot{r}$ is likely negative.
d. *Lead (L)* situation. When engaged in this situation, it is likely for σ and ε to have high values while ζ and r are likely to be low. The value of $\dot{r}$ is likely positive.
e. *Overshoot (O)* situation. When engaged in this situation, it is likely for σ and ε to have high values while ζ and r are likely to be low. The value of $\dot{r}$ is likely negative.
f. *Frontal (F)* situation. When engaged in this situation, it is likely for σ and r to have low values while ζ and ε are likely to be high. The value of $\dot{r}$ is likely positive.

This expansion, particularly involving derivation of *Chase, Undershoot, Lead* and *Overshoot* from *Advantage* and *Disadvantage*, was made possible by introducing $\dot{r}$ as one of the utilized combat parameters. Provided below is the figure depicting all 6 possible threat assessment outcomes.

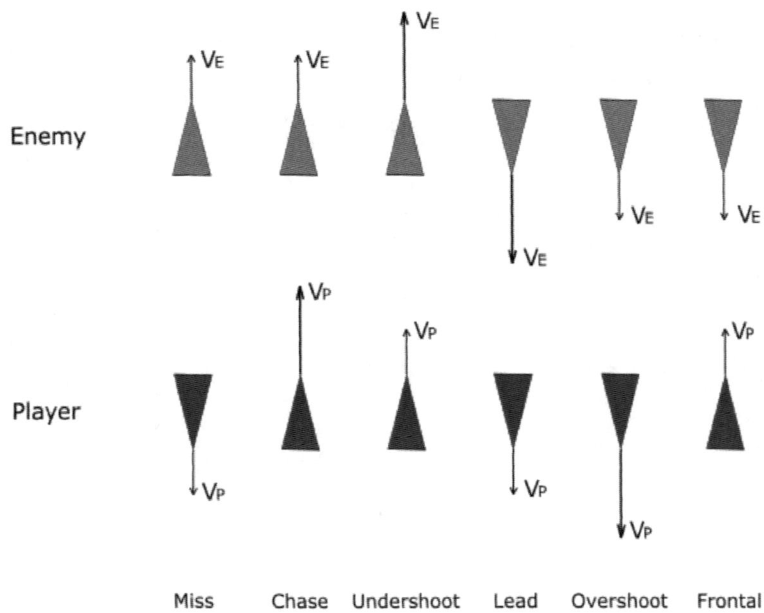

Figure 6. Depiction of the established possible threat situation outcomes

2.3 Likelihood Functions

In the designed threat assessment program, the likelihood of a situation is computed based on the likelihoods of the combat parameters having their current values while being engaged in that situation. Therefore, formulation of a set of likelihood functions involving all utilized combat parameters is necessary. To preserve each combat parameter's independency of the rest, these likelihood functions are set to be single-attributed, meaning that each function only contain one combat parameter. Last, these likelihood functions affect threat assessment outcome but not the other way around. This results in implementation of likelihood functions that are time-invariant being sufficient. The likelihood functions are set to use exponential distribution instead of linear distribution which was used by other researchers [6-7].

Formulation of the set of single-attribute time-invariant likelihood functions being used for depiction of a missile's status against the other missile are as shown in Table 1. The superscript i represents interaction of an agent with one other agent. Parameter ai acts as a constant that affects a function's gradient and maximum likelihood value. Similarly, the constant parameter kir is assigned to manipulate influence level given by r and kiṙ is assigned to regulate influence level given by ṙ. In this paper, the value of is ai set to be 1. The value of kir and kiṙ are set to be 2. In addition, the value of ai should be equal or greater than 1 to avoid any occurrence of negative likelihood value. Regardless, the values of the constants can be arbitrarily chosen [7].

Table 1. List of utilized single-attribute and time-invariant likelihood functions

Threat Situation	*Likelihood Function*	*i = BR, RB*
M, L, O	$L_\sigma^i = exp\left\{a^i\left(\frac{\sigma^i}{\pi}\right)\right\} - 1$	(8)
C, U, F	$L_\sigma^i = exp\left\{a^i\left(1 - \frac{\sigma^i}{\pi}\right)\right\} - 1$	(9)
L, O, F	$L_\zeta^i = exp\left\{a^i\left(\frac{\zeta^i}{\pi}\right)\right\} - 1$	(10)
M, C, U	$L_\zeta^i = exp\left\{a^i\left(1 - \frac{\zeta^i}{\pi}\right)\right\} - 1$	(11)
M, F	$L_\varepsilon^i = exp\left\{a^i\left(\frac{\varepsilon^i}{\pi}\right)\right\} - 1$	(12)
C, U, L, O	$L_\varepsilon^i = exp\left\{a^i\left(1 - \frac{\varepsilon^i}{\pi}\right)\right\} - 1$	(13)
M	$L_r^i = exp\left\{k_r^i a^i\left(\frac{r^i}{r_{max}^i}\right)\right\} - 1$	(14)
C, U, L, O, F	$L_r^i = exp\left\{k_r^i a^i\left(1 - \frac{r^i}{r_{max}^i}\right)\right\} - 1$	(15)
C, L, F	$L_{\dot{r}}^i = exp\{k_{\dot{r}}^i a^i(\dot{r}^i)\}$	(16)
M, U, O	$L_{\dot{r}}^i = exp\{k_{\dot{r}}^i a^i(-\dot{r}^i)\}$	(17)

2.4 Naive Bayes Algorithm

Following the referred naive Bayes algorithm, the outputs of prior likelihood L functions are then processed in the way that is expressed by Equation (18) below.

$$L(c_k^i|\Gamma_k^i = j) = L(\sigma_k^i|\Gamma_k^i = j)L(\zeta_k^i|\Gamma_k^i = j)L(\varepsilon_k^i|\Gamma_k^i = j)L(r_k^i|\Gamma_k^i = j)L(\dot{r}_k^i|\Gamma_k^i = j) \quad (18)$$

The subscript i corresponds to either B or R missile, the subscript j corresponds to a possible threat situation outcome, and the subscript k denotes corresponding stage at which the computation is performed. Equation (18) yields the likelihood of all possible outcomes given the obtained evidence. The probability P of each outcome can be obtained by evaluation formula shown in Equation (19) below.

$$P\left(c_k^i \middle| \Gamma_k^i = j\right) = \frac{L\left(c_k^i \middle| \Gamma_k^i = j\right)}{\sum_{Q=1}^{6} L\left(c_k^i \middle| \Gamma_k^i = Q\right)} \tag{19}$$

The idea of Bayesian approach in general is that update of assumption or belief is bound to happen given the accounted evidence [8]. This implies that the calculation process requires the existence of prior belief or assumption, which is formally defined as prior probabilities. Calculation process yields updated belief or assumption, formally defined as posterior probabilities. This procedure requires initialization of belief. In this paper, the initial belief vector represented by P is set to be [1/6 1/6 1/6 1/6 1/6 1/6] for both missiles.

3. Results and Discussions

Two scenarios are presented in this section. Both scenarios were given similar initial setup. The difference between the two are the maneuvers assigned to both missiles. The target is assigned with maneuvers too, but results involving the target are not recorded aside from its trajectory since this paper focuses only on the interactions between the involved missiles. Initial setup of the scenarios is as shown in Table 2. Simulations are performed numerically using MATLAB as the numerical platform.

Table 2. Initial setup of simulated scenarios

Initial Parameter	Parameter Description	Blue Missile (B)	Red Missile (R)	Target (T)
$x_e\,(m)$	position in x axis	3000	2000	5000
$y_e\,(m)$	position in y axis	4000	4000	1000
$h_e\,(m)$	altitude	3000	2000	0
γ *(deg)*	pitch angle	0	0	0
χ *(deg)*	heading angle	0	0	0
V (m/s)	velocity	250	280	100

3.1. 1st Scenario

In this simulation, missile B is assigned to engage a flight path with constant radius and altitude. Missile R and target T are assigned to maintain their respective directions, velocities, and altitudes. The performed simulation consists of 40 stages with Δt between stages value being 1 second. The visualization of the scenario's result is as follows:

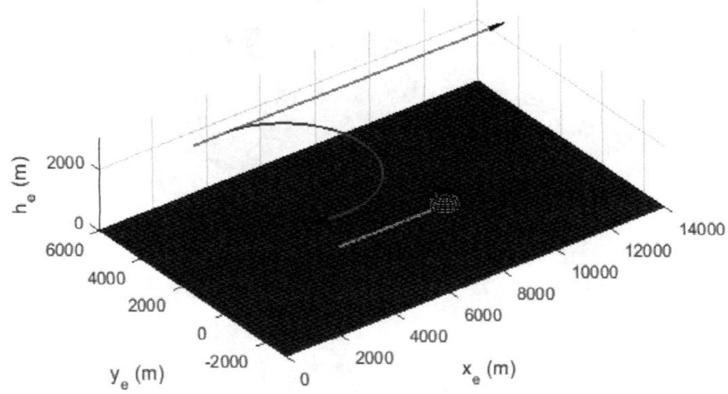

Figure 7. Scenario 1 trajectory visualization

The obtained probability distribution plots are as follows:

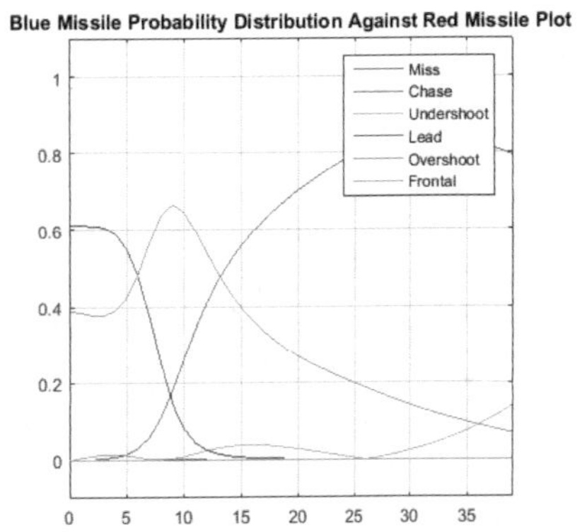

Figure 8. Blue missile's threat assessment outcomes against red missile, Scenario 1

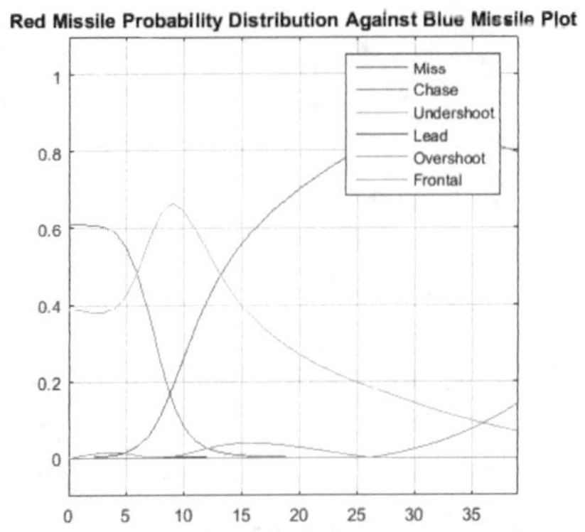

Figure 9. Red missile's threat assessment outcomes against blue missile, Scenario 1

From Figure 7, missile B experienced a change in situation with respect to missile R and vice versa. At early stages of the simulation, missile B happened to engage in *Lead* situation as missile R engaged a *Chase* situation. At latter stages of the simulation, both missiles ended up in a *Miss* situation. This interpretation is supported by Figure 8. Interestingly, Figure 8 captures the threat assessment system's interpretation of a phase of missile B's transition from *Lead* to *Miss* as an *Overshoot* situation. Figure 9 indicates that a similar case happened to missile R with the only difference being missile R's transitional phase was interpreted as a case of *Undershoot*. Both recorded interpretation of the transition phase was due to the exponential likelihood functions being used.

3.2. 2nd Scenario

In this simulation, missile B is assigned to engage a flight path with varying curvature and decreasing altitude, target T is assigned to engage a path with constant radius and altitude, and missile R is assigned to engage a flight path with varying curvature and increasing velocity. The performed simulation consists of 40 stages with Δt between stages value being 1 second. The visualization of the scenario's result is as follows:

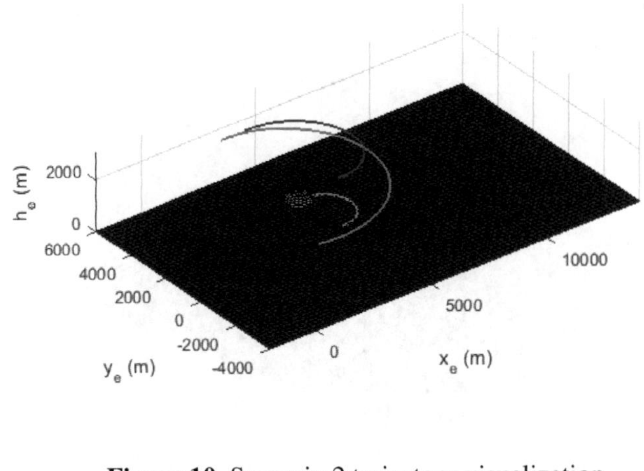

Figure 10. Scenario 2 trajectory visualization

The obtained probability distribution plots are as follows:

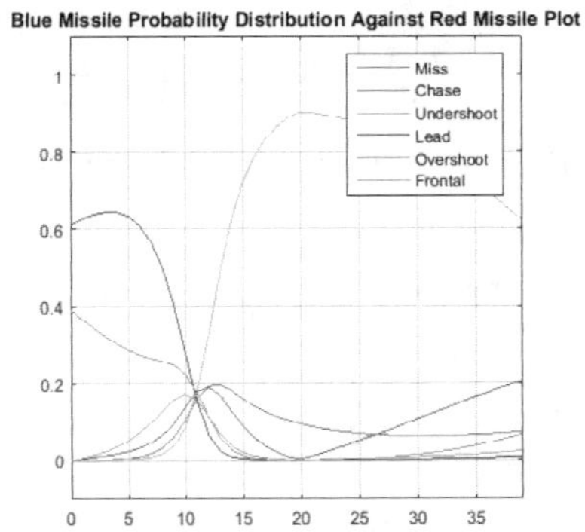

Figure 11. Blue missile's threat assessment outcomes against red missile, Scenario 2

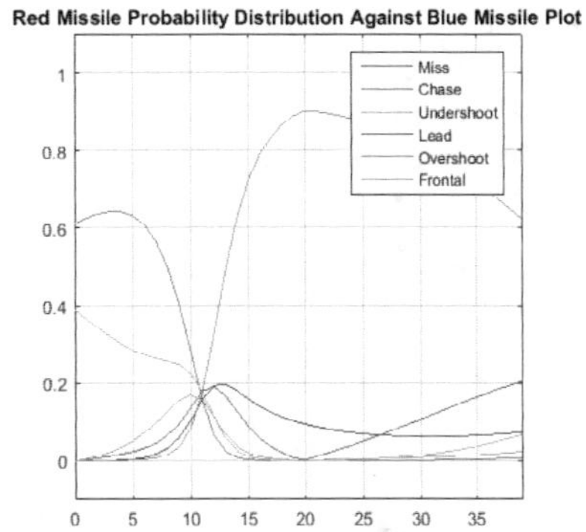

Figure 12. Red missile's threat assessment outcomes against blue missile, Scenario 2

At early stages of the simulation, missile B happened to engage in *Lead* situation as missile R engaged a *Chase* situation. This also happened in Scenario 1. However, at latter stages of Scenario 2's simulation, missile B ended up in an *Undershoot* situation while missile R ended up in an *Overshoot* situation. This interpretation is supported by Figure 11. Interestingly, unlike Scenario 1, no specific threat status visibly dominating the rest is present in Figure 11. Figure 12 indicates that a similar case happened to missile R. However, the cause of the overlapping probabilities during the transition is the same as in Scenario 1, which is the utilization of exponential likelihood functions.

4. Conclusions

In this paper, the threat assessment system is developed based on naive Bayes classifier approach. Five combat parameters in the form of line of sight, aspect angle, heading-crossing angle, range, and closure are defined to act in unison as a bag of evidence to be processed by the predefined exponential, single-attribute, and time-invariant likelihood functions. The obtained set of single-attribute likelihoods are then utilized by the naive Bayes algorithm to transform a missile's probability distribution from an arbitrarily assumed distribution to a distribution given the evidence the system collected. Both scenarios provide threat assessment results that are aligned with the established possible threat status classes. However, due to the usage of exponential likelihood functions, interesting distribution are recorded during phases that are considered transitional. Regardless, the developed threat assessment system works properly.

References

[1] Virtanen, K., Karelahti, J., Raivio, T., and Hämäläinen, R. P. (2006): *Modeling Air Combat by a Moving Horizon Influence Diagram Game*, Helsinki University of Technology, Finland.

[2] Koçkesen, L., and Ok, E. A. (2007): *An Introduction to Game Theory* (lecture notes for Koç University and New York University).

[3] Mitchell, T. M. (1997): *Decision Tree Learning*, McGraw-Hill Companies, Inc., USA.

[4] Ummels, M. (2010): *Stochastic Multiplayer Games: Theory and Algorithms*, Pallas Publication, Amsterdam.

[5] Sugiyanto, M. (2014): *Pengembangan Perangkat Lunak Analisis Probability of Kill (PK) dan Supremacy pada Air Combat Pesawat Tempur* (Institut Teknologi Bandung studies in Aerospace Engineering, Thesis), Institut Teknologi Bandung, Indonesia.

[6] Bergdahl, C. (2013): *Modeling Air Combat with Influence Diagram* (Linköpings universitet studies in Department of Computer and Information Science, Dissertation), Linköpings universitet, Sweden.

[7] Suseno, P. A. P. (2017): *Pemodelan dan Simulasi Pertempuran Udara Satu Lawan Satu Dalam Ruang Tiga Dimensi* (Institut Teknologi Bandung studies in Aerospace Engineering, Thesis), Institut Teknologi Bandung, Indonesia.
[8] Berrar, D. (2018): *Bayes' Theorem and Naive Bayes Classifier*, Tokyo Institute of Technology, Japan.

Sensitivity study and structural optimization of an aircraft wing in the maximum-stress and flutter constraints

M Kusni[1,*], M Widiramdhani[2], and B K Hadi[1]

[1] Department of Aerospace Engineering, Bandung Institute of Technology, Jalan Ganesha No. 10 Bandung 40132, Indonesia

[2] Garuda Maintenance Facility of Indonesia

*Corresponding E-mail: muhammadkusni@gmail.com

Abstract. Optimization process is very important to be done since the preliminary design step, not only for the static load aspect but also dynamic load as well. In the past, aeroelastic behavior aspect is often neglected in this stage and consequently becomes a problem at the later stage of design. This paper contains multi-constraints design sensitivity and optimization which include both maximum stress constraint and damping flutter constraint. Sensitivity study was performed to assist optimization process and makes the process more efficient. Wing structure was modeled as skins, ribs, spar-webs, and spar-caps. Optimization process was done by changing thickness of each component and beam dimension. The methodology used in this paper is a gradient-based optimization developed by MSC Nastran. The final cycle of this optimization has successfully reduced structural weight up to 33% without violating static and flutter constraint.

Keywords: sensitivity study, structural optimization, flutter constraint

1. Introduction

In the design of aerospace structures, it is important to make a structure to be save and as light as possible. A lightweight structure is not only beneficial for a structural aspect but also for other aspects such aerodynamic, cost and aircraft operations. Optimization is needed since the preliminary design stage in order to make an efficient configuration of structure that is as light as possible while satisfy safety requirement.

In order to conduct optimization process for an aircraft structures, a designer should not think for only one aspect, for example, only to satisfy aeroelastic requirement, or static requirement; since each aspect may produce different optimization result. In the past, aeroelastic aspect often neglected in the preliminary design stage. An aircraft designer just concentrates on how to make the aircraft safe from static loading. As a result, for several cases, aeroelastic behavior becomes a problem later. In order to correct this problem, engineers had to make significant additional effort to find feasible solutions. Those solutions are usually non-optional, expensive repair solutions, which must be introduced fairly late in the design process.

An optimization problem consists of maximizing or minimizing a real function by systematically choosing input values from within an allowed set and computing the value of the function. More

IOP Conf. Series: Materials Science and Engineering 1173 (2021) 012058 doi:10.1088/1757-899X/1173/1/012058

generally, optimization includes finding "best available" values of some objective function given a defined domain (or a set of constraints), including a variety of different types of objectives functions and different types domains. A basic goal of design optimization is to automate the design process by using a rational, mathematical approach to yield improved designs. It is useful to produce more efficient designs while having maximum margins of safety, to perform trade-off or feasibility studies, to assist in design sensitivities studies, or to correlate test data and analysis results (model matching).

Optimization studies have been around since a long time. A lot of academic publications have been conducted to get better optimization method. Multidisciplinary objective for design and optimization has been considered to get a comprehensive and accurate result considering many aspects as constraints [1].

It has been presented elsewhere about the development of the structural optimization that begins with the first analytical work in structural optimization [1,2]. Also, other provides theoretical lower bounds on the weight of trusses and although highly idealized, offer considerable insight into the structural optimization problem and the design process [2]. The 1940s and early 1950s is the era of development of component optimization, such as the works of Shanley [3]. During this period, the availability of the digital computer led to the application of linear programming techniques for plastic design of frames. Digital computers were becoming commonly available, and the finite element method was offering a designer a tool for analysis of increasingly complex structures.

Nowadays, MSC Nastran [4] has developed optimization program using gradient-based method to answer structural designer expectation. Nastran SOL 200 has provided optimization code for many applications. This research will study optimization of wing structures for static strength and flutter constraints. This research focus on optimizing a wing structure to make it as-light-as-possible while satisfying static strength and flutter requirements using gradient-based optimization method.

2. Wing-box model

Figure 1 shows wing-box model studied in this paper. It contains skins, spar (spar-web and spar-cap), and ribs. The engine is modeled as a concentrated mass and is positioned in its center of gravity. Fuel is modeled as distributed mass that attached in 24 nodes connected to spar.

In a finite element model, skins, spar-webs and ribs are modeled as shell properties, while spar-caps are modeled as beam properties. The material for all the components is Al 2024-T3. Table 1 shows material properties of Al 2024-T3. For convenient, skins are divided into 5 areas as shown in figure 2; ribs into 4 areas as shown in figure 3 and spar-web is just divided into front and rear spar-web as shown in figure 4. Upper spar-caps are modeled as 'T' beam as shown in figure 5 (left) and lower spar-caps are modeled as inverted 'T' beam.

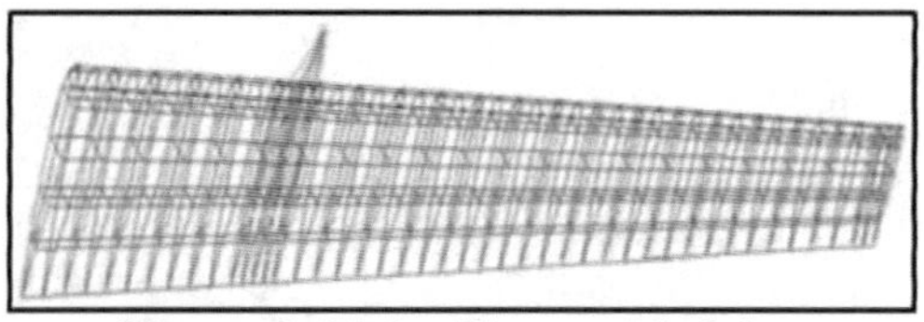

Figure 1. Finite Element Model

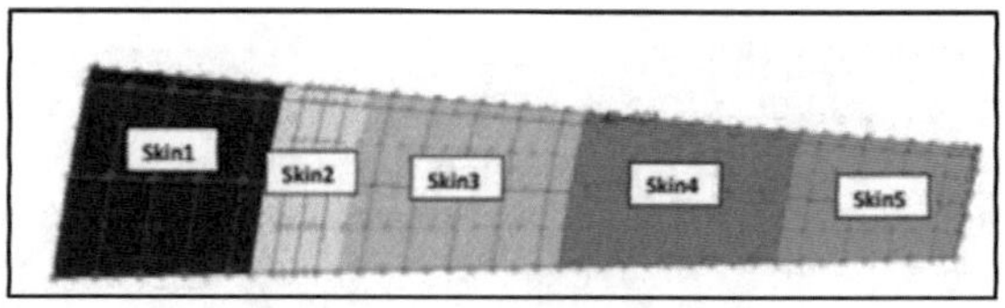

Figure 2. Skin Properties

Table 1. Material Properties

Al 2024-T3	
Density	2.78 g/cc
Modulus of elasticity	73.1 GPa
Poisson's ratio	0.33
Ultimate tensile strength	469 Mpa

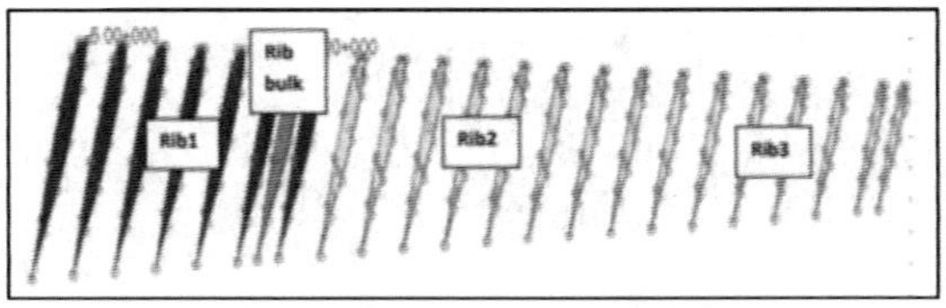

Figure 3. Finite Element Model

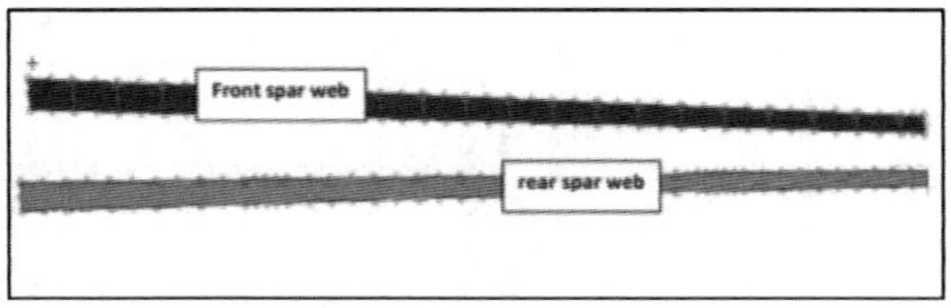

Figure 4. Spar-webs Properties

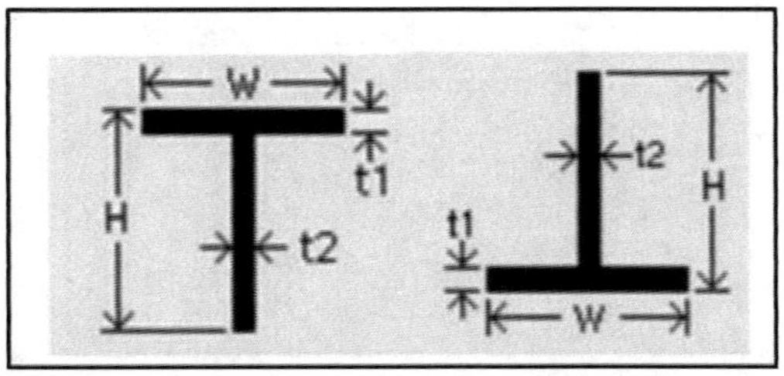

Figure 5. Spar-caps Beam Upper (left) and lower (right)

3. Design Variables and Constraints

The objective of this research is to conduct optimization process in order to minimize the weight of the wing structure while satisfying safety requirement. Optimization process is done by changing thickness of skins, spar-webs, and rib. For the case of spar-caps, optimization process is done by changing the beam properties as shown in figure 5; except the length of the beam. Table 2 consists of initial value, minimum values and maximum values of optimization process for each variable.

There are two constraints in this research. The first one is the stress constraints. Al-2024 has an ultimate tensile strength of 469 MPa; this value become stress constraint to limit the stress value. The second constraint is flutter. In order to avoid flutter occurrence in certain speed, the damping of the structures should be 0.3 at 500 m/s.

Table 2. Design variable

Variabel		Initial (mm)	Min (mm)	Max (mm)
Skin 1/2/3/4/5		4.5	0.8	10
Ribs 1		2.2	1	10
Ribs 2		2	1	10
Ribs 3		1.5	1	10
Ribs Bulk		3	1	10
Spar-web	Front	8.5	1	10
	Rear	5.6	1	10
Upper Spar-cap (Front/Rear)	H	20	10	20
	W	20	10	20
	T1	5	5	10
	T2	5	5	10
Lower Spar-cap (Front/Rear)	H	20	10	20
	W	20	10	20
	T1	5	5	10
	T2	5	5	10

4. Optimization Process

4.1 Basic Optimization Problem

Nastran uses basic optimization problem statement to optimize the structure. A basic optimization problem statement is defined as follows:

Find X to minimize (or maximize) F (X) objective, subjected to Eq. 1 – 4.

$$g_j(X) \leq 0 \qquad j = 1, \ldots\ldots, n_g \qquad \text{inequality constraints} \tag{1}$$
$$h_k(0) = 0 \qquad k = 1, \ldots\ldots, n_h \qquad \text{equality constraints} \tag{2}$$
$$x_i^l \leq x_i \leq x_i^u \qquad i = 1, \ldots\ldots, x_n \qquad \text{side constraints} \tag{3}$$
$$X = \{x_1, x_2, \ldots\ldots\ldots, x_n\} \qquad \text{design variables} \tag{4}$$

The objective function is a scalar quantity to be minimized. Side constraints are placed on design variables to limit the region of search. The inequality constraints are expressed in a less than or equal to zero form by convention. Equality constraints if present must be satisfied exactly at optimal design.

4.2 Methodology

In general, Nastran's methodology can be summarized in figure 6.

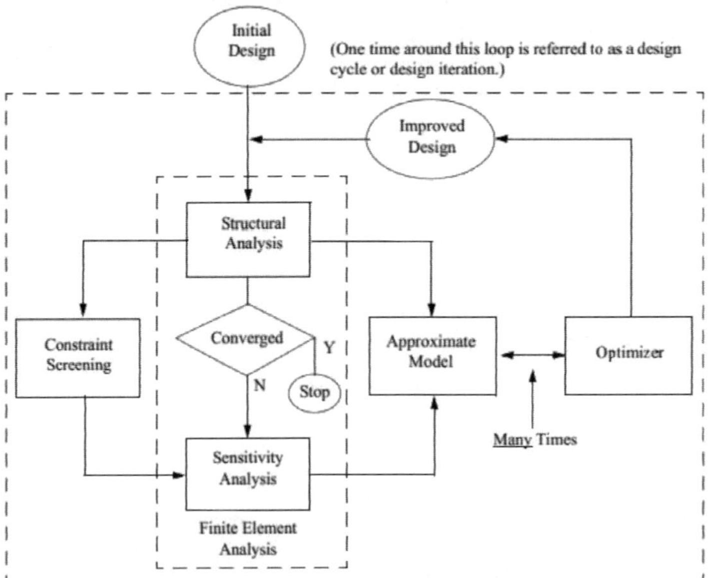

Figure 6. Nastran's Optimization Methodology

It begins with initial design that was used in the first analysis. In the structural analysis step, it can perform several analyses. In this paper, structural analysis consists of not only static load analysis but also flutter analysis. This paper uses two constraints; damping constraint for flutter and stress constraint for static load. Constraint screening is used to identify which constraint that is likely to drive the redesign process. If the result of analysis has not satisfied convergent criteria, a sensitivity analysis must be performed together with constraint screen result to make approximate model. Optimization is done with approximation model to improve efficiency. Based on the result from the optimizer, the design is updated. The cycle keeps going on until convergent criteria was satisfied.

4.3 Nastran Optimization

The optimization algorithms in MSC Nastran belong to the family of methods generally referred to as "gradient-based". This is because, in addition to function values, they use function gradients to assist in the numerical search for an optimum. To begin the optimization process, the direction of the search must be known. In general, the gradient of our objective function should be known as well as some of the constraint functions. First, forward finite difference approximation of a derivative is used to determine the gradient. For a single independent variable, the first-forward difference is given by Eq. 5.

$$\frac{df(x)}{dx} \cong \frac{f(x+\Delta x)-f(x)}{\Delta x} \tag{5}$$

For most practical design task, a vector of design variables, the resultant vector of partial derivatives, or gradient of the function $\nabla F(\mathrm{X})$ should be the main concerns. Note that this requires a number of steps in this given direction, which is equivalent to a number of function evaluations in numerical optimization for a search direction S and a vector of design variables X, the new design at the conclusion of our search in this direction can be written as in Eq. 6.

$$\boldsymbol{X}^1 = \boldsymbol{X}^0 + \alpha^* \boldsymbol{S}^1 \tag{6}$$

This relation allows updating a potentially huge number of design variables by varying the single parameter α. The new objective and constraints can now be expressed as in Eq. 7 and Eq. 8.

$$F^1 = F(\boldsymbol{X}^0 + \alpha^* \boldsymbol{S}^1) \tag{7}$$

$$g_j^1 = g_j(\boldsymbol{X}^0 + \alpha^* \boldsymbol{S}^1) \qquad j = 1, \dots, n_g \tag{8}$$

From this new point in the design space, the new gradients can be computed and establish another search direction based on this information. The process is repeated until the objectives are met.

4.4 Sensitivities Study

Design sensitivity coefficient is defined as the rate of change of a particular response quantity r with respect to a change in a design variable x or ∂x . These coefficients are evaluated at a design characterized by the vector of design variables, giving in Eq. 9:

$$\lambda_{ij} = \left.\frac{\partial r_j}{\partial x_i}\right|_{X^0} \tag{9}$$

Static response stress or strain are determined from the displacement solution. For an arbitrary response, r, the functional dependency on the displacement is written as in Eq. 10:

$$r = r(U, X) \tag{10}$$

The sensitivities of the responses with respect to changes in the design variables are approximated using first forward finite differences as below to obtain the required sensitivities (Eq. 11)

$$\frac{\partial r_j}{\partial x_i} \cong \frac{\Delta r_j}{\Delta x_i} = \frac{r_j(\boldsymbol{X}^0 + \Delta x_b \boldsymbol{U} + \Delta \boldsymbol{U}) - r_j(\boldsymbol{X}^o, \boldsymbol{U}^o)}{\Delta x_i} \tag{11}$$

Weight sensitivities are calculated using matrix $[\Delta M]$ and matrix $[DR_g]$. The latter matrix provides the displacement at all the degrees of freedom in the model based on unit displacement at the six degrees-of freedom, Eq. 12:

$$[\Delta W] = [DR_g]^T [\Delta M][DR_g] \tag{12}$$

Flutter sensitivities analysis begin with flutter equation given by Eq. 13:

$$\left[M_{hh}p^2 + \left(B_{hh} - \frac{1}{4}\rho\bar{c}V\frac{Q_{hh}^1}{k}\right)p + \left(K_{hh} - \frac{1}{2}\rho V^2 Q_{hh}^R\right)\right]\{u_h\} = 0 \quad (13)$$

Flutter sensitivity computes the rates of change of the transient decay rate coefficient γ with respect to changes in the design variables. γ is defined in connection with the complex eigenvalue p, by Eq. 14:

$$p = \omega(\gamma + i) = p_R + p_I \quad (14)$$

The sensitivity calculation for $\Delta\gamma$ is expressed in terms of the sensitivities of the eigenvalue as in equation 15:

$$\frac{\partial\gamma}{\partial x} = \frac{1}{p_1}\left(\frac{\partial p_R}{\partial x} - \frac{\gamma\partial p_I}{\partial x}\right) \quad (15)$$

5 Result and Discussion

5.1 Sensitivities Study

In order to make optimization process more efficient, Nastran does sensitivities study first. Equation 9 gives gradient equation for each variable and response. In this paper, sensitivities study was done in eight elements in the root area that give critical results. Each element gives different sensitivity coefficient for every response and design variable. For every unit dimension of design variable, the response will change according to sensitivity coefficients.

Figure 7 shows sensitivity coefficients with respect to weight response. It shows that skin 1 mostly gives highest sensitivity coefficient. Figure 8 also shows that skin1 gives highest sensitivity coefficients for almost every point according to maximum stress response. It makes sense, because skin 1 has the widest area, therefore, the weight will change significantly with the change of skin thickness. Skin 1 also placed in root area where the stresses are maximum. Therefore, skin 1 affects maximum stress significantly. Sensitivity coefficient for different response has different value. Nastran picks the most significant value to determine the direction of the search. Later it can be seen that this sensitivity will strongly decide the design optimization.

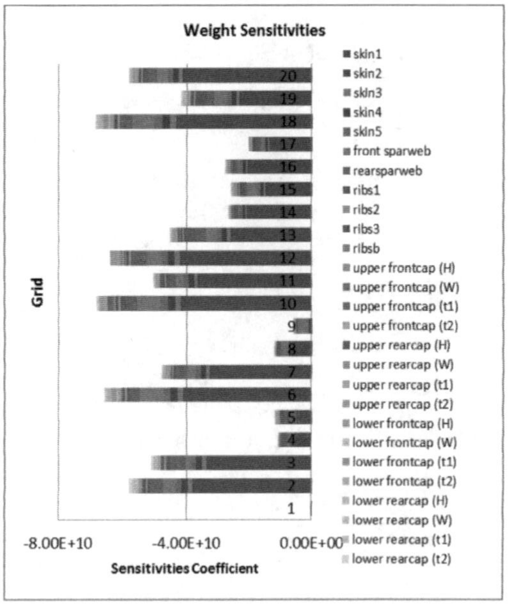

Figure 7. Weight Sensitivities

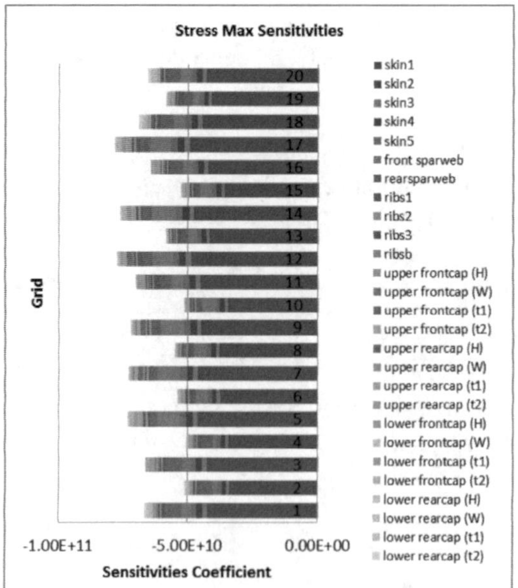

Figure 8. Maximum Stress Sensitivities

Figure 9 shows sensitivity coefficients according to flutter response. According to weight and maximum stress response, the most significant coefficient is skin1, follows by front spar-web. According to flutter sensitivities study, the most significant variables are skins not only for skin 1 but also skin 4 and skin 5. Although skin 4 and skin 5 sensitivities coefficients are large enough, but if it is compared on the value of flutter response coefficients and weight or maximum stress response coefficients, flutter sensitivities has far smaller value than maximum stress or weight sensitivities so that maximum stress constraint affects the design variable more significantly

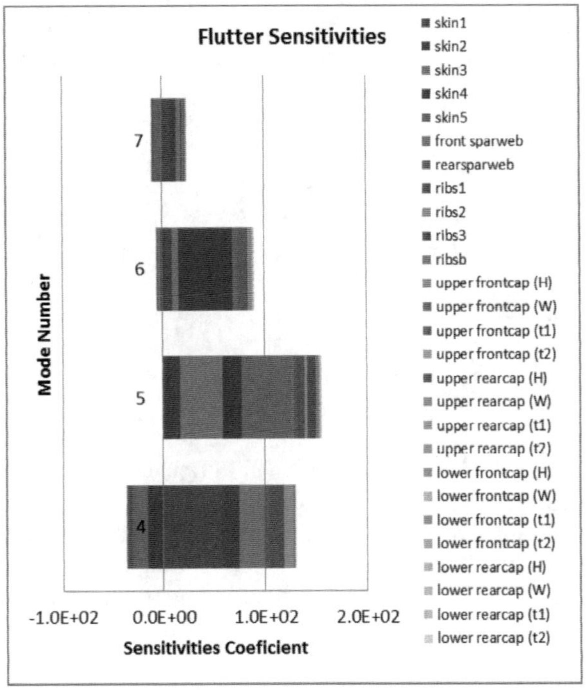

Figure 9. Flutter Sensitivities

5.2 Design Variable Histories

From design sensitivities it was found that skin 1 has largest sensitivity coefficient. From Figure 10, skin 1 is the thickest compared with other skins.

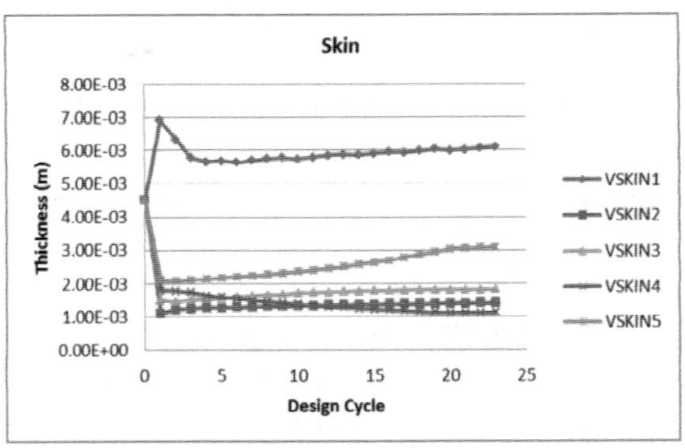

Figure 10. Skin Thickness History

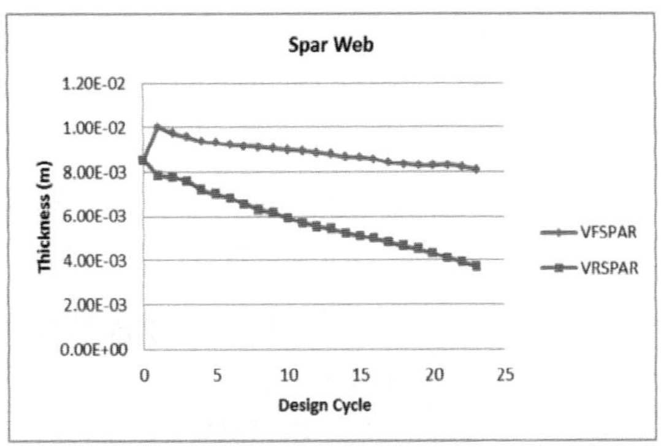

Figure 11. Spar-web Thickness History

Figure 8 also shows that front spar web has highest sensitivity coefficient after skin1. Therefore, in figure 11 for the first design cycle, front spar-web gradient is positive.

Sensitivity coefficients for ribs are small compared to other design variable, Therefore, all rib's thickness is kept to minimum except for ribs bulk. Although the direction kept decreasing, the thickness of ribs bulk still the largest between other ribs. It is rational, since ribs bulk is the place to mount the engine.

Figure 12 and 13 shows ribs thickness and spar-caps inertia history. From figure 13 we know that rear spar-cap gives higher inertia than front spar-cap.

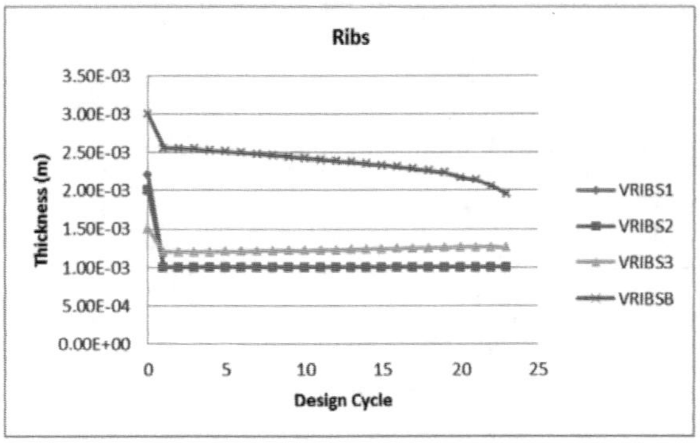

Figure 12. Ribs Thickness History

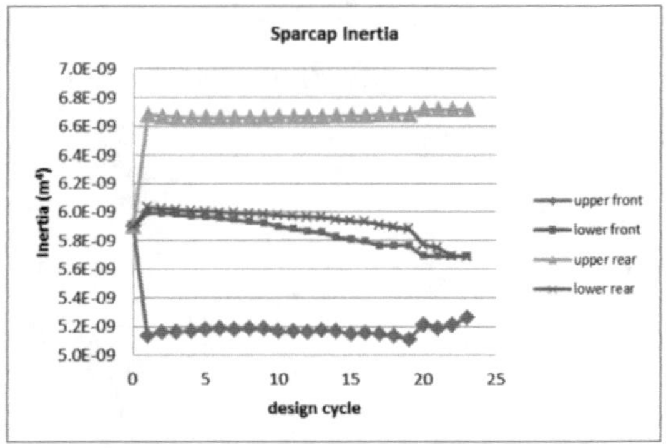

Figure 13. Spar-cap Inertia History

5.3 Weight Objective

Figure 14 shows weight history from the initial design until final design. It shows that sensitivities study is very important. It makes optimization process becomes far more efficient. Since the initial design, design sensitivities had determined the direction of the search, then optimization process continued until reach convergence criteria. For only one cycle, reduction of weight had reached 191 kg or 25% of the structural weight. For the final design, optimization process had reduced 255 kg or 33% of the structural weight.

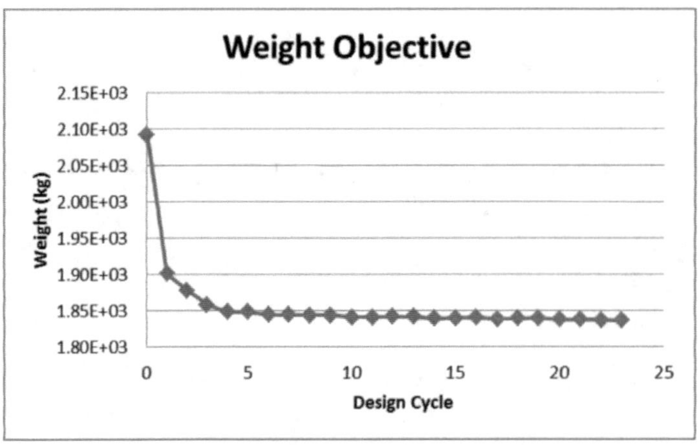

Figure 14. Weight History

Table 3. Initial-Final Constraints Comparison

	Initial	Final	Constraint
Flutter speed (m/s)	850	500	500
Max stress (Pa)	1.78E+05	2.93E+08	3.13+08

Table 3 shows comparison between initial design and final design constraint. The initial design has very large flutter speed (850 m/s), optimization process had limited the flutter speed to 500 m/s when exceeding structural damping (0.3) so the final flutter speed becomes 500 m/s. The initial design had maximum stress 178 MPa whereas the material can sustain stress until 313 MPa, after optimization process structure become more efficient, the maximum stress becomes 293 MPa.

6. Conclusions

Design sensitivities and optimization process had been applied to the wing box model using MSC Nastran software. The sensitivities study for each design variable of weight, maximum stress, and flutter response. Also, it shows that the design-variable change from initial to final cycle. Structural weight of wing had been reduced successfully until 33% without violating static and flutter constraints. Further sensitivity study, the static-load-subcase gives more critical structure condition than flutter-subcase shown by sensitivity coefficients of flutter response far smaller than maximum-stress-response sensitivity-coefficients.

References

[1] Vanderplaats G N 1981 *Structural Optimization – Past, Present, and Future* Vol. 20 No.7 (AIAA Jurnal)
[2] Maxwell C 1869 *Scientific Papers* (Dover Publications, 2, 175-177).
[3] Shanley 1952 *Weight-Strength Analysis of Aircraft Structures* (Dover; 2nd edition)
[4] MSC Software Corporation 2010 *MD Nastran 2010: Design Sensitivity and Optimization User's Guide* (Santa Ana: MSC Software Corporation)
[5] MSC Software Corporation 2010 *MSC Software Corporation MD Nastran 2010: Getting Started with MD Nastran User's Guide* (Santa Ana: MSC Software Corporation).
[6] Powell M J 1981 *Optimization Algorithms in 1979 Committee on Algorithms Newsletter 5* (Mathematical Programming Society 2-16)

Developing a non-routine maintenance load forecasting procedure in maintenance, repair and overhaul (MRO) XYZ company: A case study of B737NG aircraft

D Graciova, and E Suwondo

Department of Aerospace Engineering, Bandung Institute of Technology, Jalan Ganesha No. 10 Bandung 40132, Indonesia

*Corresponding E-mail: davegraciovaf@gmail.com, esuwondo@gmail.com

Abstract. Turn Around Time (TAT) is one of the most important performance indicators in a Maintenance, Repair, and Overhaul (MRO) company. MRO companies need a high percentage of on-time TAT to compete in the industry. In 2019, there is a 29% difference between the planned and the actual TAT in MRO XYZ. Based on observations in MRO XYZ, there is no planning to perform non-routine maintenance. 54% of the total maintenance loads is a non-routine maintenance, therefore, it needs to be planned. The first step of the research is to identify routine maintenance tasks that dominate the non-routine maintenance loads based on ATA Chapter and CAMP Number using Pareto Analysis. The next step is to develop a procedure which determines the variables and the mathematical model to estimate the Non-Routine Ratio (NRR) workloads. The last step is to implement the procedure to obtain the NRR estimation model, which is also as the validation of the developed procedure. Several routine maintenance tasks dominate Non-routine maintenance loads are categorized by ATA Chapter, CAMP Number, and task types. Dominant ATA Chapters are ATA 53, 25, and 57. Dominant CAMP numbers are 53-140-00, 53-800-00, and 53-866-00. Dominant task categories are DVI for the system, internal GVI for structure, and external inspection for zonal. The NRR forecasting model for B737NG C-Check is composed of C-Check number, aircraft's age, the ratio of ATA 53, the ratio of ATA 25, and the ratio of ATA 57. The NRR forecast model can be improved by adding some variables, such as Flight Hours (FH) and Flight Cycles (FC).

Keyword: forecasting, turn-around time, maintenance-repair-overhaul (MRO), flight cycles

1. Introduction

Turn Around Time (TAT) is an essential performance indicator in the Aircraft Maintenance, Repair, and Overhaul (MRO) industry. In this research, TAT's definition is the aircraft's period to enter the MRO facility until the aircraft returns to service. MRO is expected to have a high percentage of on-time TAT to compete in the industry and gain customer trust. In 2019, there are 918 days (28,8%) of differences in Planned TAT and Actual TAT at one of the MRO in Indonesia for B737NG. B737NG is the combined type of Boeing 737-700/-800/-900. There are additional costs for both airline and MRO if the on-time TAT isn't achieved. If there is an additional TAT, the airline cannot utilize the aircraft for commercial activities, and the MRO cannot use the slot to maintain another customer's aircraft. One of many causes of the difference TAT is non-routine maintenance works that are regularly found in routine maintenance. The total of non-routine maintenance works is as much as 54% of the routine maintenance. However, in MRO XYZ, there is still no procedure yet to forecast the non-routine maintenance workload. This condition is causing the non-routine maintenance cannot be well-prepared such as the material, manpower, and equipment. Therefore, it is needed to develop a procedure to forecast the non-routine maintenance load based on historical data.

The purpose of this research is to identify the routine maintenance tasks that dominate the non-routine maintenance works for C-Check B737NG, is to develop a procedure which determines the variables and the mathematical model to estimate the Non-Routine Ratio (NRR) workloads for C-Check B737NG, and the last is to implement the procedure as a part of the validation of the developed procedure. The methodology used in this research is formulating the research background and the problem, conducting a literature review about data science related to the problem, then preparing the data with Pareto Analysis, developing a non-routine maintenance forecast procedure, and testing the forecast procedure.

2. Materials and Methods

In this research, several literatures will be used to develop a procedure for forecasting non-routine maintenance load. This section will discuss aircraft maintenance type, a regression model with Bayesian Inference, and the tests that will be used.

2.1. Aircraft Maintenance

Routine maintenance is performed at regular intervals based on the Continuous Airworthiness Maintenance Program (CAMP). A CAMP is a list of all maintenance tasks belongs to an airline which is developed based on the Maintenance Review Board Document (MRB Doc) and has been approved by the regulatory authority [1]. Aircraft maintenance program is divided into three groups, namely the Systems, Structures, and Zonal Inspections.

Aircraft maintenance program consists of several task's types, for example Lubrication, Servicing, General Visual Inspection (GVI), Detailed Visual Inspection (DVI), etc. The interval of the maintenance tasks is clustered for the easiness of execution. Common clusters are the Daily Check, Pre-flight Check, Weekly Check, A-Check, C-Check, and D-Check. C-Check is categorized into heavy maintenance that the aircraft needs to go into a hangar and typically scheduled every 24 months.

2.2. Machine Learning

Machine Learning (ML) is one of the applications of Artificial Intelligence (AI). ML is an intersection of many fields of study such as computer science, engineering, statistics and other disciplines. Based on the learning process, ML is divided into two categories namely the Supervised and Unsupervised Learning. There are seven steps of machine learning, namely Collecting Data, Preparing Input Data, Analysing Input Data, Validating Data, Algorithm Learning, Algorithm Testing, and Applying knowledge. [3].

2.3. Multiple Linear Regression

Multiple Linear Regression (MLR) model is used to modelling the non-routine maintenance load. Multiple Linear Regression is used to forecast the relation between the dependent variable with more than one independent variable [4]. There are at least three objectives of developing a regression model:

a. Understanding the behaviour of y, given x,
b. Predicting y, given x and
c. Predicting how y would change if x were changed.

2.4. Bayesian Inference

Bayesian Inference is one of the statistical inference methods which combine the observation data with the sample information. There are three main parts of the Bayesian Inference model, namely the Prior, Likelihood, and Posterior model. The prior distribution is the initial information about the parameters, Likelihood distribution is the probability of observed data given the parameters, and Posterior distribution is the probability of the parameters given the data. From Eq. (1), the results of a Bayesian Inference are Posterior proportional to the Likelihood, multiplied by the Prior. Therefore, the results of a Bayesian Inference are not an exact value but in a form of distribution.

$$f(\theta|y) \propto f(y|\theta)f(\theta) \quad (1)$$

2.5. Tests Performed

A Normality test is used to determine if the normal distribution can model the sample data well. Two methods to perform the Normality Test are Kolmogorov-Smirnov and Chi-Square [5]. A model is considered has a normal distribution if the Kolmogorov-Smirnov or Chi-Square value less than a specific value. A Multicollinearity Test is used to evaluate whether the data have a strong correlation. The Multicollinearity Test is performed by checking the value of Variance Inflation Factors (VIF). If VIF equals 1, then there is no correlation, but if VIF's value is more than five, then there is a strong correlation between the dependent variables [6]. There are many effects due to multicollinearity, such as standard errors for the estimated coefficient will be extremely large and cannot measure the variable precisely. Information Criterion is a value to estimate the quality of the model. Widely Applicable Information Criterion (WAIC) is used in the Bayesian Analysis. The chosen model is the one with the lowest value of WAIC [4].

2.6. Markov Chain Monte Carlo (MCMC)

MCMC is a method to approximate the posterior distribution by random sampling in a probabilistic space. Monte Carlo method is an algorithm that uses random sampling to simulate a process. Markov Chain method is a stochastic process that assumes the future states depend only on the current state. No-U-Turn-Sampler (NUTS) is used to perform random samplings in MCMC. The NUTS method is a development of Hamiltonian Monte Carlo [4], therefore, it is more effective and efficient. Convergence condition from random sampling can be identified by conducting several tests. The convergence tests methods are the Autocorrelation, Monte Carlo Standard Error (MCSE), Effective Sample Size (ESS), and Gelman-Rubin Criterion. Autocorrelation should equal to zero to determine its convergence. The mean of MCSE should be less than the standard deviation of each variable. The ESS should be around 1000 and 2000. The Gelman-Rubin Criterion's approximately 1 [4]. A programming tool used in this research is Python with several libraries, such as NumPy, SciPy, Scikit-Learn, Matplotlib, PyMC3, Theano, and ArViz.

3. Results and Discussions

This section discusses the evaluation in the non-routine maintenance work loads, the development of the non-routine maintenance load forecasting procedure, and the results of applying the load forecasting procedure.

3.1. Non-Routine Maintenance Evaluation

In 2019, MRO XYZ performed C-Check for 57 B737NG aircraft, as 42% of the maintenance activities. The other 58% is aircraft maintenance such as A-Check, D-Check, Structural Check, Interval Check, and Repainting. One of the objectives in this research is to identify the routine maintenance tasks that dominate the non-routine maintenance loads based on the frequency and manhours needed. Based on ATA Chapter, the Pareto Analysis [2] identifies that ATA 53 (35%), 25 (18%), and 57(16%) dominated the non-routine maintenance loads. ATA 53, 25, and 57 are about Fuselage Structure, Equipment & Furnishing, and Main Wing Structure, respectively. Based on MPD number, the Pareto Analysis identifies that MPD number 53-140-00, 53-800-00, and 53-866-00 as the main sources of non-routine maintenance. These 3 MPD number are about the General Visual Inspection of the floor structure in forward and aft cargo compartment, General Visual Inspection of the external lower fuselage, and General Visual Inspection of the internal passenger compartment section 41, respectively.

The Maintenance Programs are divided into three groups, namely the Systems, Structure, and Zonal Inspections. The Pareto Analysis identifies that Zonal Inspection is the dominant source of the non-routine maintenance loads based on the total task frequency of 1726 times, while Structure was dominating based on the required total manhours. Based on the task types, the Pareto Analysis identifies that Detailed Visual Inspection (DVI) was dominating source of the non-routine maintenance loads of

Aircraft Systems, with a total frequency of 159 times (40%). Internal General Visual Inspection (GVI) was dominating in Aircraft Structure with a total frequency of 519 times (71%). External Inspections and Internal Inspections were dominating in Aircraft Zonal Inspections based on the total frequency and the total manhours, respectively.

3.2. NRR Forecasting Procedure

In this forecast procedure, the WAIC and Multicollinearity theory will be used to obtain a high accuracy and a high precision, respectively. Figure 1 shows the developed NRR forecast procedure.

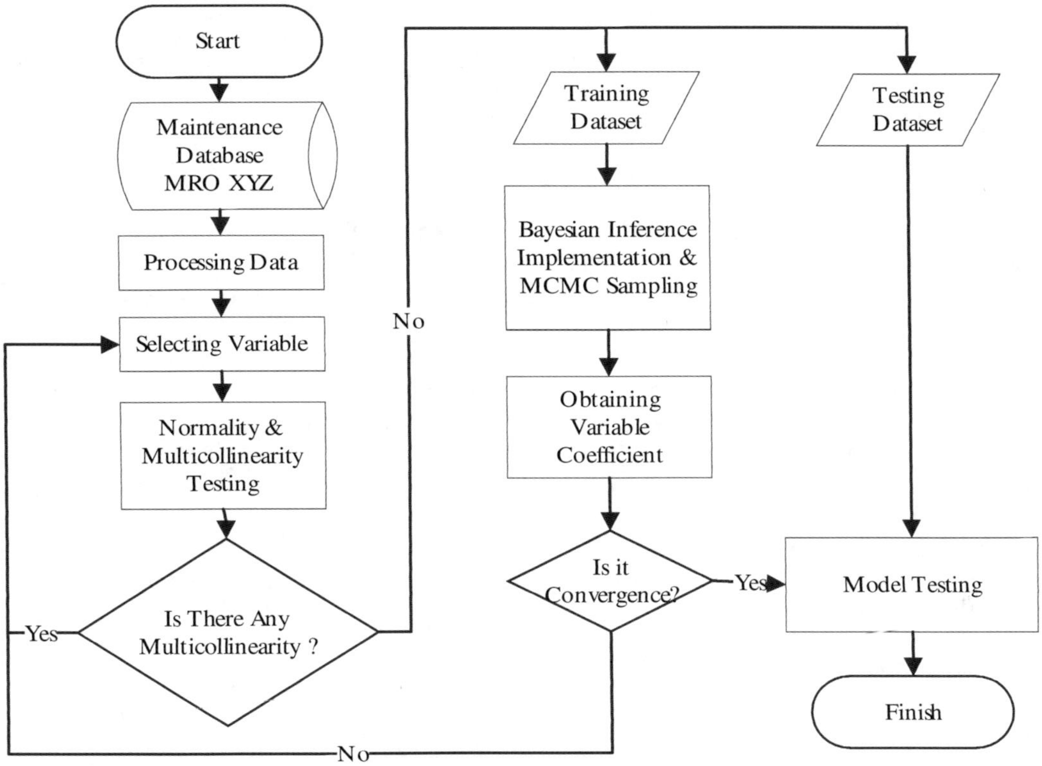

Figure 1. NRR Forecast Procedure

NRR Forecast Procedure starts with processing the maintenance data which come from various sources and in different formats. The process eliminates the missing data, sort out the inconsistent values, and give variable label. The normality and multicollinearity test are then conducted to evaluate whether the data can be modelled as normal distribution and check the correlation between the independent variables. If the data can be modelled as a normal distribution, and there is no strong correlation between independence variables, then the data divided into a training dataset and testing dataset. The ratio of the training and testing dataset is 9:1. Then, to the training dataset is implemented by the Bayesian Inference and MCMC sampling. Convergence tests are conducted to evaluate whether the result is convergence. If the results are already in the convergence state, then a model testing is conducted by using the testing dataset.

3.3. NRR Forecasting Procedure

From the 2019's maintenance database, the available variables are the C-Check Number, Additional Maintenance, Aircraft's Age, Ratio of ATA 53, 25, and 57 to the total maintenance. C-Check Number is the Number of C-Check maintenance from C01 until C07. The aircraft's age is the aircraft's length of time since its first flight; the ATA ratio is the total maintenance loads of the ATA Chapter divided by the total routine maintenance loads. The statistical summary of the maintenance database in the year 2019 is shown in Table 1.

Table 1. Statistical Summary of Maintenance Database in 2019

	C-Check Number	Aircraft Age	ATA 53	ATA 25	ATA 57	NRR
Count	50	50	50	50	50	50
Mean	3	11.34	0.087	0.025	0.147	0.476
Std	0.89	3.28	0.013	0.008	0.057	0.074
Min	1	6.37	0.051	0.009	0.088	0.306
25%	2	8.58	0.077	0.019	0.110	0.429
50%	3	10.87	0.085	0.023	0.131	0.476
75%	3	13.58	0.097	0.032	0.160	0.523
Max	4	18.67	0.121	0.043	0.368	0.644

Variables are selected from the model with the lowest WAIC. The model with the lowest WAIC is the one with C-Check Number, Aircraft's Age, Ratio of ATA 53, 25, and 57. From the normality test, all variables can be modelled as normal distribution except ATA 25 and ATA 57. However, in this case, ATA 25 and 57 still will be modelled as the normal distribution based on the Central Limit Theorem's assumption. There is also no VIF value larger than 5 in each variable, it means that there is no strong correlation between independent variables. The Bayesian Inference method is used to obtain the NRR forecast distribution. The Bayesian Inference implementation was carried out using the PyMC3 library in Python. In this research, the MCMC algorithm used 2000 samples and two chains. PyMC3 used the No-U-Turn-Sampler (NUTS), the developed Hamiltonian Monte Carlo, that is more efficient to use. From all of the convergence test, the results are all variables which are already in a convergence state. The convergence condition can also be seen from the trace plot. If the chains are mixed, then it can be considered that the model is already convergent. Figure 2 shows the trace plot example of the variable intercept, the ratio of ATA 57, and 53.

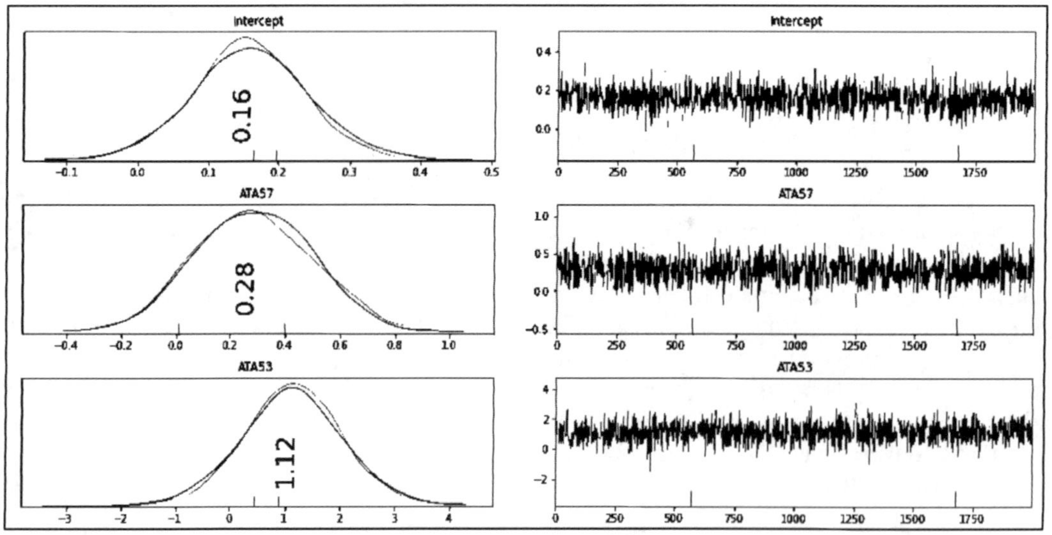

Figure 2. Example of the MCMC Trace Plot

The model to forecast the non-routine maintenance ratio, NRR = 0,157 + 0,018*(C-Check Number) + 0,005*(Aircraft's Age) + 1,116*(Ratio of ATA53) + 2,848*(Ratio of ATA25) + 0,280*(Ratio of ATA57) ± 0,064. Figure 3 shows the example of the procedure application with the data from testing dataset. The model's result using 2019 maintenance data, the actual NRR, and the Credible Interval can be seen in Figure 4. Figure 4 shows that there are only two points that out of the 95% Credible Interval, therefore, 96% of the data fit with the model's Credible Interval.

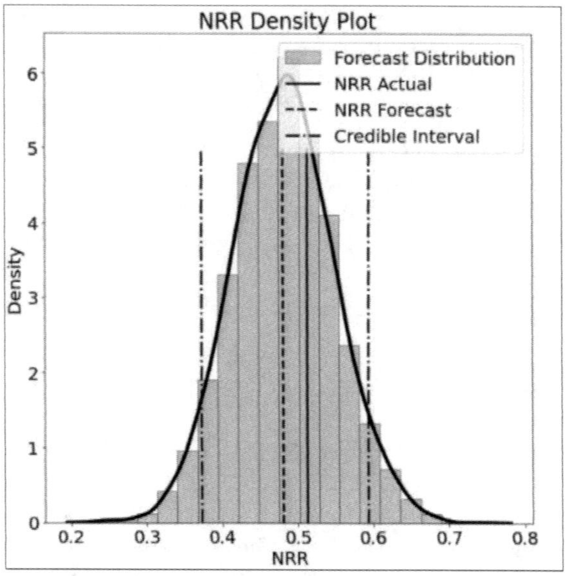

Figure 3. Example from the Testing Dataset

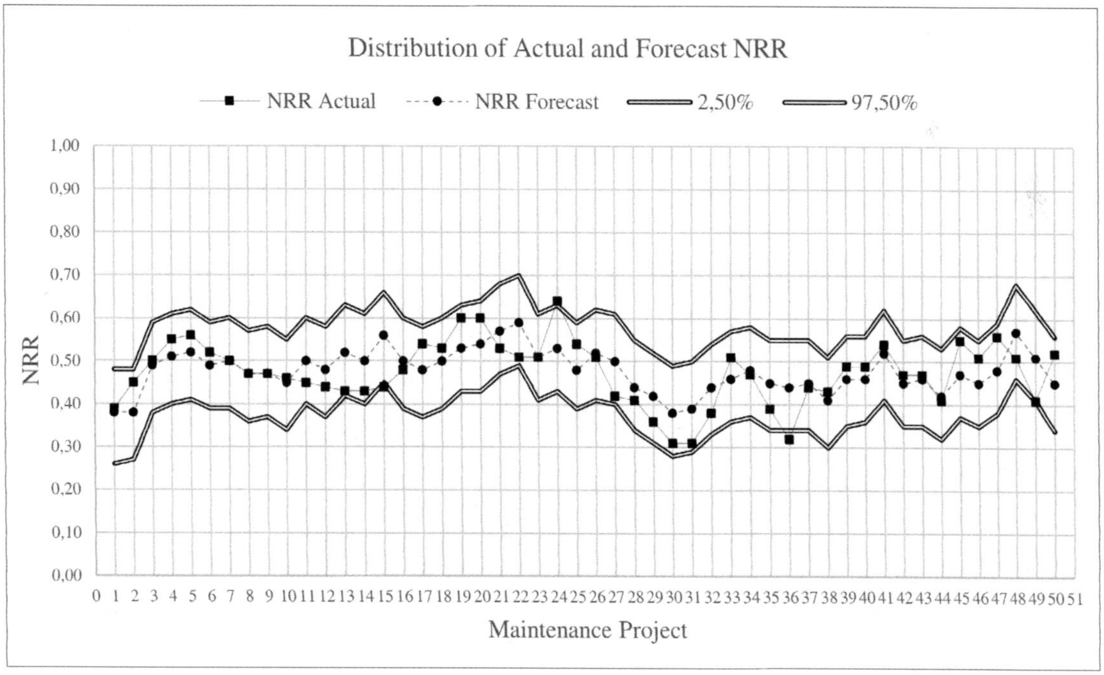

Figure 4. Distribution of Actual and Forecast NRR

The variables used in the NRR forecast are then analysed individually to evaluate its effect on NRR. The analysis shows that the C-Check Number, Aircraft's Age, Ratio of ATA 53, 25, and 57 do not clearly affect the increasing or decreasing of NRR. However, the increasing value of the C-Check Number, Aircraft's Age, Ratio of ATA 53, or 25 increases the NRR value.

4. Conclusions

Several routine maintenance tasks are dominant source of non-routine maintenance loads and categorized as ATA Chapter: ATA 53 (Fuselage), 25 (Equipment & Furnishing), and 57 (Wings); by CAMP Number: 53-140-00, 53-800-00, and 53-866-00; and by task type: DVI for System, Interval GVI for Structural, and External Inspection for Zonal Inspection. NRR forecast procedure for C-Check uses variables: C-Check Number, Aircraft's Age, Ratio of ATA 53, ATA 25, and ATA 57. The mathematical model as the result of NRR forecast procedure implementation for 737NG C-Check is: NRR = 0,157 + 0,280 x (Ration of ATA 57) $\pm$ 0,064. It is recommended that for MRO XYZ is to standardize the maintenance report, especially the maintenance activity description, the CAMP number reference, and manhours data to perform the routine and non-routine maintenance. For future research, it is necessary to add some variables, such as Flight Hours (FH) and Flight Cycles (FC), to obtain a more accurate and precise NRR forecast model.

References

[1] Mora M Telaahan Literatur Tentang Program Perawatan Pesawat Udara 2012 *Jurnal Penelitian Perhubungan Udara* **38** 356-372.
[2] O'Neill K S 2018 Applying the Pareto Principle *AABRI* **34** 1-12
[3] Harrington P 2012 *Machine Learning in Action* (New York: Manning Publications)
[4] Martin O 2018 *Bayesian Analysis with Python* (Birmingham: Packt Publishing)
[5] Ghasemi A 2012 Normality Test for Statistical Analysis: A Guide for Non-Statisticians *International Journal of Endocrinology Metabolism* **10** pp. 486-489
[6] Jamal Daoud I 2017 Multicollinearity and Regression Analysis *Journal of Physics: Conference Series*

IOP Conf. Series: Materials Science and Engineering 1173 (2021) 012060 doi:10.1088/1757-899X/1173/1/012060

V-Tail flutter analysis of wing-in-surface-effect (WISE) aircraft using a structural analysis software

M Kusni*, A Taufiqurrahman, and L Gunawan

Lightweight Structure Research Group, Faculty of Mechanical and Aerospace Engineering, Institut Teknologi Bandung; Jl. Ganesha 10, Bandung 40132, Indonesia

*Corresponding e-mail: kusni.ae.itb.ac.id

Abstract. Flutter is a dynamic aeroelastic instability that may cause structural failure and limits flight envelope of an aircraft. A passenger aircraft is required to be free from flutter and other aeroelastic instability phenomena, as stated in the regulations such as FAR 25. 33 / FAR 23.33. This paper presents the flutter analyses of the WISE aircraft using MSC Nastran. The analyses were carried out to the V-tail, one of the component where flutter might occur, by assuming rigid fuselage. Results of each analysis, in the form of velocity-damping and velocity-frequency curves, were evaluated to determine the critical flutter speed and frequency. First the analysis were conducted for sea level operation using KE, PK and PKNL methods which predict respectively the flutter speed of 1036 knot, 1037 knot, and 1037 knot. The three methods also consistently predict that the mode shapes involved in the flutter are the 5th mode and the 3rd mode. Then by using the PK-method, the analysis were repeated for air density variation. It is shown that the lower the air density, the higher the flutter speed is. It is concluded that tail flutter does not occur during the operation of the WISE craft with max operating speed of 80 knot.

Keywords: flutter, aeroelastic, V-Tail, wing in surface effect (WISE)

1. Introduction

Passenger aircraft must be safe from aeroelastic failures, that is, safe against divergence, control reversal, and flutter such as stated in FAR 25. 33 / FAR 23.33. Flutter occurs due to the interaction between structure, inertia and aerodynamic forces and is indicated with vibration with increasing amplitude. This phenomenon may cause structure failure and hence must be avoided.

This paper presents the flutter analysis of a WISE aircraft. The analysis was focused on the V-tail flutter case. This analysis was limited to flexible tail and rigid fuselage. The tail structure comprising two vertical tail and one horizontal tail, as shown in Figure 1, is modeled as stick beams with concentrated mass distributed along the length of the beam at certain offsets.

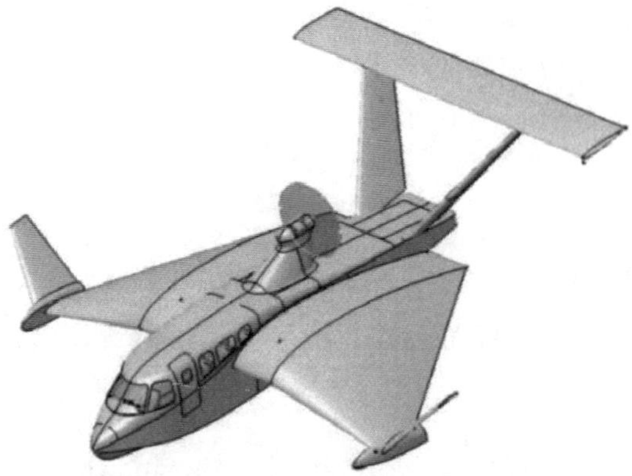

Figure 1. WISE-craft evaluated in this study [1]

2. Theory

2.1 Aeroelastic Equations of Motion

Aeroelastic equation of motion can be written in the form of the matrix equation [2-5]:

$$[M]\{\ddot{z}\} + [B]\{\dot{z}\} + [K]\{z\} = \{A\}(geometry, \rho, \mu, U, t), \tag{1}$$

with:

$[m]$ = mass matrix
$[c]$ = damping matrix
$[k]$ = stiffness matrix
$\{A\}$ = aerodynamic forces vector
$\{z\}$ = displacements vector

The aerodynamic forces comprises external and motion dependent aerodynamic forces. For instability analysis, only the latter is considered and can be written as:

$$\{A\} = \tfrac{1}{2}\rho U^2[f]\{z\} \tag{2}$$

Hence the aeroelastic equation of motion for flutter analysis is:

$$[M]\{\ddot{z}\} + [B]\{\dot{z}\} + [K]\{z\} = \tfrac{1}{2}\rho U^2[f]\{z\} \tag{3}$$

2.2 Modal Transformation

The analysis is carried out by analysing the undamped free vibration of the aircraft to obtain the natural frequencies and the corresponding mode shapes. Then, by using the modal matrix, equation 3 is transformed into the equation of motion in modal coordinates by assuming that the displacement $\{z\}$ is the superposition of displacement in each mode shape:

$$\{z\} = [\varphi]\{q\} \tag{4}$$

where $[\varphi]$ is the modal matrix and $\{q\}$ is the modal coordinates vector.

By substituting equation 4 into equation 3, and pre multiplying the result with the transposed modal matrix, the following results is obtained:

$$M_{hh}\{\ddot{q}\} + B_{hh}\{\dot{q}\} + K_{hh}\{q\} = \tfrac{1}{2}\rho U^2 Q_{hh}\{q\} \tag{5}$$

where:

$M_{hh} = [\varphi]^T[M][\varphi]$: modal mass matrix
$B_{hh} = [\varphi]^T[B][\varphi]$: modal damping matrix
$K_{hh} = [\varphi]^T[K][\varphi]$: modal stiffness matrix
$Q_{hh} = [\varphi]^T[f][\varphi]$: modal aerodynamic matrix

The modal mass and stiffness matrices are diagonal with dimensions $[n \times n]$, where n is the number of mode shapes used in the modal matrix.

2.3 Flutter Analysis Method
Several parameters are needed to find the solution of flutter equation. Several methods have been developed and implemented in NASTRAN, such as KE-method, the PK-method, and the PKNL-method.

2.3.1 KE-method
KE-method is special case of K-method where viscous damping is neglected. This basic principle of this method is based on the assumption that the solution of the flutter equation is harmonic motion, $\{q\} = \{\hat{q}\}e^{i\omega t}$, with artificial structural damping being added to the system. The equation for flutter analysis is [6]:

$$\left[-\left[M_{hh} + \frac{\rho}{2}\left(\frac{c}{2k}\right)^2 Q_{hh}(M_\infty, k)\right]\frac{\omega^2}{1+ig} + K_{hh}\right]\{\hat{q}\} = 0 \tag{6}$$

This method compute flutter roots ($\boldsymbol{\omega^2/(1+g)}$) for specified values of density ($\boldsymbol{\rho}$), Mach number ($\boldsymbol{M_\infty}$), reduced frequency ($\boldsymbol{k = \omega c/(2U)}$), where $\boldsymbol{c}$ is reference length. From the roots and the inputs, the value of $\boldsymbol{U}$, $\boldsymbol{g}$, and $\boldsymbol{\omega}$ are determined. Flutter speed is obtained from values of $\boldsymbol{\rho}$, $\boldsymbol{M_\infty}$, and $\boldsymbol{k}$ for which $\boldsymbol{g = 0}$.

2.3.2 PK-method
In PK-method, the motion is harmonic with exponentially varying amplitude, $\{q\} = \{\hat{q}\}e^{pt}; p = \omega(\gamma \pm i)$, and the eigenvalues of flutter equations is complex. The unsteady aerodynamic forces depends only on the reduced frequency $k = \omega c/(2U)$. The basic equation for flutter analysis is [6]:

$$\left[M_{hh}\, p^2 + \left(B_{hh} - \frac{1}{4}\rho c U Q_{hh}^I/k\right)p + \left(K_{hh} - \frac{1}{2}\rho U^2 Q_{hh}^R\right)\right]\{\hat{q}\} = 0 \tag{7}$$

where:

Q_{hh}^I = modal aerodynamic damping matrix, a function of M_∞ and k.
Q_{hh}^R = modal aerodynamic stiffness matrix, a function of M_∞ and k.
p = eigenvalue = $\omega(\gamma \pm i)$
γ = transient decay rate coefficient.

The PK-method extracts these roots for user-specified values of ρ, M_∞, and U. From these roots, the frequency and damping can be determined.

2.3.3 PKNL-method
PKNL or PK No Looping method is a variation of the PK-method that does not loop on all combinations of users input ρ, M_∞, and U [7], but only for combination of ρ and matching pairs of M_∞, and U. Hence this method is more efficient than the PK-method.

3. Aeroelastic Analysis using MSC Nastran
Aeroelastic analysis using MSC Nastran is divided into two steps, the free vibration analysis and flutter analysis. Free vibration analysis requires stiffness and mass model of the structure, and flutter analysis requires the aerodynamic model and relation between structure and aerodynamic motion.

3.1 Free Vibration Analysis
Free vibration analysis is performed using SOL103 to obtain the dynamic characteristics of the V-tail structure, i.e. the natural frequencies and the corresponding mode shapes. This analysis needs structure data to create the $[K]$ and $[M]$ in equation 3, and the results are used as the basis of $[\varphi]$ in equation 4.

In this work, the vertical and horizontal tail structure are modeled using stick beams along the elastic axis of each component. The mass of the structure is lumped at the c.g. of that section and connected to the node of the beam by a rigid element.

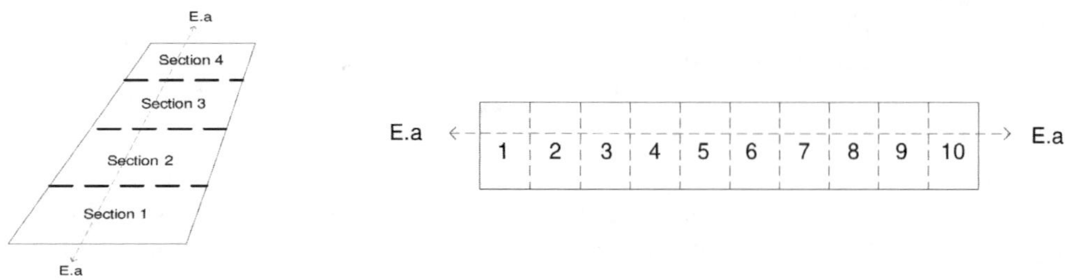

Figure 2. The division of the (a) VT structure and (b) HT structure

3.1.1. Vertical Tail Model

The vertical structure of the tail is divided into four sections based on the number of ribs, as shown in figure 2.a. Data of vertical tail dimensions: root chord = 1700 mm, tip chord = 1040 mm, mean chord = 1400 mm. For each section, the area and stiffness are shown in table 1 and the inertial properties are shown in table 2.

Table 1. Area and stiffness of vertical tail

Section	Area (mm^2)	Area moment of inertia (mm^4)			E (KN/m^2)	G (KN/m^2)
		I_{11}	I_{22}	I_{33}		
1	5980.17	2.59E+7	7.63E+8	2.74E+8	1.630E+7	6.52E+6
2	5516.68	2.04E+7	5.91E+8	2.1E+8		
3	4682.61	1.33E+7	3.48E+8	1.27E+8		
4	4256.95	9.2E+6	2.62E+8	8.97E+7		

Table 2. Inertias of vertical tail

Section	Mass (kg)	Xcg (mm)	Ycg (mm)	Zcg (mm)	J_{11} ($kg.mm^2$)	J_{22} ($kg.mm^2$)	J_{33} ($kg.mm^2$)
1	17.95	13640	-1182.2	1754.6	2.096E+7	1.053E+7	2.485E+7
2	11.59	13908	-1435.2	2457.6	2.452E+7	8.789E+6	2.989E+7
3	10.14	14848	-2142.9	3113.6	4.098E+7	6.785E+6	4.528E+7
4	2.69	15348	-2465.8	3800.4	1.355E+7	8.58+E5	1.423E+7

Each section is modeled with 3 beam elements on the elastic axis of the structure. Hence, model of each vertical tail (right and left) consists of 12 beam elements.

3.1.2. Horizontal Tail Model

The horizontal structure of the tail is divided into 10 sections based on ribs, divided as shown in figure 2.b. For each section, the area and stiffness are shown in table 3 and the inertial properties are shown in table 4. Each section is modeled with one beam element on the elastic axis of the structure. Overall, the horizontal tail structure is modeled with ten beam elements.

Table 3. Area and stiffness of horizontal tail

Section	Area (mm^2)	I_{11} (mm^4)	I_{22} (mm^4)	J (mm^4)	E (KN/m^2)	G (KN/m^2)
All	10088.23	1.15E+09	2.00E+09	1.12E+09	1.41E+7	5.24E+6

Table 4. Inertia properties of half HT (symmetric to xz plane)

Section	Mass (kg)	Xcg (mm)	Ycg (mm)	Zcg (mm)	I_{11} ($kg.mm^2$)	I_{22} ($kg.mm^2$)	I_{33} ($kg.mm^2$)
1	4.87	15850	4900	3780	1.2983E+7	1.148E+6	1.355E+7
2	9.30	15914	3640	3756	1.2983E+7	1.148E+6	1.355E+7
3	8.14	15999	2373	3788	2.6086E+7	2.255E+6	2.711E+7
4	7.18	16042	1510	3840	2.6086E+7	2.255E+6	2.711E+7
5	7.74	16043	510	3840	2.2742E+7	2.914E+6	2.330E+7

3.1.3. V-Tail Model

Figure 3a shows the structural model of the V-tail where rigid elements are used to connect the vertical to the horizontal tail. The vertical tails are rigidly clamped at the bottom ends. Thin plate elements with zero properties are added to the beam elements using rigid elements to visualize the mode shapes of the tail, as shown in figure 3.b.

Figure 3. Structural model of V-tail (a) beams (b) beams with plates elements

3.2 Flutter Analysis

Flutter analysis is carried out using Nastran SOL 145. In this analysis, aerodynamic model is prepared to create the aerodynamic matrix Q_{hh} in equation 5 using the Doublet Lattice method. Four lift surfaces were defined, namely, the right and the left vertical tail, and the left half and the right half of the horizontal tail. Each lift surface was divided into aerodynamic panels using CAERO1 card.

3.2.1. Aerodynamic Model of Vertical Tail

Each vertical tail was defined as one lift surface which was divided into 96 aerodynamic panels (24 spanwise and 4 chordwise). With this division, each panel has an aspect ratio of 0.35. Figure 4.a shows the aerodynamic panels of the vertical tail.

3.2.2. Aerodynamic Model of Horizontal Tail

Two lift surfaces were defined for the horizontal tail, left and right half of the horizontal tail. Each lift surface was divided into 32 aerodynamic panels (8 spanwise and 4 chordwise). The aerodynamic panel has an aspect ratio of 0.683. Figure 4.b shows the 64 aerodynamic panels of the horizontal tail.

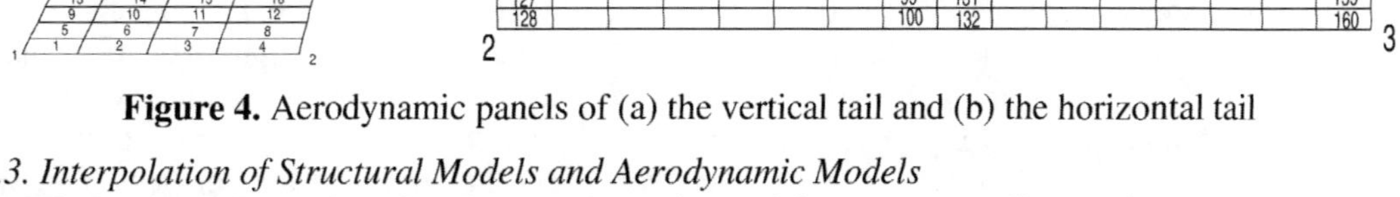

Figure 4. Aerodynamic panels of (a) the vertical tail and (b) the horizontal tail

3.2.3. Interpolation of Structural Models and Aerodynamic Models

Interpolation is performed so that the aerodynamic model moves according to the structural motion. SPLINE2 card was used for this purpose. Since four lift surfaces were defined for the V-Tail, four cards were used to interpolate motion of the panels to that of the structure.

3.2.4. Flutter Analysis Parameters

The parameters for flutter analysis are defined using FLUTTER card such as the methods, air density, Mach number, reduced frequency or velocity, number of eigenvalues. From the results displayed as v-g and v-f graphs, flutter characteristics are determined.

4. Results and Discussions

First, the dynamic characteristics of the V-Tail was determined. Then flutter characteristics at sea level, which practically has the same air density as that in operating altitude of 3 m, was analyzed using the KE, PK, and PKNL methods. Finally, sensitivity of the flutter characteristics of V-Tail to (altitude/air density) was evaluated using the PK method.

4.1. Structural Dynamic Characteristics

Undamped free vibration analysis was carried out to determine the natural frequencies and the corresponding mode shapes of the V-Tail structure. Table 3 shows the lowest 14 frequencies and mode shapes of the tail, and figure 14 depicts the first 8 mode shapes. It can be seen that mode shapes at lower frequencies are dominated by bending of vertical tail in symmetric or anti-symmetric configuration. At higher frequencies the mode shapes involve the torsion of horizontal tail. At even higher frequencies, the mode shapes involve bending and torsion of both vertical and horizontal tail.

4.2. Flutter Characteristics

Flutter analysis at sea level was carried out by using the 14 mode shapes for M = 0.2 and a flight altitude of 0 m as the air density at 3 m. Figure 6.a shows the U-g curves, where damping curve that first intersects with intersects the damping axis (g) = 0% first is that of the 5th mode shape at 1037 knot. The U-f curves in figure 6.b shows that the frequencies of the 5th mode at 1037 knot is 11.37 Hz. Figure 6.b also shows that the frequencies of the 5th and the 3rd modes are getting closer, as the air speed increases. This shows that the flutter occurs from the coupling between the 5th and the 3rd modes. From this information, it can be concluded that flutter at sea level occurs first for 5th mode at velocity of 1037 knot and frequency of 13.37 Hz.

In addition to the PK method analysis, flutter analysis at sea level was also conducted using the KE and PKNL methods. Figure 7 shows the U-g curves obtained from the three methods which shows that the three methods predict the flutter speed almost close to each other. The three methods consistently predict that the mode shapes involved in the flutter are the 3rd and the 5th mode shapes.

4.3. Effect of Air Density on the Flutter Speed

To observe the effect of air density on the flutter velocity, additional flutter analyses were conducted with air densities of 1.112 kg/m^3, 1.007 kg/m^3, and 0.9093 kg/m^3 using PK-method.

Figure 8.a shows results of analysis for air density of 1.007 kg/m^3, where the damping curve that intersects the damping axis = 0% is that of the 5th mode at 1087 knot. The U-f curves in figure 8.b shows that the frequencies of the 5th mode at 1087 knot is 11.37 Hz. Figure 8.b also shows that the frequencies of the 5th and the 3rd modes are getting closer, as the air speed increases. This shows that the flutter occurs from the coupling between the 5th and the 3rd modes. From this information, it can be concluded that flutter at air density of 1.112 kg/m^3 occurs first for 5th mode at velocity of 1037 knot and frequency of 13.37 Hz.

Figure 9.a shows results of analysis for air density of 1.007 kg/m^3 where the damping curve of 12th mode intersects the damping axis = 0% first at 1131 knot. The U-f curves in figure 9.b shows that the frequencies of the 12th mode at 1131 knot is 36.63 Hz. Figure 8.b also shows that the frequencies of the 12th and the 13th modes are getting closer, as the air speed increases. This shows that the flutter occurs from the coupling between the 12th and the 13th modes. In the flutter analysis at air density of 1.007 kg/m^3, the flutter occurs at different mode than at lower densities.

Figure 10.a and 10.b show results of analysis for air density of 0.9093 kg/m^3. In this case, flutter occurs similar to that at air density of 1.007 kg/m^3. Flutter mode is the 12-th mode at a frequency of 36.54 Hz at speed of 1173 knot. Table 4 compiles the flutter characteristics of the V-tail at several air densities. The values of flutter speed is increased with lower air density. This result is in agreement with statement of Bisplinghoff [2].

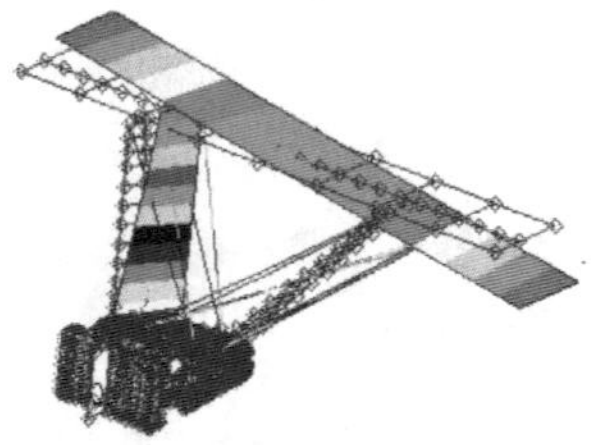
1st mode shape, 4.76 Hz

2nd mode shape, 9.47 Hz

3rd mode shape, 11.39 Hz

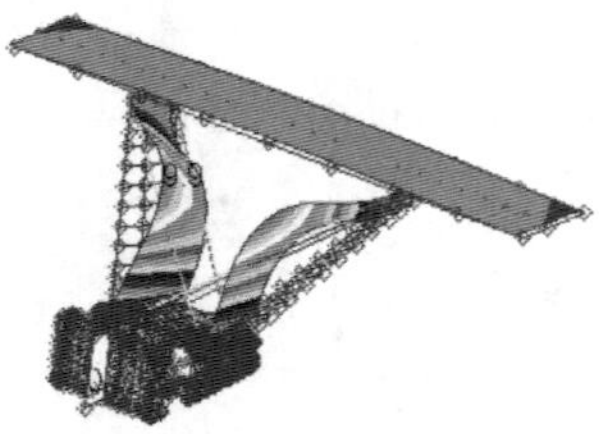
4th mode shape, 15.78 Hz

5th mode shape, 16.17 Hz

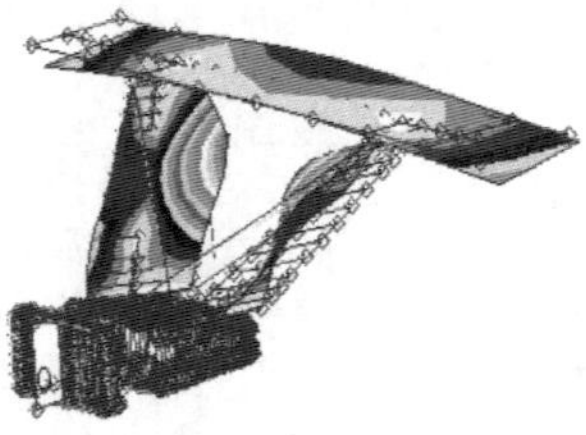
6th mode shape, 22.14Hz

7th mode shape, 23.10 Hz

8th mode shape, 29.16 Hz

9th mode shape, 31.49 Hz

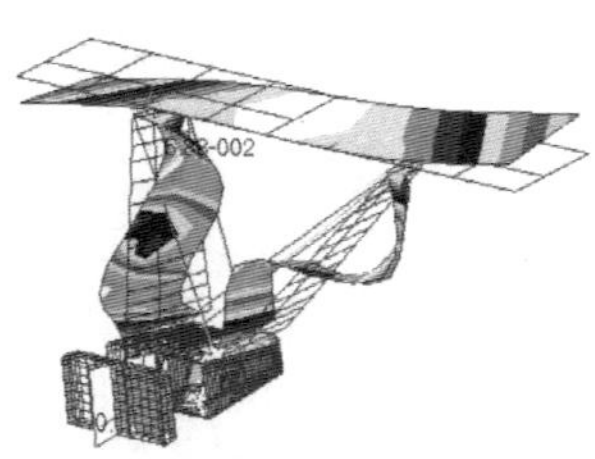
10th mode shape, 32.24 Hz

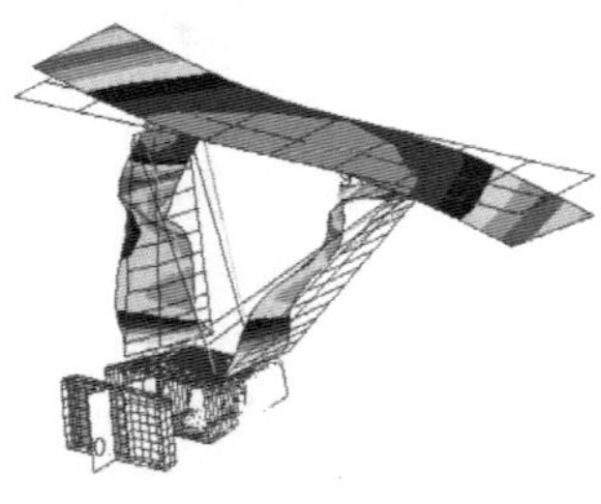
11th mode shape, 33.35 Hz

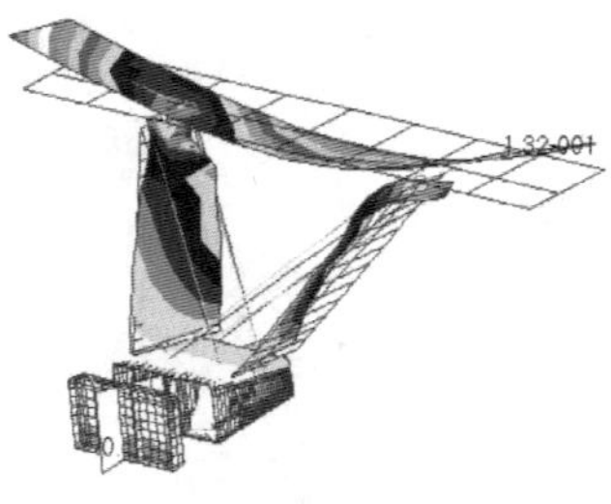
12th mode shape, 36.58 Hz

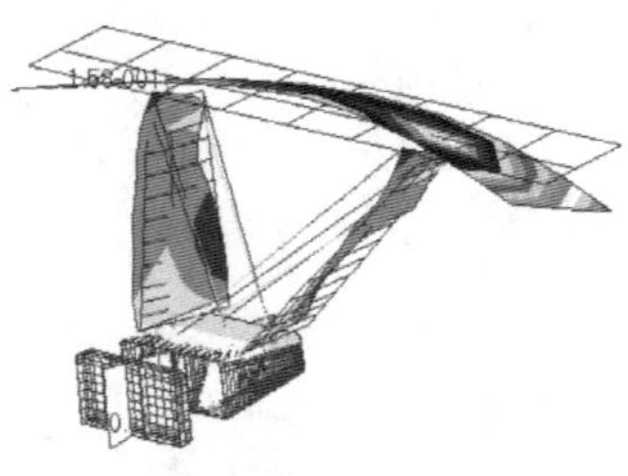
13th mode shape, 42.60 Hz

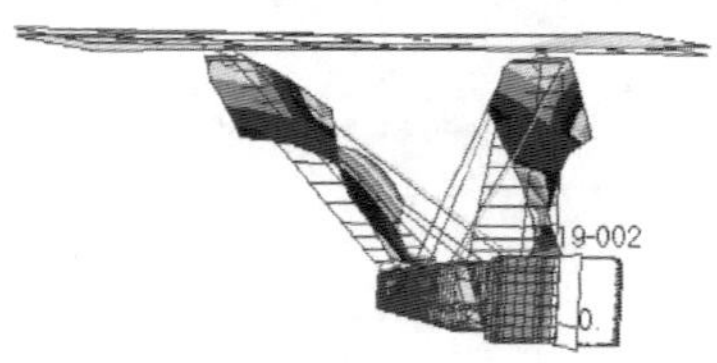
14th mode shape, 43.50 Hz

Figure 5. Mode shapes and natural frequencies of the V-Tail structure

Table 5. Mode shapes and natural frequencies of the V-tail structure

Modes Shape	Natural Frequency (Hz)	Mode Shape
1-st	4.76	Bending of vertical tail
2-nd	9.47	Bending of vertical tail
3-rd	11.39	Bending of vertical tail and torsion of horizontal tail
4-th	15.78	Bending of vertical tail
5-th	16.17	Bending and torsion of vertical tail
6-th	22.14	Bending and torsion of vertical tail, bending of horizontal tail
7-th	23.10	Bending and torsion of vertical tail, torsion of horizontal tail
8-th	29.16	Bending of vertical tail, bending of horizontal tail
9-th	31.49	Bending and torsion of vertical tail
10-th	32.24	Bending and torsion of vertical tail, bending and torsion of horizontal tail
11-th	33.35	Bending and torsion of vertical tail and horizontal tail
12-th	36.58	Bending and torsion of vertical tail and horizontal tail
13-th	42.60	Bending of vertical tail, bending of horizontal tail
14-th	43.50	Bending and torsion of vertical tail

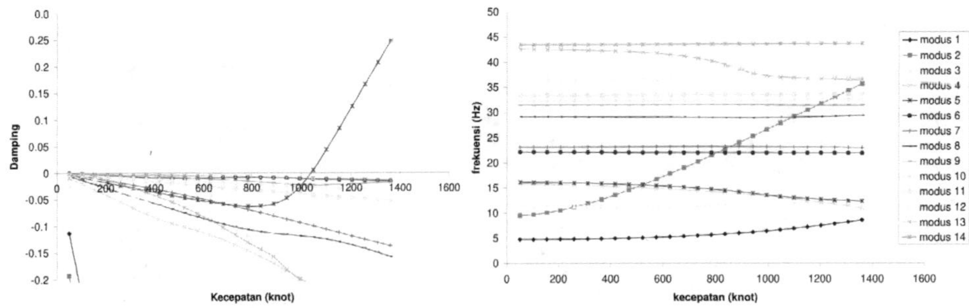

Figure 6. (a) U-g and (b) U-f curves at sea level

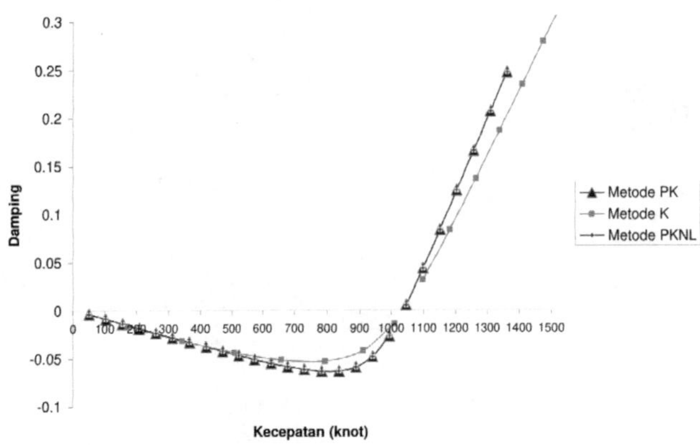

Methods	V_F (knot)	ω_F (Hz)
KE	1036	13.38
PK	1037	13.37
PKNL	1037	13.37

Figure 7. U-g curves of V-tail at sea level calculated using PK, K, and PKNL methods

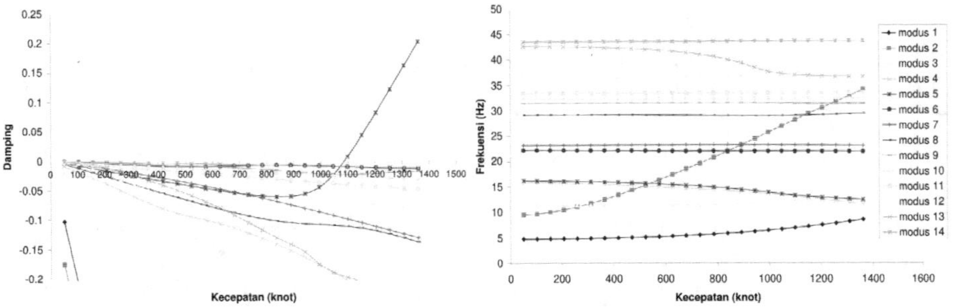

Figure 8. The U-g and U-f curves at air density of 1.112 kg/m^3

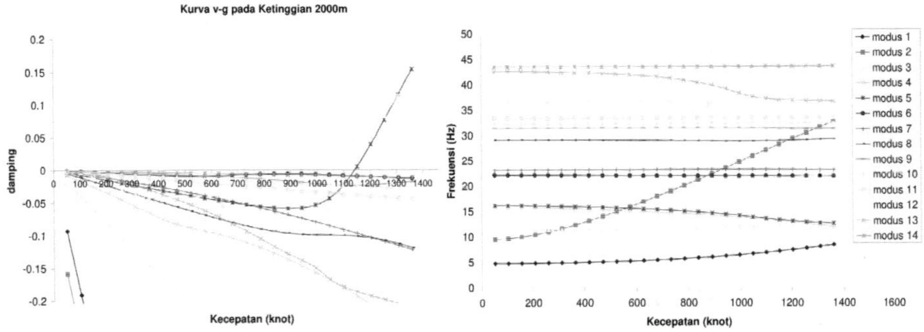

Figure 9. The U-g and U-f curve at air density of 1.007 kg/m^3

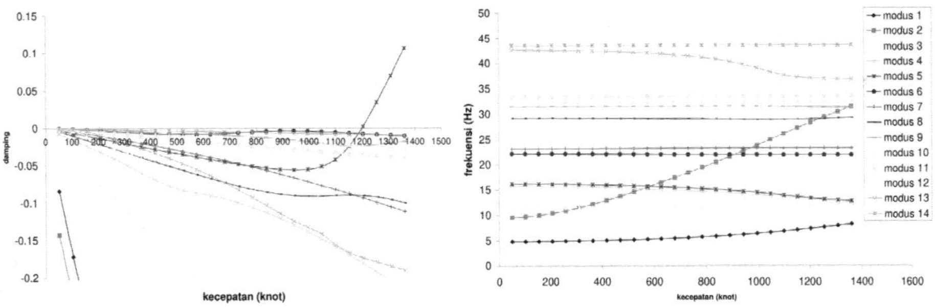

Figure 10. The U-g and U-f curves at air density of 0.9093 kg/m^3

Table 6. Flutter characteristics of V-Tail at several air density

Air density (kg/m^3)	V_F (knot)	ω_F (Hz)	Flutter mode
1.225	1037	13.37	5th mode, bending and torsion of vertical tail
1.112	1087	13.37	5th mode, bending and torsion of vertical tail
1.007	1131	36.63	12th mode, bending and torsion of vertical tail and horizontal tail
0.9093	1173	36.54	12th mode, bending and torsion of vertical tail and horizontal tail

5. Conclusions

From the results of the flutter analysis that have been performed, it can be concluded that the V-tail structure of the WISE aircraft is safe from flutter for its max operating speed at 80 knots and an altitude of 3 m above sea level. However, further study is needed by using the full structure configuration of the WISE craft to obtain more realistic prediction.

IOP Conf. Series: Materials Science and Engineering 1173 (2021) 012060 doi:10.1088/1757-899X/1173/1/012060

References

[1] Jatiningrum D, Jenie Y I, Muhammad H, Pasaribu HM 2010, *Verification of Sail-Flight Testing Procedures of Wing-in-Surface Effect Craft on Engineering Flight Simulator*, ICAS 27th Conference, 19-24 Sept 2010, Nice, France.

[2] Bisplinghoff R L, H. Ashley, and R L Halfman 1955 *Aeroelasticity*, Addison-Wesley Publishing Company, Inc., Massachusetts.

[3] Dowell E H, Curttiss Jr. H C, Scanlan R H, and Sisto F 1980, *A Modern Course in Aeroelasticity*, Sijthof and Noordhoff, Netherland.

[4] Fung Y C 1996 *An Introduction to the Theory of Aeroelasticity*, Dover Publications, Inc., New York.

[5] Scanlan R H and Rosenbaum R 1968 *Introduction to the Study of Aircraft Vibration and Flutter*, Dover Publications, Inc., New York.

[6] *Aeroelastic Analysis User's Guide 2004*, MSC.Nastran Version 68, MSC Software Corporation, USA.

[7] Johnson E H 1997, *MSC Developments in Aeroelasticity*, MSC Aerospace Users' Conference, The MacNeal-Schwendler Corporation, California, USA.

[8] Dasril H 2007, Pemodelan dan Analisis Dinamika Struktur Pesawat WISE dengan Metode Elemen Hingga, Undergraduate Thesis (unpublished)), Dept. of Aerospace Engineering ITB, Indonesia.

[9] Taufiqurrahman A 2007, Pemodelan dan Analisis Dinamika Struktur Pesawat WISE dengan Metode Elemen Hingga, Undergraduate Thesis (unpublished), Dept. of Aerospace Engineering ITB, Indonesia.

IOP Conf. Series: Materials Science and Engineering 1173 (2021) 012061 doi:10.1088/1757-899X/1173/1/012061

Corrosion of Carbon Steel and alloy Steel: effect of humidity and hydrochloric acid

Abdulkader A. Muhammad[1], Thamir A. D. M. S. Almula [1,*], Dyhia A. Sultan[1]

[1] Engineering Technical College / Mosul, Northern Technical University NTU, 41002 Mosul, Iraq

* Email: thamir_own@ntu.edu.iq, thamir_own@yahoo.com

Abstract. The corrosion performance of hypo-eutectoid carbon steel, and alloy steel used in steam boilers at Al-Najibiyah Thermal Power Station in the Iraqi city of Basra was investigated. These metals were exposed to different media such as moisture and acids for (30, 45, and 60) days, then the rate of corrosion was calculated in a way of weight loss. The influence of carbon content, corrosive media and duration of immersion on samples of hypo-eutectic carbon steel and alloy steel were studied. The results showed an increase in the percentage of corrosion as a result of the increase in carbon content, and alloy steel has a high resistance to corrosion than carbon steel in the different corrosive media used in this study.

Keywords: Corrosion, Carbon steel, Alloy steel, Moisture, Hydraulic acid, Steam boilers

1. Introduction

The major technical challenges and the great cost directly correlated to corrosion, which is provide strong motivations for researchers and engineers and other technical scientist staff to improve a firm grasp development on the essential bases of corrosion. Adopting the fundamentals of corrosion is essential not only for recognizing corrosion mechanisms (an important achievement by itself), but also for avoiding corrosion by proper corrosion protection means as well as predicting the corrosion performance of metallic material within service conditions. Understanding the corrosion mechanisms is important for developing the design knowledge of corrosion resistance for alloys and to estimate the long-term performance of metallic material in corrosive environments [1].

Corrosion is defined in several forms, the most important of which are: metal decay caused by its interaction with the corrosion medium to which it is exposed; it is also known as metal damage as a result of chemical or electrochemical interaction with the atmosphere. Or the reverse procedure of extracting the raw materials from metal, i.e., restoring the metal to its original, free, natural form before it was extracted, or an electrochemical reaction of metals surrounded by soil or water chemicals. These conduction cells generate electricity as a result of the natural voltage difference between metals which results degradation of the more active metals [2-5].

The microstructure of the carbon steel which has been used in the current research consists of ferrite and pearlite, where the ferrite represents a single phase, while the pearlite represents two different phases, namely the ferrite and cement. The two different phases give distinct mechanical properties but are more susceptible to corrosion than the single-phase microstructure. The micro-galvanic corrosion cell is formed because of the expected voltage difference between the two microstructure phases [6].

The corrosion that will study in this research is wet corrosion, which includes electrochemical reactions, and this type of corrosion occurs only when the medium is an electrolyte, i.e. electricity conductor. This type of corrosion has been focused on which attack metallic parts, for example, corrosion of carbon steel in the presence of water, this type of corrosion is called galvanic corrosion attributed to the discovery of the Italian scientist Galvani. The Galvanic corrosion cell consists of two poles, one of which is the anode, and the other represents the cathode, and the voltage difference between the anode and the cathode can occur if the two poles of the cell are of different metals or of the same metal. In this research, the focus will be on the corrosion of the one metal, so that the single metal becomes two poles, one of which represents the anode, while the other represents the cathode [7].

Numerous researches and studies have been conducted in the field of metal corrosion, the researcher Raja and his colleagues [8] studied effect of carbon percentage on sulfuric acid corrosion rate (5% H_2SO_4) as Corrosion Medium, and concluded that corrosion rate increases with increasing carbon percentage in steel due to increased surface area of micro-galvanic corrosion cell. Batis and Rakanta [9] studied the effect of carbon percentage on corrosion resistance in the atmosphere and found that steel containing 0.08% C more corrosion-resistant than steel containing 0.30% C, and that is due to decreasing the area of corrosion cells and the two researchers showed that corrosion resistance in carbon steel decreases with increasing carbon content. Takasaki and his colleagues [10] investigated the influence of corrosion exposure period on corrosion rate for mild steel and concluded that increasing the time of exposure reduces corrosion rates, and studied the effect of SO_4^{-2}, CL^{-1} ions on corrosion rates in mild carbon steel and showed that the corrosion rates rise with the concentration of ions.

Wall F. D and his colleagues [11] also studied the effect of corrosion exposure time on corrosion rate, where they found a decrease in corrosion rate through time, they explained that by the presence of rust and corrosion layers that tend to decrease the rate of corrosion, and these layers may form protective membranes that reduce the interaction of the metal surface with the corrosion medium. Corvo F. and his colleagues [12] concluded that the cause of variation in the corrosion rate of carbon steel in the air media is the difference in the chloride ions Cl^- concentration that cause accelerated corrosion, especially in areas exposed to rain, for continuous periods or in marine areas, the increasing of chloride ions concentration rises the conductivity, which increases the reaction rate of the electrochemical, as a result it increases the rate of corrosion.

Increasing the concentration of chloride ions Cl^- would maximize the weight loss of carbon steel and increase conductivity as well, demonstrating that increased concentration of NaCl leads to increased corrosion rates. Corrosion is closely related to the quality of metal exposed to corrosion and to the quality of the corrosive medium, there are several types of corrosive media that produce different types of corrosion, and the most important of these are the atmosphere, water, soil, acids and salts. When conducting tests and examinations related to corrosion phenomenon in a certain media and for a particular metal, it is necessary to carefully analyze both the medium and the metal, in order to determine the type of corrosion which will result, and consequently to conclude the methods to protect it [13]. The researcher R. M. Azoor and his colleagues provided the foundation for the presence of the optimal moisture level, which results to relatively great corrosion rates in soil. the corrosion rates of cast iron in silt, sand and clay were observed at various degrees of saturation [14].

Previous research shows that the researchers compared the corrosion of different types of carbon steel under different conditions of humid atmosphere and acidic media such as sulfuric acid and also studied the impact of the duration of corrosion exposure on the rate of corrosion.

In the present study, corrosion of low, medium and high carbon steels in wet and acidic media will be studied and compared with the corrosion of alloy steel used in the steam boilers at Al-Najibiyah Thermal Power Station in the Iraqi city of Basra under the same conditions.

2. Experimental work

2.1 Material

In this study, three types of hypo-eutectoid steel were used, which are characterized by the wide range of usage in engineering and industrial applications, these types of steel contain different carbon content, as well as a sample of alloy steel, which is used in steam boiler at the Najibiyah Thermal Power Station in the Iraqi city of Basra was taken, and is shown in figure 1. The chemical analysis of the specimens used in this research was carried out using a metal spectroscopy device shown in figure 2(a, b), to determine the carbon content and other chemical constituents, in order to obtain reliable results, three readings were taken and the average of these readings was adopted. Tables 1 and 2 show the chemical composition of carbon steel and alloy steel, respectively.

Table 1. Ultimate tensile strength values and elongation to fracture.

Type	Chemical composition (%)									
	C	Si	Mn	p	S	Cr	Ni	Mo	AL	Fe %
S1	0.11	0.14	0.42	0.008	0.012	0.112	0.041	0.014	0.006	Ba
S2	0.385	2.02	0.92	0.015	0.037	0.231	0.22	0.072	0.017	Ba
S3	0.64	0.21	0.82	0.002	0.04	0.027	0.083	0.043	0.041	Ba

Table 2. The chemical composition of alloy steel studied.

S4	Mn	Cu	Ni	Mo	Fe %
Alloy steel	0.54	0.16	0.1	0.01	Ba

Figure 1. Steam boiler

Figure 2. Steam boiler (zoom in view)

2.2 Heat treatment of specimens

Normalizing is a heat treatment process apply to carbon steel to restore the metal to its equilibrium state. This process was performed on the carbon steel specimens to restore them to their equilibrium state, namely, ferrite and the Pearlite [15]. To investigate the effect of the microstructure of carbon steel on corrosion resistance in different media, normalizing process was carried out for carbon steel specimens used in the research by heating them to a temperature above the crystallization temperature by 30-50 °C, and then maintained at this temperature depending on the thickness of the specimen [15]. The specimens were then removed from the oven and left to cool in still air until room temperature. Table 3 shows the retention time and temperature of the normalizing process performed on the carbon steel specimens used in this research.

Table 3. The retention time and temperature of normalizing process performed on carbon steel.

Retention time (min.)	Normalizing temperature (°C)	Carbon content
30	900	0.11
30	850	0.385
30	800	0.64

2.3. Heat treatment of specimens

Carbon steel specimens were prepared for microstructure test by conventional method including: grinding and smoothing by using sandpaper of SiC with (320, 600, 800, 1000, 1200) degree and electrical polishing followed by the etching process using the nital solution, which consists of the following percentages: 2% concentrated nitric acid and 98% methanol. Figure 3 shows the microstructure of carbon steel specimens that have been heat treated by normalizing process.

(a) Carbon steel with 0.11% carbon content (b) Carbon steel with a carbon content of 0.358% (c) Carbon steel with a carbon content of 0.64%

Figure 3. Microstructure of carbon steel specimens that have been heat treated by normalizing process

3. Corrosion tests

3.1. Exposure period to corrosive medium

First period: 30 days (720 hours)
Second Period: 45 days (1080 hours)
The third time: 60 days (1440 hours)

3.2. Corrosive media

Saline solution: saline water was used, 5% g of sodium chloride (NaCl) added to (95% g) of distilled water, where it was ascertained that sodium chloride was completely dissolved in water before immersion of the specimens. Acid solution: diluted hydrochloric acid was used (5% Vol HCl).

3.3 Calculating the rate of corrosion in a way of weight loss

3.3.1 Calculating the weight loss

Weight loss was calculated as follows:

- The specimens were cleaned before the weight was measured with diluted HCL 10% concentration, then swap with filtered water then alcohol then finally dried.

IOP Conf. Series: Materials Science and Engineering 1173 (2021) 012061 doi:10.1088/1757-899X/1173/1/012061

- The weight of each specimen was accurately found by using a sensitive electronic balance (Sartorius) Germany-made, with a precision of 0.1 mg, where the original weight was recorded.
- The specimens are placed in the corrosive media so that they are suspended with insulating filaments and completely immersed in the glass basin containing the corrosive media.
- After the exposure period is finished, the specimens are removed from the corrosive medium and cleaned from resulting rust layers by using a soft brush and water, and then cleaned the eroded layers using a diluted hydrochloric acid solution (HCl) without affecting the metal. Then the specimens are washed with distilled water and alcohol, then dried and each weight is measured to find the final weight. In order to calculate the change in weight (Δw) which is the difference between the original weight (the weight of the specimen before the corrosion test) and the final weight (the weight of the specimen after the corrosion test and removal of corrosion residues).

3.3.2 Calculating the rate of corrosion

Corrosion rate was measured in Mils per year (mpy) depending on the main function, which is the weight loss (Δw) during the period of the test (T). The test constants in this study are: density of steel (D), surface area exposed to corrosion (A). The following equation has been adopted to compute the rate of corrosion:

$$CR(mpy) = (k \times \Delta w)/(D \times A \times T) \qquad (1)$$

where CR is the corrosion rate, and K is a constant value equal ⟦3.4*10⟧^6 .

The corrosion rate can also be calculated in millimeters per year and is usually called the corrosion penetration rate.

Corrosion penetration rate is given as follows:

$$CR\ (mm/y) = (87.6\ \Delta w)/(D \times A \times T) \qquad (2)$$

4. Results and discussion

The main objective of this research is to study the effect of moisture and acid solution on the corrosion of carbon steel and alloy steel at different time intervals (30, 45, 60 days). Figures 4, 5, 6, 7, 8, 9, and 10 show the results of corrosion tests. The results in figure 4 show the extent to which the carbon content affects the corrosion rates of hypo-eutectoid steel in the corrosive media during exposed period (30 days). It has been shown that when the carbon content increases in the three types of carbon steel which are adopted in the current research, the corrosion rates increase, this increase is gradual and somewhat similar to the linear relationship.

The Carbon steel corrosion rate of 0.64% C is higher than that of carbon steel with the lowest carbon content of 0.112% C, which means that the corrosion rate rises with the increase in carbon content. Corrosion rates depend on the type of carbon steel in general and on its microstructure in particular, where the change in carbon content affect the microstructure of steel. The microstructure of the hypo eutectoid steel composed of alpha ferrite and pearlite with different percentage depending on carbon content.

The increase in corrosion rates, with the increase in carbon content, is caused by the fact that increasing the carbon content leads to an increase in the proportion of pearlite, leading to an increase in the number of galvanic corrosion cells. The pearlite is a two-phase microstructure composed of alpha ferrite and cementite (⟦Fe⟧_3 C), Which reflects the best installation of the galvanized cells represented by the anode, which is the cementite and the cathode, which is the alpha ferrite, and then the pearlite represents the galvanic corrosion cell because of the voltage difference between the cathode and the anode, between the ferrite and cementite when they exist together and when the electrolyte is available. As a result, with the increase in the carbon content in the hypo-eutectoid steel, the anodic area (the area of the galvanic corrosion cells) represented by the pearlite, leading to maximize in the rate of corrosion.

Figure 5 shows the rates of corrosion of hypo-eutectoid steel in the corrosive media during 45 days. The results show that corrosion rates are increasing as carbon content increases, and this increase is also gradual. Figure 6 shows the rates of corrosion of hypo-eutectoid steel in the corrosive media during 60 days. The results show that when the carbon content of hypo-eutectoid steel increases, the corrosion rate increases, and this increase is gradual. The highest carbon content of 0.64% C is the highest in corrosion rate compared with the other carbon steel used, and the lowest corrosion rate occurred for carbon content of 0.11% C.

The corrosion rate increases with the increase in carbon content due to the change in the microstructure, which leads to the increase of the cementite phase and the decrease of the ferrite phase, thus increasing the anode area with the increase of the carbon content that leads to the increase of the corrosion rates. The results show that there is a variance in the corrosion rates of hypo-eutectoid steel by changing the corrosive medium, which indicates the important role played by the corrosive medium that affect the corrosion rates.

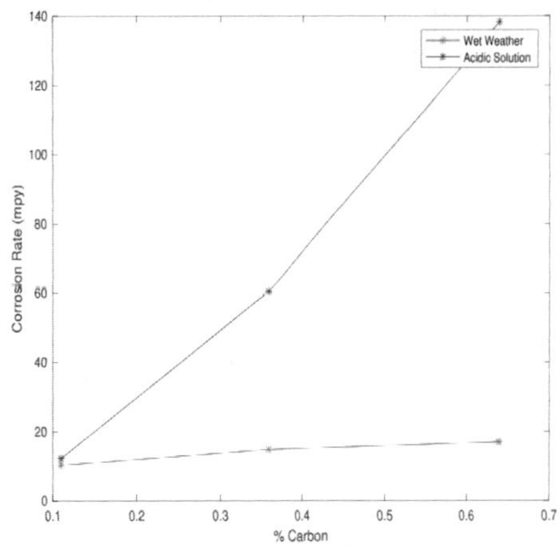

Figure 4. Effect of carbon content on corrosion rates during 30 days.

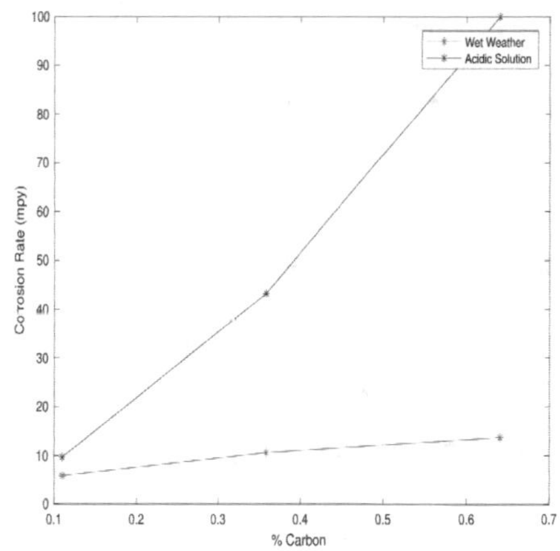

Figure 5. Effect of carbon content on corrosion rates during 45 days.

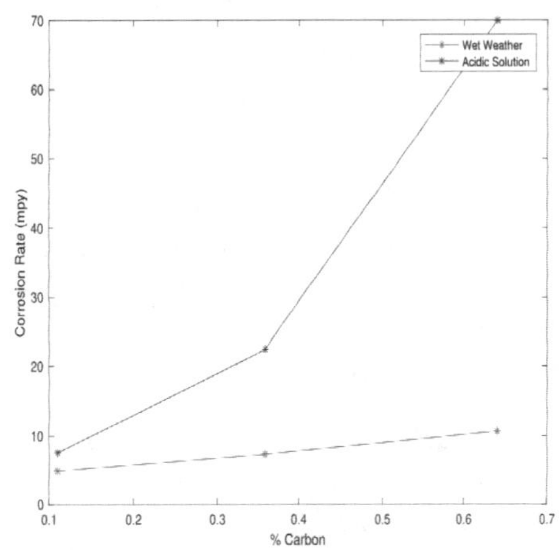

Figure 6. Effect of carbon content on corrosion rates during 60 days.

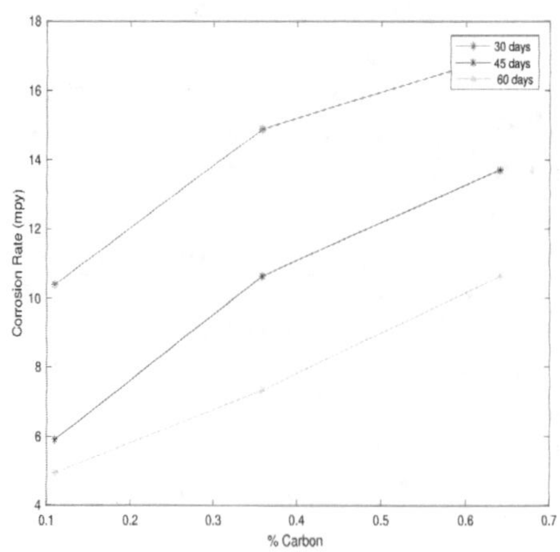

Figure 7. Effect of wet weather exposure period on the corrosion rate of carbon steel.

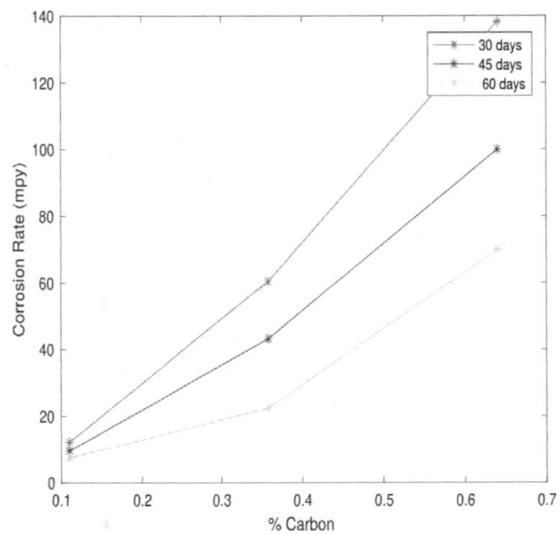

Figure 8. Effect of acidic solution exposure period on the corrosion rate of carbon steel

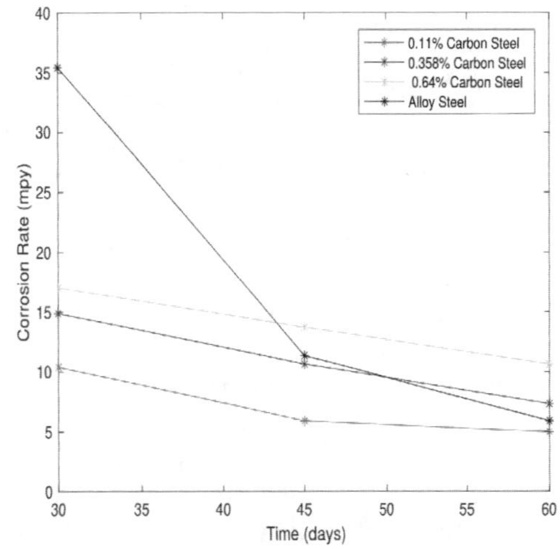

Figure 9. Corrosion rates for carbon steel and alloy steel in a wet weather during different time periods

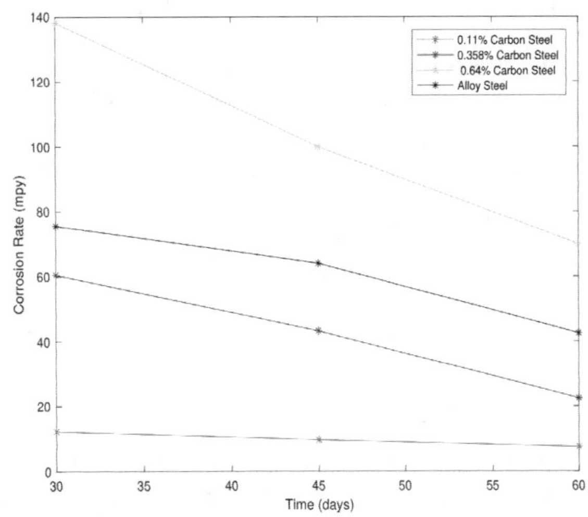

Figure 10. Corrosion rates for carbon steel and alloy steel in an acidic solution during different time periods

All the results shown in Figures 4, 5, and 6 indicate that corrosion rates of carbon steel in acidic media were higher than those of wet media, the results show that corrosion rates vary from one medium to another depending on the nature of the medium and the severity of its effect. Figure 7 shows the effect of wet weather exposure period on the corrosion rate of carbon steel. Figure 7 show the results that corrosion rates decrease gradually with increased exposure period in wet weather. It seems that the lowest corrosion rates were obtained within a 60-day exposure period, while the highest corrosion rates were obtained during a period of 30 days.

The reason for the decrease in corrosion rates with increasing in exposure period is due to the slow rate of electrochemical reaction caused by the formation of a membrane or layers of corrosion residues on the surface of carbon steel that impedes the persistence of electrochemical corrosion, thus reducing corrosion rates. And due to the depletion of dissolved oxygen in corrosive medium as well. At the beginning of the corrosion test, the electrochemical reaction occurs rapidly resulting in high corrosion rates, but over time the dissolved oxygen is reduced, increasing the concentration of hydrogen ions that accumulate at the anode, which inhibits corrosion reactions.

This is called polarization and the value of acidic function (pH) is also affected by the concentration of dissolved oxygen, since the acidity and the base of the medium are affected by the value of PH. If (pH< 6) the medium is acidic and have a very high effect on the metal, if the value of (pH=7) the medium will be neutral, and the medium will be alkaline if (pH>8) which is the least influential medium in steel.

Figure 8 shows the effect of corrosion exposure period in acidic solution on the corrosion rates of the hypo-eutectoid steel. It is clear from the figure that the corrosion rates gradually decrease with increasing exposure period to corrosive medium, it seems that the lowest corrosion rates were obtained during a 60-days exposure period, and the highest corrosion rates were obtained during a 30-days exposure period. Also, as the case in figure 4, the reason for the decrease in corrosion rates with increasing in exposure period is due to the slow rate of electrochemical reaction caused by the formation of a membrane or layers of corrosion residues on the surface of carbon steel that impedes the persistence of electrochemical corrosion, thus reducing corrosion rates.

It is therefore possible to conclude that the longer the time for immersing, the greater the carbon steel resistance to corrosion and rust, because increasing the time of the immersion provides an opportunity to complete the interaction between the immersion solution (corrosive medium) and the surface of the metal, and this helps to form a layer of corrosion residues on the surface of the metal then increase its thickness and form an obstacle to the persistence of electrochemical corrosion.

Figure 9 shows that the corrosion rate of alloy steel is greater than that of the three types of carbon steel during the 30-days exposure period in wet weather. While in the 45-days exposure period, the corrosion rate of alloy steel is higher compared with carbon steel of 0.11% C and 0.358% C, and less than that of 0.64% C. In the 60-day exposure period, the corrosion rate of alloy steel is lower than the corrosion rate of carbon steel of 0.358% C and 0.64% C, and higher than the corrosion rate of carbon steel of 0.11% C. It is concluded that the corrosion rate of alloy steel in wet weather decreases at a greater rate with increasing exposure period compared with the reduction in the corrosion rate of carbon steels.

Figure 10 observes the corrosion rates of carbon steels and alloy steel in an acid solution at different time periods. The figure shows that the corrosion rate of alloy steels is lesser than the corrosion rate of carbon steel 0.64% C, unlike the other two types of carbon steel 0.11% C and 0.358% C, the corrosion rate of alloy steel is greater than that of these two types of carbon steel.

5. Conclusion

From the previous results obtained from this study, it can be concluded that the low carbon steel (0.11%C) has higher corrosion resistance than that of medium carbon steel (0.358%C) and high carbon steel (0.64%C) in the wet and acid corrosive media, this indicates that the corrosion resistance decreases with increasing carbon content, however, this decrease in corrosion resistance is more pronounced in the case of acid medium. It is also noticed that the corrosion rate decreases with increasing the exposure duration to wet and acid corrosive media for the three types of carbon steels. In the wet medium, it is noticed that the corrosion rate of alloy steel which is used in the steam boilers at Al-Najibiyah Thermal Power Station in the Iraqi city of Basra is higher than that of the three types of carbon steels after 30 days, but begins to decrease whenever the exposure duration exceeds 30 days. So, after 45 days the corrosion rate of alloy steel is less than that of high carbon steel, and it is nearly equal to that of medium carbon steel. After 60 days, it is noticed that the corrosion rate of alloy steel is lower than that of high carbon steel and medium carbon steel, but slightly more than that of low carbon steel. In the acid medium, it is noticed that the corrosion rate of alloy steel is lower than that of high carbon steel, but still higher than that of medium carbon steel and low carbon steel for the three-exposure duration 30, 45, and 60 days.

References

[1] Shaw, B. (2003). ASM handbook volume 13a: corrosion: fundamentals, testing and protection. D. Stephen, ASM International, Materials Park, Ohio, USA. Singh, AK, & Quraishi, MA (2010). Effect of Cefazolin on the corrosion of mild steel in HCl solution. Corrosion Science, 52(1), 152-160.
[2] Fontana, M. G. (2005). Corrosion engineering: Tata McGraw-Hill Education.
[3] Hou, X., Gao, L., Cui, Z., & Yin, J. (2018). Corrosion and protection of metal in the seawater desalination. Earth Environ Sci, 108, 022037.
[4] Takasaki, S., & Yamada, Y. (2007). Effects of temperature and aggressive anions on corrosion of carbon steel in potable water. Corrosion Science, 49(1), 240-247.
[5] Stephen, C. (2003). Galvanic Corrosion. University of Delaware, USA.
[6] Shreir, L. L. (2013). Corrosion: corrosion control: Newnes.
[7] Roberge, P. R. (2000). Handbook of corrosion engineering: McGraw-Hill.
[8] Raja, V., Baligidad, R., & Shankar Rao, V. (2002). Effect of Carbon on Corrosion Behavior of Fe3 Al Inter-metallic's in 0.5 N Sulfuric acid. Journal of Corrosion Science, 33, 521-533.
[9] Batis, G., & Rakanta, E. (2005). Corrosion of steel reinforcement due to atmospheric pollution. Cement and concrete composites, 27(2), 269-275.
[10] Trethewey, K. R., & Chamberlain, J. (1995). Corrosion for science and engineering.
[11] Wall, F., Martinez, M., Missert, N., Copeland, R., & Kilgo, A. (2005). Characterizing corrosion behavior under atmospheric conditions using electrochemical techniques. Corrosion Science, 47(1), 17-32.
[12] Corvo, F., Minotas, J., Delgado, J., & Arroyave, C. (2005). Changes in atmospheric corrosion rate caused by chloride ions depending on rain regime. Corrosion Science, 47(4), 883-892.
[13] García, K., Barrero, C., Morales, A., & Greneche, J. (2008). Lost iron and iron converted into rust in steels submitted to dry–wet corrosion process. Corrosion Science, 50(3), 763-772.
[14] Azoor, R. M., Deo, R. N., Birbilis, N., & Kodikara, J. (2019). On the optimum soil moisture for underground corrosion in different soil types. Corrosion Science, 159, 108116. doi:https://doi.org/10.1016/j.corsci.2019.108116
[15] Brooks, C. R. (1996). Principles of the heat treatment of plain carbon and low alloy steels: ASM international.

Essential characterization of etal owder for dditive anufacturing

S M Q M Ramli[1,*], N A Fadzil[1], H Ghazali[1], P Viklund[2], W F F W Ali[1,3*]

[1] School of Mechanical Engineering, Faculty of Engineering, Universiti Teknologi Malaysia, 81310 Johor Bahru, Johor, Malaysia.

[2] GKN Aerospace Engines, 6031 Connection Drive, Suite 100 Irving, Texas USA 75039

[3] Centre of Excellence for Frontier Materials Research (FrontMate), Universiti Malaysia Perlis, 02600 Arau, Perlis, Malaysia.

* Email: wan_fahmin@utm.my, mazarinaramli@gmail.com

Abstract. The aerospace industry is mostly focused on complex, low production volume as the components are often known with complicated internal geometries and dimensions. Heavy market competitions and restrictions in high manufacturing cost could be overcome with additive manufacturing (AM). AM has received notable attention in its ability to produce customized metal parts and exhibiting top-notch capabilities that would lead the AM industry in the future. Nonetheless, in producing trusted, high-quality products, characteristics of metal powders should be accounted so it would meet proper quality such as powder morphology, particle size distribution, porosity, and powder's ability to flow. SEM, OM, laser diffraction and rheometer are suggested in studying the powder's characteristics. All characterization methods found are used practically. These characteristics are important to assess AM quality. Comparison in different aerospace materials is studied to get the actual material combination in getting an optimum metallurgical bond between the substrate and molten pool of the powders. Therefore, characterization methods of 3D metal additive and characteristics of aerospace materials are evaluated and presented systematically.

Keywords: Additive manufacturing, 3D metal printing, characterizations, metal powders

1. Introduction

Additive manufacturing (AM) successes have received significant media attention in the media press in the last couple of years, where it involved in joining metal powders layer upon layer with no or minimal pre-processing and post-processing requirements [1-3] This approach can give industry new design flexibility, reduce energy use and shorten time to market when compared with conventional machining methods which are time-consuming and uneconomical [4-6]. One of the assuring AM technologies is direct laser metal deposition (DLMD) where it derived from micro-welding and laser technologies to generate molten pool and solidified after to form a complex structure. DLMD also has the ability to repair existing worn-out parts that have not been repairable in the past, including both geometrical and structural restoration [7-13].

There are two main parameters for metal AM which are a type of input raw material and energy source used to form the desired part [14]. Input raw materials, as shown in Figure 1, play a significant role in producing a high-quality printed aerospace component and repairing existing component. For example, Ti-6Al-4V is commonly used as fans, compressor disc/blade and load-carrying components due to its strength and thermal stability at high-temperature environment [9] meanwhile SS316 has resistance towards corrosive environment [6]. It has been found that SS316 produced by AM have better yield strength (649.3 MPa) and hardness (235 HV) [16, 17]. Inconel 718 usually used in engine part due to excellent wear resistance and high tensile-creep rupture properties at temperature of 1300 °F [18, 19]. However, these materials are hard to machine due to high work hardening rates make AM is favorable over machining process [20].

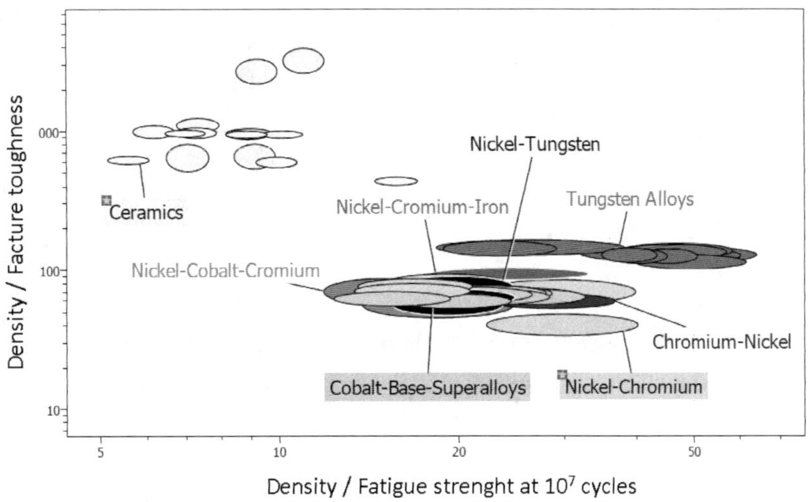

Figure 1. The density of different materials in 900 C temperature [1]

In general, DLMD in AM requires metal powder diameters in the range of 50 µm to 150 µm [21, 22]. Theoretically, the smaller and more spherical particle sizes are preferable because it is easy to infuse fine powder into a narrow stream. In ensuring the repeatable output of metal parts, these characteristics give a balance of good flowability, fair powder consumption efficiency, and smooth surface finish [23-27]. It is worth noting that the optimal particle size may also differ depending on the properties of the alloy. For example, light alloys such Al-Si -10Mg flow at best in a coarser shape, whereas dense alloys like Ti-6Al-4V work better in a more refined form [1,28]. However, the wide variety of powder manufacturers on the market has hampered the establishment of universal standards for powder properties [29]. Thus, determining the powder characteristic for metal-additive manufacturing is essential for industry to produce consistent parts with predictable properties. Otherwise, it may resulting in high degree of variation in process outcomes. Hence, characterizing raw metal powders for better output will be discussed.

2. Powder Characterization/Sampling

Consistent powder characteristics are crucial for repeatable manufacture so that printed component has desired mechanical properties. Thus, a deeper understanding and knowledge about powder characteristics and its impact on AM is crucial which affect the development, or quality of the printed parts [30]. In addition, powder sampling e.g., scoop sampling, chute splitting or spin rifling also must be done carefully to ensure the accuracy of the results. Furthermore, the effect of powder properties on the performance and processing of the final component could be done by through the characterization of the mechanical, chemical and physical properties of a powder [31, 32].

2.1 Particle Density/Bulk Properties

Particle density is an essential powder characterization to estimate the built-in porosity or related to a powder's ability to pack which concerning die packing and spreading mechanism during the deposition process. This property is measured based on ideal gas law derived from primitive density equation (mass/volume). In this method, the metal powder is placed in a pycnometer under controlled temperature (15-30 °C), and atmospheric pressure (0 – 0.7 MPa), type of gas (helium) [33, 34]. There are many ways to measure powder density; in particular, these methods include measurement of apparent density, tapped density, actual density, and powder-bed density. The apparent density simulates the loss of powder packing state following the ASTM B212-09 and ASTM B707-10 standards. It shows the correlation of a powders mass to freely fill a hollow space [35]. The tapped density indicates that the stress-packing state of self-powder is determined using B527 [35]. The actual density represents the bulk density of a powder using pycnometer as in ASTM B923 [35].

Thejane et al. [36, 37] indicated that the density of Ti-6Al-4V measured using pycnometer (4.0-4.42 g/m3) was close to the theoretical value (4.43 g/m3). Cordova et al. [21] found Inconel 718 has highest density (8.26 g/cm3), followed by Ti-6Al-4V (4.38 g/cm3), Scalmalloy (2.68 g/cm3) and Al-Si-10Mg (2.65 g/cm3), respectively [21]. The apparent and taped density of Inconel 718 virgin powder is ten times higher than reused IN718 powder but vice versa for true density [35].

2.2 Particle Size Distribution (PSD)

It is important to accurately characterize particle size distribution (PSD) in delivering high quality of AM product defined by (D_v90-D_v10)/D_v50, where Dv = $(\frac{6}{\pi}Vp)^{1/3}$ respectively [36]. Noted that, Dv is diameter of volume-equivalent sphere and Vp is the volume of particles. Through PSD, the powders ability to flow, spread and packing during deposition process can be illustrated using histogram. Using smaller particle size is likely induce coagulation inside nozzle subsequently affect the mass flow. There are several methods used to determine PSD of the metal powders such as dry sieving (DS) and laser diffraction (LD). For DS, ASTM B214-07 or ASTM E161-00 is needed. The standard suggests sieve need to be arranged in consecutive order between 5 µm to 1 mm and agitated for 15 minutes. Compared to DS, LD provides real-time monitoring to facilitate the distribution of particle size during production and, at the same time, provides immediate feedback to improve the production process [37]. In LD, at wider angles, small particles scatter weaklier while large particles scatter light strongly at smaller angles [37].

However, there are some limitations of this method in which it assumed that all powders are in spherical structure, although irregular structures are presented [31,38] which give "ghosts" diffractions interpretation [39]. It has been found that defect formation can be reduced when the PSD of metal powder is within the range of 20µm to 50µm [22]. Author [54] in their finding, indicated that Inconel 625 with 60µm - 80µm provide excellent mechanical strength [35].

2.3 Particle Morphology/Surface Roughness

The morphology of metal powders and uniformity (e.g., spherical, angular, dendritic, dendritic, dish-shaped, circular, or quasi-quantitatively as established in ASTM B243-11) are crucial elements in determining powder flowability in AM [40,41] besides packing density [21]. It can be estimated using $\frac{Ss}{Sp}$ in which Sp is the surface area of a given volume while Ss is equivalent surface area. However, these shapes are highly dependent on the type of powder processing method [42]. For example, the atomization process yields a smooth and spherical shape but exhibit leaf-like and irregular structure under precipitation or comminution [42]. Ideally, spherical particles provide better densification than non-spherical particles [43, 56]. There are several methods to determine the size and morphology of particle properties as such adsorption, sieve, laser, and microscopy analyses [44]. Barret [51] offers a comprehensive review in expressing dimensional parameters for metal powders as depicted in Table 1.

Table 1: Dimensional parameters to characterize particle morphology [51]

No	Formula	Description	Range
1	$\frac{L+i}{2S}$	Flatness index	1 to ∞
2	$\frac{i}{L}, \frac{S}{i}$	Ordinate and Abscissa for plot to evaluate shape	0 to 1
3	$\frac{i.100}{L}$	Elongation	0 to 100
4	$\frac{S.100}{L}$	Flatness	0 to 100
5	$\frac{S}{L}$	Flatness	0 to 1

L = long axis, S = short axis, i = intermediate axis

Undeniably, different microscopy methods have been extensively used among researchers to study surface morphology. As much useful as the method is, there is a setback too. The number of particles analysed in each image is limited to 100 particles or lesser [44]. Chen et al. [45] investigated the characteristics of 15Cr-13Mo-Y using the scanning electron microscope (SEM) and metallographic microscope. Cleary and Sawley investigated powders' aspect ratios discharge from hoppers have affected the mass flow rates [46]. The flow rate decreases when the powder shape changes from equiaxed to elongated [46].

2.4 Chemical Composition

In practice, raw metal powders used in AM are not 100% pure and might contain other unwanted materials. Thus, to ensure fidelity of powders require (i) micro-analysis methods such as energy dispersive spectroscopy (EDS), (ii) surface-analysis methods e.g., atomic emission spectroscopy (AES), X-ray photoelectron spectroscopy (XPS), secondary ion mass spectrometry (SIMS), and (iii) bulk-analysis methods e.g., X-ray diffraction (XRD) and X-ray fluorescence (XRF) by identifying the chemical composition of metal powders. The presence of unwanted phases or secondary phases significantly affects the diffusion process during micro-welding, affecting the printed part's mechanical properties [42]. For example, a small amount of oxygen and nitrogen gases has negatively influenced AM product's strength and ductility [42].

2.5 Porosity

The porosity of the powders either in interconnected or particles' surface is unwanted characteristic that need to be avoided. It has been established that powders' porosity might be transported to the melt pool hence lowering the mechanical properties of deposited structures or printed parts [47]. This characteristic is highly dependent on type of powder processing. For example, although atomization process able to produce uniform-spherical particles however it induced higher porosity than precipitation method due to entrapped gas/air inside the particles. The most appropriate approach in visualizing the metallic powder's porosity is using X-ray computed tomography (XCT) [48, 49]. However, metallography techniques (viewed under optical microscope) can still be used [36]. Author [35] used the optical microscopy (OM) technique to observe the powders' internal porosity, defects, and cross-sections [26].

2.6 Flowability/Rheology

The metal powder's flowability significantly affects the smoothness of the powder deposition and equable powder feeding in AM systems [39] in which ultimately affect the production rate and packing ability. This property is very much affected by particle size and distribution. It has been portrayed that powder with less than 10μm diameter has poor flowability than large particles due to higher surface fiction or cohesive force between the particles [39]. In addition, smaller particle size may also result in agglomeration. Among standardized methods to measure metal powder's flowability or rheological behavior (e.g., dynamic flow properties) is using FTF *Freeman* rheometer and *Hall & Gustavson* flowmeter based on ASTM B213-11 and ASTM B964-09, respectively. However, in term of sensitivity and accuracy, rheometer provides a finer evaluation of powder rheological behavior than flow meter [50].

2.7 Thermal Analysis / Conductivity

Thermal properties of metal powders are also extremely important to AM. It measured the ability of metal powders to absorb and transfer heat (thermophysical property) during deposition process [52]. Usually, differential scanning calorimetry (DSC) and laser flash experiment are utilized to determine the thermal characteristic (thermal conductivity, K) and calculating the kinetic parameters of metal powders. Other methods as such photoacoustic, crenel heating excitation and photopyroelectric techniques are also employed [53-56]. This property, K can be mathematically represented by $\frac{Z^2 \rho C_p}{6T_{max}}$ (where z is distance from powder surface to thermocouple junction, ρ is bulk density and C_p is specific heat capacity at time at maximum voltage (T_{max}) will elucidate the heat-dissipation effect during consolidation process [53]. However, this equation is valid when $\frac{Z}{a} > 3.87$ [54]. However, thermal conductivity for mixed powder with different particles size are varies.

3. Conclusion

To ensure the integrity of the printed products in AM, feedstock or metal powders' characteristic needs to be understood. Density, particle size distribution, morphology, chemical composition, porosity, and flowability are essential characteristics in ensuring the process's workability. Density will determine the spreading and packing density during deposition meanwhile particle size distribution determined the flow rate of the process. Smaller particle size induced coagulation. Other properties as such porosity need to be minimized since it affects strength of the printed part. Flowability is highly dependent on the particle morphology and density of metal powders. Although it may seems the characterization of metal powders are readily applicable however, standardizing methods used might be more challenging.

Acknowledgment

The authors would like to express their gratitude to the Universiti Teknologi Malaysia (UTM) for financial support under UTMICONIC research grant (Vot no. 09G59).

The framework of the work was established by Mazarina MR to meet study objective related with characterization of metallic powders based on the state of the art and industrial partners from GKN Aerospace needs. The authors declare not conflict of interest.

References

[1] Gu, D., & Shen, Y. 2008. *Mater. Sci. Eng.* A, **489**(1-2), 169–177.

[2] El Cheikh, H., Courant, B., Branchu, S., Huang, X., Hascoët, J.-Y., & Guillén, R. 2012. *Optics Lasers Eng.,* **50**(12), 1779–1784.

[3] Duda, T., & Raghavan, L. V. 2016. *IFAC-PapersOnLine*, **49**(29), 103–110.

[4] Najmon, J. C., Raeisi, S., & Tovar, A. 2019. Review of additive manufacturing technologies and applications in the aerospace industry. In Additive Manufacturing for the Aerospace Industry. Elsevier Inc

[5] Ngo, T. D., Kashani, A., Imbalzano, G., Nguyen, K. T. Q., & Hui, D. 2018. *Compo.Part B: Eng.,* **143**, 172–196.

[6[Graf, B., Gumenyuk, A., & Rethmeier, M. 2012. *Physics Procedia*, **39**, 376–381.

[7] Leino, M., Pekkarinen, J., & Soukka, R. 2016, *9th Int. Conf. Photon. Tech.*, **83**, 752–760.

[8] Liu, Q., & Hinton, B. 2008. *Int. J. Struc. Integrity*, **2**(3), 314–331.

[9] Pinkerton, A. J., Wang, W., & Li, L. 2008. *J. Eng. Manufact.* **222**, 827–836.

[10] Plourde, R. 2003. Laser-Based Repair System Reclaims High Value Military Components. June, 7–9.

[11] Sexton, L., Lavin, S., Byrne, G., & Kennedy, A. 2002. J. *Mater. Process. Tech.*, **122**, 63–68

[12] Bergan, P. 2000. Implementation of laser repair processes for navy aluminium components. In: Proceeding of diminishing manufacturing sources and material shortages conference (DMSMS).

[13] Barile, C., Casavola, C., & De Cillis, F. 2019. *Compo. Part B: Eng.,* **162**, 122–128.

[14] Li, P. H., Guo, W. G., Huang, W. D., Su, Y., Lin, X., & Yuan, K. B. 2015 *Mater. Sci. Eng. A,* **647**, 34–42

[15] Greco, S. 2020. Selective laser melting (SLM) of AISI 316L - impact of laser power, layer thickness, and hatch spacing on roughness, density, and microhardness at constant input energy density. 1551–1562.

[16] Tolosa, I., Garciandía, F., & Zubiri, F. 2010. Study of mechanical properties of AISI 316 stainless steel processed by "selective laser melting", following different manufacturing strategies. 639–647.

[17] High Temp Metals: Inconel 718 Technical Data. 2017. http://www.hightempmetals.com/techdata/hitempInconel718data.php

[18] Special Metals Corporation: INCONEL® Alloy 718. (2007).https://www.specialmetals.com/assets/smc/documents/inconelalloy718.pdf

[19] Junker, D., Hentschel, O., Schmidt, M., & Merklein, M. 2001. *Metals*, **8**(10), 1–13.

[20] Cordova, L., Campos, M., & Tinga, T. 2017. Paper presented at VI Congreso Nacional de i Congreso Iberoamericano de Pulvimetalurgia, **6**(3), 1-6

[21] Masaylo, D., Igoshin, S., Popovich, A., & Popovich, V. 2020. *Mater. Today: Proceedings*, **30**, 665–671.

[22] Dietrich, S., Wunderer, M., Huissel, A., & Zaeh, M. F. 2016. *Procedia Manufac.*, 6, 88–95.

[23] Zhong, C., Chen, J., Linnenbrink, S., Gasser, A., Sui, S., & Poprawe, R. 2016. *Mater. Design*, **107**, 386–392.

[24] Averyanova, M., Bertrand, P., & Verquin, B. 2010. *Proceedings Inter.DAAAM Sym.*, **21**(1), 1531–1532.

[25] Iebba, M., Astarita, A., Mistretta, D., Colonna, I., Liberini, M., Scherillo, F., Pirozzi, C., Borrelli, R., Franchitti, S., & Squillace, A. 2017. *J. Mater. Eng Perform.,* **26**(8), 4138–4147.

[26] Clayton, J., & Deffley, R. 2014. *Metal Powder Report*, **69**(5), 14–17.

[27] Thomas, M., & Drawin, S. (2016). Role of metal powder characteristics in additive manufacturing. World PM 2016 Congress and Exhibition.

[28] Hofmann, D. C., Kolodziejska, J., Roberts, S., Otis, R., Dillon, R. P., Suh, J.-O., Borgonia, J.-P. 2014. *J. Mater. Res*, **29**(17), 1899–1910.

[29] Slotwinski, J. A., Garboczi, E. J., Stutzman, P. E., Ferraris, C. F., Watson, S. S., & Peltz, M. A. 2014. *J. Res. Nat. Inst. Standards Tech*, **119**, 460–493

[30] Slotwinski, J. A. & Garboczi, E. J. 2015. *J. Mineral, Metals Soc.,* **67**(3), 538-543.

[31] Staab, T. E. M., Krause-Rehberg, R., Vetter, B., & Kieback, B. 1999. *J. Phys. Condensed Matter*, **11**(7), 1807–1822

[32] Sheehan,C.(n.d.). Density of solids. U.S. Pharmacopeia. http://www.pharmacopeia.cn/v29240/usp29nf24s0_c699.html#:~:text=The test gas is helium,the course of the measurement .
[33] Sereno, A. M., Silva, M. A., & Mayor, L. 2007. Int. J. Food Prop., **10**(3), 455–469.
[34] Nguyen, Q. B., Nai, M. L. S., Zhu, Z., Sun, C.-N., Wei, J., & Zhou, W. 2017. *Eng.*, **3**(5), 695–700.
[35] Jansson, P. 2018. Investigation and Evaluation of Metal Powder Characterization Techniques currently available at the KTH Department of Material Science and Engineering.
[36] Levoguer, C. 2013. *Metal Powder Report*, **68**(3), 15–18.
[37] Ferraris, C.F., Guthrie, W., Aviles, A.I., Peltz, M.A., Haupt, R. & MacDonald, B.S. (2006).
[38] Certification of SRM 114q: Part II(Particle Size Distribution), NIST Special Publication, 260-166.
[39] Popovich, A., & Sufiiarov, V. (2016). Metal Powder Additive Manufacturing. Intech, tourism, 13. https://www.intechopen.com/books/advanced-biometric-technologies/liveness-detection-in-biometrics
[40] Dawes, J. (2017). Optimizing Metal Powders for Additive Manufacturing. Additive Manufacturing. https://www.additivemanufacturing.media/blog/post/optimizing-metal-powders-for-additive-manufacturing-exploring-the-impact-of-particle-morphology-and-powder-flowability
[41] Mikli, V., Kaerdi, H., Kulu, P., & Besterci, M. 2001. *Eng.* **7**(1), 22–34.
[42] Benson, J. M., & Snyders, E. 2015. *South African J. Indust. Eng.*, **26**;104–114.
[43] Micromeritics. 2016. Characterization of Metal Powders for Additive Manufacturing. 1–2.
[44] Van Der Wiel, D., & Operations Manager, A. (2019). Advanced Powder Characterization for 3D Printing. 22–26.
[45] Chen, X., Chen, S., Liang, J., Cui, T., Liu, C., & Mei, W. 2019. *Powder Metal.*, **62**(4), 218–228.
[46] Cleary, P. W., & Sawley, M. L. 2002. *Appl. Math. Model.*, 26(2), 89–111.
[47] Cunningham, R., Nicolas, A., Madsen, J., Fodran, E., Anagnostou, E., Sangid, M. D., & Rollett, A. D. 2017. *Mater. Res. Lett.,* **5**(7), 516–525.
[48] Heim, K., Bernier, F., Pelletier, R., & Lefebvre, L. P. (2016). High resolution pore size analysis in metallic powders by X-ray tomography. Case Studies in Nondestructive Testing and Evaluation, 6, 45–52.
[49] Léonard, F., Tammas-Williams, S., & Todd, I. 2016. CT For Additive Manufacturing Process Characterisation: Assessment of Melt Strategies on Defect Population. (Accessed November 8, 2020) www.3dct.at.
[50] Clayton, J., & Deffley, R. 2014. *Metal Powder Report*, **69**(5), 14–17.
[51] Salah E.B., Morgan L., Christopher A.D., and Vladimir B., 2020, *Addit. Manuf.* **31**(1-16)
[52] Barret P.J.,1980, *Sedimentology*. **27**(3) 291-303
[53] Davoud J., Wessel W.W., and Bernard J.G (2018) *Appl. Therm. Eng.* **143**(403-414)
[54] Rizal MA., Tatsuaki F., Takashi E., Akira H., Ryutara T., and Sanusi AA. 2012, *Key Eng. Mater.* **523-524**(244-249)
[55] Yun B., Grady W., and Cristopher B. W (2017) *J. Manuf. Sci. Eng.* **13**(1-6)
[56] Jacob G., Donmez., Slotwinski and Moylan (2016) *Meas. Sci. Technol.* **27**(1-12)

Micro-electro discharge machining (Micro-EDM) models for conductive and nonconductive materials: A review

M Y Ali[1*], A Sabur[2], S R Ya'akub[1], A D A Hamid[1], A R Alao[1], and R Singh[1]

[1]Mechanical Engineering Programme Area, Faculty of Engineering, Universiti Teknologi Brunei, Tungku Highway, Gadong BE1410, Brunei Darussalam

[2]Ministry of Fishers & Livestock, Bangladesh Secretariat, Abdul Gani Road, Dhaka 1000, Bangladesh

* Contact Email: yeakub.ali@utb.edu.bn

Abstract. Models are useful in controlling the process effectively and efficiently in micro electro discharge machining (Micro-EDM). There are two types of models which are empirical models and theoretical models. Most of the models are basically formulated for conductive work materials. However, nowadays there is a trend to cut nonconductive materials using micro-EDM where the models developed for conductive materials are not applicable. There are only few models that are developed to cut nonconductive materials with micro EDM but these models have limitations. In this article, models for cutting conductive materials and nonconductive materials using micro EDM are compared in view of their applications.

Keywords: Micro EDM, Modelling, nonconductive materials, ceramics, aluminium oxide

1. Introduction

Micro-electro discharge machining (Micro-EDM) is a special application of electro discharge machining to produce micro sized products with metallic materials. The material removal process in micro-EDM is similar with general EDM process which is based on a series of electrical sparks between the electrode and workpiece [1]. Any materials to be cut using micro-EDM must have electrical conductivity more than 0.1 Scm^{-1}. The material hardness and brittleness do not affect the cutting process [2-3]. In micro-EDM, thermal energy evolved from the electrical sparks flows into the electrode, workpiece and dielectric fluid. A fraction of spark energy or heat conducted into a highly localized area on workpiece causes temperature rise, melting and vaporization which eventually creates a micro crater [4]. As such most of the theoretical models were developed based on single spark only although that is not the real case in practice where millions of sparks were used to create desired shape [5-6]. In addition, the process becomes more complicated when applied for cutting nonconductive materials (electrical conductivity less than 0.1 Scm^{-1}) as it involves assisted electrode to create initial spark between real electrode and workpiece. In addition to melting and vaporization, there exist a complex and uncountable phenomena call "spalling" in material removal process [7].

Models are useful in controlling the process effectively and efficiently. Two approaches are widely used for modelling in micro-EDM. The first one is empirical models that are developed on the basis of experimental investigations. The second one is theoretical models that are developed using the fundamentals of the electrical circuit, theories of thermodynamics and heat transfer. In theoretical modelling, it is presumed that the electrical energy is converted to thermal energy and causes the material removal of the workpiece. Theoretical models can be categorized based upon the assumption of heat flux whether it is uniform (disc), point heat source, or cylindrical plasma [8]. The models can be further divided into three categories: anode erosion, cathode erosion and plasma channel modelling. Each of the models has its own characteristics according to the variation of hypothesis and development process. Although there are mathematical models for to estimate micro EDM characteristics for conductive materials are of plenty, there are scarcity of models for nonconductive materials. In this article a comparison of EDM models for conductive and nonconductive materials are compared.

2. Micro-EDM Models for Conductive Materials

2.1 Dibitonto's model

Functional relationship between the input parameters and crater radius in EDM has been assuming that the plasma radius at the cathode is smaller than the anode [9]. A schematic diagram of an EDM crater as assumed in this model is shown in Figure 1. Point heat source was approximated with spherical symmetry of melt front. As a result, the crater profile was assumed hemispherical. The cathode was assumed to consume 18% of the total energy. In this modelling, the crater radius was expressed as the function of heat energy. The temperature distribution during EDM is expressed by:

$$T_m = T_0 + \frac{qr}{K_t} erfc\left(\frac{r}{2\sqrt{\alpha t}} \right) \quad (1)$$

where T_m = melting temperature of the material, r = crater radius, q = power, t = discharge duration, T_o = room temperature, K_t = workpiece heat conductivity, α = workpiece heat diffusivity ($K_t/\rho C_p$), ρ = workpiece density, and C_p = workpiece specific heat.

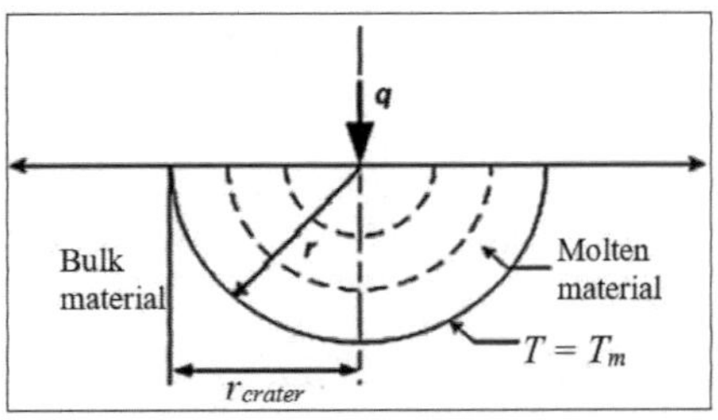

Figure 1. Schematic diagram of an EDM crater [9]

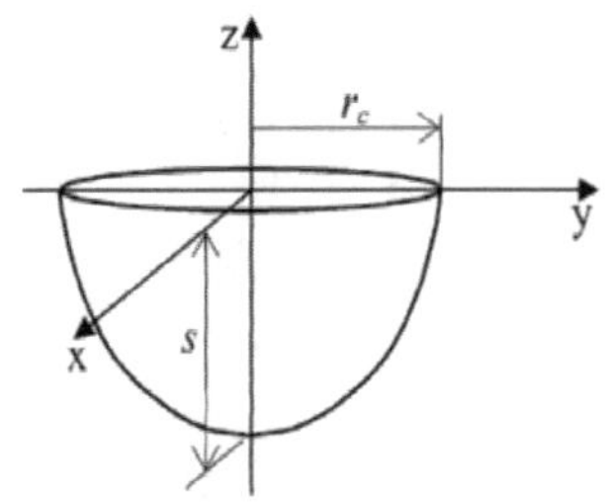

Figure 2. Schematic diagram of paraboloid crater geometry [8]

2.2 Salonitis' Model

MRR and R_a models have been developed in EDM using three dimensional heat conduction equation. In the modelling, uniform heat source and craters of circular paraboloid geometry with no recast layer were assumed. It was also postulated that the idling time is insignificant compared with the discharge duration. The crater geometries are as shown in Figure 2. The crater diameter on the surface has been determined from empirical relations. The developed models in terms of the *MRR* and R_a are expressed by:

$$MRR = \frac{\pi s r_c^2 \times N_s}{2 \times t} \quad (2)$$

$$R_a = \frac{1}{4}\left[\frac{r_c + r_s}{r_c} \right]^2 s \quad (3)$$

where s = crater depth, r_c = crater radius, r_s = heat input radius (μm), N_s = number of sparks in unit time, and t = time. However, s and r_c can be presented as,

$$s = \frac{q_w t}{\rho\left[L_v + c_p(T_s - T_0)\right]} \tag{4}$$

where q_w = heat source intensity distribution, ρ = density, L_v = latent heat of vaporization, c_p = specific heat, T_s = erosion front temperature, and T_0 = ambient temperature. Similarly, r_c can be presented as,

$$r_c = \frac{d}{2} = M \times (I \times t)^N \tag{5}$$

In Equation (5), M and N are material dependent properties which are determined from experimental investigations.

3. Micro-EDM Models for Nonconductive Ceramics

Theoretical models which could be applied in micro-EDM of nonconductive ceramics are not available in the literature. Therefore, some models developed based on experimental investigations have been reviewed in this section. The effect of tool polarity, voltage, feed rate and rotational speed of electrode on *MRR* and surface roughness in ED milling of insulating Al_2O_3 has been investigated with a very thin copper sheet AE, steel wheel tool electrode and water based emulsion dielectric. AE was fed continuously in the machining zone which was fed back immediately after the spark occurrence [10]. Therefore, heat was absorbed by Al_2O_3 workpiece instead of AE [11]. The overall findings of the investigation on the machining characteristics in ED milling of Al_2O_3 are summarized in Table 1.

The effects of emulsion concentration, $NaNO_3$ concentration, polyvinyl alcohol concentration and flow velocity of the machining fluid on the characteristics in ED milling of alumina has been investigated using water-based emulsion dielectric fluid [10]. Other models of EDM with powder addition in dielectric fluid [12] are considered out of the scope of this paper. The study showed that a higher *MRR* and good surface quality can be obtained using suitable emulsion concentration. It was also observed that the *MRR* increases with higher flow velocity of fluid, with negligible change in surface roughness.

Empirical models of the *MRR* and surface roughness in EDM of nonconductive ZrO_2 have been developed by the Taguchi method [13]. Table 2 shows the optimal combination of the machining parameters for *MRR* and surface roughness. It was observed in the investigation that the main parameters of material removal in EDM of ZrO_2 are peak current (I_p) and pulse-on time (t_p). Higher I_p and t_p increase *MRR* due to enlarged energy in a single pulse. The machined surface becomes coarser at higher I_p and t_p i.e. surface roughness is increased. Effect of tool materials on output characteristics *MRR* and surface roughness has been investigated in EDM of Al_2O_3 ceramic material using colloidal graphite solution AE and electrodes of Cu, graphite (Poco EDM-3) and Cu-infiltrated-graphite (PocoEDM-C3) [14]. This research identified the effect of electrode material and polarities on EDM characteristics viz. *MRR* and surface roughness.

Table 1. Effect of parameters on characteristics in ED milling of Al_2O_3 [10]

Parameter	*MRR*	Surface roughness
Tool polarity (TP)	2-4 times higher in +ve	Higher in +ve
Peak voltage (PV)	Rises with an increase in PV	Rises initially with the increase in PV and then decreases with the increase in PV
Rotational speed (RS) of the tool electrode	Initially increases faster with the increase of RS and then increases slowly with an increase in RS	Initially increases with the increase of RS and then increases slowly with an increase in RS
Feed speed (FS) of the workpiece	Initially increases faster with the increase of FS and then decreases with an increase in FS	Initially decreases with an increase of FS and then increases slowly with an increase in FS

Table 2. Optimal combination of the machining parameters for EDM on ZrO_2 ceramic [13]

Process characteristics	Optimized process parameters					
	AE	Peak current, I_p (A)	Auxiliary current with high voltage, I_H (A)	Pulse duration, t_P (µs)	Electrode jumping interval, *EJI* (s)	Servo reference voltage, S_v (V)
Maximum *MRR* (0.2869 mm^3/min)	Cu foil	4	0.4	200	4	55
Minimum *TWR* (0.0026 mm^3/min)	Al foil	2	0.9	50	3	70
Minimum surface roughness (3.37 µm)	Al foil	2	0.9	50	3	40

Table 3: Surface resistivity of conductive layer generated with various electrodes [14]

Electrode material	Electrical resistivity (Ω cm)	Electrical conductivity (Scm^{-1})	Negative polarity		Positive polarity	
			Thickness (µm)	Surface resistivity (Ω/cm^2)	Thickness (µm)	Surface resistivity (Ω/cm^2)
Pure copper	1.018×10^1	9.80×10^{-2}	25	4.071×10^3	0	–
Cu-infiltrated graphite	4.924×10^0	2.03×10^{-1}	90	5.471×10^2	60	8.206×10^2
Graphite	1.491×10^1	6.70×10^{-2}	30	4.971×10^3	10	1.491×10^4

About 40% increase in process speed and 25% increase in bore depth were observed with electrode vibration or rotation. The characteristics of voltage and current signals in micro-EDM of ZrO_2 were compared with micro-EDM of steel in their investigation. The result shows that current and voltage waveforms are not influenced by the machining depth and flushing conditions in micro-EDM of steel. In nonconductive ceramics, similar discharge shapes to those in metal are observed during the machining of starting AE. But it becomes short-peak-long-hold type during the machining of nonconductive ceramics which is named ceramic discharge. Peak current is also decreased with the increase of machining depth in ceramics [16].

4. Comparison of Models for Conductive and Nonconductive Materials

From the above review it can be concluded that there is some specific difference between the modelling of EDM of conductive materials and nonconductive ceramic materials. A comparison between EDM modelling for conductive and nonconductive ceramics is given in Table 4. It is observed that the models of micro-EDM for conductive materials have been developed on the basis of melting and vaporization phenomena, assuming that the energy supplied to the workpiece surface increases the temperature to the boiling point and is removed by vaporization immediately. In micro-EDM, about 90% of the molten material is removed by vaporization and the material beneath the surface gets less energy which is removed by melting [17]. Melting and vaporization are the usual mechanisms in EDM and most of the theoretical models have been developed considering fundamental heat transfer theories.

Micro-EDM shows some differences from macro-EDM with respect to the level of energy, crater diameter and power density, although the material removal mechanism is the same. Spalling is assumed negligible and insignificant in micro-EDM of conductive materials. But in micro-EDM of nonconductive ceramics, the main material removal mechanism is spalling, although melting and vaporization occur depending on the type of ceramic. A comparison between EDM modelling for conductive and nonconductive ceramics is given in Table 4.

Table 4. Comparison between EDM modelling for conductive materials and nonconductive ceramics

Criterion	Workpiece		References
	Conductive material	Nonconductive ceramics	
Need of AE	No	Yes	
Dielectric	Any type	Carbonic for EDM or micro-EDM Any type for ED milling	[2] [10]
Main material removal mechanism	Melting and vaporization	Spalling	[17]
Crater geometry	Regular shape	No regular shape	[11]
Voltage and current shapes	Follow the setup value	Different pattern than setup value	[14,16]
Percentage of effective discharge	Higher	Lower	[16]
Material removal rate	Higher	Lower	
Surface roughness	Lower	Higher	
Models available	Both theoretical and empirical for EDM and micro-EDM	Only empirical for EDM	

5. Summary

In this article the recent development in the field of micro-EDM for cutting nonconductive ceramic materials has been discussed. Several micromachining process have been discussed to have an overview of micro metal cutting to justify the importance of micro-EDM process usually used for cutting conductive materials. The main focus of the article was material removal rate which eventually involves process parameters and special experimental setup. Use of assisting electrode to initiate the spark and to form pyrolytic carbon for subsequent spark are the key to cut nonconductive materials. However, accurate theoretical model for cutting conductive materials is hardly possible which is almost impossible for the case of nonconductive materials. Most of the models are of combination of theoretical and experimental in nature. Although models are not perfect, a comparison has been presented for the ease of selection for any specific purpose. This review showed that there exists a huge need to conduct research to cut nonconductive materials using micro-EDM.

References

[1] Ali M Y, Banu A, Salehan M, Adesta EYT, Hazza, M and Shaffiq M 2018 Dimensional Accuracy in Dry Micro Wire Electrical Discharge Machining *J. Mech. Eng. and Sci.* 12(1) 3321-3329

[2] Hösel T, Müller, C and Reinecke H 2011 Spark erosive structuring of electrically nonconductive zirconia with an assisting electrode *Int. J. Prec. Eng. Manuf.* 11(4) 629-632

[3] Banu A, and Ali M Y 2016 Electrical Discharge Machining (EDM): A Review. *Int. J. Eng. Mater. Manuf. 1*(1) 3-10. https://doi.org/10.26776/ijemm.01.01.2016.02

[4] Yoo B H, Min B-K and Lee S J 2010 Analysis of the machining characteristics of EDM as functions of the mobilities of electrons and ions *Int. J. Prec. Eng. Manuf.* 11(4) 629-632

[5] Dhanik S, and Joshi S S 2005 Modeling of a single resistance capacitance pulse discharge in micro-electro discharge machining *J. Manuf. Sci. Eng.* 127(4) 759-767

[6] Ali M Y, Sabur A, Banu A, Maleque A and Adesta, E Y T 2018 Micro Electro Discharge Machining for Nonconductive Ceramic Materials *Int. J. Eng. Mater. Manuf. 3*(1) 55-62 https://doi.org/10.26776/ijemm.03.01.2018.07

[7] Liu Y, Yu L, Xu Y, Ji R and Li Q 2009 Numerical simulation of single pulse discharge machining insulating Al_2O_3 ceramic *J. Eng. Manuf.* 223(1) 55-62

[8] Salonitis K, Stournaras A, Stavropoulos P and Chryssolouris G 2009 Thermal modeling of the material removal rate and surface roughness for die-sinking EDM *Int. J. Adv. Manuf. Technol.* 40(3-4) 316-323

[9] Dibitonto D D, Eubank P T, Patel M R and Barrufet, M A 1989 Theoretical models of the electrical discharge machining process I. A simple cathode erosion model *J. Appl Phys* 66(9) 4095–4103
[10] Liu Y, Li X, Ji R, Yu L, Zhang H and Li Q 2008 Effect of technological parameter on the process performance for electric discharge milling of insulating Al_2O_3 ceramic J. *Mater. Process Technol.* 208(1) 245-250
[11] Ji R, Liu Y, Zhang Y, Wang F, Cai B and Fu X 2012 Single discharge machining insulating Al_2O_3 ceramic with high instantaneous pulse energy in kerosene *Mater Manuf Process* 27(6) 676-682
[12] Jamil H, N A B, Lajis, M A B and Idris M R B 2018 Modelling and optimization of Chromium Powder Mixed EDM Parameter Effect Over the Surface Characteristics by Response Surface Methodology Approach *Int. J. Eng. Mater. Manuf.* *3*(2) 78-86 https://doi.org/10.26776/ijemm.03.02.2018.02
[13] Chen Y F, Lin Y J, Lin Y C, Chen S L and Hsu L R 2010 Optimization of electrodischarge machining parameters on ZrO_2 ceramic using the Taguchi method *J. Eng. Manuf.* 224(2) 195-205
[14] Muttamara A, Fukuzawa Y, Mohri N and Tani T 2009 Effect of electrode material on electrical discharge machining of alumina. *J. Mater. Process. Technol.* 209(5) 2545-2552
[15] Schubert A and Zeidler H 2009 Machining of nonconductive ZrO_2 ceramics with micro-EDM. Proc. 9th Int. Conf. Europe Soc. Preci Eng & Nanotechnology, San Sebastian, Spain, 2, 6-9
[16] Schubert A, Zeidler H, Wolf N and Hackert M 2011 Micro electro discharge machining of electrically nonconductive ceramics *AIP Conf. Proc.* 1353 1303
[17] Zahiruddin M and Kunieda M 2012 Comparison of energy and removal efficiencies between micro and macro EDM *CIRP Ann. Manuf. Technol.* 61(1) 187-190

IOP Conf. Series: Materials Science and Engineering 1173 (2021) 012064 doi:10.1088/1757-899X/1173/1/012064

Modelling and design of flight control for quadcopter in ballistic airdrop mission under wind perturbation

C Liuswanto [1*], and Y I Jenie [1]

[1] Department of Aerospace Engineering, Bandung Institute of Technology, Jalan Ganesha No. 10 Bandung 40132, Indonesia

*Corresponding E-mail: chandraliuswanto@students.itb.ac.id

Abstract. Airdrop mission are often carried for areas that are difficult to reach using land route and do not have the infrastructure for manned aircraft. The benefits of using Unmanned Aerial Vehicle (UAV) for airdrop mission are getting much more attention with the rapid development of unmanned technology since last decade. The purpose of this paper is to develop a mathematical model of a quadcopter to carry out ballistic airdrop mission in a windy environment, modelled in the MATLAB/Simulink software. The quadcopter based on nonlinear dynamics model which is obtained by using Newton-Euler equation, equipped with sensor model, and free-fall payload model. The sensor consists of Inertial Measurement Unit (IMU), barometer, and Global Navigation Satellite System (GNSS). Sensor fusion algorithm is developed to estimate attitude and position of quadcopter using Kalman filter, as well as to estimate the wind speed. A Proportional, Integral, and Derivative (PID) based controller is designed to obtain the attitude and position control of quadcopter using estimated state of quadcopter. Then, the quadcopter will carry out an airdrop mission by following a simple automated waypoint navigated system. The results of the ballistic airdrop show good precision and accuracy in a high speed, but emphasize the importance of validation model.

Keywords: ballistic airdrop, UAV, quadcopter, sensor, control system

1. Introduction

Unmanned Aerial Vehicle (UAV) has been growing rapidly in various aspect [1]. UAV originated from military ground, now UAV has been used in many fields, particularly in cargo dropping. Cargo dropping mission are often carried for areas that are difficult to reach using land route and do not have the infrastructure for manned aircraft. The benefits of using Unmanned Aerial Vehicle (UAV) for airdrop mission are getting much more attention with the rapid development of unmanned technology since last decades. There are two type most popular UAV in the market, especially in cargo dropping is fixed wing and multicopter. In general, fixed wing UAV is more efficient while cruising [2], therefore capable to fly further than any other UAV type [3] but need large area to take-off and landing.

In the other hand, multicopter have worst efficiency in cruising, but advantage of multicopter is capable to take-off and landing in vertical, therefore there is no need large area for take-off and landing, and capable to hover. With those advantage, particularly in cargo dropping, multicopter capable to operate from small area, and higher accuracy for payload dropping [4].

There are many types of multicopter and divided by number of rotors. In this paper, quadcopter is chosen with X-frame due to simplicity and more stable [5]. With cargo dropping mission will operate in rural area, there will be unpredicted environment [6], and need to operate at lower price than current method, which is using a helicopter; therefore, cargo dropping using quadcopter need an adequate control system, and sensor system. This paper deal with simulation cargo dropping using quadcopter with control system and sensor system in MATLAB/Simulink with Control System Toolbox.

2. Mathematical Model

2.1. Quadcopter

Mathematical dynamic model of the quadcopter is derived and can be used to simulated quadcopter by using simulation programs MATLAB/Simulink version 2018a. Quadcopter has four independently controllable rotor and assemble in cross configuration with two rotors are rotating in clockwise direction while the other two are rotate in anti-clockwise direction, as shown in **Figure 1**. These rotating rotors create upward thrust force and the difference of their rotating speed cause unbalanced torque applied to the centre of mass, and attitude can be change consequently. By assigning the rotating speed separately, total thrust and torque about body axis frame of the quadcopter can be obtained, thus control attitude and position of quadcopter in three-dimension space could controllable.

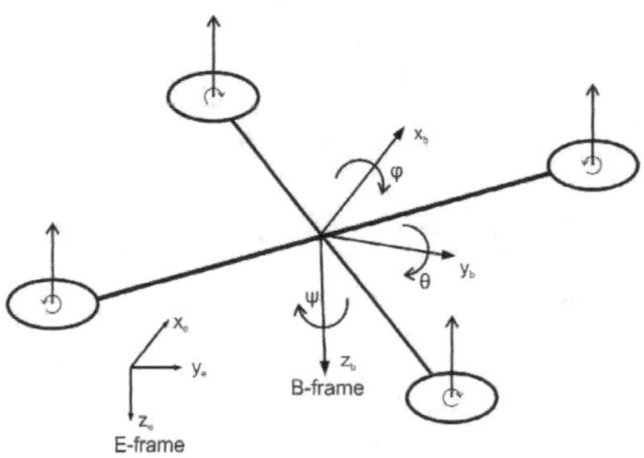

Figure 1. Quadcopter cross configuration with coordinate frame in three-degree freedom

Equation of motion of quadcopter is derived from Equation 1 to Equation 9 using Newton-Euler equation in body frame with assumption that the quadcopter is rigid, where (x_b, y_b, z_b) is position in body frame, (p, q, r) is angular rate, (ϕ, θ, ψ) is Euler angle using Z-Y-X rotation, (L, M, N) is moments due to propulsion system, (D_x, D_y, D_z) is forces due to aerodynamics drag, (I_{xx}, I_{yy}, I_{zz}) is moments of inertia of quadcopter, (c, s) denote cos and sin, respectively, and (g) is gravitational acceleration. Position quadcopter in local horizon frame will develop in Equation 10 to Equation 12 using directional cosines matrix from position quadcopter in body frame and Euler angle.

$$\ddot{x_b} = r\dot{y_b} - q\dot{z_b} - g\, s\theta - D_x/m \tag{1}$$

$$\ddot{y_b} = p\dot{z_b} - r\dot{x_b} + g\, s\phi c\theta - D_y/m \tag{2}$$

$$\ddot{z_b} = q\dot{x_b} - p\dot{y_b} + g\, c\phi c\theta - (D_z + T)/m \tag{3}$$

$$\dot{p} = (qr(I_{yy} - I_{zz}) + L)/I_{xx} \tag{4}$$

$$\dot{q} = (pr(I_{zz} - I_{xx}) + M)/I_{yy} \tag{5}$$

$$\dot{r} = (pq(I_{xx} - I_{yy}) + N)/I_{zz} \tag{6}$$

$$\dot{\phi} = p + q\, s\phi s\theta/c\theta + r\, c\phi s\theta/c\theta \tag{7}$$

$$\dot{\theta} = q\, c\phi - rs\phi \tag{8}$$

$$\dot{\psi} = q\, c\phi/c\theta + r\, c\phi/c\theta \tag{9}$$

$$\dot{x_e} = \dot{x_b}(c\theta c\psi) + \dot{y_b}(s\phi s\theta c\psi) + \dot{z_b}(c\phi s\theta c\psi + s\phi s\psi) \tag{10}$$

$$\dot{y_e} = \dot{x_b}(c\theta s\psi) + \dot{y_b}(s\phi s\theta s\psi + c\phi c\psi) + \dot{z_b}(c\phi s\theta s\psi - s\phi s\psi) \quad (11)$$

$$\dot{z_e} = \dot{x_b}(-s\theta) + \dot{y_b}(s\phi c\theta) + \dot{z_b}(c\phi c\theta) \quad (12)$$

Controlling quadcopter by changing each rotor thrust is not intuitive, hence propulsion system will be modelled as ratio of thrust and using motor mixing. Motor mixing will calculate each rotor based on attitude reference of quadcopter. The propulsion system is modelled from Equation 13 to Equation 16 where u_x is thrust force and moment ratio of rotor x, F_{max} is maximum thrust of each rotor, M_{max} is maximum torque of rotor, l is diagonally length from centre of mass to rotor, and θ_f is angle from X_b to rotor 1 or rotor 2. Motor mixing is modelled from Equation 17, cmd subscript stand for command.

$$T = (-u_1 - u_2 - u_3 - u_4)F_{max} \quad (13)$$

$$L = (u_1 - u_2 + u_3 - u_4)F_{max} l \sin \theta_f \quad (14)$$

$$M = (u_1 + u_2 - u_3 - u_4)F_{max} l \cos \theta_f \quad (15)$$

$$N = (u_1 - u_2 - u_3 + u_4)M_{max} \quad (16)$$

$$U = \begin{bmatrix} roll_{cmd} \\ pitch_{cmd} \\ yaw_{cmd} \\ alt_{cmd} \end{bmatrix} = \begin{bmatrix} 1 & 1 & -1 & -1 \\ 1 & -1 & 1 & -1 \\ 1 & -1 & 1 & -1 \\ 1 & 1 & 1 & 1 \end{bmatrix} \begin{bmatrix} u_1 \\ u_2 \\ u_3 \\ u_4 \end{bmatrix} \quad (17)$$

Aerodynamics drag is derived utilizing drag model [7] and shown in Equation 18. Aerodynamics drag calculated in body frame, and proportional to the square of sum of quadcopter speed in body frame and wind speed in body frame. External wind will be modelled constant and parallel to XY plane earth frame. External wind will transform to body frame in Equation 19 to Equation 21. Where (D_x, D_y, D_z) is aerodynamics drag in body frame, (w_{x_e}, w_{y_e}) is wind speed in earth frame and $(w_{x_b}, w_{y_b}, w_{z_b})$ is wind speed in body frame.

$$\vec{D} = \begin{bmatrix} D_x \\ D_y \\ D_z \end{bmatrix} = \frac{1}{2}\rho C_d \begin{bmatrix} (\dot{x_b} + w_{x_b})^2 A_x \\ (\dot{y_b} + w_{y_b})^2 A_y \\ (\dot{z_b} + w_{z_b})^2 A_y \end{bmatrix} \quad (18)$$

$$w_{x_b} = w_{x_e}(c\theta c\psi) + w_{y_e}(s\phi s\theta c\psi) \quad (19)$$

$$w_{y_b} = w_{x_e}(c\theta s\psi) + w_{y_e}(s\phi s\theta s\psi + c\phi c\psi) \quad (20)$$

$$w_{z_b} = w_{x_e}(-s\theta) + w_{y_e}(s\phi c\theta) \quad (21)$$

The quadcopter model is based on DJI M100. The initial parameter of quadcopter is listed in table 1.

Table 1. Specification of DJI M100 [8-9]

Parameter	Symbol	Value	Unit
Empty Weight	m	2.355	kg
Maximum Take-off Weight	MTOW	3.6	kg
Moment Inertia – X	I_{xx}	0.0554	kgm^2
Moment Inertia – Y	I_{yy}	0.0578	kgm^2
Moment Inertia – Z	I_{zz}	0.1185	kgm^2
Rotor length	l	650	mm
Rotor angle	θ_f	45	deg
Area – X	A_x	0.0309	m^2
Area – Y	A_y	0.0309	m^2
Area – Z	A_z	0.0848	m^2
Aerodynamic drag coefficient	C_d	0.9	-
Maximum propulsion thrust	F_{max}	10.5	N
Maximum propulsion torque	M_{max}	0.168	Nm

2.2. Airdrop Payload

Airdrop is a method to deliver a cargo by dropping from aircraft while still on the air and can be used to deliver payload to remote area. There are three types of airdrop, free-drop, high-velocity, and low-velocity [10]. This research will be focused on free-drop type to save payload weight by not using parachute or any device to slow the motion of free-fall payload. Quadcopter is suitable for dropping using airdrop type due to capability of quadcopter to hover.

Payload mathematical model is derived from [10] where payload modelled as free-falling object with aerodynamics drag and assuming the moment from drag is be small enough to ignore. Payload model is in Equation 22.

$$\begin{bmatrix}\ddot{x_p}\\ \ddot{y_p}\\ \ddot{z_p}\end{bmatrix} = \begin{cases}\begin{bmatrix}\ddot{x_e}\\ \ddot{y_e}\\ \ddot{z_e}\end{bmatrix} & if\ release = false\\ \begin{bmatrix}0\\0\\g\end{bmatrix} - \frac{1}{2}\rho C_{d_p}\begin{bmatrix}(\dot{x_p}+w_{x_e})^2 A_{x_p}\\ (\dot{y_p}+w_{y_e})^2 A_{y_p}\\ (\dot{z_p}+w_{z_e})^2 A_{z_p}\end{bmatrix} & if\ release = true\end{cases} \tag{22}$$

Where (x_p, y_p, z_p) is position payload in earth frame, $(A_{x_p}, A_{y_p}, A_{z_p})$ is cross section area of payload and $C_{d_p}$is payload aerodynamics drag. Payload position will follow quadcopter position until release flag is true. Payload geometry used model adapted from [11] and listed in Table 2

Table 2. Geometry specification of the payload [10]

Parameter	Symbol	Value	Unit
Area – X	A_{x_p}	0.035216	m^2
Area – Y	A_{y_p}	0.011408	m^2
Area – Z	A_{z_p}	0.006532	m^2
Aerodynamic drag coefficient	C_{d_p}	0.8	-

Inertia of payload will assume as negligible due to smaller size than quadcopter and aerodynamics drag coefficient assume as 0.8 due to shape of payload resemble of cuboid.

3. Sensor

In this section, sensor model and state estimator of quadcopter is developed to estimate state of quadcopter. Sensor is necessary to estimate the vehicle state, i.e. its altitude and attitude. The most common sensor on board a quadcopter is Inertial Measurement Unit (IMU), Barometer and Global Navigation Satellite System (GNSS). IMU include a 3-axis accelerometer, gyroscope, and magnetometer. IMU modelled as analog device, where sensing state is converted to 16-bit data using simple gain and is called as an Analog-to-Digital Converter (ADC). Output from ADC then interpreted to measurement data using the invert gain of ADC and both gain will eliminate each other but with added random measurement noise. Data rate of all sensor but GNSS is 100 Hz and GNSS is 10 Hz. Sensor specification and modules are shown in **Table 3**.

3.1. Accelerometer

Accelerometer in a IMU is an electromechanical device that measure linear acceleration in body frame due to external forces and static acceleration by sensing changes in capacitance. Placement of acceleration is crucial, due to there will be centripetal forces if the accelerometer is not on centre of gravity. Accelerometer is described by the following Equation 23.

$$[\tilde{a}_x \quad \tilde{a}_y \quad \tilde{a}_z]^T = C_h^b\, g\vec{z} + [\ddot{x_b} \quad \ddot{y_b} \quad \ddot{z_b}]^T + \eta_a \tag{23}$$

where $(\tilde{a}_x, \tilde{a}_y, \tilde{a}_z)$ is acceleration measurement by accelerometer in body frame, $C_h^b$is rotation matrix from earth frame to body frame and η_a is measurement noise of accelerometer. Acceleration measurement is measuring acceleration of motion and gravity acceleration in body frame. Accelerometer data will be being used to estimate position, attitude of quadcopter, and external wind velocity.

Table 3. Gyroscope, Accelerometer, Magnetometer, Barometer, and GNSS module and specification

Sensor	Parameter	Value	Unit
ST Micro L3GD20 Gyroscope	Scale	± 250	dps
	Update rate	100	Hz
	Rate noise density	0.03	dps/$\sqrt{\text{Hz}}$
ST Micro LSM303D Accelerometer	Scale	± 16	g
	Update rate	100	Hz
	Rate noise density	150	μg/$\sqrt{\text{Hz}}$
ST Micro LSM303D Magnetometer	Scale	± 2	gauss
	Update rate	100	Hz
	Rate noise density	5	mgauss/$\sqrt{\text{Hz}}$
MEAS MS5611 Barometer	Update rate	100	Hz
	Vertical position accuracy	0.1	m
NEO-M8N GNSS modules	Update rate	10	Hz
	Position accuracy	2.5	m

3.2. Gyroscope

Gyroscope is similar to accelerometer that is an electromechanical device but measuring angular velocity of an object. Gyroscope is the most relevant for air vehicle navigation and control, along with accelerometer [7]. Gyroscope is described by the following equation below.

$$[\tilde{p} \quad \tilde{q} \quad \tilde{r}]^T = [p \quad q \quad r]^T + \eta_g \tag{24}$$

where $(\tilde{p}, \tilde{q}, \tilde{r})$ is gyroscope measurement by IMU and η_g is measurement noise of gyroscope. Data form gyroscope will be used for estimate position and attitude of quadcopter.

3.3. Magnetometer

Magnetometer is a sensor that can measure the Earth magnetic field vector. Magnetometer is used to estimate the magnetic field by following equation below

$$\begin{bmatrix}\tilde{b}_x & \tilde{b}_y & \tilde{b}_z\end{bmatrix}^T = C_h^b\,[b_x \quad b_y \quad b_z]^T + \eta_m \tag{25}$$

where (b_x, b_y, b_z) is Earth magnetic field vector, $(\tilde{b}_x, \tilde{b}_y, \tilde{b}_z)$ is magnetic measurement from the magnetometer, and η_m is measurement noise of magnetometer. Earth magnetic is different for each location on earth and will be assume constant, due to small movement in Earth position.

3.4. Barometer

Barometer is a sensor that can measure atmospheric pressure. Measurement data from barometer can be used to determining relative height using hypsometric equation, where increasing altitude will decrease atmospheric pressure. Equation that estimates relative altitude takes the following form below

$$\tilde{z}_e = \frac{RT}{g}\log\left(\frac{p_1}{p_0}\right) + h_0 + \eta_b \tag{26}$$

where $\tilde{z}_e$ is estimate altitude on Earth frame, R is air specific gas constant, T is temperature in reference altitude, h_0 is reference altitude, p_0 is reference atmospheric pressure, p_1 is measurement atmospheric pressure, and η_b is measurement noise of barometer. Reference altitude is chosen at take-off point.

3.5. Global Navigation Satellite System (GNSS)

GNSS is a satellite navigation system that provide geo-spatial position, but GNSS modules have lower update rate. GNSS can be used to determine absolute position quadcopter in Earth frame. Typical GNSS modules have lower vertical accuracy than horizontal accuracy, therefore barometer is used to altitude estimation. GNSS modules is described by the following equation

$$[\tilde{x}_e \quad \tilde{y}_e\,]^T = [x_e \quad y_e\,]^T + \eta_{GNSS} \tag{27}$$

where $(\tilde{x}_e \quad \tilde{y}_e\,)$ is horizontal position output from GNSS modules in Earth frame, and η_{GNSS} is measurement noise of GNSS modules.

3.6. Attitude estimation

Attitude estimation of quadcopter will be based on Attitude and Heading Reference System (AHRS) from [12], where using Kalman filter to estimate Euler angle from IMU. AHRS algorithm combine gyroscope data to traditional attitude determination to lessen oscillation of estimation and drifting from using gyroscope measurement. AHRS will provide optimal sensor fusion to estimate correct vehicle attitude.

AHRS algorithm will fusion measurement sensor data with Kalman filter to estimate Euler angle. AHRS contain six state of quadcopter which is roll, pitch, yaw, drift roll, drift pitch, and drift yaw $(\phi\ \theta\ \psi\ d\hat{\phi}\ d\hat{\theta}\ d\hat{\psi})$. State vector is initialized before Kalman filter begins to operate, which is assume quadcopter not rotating and at level. AHRS algorithm is represent at Equation 28, where A is state matrix to ensure the angle from previous estimation and removes impact of the drift from angle prediction, and B is distribution matrix with allows for multiplication of gyroscope measurement by the time difference from previous measurement.

$$A = \begin{bmatrix} I_{3x3} & -dt\, I_{3x3} \\ 0_{3x3} & I_{3x3} \end{bmatrix} \qquad B = \begin{bmatrix} -dt\, I_{3x3} \\ 0_{3x3} \end{bmatrix} \tag{28}$$

where I_{3x3} is identity matrix with size 3x3 and 0_{3x3} is zero matrix with size 3x3. In measurement vector only Euler angle is used due to drift Euler angle is cannot be obtain from the sensor. Measurement vector input is estimate using traditional attitude determination, where using accelerometer and magnetometer to estimate Euler angle. Equation below is represented measurement matrix of AHRS.

$$z = \begin{bmatrix} \tan^{-1}\left(\dfrac{-\tilde{a}_x}{\sqrt{\tilde{a}_x{}^2 + \tilde{a}_z{}^2}}\right) \\ \tan^{-1}\left(\dfrac{-\tilde{a}_x}{\tilde{a}_z}\right) \\ \tan^{-1}\left(\dfrac{\tilde{b}_z \sin\phi - \tilde{b}_y \sin\phi}{\tilde{b}_x \sin\theta + \tilde{b}_y \sin\theta \sin\phi + \tilde{b}_z \cos\phi \sin\theta}\right) \end{bmatrix} \tag{29}$$

Value of covariance of measurement and noise matrix is obtained by trial and error and have the following form

$$Q = \begin{bmatrix} 2.56e^{-14}\, I_{3x3} & 0_{3x3} \\ 0_{3x3} & 4e^{-11}\, I_{3x3} \end{bmatrix} \qquad R = 1e^{-15} * [I_{6x6}] \tag{30}$$

3.7. Position estimation

Position estimation will fusion Inertia Navigation System (INS) and GNSS using Kalman filter. INS calculate position by dead-reckoning the position and velocity object without external references but will suffer from drift issue. By providing GNSS data and combining information using Kalman filter, the error in position estimation are stable. INS can be short fallback, if GNSS is unavailable.

Position estimation model is adapted from [13]. Kalman filter state is position and velocity quadcopter in Earth frame $(\hat{x}_e, \hat{y}_e, \hat{z}_e, \dot{\hat{x}}_e, \dot{\hat{y}}_e, \dot{\hat{z}}_e)$, with input as acceleration in Earth frame without gravitational acceleration. Position estimation Kalman filter is represent in Equation 30, where A is state matrix that integrated previous state matrix into current state, and B is input model where integrated acceleration from acceleration into estimated state.

$$A = \begin{bmatrix} I_{3x3} & dt\, I_{3x3} \\ 0_{3x3} & I_{3x3} \end{bmatrix} \quad B = \begin{bmatrix} 0.5\, dt^2\, I_{3x3} \\ dt\, I_{3x3} \end{bmatrix} \tag{31}$$

In measurement vector, only position is used as state and represent as Equation 31. Where $\tilde{x}_e$ and $\tilde{y}_e$ is obtain from GNSS and $\tilde{z}_e$ from barometer.

$$z = \begin{bmatrix} \tilde{x}_e \\ \tilde{y}_e \\ \tilde{z}_e \end{bmatrix} \tag{32}$$

Covariance matrix of noise and process is obtained by trial and error, and represent as Equation 32

$$Q = \begin{bmatrix} 1e^{-10}\, I_{3x3} & 0_{3x3} \\ 0_{3x3} & 1e^{-7}\, I_{3x3} \end{bmatrix} \quad R = \begin{bmatrix} 0.01\, I_{3x3} & 0_{3x3} \\ 0_{3x3} & 0.001\, I_{3x3} \end{bmatrix} \tag{33}$$

3.8. Airspeed estimation

Airspeed estimation of quadcopter can be calculated using accelerometer and derived from [7]. When quadcopter stay while in hover, total thrust can be estimated by Euler angle. By calculated equilibrium forces with estimated total thrust, aerodynamics drag can be estimated and is correlated to airspeed. Total thrust estimation derived from translation dynamics of quadcopter in Equation 34, by assuming quadcopter not moving at Z axis Earth frame, so the aerodynamics drag in z_e is negligible

$$\hat{T} = -\frac{mg}{\cos\hat{\theta}\cos\hat{\phi}} \tag{34}$$

where $\hat{T}$ is estimated total thrust in body frame, and Euler angle is estimated earlier from AHRS algorithm. Subsequently, total thrust is used for estimated aerodynamics drag by using translation dynamics of quadcopter. Estimated wind speed is represent at Equation 35 to Equation 37

$$\hat{w}_x = \sqrt{2\frac{-(\cos\hat{\phi}\sin\hat{\theta}\cos\hat{\psi} + \sin\hat{\phi}\sin\hat{\psi})\hat{T} - m\ddot{\tilde{x}}_e}{\rho\, A_x C_d}} \tag{35}$$

$$\hat{w}_y = \sqrt{2\frac{-(\sin\phi\sin\theta\cos\psi - \cos\phi\sin\psi)\hat{T} - m\ddot{\tilde{y}}_e}{\rho\, A_y C_d}} \tag{36}$$

$$[\ddot{\tilde{x}}_e \quad \ddot{\tilde{y}}_e \quad \ddot{\tilde{z}}_e]^T = C_b^h[\tilde{a}_x \quad \tilde{a}_y \quad \tilde{a}_z]^T \tag{37}$$

Where $(\ddot{\tilde{x}}_e, \ddot{y}_e, \ddot{z}_e)$ is acceleration measurement with transformation to Earth frame.

4. Controller Design

Quadcopter system input signal to each rotor by using motor mixing, where each rotor rotation speed can be controlled independently. Single Input Single Output (SISO) controller is designed using

Proportional, Integral and Derivative (PID) feed forward controller. PID controller is chosen due to simplicity and easy to implement and will be used parallel form, where represent in Laplace domain at Equation 38.

$$G(s) = K_p + \frac{K_i}{s} + K_d \frac{s\,N}{s+N} = \frac{(K_p + K_d N)s^2 + (K_p N + K_i)s + K_d}{s(s+N)} \tag{38}$$

where K_p is proportional gain, K_i is integrator gain, K_d is derivative gain, and N is derivative filter. Since quadcopter is a system with degree of freedom higher than control available, controller design will use two loops, which is inner loop, and outer loop. Inner loop will feedback Euler angle of quadcopter and outer loop will feedback position of quadcopter. With quadcopter dynamics control in body frame, there will be a transformation to change Earth frame in outer loop to Body frame. This control loop is derived from [14] and [15] and is illustrated in Figure 2.

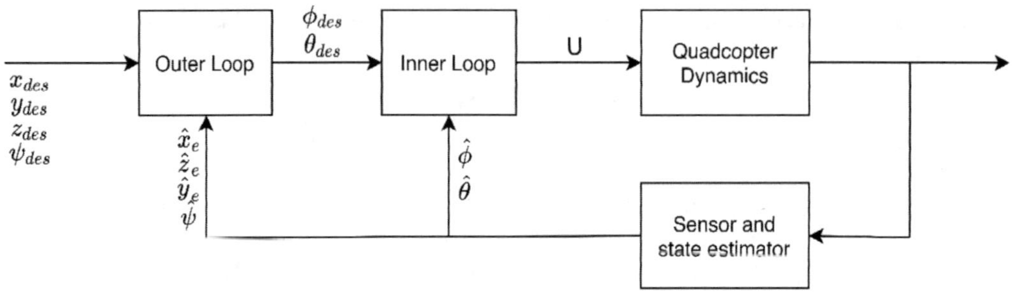

Figure 2. Quadcopter control diagram

4.1. PID value of quadcopter

Linear quadcopter will be used to determine PID value. Linear quadcopter model derives from nonlinear quadcopter equation of motion while quadcopter have empty weight at hover with an ideal sensor. Therefore, below are open loop linear quadcopter model in Laplace domain.

$$\frac{\phi(s)}{roll_{cmd}(s)} = \frac{209.7}{s^2} \tag{39}$$

$$\frac{\theta(s)}{pitch_{cmd}(s)} = -\frac{124}{s^2} \tag{40}$$

$$\frac{\psi(s)}{yaw_{cmd}s)} = \frac{6.3}{s^2} \tag{41}$$

$$\frac{z_e(s)}{alt_{cmd}(s)} = -\frac{11.67}{s^2} \tag{42}$$

Quadcopter model shows that the open loop system has double integrator, where there are no zeros and two poles at origin. Due to double integrator, using constant gain only change poles to up and down, therefore feedback quadcopter will need different form from constant gain.

Design Requirement and Objective (DRO) of PID value derive from the mission of quadcopter, which is dropping payload. Quadcopter need to stable in position only when dropping payload, so PID value is achieve by feedback the open loop system with PID in Laplace domain until control performance of the quadcopter near or matches the DRO mission and it is in **Table 4**. The DRO is overshoot under 30%, 2 second or faster rise time and 10 second or faster settling time.

Table 4. PID value from outer loop and inner loop

Feedback	P	I	D	N
Roll (ϕ)	0.466	0.500	0.086	50
Pitch (θ)	-0.789	-0.845	-0.145	50
Yaw (ψ)	0.115	0.010	0.277	5
Altitude (z_e)	0.111	0.015	0.176	10
Position X (x_e)	-0.069	-0.006	-0.193	10
Position Y (x_e)	0.069	0.06	0.193	10

4.2. Mission profile

With quadcopter will run autonomously to deliver airdrop without human input, mission profile and autonomous system is created. Because payload will drop ballistic at specific altitude, wind perturbation can interfere dropping point and increasing possibility of missing or broken payload, therefor using wind estimation is needed to increase accuracy dropping. Mission profile of quadcopter to ballistic airdrop is divided into 6, which are:

a) Take-off : Take-off to desired altitude
b) Cruise : Fly toward target at desired altitude
c) Hover : Hover for a few second to estimate wind speed at target with desired altitude
d) Correction : Fly to correction location for quadcopter to compensation wind speed
e) Drop and Cruise : Airdrop is drop and quadcopter fly to origin position at desired altitude
f) Landing : Reduce altitude at constant speed

In simulation, mission profile is run sequentially with if statement and using waypoint model as reference. In mission 1 to 3 and 5 to 6, final position is not really mater to accuracy payload, so if the error position is under 2.5m from the target, quadcopter will run next mission. With mission 4, which is correction, dropping accuracy is crucial, so final position before dropping is set to under 1m and absolute velocity under 0.1m/s.

5. Results and Discussions

Simulation is conducted using MATLAB/Simulink R2018a using ode45 as solver, zero order hold for discrete and estimation every 100 ms. Model in Simulink is shown in **Figure 3**. Simulation will conduct and will focus to attitude quadcopter while releasing payload, full mission, and accuracy of airdrop.

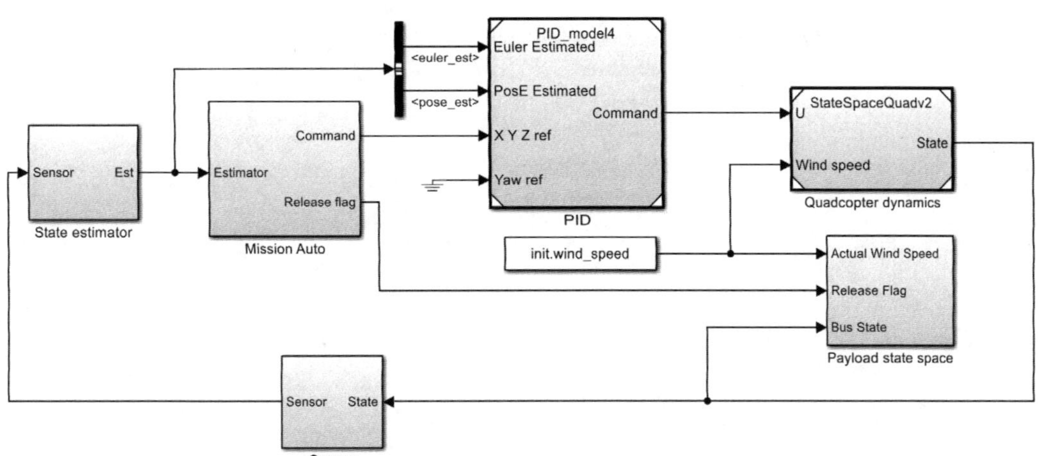

Figure 3. Simulink model for autonomous airdrop quadcopter

5.1. Attitude quadcopter

When quadcopter release payload, there will be a movement due to different weight in quadcopter. With configuration of payload of 1.245 kg, wind speed of 20 m/s and releasing payload at 5 second, simulation result is shown in **Figure 4**.

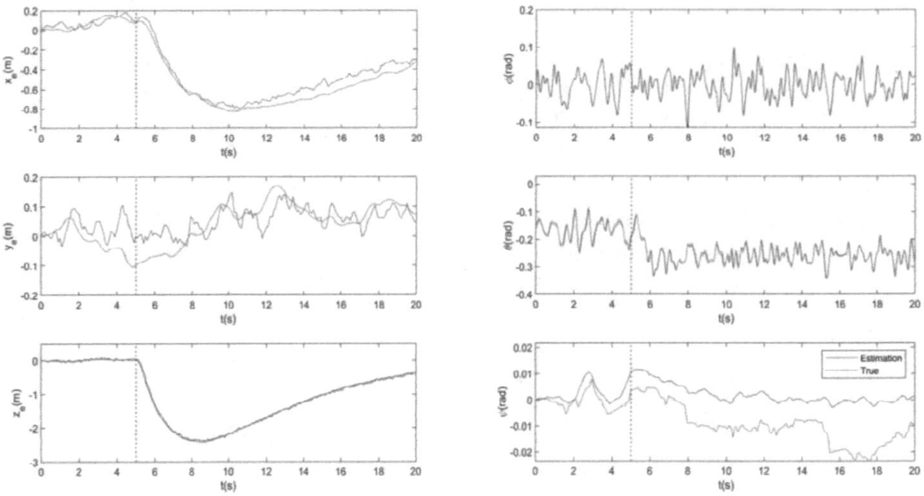

Figure 4. Quadcopter position and attitude simulation result while releasing airdrop with payload 1.245kg, and wind speed from x_e of 20 m/s

It shows that sensor estimation and control design works, quadcopter can hover, but with some oscillation at attitude. After releasing payload, quadcopter altitude increasing by 2 m, drift by the wind at 0.8m, and pitch angle increase by 0.1 rad. With lower mass, quadcopter need to increase angle to hover at same position due to lower thrust produce to stay level.

5.2. Full airdrop mission

Airdrop is simulated using speed wind from x_e of 20m/s, payload of 1.245kg, origin at (0,0,0)m, and target airdrop is (50,20,0)m. Trajectory of quadcopter and payload is shown in **Figure 5** and **Figure 6**.

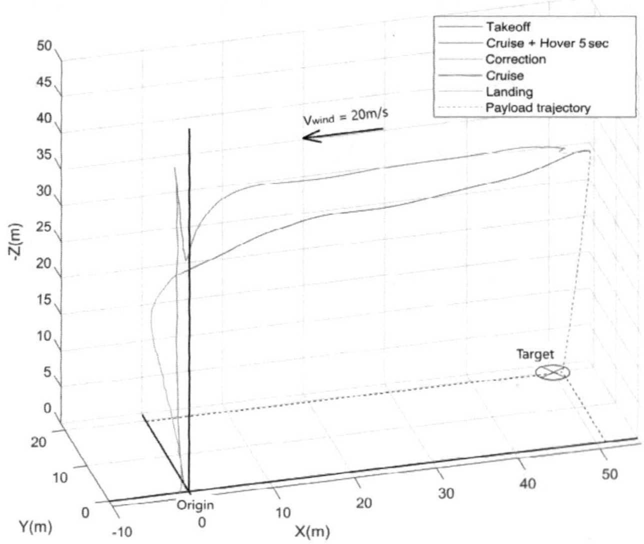

Figure 5. Full mission quadcopter and payload trajectory in cargo dropping mission

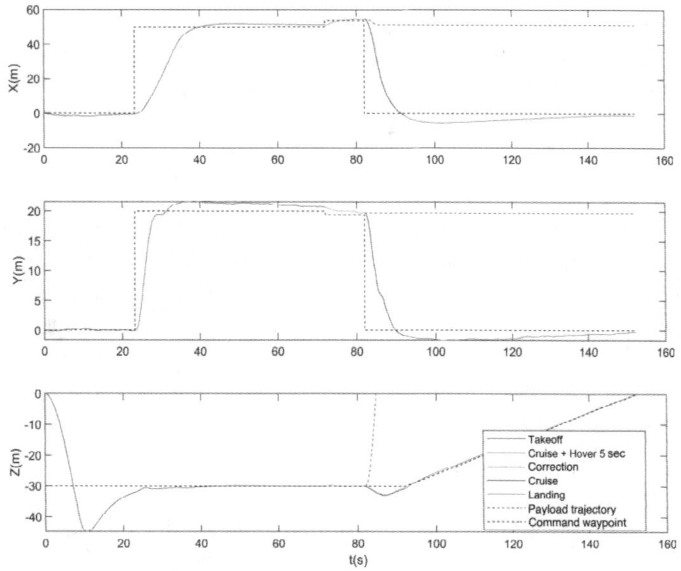

Figure 6. Full mission quadcopter and payload trajectory vs time in cargo dropping mission

Result of simulation is quadcopter capable to perform mission properly with payload drop at (51.11,19.74,0)m and quadcopter landing at (-1.20,0.25,0)m. Mission profile of each mission is distinguishable.

5.3. *Accuracy of airdrop*

To find out the accuracy of the system payload dropping, the full airdrop mission will be simulated 40 times with same conditions as the previous mission but with moving the dropping location to (50,0,0)m and vary wind speed to a single axis. Wind speed, location and origin is straight; therefore, accuracy of each axis can be determined. Result of the simulation of 40 times is shown in Fig. 7.

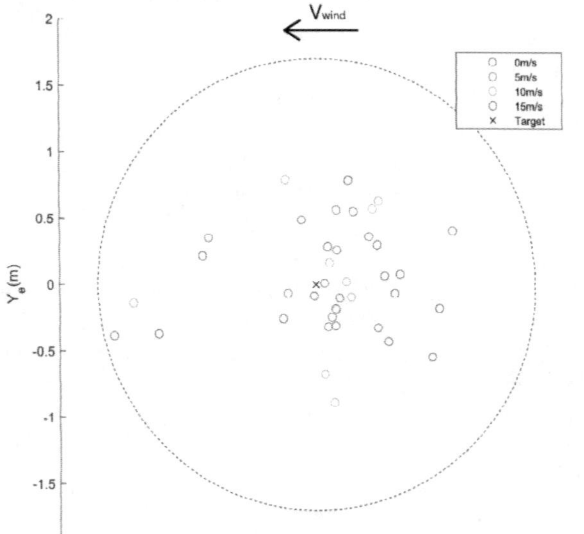

Figure 7. Airdrop location with payload of 1.245 kg and various wind speed

The result of simulation shows that payload dropping is quite accurate, with mean error of 0.57m, standard deviation of 0.371, and maximum error of 1.62m. Accuracy and precision with low wind is lower than with high wind, due to wind speed estimation is works worse with low wind [7].

6. Conclusion

This paper presented that with adequate sensor and control systems, regular quadcopter capable to airdrop payload using MATLAB/Simulink R2018a. Sensor systems consist of IMU, barometer, GNSS modules, airspeed estimation, attitude estimation, and position estimation. Control systems consist of UAV feedback using PID, and waypoint model derive from airdrop mission profile. Result of cargo dropping shows that the quadcopter capable to do cargo dropping in wind condition, there is no indication of control failure, both in windy conditions and without wind. Accuracy and precision of cargo dropping shows that quite good, with payload fell within a radius of less than 1.9 m after dropped at altitude of 30 m in various wind.

References

[1] Artale V, Milazzo C L R and Ricciardello A 2013 Mathematical modeling of hexacopter ams 7 4805–11
[2] Gunarathna J kalpa and Munasinghe R 2018 Development of a Quad-rotor Fixed-wing Hybrid Unmanned Aerial Vehicle 2018 Moratuwa Engineering Research Conference (MERCon) 2018 Moratuwa Engineering Research Conference (MERCon) (IEEE)
[3] Mathisen S H, Grindheim V and Johansen T A 2017 Approach Methods for Autonomous Precision Aerial Drop from a Small Unmanned Aerial Vehicle IFAC-PapersOnLine 50 3566–73
[4] Bouabdallah S 2007 Design and control of quadrotors with application to autonomous flying M.S. thesis Université Aboubekr Belkaid
[5] Ye H 2018 Control of quadcopter UAV by nonlinear feedback M.S. thesis Case Western Reserve University
[6] Safi'i I, Fariz A, Rahman M F, Arifianto O, Sasongko R A and Jenie Y I 2018 The development of operational concept and Design Requirements and Objectives (DRO) of medical transport drone in Liukang Tangaya Archipelago J. Phys.: Conf. Ser. 1130 012034
[7] Moyano Cano J 2012 Quadrotor UAV for wind profile characterization M.S. thesis Universidad Carlos III de Madrid
[8] Park S 2019 Coupled analysis for 6-DOF dynamics of UAV and modified blade element momentum theory considering gusst and flight conditions M.S. thesis Seoul National University
[9] Jeaong H, Jo S, Kim S, Suk J, and Lee Y G 2018 Simulation and flight experiment of a quadrotor using disturbance observer based control *10th international micro air vehicle competition and conference*
[10] Mathisen S G, Leira F S, Helgesen H H, Gryte K and Johansen T A 2020 Autonomous ballistic airdrop of objects from a small fixed-wing unmanned aerial vehicle *Auton Robot* 44 859–75
[11] Iqbal U, Shah S I A and Sadiq M S 2016 Servo actuated payload carry and drop mechanism for unmanned helicopter 2016 International Conference on Emerging Technologies (ICET) 2016 International Conference on Emerging Technologies (ICET) (IEEE)
[12] Tomaszewski D, Rapiński J and Pelc-Mieczkowska R 2017 Concept of AHRS algorithm designed for platform independent IMU attitude alignment Reports on Geodesy and Geoinformatics 104 33–47
[13] Al-Sinbol G, Perhinschi M G and Wilburn B K 2015 Simplified GPS model for UAV fault tolerant control laws design Int Jnl of Intel Unmanned Syst 3 39–60
[14] Mukarram A, Fiaz U A and Khan U I 2015 Altitude control of a quadcopter M.S. thesis Pakistan Institute of Engineering & Applied Sciences
[15] Kotarski D, Benic Z and Krznar M 2016 Control design for unmanned aerial vehicles with four rotors: Interdiscip. Descr. Complex Syst. 14 236–45.

Using FC-CLAHE-ADF to enhance digital mammograms for detecting breast lesions

S H Suradi*, and K A Abdullah

School of Medical Imaging, Faculty of Health Sciences, Universiti Sultan Zainal Abidin, 21300, Kuala Nerus, Terengganu, Malaysia

* Email: harithsuradi@gmail.com

Abstract. Digital mammograms are commonly used for breast screening, especially to aid the detection of cancer. However, digital mammograms suffer with low contrast images due to the low exposure factors used. This paper presents a novel image enhancement technique for digital mammogram images known as Fuzzy Clipped Contrast-Limited Adaptive Histogram Equalization with Anisotropic Diffusion Filter (FC-CLAHE-ADF). This proposed FC-CLAHE-ADF has adopted a fuzzy-based and histogram-based image enhancement technique where it can further reduce the noise of the digital mammograms while preserving the contrast and brightness. A total of six digital mammograms were retrieved from Mammographic Image Analysis Society (MIAS) open-source database. The performance of FC-CLAHE-ADF was compared to Recursive Mean-Separate Histogram Equalization (RMSHE), Fast Discrete Curvelet Transform via Unequally Spaced Fast Fourier Transform (FDCT-USFFT), and FC-CLAHE only. In summary, this novel FC-CLAHE-ADF has provided the most superior results, among other selected enhancement techniques. The resulting images have been able to demonstrate breast lesions better.

Keywords: digital image processing, fuzzy inference system, histogram equalization, mammography

1. Introduction

One of the main causes of death among women is breast cancer [1,2]. Digital mammograms are used at an early stage to screen and diagnose breast cancer [3]. In previous studies, they have shown that the early screening of breast cancer improves breast cancer survival [4–6]. However, digital mammograms suffer with low contrast images due to the low exposure factors used. Therefore, the image enhancement technique is required to improve the current image quality.

Image enhancement techniques are used to highlight the details and enhance any suspected abnormalities in the images [7]. These techniques involve manipulation of contrast and intensity appearances. In previous, various mammographic image enhancement techniques have been introduced, such as histogram-based, frequency-based, fuzzy-based, and filter-based. Examples of those techniques are Histogram Equalization (HE) [8], Contrast Limited Adaptive Histogram Equalization (CLAHE) [9,10], Recursive Mean-Separate Histogram Equalization (RMSHE) [11], fuzzy logic based enhancement [10,12–14], Fast Discrete Curvelet Transform via Unequally Spaced Fast Fourier

Transform (FDCT-USFFT) [15] and nonlinear unsharp masking [16]. However, these image enhancement techniques still have limitations such as high computational burdens and image noise [17].

A novel image enhancing technique has been proposed in this paper for digital mammograms known as Fuzzy Clipped Contrast-Limited Adaptive Histogram Equalization with Anisotropic Diffusion Filter (FC-CLAHE-ADF).

2. Materials and methods

2.1. Image database

A total of six digital mammograms were retrieved from Mammographic Image Analysis Society (MIAS) open-source database [18]. The image size of each image datasets is 1024 X 1024 pixels. Table 1 shows digital mammograms with various background tissues and abnormalities.

Table 1. Digital mammograms with various background tissues and abnormalities.

Condition	Image	Background tissue	Abnormality
Normal	1	Fatty	Normal
	2	Fatty	Normal
Benign	3	Fatty	Architectural distortion
	4	Fatty-glandular	Asymmetry
Malignant	5	Fatty	Spiculated masses
	6	Fatty-glandular	Circumscribed masses

2.2. Fuzzy Clipped Contrast-Limited Adaptive Histogram Equalization with Anisotropic Diffusion Filter (FC-CLAHE-ADF)

Contrast-Limited Adaptive Histogram Equalization (CLAHE) is one of the derivation methods in Histogram Equalization (HE)-based. CLAHE divides the input images into several non-overlapping sub-images and clips them to limit the amount of image enhancement. The clipped histogram changes the structure of the input images by reducing or increasing the values with reference to the threshold or the clipping limit before the equalization. The input images are clipped and calculated by using Equation (1).

$$Clip\ Limit = \left[\frac{\varphi}{256}\right] + \left[\beta.\left(\varphi - \left[\frac{\varphi}{256}\right]\right)\right] \tag{1}$$

Where β is the clipping enhancement parameter and [.] denotes the truncating value to the nearest integer. The φ is a product of block size. The value of 256 denotes the pixel intensity values range from 0 to 255.

FC-CLAHE-ADF is derived from the CLAHE technique. This FC-CLAHE-ADF technique consists of a fuzzy clipped inference system, which automates clip-limit selection. ADF is used to reduce the noise of the mammographic images while preserving the details. The flowchart of the proposed FC-CLAHE-ADF enhancement technique is shown in Figure 1.

Fuzzy Clipped (FC) Interface System has been set up with two input measures, which include the contrast (C) and the discrete entropy (E). The output measure of FC is the fuzzy clipping enhancement parameter (fβ). The fuzzy sets of (C) are Low (V1), Medium (V2), and High (V3), while (E) are Small (W1) and Big (W1). The values for V1, V2, V3, W1, and W2 were fixed using ground truth or knowledge base benchmark datasets. Triangular membership function was used to construct the rules of FC and classify the f β into clip limit low (CL1) and clip limit high (CL2). The value of each rule for overall FC output was aggregated into a single fuzzy set. The result of the defuzzification of FC is 0.0191. As a result, the $f\beta$ is 0.0119 for the given values of (C =0.4) and (E =0.4). The $f\beta$ varies between 0 and 0.1. The FC-CLAHE method uses Equation (2) to calculate the new clip limit that depends on the C and E of the input images.

$$Fuzzy\ Clip\ Limit = \left[\frac{\varphi}{256}\right] + \left[f\beta.\left(\varphi - \left[\frac{\varphi}{256}\right]\right)\right] \quad (2)$$

Where $f\beta$ is the fuzzy clipping enhancement parameter (ranges from 0 to 0.1) and [.] denotes the truncating value to the nearest integer. φ is the product of block size. The value of 256 denotes the pixel intensity values range from 0 to 255.

Anisotropic diffusion filter (ADF) was used to remove the noise and preserve the details by using a partial differential equation. The model of ADF is introduced by Perona and Malik [19]. The anisotropic coefficient is used to stop the diffusion over the edges of the image [20]. Equation (3) describes the ADF model.

$$\frac{\partial I(x,y)}{\partial t} = div\big(g(|\nabla I(x,y)|)\nabla I(x,y)\big) \quad (3)$$

Where $div(.)$ and ∇ represent the divergence and the gradient operators respectively, while $g(|\nabla I(x,y)|)$ is a reducing function known as an edge stop mechanism, which plays an important role in regulating the mechanism of diffusion. Another two decreasing functions can be described in Equation (4).

$$g'(|\nabla I(x,y)|) = \frac{1}{1+\frac{|\nabla I(x,y)|^2}{k^2}} \quad (4)$$

$$g\text{^}''\ (|\nabla I(x,y)|) = \exp\ (-|\nabla I(x,y)|\text{^}2/k\text{^}2$$

Where k is a parameter with a threshold role. If $|\nabla I| > k$ so the pixels are considered edges and will be less blurred. On the other hand, if $|\nabla I| \leq k$, the pixels are considered as the interior region and will be highly smoothed.

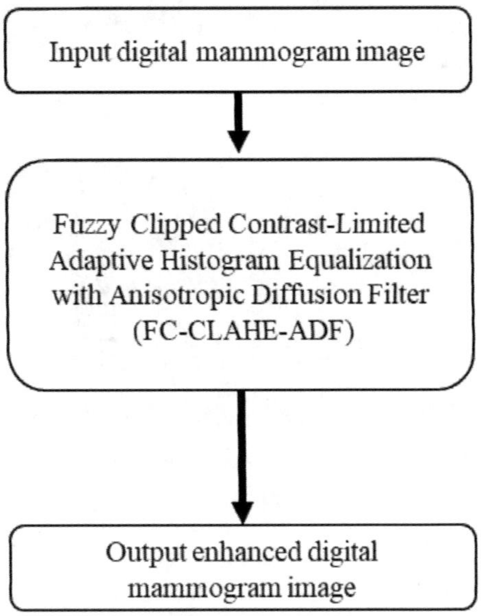

Figure 1. Flowchart of the proposed FC-CLAHE-ADF enhancement technique.

3. Results and Discussion

The resulting images of various image enhancement techniques are shown in Figure 2. The enhanced digital mammograms have shown superior image quality than the original images. The enhanced digital mammograms have preserved the details for the interpretation of breast lesions.

Figure 2 (a) has shown the original image of digital mammograms with low contrast features. The details of dense tissue in the breast are reduced in the images. The details, such as breast textures, edges, and lines, are blurred and grainy. The low contrast and low intensity could affect the detection of potential tumors as they could reside within dense tissues. Figure 2 (b) and (d) have shown the enhanced images by RMSHE and FC-CLAHE, respectively. Both of these techniques have shown improvement but are overly enhanced, which leads to noise amplification.

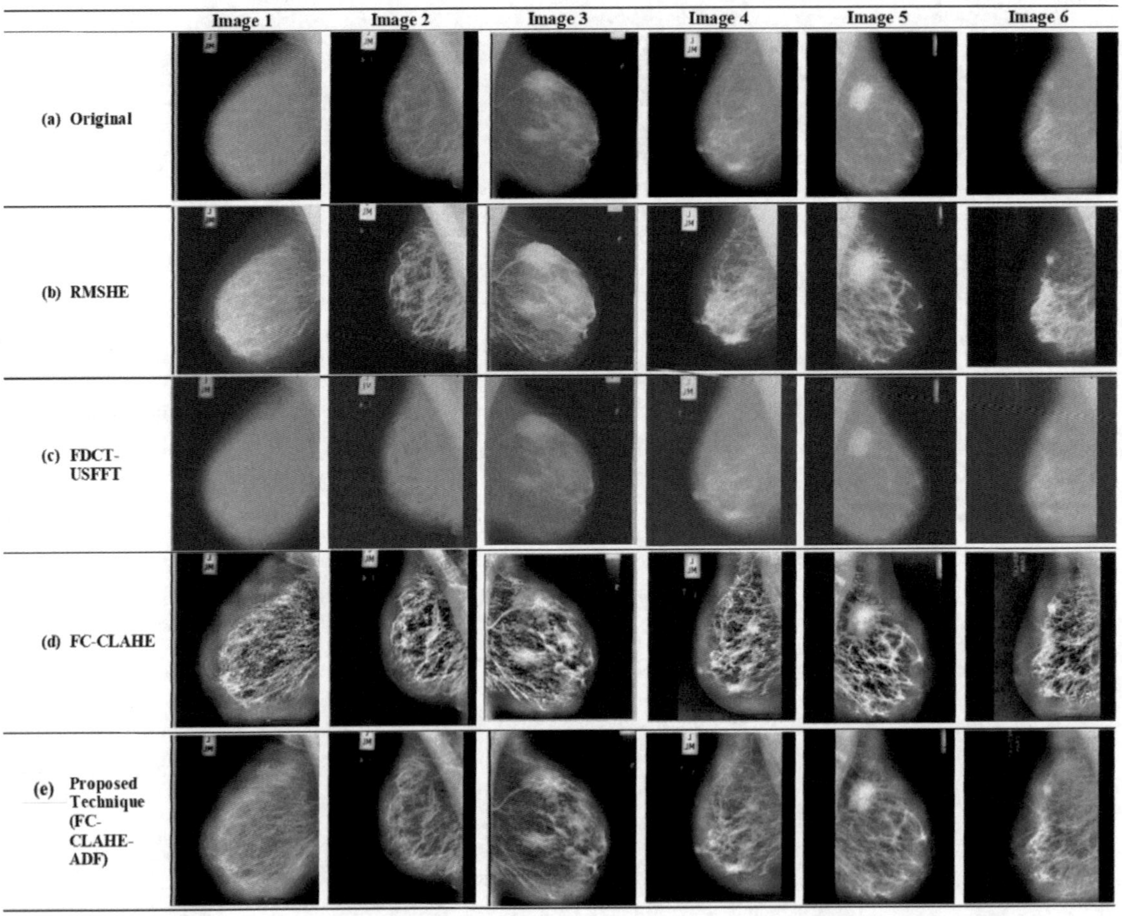

Figure 2. Comparison of varied image enhancement techniques using different digital mammograms: (a) original image, (b) RMSHE, (c) FDCT-USFFT, (d) FC-CLAHE and (e) proposed technique (FC-CLAHE-ADF).

The enhanced images produced by FDCT-USFFT in Figure 2 (c) have shown no improvement even after the enhancement process. The proposed FC-CLAHE-ADF in Figure 2 (e) has shown appropriate noise reduction while preserving the details. FC-CLAHE-ADF has produced digital mammograms without overly enhanced as compared to other techniques. Therefore, the proposed FC-CLAHE-ADF could assist in the identification of breast lesions, especially for early breast cancer detection.

IOP Conf. Series: Materials Science and Engineering 1173 (2021) 012065 doi:10.1088/1757-899X/1173/1/012065

4. Conclusion

A novel of FC-CLAHE-ADF has been proposed for digital mammograms. The overall quality of enhanced images has been improved as compared to the other techniques such as RMSHE, FDCT-USFFT, and FC-CLAHE only.

References

[1] Suppaya K, Nasir F M and Ghani A A 2020 Variations of Bi-Rads 5 in Mammography by Age , Ethnicity , and Breast Density : A Retrospective Study in University Malaya Medical Centre *Asian J. Med. Biomed.* **4** 11–6

[2] Tripathy S and Swarnkar T 2020 Unified Preprocessing and Enhancement Technique for Mammogram Images *Procedia Comput. Sci.* **167** 285–92

[3] Brooks A D, Nover A B, Jagtap S, Anjum W, Yegingil H, Shih W Y and Shih W H 2009 Modern breast cancer detection: A technological review *Int. J. Biomed. Imaging* **2009**

[4] Berg W A, Blume J D, Cormack J B, Mendelson E B, Lehrer D, Böhm-Vélez M, Pisano E D, Jong R A, Evans W P, Morton M J, Mahoney M C, Larsen L H, Barr R G, Farria D M, Marques H S and Boparai K 2008 Combined screening with ultrasound and mammography vs mammography alone in women at elevated risk of breast cancer *JAMA - J. Am. Med. Assoc.* **299** 2151–63

[5] Black E and Richmond R 2019 Improving early detection of breast cancer in sub-Saharan Africa: Why mammography may not be the way forward *Global. Health* **15** 3

[6] Wang L 2017 Early diagnosis of breast cancer *Sensors (Switzerland)* **17**

[7] Quintanilla-Dominguez J, Ojeda-Magañ B, Cortina-Januchs M G, Ruelas R, Vega-Corona A and Andina D 2011 Image segmentation by fuzzy and possibilistic clustering algorithms for the identification of microcalcifications *Sci. Iran.* **18** 580–9

[8] Gonzalez R C, Woods R E and Masters B R 2009 Digital Image Processing, Third Edition *J. Biomed. Opt.* **14** 029901

[9] Pisano E D, Zong S, Hemminger B M, DeLuca M, Johnston R E, Muller K, Braeuning M P and Pizer S M 1998 Contrast limited adaptive histogram equalization image processing to improve the detection of simulated spiculations in dense mammograms *J. Digit. Imaging* **11** 193–200

[10] Sondele S and Saini I 2013 Classification of mammograms using bidimensional empirical mode decomposition based features and artificial neural network *Int. J. Bio-Science Bio-Technology* **5** 171–80

[11] Akila K, Jayashree L S and Vasuki A 2015 Mammographic image enhancement using indirect contrast enhancement techniques - A comparative study *Procedia Comput. Sci.* **47** 255–61

[12] Magudeeswaran V and Ravichandran C G 2013 Fuzzy logic-based histogram equalization for image contrast enhancement *Math. Probl. Eng.* **2013**

[13] Jenifer S, Parasuraman S and Kadirvelu A 2016 Contrast enhancement and brightness preserving of digital mammograms using fuzzy clipped contrast-limited adaptive histogram equalization algorithm *Appl. Soft Comput. J.* **42** 167–77

[14] Chan N H 2019 An Improved Enhancement Technique for Mammogram Image Analysis : A Fuzzy Rule-Based Approach of Contrast Enhancement *2019 IEEE 15th Int. Colloq. Signal Process. Its Appl.* 202–6

[15] Senthilkumar B and Umamaheswari G 2015 Breast cancer detection using combined curvelet based enhancement and a novel segmentation methods *Biomed. Pap.* **159** 83–6

[16] Panetta K, Yicong Zhou, Agaian S and Hongwei Jia 2011 Nonlinear Unsharp Masking for Mammogram Enhancement *IEEE Trans. Inf. Technol. Biomed.* **15** 918–28

[17] Wu S, Zhu Q, Yang Y and Xie Y 2013 Feature and contrast enhancement of mammographic image based on multiscale analysis and morphology *2013 IEEE Int. Conf. Inf. Autom. ICIA 2013* **2013** 521–6

[18] Clark A F 2012 The mini-MIAS database of mammograms.

[19] Perona P and Malik J 1990 Scale-Space and Edge Detection Using Anisotropic Diffusion *IEEE Trans. Pattern Anal. Mach. Intell.* **12** 629–39
[20] Kurt B, Nabiyev V V. and Turhan K 2012 Medical images enhancement by using anisotropic filter and CLAHE *INISTA 2012 - Int. Symp. Innov. Intell. Syst. Appl.* 1–4

Numerical simulation on the permeability of non-crimp Carbon fiber epoxy using out-of-autoclave manufacturing method

A Levino, B K Hadi*, and A Kuswoyo

Lightweight Structure Laboratory, Faculty of Mechanical and Aerospace Engineering Bandung Institute of Technology (ITB), Jl. Ganesha No.10, Lb. Siliwangi, Kecamatan Coblong, Kota Bandung, Jawa Barat 40132, Indonesia

* Email: bkhadi@ae.itb.ac.id

Abstract. Composites made using Resin Transfer Molding (RTM) processes have attractive properties to the aerospace and automotive industries because they can produce complex composite parts with good surface finishing. However, it still has a problem filling a large part with high fiber content at low injection pressure. Therefore, during the experiment, resin flow has to be the primary concern. This work presents a resin flow simulation of two types of mold where it will visualize the manufacturing process without actually doing it. Numerical simulation for the RTM manufacturing process is attempted by using Fluent-Ansys. It allows us to simulate the resin flow through the porous medium and obtain the total mold filling time and volumetric flow rate. It will lead to an explanation of the strength of the composite produced. Several assumptions and techniques are used to simplify the problem. Some variables, such as porosity and pressure injection, will be varied to see the effect of those variables on the RTM manufacturing process. According to the simulation result, porosity and pressure injection affect the RTM manufacturing process, especially for resin filling time and volumetric flow rate. The increase of porosity will increase the flow time, while pressure injection will decrease the flow. It happens due to how easy the resin flows through the porous medium, and the total area needs to be filled by the resin. This work is validated by comparing the obtained results with analytic solutions for a specific mold of Darcy's law.

Keywords: Resin transfer molding, porosity, pressure injection, analytic solution, Darcy's law

1. Introduction

Since the birth of aviation, designers have continuously endeavored to improve the lift to weight ratios of aircraft. Composite materials have played a significant role in weight reduction. Therefore, the development of lightweight material has been attracting attention. The use of composite is increasing to develop lightweight and high-strength vehicle components as these enable designers to overcome the barriers created by using metals. Many companies are competing against each other to offer the best composite material. Producing a composite that has good quality, especially for the aerospace industry, is usually done by the autoclave method. However, this method needs a higher cost than it needs to be [1]. Thus, many innovations are made to produce a composite with a low cost. One of them is called Out-of-Autoclave. The example of out-of-autoclave is Resin Transfer Molding (RTM) and Vacuum Assisted Resin Transfer Molding (VARTM).

The resin transfer molding (RTM) process has been the subject of a great deal of practical and theoretical development for aerospace applications since the early 1980s. For instance, Mitsubishi Regional Jet (MRJ) has been used composite produced by the RTM manufacturing process, where it offers impressive operating costs with good efficiency. RTM suits to manufacture high quality of complex products, and the products have high fiber content. In the forming process, there are less volatile components and little environmental pollution. With less investment and high production efficiency, the production automation is adaptable for this process [2] The strength of the composite produced by RTM is dependent on the impregnation of the resin flow through the Fiber. Permeability defines the resistance to the resin flow. Permeability is a reinforced medium property used to measure (quantify) how easy a fluid can pass through a porous medium [3]. Therefore, Permeability is a property that needs to be the primary concern.

In this paper, the resin flow process in the RTM manufacturing process has been simulated using FLUENT and the outputs are focused on investigating the effect of variation of porosity and pressure injection to the total mold filling time and pressure distribution that occur during the process. Filling time is obtained and after that, the result will be compared to the result of the analytical method.

2. Analytical Models

To simulate the RTM process; several assumptions are used to simplify the process, such as the resin's density remains constant, assuming the resin as a kind of Newton fluid, viscous force, and surface tension are neglected, type of flow is rectilinear, and curing process is neglected by assuming after full filling, the resin begins to cure.

The basic theory of prediction of the flowing resin is generally modeled as a Newtonian fluid with constant density and viscosity, so the Navier-Stokes equation is valid. However, the common equation used to model the flow of resin in Out-of-Autoclave is Darcy's law [4]. It is a simplification of the Navier-Stokes equation. Newtonian fluid flow in porous media is often modeled using a generalized version of the full non-linear Navier-Stokes equations that include additional terms describing the resistance to flow due to the porous matrix. Darcy's law can be expressed as:

$$q = -\left(\frac{K}{\mu}\right)(\Delta P) \tag{1}$$

where q is the flow rate, K is the permeability of porous medium, P is the pressure gradient, and μ is the viscosity. It assumed that there is no movement perform during the flow process, therefore continuity condition is applied.

$$q = \frac{dx}{dt} \tag{2}$$

since,

$$v = \frac{q}{\emptyset} \tag{3}$$

where v is the velocity of the fluid and ϕ is the porosity. Therefore,

$$v = \frac{dx}{\emptyset\, dt} \tag{4}$$

Additional assumption like constant pressure is applied due to the constant flow rate of the resin supply system.

$$\nabla P = \frac{\partial P}{\partial x} = \frac{P}{x} \tag{5}$$

By substituting and doing the integration to Darcy's law, the formulation of the total filling time, in this case, can be obtained as follow:

$$t = \frac{x^2\, \mu}{2\, KP} \tag{6}$$

3. Simulation Model

The RTM process is carried out on two types of two-dimensional mold such as rectangular and rectangular with hole. several variables are varied Several variables are varied, such as porosity (0.4, 0.6, and 0.8) and pressure injection (70000 Pa, 100000 Pa, and 130000 Pa). Other variables, like permeability, air properties, and resin properties, will remain constants [5].

Table 1. Fixed-Parameter

Fixed Variable	Magnitude	Unit
Air density	1.225	Kg/m^3
Air viscosity	1.7894×10^{-5}	Kg/ms
Permeability	3.89×10^{-9}	m^2
Resin density	916	Kg/m^3
Resin viscosity	0.7115	Kg/ms

There are 2 types of process such as mechanical modelling and FLUENT. The mechanical model comprises of engineering data, geometry, and model while Fluent comprises of problem setup, solution, and result.

3.1. Mechanical Model

This process allows us to create the mold's geometry, determine the type of material of the mold, and perform the meshing. Rectangular and rectangular with hole mold are built with dimension 1 x 0.5 m and a radius of 0.1 m. Non-crimp carbon fiber is applied as the mold's material and there are three meshing sets such as general mesh, edge mesh, and face mesh with size 5 mm.

(a) (b)

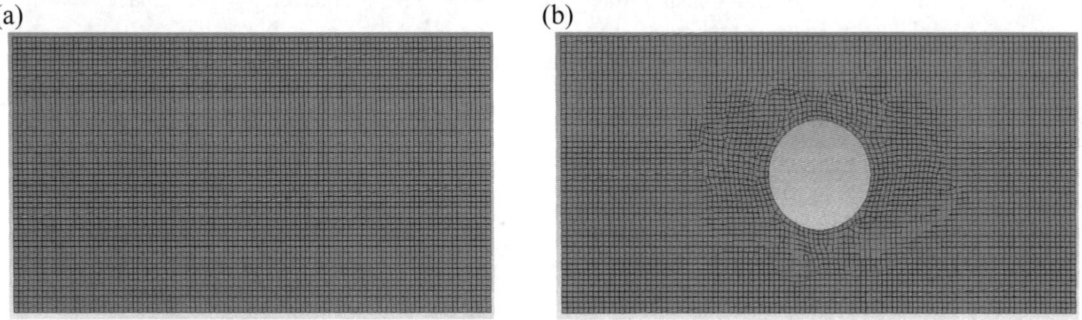

Figure 1. Mold geometry models (a) Rectangular (b) Rectangular with Hole

3.2. Fluent program

The FLUENT is set as 2D double precision solver. The specific settings in FLUENT are as follows:

a. The boundary condition is defined as follows, a pressure inlet condition is defined at the inlet surface, and the value of pressure inlet defines the pressure injection will be used while a pressure outlet at the outlet surface.
b. The Reference Values task page is set as default.
c. The pressure coupled equations of half implicit method (i.e., SIMPLE algorithm) are adapted to the solution. The under-relaxation factor and discrete format are the same as the default value.
d. The convergence accuracy during the calculation is 0.0001 up to 0.000001.
e. The time step and iteration are 0.1 and 10, respectively.
f. The gravitational effect is neglected.
g. The number of steps will be varied according to the porosity and pressure injection used.
h. Input the process conditions, including the viscosity and density of resin, the porosity of porous media and its viscous resistance and inertia resistance to resin, and the boundary conditions. Then initialized, and set a solution animation to monitor the volume fraction of resin, calculated at last.

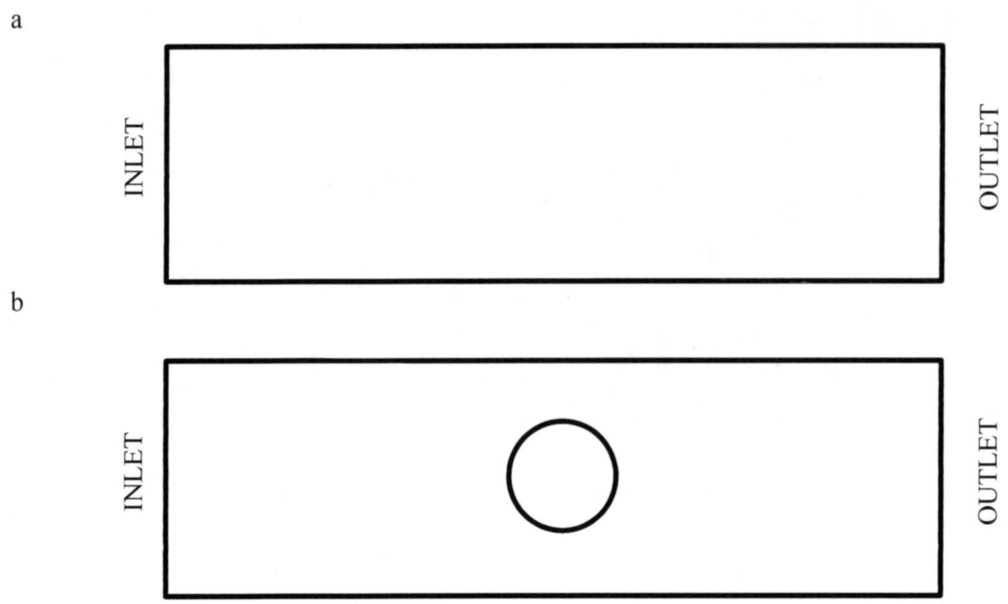

Figure 2. Simulation sketch modelling (a) Rectangular (b) Rectangular with hole

In this case, the resin is injected from the left boundary side of the mold (Figure 2), then resin (epoxy) is injected into the mold under a constant injection pressure. The interaction between phases like surface tension is neglected to simplify the problem. The whole zone of the mold is defined as fluid.

Next, input others variables such as viscosity and density of the resin, porosity, permeability, and boundary condition. In this case, pressure injection at the inlet is varied as mention previously while at the outlet, the pressure is set as a constant 0 Pa while the other boundaries are set as atmospheric pressure. Then do the initialization and set a solution animation to monitor the volume fraction of resin.

VOF method is used to simulate the RTM process where has been frequently used to solve moving boundary problems in a fixed mesh. Furthermore, it is representative of the boundary tracking method. For the RTM process's numerical simulation, the resin flow front position in the mold is obtained by the VOF method [6].

4. Results dan Discussions

4.1. Simulation results

4.1.1. Rectangular mold

Figure 3 shows the resin filling process with porosity 0.4 and pressure injection 70000 Pa done by FLUENT. It took 520.2 seconds to fill the whole mold with the resin. There is no difference in the visualization of the resin filling process for other pressure injection and porosity configuration, except for the total time for the resin to completely fill the mold. Table 2 below shows the results of the resin filling time for each configuration.

IOP Conf. Series: Materials Science and Engineering 1173 (2021) 012066 doi:10.1088/1757-899X/1173/1/012066

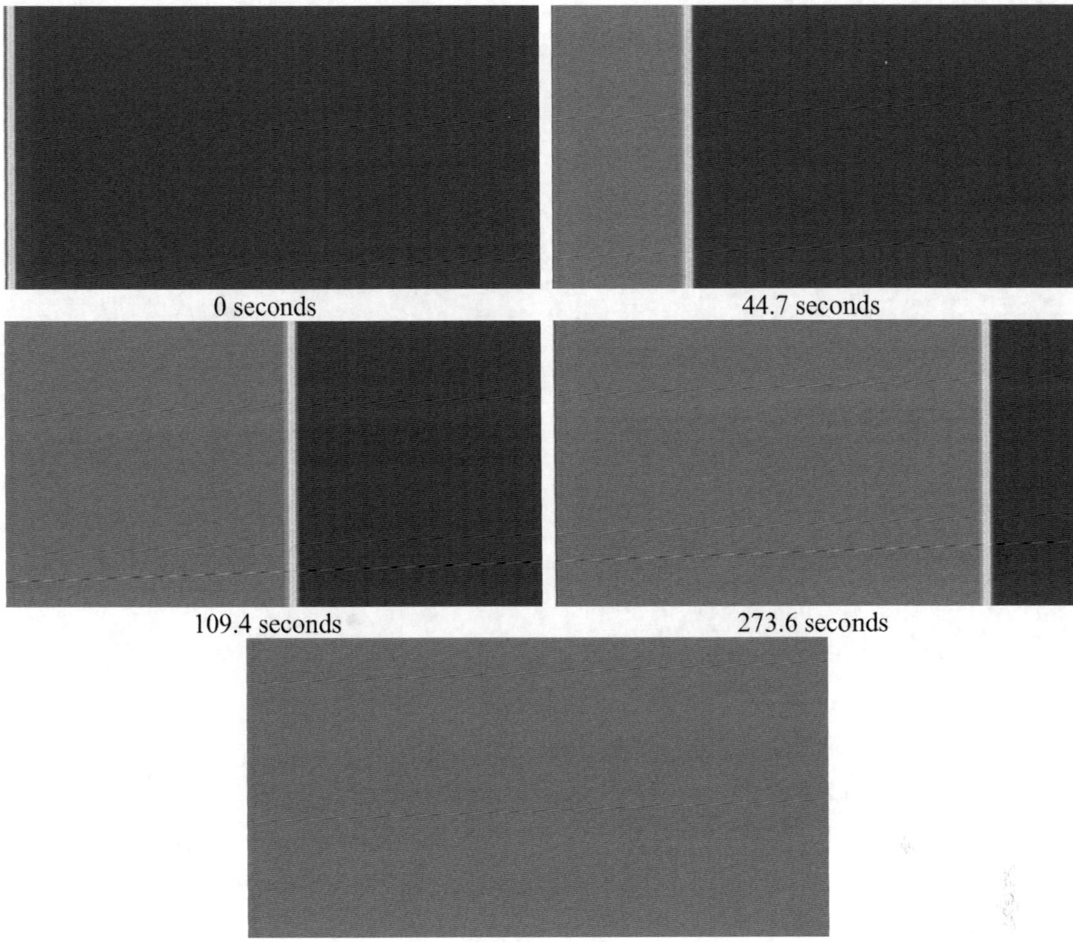

Figure 3. Resin filling simulation of porosity 0.4 and pressure injection 70000 Pa

Table 2. Rectangular total resin filling time

Rectangular Mold	Total resin filling time in simulation (s)		
Pressure Injection (Pa) / Porosity	70000	100000	130000
0.4	520.5	381.7	346
0.6	785.8	559.8	454.7
0.8	1044.6	735.8	544.7

Moreover, the figure below shows the effect of porosity and pressure injection difference on the total resin filling time.

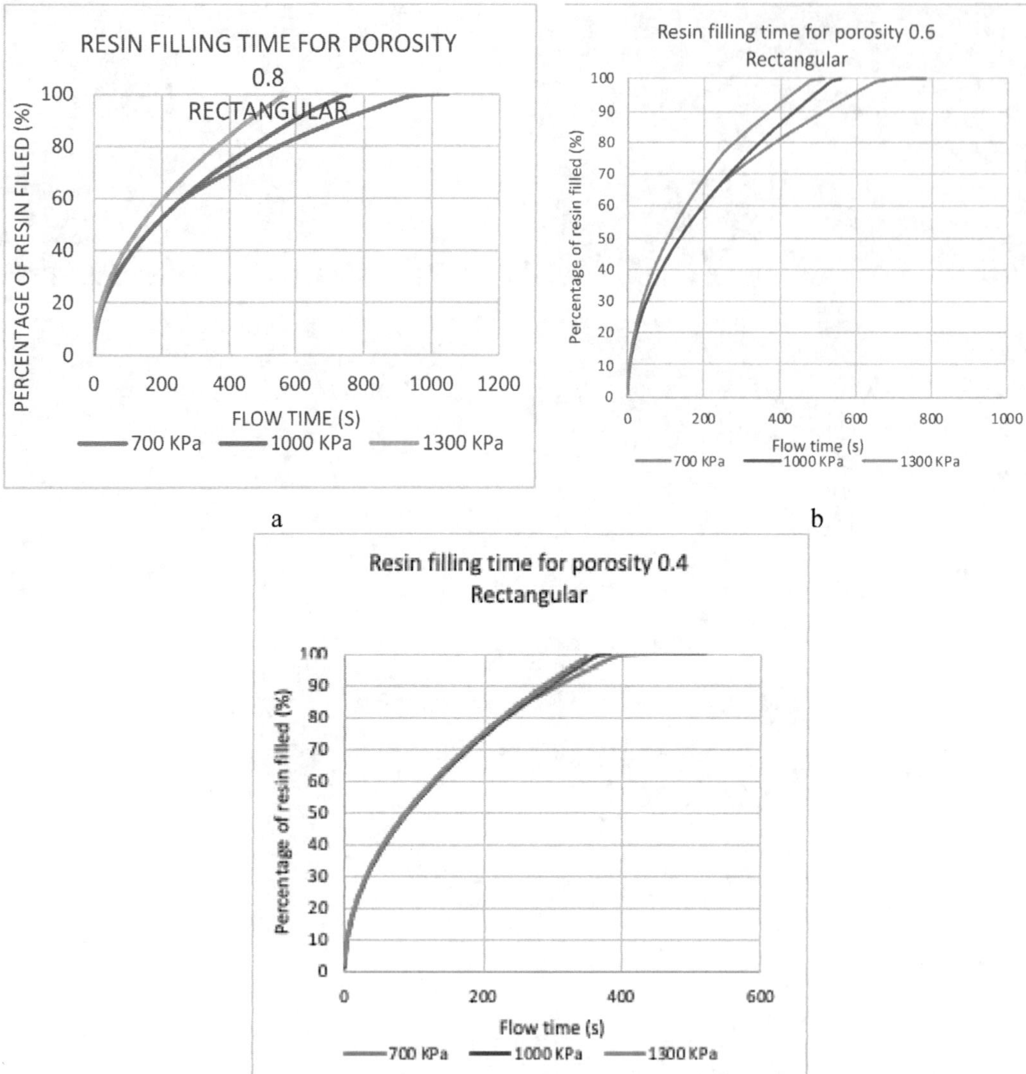

c

Figure 4. Rectangular resin filling time (a) 0.8 (b) 0.6 (c) 0.4

Other output results can be obtained through this simulation is volumetric flow rate. Here's the graphs show the volumetric flow rate for each pressure injection.

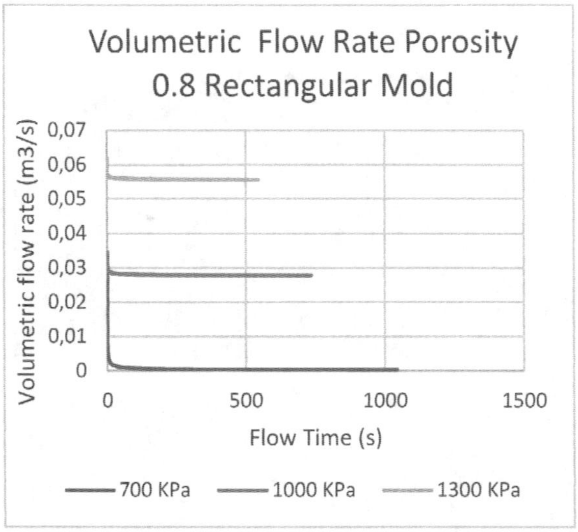

a

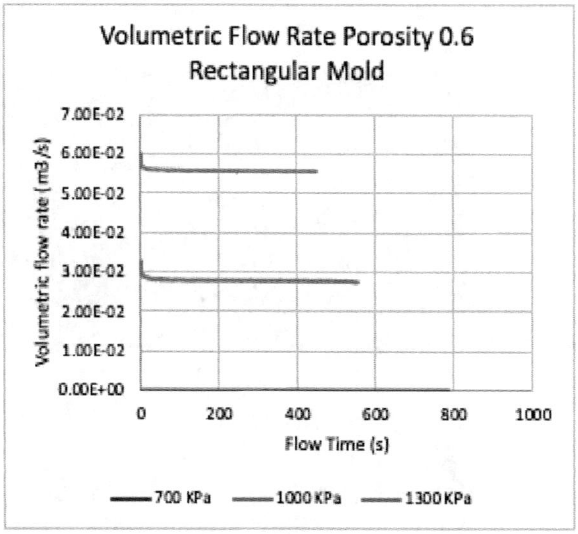

b

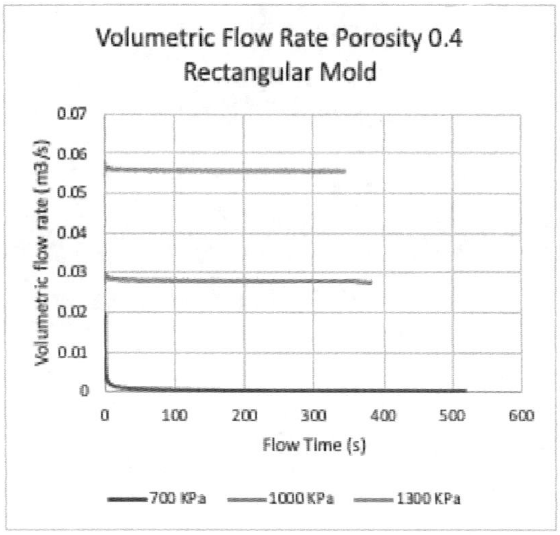

c

Figure 5. rectangular volumetric flow rate (a) 0.8 (b) 0.6 (c) 0.4

According to figure 4, it shows the effect of porosity and pressure injection on resin filling time. Increasing the porosity took a long time for the resin to fill the whole mold while increasing the pressure injection shortens the total resin filling time. At porosity 0.4, overall, the resin filling processes do not give a significant effect (the lines almost coincide) if it is compared with porosity 0.8. however, the total resin filling time is still affected by it. It is due to the increasing area needs to be covered by the resin. Porosity is the ratio between void area and total mold area. Since the total mold area is constant, the porosity value here will affect the void area. By increasing the porosity in this case, the total void area is increasing.

Figure 5 explains the reason why increasing pressure injection shorten the total resin filling time. It is due to the increasing volumetric flow rate as the effect of increasing pressure injection where it will lead to the increasing of the pressure gradient. Therefore, it shortens the resin filling time process.

4.1.2. Rectangular with hole mold

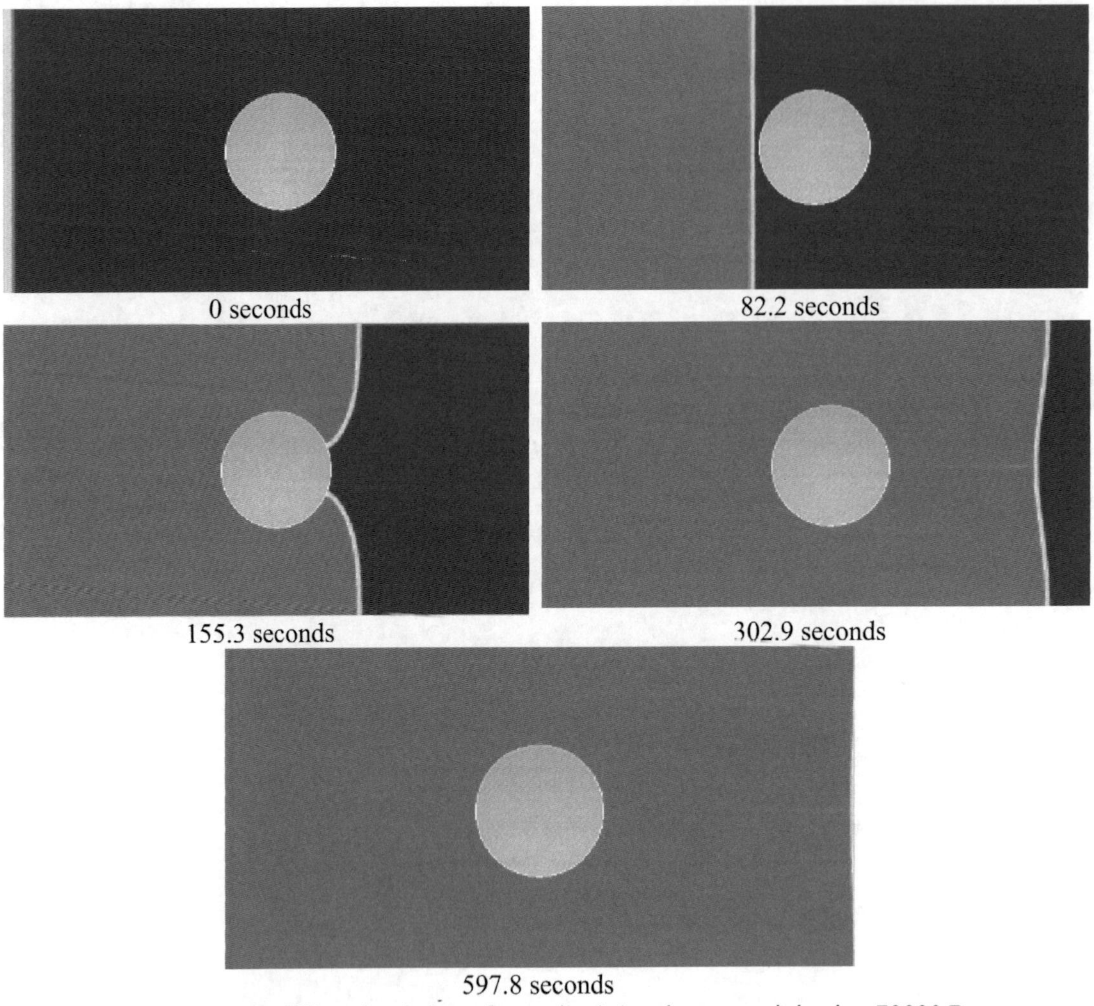

Figure 6. Resin filling simulation of porosity 0.4 and pressure injection 70000 Pa

Figure 6 rectangular with hole shows the resin filling process with porosity 0.4 and pressure injection 70000 Pa done by FLUENT. It took 597.8 seconds to fill the whole mold with the resin. Same as rectangular mold, there is no difference in the visualization of the resin filling process for other pressure injection and porosity configuration, except for the total time for the resin to completely fill the mold. Table 3 below shows the results of the resin filling time for each configuration.

Table 3. Rectangular with hole total resin filling time

	RECTANGULAR HOLE		
Pressure Injection / Porosity	700 KPa	1000 KPa	1300 KPa
0.4	597.8	456.2	402.3
0.6	842.3	762.2	776.6
0.8	1492.4	1107.3	1125.1

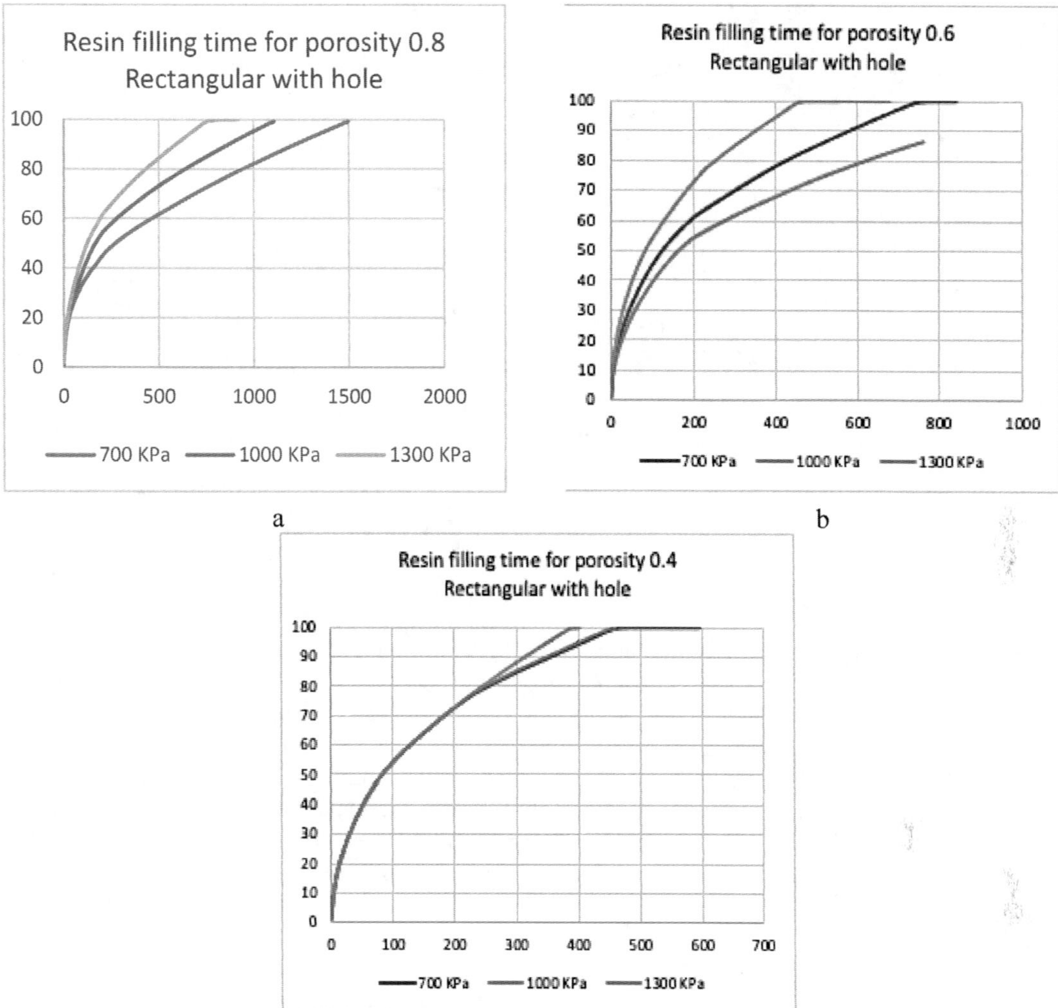

a b

c

Figure 7. Rectangular with hole resin filling time (a) 0.8 (b) 0.6 (c) 0.4

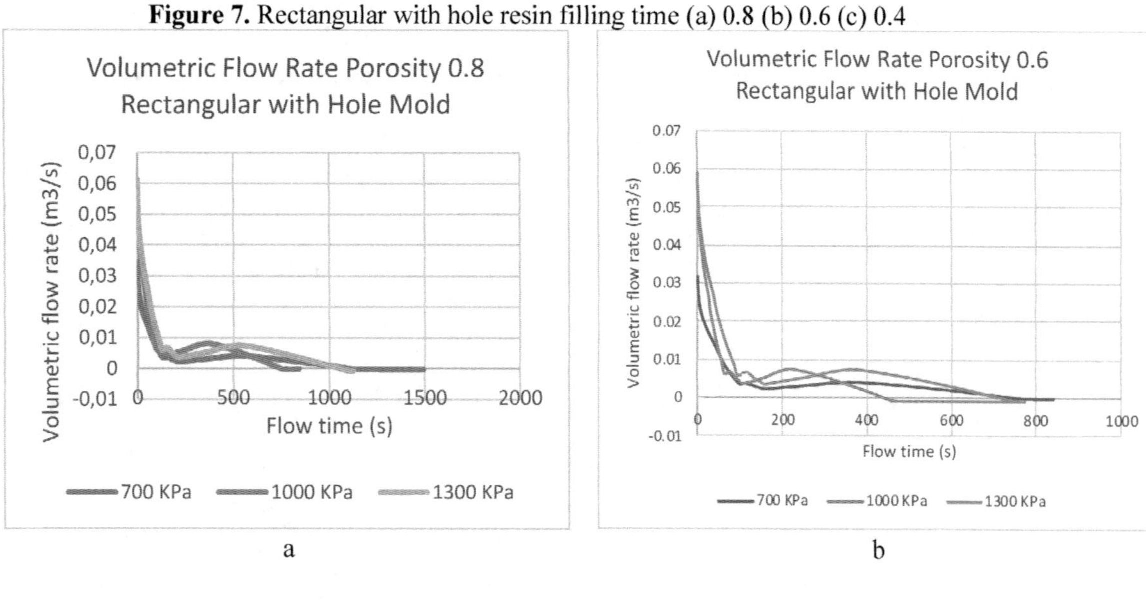

a b

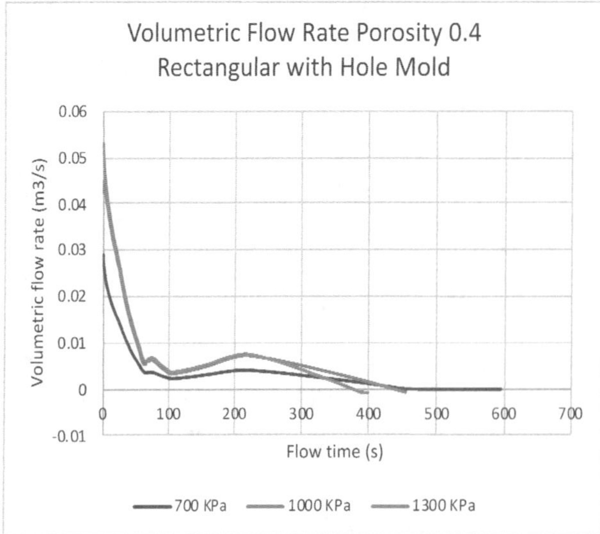

c

Figure 8. Rectangular with hole resin filling time (a) 0.8 (b) 0.6 (c) 0.4

Figure 7 shows the effect of pressure injection and porosity on rectangular with hole on resin filling time. When the pressure injections are varied into 70000, 100000, and 130000 Pa, the total resin filling time reduces. It significantly reduces at porosity 0.8 compared with other porosities. When the porosity is varied into 0.4, 0,6, and 0.8, the total resin filling time increases the same as the rectangular mold.

Figure 8 shows that the volumetric flow rate of rectangular with a hole is fluctuated due to the separation flow when the resin flow reaches the hole. Same as the rectangular, the volumetric flow rate escalates by increasing the pressure injection. The porosity difference does not significantly affect the volumetric flow rate for the same pressure injection, according to figure 8.

Not only porosity and pressure injection affect the RTM process. In this case, other variables, like the geometry of the mold, also affect the RTM process. Comparing data obtained from rectangular and rectangular with a hole can be concluded that different mold will increase the total resin filling time at the same pressure injection. It is due to the complexification shape where related to the type and the number of flows used in the process. More complex, there will be more flow definitions that need to be applied. In this case, rectangular mold is a rectilinear flow while rectangular with hole mold is rectilinear initially, but when it reaches the hole, the flow will separate into two flow.

As a consequence, the type of flow is changing. There will be flow in the y-direction. Therefore, it will affect the resin filling time. Moreover, the effect of mold variation on the volumetric flow rate is for the rectangular mold, the volumetric flow rate is like a curve, while for the rectangular with hole mold, fluctuated appears. It is due to the separation flow occurs when the flow reaches the hole. The effect of porosity difference does not significantly affect the volumetric flow rate for both molds. In contrast, the effect of pressure injection difference significantly affects the volumetric flow rate. On the other hand, the volumetric flow rate for rectangular with a hole is not significantly affected by the pressure injection difference.

4.2. Analytical results

Analytical result is done to validate the simulation done by FLUENT. this method is limit only for rectangular mold due assumption that used.

Table 4. Rectangular analytical result

Rectangular Mold		Total resin filling time of analytical solution (s)		
Pressure Injection (Pa)	Porosity	70000	100000	130000
	0.4	522.6	365.8	281.4
	0.6	783.9	548.7	422.1
	0.8	1045.2	731.6	562.8

Table 5. Rectangular comparison error result

Rectangular		Error (%)		
Pressure Injection (Pa)	Porosity	70000	100000	1300000
	0.4	0.40	4.16	18.17
	0.6	0.25	1.98	7.17
	0.8	0.05	0.57	3.32

From table 4 and 5, it shows the comparison between analytical and numerical solution. For pressure injection 70000 and 100000 Pa, the errors are still low (below 10%), while at pressure injection 130000 Pa at porosity 0.4, the error exceeds 10%. Many things have been done to reduce this error, but it showed the same result. The error possible occurs due to the boundary condition setup. Boundary condition setup is used to guess the initial pressure and velocity at the inlet. Boundary condition setup has many possibilities; it is due to the pressure injection and operating pressure. Since in this experiment, the operating pressure is set as a constant (atmospheric condition). Therefore, the initial pressure and velocity guessing by boundary conditions are incorrect because operating pressure is set up in Fluent is not applicable or correct for in this case.

5. Conclusions

In this research, current work is mainly focused on the effect of pressure injection, porosity, and type of mold on the RTM manufacturing process. Several parameters are still constant to simplify the problem and simulation procedure. For instance, the value of permeability is set as constant to simplify the problem, and that is why non-crimp carbon fiber is used. Future work involves presenting all the variables that might affect the process, such as permeability is as a function, use other types of flow, and experiment has to be done to validate the simulation results.

Acknowledge

This work was supported by the Lightweight Structure Research Laboratory Faculty of Mechanical and Aerospace Engineering Bandung institute of technology. Gratitude also goes to Mr. Bambang Kismono Hadi and Mr. Andi Kuswoyo as an advisor in this research.

References

[1] R. Alagirusamy, "Hybrid yarns for thermoplastic composites," Indian Institute of Technology., New Delhi , 2010.

[2] Yinfei Dong, Xuejian Jiao,, "The Simulation of RTM Based On FLUENT," *Advanced Materials Research,* Vols. 538-541, pp. 873-876, 2012.

[3] Yang B, Kozey V, S. Adanur, Kumar S., "“Composite : Part B,” Bending, Compression and Shear Behaviour of Woven glass fiber-epoxy composites.," vol. 31, pp. 715-721, 2000.

[4] Dinh Giap Ta, Dinh Quy Vu, Xuan Tung Pham, Quoc Huy Vu, "Dinh Giap Ta, Dinh Quy Vu, Xuan Tung Pham, Quoc Huy Vu," 2013.

[5] J. da S. Portoa, M. Letzow, E. D. dos Santosa, S. C. Amicob, J. A. Souzaa, L. A. Isoldia, "COMPUTATIONAL MODELING OF RTM AND LRTM PROCESSES APPLIED TO COMPLEX GEOMETRIES," 2013.

[6] Cyril W. Hirt, Benjamin Nichols, "Volume of Fluid (VOF) MeThod for the Dynamics of Free Boundaries.," 1981.

Flexural behaviour of jute, glass, and Carbon fibre reinforced Polyester hybrid composites

A Murdani*, and U S Amrullah

Department of Mechanical Engineering, Politeknik Negeri Malang, Jalan Sukarno Hatta No. 9, Malang 65141, Indonesia

* Corresponding e-mail: anggitm@polinema.ac.id

Abstract. This study is a part of a project in developing natural fibre reinforced composite for wind turbine blade. The objective of this study is to evaluate the effect of fibre stacking sequence on flexural behaviour of the natural fibre reinforced composite hybridized with synthetic fibre. The flexural behaviour of the composite becomes an important consideration in deciding stacking sequence of the fibres. Material used in this study is polyester resin as the matrix. Jute, glass, and carbon fibres are used as the reinforcement. The composites were fabricated using vacuum bagging method cured at room temperature. The combinations of fibre as the reinforcement are configured by arrange two and three types of fibre. As a natural fibre, jute exists in all kind of combination. Flexural behaviour was observed using 3-point-bending test. The specimens were tested in accordance to the fibre stacking sequence. The result shows that stacking sequence of the fibres give significant different flexural behaviour regardless the effect of the fibre strength.

Keywords: natural fibre composites, wind turbine, blade, green materials, renewable energy

1. Introduction

Low speed wind turbines need the appropriate blade properties, such as strength, stiffness, and lightweight [1-4]. The materials used for the blade are composites consisting of resin and reinforcing fibres [5]. The reinforcing fibre used is generally glass fibre.

The problem arose when a huge amount of blade waste will occur in the next 30 years [5]. This blade material waste will not easily degrade naturally and has a long-term impact on the environment. Therefore, this research is looking for solutions to reduce non-degradable waste. The wind turbine blade will be made of partially natural material, i.e., natural fibre as reinforcing composite.

To meet the properties required by low-speed wind turbine blade, natural fibre is hybridized by synthetic fibres, i.e., glass and carbon fibre, as reinforcement of the composite [6-10]. Characterization of composite is carried out in order to obtain the best configuration to meet the demand of low-speed wind turbine blades characteristic.

2. Materials and Method

The material used in this research consists of matrix and fibres. The matrix material is polyester with the product name Yucalac 157 BTQN-EX, licensed by Showa Highpolymer Co., Ltd (Japan) and Polynt

SpA (Italy). At normal temperature, this type of polyester is resistant to water and weak acids. This polyester is widely used in shipbuilding, tanks, sanitary ware, and FRP Pipe.

As the reinforcement, good quality of jute (*corchorus capsularis*) fibre cloth obtained from local supplier is used. The jute fibre is in the form of plain weave type. The jute fibre cloth obtained from the supplier is dried prior to composite fabrication. Drying is carried out using an electric heater with a temperature of 150°C until the weight of the jute fibre cloth is no longer reduced. The final weight of the dried jute fibre cloth is 25 g/m^2. For hybridizing, S-glass type fibres cloth with areal weight 220 g/m^2 in plain weave woven and 2x2 plain weave carbon fibre cloth with areal weight 220 g/m^2 are used.

Figure 1 shows the experiment design in this research. The hybrid composite, in the combination of jute, glass, and carbon fibres, arranged according to their pair with the notation of jute fibre (J), glass fibre (G), and carbon fibre (C). The hybrid combination considers the aspects of mechanical properties, thickness, and economical cost. As a natural fibre, jute exists in all kind of combination. The configurations of the specimens are GJJ, GJG, CJJ, and CJG. It can be seen that the role of glass and carbon fibres as the outermost layer or skin. Specimens were made using vacuum bagging technique.

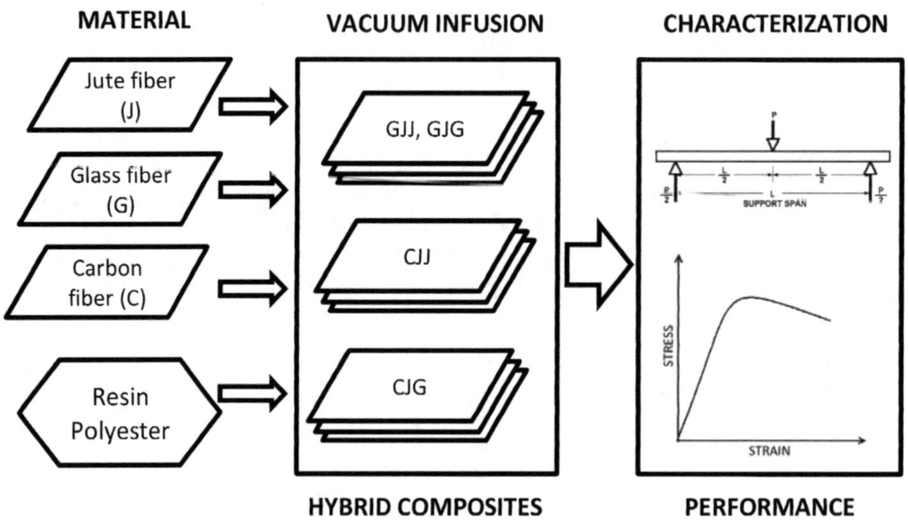

Figure 1. Experimental configuration.

Three-point flexural tests were carried out according to ASTM D7264. Each specimen configuration is tested in two stacking sequence. For example, GJJ specimen is tested in GJJ and JJG sequences. The first letter is the surface that contacted to the loading nose, i.e., the compression surface. The last letter is the tension surface.

3. Results and Discussions

The test results are presented in Figures 2 - 5. Figure 2 shows the test result of GJJ specimens. JJG specimen show higher flexural strength and larger deflection than GJJ specimen does. In the case of JJG specimen, glass fibre cloth is in the tension surface of the specimen. Therefore, glass fibre, which has higher strength than jute fibre, contributes to the flexural strength of the specimen [11-12]. Also, since glass fibre has good elongation, the total deflection of the JJG specimen shows larger than GJJ specimen. The flexural strength of JJG specimen (140 N/mm^2) is more than two times of GJJ specimen (65 N/mm^2).

Figure 3 shows the test result of GJG specimen. Since the both side of the specimen is the same fibre cloth layer, the test result shows a single curve. GJG specimen shows a similar flexural behaviour to that of JJG specimen in Figure 2. Although in the case of GJG specimen consists of two layers of glass fibre cloth, the first layer, the compression surface, has no obvious effect on the flexural strength [13].

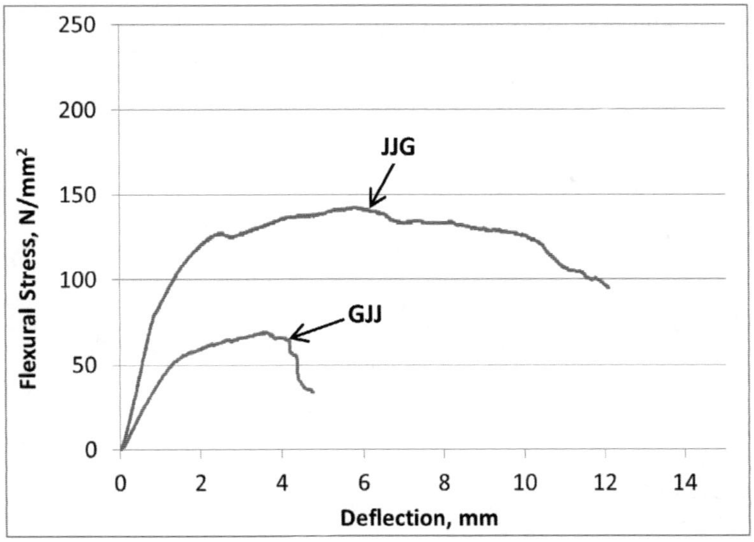

Figure 2. Specimen with stacking sequence GJJ and JJG.

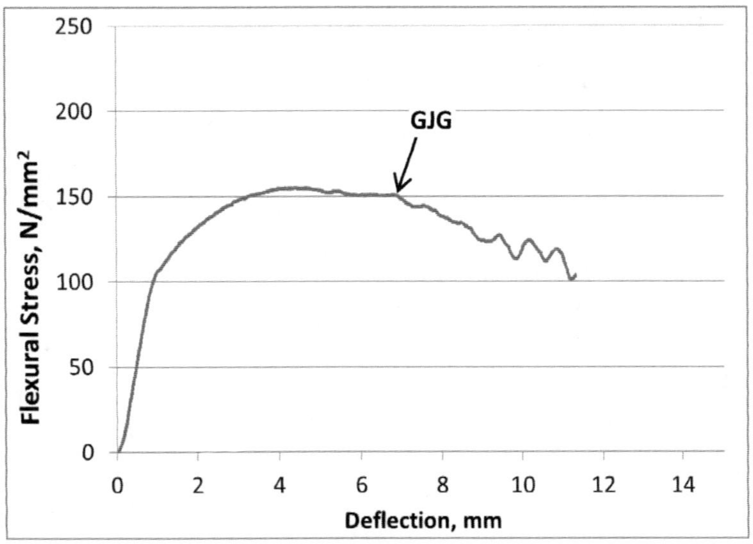

Figure 3. Specimen with stacking sequence GJG.

In Figure 4 and 5, carbon fibre cloth gives significant effect on the flexural strength in JJC specimen of 158 N/mm^2 compared to CJJ specimen of 93 N/mm^2 (Figure 4) and GJC specimen of 209 N/mm^2 compared to CJG specimen of 162 N/mm^2 (Figure 5). In these cases, carbon fibre cloth is in tension surface. Therefore, the flexural strength of the specimen is mainly affected by carbon fibre cloth. In addition, it can be seen in Figure 4, specimen with jute fibre cloth in the tension surface shows a smaller total deflection than that of with other fibre cloth. In the case of GJC and CGJ specimens, both specimens have the highest flexural strength among other specimens with corresponding tension surface of carbon fibre (JJC) and glass fibre (JJG, GJG).

Jute fibre cloth exists in all cases of specimens. In the configuration of jute and glass fibre, there are two options of combination, i.e., JJG and GJG. According to Figure 2 and 3, the flexural strength of both configurations is almost the same. Therefore, based on the goal to reduce the use of the synthetic fibre, JJG specimen is the best choice for building wind turbine skin.

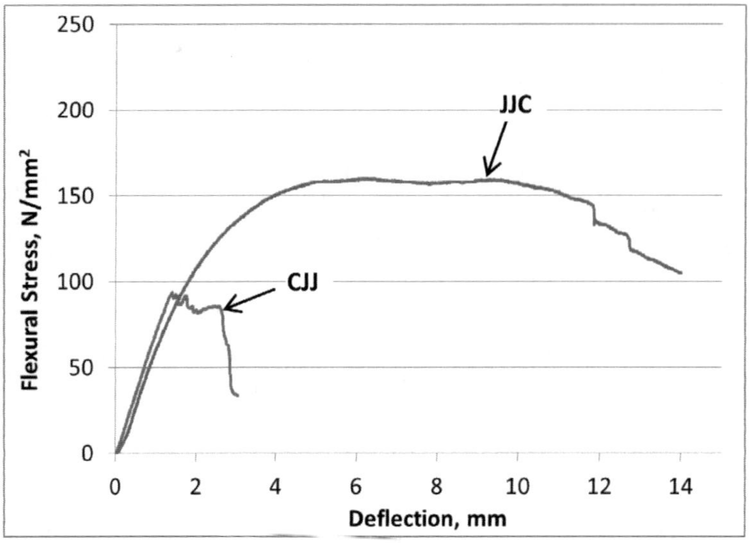

Figure 4. Specimen with stacking sequence CJJ and JJC.

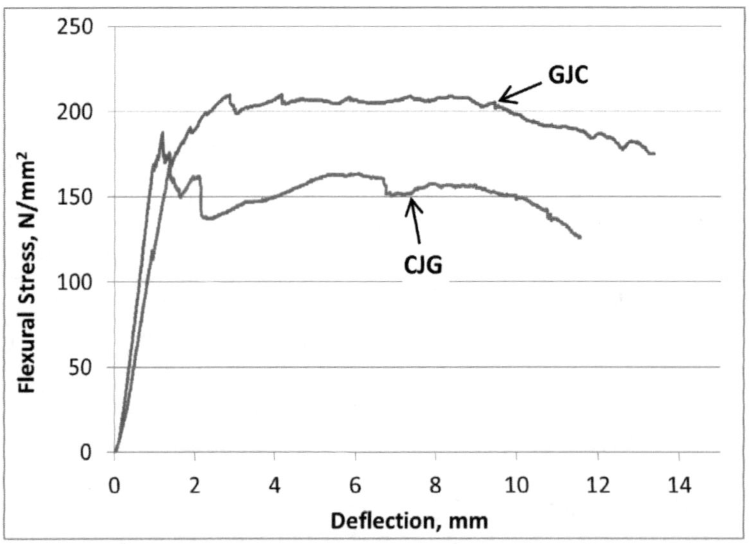

Figure 5. Specimen with stacking sequence CJG and GJC.

Figure 6 - 9 are the examples of fractured specimen of configuration of CJG and GJC. Figure 6 and 8 show the fractured surface of CJG specimen in which the tension surface is glass fibre. It can be seen in figure 8, tension surface shows less damage since the matrix remain tightly bonded to glass fibre [14]. On the other hand, in Figure 7 and 9, in tension surface, part of matrix is broken in the brittle manner since the carbon fibre cloth shows the high stress and large elongation. The total elongation of the specimen GJC is longer than that of CJG specimen.

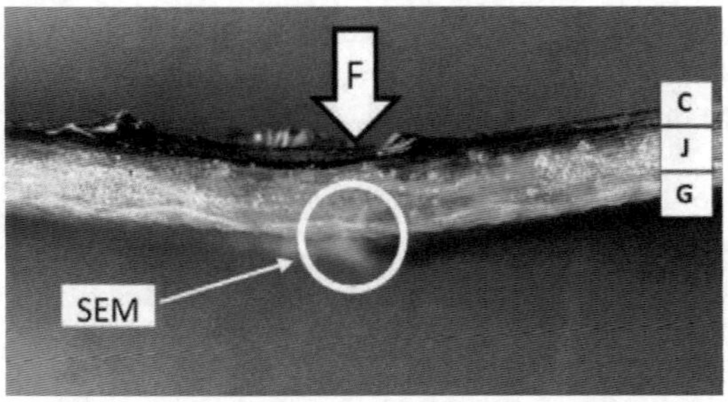

Figure 6. Example of fracture of CJG

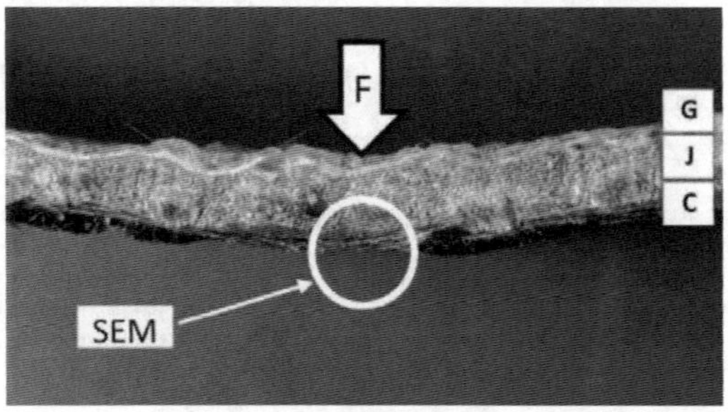

Figure 7. Example of fracture of GJC

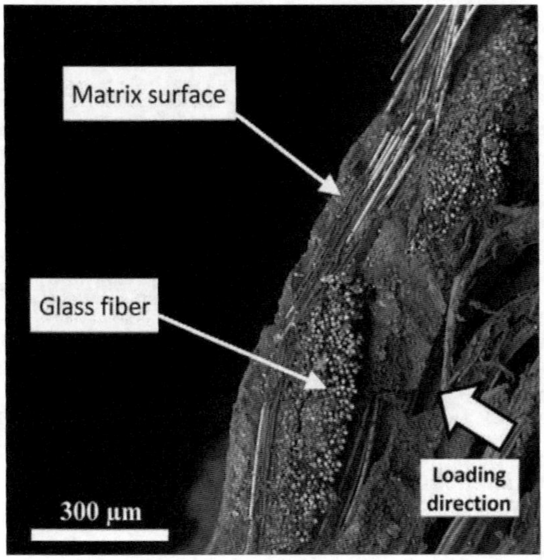

Figure 8. SEM Image of fracture of CJG

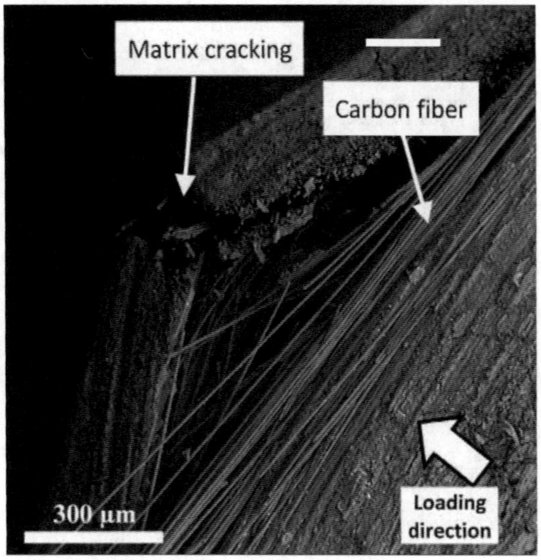

Figure 9. SEM Image of fracture of GJC

4. Conclusion

Hybrid polymer composite reinforced with jute and glass fibre has properties: good strength and stiffness, where the glass fibre is the skin (outer surface) of wind turbine blades. This configuration is also considered economical cost for common application. The result also shows that stacking sequence of the fibres give significant different flexural behaviour regardless the effect of the fibre strength.

References

[1] Kumar A, Dwivedi A, Paliwal V, and Patil PP 2014 Free Vibration Analysis of Al 2024 Wind Turbine Blade Designed for Uttarakhand Region Based on FEA, *Procedia Technology* 14, 336 – 347

[2] Rocha IBCM, Raijmaekers S, Nijssen RPL, Van der Meer FP, and Sluys LJ 2017 Hygrothermal ageing behaviour of a glass/epoxy composite used in wind turbine blades *Composite Structures* 174, 110–122

[3] Aceves CM, Sutcliffe MPF, Ashby MF, Skordos AA, and Román CR 2012 Design methodology for composite structures: A small low air-speed wind turbine blade case study. *Materials and Design* 36: 296–305, Elsevier
[4] Sudarsono 2013. Optimasi Rancangan Kincir Angin Modifikasi Standar Naca 4415 Menggunakan Serat Rami (Boehmeria Nivea) Dengan Core Kayu Sengon Laut (Albizia Falcata)Yang Berkelanjutan. Disertasi, *Program Doktor Ilmu Lingkungan Program Pascasarjana Universitas Diponegoro Semarang*
[5] Liu L and Barlow CY 2017 Wind turbine blade waste in 2050 *Waste Management* 62, 229–240
[6] Dalbehera S and Acharya SK 2016 Effect of cenosphere addition on the mechanical properties of jute-glass fiber hybrid epoxy composites *Journal of Industrial Textiles* Vol. 46(1) 177–188
[7] Johnson S., Kang L., and Akil H. Md. 2016. Mechanical behavior of jute hybrid bio-composites Composites Part B 91, 83-93
[8] Jothibasu S, Mohanamurugan S, Vijay R, Singaravelu DL, Vinod A, and Sanjay MR 2018 Investigation on themechanical behavior of areca sheath fibers/jute fibers/glass fabrics reinforced hybrid composite for light weight applications *Journal of Industrial Textiles* 0 (00) 1–25
[9] Murdani A, Hadi S, and Amrullah US 2017 Flexural properties and vibration behavior of jute/glass/carbon fiber reinforced unsaturated polyester hybrid composites for wind turbine blade *Key Engineering Materials* 748 62-68
[10] Sezgin H and Berkalp OB 2016 The effect of hybridization on significant characteristics of jute/glass and jute/carbon-reinforced composites *Journal of Industrial Textiles* 0 (00) 1–14
[11] Khan MZR, Srivastava SK, and Gupta MK 2018 Tensile and flexural properties of natural fiber reinforced polymer composites: A review *Journal of Reinforced Plastics and Composites* Vol. 37(24) 1435–1455
[12] Haghdan S and Smith GD 2015 Natural fiber reinforced polyester composites: A literature review *Journal of Reinforced Plastics and Composites* Vol. 34(14) 1179–1190
[13] Sushmita, Rajesh M, Hemanth K, Prabhu R, Meloth SC, and Binu KG 2016 Processing and Testing of Hybrid Sandwich Composites for Vibration Damping and Mechanical Properties *Journal of Mechanical Engineering and Automation* 6(5A): 22-27
[14] Rahman AS 2019 Design of cost-effective and efficient fiber-reinforced composite blades for wind turbin *Reinforced Plastics* Volume 63, Number 1

Comparison analysis of abaca fiber/polyester and abaca-e-glass/polyester hybrid composites to impact strength and its application to ballistic

L Banowati[1,*], P Sari[1], and B K Hadi[2]

[1]Department of Aviation Engineering, Faculty of Engineering, Nurtanio University Jl. Pajajaran No. 219, Husein Sastranegara, Bandung, Indonesia

[2]Faculty of Mechanical and Aerospace Engineering, Bandung Institute of Technology, Indonesia Jl. Ganesha 10, Bandung, Indonesia

* E-mail: liesbano@gmail.com

Abstract. The aim of this study consentrates of mechanical properties of abaca fiber and compares with the hybrid fiber composite on using abaca-e-glass fiber as reinforcement for composites. The fibers were arranged to make a bidirectional ±45° fiber arrangements by using polyester as a matrix with hand lay-up method is used to manufacture the composite. Based on the impact test results showed that the abaca-E-glass/polyester has a higher strength than the abaca/polyester specimen. Therefore, the abaca-E-glass/polyester hybrid composite has a better ability to ballistic test.

Keywords: abaca fibre, polyester, hybrid composite, ballistic, e-glass/polyester

1. Introduction

Natural fibres have attracting the interest to engineers, researchers, professionals and scientists all over the world as an alternative reinforcement for fibre reinforced polymer composites, because of its superior properties such as high specific strength, low weight, low cost, fairly good mechanical properties, non-abrasive, eco-friendly and bio-degradable characteristics [1].

Hybridization of natural/glass fiber reinforced polymer composites have been developing to build their applications in the field of engineering and technology [2]. Glass Fibre Reinforced Polymers are mixing with natural fibres to increase Engineering and Technology applications [1]. Abaca, common name for Musa textilis, is a species of banana native to the Philippines, grown widely as well in Borneo and Sumatra [3]. Abaca fiber is known as one of the strongest natural fibers native to the Philippines and widely distributed in the humid tropic's countries including Indonesia. In the last years, natural fibers reinforced composites have received high attention due to their low density, excellent thermal properties, low cost, biodegradability, availability, non-toxicity and absorbing CO2 during their growth [4-5]. It has been reported that Abaca fiber is resistant to rotting and has a high tensile strength, and a specific flexural strength comparable to that of glass fiber [6]. Therefore, abaca fibers can be used as reinforcement in thermoset polymer composites. Thermoset polymer was used extensively in natural composites, whether in the form of ramie/epoxy composite [7], flax/epoxy composites [8-9] and natural fiber/vinyl-ester composites [10]. Therefore, in this paper, we used the polyester thermosetting polymer which was then mixed with alkali treatment abaca and e-glass fibers. The fibers were arranged to make a bidirectional ±45° fiber arrangements. The processing method was using a hand lay-up.The produced specimens were prepared for the impact tests according to ASTM D 6110-04 standards.

Therefore, the abaca/polyester and abaca-E-glass/polyester hybrid composites applied to ballistic respectively. Ballistics aims to see the phenomenon of bullets fired at the target plate so that they can know the characteristics of bullets. Then can be developed bulletproof material that can reduce energy from the shooting process as much as possible [11]. Furthermore, in 1965 Stephanie Kwolek synthesized Kevlar synthesis fibers which could be used as ballistic materials, but Kevlar was expensive.

Research on ballistic testing has been carried out by many researchers in the world, and at this time ballistic testing is still being developed, but very little has been published because the capability of making ballistics is very closely related to a country's defense forces. The following are some studies on ballistics that have been carried out. Economical bullet resistant shirts made from polymer composites reinforced with silicon carbide granules and carbon fiber capable of holding level II bullets by using special magnum 38 revolver guns at a range of 10 meters [12].

Engineering raw materials and cocoon waste have been made bullet proof vests and are able to withstand the rate of bullets from the M16 rifle at a distance of 100 mm [13].Study of the characteristics of composite leather and hemp skin with an epoxy resin matrix as an impact resistant material and ballistic test results of all bullet translucent samples [14].Analysis of the effect of the angle of attack on the ballistic impact on composite matrix resin materials with a reinforcement of water hyacinth (eichhornia crassipes) woven from the test results at a firing angle of 0°, 15°, 30°, 45° at a distance of 5 meters impermeable bullet [15].

Polymer composites as ballistic resistant materials are able to provide better performance compared to metal materials, especially in enhancing the mobility capabilities of military defense personnel [16]. In the present study, based on impact test results, this composite was applied to the manufacture of ballistic test plates. The next step is to perform a ballistic test to determine the penetration of SS2-V4 weapons and tanfoglio guns with a caliber of 5.6 and 9 mm respectively. It is expected to produce bullet resistant composite materials that are easy to make, inexpensive, and have the strength to withstand bullets.

2. Research Methodology

Figure 1. shows the alkali treatment of abaca fibers which is soaked using 5% NaOH solution for 2 hours, then Figure 2. washed the fiber with clean water and dried at room temperature.

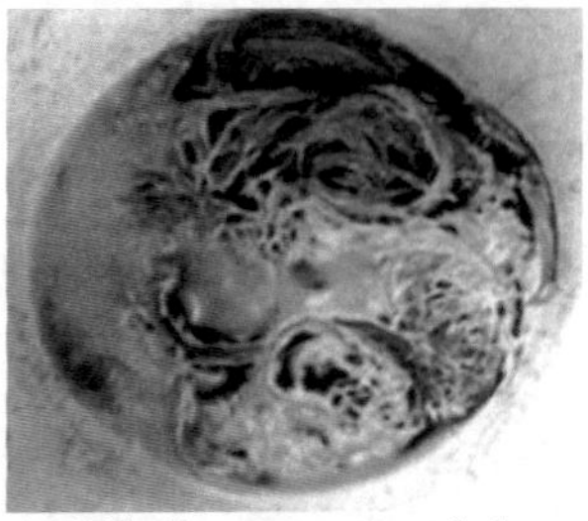

Figure 1. Alkali treatment of abaca fibers

Figure 2. Drying the abaca fibers

The fiber was carefully arranged that eventually will produce with ±45° fiber orientation for both abaca and E-glass fiber respectively and placed into the molding, were then mixed with polyester resins as a matrix using hand lay-up method.

The polyester resins and fiber were measured in order to achieve 50% abaca and abaca-E-glass hybrid fiber volume fractions respectively. Figure 3. shows the final specimens for impact tests according to ASTM D 6110 – 04 standards.

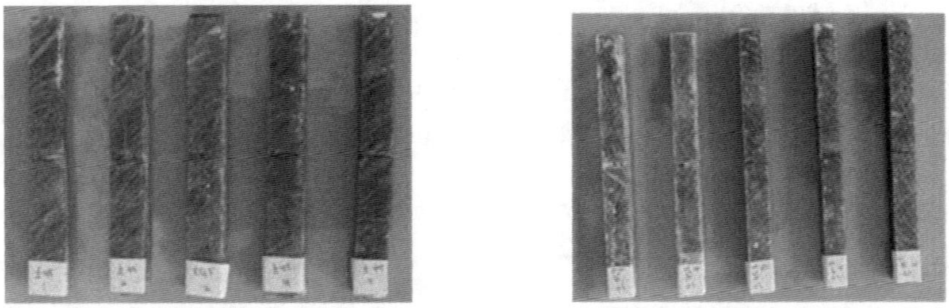

(a) (b)

Figure 3. Tests Specimens: (a) Abaca/polyester Composites, (b) Abaca-E-glass/ polyester Hybrid Composites

Furthermore, making ballistic testing to determine the two types of composites is able to withstand bullets. Figure 4. shows the ballistic testing sketch.

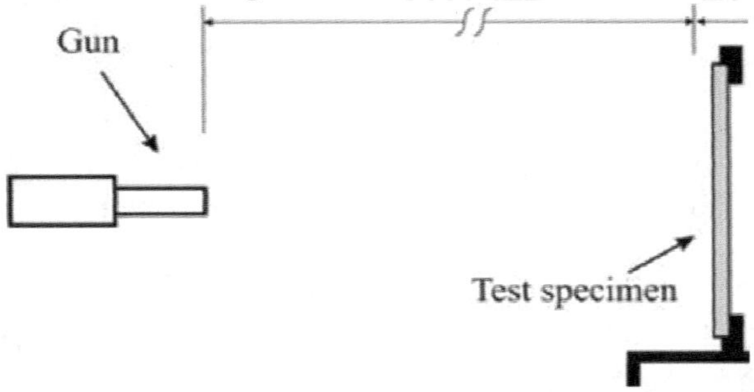

Figure 4. The Ballistic Testing Sketch

The ballistic specimens placed on the support and then the specimen was subjected to shooting at a distance 100 meters and 10 meters by using SS2-v4 weapons and tanfoglio gun respectively as shown in Figure 5.

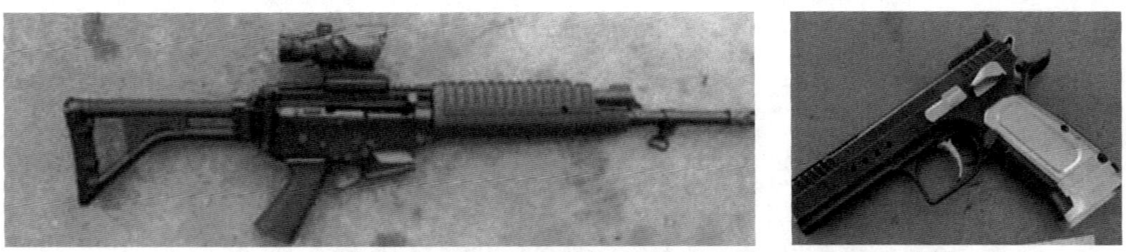

(a) (b)

Figure 5. (a) SS 2 - V4 Weapons, (b) Tanfoglio Guns

3. Results and Discussion

Impact tests was carried out according to ASTM D 6110 – 04. The analysis was done for all fiber composite configurations for both abaca/polyester and abaca-E-glass/polyester hybrid composites is given in Table 1.

Table 1. Present Results on The Impact Strengths of Abaca/polyester and Hybrid Abaca-E-glass/polyester

Composite	Impact Strength (J)
Abaca/polyester	3.36
Hybrid abaca-E-glass/polyester	16.5

Table 1. shows that the value of impact strength for hybrid abaca-E- glass/polyester specimen is higher than abaca/polyester. It is important also to compare the results of the current study with similar impact strength of natural fiber and hybrid natural-glass fiber composites available in the literature. Table 2. and 3. shows the comparison with other natural fiber composite and other natural-glass fiber hybrid composite systems. It shows that the mechanical properties found in the present study are able to compete with other natural fiber and natural-glass fiber composites.

Table 2. Comparison with Other Natural Fiber Composite Systems

Composite	Impact Strength (J)	References
Kenaf/epoxy	3-9	[17]
Kenaf/polyester	4.53	[18]

Table 3. Comparison with Other Natural-Glass Fiber Hybrid Composite Systems

Composite	Impact Strength (J)	References
Nem/kenaf/E-glass/polyester Kenaf/polyester	11.26	[17]
Kenaf–glass fibre reinforced polymer composites	6	[19]

Therefore, it can be concluded that the abaca-E-glass/polyester hybrid composites has a good potential to be developed further for wide application, whether in the automotive industries as well as in the ballistic equipment.

The impact strength of this composite is needed to determine which type of composite is suitable for ballistic tests. The greatest impact strength was selected for ballistic testing. The impact toughness value is influenced by the absorbed energy value and the cross-sectional area of the test specimen, and the addition of the number of E-glass fiber layers causes an increase in the resistance of the composite material to shock loads. Based on the impact test results show the abaca-E-glass/polyester fiber will be used as a bullet proof material.

Composite ballistic testing aims to observe the behaviour and the impact of the two composite specimens on the fired bullets. Figure 6. shows the ballistic specimen test result scheme that the abaca-E-Glass/polyester ballistic specimen test can withstand bullets with a depth of 9 mm by using tanfoglio gun, whereas by using SS2-V4 weapon can be penetrated by bullets.

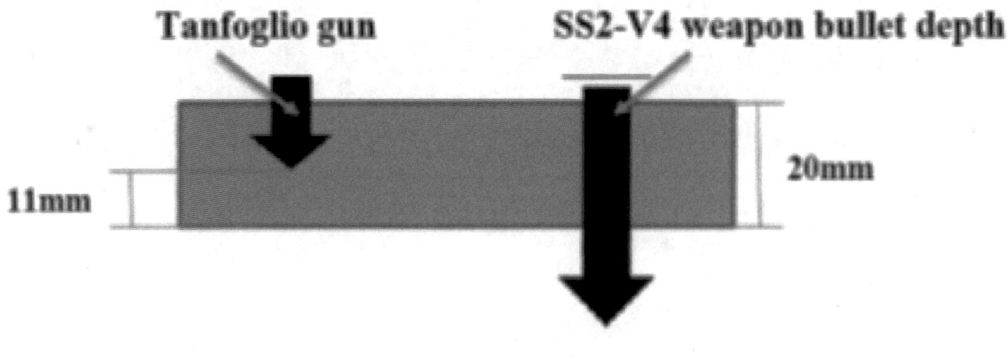

Figure 6. Ballistic Specimen Test Result Scheme

Figure 7. shows the abaca-E-Glass/polyester ballistic test results by using a SS2-V4 weapons.

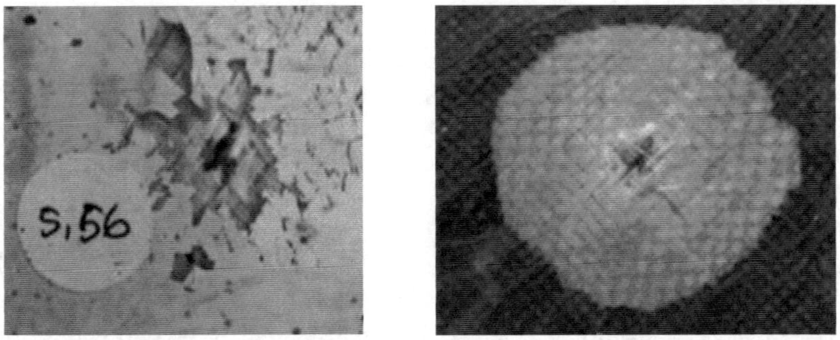

(a) (b)

Figure 7. Abaca-E-glass /polyester Ballistic Test Specimens by Using SS2-V4 Weapons: (a) Upper surface, (b) Lower surface

In the figure above shows that a SS2-V4 weapons using 5.56 x 45 mm caliber with a distance of 100 meters can penetrate the specimen. The bullet damaged the target. Whereas, in Figure 8. shows the abaca-E-Glass/polyester ballistic test results by using a tanfoglio guns.

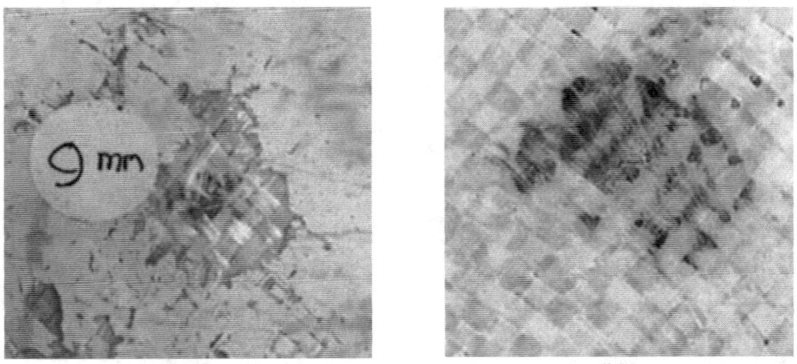

(a) (b)

Figure 8. Abaca-E-Glass/polyester Ballistic Test Specimens by Using a Tanfoglio Gun: (a) Upper surface, (b) Lower surface

In the figure shows that a bullet fired from tanfoglio gun cannot penetrate the specimen. The bullet damaged the target. In Figure 8. shows that the bullet cannot penetrate the specimen and the speed of the 9 mm caliber bullet fired from the tanfoglio gun was lower than the 5,56-caliber bullet fired from the SS2-V4 weapons, because it is held back by fiber. Delamination occurs at the lower surface on the inner layer of the panel due to adhesive failure.

4. Conclusion

The present study concludes that a hybrid abaca-E-glass/polyester composite have a good mechanical property compared with other natural-glass fiber hybrid composite systems. Despite the simple manufacturing processes, the specimens produce good qualities with moderate impact strengths. Therefore, hybrid abaca-E-glass/polyester composite has a good potential to be developed further, for applied in ballistic equipment.

References

[1] Study on Mechanical Properties of Natural - Glass Fibre Reinforced Polymer Hybrid Composites: A Review Author links open overlay panel M.R.SanjayG.R.Arpitha B.Yogesha Another reference https://doi.org/10.1016/j.matpr.2015.07.264 Selection and peer-review under responsibility of the conference committee members of the 4th International conference on Materials Processing and Characterization. Copyright © 2015 Published by Elsevier Ltd.

[2] Studies on Natural/Glass Fiber Reinforced Polymer Hybrid Composites: An Evolution Author links open overlay panel MR Sanjay B Yogesha https://doi.org/10.1016/j.matpr.2017.02.151
[3] Reynaldo M. P. Natural Organic FiberMeshes as Reinforcements in Cement Mortar Matrix, Proceedings of The 2011 IAJC-ASEE International Conference.
[4] Faruk, O., Bledzky, A. K., Fink, H. P., and Sain, M. (2012). "Biocomposites reinforced with natural fibers: 2000-2010." ProgPolymSci, Vol. 37(11), 1552-1596.
[5] Saheb, D. N., and Jog, J. P. (1999). "Natural fiber polymer composites;areview."AdvPolym Technol, Vol. 18(4), 351-363.
[6] Knothe, J., Rebstock, K., and Schloesser, T. (2000). "Natural fiber reinforced plastics in Automotive exterior applications." In: The 3rd international wood and natural fiber composites "symposium. Kassel, Germany, 1-12.
[7] H. Kishi, A. Fujita, Env Eng and Management J. Vol 7 No 5 (Oct 2008), pp. 517-523.
[8] H. Bos, The Potential of Flax Fibers as Reinforcement for Composite Materials, (Dissertation, Technische Universiteit Eindhoven, 2004), ISBN: 90-386-3005-0.
[9] E. Sparning, Mechanical Properties of Flax Fibers and Their Composites, (Licentiate Thesis,
[10] Lulea University of Technology, 2006), ISSN: 1402-1757.A. Alhuthali, I.M. Low: J Mater Sci Vol 48 (2013), pp. 6331-6340.
[11] Prasetyo Adi Prabowo, Ismoyo Haryanto 920140. "Analisa Kecepatan Sisa Proyektil Berhidung Tumpul Yang Ditembakkan Pada Baja Weldox 460 E", Jurnal Teknik Mesin, http://http/ejournal-s1.undip.ac.id/index.php/jtm, Universitas Diponegoro, Indonesia
[12] Nindhia, Tjokorda Gde Tirta dan Daud Simon Anakottapary. 2011. Baju Tahan Peluru Ekonomis Terbuat Dari Komposit Polimer Diperkuat Butiran Silikon Karbida dan Serat Karbon. Universitas Udayana .
[13] Moeliono M dan Siregar Y. 2012. Rekayasa Bahan Baku Sutera Dan Limbah Kokon Untuk Rompi Tahan Peluru (The Engineering Of Silk Material And Cocoon Waste For Ballistic Proof. Jurnal Riset Industri, Vol. 06, Balai Besar Tekstil, Bandung.
[14] Budianto Aris dan Heru Santoso. 2014. Studi Karakteristik Kulit Kras Dan Rami Dengan Matrik Resin Epoksi Sebgai Bahan Tahan Impak. Universitas Gadjah Mada Yogyakarta
[15] Latif Abdul dan Helmy Purwanto. 2017. Analisa Pengaruh Sudut Tembak (Angle Attack) Terhadap Dampak Balistik Pada Material Komposit Mtarik Resin Dengan Penguat Kulit Eceng Gondok (eichhornia Crassipes) Yang Dianyam. Vol.13, No.2, Hal.42-48. Universitas Wahid Hasyim.
[16] Mardiyanti. 2018. Komposit Polimer Sebgai Material Tahan Balistik. Vol.1, No.1. Institut Teknologi Bandung.
[17] Alavudeen, A.; Rajini, N.; Karthikeyan,S.;Thiruchitrambalam, M.; Venkateshwaren, N. Mechanical properties of banana/kenaf fiber-reinforced hybrid polyester composites: Effect of woven fabric and random orientation. Mater. Des. 2015, 66, 246–257
[18] Wan Busu, W.N.; Anuar, H.; Ahmad, S.H.; Rasid, R.; Jamal, N.A. The mechanical and physical properties of thermoplastic natural rubber hybrid composites reinforced with Hibiscus cannabinus, L and short glass fiber. Polym. Plast. Technol. Eng. 2010, 49, 1315–1322.
[19] Ramesh, M.; Nijanthan, S. Mechanical property analysis of kenaf–glass fibre reinforced polymer composites using finite element analysis. Bull. Mater. Sci. 2016, 39, 147–1

Design, development and testing of Universiti Malaysia Sabah (UMS) eco-solar dryer

B C Lim[1], A M Wash[1], S S Arifin[1], N S Kassim[1], A S T Tan[1], A Abdullah[2, a], W A A Zal[3], and J Janaun[1, b]

[1] Industrial Dryer and Palm Oil Recovery Research Group, Chemical Engineering Program, Universiti Malaysia Sabah, Jalan UMS, 88400 Kota Kinabalu, Sabah, Malaysia.

[2] Chancellery Office, Universiti Malaysia Kelantan, 16300 Bachok, Kelantan, MALAYSIA

[3] Institute for Poverty Research and Management (InsPeK), Universiti Malaysia Kelantan, 16300 Bachok, Kelantan, MALAYSIA

* Corresponding Authors: [a]arham.a@umk.edu.my; [b]jidon@ums.edu.my

Abstract. Utilization of solar drying for food preservation would be a step towards sustainable food processing. Even today, most of the agricultural and marine products are dried under open sun-drying. However, this method is associated with several drawbacks, which then leads to the use of solar dryer. Universiti Malaysia Sabah (UMS) has developed a solar drying system with loading capacity of 50kg to enhance the drying process of various food products. Design parameters particularly focused in this project includes the air movement pattern in drying chamber, heating system employed, air circulation ventilation and the airflow configuration in solar heat collector. Results revealed that solar dried products attained desired moisture content in shorter drying time. Temperature profile in updraft and downdraft drying mode indicated that no temperature fluctuation occurred when switching in between. Maximum temperature achieved under complete solar mode were 65°C and 83°C for single-pass and multi-pass solar heat collector, respectively. Under solar-hybrid mode, higher temperature was attained under intermittent ventilation mode. It can be concluded that UMS Eco-Solar Dryer is an environmentally and economically efficient drying system for the food processing industry.

Keywords: Downdraft, updraft, hybrid solar dryer, solar heat collector, efficiency

1. Introduction

Utilization of solar drying for food preservation would be a step towards sustainable food processing. Even today, most of the agricultural and marine products are dried under open sun drying. Due to its free and renewable energy source from the sun, it is still an extensively used method to preserve perishable products in most tropical and subtropical countries [1]. However, this method is associated with several drawbacks in term of hygienic, drying efficiency and quality characteristics of end products. Hence, the revolution of today's technology has come up with solar dryer as a solution to overcome such problem encountered during open sun drying. To ensure maximum performance is achieved during the drying process, continuous efforts in optimizing the performance of solar dryer are essential.

There are several drawbacks encountered when employing solar drying in food drying, and such include uncontrollable drying condition in the dryer due to the periodic characteristics of solar radiation. Although this could be easily overcome by installing electrical or photovoltaic powered auxiliary devices, higher electrical energy consumption and expenses are required for the operation. This could be a limitation, particularly for the application of small-scale farmers in the rural area. Hence, different approaches must be taken to enhance the performance of a solar dryer.

In a natural convection (passive) solar dryer, air movement in the dryer is voluntarily circulated by natural buoyancy force action [2]. Since no additional auxiliary system required to operate, passive solar dryers are extensively utilized in smallholders and rural communities. In an updraft drying system, low density heated air flowing into the drying chamber will flow in an upright motion across each drying tray for drying purposes. Without adequate ventilation, drying air will eventually lose its momentum to progressively rise up while carrying moisture removed from the commodities and consequently reduced the drying efficiency. On the other hand, using downdraft solar dryer will result in less efficient drying when heated air begins gaining less moisture from the commodities or when the moisture level is low. These drawbacks encountered in natural convection can be overcome using a forced convection solar dryer. In this project, Universiti Malaysia Sabah has developed a solar dryer classified as an indirect tray-type cabinet solar dryer, namely UMS Eco-Solar Dryer to overcome these limitations in drying food products.

2. Methodology

2.1. Design consideration

Many types of solar dryer have been introduced for domestic and industrial application in food dryings such as cabinet dryers, greenhouse dryers and chamber-type dryers. Among these dryers, solar dryer with multiple tray arrangement is most extensively used due to its capability to dry the high volume of products without complex design. The drying process in a tray-type dryer takes place by forcing heated drying air stream to flow across each level of drying tray either from the uppermost or lowermost level depending on the location of drying air inlet. In a tray-type dryer, an important parameter that defines the drying performance is the uniformity of airflow distribution across the food products [3].

Hence, in this project, a solar dryer system that offers a flexible feature in switching between updraft and downdraft drying mode is proposed. With this, extra manpower to manually turn over the food products can be avoided, and the basic problem of updraft and downdraft drying system can be resolved. In an attempt to ensure adequate air circulation in the dryer, wind ventilated system is employed. Auxiliary heating source is installed to ensure continuous heating throughout unfavorable weather and during the off-sunshine period.

2.1.1. Solar heat collector

The solar heat collector was designed to achieve drying temperature T_p, of 50 °C to dry 50 kg of palm fruitlet with the initial moisture content of 35% to 15% within 16 hours drying time. To enhance heat absorptivity of the collector, black painted aluminum cans were used for the construction of absorber. The frame was made up of plywood due to its high insulation performance (k = 0.02675 W/mK). Furthermore, heat intensity was enhanced by installing acrylic cover with 5 cm air gap from the absorber to produce the greenhouse effect in the collector.

Two solar heat collectors that provide different airflow configuration were proposed – single-pass and multi-pass. Flow configuration in the single pass solar heat collector is shown in Figure 1. This design will be used for the commissioning of solar dryer under non-loaded and loaded condition. The latter design offers longer drying air residence time in the air channel to attain higher drying air temperature than that of single-pass solar heat collector. This approach was taken to ensure the highest possible drying temperature attained in the drying system with electrical assisted devices. Drying condition of the system can then be regulated through the air ventilation system. The area of solar heat collector was determined from equation (1) [4].

$$Q = A_c((I_t\delta_\alpha - U_L(T_p - T_a)F_R) \quad (1)$$

Where A_c is the area of solar heat collector, I_t is the solar radiation, δ_α is the transmissivity, U_L is the heat coefficient, F_R and T_a are the heat removed factor and ambient temperature, respectively.

2.1.2. Drying chamber

The drying chamber was constructed to have a closed system with an airtight environment which helps to build up pressure inside the chamber. The enclosed chamber was covered with glass wool fiber for insulation purpose. The updraft and downdraft system windows are hinged on the drying chamber inner wall to enable easy opening and closing. Figure 2 demonstrates the airflow configuration for different drying mode. Drying area in the chamber was fixed at 0.64 m^2 with five drying trays arranged in 0.1m interval space. Minimum height drying tray was determined from equation (2) as follow:

$$H_t = \frac{H}{n} \quad (2)$$

where H_t is the minimum height of drying tray, H is the height of 50 kg palm fruit and n is the number of trays.

Six infrared heating bulbs were installed at the top of the drying chamber as an auxiliary heating source. The heating system was connected to a control panel that automatically regulates drying condition in the chamber according to the value set in the differential temperature controller.

The calculation of energy required to heat drying air in the chamber up to 80 °C is expressed in equation (3) [5].

$$Q = m \times C \times \Delta T \quad (3)$$

where Q is the heat energy required, m is the mass of air in the drying chamber, C is the heat capacity of air at 328 K, and ΔT is the temperature change from initial temperature.

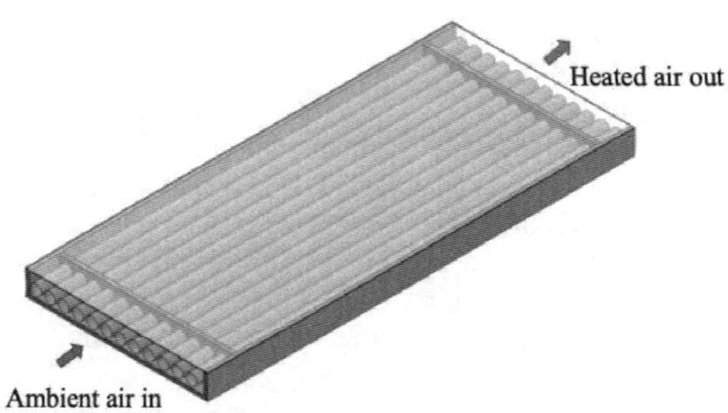

Figure 1. Airflow configuration in single-pass solar heat collector

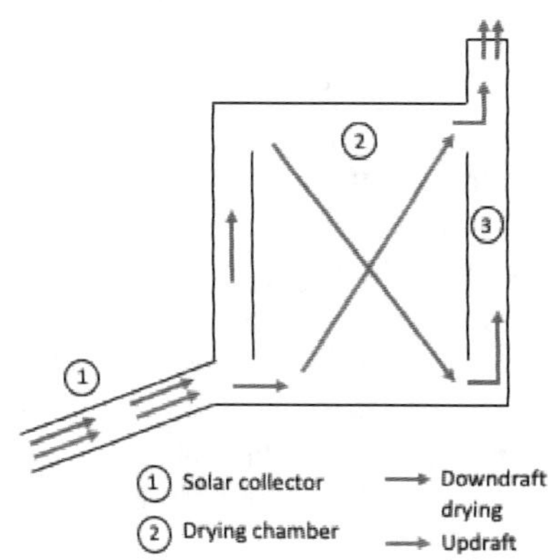

Figure 2. Air flow configuration in updraft and downdraft drying mode.

2.1.3. Chimney

A wind ventilator was installed at the top of the chimney to allow forced convection throughout the process. In this design, the controlling factor is the height of the chimney and the draft produced was determined based on equation (4) [6].

$$D_i = Hg(\rho_a - \rho_e) \tag{4}$$

where $D_i$is the pressure difference, H is the height of the chimney, ρ_a and ρ_e is the density of ambient air and exit air, respectively, and g is the gravitational force.

2.2. Experimental set-up

A series of experimental tests were performed to study the dryer performance under no-load and loaded condition. Another test with a solar dryer integrated with a multiple-pass solar heat collector was performed to determine its performance in term of drying temperature attained at the inlet of the drying chamber.

Variables such as solar radiation, ambient temperature, drying chamber temperature, relative humidity and drying chamber airflow were measured using Zonen CMP3 pyranometer, GLOBE Hygro-Thermometer, thermocouple and anemometer, respectively. The initial moisture content of drying products – Senduduk and Jasmine flower were determined by Sartorius MA35 Moisture Analyzer Test. The weight change of products was recorded every 2 hours interval.

3. Results and Discussions

3.1. Commissioning without load

The first experiment was conducted without load to verify the performance of the solar drying system proposed. Tests were conducted under complete solar mode and hybrid mode for comparison. Under complete solar mode, the variation of average drying air temperature in the chamber for both updraft (UT) and downdraft (DT) drying mode with average airflow of 0.2 m/s and 0.13 m/s, respectively are presented in Figure 3. Based on the temperature profile exhibited, updraft drying mode offers a higher drying temperature.

This is due to higher solar radiation was determined when performing updraft drying. However, minimal drying air temperature difference of 1±0.9 °C to 4±2.4 °C was observed between updraft and downdraft drying modes. Therefore, it could be deduced that both drying modes exhibits similar temperature profile, and that no temperature fluctuation would occur while switching between the two different drying modes.

Temperature variation across the drying trays under updraft and downdraft drying mode is presented in Figure 4. In both drying modes, the uppermost level drying tray received higher drying temperature than that at the lowest level drying tray. In updraft mode, hot air with low density rises up, hence resulting in a higher temperature at the top. Whereas in the downdraft mode, additional energy and longer air residence time are required for low-density hot air to reach the bottom level, resulting higher temperature observed at the uppermost tray located near the drying chamber inlet.

From this observation, it can be inferred that the solar drying system proposed could improve the air circulation in the drying chamber. When employing updraft mode during a drying process, the solar dryer will eventually encounter inadequate ventilation as hot air gains moisture from the products and loses its momentum to progressively rise up. At this point, switching to downdraft mode allows hot air from the inlet to compensate low temperature at the top. The minimum and maximum drying temperature observed from both drying modes were 31 °C and 65 °C, respectively.

In hybrid mode, the dryer was operated under two ventilation mode – continuous ventilation and intermittent ventilation. The temperature profile in the chamber was observed with the temperature set at 55 °C. Figure 5 shows temperature variation across the drying trays with continuous ventilation. Based on the graph plotted, the desired temperature was observed only at the uppermost drying tray between 10:00 AM to 12:00 PM under this condition. This was due to the heating source was installed right above the drying tray, and the temperature sensor was placed on the middle tray, hence, resulting in insufficient heating in the remaining trays. In contrast, the desired temperature was achieved at top and middle tray before 10:00 AM under intermittent ventilation mode, as shown in Figure 6.

Based on the data recorded, it can be deduced that the intermittent ventilation system offers shorter heating to achieve the desired temperature. As a comparison, Noh *et al.* suggested using intermittent ventilation to reduce the energy consumption of the drying process [7]. Based on their findings, the intermittent ventilation system offers a higher temperature in the drying chamber compared to passive ventilation and continuous active ventilation mode. However, this results in poor temperature distribution uniformity in the drying chamber than that of continuous ventilation.

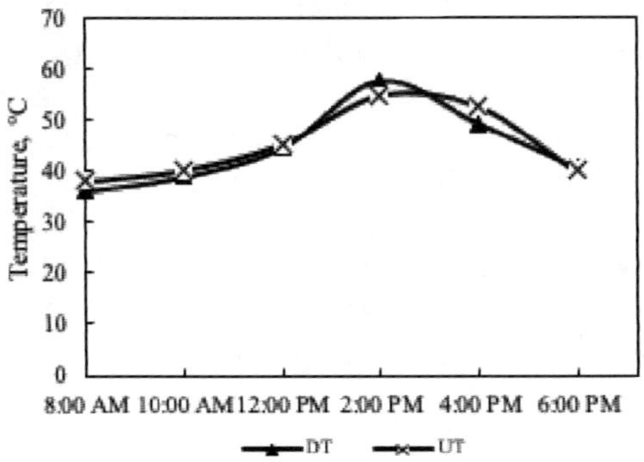

Figure 3. Temperature profile in drying chamber under updraft and downdraft drying modes.

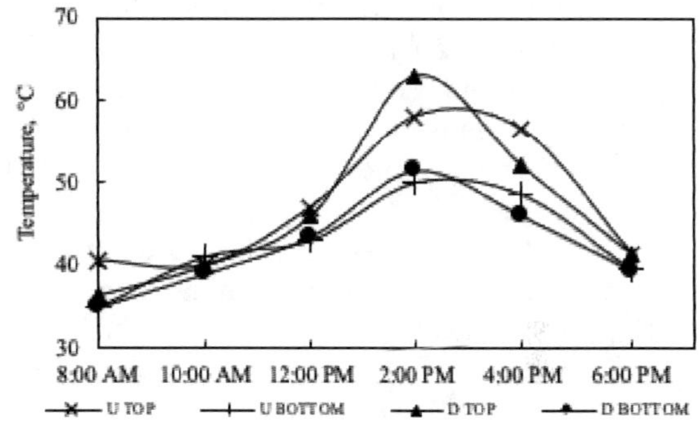

Figure 4. Temperature profile across drying trays under updraft and downdraft drying modes.

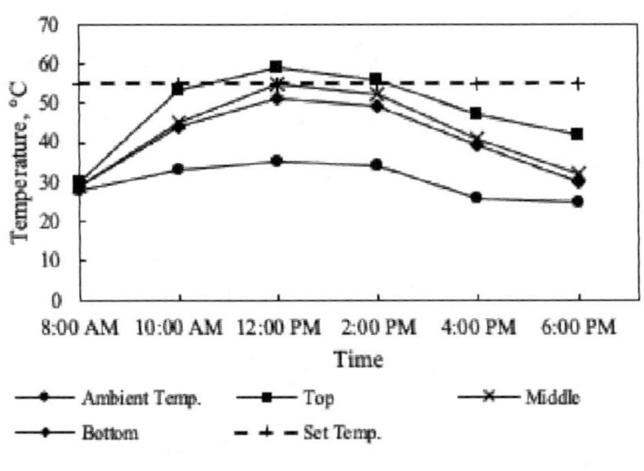

Figure 5. Temperature profile across drying trays at the temperature set at 55 °C under continuous ventilation mode.

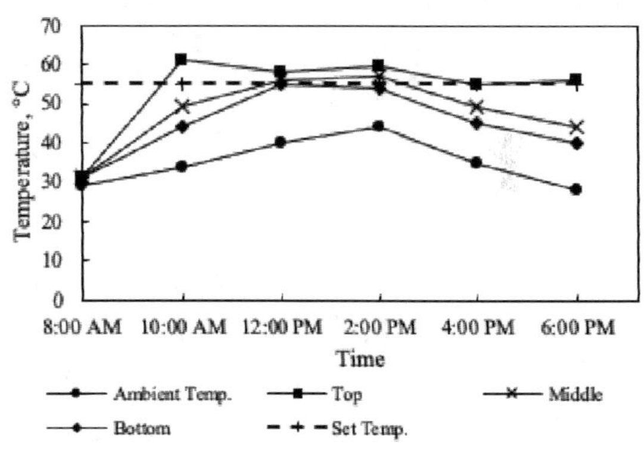

Figure 6. Temperature profile across drying trays at the temperature set at 55 °C under intermittent ventilation mode.

3.2. Commissioning with load

Drying products used to verify the performance of hybrid dryer developed were Senduduk and Jasmine flowers with the initial moisture content of 64.27 % and 74.13 %, respectively. Weight loss of about 75 % to 77% was taken as the reference line for the drying process under drying temperature of 40 °C. Figure 7 shows the weight loss of products against drying time in a solar dryer. Based on the plotted results, drying time required to achieve desired moisture loss in solar dryer are 26 hours and 32 hours for Senduduk and Jasmine, respectively. A longer drying time of more than 56 hours was required under open sun drying.

3.3. Test with a multiple-pass solar heat collector

Experiment tests to study the performance of multiple-pass solar heat collector were conducted under wind ventilated condition with airflow ranged from 0.2 to 2 m/s. Variables measured were ambient temperature, collector outlet temperature, drying chamber temperature, airflow rate and solar radiation. Performance of the collector was evaluated based on the collector efficiency calculated from these parameters. The test was performed from 9:30 AM to 4:30 PM. Figure 8 shows the temperature variation taken on 20 July 2020 that exhibits the highest outlet temperature during the one-month experimental period.

Throughout the day, maximum outlet temperature attained was 83 °C at 1:00 PM with the solar intensity of 209.8 W/m^2. Whereas the minimum temperature of 36 °C was recorded at 9:00 AM with low solar intensity. Comparing to the maximum temperature recorded from previous experiments that used single-pass solar heat collector, this result indicated that thermal enhancement was achieved by increasing the length of air channel in solar heat collector. This result is in accordance with Yeh *et al.* findings [8].

Data that shows the collector efficiency of multi-pass solar heat collector against air velocity are shown in Figure 9. Based on the graph plotted, lower efficiency was observed at low air velocity. In contrast, the efficiency increased at high air velocity, which enhanced the heat transfer rate in the solar heat collector.

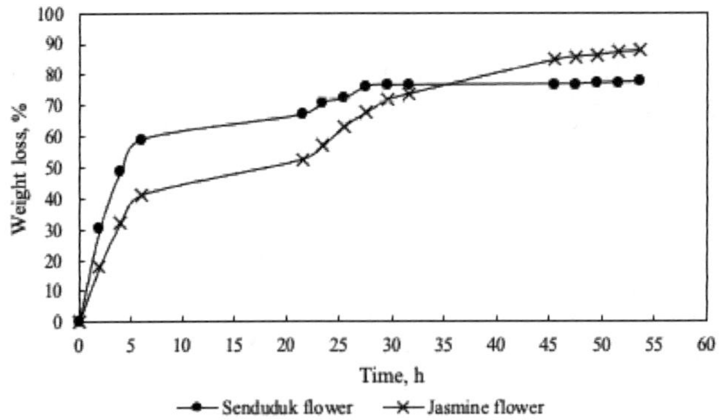

Figure 7. Weight loss of Senduduk and Jasmine flowers throughout the drying period.

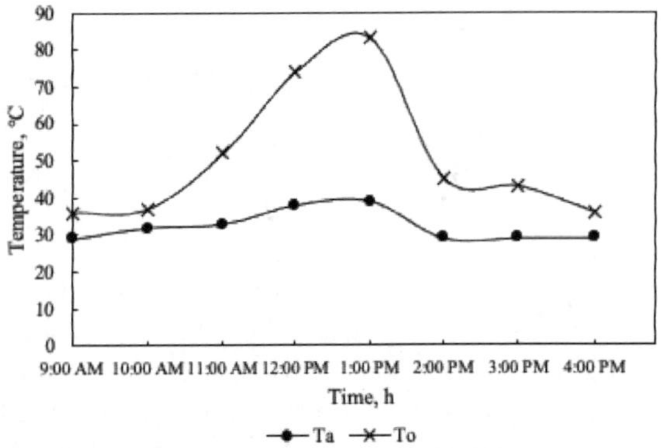

Figure 8. Temperature attained at the outlet of multiple-pass solar heat collector throughout the experimental period.

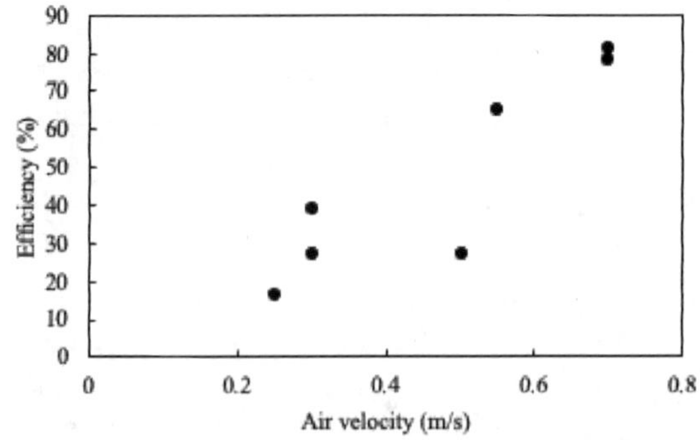

Figure 9. Collector efficiency against air velocity.

4. Conclusion

An indirect type of solar dryer with 50 kg drying capacity was developed in UMS. Uniformity of temperature distribution across the multiple tray arrangement in the chamber was enhanced by employing flexibility features to switch the drying system in between updraft and downdraft drying modes. A maximum collector outlet temperature of 65 °C was observed by using a single-pass solar heat collector. The total drying time of 26 hours and 32 hours were required to achieve 78 % weight loss in Senduduk and Jasmine flowers, respectively under this condition. By integrating the multiple-pass solar dryer, the maximum temperature at the collector outlet was increased to 83 °C, indicating thermal enhancement.

Acknowledgments

This work was primarily supported by Ministry of Higher Education Malaysia National Blue Ocean Strategy (NBOS) (AM00091) and partly by Ministry of Finance Malaysia Social Enterprise Project a collaborative project between Universiti Malaysia Kelantan and Universiti Malaysia Sabah (R/MOF/B0108/01759A/001/2020/00722).

References

[1] W. Szulmayer 1971, From sun-drying to solar dehydration: I - Methods and equipment, *Food Technology in Australia*, vol. **23**, pp. 440–443.

[2] M. A. Basunia and T. Abe 2001, Thin-layer solar drying characteristics of rough rice under natural convection, *Journal of Food Engineering*, vol. **47**, no. 4, pp. 295–301.

[3] S. Misha, S. Mat, M. H. Ruslan, K. Sopian, and E. Salleh 2013, Review on the application of a tray dryer system for agricultural products, *World Applied Sciences Journal*, vol. **22**, no. 3, pp. 424–433.

[4] S. A. ManzoorAhmad, A. Munir, and A. Wajid 2015, Design, development and performance evaluation of a small-scale solar assisted paddy dryer for on farm processing, *Journal of Food Processing & Technology*, vol. **6**, no. 9, pp. 1–6.

[5] K. Li and N. Hong 2018, Dynamic heat load calculation of a bridge anti-icing system, *Applied Thermal Engineering*, pp. 198–203.

[6] A. C. Jambhulkar, V. B. Pawar, S. B. Pawar, A. S. Dharwadkar, P. S. Pandure, and S. P. Gadewar 2017, Solar Drying Techniques And Performance Analysis: A Review, *IOSR Journal of Mechanical and Civil Engineering (IOSR-JMCE)*, pp. 35–39.

[7] A. M. Noh, S. Mat, and M. H. Ruslan 2018, CFD simulation of temperature and air flow distribution inside industrial scale solar dryer, *Journal of Advanced Research in Fluid Mechanics and Thermal Sciences*, vol. **45**, no. 1, pp. 156–164.

[8] H. M. Yeh, C. D. Ho, and C. H. Chen 1999, The effect of collector aspect ratio on the collector efficiency of sheet-and-tube solar fluid heaters, *Tamkang Journal of Science and Engineering*, vol. **2**, no. 2, pp. 61–68.

Inorganic flame retardants' efficacy of Aluminum Hydroxide, Magnesium Hydroxide in the combustion rate of intermediate Calamagrostis from Ecuador' Moorlands

M Córdova-Suárez[1,*], E M Barreno-Avila[2], D S Pozo-Álvarez[3], and J C Córdova-Suárez[3]

[1] School of Engineering, Universidad Nacional de Chimborazo, Av. Antonio Jose de Sucre, Riobamba-Ecuador

[2] Faculty of Science and Engineering, WA School of Mines: Minerals, Energy and Chemical Engineering, Kent St, Bentley WA 6102, Australia

[3]School of Food Science and Engineering, University, Universidad Técnica de Ambato, Los Chasquis Avenue and Río Payamino, 180201, Ecuador

E-mail: hesconsultores@hotmail.com

Abstract. In this research, intermediate calamagrostis vegetal fiber (FIC) combined samples were carried out in combination with polyester resin (RP), Aluminum Hydroxide (HA) and Magnesium Hydroxide (MH). The aim of this study is to determine inorganic flame-retardant effectiveness in the intermediate calamagrostis' combustion rate reduction from Ecuador's moorlands and possible forest fires' control of this endemic plant species. The flammability index was obtained following the ISO 3795 standard; In addition, the contamination degree: Carbon Monoxide (CO) was determined, following the standard EPA 40 CFR, method 60 Annex A. The composite material was evaluated in a horizontal flammability chamber. Furthermore, the experimental design considered seven tests, one control and two flame retardants' types. Aluminium Hydroxide and Magnesium Hydroxide accounted for three different concentrations at 3%, 6% and 9% percent each one. Thus, these flame retardants' effectiveness was greater for Aluminium Hydroxides at all the proposed concentrations, while for Magnesium Hydroxide; its effectiveness was positive for a concentration of 9% only. The study determines that the best combination of materials are: 70% FIC, 21% PR, 9% HA, resulting in a combustion rate of 190.98 mm / min and self-extinguishing capacity. Finally, an average decrease of 50% in the generation of CO and the amount of smoke was determined.

1. Introduction

When a fire occurs every second counts and to prevent it, or at least to slow its spread, many woods, plastics, fabrics and electronic devices are protected with fire retardants that reduce the flammability of the product by eliminating the chain reaction of a fire [1]. Depending on their nature, these compounds can act physically (by cooling, formation of a protective layer or dilution of the combustible material) or chemically (reaction in condensed or gaseous phase). Hence, they can interfere with several phenomena involved in the combustion of a polymer (heating, ignition, pyrolysis, thermal degradation, flame propagation) [2]

There are several fire retardants, inorganic ones based on halogenated compounds, phosphorus, antimony trioxide, and nitrogen; and intumescent, among others that act in different ways [3].

Halogenated Compound. The most commonly used fire retardants are those based on halogens; however, their application has been restricted due to the environmental problems they produce [4]. Halogens, mainly chlorine and bromine, are found in the chemical structure of many of the most effective and widely used flame retardant additives, such as brominated compounds [5]. These are developed by removing the H+ and OH - radicals in the gas phase of the flame. This slows down the burning process. Thus, reducing the production of heat and the generation of flammable gases [6].
On the other hand, the production of harmful and corrosive smoke and toxic gases during combustion, as well as environmental regulations and considerations, have promoted the use and development of halogen-free flame retardants [7].

Inorganic compounds (aluminum hydroxide, magnesium hydroxide). They obstruct the flame through physical processes such as the release of water or non-flammable gases that dilute those who fan the fire, cooling and producing a non-flammable and resistant layer on the surface of the material [8]. They decompose through an endothermic reaction, releasing water vapour which removes heat and reduces the formation of combustible gases, together they form a protective layer on the surface of the material which intervenes by slowing down the spread of thermal degradation [1,8]

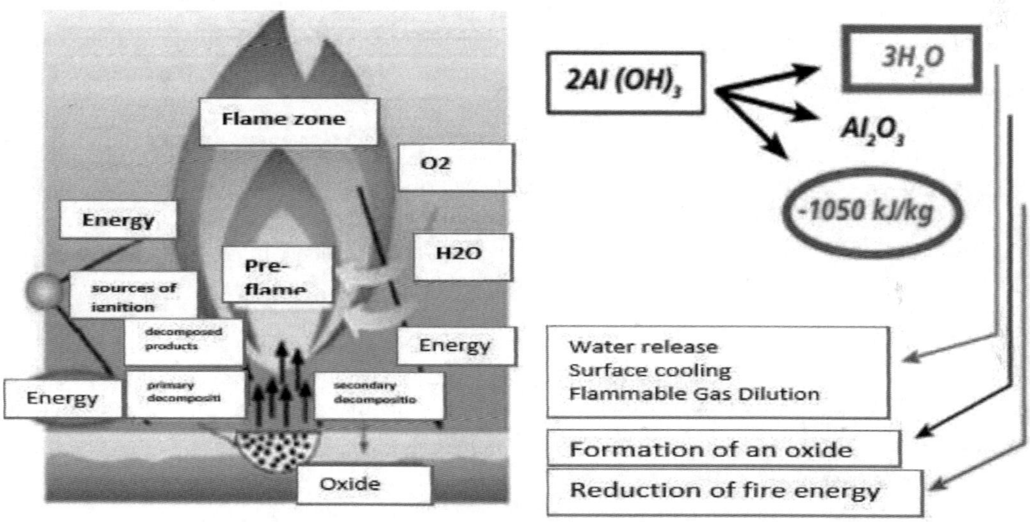

Figure 1. Processes involved in inorganic flame retardant [3]

Aluminium hydroxide and magnesium hydroxide flame retardants are considered non-toxic and therefore environmentally friendly, in addition to producing low smoke emissions [9]. Aluminum and magnesium hydroxides (M(OH)x) decompose endothermically and this begins at temperatures close to 200 °C and 300 °C respectively, releasing water in the form of vapour and thanks to this phenomenon this form of metal oxide (M2Ox) [7].

Phosphorous Base. The phosphorus base plays an efficient role in the solid phase of the materials being burned. When heated, the phosphorus reacts to give a polymeric form to phosphoric acid, forming a crystalline layer that inhibits the process of pyrolysis and release of flammable gases, which are needed to maintain the flame. By this mechanism the amount of fuel is significantly compressed because more carbonized zone is formed than fuel gas [6].

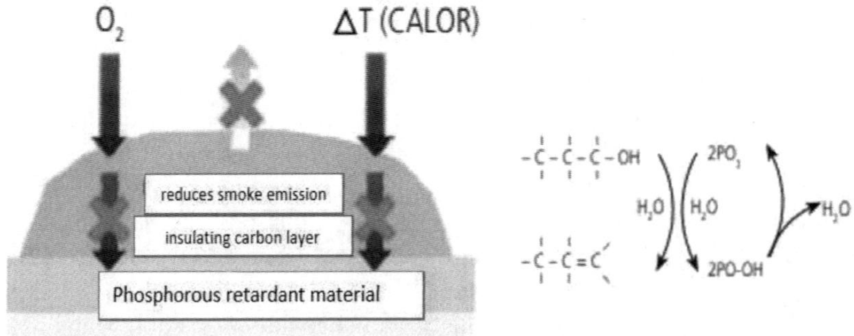

Figure 2. Processes involved in Phosphorous-based fire retardants [3]

Intumescent Coatings. Intumescent coatings are used to protect materials such as wood and plastics from flame and prevent incineration. They are also used to safeguard steel and other materials, retarding flame and structural damage during fire. They are produced from a mixture of products that are used on the surface like a paint. They are intended to disperse to form an insulating and fire-resistant layer covering the material exposed to the heat [6].

2. Materials and Methods

2.1. Raw material collection.

- For Calamagrostis Intermedia sample; the land was identified, where the sampling was done, and this was collected in the province of Chimborazo with the coordinates 1°46'23.3 "S 78°44'53.3 "W
- Aluminium hydroxide and magnesium hydroxide were collected from the laboratories of the Faculty of Food Science and Engineering and Biotechnology of the Technical University of Ambato.
- The polyester resin was obtained from the Master Fiber factory located in the city of Ambato with coordinates 1°12'43.5 "S 78°35'42.5 "W

2.2. Raw material cleaning.

- The collected sample of intermediate calamagrostis was exposed to the sun for complete drying of the fibre.
- The sample was cut to a length of 400mm without roots.
- Any remaining dust particles were removed.

2.3. Elaboration of the mold for specimens (test material).

- The mould for making the specimens (test material) was made of wood.
- The dimensions were 400 mm long by 400 mm wide, as 5 replicas were made for each concentration.
- The specimens (test material) were then removed from the mould and cut to the dimensions according to ISO 3795.

2.4. Cleaning the specimen mold (test material).

- The wooden mould was cleaned with paint thinner the first time, from the second time onwards, any resin residue was cleaned with a spatula.
- Mould release wax was applied to the entire mould, with a greater proportion at the corners to prevent the resin from sticking to the mould.

2.5. Production of test specimens (test material).

- The vegetable fibre was subjected to a heat press for complete drying.
- For specimens dimensioning, vegetable fibre layer thickness was represented approximately 3 layers to have the desired thickness, since the minimum thickness to have is 3 mm from the ISO 3795 standard, but the thickness with which it was worked was 10 mm.

- The 3 layers of vegetable fibre were weighed, this represented 70% of the volumetric fraction of the composite material, it should be stressed that this was done for each combination.
- 30 % was distributed between the resin and the fireproofing with the following configuration: (27% - 3%), (24% - 6%), (21% - 9%) respectively A rule of three was used to calculate these percentages.

Taking for example the combination A of the study, with a configuration: 27% polyester resin, 70% reinforcement and 3% fireproof (Aluminium Hydroxide), the following is obtained. Weight of the 3 layers of vegetable fiber = 0.925 kg as follow:

- Reinforcement = Plant fibre = 70% of the volumetric fraction
- Matrix = Polyester resin = 27 % of the volume fraction
- Weight of polyester resin = X
- Fireproof = 3% of the volumetric fraction
- Weight of the fire retardant = Z

$$0{,}925\ Kg = 70\% \text{ and } X = 27\%$$

$$X = (0{,}925\ Kg * 27\%) / 70\% \quad (1)$$

$$X = 0{,}356\ Kg$$

$$0{,}925\ Kg = 70\% \text{ and } Z = 3\%$$

$$Z = (0{,}925\ Kg * 3\%) / 70\% \quad (2)$$

$$Z = 0{,}039 Kg$$

In other words, for the 5 replicas of each specimen was needed 0.356 Kg of polyester resin and 0.039Kg of fireproof (aluminium hydroxide), because the 400 mm long by 400 mm wide mould was used. Then make the individual cut of each specimen with its due dimensioning according to ISO 3795, this means that for each individual specimen, the proportion was given according to Tab. 1 and 2.

Table 1. Quantity of kilograms to be used to make the specimens (test material) with polyester resin and aluminium hydroxide

Combination	70% FCI[a]	27 % RP[b]	3 % HA[c]
A	0,185 kg	0,0713 kg	0,0079 kg
Combination	70% FCI[a]	24 % RP[b]	6 % HA[c]
B	0,185 kg	0,0634 kg	0,0158 kg
Combination	70% FCI[a]	21 % RP[b]	9 % HA[c]
C	0,185 kg	0,0555 kg	0,0237 kg

[a] FCI= Intermediate calamagrostis fiber ; [b] RP= Polyester resin ; [c] HA= Aluminium Hydroxide

Table 2. Quantity of kilograms to be used to make the specimens (test material) with polyester resin and Magnesium hydroxide.

Combination	70% FCI[a]	27 % RP[b]	3 % HM[c]
D	0,185 kg	0,0713 kg	0,0079 kg
Combination	70% FCI[a]	24 % RP[b]	6 % HM[c]
E	0,185 kg	0,0634 kg	0,0158 kg
Combination	70% FCI[a]	21 % RP[b]	9 % HM[c]
F	0,185 kg	0,0555 kg	0,0237 kg

[a] FCI= Intermediate calamagrostis fiber ; [b] RP= Polyester resin ; [c] HA= Aluminium Hydroxide

- To these combinations a Methyl Ethyl Keton Peroxide (PMEC) catalyst was added to steel the drying process of the samples for which it was calculated, 20% of the weight of the resin to be used.
- For the freezing process, the sample containing aluminium hydroxide was exposed to room temperature for 15 to 30 minutes, while the one containing magnesium hydroxide was subjected to a time of 60 to 90 minutes.
- For the curing time, it was left to dry for a period of approximately 3 hours, for the specimens (test material) containing aluminum hydroxide and approximately 7 hours, for the specimens (test material) containing magnesium hydroxide..
- Once the material has cured, the specimens (test material) were removed from the mould and then cut, sanded and measured to check their dimensions and compliance with the standard.

2.6. Specimen sizing (test material)

The ISO 3795 standard was used to size the specimens (test material) and it was established that 5 specimens should be made for the test, complying with the following dimensions, length: 356 mm, width: 60 mm and in the case of the thickness it was 10 mm. It is recommended 2 layers to be used in the study [10], but in this case, 3 layers were used to achieve the desired thickness.

Table 3. Averages of the thicknesses of each combination.

Combination	Composition	Average thickness (mm)
A	70 % FCI[a] + 27 % RP[b] + 3% HA[c]	10
B	70 % FCI[a] + 24% RP[b] + 6% HA[c]	10
C	70 % FCI[a] + 21% RP[b] + 9% HA[c]	10
D	70 % FCI[a] + 27% RP[b] + 3% HM[d]	10
E	70 % FCI[a] + 24% RP[b] + 6% HM[d]	10
F	70 % FCI[a] + 21% RP[b] + 9% HM[d]	10

[a] FCI= Intermediate calamagrostis fiber
[b] RP= Polyester resin
[c] HA= Aluminium Hydroxide
[d] HM= Magnesium Hydroxide

2.7. Flammability Index Test.

Corresponding to the ISO 3795 standard, the flammability test was carried out on five samples of each material, each of which met the following dimensions: 60 mm wide by 356 mm long and with a minimum thickness of 3 mm and no more than 13 mm, as the method is not applicable outside that range. Fourteen marker threads were drawn and fixed perpendicularly to the sample at the indicated points.

- The first point was at 40 mm,
- From the second point to the thirteenth point it was 20 mm,
- Finishing the fourteenth point with 14 mm

Having a total of 274 mm, according to the standard.

1. This test was carried out inside a gas extraction chamber. Inside the combustion chamber, the individual specimens (test material) were placed horizontally on their respective U-shaped support (Svebrant, 2011).
2. A bunsen burner and LPG were used to generate the flame.
3. The samples were exposed to an initial flame for 15 seconds inside the combustion chamber so that the flameacted on the free edge of the sample.
4. After a series of tests, it was allowed to cool down, so that the temperature of the combustion chamber and thesample holder was below 30°C before starting the next test (Serrano Aguiar & Padilla Porras, 2013).
5. From the following equation Eq. 3, the burning rate (B) was determined in mm/min.

Combustion rate

$$B = 60\frac{d}{t} \quad (3)$$

where: d = length of distance burned (mm) and t = time it takes for that distance to burn. (s)

2.8. Test to determine the presence of carbon monoxide (CO).

The test to determine the presence of carbon monoxide was carried out since when making combustion toxic gases are released to the environment what was wanted to verify with this test was that the flame-retardants used in the fiber do not generate these polluting gases for what it was resorted to the taking of samples to then check indicators, with tables established by the Ministry of the Environment of Ecuador.

The gas analyser FYRITE INSIGHT PLUS was used, as follow:

1. The sampling probe was placed at the first measuring point in such a way as to allow at least three measurements.
2. Measurement parameters were set.
3. Extraction systems were switched on.
4. Flammability test started.
5. The combustion test was carried out, taking CO measurements every 80 seconds once the test piece had been lit.
6. Finally, the instrument was purged for the next measurements [11].

3. Results and Discussions

3.1 Flammability test.

The flammability test was carried out to help monitor combustion processes of the specimens (test material) effectively to measure the speed of combustion, as well as the production of carbon monoxide.

The material was evaluated in the horizontal flammability chamber of the Faculty of Mechanical Engineering of the Technical University of Ambato, 7 tests were carried out, one control and two types of fire retardants (Aluminum Hydroxide and Magnesium Hydroxide), each with 3 different concentrations at 3, 6 and 9 percent, these values established according to other results [12].

The combustion process took place by deflagration since the flame front moved at a constant speed, this combustion took place by the heating in the form of conduction-convection of the specimen and the chain reactions that advanced the process of each of the specimens (test material) [13]. Table 4 displays the summary of results of the flammability index of the 7 combinations, Test Bench.

Table 4: Summary of flammability index results

Combination	Flammability Index Average (mm/min)
No fire-retardant	302,398
A [a]	259,052
B [b]	253,728
C [c]	190,98
D [d]	365,688
E [f]	315,192
F [g]	234,178

[a] **Combination A:**70% FCI, 27% RP, 3% HA;
[b] **Combination B:**70% FCI, 24% RP, 6% HA.;
[c] **Combination C:** 70% FCI, 21% RP, 9% HA.;
[d] **Combination D:**70% FCI, 27% RP, 3% HM.;
[e] **Combination E:** 70% FCI, 24% RP, 6% HM.;
[f] **Combination F:** 70% FCI, 21% RP, 9% HM.

From Figure 3, it can be determined that the surface of each bar is proportional to the frequency of thevalues represented, thus obtaining a panoramic view of the distribution of the samples. The flame retardants that are working effectively are aluminium hydroxide at 3, 6 and 9 percent, while magnesium hydroxide is only effective at 9 percent, since this hydroxide at the concentrations of 3 and 6 percent has a combustion rate greater than that of the control, which means that they are not effective.

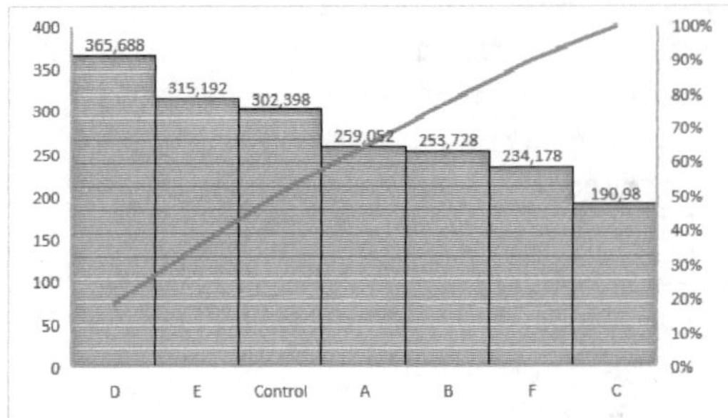

Figure 3. Histogram of flammability index results

It can also be determined that the best concentration to work with these fire retardants is the aluminium hydroxide at 9 percent since it presents a self-extinguishing behaviour, which makes it ideal and in comparison, to a study made by [12] these hydroxides present similar characteristics being the aluminium hydroxide at 9 percent the best combination.

3.2 Statistical analysis

Null hypothesis = All averages are equal

Alternate hypothesis= At least one average is

differentSignificance level α = 0.05

Table 5. ANOVA: Combustion rate B(mm/min) vs. Concentration

	GL	SC Adjusted.	MC Adjusted	F-value	p-value
Concentrations	6	100105	16684.1	45.85	0.000
Error	28	10190	363.9		
Total	34	110295			

Since the p value is less than the significance level (0.05), the null hypothesis is rejected so at least onemean is different, as presented on Table 6.

Table 6. Averages of Burning Rate B(mm/min) vs. Concentration

Concentration	N	Mean	Standard Deviation	IC 95%
AL 3	5	259.1	45.2	(241.6; 276.5)
AL 6	5	253.728	1.291	(236.252; 271.204)
AL 9	5	190.98	2.32	(173.50; 208.46)
Control	5	302.40	8.82	(284.92; 319.87)
MG 6	5	315.19	14.34	(297.72; 332.67)
MG 9	5	234.18	9.08	(216.70; 251.65)
MG3	5	365.69	11.56	(348.21; 383.16)
Grouped data standard deviation			= 19.0767	

Figure 4 shows that, in relation to the control, the combination that works more efficiently is aluminiumhydroxide at 9 percent, which corroborates the regression analysis, and previous studies have shown similar results, as mentioned above [14].

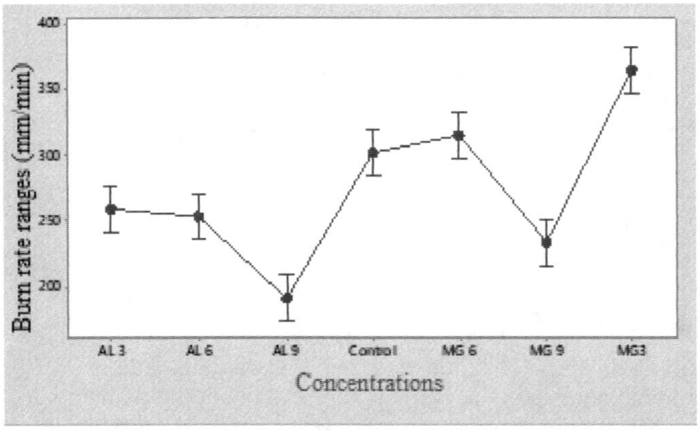

Figure 4. Burn rate ranges (mm/min) vs. Concentrations

The fire retardants that are found above the control flaps have a higher combustion rate, so the fire retardant does not work effectively and therefore the use of these would not be appropriate [3].

3.3 Multiple Dunnet comparisons with a control

Table 7. Multiple Dunnet comparisons with a control

Concentration	N	Mean	Group
Control	5	302.40	A
MG 3	5	365.69	
MG 6	5	315.19	A
AL 3	5	259.1	
AL 6	5	253.728	
MG 9	5	234.18	
AL9	5	190.92	

Averages not labelled with the letter A are significantly different from the average of the control level, indicating once again that Mg 6 has a significant similarity to the control average and does not function properly, so it has no effect when used. In addition, Mg 3 is not labelled with the letter A as this is significantly different but compared to the control it has a much higher combustion rate so it does not function effectively either.

Table 8. Simultaneous Dunnett tests for the average level - Control average

Level difference	Difference from averages	EE difference	IC 95%	T Value	Adjusted p-value
AL 3 - Control	-43.3	12.1	(-76.3; -10.4)	-3.59	0.006
AL 6 - Control	-48.7	12.1	(-81.6; -15.7)	-4.03	0.002
AL 9 - Control	-111.4	12.1	(-144.4; -78.5)	-9.23	0.000
MG 6 - Control	12.8	12.1	(-20.2; 45.7)	1.06	0.782
MG 9 - Control	-68.2	12.1	(-101.2; -35.3)	-5.65	0.000
MG3 - Control	63.3	12.1	(30.3; 96.2)	5.25	0.000

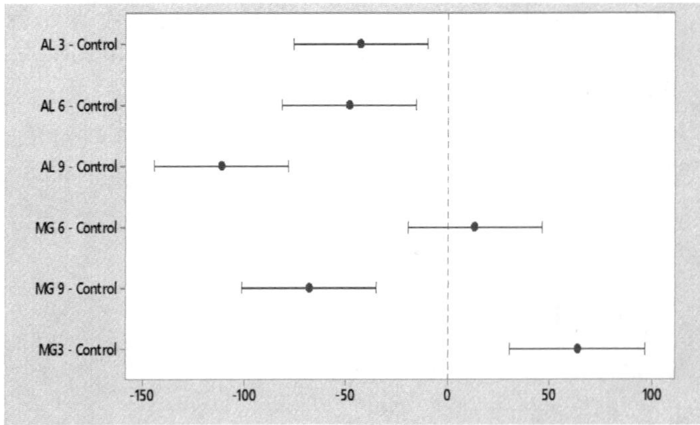

Figure 5. Dunnett's 95% simultaneous control intervals

If an interval does not contain zero, the corresponding mean is significantly different from the control mean, yielding that the best combination is 9 percent Aluminum Hydroxide. This compound presents a self-extinguishing behavior which makes it ideal. The fireproofing action of aluminum hydroxide is based on the thermal elimination of chemically bonded water in case of fire at temperatures in the range of 200 to 400 ° C. These compounds decompose through an endothermic reaction, releasing water vapour that removes heat and reduces the formation of combustible gases, as well as forming a protective layer on the surface of the material that acts by preventing the thermal degradation from being prolonged. [1].

The flame propagation in the analysed samples was symmetrical and kept within the limits indicated by the standard. This endothermic decomposition of the aluminium hydroxide consumes energy, due to the cooling of the polymer surface. [15] Aluminium hydroxide acts essentially by forming a barrier containing the degradation products and diluting the concentration of combustible gases.[16]

One of the reasons why the 3 and 6 per cent Magnesium Hydroxide did not present a positive efficiency is due to the fact that when the combination with the polyester resin was made, it had a high viscosity and when the sample was homogenized it was difficult, since its molecular weight is 58.3197 g/mol which makes it very light, producing lumps, thus causing an incorrect adhesion between them. [17] Furthermore, that a higher percentage of the catalyst Methyl Ethyl Keton Peroxide was used, thus using not 20% of the weight of the resin used, but rather 30% of it, this was because the curing time was too long and by means of this catalyst the reaction was accelerated. Another reason why I do not present a positive efficiency, was because this type of fireproofing requires a high amount of concentration as mentioned [18], so it is determined that at 9 percent this Hydroxide has a positive response, besides being the best combination, since the concentration is higher and provides the fireproofing properties that are required.

3.4 Test to determine the presence of carbon monoxide (CO).

The gas analyser FYRITE INSIGHT PLUS was used because it analyses gases that are products of a combustion, its analysis is limited to the determination of CO, CO2, O2, ambient temperature, gas temperature, its adaptability with the EPA 40 CFR method 60 Annex A, makes it the ideal equipment for the analysis of the combustion processes that are generated in the flammability chamber, as on Table 9 below.

Table 9. CO analysis in combustion

Sample	Control (PPM)	A (PPM)	B (PPM)	C (PPM)	D (PPM)	E (PPM)	F (PPM)
1	2	6	4	1	2	1	12
1	2	8	5	1	8	5	20
1	3	8	6	2	13	5	27
2	2	7	1	1	5	1	0
2	2	8	0	1	4	5	0
2	2	8	0	2	9	4	0
3	3	6	3	1	3	3	1
3	4	7	3	0	6	8	3
3	5	8	3	0	2	8	3
4	2	5	1	1	5	1	10
4	3	18	1	2	6	5	12
4	4	19	1	2	8	6	15
5	3	6	3	0	6	2	12
5	4	6	4	1	8	5	15
5	4	7	4	1	9	5	22
Average	3	8	3	1	6	4	10

Note: FCI= Fiber of intermediate calamagrostis; RP= Polyester resin; HA= Aluminum hydroxide; HM= Magnesium hydroxide. **Combination A**: 70% FCI, 27% RP, 3% HA; **Combination B**: 70% FCI, 24% RP, 6% HA; **Combination** C: 70% FCI, 21% RP, 9% HA; **Combination** D: 70% FCI, 27% RP, 3% HM; **Combination** E: 70% FCI, 24% RP, 6% HM; **Combination** F: 70% FCI, 21% RP, 9% HM.

The gas analyser FYRITE INSIGHT PLUS was used because it analyses gases that are products of a combustion, its analysis is limited to the determination of CO, CO2, O2, ambient temperature, gas temperature, its adaptability with the EPA 40 CFR method 60 Annex A, makes it the ideal equipment for the analysis of the combustion processes that are generated in the flammability chamber.

The percentage of acceptance or permissible error for the reading according to the analyser is 10% for readings between 0 and 200 PPM. As it is observed in figure 6 the combination that presents a smaller value of CO is the aluminum hydroxide to 9 percent, being this ideal to reduce the environmental impact that this one has since the aluminum oxide that remains as residue has a high specific surface area and absorbs compounds of polycyclic and aromatic hydrocarbons formed in the combustion of the polymer. As a result, these compounds are removed from the combustion process. Since polycyclic and aromatic hydrocarbon compounds are constituents of black smoke from a fire, aluminium hydroxide also contributes to reducing the density of the smoke [15].

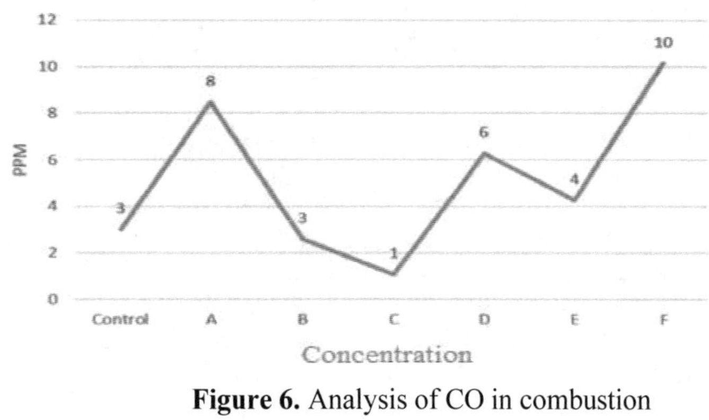

Figure 6. Analysis of CO in combustion

These CO values in the combustion of intermediate calamagrostis, do not exceed the permissible limits and are therefore considered environmentally friendly, as the maximum concentration in 1 hour of carbon monoxide must not exceed thirty thousand micrograms per cubic metre 30 000 µg/m3[19].

The maximum concentration of the combinations was 9 percent magnesium hydroxide with a value of 11432.7 µg/m3, this is well below the maximum value, without the need to implement treatment protocols or improvement in the quality of air emitted into the atmosphere [20]

4. Conclusions

The flammability analysis was carried out using the ISO 3795 standard, which is the best procedure for preparing the test material, obtaining the appropriate sizing, so that the specimens can be burnt. Out of 30 test pieces, the efficiency of inorganic fire retardants applied to the combustion speed of intermediate calamagrost is in the Ecuador's moorlands was determined. Statistical calculations showed that the best combination was C (70% ICF, 21% PR, 9% AH), resulting in a combustion rate of 190.98 mm / min and self-extinguishing capacity. In addition, a positive answer to the effectiveness was verified for Aluminium Hydroxides to all the proposed concentrations, whereas for Magnesium Hydroxide the effectiveness was positive only for the combination F. These results were satisfactory when delaying the combustion flame in the test tubes that were the material of test, this was given by means of the use of the horizontal combustion chamber to controlled environmental conditions under the norm ISO 3795. The results of the carbon monoxide determination test were evaluated for the 35 prepared samples from the moorlands of Chimborazo, Ecuador. Combination C and combination F have a gray-dark coloring in the gas product of combustion. These results were obtained by using the horizontal combustion chamber under controlled environmental conditions and the FYRITE INSIGHT PLUS gas analysis measurement equipment. Finally, an average decrease of 50% in CO generation and the amount of smoke was determined.

References

[1] Tovar Y M, "Desarrollo de un compuesto de polietileno reticulable con retardante de llama libre dehalógeno y baja emisión de humo como material de aislamiento en la fabricación de cables para bajatensión en la empresa Centelsa (XLPE-LSHFFR)," Universidad del Valle, 2010.

[2] Jouyandeh M *et al.*, "Properties of nano-Fe3O4 incorporated epoxy coatings from Cure Index perspective," *Prog. Org. Coatings*, vol. 133, pp. 220–228, Aug. 2019, doi: 10.1016/j.porgcoat.2019.04.034.

[3] Castro Miniguano C B, "Estudio del comportamiento frente al fuego de los materiales utilizados enla fabricación del interior de carrocerías y su incidencia en la homologación de modelos y competitividad de la empresa carrocera Industria Metálica Cepeda," 2014, Accessed: Oct. 15, 2020.[Online]. Available: https://repositorio.uta.edu.ec:8443/jspui/handle/123456789/6539.

[4] Torra R, "Nueva generación de polímeros termoestables ignífugos basados en residuos de la agroindustria," nstituto de Desarrollo Tecnológico para la Industria Química INTEC-UNL-CONICET, 2016.

[5] Huang G *et al.*, "Realizing simultaneous improvements in mechanical strength, flame retardancy and smoke suppression of ABS nanocomposites from multifunctional graphene," *Compos. Part B Eng.*, vol. 177, p. 107377, Nov. 2019, doi: 10.1016/j.compositesb.2019.107377.

[6] Sánchez Terán L F, R. S. Roux Gutiérrez, and J. A. Espuna Mujica, "Estudio calorimétrico con el uso de termopares en aislante hecho de mezclilla de desecho para viviendas en Saltillo, Coahuila," *Nov. Sci.*, vol. 10, no. 20, p. 280, May 2018, doi: 10.21640/ns.v10i20.1319.

[7] Yang H, Yu B, Song P, Maluk C, and Wang H, "Surface-coating engineering for flame retardant flexible polyurethane foams: A critical review," *Composites Part B: Engineering*, vol. 176. ElsevierLtd, p. 107185, Nov. 01, 2019, doi: 10.1016/j.compositesb.2019.107185.

[8] Johannes Karl Fink, *Flame Retardants*, 1st ed., vol. 1. Beverly, USA: Scrivener Publishing LLC,

2020.

[9] Beltran Ramirez F I, "Estudio de la influencia del tamaño de partícula de hidróxidos metálicos de Aly Mg (micro y nano) y el efecto sinérgico de sus mezclas en las propiedades mecánicas y de retardancia a la flama de compuestos a base de polietileno de alta densidad," CENTRO DE INVESTIGACIÓN EN QUÍMICA APLICADA, Saltillo, 2012.

[10] Arias M F, "Fabricación de tejido composite con fibras de polímeros termoplásticos y fibras naturalespara aumentar sus propiedades de retardancia a la llama," 2018, Accessed: Oct. 15, 2020. [Online]. Available: https://riunet.upv.es/handle/10251/94726.

[11] Gualpa Caisachana A M, "Diseño y construcción de un sistema de ventilación y medición de gases producidos en la cámara de inflamabilidad del Laboratorio de Ingeniería Mecánica de la UniversidadTécnica de Ambato," Universidad Técnica de Ambato. Facultad de Ingeniería Civil y Mecánica, Carrera de Ingeniería Mecánica, Ambato, 2020.

[12] Paredes Villacis D M, "Análisis de la aplicación de retardantes de llama (hidróxido de aluminio, hidróxido de magnesio) en un material compuesto de fibra de vidrio con matriz polimérica utilizadoen la fabricación del interior de carrocerías metálicas y su influencia en el índice de inflamabilidad,"Universidad Técnica de Ambato. Facultad de Ingeniería Civil y Mecánica, Carrera de Ingeniería Mecánica, 2019.

[13] Cabrera Valencia A F, "Diseño y construcción de un banco de pruebas para ensayos de inflamabilidad vertical de materiales utilizados en el interior de las carrocerías," Universidad Técnicade Ambato. Facultad de Ingeniería Civil y Mecánica. Carrera de Ingeniería Mecánica, Ambato, 2018.

[14] Villena Portero J L, "Análisis de la velocidad de combustión y su incidencia en los materiales utilizados en el interior de carrocerías para vehículos," Universidad Técnica de Ambato. Facultad deIngeniería Civil y Mecánica, Carrera de Ingeniería Mecánica, Ambato, 2019.

[15] Alfred Reimer C B, Carsten Ihmels, "(12) United States Patent45) Date of Patent: (54) PROCESS FOR PREPARING ALUMINIUM S371 (c)(1), (2), (4) Date (30) Foreign Application Priority Data," Aug. 2015. Accessed: Oct. 15, 2020. [Online]. Available: www.hubermaterials.com,

[16] Laoutid F, Duriez V, Brison L, Aouadi S, Vahabi H, and Dubois P, "Synergistic flame-retardant effect between lignin and magnesium hydroxide in poly(ethylene-co-vinyl acetate)," *Flame Retard.Therm. Stab. Mater.*, vol. 2, no. 1, pp. 9–18, Jul. 2019, doi: 10.1515/flret-2019-0002.

[17] Freire O F , "Análisis de la incidencia de retardantes de llama (hidróxido de aluminio, hidróxido de magnesio) en las propiedades mecánicas de un material compuesto de matriz polimérica reforzado con fibra de vidrio utilizado en la fabricación del interior de carrocerías metalmecánicas," Universidad Técnica de Ambato. Facultad de Ingeniería Civil y Mecánica, Carrera de Ingeniería Mecánica, Ambato, 2019.

[18] Quiñones López D D and Quiñones López D A, "Obtención de un material compuestos de polipropileno/fibra de vidrio modificada con silanoles silsesquioxanos poliedricos oligoméricos (POSS)," IT Ciudad Madero, 2018.

[19] "Norma Ecuatoriana de la Calidad del Aire." http://www.quitoambiente.gob.ec/ambiente/index.php/norma-ecuatoriana-de-la-calidad-del-aire (accessed Oct. 15, 2020).

[20] "Decreto ejecutivo 3516 | Ecuador - Guía Oficial de Trámites y Servicios."https://www.gob.ec/regulaciones/decreto-ejecutivo-3516 (accessed Oct. 15, 2020).

Fusion deposition modeling (FDM) 3D printing parameters correlation: An analysis of different polymers surface roughness

A F Barreno-Avila[1], M Monar-Naranjo[2*], E M Barreno-Avila[1]

[1] Faculty of Civil and Mechanical Engineering, Universidad Técnica de Ambato, Los Chasquis Avenue and Río Payamino, 180201, Ecuador

[2] Faculty of Design and Architecture, Universidad Técnica de Ambato, Los Chasquis Avenue and Río Payamino, 180201, Ecuador

*Corresponding E-mail: mb.monar@uta.edu.ec

Abstract. One of the most popular methods for manufacturing new products is FDM (Fusion Deposition Modeling). In which two segments are established: industrial and desktop. The latter group has had notable growth in recent years. One of the disadvantages found in this method is its surface finish, which presents a remarkable roughness. In this research work, the correlation that exists in the printing configuration' parameters and the resulting roughness analyzed by means of the ISO 4287-1997 standard will be analyzed. The configuration parameters to take into account are: layer height, printing speed. The materials to be analyzed are: PLA, PET-G, ASA, Wood Filament, TPU. Besides, the equipment used is a Prusa brand desktop printer with the manufacturer's software. What is more, the acceleration, extruder temperature and bed recommended in the software are used for each material. Hence, the constant printing characteristics are: perimeter thickness, 10% infill, extrusion diameter 0.4 mm. The objects to be evaluated are printed and the roughness on the walls of the Z axis, base plane and upper plane is measured. This analysis was carried out by means of the correlation by the use of the analysis of variance ANOVA. The printing speed does not affect the resulting surface roughness in any plane, while the layer height only shows a significant correlation (<0.05) with the roughness of the lateral plane of the samples

Keywords: 3D printing, surface roughness, polymers, correlation, Anova.

1. Introduction

3D printing has made great strides in the manufacturing industry, making it ideal for low production processes or prototyping machining [1]. Additive manufacturing challenges the production industry, making possible to produce designs that cannot be made on CNC. New design methods and new engineering needs are supported by these new manufacturing methods [2].

The 3D printer revolution can reach not only factories, but also homes. The desktop printer has become popular in recent years allowing people to buy an affordable machine and use it in their house.

Desktop printers have an annual usage increase of 170% [3]. In the desktop printers the more popular are the FDM (Fusion deposited modelling). The most popular desktop printers are using the FDM (Fusion deposited modelling) method, which uses polymers as a deposition material, generally.

The new modular concept for manufacturing initiated with the desktop printers can be used not only in the home, industries, laboratories and academia use them. Although all the advantages 3D printers bring to users, there are limitations in novel design production. Thus, the manufacturing process is slow, the price increases with each replica, the mechanical aspects are not excellent, as it is done by the adhesion of polymers that can break depending on the mechanical stress that can be applied to it.

Another weakness of 3D printing manufacturing based in FDM is the superficial roughness. The research in other states that some of the studies show improvements in the post processing of the product, otherwise some research find out the correct setup in order to decrease the roughness [4]. The superficial roughness is important in many aspects, such as a professional finish, molding procedures, the friction than can be caused if the new product is used into a mechanism.

In order to take advantage of the infinitive applications for 3D printing, the user needs to know how to setup optimal parameters for every material, so the roughness, strengthens, size will be the predicted. In addition, is important to lower the time, energy consumed and the material wasted.

2. Materials and Methods

2.1. Investigation methods

The investigation is both quantitative and deductive. Further, the strategy used is experimental and transversal [5]. The data collecting method was done through laboratory testing and it was used bibliographic research, too.

2.2. Materials and parameters

The materials selected for the research were TPU, PLA, Wood Filament, ASA, PET-G. The group of materials was selected under three parameters: bibliographic research, stock in the market and usability. These materials were the most stable and suitable for non-enclosure printers [6, 7].

The constant printing characteristics are perimeter thickness, 10% infill, extrusion diameter 0.4 mm. The print speed selected was 30 mm per second and 60 mm per second. That speed was used in each sample for: speed inside, outside and filled [3, 8]. The other variable was layer height, which depends on the machine capacity. This could be 0.05 mm up to 80% of the nozzle size. In this investigation, a common size of 0.4 mm nozzle was used. The height of layer used was 0.1 and 0.2 mm. In Table 1, it was described the information and parameters of printing configuration recommended by the users experience and the machine manual.

Table 1. Printing parameters

Material	Speed min	Speed max	Layer height min.	Layer height max.
TPU	30 mm/s	60 mm/s	0.1 mm	0.2 mm
PLA	30 mm/s	60 mm/s	0.1 mm	0.2 mm
Wood Filament	30 mm/s	60 mm/s	0.1 mm	0.2 mm
ASA	30 mm/s	60 mm/s	0.1 mm	0.2 mm
PET-G	30 mm/s	60 mm/s	0.1 mm	0.2 mm

2.3. Samples dimensions

The samples were cubes of 4*4*4 cm, model based in [8,9]. The authors mention that some inclination or shape in the sample should create a different measure in the superficial roughness. The focus in this investigation is the setup parameters and their incidence in the roughness, for that reason the samples are made with fully vertical and horizontal planes.

2.4. Null hypothesis

There is no relation between variables in 3D printing process (speed and layer height) and the superficial roughness (measured in the three planes first layer, last layer and height layer).

2.5. Surface roughness testing

Mitutoyo SJ 210 equipment shown in figure 1 was used in order to comply with ISO 4287-1997. The roughness test is carried out on the specimens of the 5 different materials in each of the axes; base, height and top layer. Once the tests were performed on the 14 samples, the graphs and the values of Ra, Rz and Rp (as shown in Figure 1) were obtained, which describe the average roughness, the roughness between the highest peaks and the roughness in the valleys.

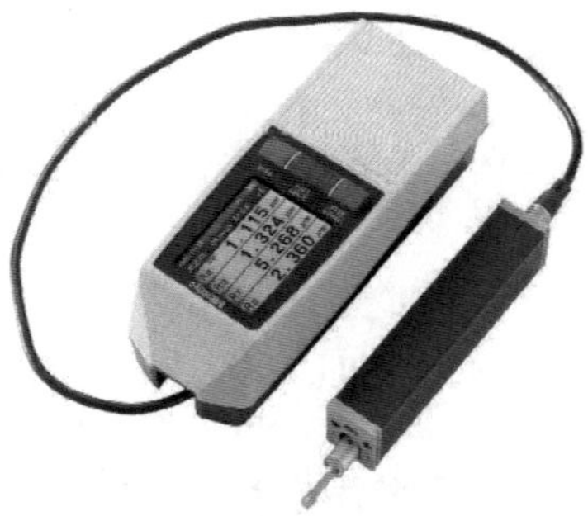

Figure 1. Mitutoyo SJ 210 roughness meter

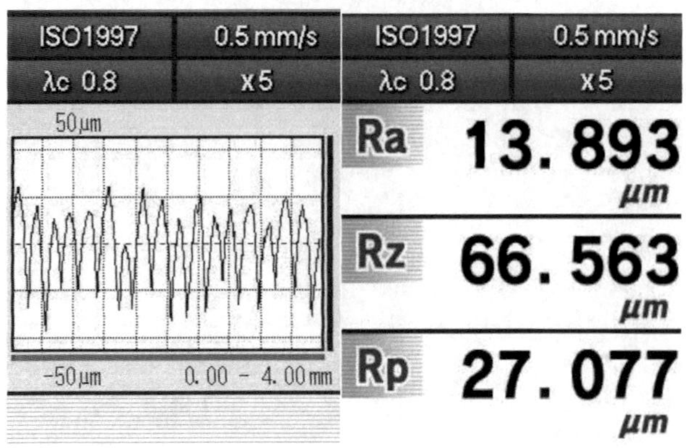

Figure 2. Graph and values shown by the roughness meter

In table 2 the average values for each sample were tabulated.

2.6. Usage of data and statistical test

All the Ra values were tested to prove that the group meets the conditions of normality. Then, each parameter (Layer height and speed) was compared with every result of Ra in the three planes. ANOVA test was selected to establish correlation between the data.

Table 2. Average measures per each sample

				Ra (Roughness Average)		
Material	**N° sample**	**Layer height (mm)**	**Speed (mm/s)**	**Top plane**	**Lower plane**	**Lateral plane**
PLA	1	0,2	60	6,45	11,87	12,79
	2	0,1	60	2,83	16,48	8,22
	3	0,2	30	5,03	15,67	12,15
	4	0,1	30	3,46	15,83	6,94
Woodfilament	5	0,2	60	3,74	9,77	13,55
	6	0,1	60	5,18	9,8	8,61
	7	0,1	30	6,11	9,84	8,99
	8	0,2	30	6,82	8,96	12,84
PETG	9	0,1	30	5	17,31	6,98
	10	0,2	30	6,55	17,26	12,35
ASA	11	0,1	60	4,51	14,88	7,6
	12	0,2	60	3,76	15,81	13,09
TPU	13	0,1	30	1,15	6,3	7,64
	14	0,2	60	1,72	7,05	14,51

3. Results and Discussion

Table 3 shows the results from Anova test using the relationship between Layer height and Ra Values. In addition, Table 4 shows the results from Anova test using the relationship between Speed and Ra Values.

Table 3. Anova test comparing layer height and Ra

		Sum of Squares	df	Mean Square	F	Sig.
Ra Top plane	Between Groups	2,428	1	2,428	0,758	0,401
	Within Groups	38,458	12	3,205		
	Total	40,885	13			
Ra Lower plane	Between Groups	1,172	1	1,172	0,070	0,796
	Within Groups	200,871	12	16,739		
	Total	202,043	13			
Ra Lateral plane	Between Groups	94,121	1	94,121	150,598	0,000
	Within Groups	7,500	12	0,625		
	Total	101,620	13			

Table 4. Anova test comparing speed and Ra

		Sum of Squares	df	Mean Square	F	Sig.
Ra Top plane	Between Groups	2,512	1	2,512	0,785	0,393
	Within Groups	38,374	12	3,198		
	Total	40,885	13			
Ra Lower plane	Between Groups	2,169	1	2,169	0,130	0,725
	Within Groups	199,874	12	16,656		
	Total	202,043	13			
Ra Lateral plane	Between Groups	7,845	1	7,845	1,004	0,336
	Within Groups	93,775	12	7,815		
	Total	101,620	13			

Although there are differences between the Ra values in the three planes analyzed, no significant difference is shown in any of the materials tested in their upper and lower planes. The variation of the parameters layer height and speed do not affect the surface roughness of the upper and lower planes. In the lateral planes (Z axis) it is shown that there is a significant correlation between the variation of the layer height and the surface roughness of the sample, regardless of whether it is in one material or another. On the other hand, the speed parameter does not affect the surface roughness of the lateral planes in any material.

Refer to other results, on print roughness analysis by the FDM method, was carried out a study with variables such as speed, nozzle type and layer height. The roughness analysis is done at 0° to 45° and 90° [4]. The material used is PLA. The researcher obtains that the best characteristic is given in a nozzle width of 0.3 mm and a layer height of 0.2 mm and the roughest element was formed with a nozzle of 0.5 mm and a layer height of 0.1 mm. This is because the nozzle expels more material, because of its thickness, and the space in which it is distributed layer by layer is smaller, allowing tiny folds of plastic to form and these generate more roughness to the object.

Also, parameters such as nozzle temperature, printing speed and layer height are considered. The material used in the study is PLA [9]. It is determined that the surface roughness is defined at 66% by the value given to the height of layer, while the other two variables such as speed (measured from 30 to 60 mm / s) provides 24% and temperature (measured from 190 ° C to 215 ° C) with 6.6%. The parameters were measured up to a layer height of 0.3 mm, the higher this number the wider the roughness.

4. Conclusion

The present study found that the best setup for PLA was 0.1 mm layer height and 30mm/s in order to achieve the lowest surface roughness on the lateral planes of the cubes used as samples. Further research is needed comparing other parameters as nozzle temperature, printing time and extrusion diameter will give deeper understanding about the behaviour of the surface roughness. In addition, vary the planes angles should be tested in order to define more interactions between the results

References

[1] N. Hopkinson and P. Dicknes, "Analysis of rapid manufacturing—using layer manufacturing processes for production," Proceedings of the Institution of Mechanical Engineers, Part C: Journal of Mechanical Engineering Science, vol. 217, no. 1, pp. 31-39, 2003.

[2] M. Simons, "Additive manufacturing—a revolution in progress? Insights from a multiple case study," The International Journal of Advanced Manufacturing Technology, vol. 96, no. 1-4, pp. 735-749, 2018.

[3] Y. Li, B. S. Linke, H. Voet, B. Falk, R. Schmitt, and M. Lam, "Cost, sustainability and surface roughness quality–A comprehensive analysis of products made with personal 3D printers," CIRP Journal of Manufacturing Science and Technology, vol. 16, pp. 1-11, 2017.
[4] A. Srivastava and J. Bhaskar, "Experimental Investigations of Printing Parameters of Fused Deposition Modeling-Based 3D Printers for Average Surface Roughness," in Advances in Additive Manufacturing and Joining: Springer, 2020, pp. 253-265.
[5] K. Ranjit, "Research Methodology 3rd ed," ed: Sage, 2011.
[6] M. Kariz, M. Sernek, M. Obućina, and M. K. Kuzman, "Effect of wood content in FDM filament on properties of 3D printed parts," Materials Today Communications, vol. 14, pp. 135-140, 2018.
[7] V. Mazzanti, L. Malagutti, and F. Mollica, "FDM 3D printing of polymers containing natural fillers: A review of their mechanical properties," Polymers, vol. 11, no. 7, p. 1094, 2019.
[8] B. N. Turner and S. A. Gold, "A review of melt extrusion additive manufacturing processes: II. Materials, dimensional accuracy, and surface roughness," Rapid Prototyping Journal, 2015.
[9] M. S. Alsoufi and A. E. Elsayed, "How surface roughness performance of printed parts manufactured by desktop FDM 3D printer with PLA+ is influenced by measuring direction," American Journal of Mechanical Engineering, vol. 5, no. 5, pp. 211-222, 2017.

Influence of Textured Nickel/Palladium/Gold-Silver (Ni/Pd/Au-Ag) Layer on Pre-plated Leadframe for Automotive Applications

A Suhaimi[1,*], G Omar[1,2], and M T Asmah[3]

[1] Fakulti Kejuruteraan Mekanikal, Universiti Teknikal Malaysia Melaka, Hang Tuah Jaya, 76100 Durian Tunggal, Melaka, Malaysia

[2] Advanced Manufacturing Centre, Universiti Teknikal Malaysia Melaka, Hang Tuah Jaya, 76100 Durian Tunggal, Melaka, Malaysia

[3] Fakulti Kejuruteraan Elektrik dan Elektronik, Universiti Tun Hussein Onn Malaysia, Parit Raja, 86400 Batu Pahat, Johor, Malaysia

* Email: suhaimi.azizan83@gmail.com

Abstract. High quality and reliability of integrated circuit packaging is required for automotive applications due to the safety concerns. Therefore, elimination and controlling the package defects which can lead to reliability failure is essential. One of the defects that leads to reliability issue is delamination. This study presents the influence of textured leadframe surfaces on the surface wettability and delamination occurrence in automotive package unit. Standard bare Cu leadframe was used as comparison. The leadframe surface was textured with Ni/Pd/Au-Ag using electroplating technique. The surface morphologies of both bare and textured leadframe were characterized using field-emission scanning electron microscope and optical profiler. The energy dispersive spectroscopy was used to quantify the elemental composition of the samples. All samples went through the wettability test under two conditions, which were room temperature and 175°C. The reliability test of complete package was subjected to moisture sensitivity preconditioning level 1 according to IPC/JEDEC STD 020 and the baking process was performed at 125°C for 24hrs followed by soaking process under the condition of 85°C and 85 of relative humidity. Scanning acoustic tomography according to IPC/JEDEC STD 035 was performed after three cycles of infrared reflow at 260°C. Higher surface roughness was observed for pre-plated leadframe (PPF) compared with bare leadframe. The lowest value of average contact angle (~55°) was observed for PPF at 175°C indicated better wettability feature. Furthermore, no delamination occurrence was observed for PPF sample compared with bare copper leadframe. The obtained result reveals that leadframe with textured surface of Ni/Pd/Au-Ag improved interfacial surface adhesion which eliminated the delamination defect and thus improved the package robustness and reliability.

Keywords: Ni/Pd/Au-Ag, Texturing, Wettability, Delamination, Automotive

1. Introduction

A reliable and full capability of integrated circuit (IC) device is obligatory in automotive applications. This device requires high quality and robust packaging solution as any functionality impairment will lead to reliability and safety concerns at field application. The device used in automotive system application such as air bag, door lock and tire pressure strongly requires high performance and without failures. Therefore, improvement in package reliability that satisfied the current automotive standard is vital. One of the common failures in IC packaging is delamination that affects the product quality. Delamination is defined as weak interfacial adhesion (loss of adhesion) of a film from a substrate. This failure caused by coefficient of thermal expansion (CTE) mismatch between different types of dissimilar material interfaces at elevated temperature. In IC packaging, delamination occurrence is commonly observed between mold compound and metal substrate interfacial [1-3].

Metal substrate or copper-based (Cu-based) leadframe is widely used in IC packaging for chips mounting and electrical connection between IC chips and external leads of the package [4-5]. This day, pre-plated leadframe has gain popularity and extensively used due to its advantages, ability to improve the leadframe solderability and also improve the die attach, wire bonding and mold compound adhesion to the base substrate [6]. This PPF is also a reliable alternative to the typical Ag-plated Cu leadframe that suffered with tin (Sn)-whisker issue and toxic lead-based plating. Thus, NiPd-based pre-plated leadframes, NiPdAu, NiPdAu-Ag and NiPdAu-Pd have been introduced in order to remove certain hazardous materials in the process and at the same time meets the environmental regulations. In addition, the application of multi-layer plating on the Cu leadframe helps to eliminate the post-plating process of the final package's leads [7-9].The multi-layer plating plays different role to the Cu leadframe. The Ni layer which is plated on the Cu surface provides corrosion and oxidation resistance to the leadframe. This layer also prevents Cu (from the leadframe) diffusion into the Pd layer. The Pd layer which plated on the Ni layer hinders the underlying Ni surface from being oxidized and prevents diffusion from the top layer to the Ni layer. The top layer, Au or Au alloy prevents the Pd layer from being oxidized and facilitates the wire bonding and soldering process. This layer also helps to prevent outgas absorbed by Pd layer. In addition, surface hardness can be reduced by adding with Ag for better bondability [4,10].

Excellent interfacial adhesion between mold compound and Cu leadframe is in demand in order to overcome the delamination issue. One of the techniques that can be used to improve the interfacial adhesion is modification of the plated surface. The modification or roughening process can be performed on the top of copper lead frame surfaces followed by Ni layer prior to Pd plating process. This modification improves the surface energy and high mechanical interlocking between Cu leadframe surface and mold compound can be achieved. A rough surface that produced by this modification is not only helps to improve adhesion but also effective with downbonds [6-7,11]. However, optimum surface roughness is essential to ensure interfacial adhesion. This is because a very rough surface is unfavourable from the adhesion standpoint while a smooth substrate tends to reduce the adhesion degree [2].

This research studied the influence of textured (roughened) PPF surface which comprises of Ni/Pd/Au-Ag layer on the wettability property and surface delamination for automotive device applications.

2. Material and Method

2.1. Substrate Preparation

A commercial Cu alloy leadframe (C194) was used as a substrate in this study. These materials were obtained from one of the leadframe supplier. The dimension of the substrate used was 219.6mm x 31.5mm x 0.127mm (Length x Width x Thickness). Prior to electroplating process for the PPF preparation, the Cu substrate was cleaned using anodic cleaning process in a detergent solution at certain current density for a certain time. The substrate was then rinsed with de-ionized (DI) water. The substrate went through an acid cleaning process and finally rinsed with DI water. The cleaned Cu substrate underwent roughening treatment based on the electroplating technique. The Cu substrate was submerged into Ni solution in order to form an intermediate layer. This process was associated with an application of modulated current to the plating solution and metal substrate. The final thickness of plated Ni was ~0.5μm. The second layer of Pd was electroplated on the underlying Ni layer with the thickness of ~12nm. The top layer which was Au alloy (Au-Ag) layer was then electroplated on the Pd layer with the thickness of ~5nm. The schematic diagram of the prepared PPF is depicted in figure 1. In addition, bare Cu leadframe from the same supplier was used as comparison.

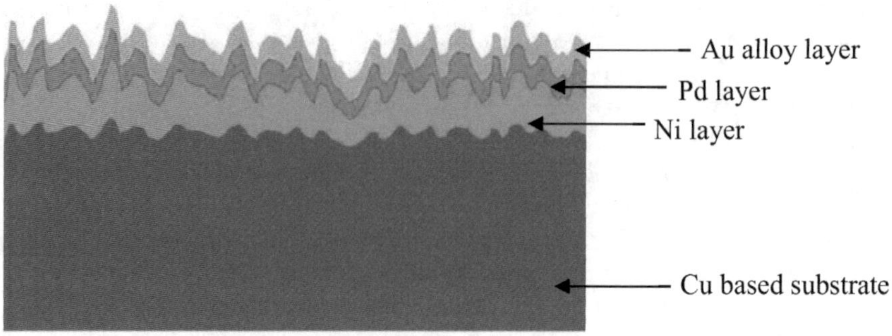

Figure 1. Schematic diagram of pre-plated leadframe.

2.2. *Surface Characterization*
The surface morphology of the sample was characterized using digital microscope (Keyence VHX-6000) and field-emission scanning electron microscope, FESEM (Zeiss Merlin). In addition, the elemental compositions of the sample were measured using energy dispersive spectroscopy, EDS (Bruker Quantax System) and the sample surface roughness was characterized using optical profiler (Bruker ContourGT).

2.3. *Wettability Test*
The surface wettability of metal substrate surface was measured by self-fabricated contact angle equipment. Two samples were prepared for this test. One of the samples was exposed to air at room temperature while another sample was exposed to 175°C for 120s in order to simulate the manufacturing molding process. DI water was then dropped onto both samples surface for wettability test.

2.4. *Reliability Test*
Prior to reliability test, the PPF and bare Cu leadframe underwent all the manufacturing assembly processes to become a complete IC package. This finish part was subjected to moisture sensitivity preconditioning level 1 (PC-MSL 1) which the baking process took 24hrs at 125°C. A soaking process took place in moisture and temperature chamber under the condition of 85°C/85RH (relative humidity). The sample then went through an infrared reflow at 260°C for three cycles. This reliability test was followed the IPC/JEDEC STD 020 (Moisture/Reflow Sensitivity Classification for Non-Hermetic Solid State Surface Mount Devices) standard. Any delamination failure of the sample could be detected after this process. In addition, scanning acoustic tomography, SAT (Sonix Echo) was used to measure the delamination of the sample. The SAT frequency was set to 75MHz C-SCAN, 23.3uS water path and 55dB of attenuation. This SAT testing was followed IPC/JEDEC STD 035 (Acoustic Microscopy for Non-Hermetic Encapsulated Electronic Components) standard.

3. Results and Discussion

3.1. Surface Morphology

Microscope and FESEM images of the leadframe surface for bare Cu and PPF are presented in figure 2 and figure 3, respectively. The difference of the surface morphology for bare Cu leadframe and PPF can be clearly seen from both Microscope and FESEM images. Figure 3(a) depicts that the surface morphology of bare Cu leadframe is relatively smooth compared with PPF surface (figure 3(b)). A uniform grain size and shape can be observed on the roughened PPF surface (figure 3(b-ii)). Similar observation of PPF surface structure has been published and reported, previously [7]. In addition, the average surface roughness for bare Cu surface and PPF was about 150nm and 275nm, respectively. This result shows that the PPF surface is more roughen than bare Cu surface. Furthermore, the range of grain size thickness for PPF was between 0.8μm and 1.5μm. Basically, rough metal surface increases the leadframe surface area and provides mechanical interlocking between leadframe surface and mold compound. Therefore, the interfacial adhesion of dissimilar surface materials can be improved. This improvement helps to control the delamination occurrence and at the same time improve the package reliability. Furthermore, the rough surface keeps the stress on the ball and stitch bonds low. This function hinders broken stitch bonds and lifted ball. On top of that, rough leadframe surface also preserves the thermal path [6-7,12].

Elemental compositions for both bare Cu leadframe and PPF leadframe are exhibited in figure 4, respectively. No foreign elemental has been observed for bare Cu leadframe (figure 4(a)). As expected, the main four electroplating elements can be clearly seen for PPF leadframe surface, which were Ni, Pd, Au and Ag (figure 4(b)).

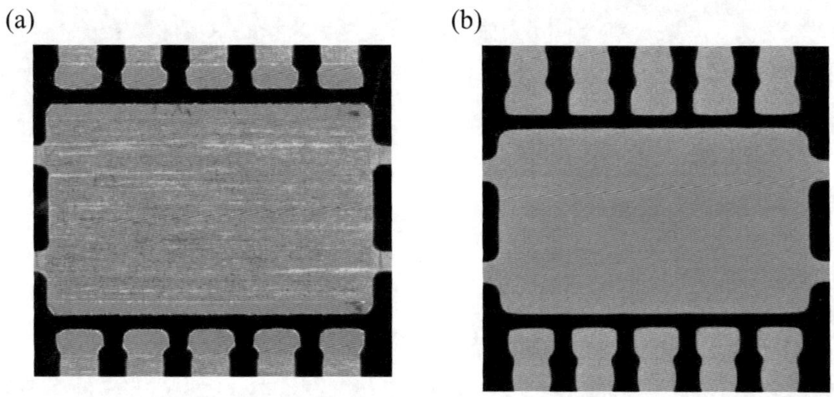

Figure 2. Microscope images of the surface (a) bare Cu leadframe sample, (b) PPF leadframe sample.

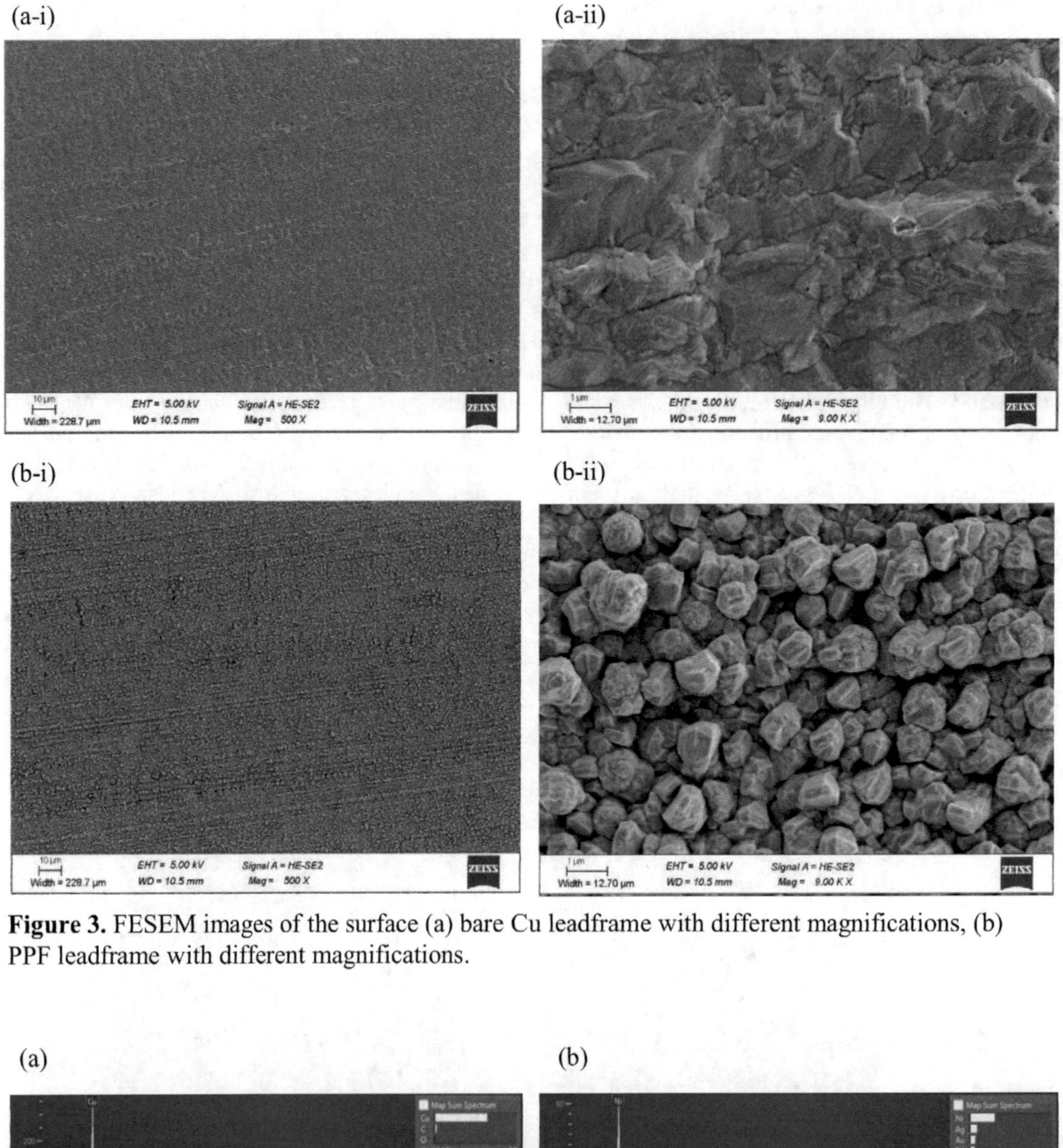

Figure 3. FESEM images of the surface (a) bare Cu leadframe with different magnifications, (b) PPF leadframe with different magnifications.

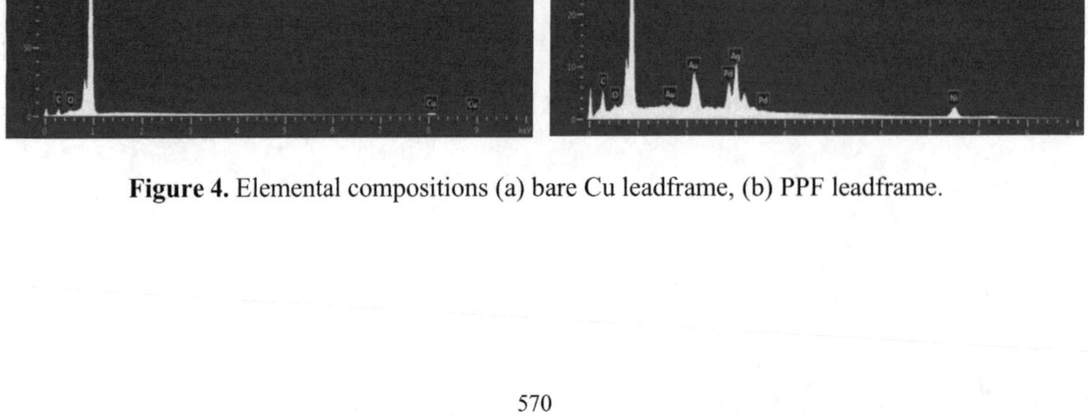

Figure 4. Elemental compositions (a) bare Cu leadframe, (b) PPF leadframe.

3.2. Wettability Test

A comparison of wettability result for both bare Cu and PPF lead frame is presented in table 1. Lower average contact angle was observed for bare Cu leadframe (~58°) compared with PPF (~71°) when both samples were exposed to air at room temperature. However, the lowest average contact angle was obtained for PPF (~55°) after exposure to 175°C for 120s. This result shows that PPF possesses better wettability feature at molding compound temperature compared with bare Cu leadframe. In addition, good wettability and clean surface is demonstrated by a small contact angle value [13].

Table 1. Comparison of contact angle result for both bare Cu leadframe and PPF leadframe.

Conditions	Bare Cu Leadframe		PPF Leadframe	
	Average Contact Angle (°)	Image	Average Contact Angle (°)	Image
Room Temp.	58.02		71.23	
175°C	66.28		55.16	

3.3. Reliability Test

The delamination defect of package unit can be observed after SAT was performed. Figure 5 shows the SAT images for both bare Cu leadframe and PPF leadframe. The delamination defect is indicated by the red. Therefore, each package unit from bare Cu leadframe sample was suffered from delamination. However, no delamination was observed for all packages from PPF leadframe sample. This result shows significant improvement in adhesion between mold compound and roughened leadframe surface with no delamination occurrence was observed. This observation was also reported by previous reports [1,6]. Hence, roughen the leadframe surface helps to improve interfacial adhesion strength and possible to reduce and eliminate the delamination problem between mold compound and leadframe surface.

Figure 5. SAT images (a) bare Cu leadframe, (b) PPF leadframe.

4. Conclusion

Dissimilar material interfaces caused by CTE mismatch leads to the lack of interfacial adhesion between mold compound and leadframe surface. This problem becomes more serious with delamination occurrence in the automotive package device which any functionality impairment will lead to safety and reliability concerns at field applications. However, the severe defect such as delamination can be reduced and eliminated with several improvements. One of the improvements that can be performed is by leadframe surface modification. This modification will create a rough leadframe surface and therefore increase the leadframe surface area. Thus improves the mechanical interlocking between leadframe surface and mold compound. An improvement in the interlocking of both surfaces leads to improve the interfacial adhesion bonding joint strength and at the same time helps to control the delamination defect of the package. Sample with roughen surface (PPF) shows better wettability feature at high temperature with the lowest average contact angle value compared with bare Cu leadframe. In addition, no delamination occurrence was observed for PPF sample compared with bare Cu leadframe sample. As conclusion, textured and roughened leadframe surface will influence the wettability feature and delamination occurrence and subsequently affect the device reliability for automotive package applications.

Acknowledgements
The authors wish to thank the Universiti Teknikal Malaysia Melaka and all the work force involved in this study for the support and guidance.

Rerefences

[1] Anan S, Kessararat U and Chalermsak S 2020 Effect of epoxy molding compound material and leadframe roughness to integrated circuit package for automotive devices *ECTI Transactions on Electrical Eng. Electronics and Communications* **18** (2) August 179-188
[2] Khanna V K 2011 Adhesion-delamination phenomena at the surfaces and interfaces in microelectronis and MEMS structures and packaged devices *J. Phys. D: Appl. Phys.* **44** 034004 1-19
[3] Alvin B D, Rod J D S Jr, Ton C P, Darwin J D L, Ivan T G C and Allen M M 2018 An alternative packaging solution in achieving moisture sensitivity level one (1) for small outline integrated circuit (SOIC) automotive package *38th Int. Electronics Manufacturing Technology Conf. (Melaka: IEEE)* pp 1-4

[4] Lilin L, Ran F, Deming L and Tong -Y Z 2008 Cu out-diffusion kinetics in pre-plated Cu-alloy leadframes investigated by a developed EDX-based oxidation test *Int. Conf. on Electronic Packaging Technology & High Density Packaging (Shanghai: IEEE)* pp 1-6

[5] Vivet L, Joudrier A -L, Tan K -L, Morelle J -M, Etcheberry A and Chalumeau L 2013 Influence of nickel-phosphorus surface roughness on wettability and pores formation in solder joints for high power electronic applications *Appl. Surface Sci.* **287** 13-21

[6] Christopher H 2013 *Leadframe – Part II (Albuquerque: Semitracks Inc.)* Issue 54 pp 1-11

[7] Wu -H L, Albert A, Mary G M, Nicolas T M and Raymond S C 2011 Cu wire bonding in Ni/Pd/Au-Ag and roughened Ni/Pd/Au-Ag pre-plated leadframe package *13th Electronics Packaging Technology Conf. (Singapore: IEEE)* pp 794-97

[8] Raymond S C, Joseph A M B, Wu -H L, Ariel L M and Mary G M 2013 Study of die bond on roughened NiPdAu-Ag pre-plated frame with anti-EBO *15th Electronics Packaging Technology Conf. (Singapore: IEEE)* pp 341-46

[9] Sock C T, Kok -T L and Mohd E A M 2015 Leadframes' AuAg plating thickness influencies to stitch binding of palladium coated copper wires *16th Int. Conf. on Electronic Packaging Technology (Changsha: IEEE)* pp 114-17

[10] Tok C W and Johnny Y P H 2011 Bondability of copper wires on PPF leadframe *13th Electronics Packaging Technology Conf. (Singapore: IEEE)* pp 12-9

[11] Wenjing Z, Wei L, Anmin H and Ming L 2012 Adhesion improvement of Cu-based substrate and epoxy molding compound interface by hierarchical structure preparation *Microelectronics Reliability* **52** 1157-64

[12] Dan H, Bruce L and John G 2008 Increasing IC leadframe package reliability *10th Electronics Packaging Technology Conf. (Singapore: IEEE)* pp 1209-13

[13] Sung Y, Jang -K K, Chee Y Y and Jang -H H 2000 Bonding strengths at leadframe encapsulant-gold-plated copper leadframe interface *Microelectronics Reliability* **40** 1207-14

IOP Conf. Series: Materials Science and Engineering 1173 (2021) 012073 doi:10.1088/1757-899X/1173/1/012073

Short review on heteropoly acid based catalyst for valorization of biomass waste into valueable chemicals

N L Z Z Adil[1], T S T Saharuddin[1*], L N Ozair[1] and F W Harun[1]

[1] Chemistry, Faculty of Science and Technology, Universiti Sains Islam Malaysia Bandar Baru Nilai 71800, Nilai, Negeri Sembilan, Malaysia

* Email: shafazila@usim.edu.my

Abstract. Depletion of fossil fuels resources has triggered more research on finding potential alternative sources that are more sustainable such as biomass waste. However, deconstruction of the complex biomass waste into desired products is too challenging and usually require a strong catalyst with a great hydrolysing property. Common acids such as sulphuric acid, hydrochloric acid and nitric acid were the most studies for biomass conversion. These liquid acids catalysts suffer low recovery and recyclability that can be overcome by a solid acid catalyst. Heteropoly acid (HPA) catalyst is the preferred choice to replace these common acid catalysts as it is known to have a strong Bronsted acid site with an oxidizing property that allows this catalyst to hydrolyse and oxidize in one-step reaction. Furthermore, heteropoly acid (HPA) catalyst can be modified into heterogenous type of catalyst by solidifying HPA to increase the surface area and recyclability. Future work of this research is necessary to improvise the previous method of the catalyst preparation and to suppress the by-products after the catalytic process. In this review, we summarize the use of HPA catalyst in the complex reaction process of biomass conversion to valuable chemical products.

Keywords: Heteropoly acid, heterogenous catalyst, biomass waste

1. Introduction

The key ingredients for industrial chemical production are petroleum-derived feedstocks that come from fossil fuel with the growth in demand for valuable chemicals, the industry greatly relies on the use of fossil fuels. However, the usage of fossil fuel in chemical industries is responsible for 5.5% CO_2 and 7% of greenhouse gasses emission. This emission includes process emissions during chemical transformation [1]. Increasing demand for chemical industries will cause a serious threat not only to the environment but to public health [2]. In this regard, biomass wastes are the alternative supply to greener route for chemical productions. Besides, abundant of natural resources are underutilized and usually dispose of by dumping and open burning which create additional environmental problems [3]. Hence, biomass wastes are a promising source as compared to non-renewable fossil fuel that will soon be exhausted [4].

Today, growth in the global chemical industry subsequently drive the market to utilize biomass waste as a feedstock to replace fossil fuels [6]. Biomass waste is the most abundant natural source composed by an organic substance that derived directly or indirectly from the photosynthesis process. In biomass utilization, cellulose is the major component of lignocellulosic biomass that effectively converted into various value-added chemicals [5]. Conversion of lignocellulosic biomass into chemicals will starts by breaking down cellulose into its constituents such as hexose and pentose. Efficient utilization of renewable and inedible biomass waste meets the principles of green chemistry and create a sustainable green chemical [5]. Biomass can be classified as forestry waste, animal and human residues, aquatic biomass and municipal solid waste. The chemical composition of each biomass sources is different from one another. For example, plant origin waste consists of cellulose, hemicellulose and lignin while animal waste is rich in protein [6]. Table 1.1 show the possibility of converting raw biomass into value-added chemicals.

Table 1. Possible value-added chemicals converted from raw biomass.

Raw Biomass	**Component of Biomass**	**Chemical Product(s)**	**References**
Corncob	Hemicellulose	Xylitol	[7]
Pinus Radiata	Lignin	Formic acid	[8]
Sugarcane Bagasse	Cellulose	Formic acid and acetic acid	[9]
Beech wood	Hemicellulose	Xylitol	[10]
Rice husk	Cellulose	Levulinic acid	[11]
Sugarcane bagasse	Hemicellulose	Levulinic acid, furfural, formic acid and lactic acid	[12]
Rice straw Cotton straw Barley straw Paddy straw	Hemicellulose and cellulose	Levulinic acid	[13]
Rice husk Soy bean peel	Cellulose, hemicellulose and lignin	5-hydroxymethyl furfural and furfural	[14]

However, the harsh condition is needed to induce depolymerization of lignocellulosic. The reaction involves high temperature and pressure, corrosive acid or alkali that require an improvement strategy. Therefore, mild, selective and greener process should have replaced the previous method in order to provide a green route valorization of biomass by introducing heterogenous catalyst, water-based and environmental-friendly oxidant [15]. A study reported that common methods for valorization of biomass are direct combustion, thermochemical conversion and bio-chemical transformation [16]. For thermochemical conversion, three techniques that commonly use include pyrolysis, liquefaction and gasification. All of these techniques lead to the production of bio-oils, chemicals and syngas. Furthermore, they also reported that liquefaction involves chemical reforming process of organic matters which combine the use of high hydrogen partial pressure and catalyst to improve the selectivity and quality of the products. Meanwhile, gasification is considered a cost-effective technique because gasification allows different types of biomass to be valorized. The product of gasification such as bio-syngas can be used as raw material for Fischer Tropsch Synthesis to produce green diesel and high value-added chemicals.

Therefore, increasing research efforts had been done on the usage of catalyst for biomass conversion to improve the reaction rate for the reaction to proceed effectively. Another point to emphasize is low reactivity of cellulose requires a highly reactive catalyst to access and break down the glycosidic bonds in cellulose [17]. Therefore, the suggested catalysts used for valorization of biomass for chemical productions are acid catalyst, base catalyst or photocatalyst. These catalysts have different characteristics that enable them to depolymerize lignocellulosic biomass optimally. Heteropoly acid (HPA) catalyst is one of the interests focus nowadays since this catalyst possesses Bronsted acidity and oxidizing property that works as a bifunctional catalyst such as phosphomolybdic acid [18]. According to their finding, phosphomolybdic acid not only possesses acidic properties but also provides an oxidative catalytic system that able to degrade lignin completely with a high yield of aromatic oil. Another study found that conversion of cellulose into 5-hydroxymethylfurfural by a stable type of HPA able to maintain the same yield after a few runs [19]. Therefore, HPA does not only has good stability after being reused but HPA also contains high acidity that allows its effectiveness during cellulose conversion [20]. Thus, an effort has been made to write up a short review that mainly focuses on breaking down biomass waste using HPA based catalyst.

2. Acid and base catalyse depolymerization of lignocellulosic

Acid catalysis is defined as a reaction in which the acid catalyst donates a proton to the targeted species [21]. The liquid acids that are commonly used in acid hydrolysis include HCl, HNO_3 and H_2SO_4. However, recent studies have been focusing on the usage of solid acids compared to liquid acids for cellulose hydrolysis since the solid acids give better product separation, reusability and recovery of catalyst. Hence, a comparable Brønsted acid solid catalyst as the common liquid acids had gathered a lot of attention include acidic ionic liquid [22], metal oxides such as layered niobium oxide that possess Bronsted acid site [23] and heteropoly acid catalyst [24]. There are many functions of acid catalyst that several can be used for depolymerization of cellulose [25]. An investigation was made on acid hydrolysis of cellulose into levulinic acid in which cellulose initially break down into lower molecular weight fragments by the cleavage of β-(1-4)-glycosidic bond of cellulose. Then, lower molecular weight fragments such as glucose will decomposed into hydroxymethylfurfural and further converted into formic acid [26].

Another study was done to observed variety of solid acid catalyst ranging from crystalline to amorphous structure to understand the efficiency of structure towards lignin conversion [27]. From their observation, well defined structure catalyst such as zeolite is unstable because zeolite undergoes structure alterations under high temperature compared to SiO_2-Al_2O_3 that are more amorphous but have strong Bronsted acid site. An organic electrolytes solution which is 1-butyl-3-methylimidazole chloride solution with the presence of different mineral acids such as HCl, H_2SO_4 and H_3PO_4 was experimented to catalyze the hydrolysis of cellulose [28]. From the investigations, authors reported that the high acidity of H_2SO_4 mineral acid in aqueous solution contribute to a high acidity in organic electrolyte solutions that effectively enhance the cellulose hydrolysis.

Meanwhile, base catalyzed depolymerization by Bronsted or Lewis base usually present in liquid solvent such as water. The proton withdrawal occurred due to the electron donor such as HO- that interact with the reactant. Furthermore, hydrothermal base lignin depolymerization relies mostly on Bronsted base alter the electronic structure of phenolics and other aromatic subunits which causes shift and destabilizes other linkages. Therefore, this will cause bond cleavage in lignin and lead to depolymerization [29][30][31]. To date, depolymerization of cellulose by base catalyst effectively cleave aryl-alkyl (β-O-4) in lignin but base catalyzed reaction requires high temperature which is around 340 °C [32]. Examples of base catalyst used by previous researchers include Pt-Mg-Al hydrotalcite as metal/base catalyst [33], ZrO_2-based catalysts [34], TiO_2 [35][36] and alkali-metal substituted zeolite catalyst [37].

3. Heteropoly Acid Catalyst

The general formula of keggin type heteropoly acid (HPA) is $[XM_{12}O_{40}]_{n-}$ where X is the heteroatom and M is the transition metal. Figure 1 shows the chemical structure of HPA. HPA catalyst belongs to polyoxometalates (POMs) class in which consist of transition metal-oxygen-anion clusters that contain p or d block elements. Polyoxometalates (POMs) consist of cations and polyanions clusters in which oxometal polyhedral of transition metal at their high oxidation state and heteroatom are the basic construction units of POMs [38]. The primary structure of HPA is the heteropolyanions that linked to each other creating secondary structure of solid HPA. The structure of HPA must consist of tetrahedron (XO_4) where X is the heteroatom surrounded by 12 edges and corner-sharing metal oxygen octahedra (MO_6) [39][40]. HPA is defined based on its ionic clusters with general formula $[X_xM_mO_y]_{q-}$ in which

M is metal typically molybdenum, vanadium and phosphorus and X is the heteroatom usually silicon or phosphorus. HPA contains protons that made up the counter ions of HPAs [41]. At the surface of heteropolyanions, there are abundance of O2 that can donate electron which make HPA as soft base. Meanwhile, the metal ions in HPA have unoccupied orbital that accept electrons which at the same time make HPA as Lewis acid. Therefore, their strong ability to accept and withdraw electrons indicate their available redox nature [38].

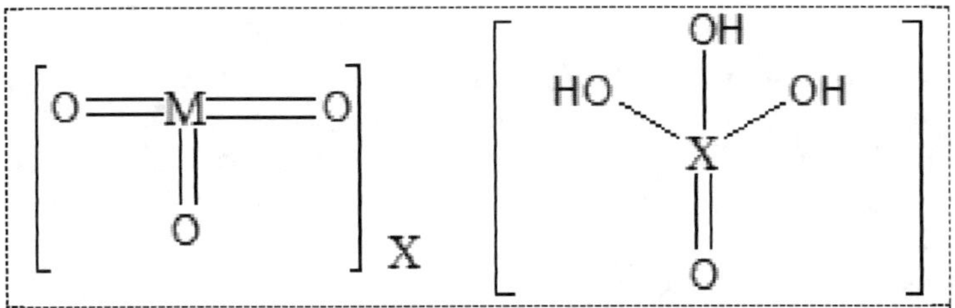

Figure 1. Chemical structure of HPA

HPA catalyst is known to be a promising catalyst due to its strong Bronsted acidity and together with the oxidizing property that allows HPA to be widely used in acid catalyzed oxidation reaction [42][43][44][45][46]. HPA exhibit stronger acidity approach compared to typical acids such as H_2SO_4, HNO_3 and HCl [18][40]. This type of catalyst also known for having a high proton mobility and stability [47] because in catalytic reaction HPAs play important role as proton donor that exhibit high catalytic activity [48]. HPA is reported to have good oxidation properties to cleave cellulose, lignin and model substrates under mild conditions [49]. Broader research areas of HPAs for the conversion of cellulose involve hydrolysis, alcoholysis, hydrolysis-hydrogenation, hydrolysis-oxygenation reaction [50]. Another advantage of HPAs is highly soluble in water which makes this catalyst greatly efficient to break down biomass in water [51]. While in water, HPA will dissociates protons and increase the amount of freely available H+ to interact with the oxygen atom in ether linkage of cellulose [52]. HPA then will continue to break the rigid structure of cellulose by attacking the intermolecular and intramolecular bonds in the structure of cellulose [53][42].

Table 2. shows the ability of HPAs to convert different types of lignocellulosic biomass into desire product.

Heteropoly Acid Catalyst	**Substrate**	**Product**	**Reaction Involved**	**References**
$K_4[PVW_{11}O_{40}]$-KPWV$_1$ and $K_6[PV_3W_9O_{40}]$-KPWV$_3$	Cellulose	Glucose	Hydrolysis	[37]
$H_4PMo_{11}VO_{40}$ and $Cs_4PMo_{11}VO_{40}$	Methacrolein	Methacrylic acid	Oxidation	[54]
$H_4SiW_{12}O_{40}$, $H_3PW_{12}O_4$, SiO_2-$H_4SiW_{12}O_{40}$ and SiO_2-$H_3PW_{12}O_4$	Cellulose	Lactic acid	Hydrolysis	[55]
$H_4SiW_{12}O_{40}$, $H_3PW_{12}O_4$ and H_2WO_4	Cellulose	Glucose	Hydrolysis	[42]
$Cs_3HSiW_{12}O_{40}$	Cellulose	Sorbitol	Hydrolysis-hydrogenation	[56]

(Cesium salt of $HSiW_{12}O_{40}$)				
$H_4SiW_{12}O_{40}$, $H_3PW_{12}O_4$ and $H_3PMo_{12}O_{40}$	Lignocellulosic biomass	Glucose	Hydrolysis	[52]

4. HPA as Bifunctional Catalyst

Single function catalyst is widely used for chemical transformation but the process is undesirable because requires numbers of isolations, purification steps and unwanted by-products. Therefore, the bifunctional catalyst will combine multiple reactions in a one-pot reaction [17][57]. A bifunctional catalyst having both acid sites and redox site can provide two successive steps to convert cellulose into products. Acid sites are used to hydrolyze β-1,4-glycosidic bond and redox site to activate O2 for the oxidation of the aldehyde group into the carboxyl group [58]. These two important characteristics can be observed in HPAs catalyst and much attention has been drawn to study the environmental route for cellulose conversion. An investigation was done on different types of HPAs as bifunctional catalyst which are $Cs_{2.5}H_{0.5}PMo_{11}VO_{40}$, $Cs_{3.5}H_{0.5}PW_{11}VO_{40}$ and $Cs_{4.5}H_{0.5}SiW_{11}VO_{40}$ in one pot hydrolysis-oxidation of starch into formic acid [56]. The results showed strong acidity of catalyst provides a route for hydrolysis reaction while vanadium ions contribute in the oxidation process.

Another finding synthesized Mo-V-P HPA that possess both acidic and oxidizing properties and yielded 66% formic acid from cellulose under mild temperature [46]. During the reaction, cellulose was hydrolyzed to glucose by the acidic sites of HPA and proceed with oxidation in which vanadium is reduced from V^{5+} to V^{4+} to oxidize glucose into formic acid.

However, there are a few types of modified HPA display a bifunctional catalyst by having a synergy Bronsted and Lewis acidity. This characteristic was observed in a research [59] using amino functionalized ionic liquid-HPA catalyst with sequential hydrogen transfer between the catalyst and glucose which leads to the dehydration of glucose into hydroxymethylfurfural. Besides that, a bifunctional molybdenum-based polyoxometalates supported on titanium was studied for conversion of lignin derived compound into alkylated phenols [60]. The pairing between Bronsted and Lewis acid for this catalyst enhance the catalytic reaction.

5. Homogenous HPA Catalyst

HPA catalyst is a versatile catalyst that can be used as homogenous, heterogenous or biphasic catalyst system. These catalytic systems showed great effectiveness in degrading lignocellulosic biomass but the limiting factor of homogenous catalyst is the difficulty for recovery since HPA is expensive compared to conventional acid catalysts [51]. However, from the investigation phospotungstic acid ($H_3PW_{12}O_{40}$) can be reused again after extraction using diethyl ether. This extraction method is cheaper for recovery compared to filtration using mesoporous ceramic membrane [61]. Moreover, diethyl ether can also be used for separation and reusable of homogenous HPA [62]. The efficiency of HPW for production of glucose from cellulose under hydrothermal conditions was tested. Results of the investigation showed that the glucose yield is 50% with 90% selectivity at 453 K. The solubility of HPW in water and its high Bronsted acidity allows large dissociation of H^+ during cellulose conversion which causes interaction between protons from HPW and oxygen atom from ether linkage of the cellulose.

Meanwhile, a study reported that silicotungstic acid as bifunctional catalyst is due to the presence of Bronsted and Lewis acidities that can enhance the depolymerization of solid cellulose into methyl levulinate [63]. This catalyst can be reused for 5 times and the yield of each cellulose conversion have no significance different. Another study who investigated four different types of homogenous HPAs such as Keggin, Wells-Dawson, Anderson and Lindqvist for oxidation of cellulose into formic acid observed that Lindqvist gave better selectivity of formic acid compared to other types of HPAs [64]. Another study reported that most of the previous studies were focusing more on strong Bronsted HPA catalyst which are 12-tungstoboric acid, silicotungstic acid and phosphotungstic acid for fractionation of cellulose to glucose, but these two catalysts have low selectivity towards glucose [65]. Hence, the efficient HPA for cellulose conversion into glucose is 12-tungstoboric acid ($H_5BW_{12}O_{40}$) catalyst. From these 3 different HPAs that have been studied, they concluded that the sequence of the reactivity is as followed $H_5BW_{12}O_{40}$> $H_4SiW_{12}O_{40}$>$H_3PW_{12}O_{40}$.

Another study was done to understand the relation of different anion charge in $H_3PMo_{12}O_{40}$ towards fatty acid yield from hydrolysis of macauba oil [39]. Higher charge of anion in $H_3PMo_{12}O_{40}$ favors hydrolysis reaction because more soluble and strong oxidizing property. An investigation on three different HPAs which are silicotungstic acid ($H_4SiW_{12}O_{40}$), phosphotungstic acid ($H_3PW_{12}O_4$) and phosphomolybdic acid ($H_3PMo_{12}O_{40}$) for hydrolysis of cellulose into glucose observed that tungsten-based HPA yield more glucose but low selectivity because $H_3PW_{12}O_4$ have stronger acidity which hydrolyze all cellulose [52]. Meanwhile, molybdenum based HPAs produce better selectivity because of their moderate acidity and oxygen activation abilities. In another experiment, two different HPAs which are $H_3PMo_{12}O_{40}$ and $H_3PW_{12}O_4$ for valorization of lignin [51]. From the observation, $H_3PMo_{12}O_{40}$ yield higher carboxylic acids which attributed to its superior potential compared to $H_3PW_{12}O_4$. Although $H_3PW_{12}O_4$ showed remarkable yield at preheating period but at the end over-oxidation occur because of the acid strength. Thus, both acidity and oxidizing property should be balance in breaking down cellulose.

Another research on three different tungsten based HPAs which are $H_3PW_{12}O_4$, $H_4SiW_{12}O_{40}$ and H_2WO_4 for cellulose hydrolysis to glucose in homogenous system [42]. $H_3PW_{12}O_4$ showed the highest yield and selectivity of glucose because of strong Bronsted acidity that allows $H_3PW_{12}O_4$ to liberate high concentration of protons in water. A comparison was made in another study to understand the effectiveness between $H_3PMo_{12}O_{40}$ and H_2SO_4 to yield higher total organic carbon and from the observation $H_3PMo_{12}O_{40}$ yield better because $H_3PMo_{12}O_{40}$ is more soluble in water to break down cellulose in water medium [66].

6. Heterogenous HPA Catalyst

A pure form of HPA usually a homogenous catalyst that is efficient in the catalytic reaction due to high solubility in water and organic solvents. However, homogenous catalyst generally suffers low recovery, low surface area and difficult to separate from the reaction medium [67]. Heterogenous HPA is the suggested solution and the proses requires solidification for better recovery and recyclability [68]. Moreover, heterogenous HPAs in solid form are water tolerant and stable in reactions involving water such as hydrolysis, oxidation and esterification [69]. There are few proposed methods to solidify HPAs such as introduction of large inorganic cations, encapsulation by metal organic framework and impregnation with support material [70][52][68]. Incorporation of inorganic cations include substitution of cations with suitable size, amount, charge and hydrophobicity that result ionic interaction with HPA [69]. Common cations used to heterogenized HPAs are Cs^+, Ag^+, Sn^{2+}, Zn^{2+}, Mn^{2+}, Co^{2+} and Cu^{2+}.

For example, a study has observed a high percentage yield of lactic acid (~93%) produced from glycerol conversion using silver-exchanged phosphomolybdic acid [71]. Their analysis showed that silver-exchanged phosphomolybdic acid is a promising catalyst for the reaction due to the strong Lewis acid property, high redox potential and water tolerant. Previous studies reported that the success of HPA for hydrolysis of cellulose is due to its strong acidity. A finding was done to compare 2 types of HPAs which are $K_4[PVW_{11}O_{40}]$-KPWV$_1$ and $K_6[PV_3W_9O_{40}]$-KPWV$_3$ for hydrolysis of cellulose into glucose to understand the effect of redox properties in metal ions of HPA by replacing tungsten atom to vanadium atom [41]. A previous finding was done to understand the ability of $CePW_{12}O_{40}xH_2O$ to degrade lignin in the presence of isopropanol as a solvent. Almost 73.4% of lignin was degraded into monomers thru cleavage of C-C and C-O bonds followed by dehydration, dealkylation and demothyxylation reaction of monomeric reaction [72].

Meanwhile, heterogenized of HPA by metal-organic framework also has attracted attention among researchers due to their unique combination of properties such as high surface area, crystalline structures and tunable pore size [67]. An example of a reported study on heterogenous silicotungstic acid (HSiW) encapsulated UiO-66 metal-organic frameworks showed efficient one-pot esterification of lauric acid to produce biodiesel which achieves 80% conversion and stable after 6 times of catalytic reaction [73].

Meanwhile, a metal organic framework supported phosphomolybdic acid catalyst was investigated for conversion of biomass-derived levulinic acid into ethyl levulinate using [74]. The metal support was synthesized using 1,3,5-benzenetricarboxylic acid and copper nitrate that provides Lewis acid site and enhance the esterification reaction.

Besides that, impregnation of HPA on support material may also provide better properties of pure HPA. Common support material that has been investigated by previous researchers include silica [75], zirconia [76], activated carbon [20] and bentonite [48]. Support play important role to provide better stability, efficiency and high dispersion of active sites for catalytic reaction. An example, HPA can be supported on inert solid matrices such as silica for easy separation from hydrolysate [52]. 5-hydroxymethylfurfural production from glucose was obtained using 2 different types of HPAs which are phosphotungstic acid ($H_3PW_{12}O_{40}$) and phosphomolybdic acid ($H_3PMo_{12}O_{40}$) on niobium (V) oxide (Nb_2O_5) and aluminium oxide (Al_2O_3) [19]. The result showed that $H_3PW_{12}O_{40}$/ Nb_2O_5 gave the highest yield (40.8%) and can be recycled four times without reducing yield due to the presence of Lewis and Bronsted acid sites on Nb_2O_5. They also stated that low yield of product observed for $H_3PW_{12}O_{40}$/ Al_2O_3 is due to only Lewis sites existed on the Al_2O_3 surface.

Meanwhile, HPW supported on activated carbon was also studied to determine the total percentage of reducing sugars by comparing with non-supported HPW [20]. The BET analysis showed supported HPW possess higher acidity compared to un-supported HPW. Besides, activated charcoal also possesses high surface area that improves the catalytic reaction. Another study examine the effectiveness of carbon foam supported HPW for cellulose conversion into glucose [73]. The product yield is 58.47% with cellulose conversion of 65.47%. The carbon foam consists of oxygen-containing functional groups that increase the affinity between cellulose and catalyst by forming a hydrogen bond with free hydroxyl groups of cellulose. Furthermore, carbon foam supported HPW is also rich with active sites that contribute to the disruption of β-1,4-glycosidic bonds of cellulose.

7. Conclusion

Several modified and unmodified heteropoly acid catalysts have been outlined in this review as a promising catalyst for valorization of biomass. The reason here is because HPAs is a bifunctional catalyst having both acid and redox site that can accomplish the conversion of biomass into valuable chemical products. Besides the preparation of the HPAs, the lignocellulosic types and the conditions of the conversion reactions also plays a crucial role in achieving a higher yield of the targeted product. Furthermore, Heterogenous HPAs based catalyst offers better properties compare to homogenous type especially in regeneration, recyclability and selectivity. However, some biomass conversion using heterogenous HPAs still obtained the undesirable products due to overoxidation of the product and sometimes the catalyst prepared is less acidic compared to the homogenous type. Thus, future works should be focused more on modifying the catalyst functionality by having a catalyst that has multiple functions in order to cater each of the limitations and fit for biomass conversion that usually involves a complex of reactions.

8. References

[1] Levi P G and Cullen J M 2018 *Environ. Sci. Technol* **52** 1725

[2] Perera F 2018 *Int. J. Environ. Res. Public Health* **15** 16

[3] Perkins G, Batalha N, Kumar A, Bhaskar T and Konarova M 2019 *Renewable and Sustainable Energy Reviews* 115 109400

[4] Güney T 2019 *International Journal of Sustainable Development & World Ecology* **26** 389

[5] Li G, Liu W, Ye C, Li X and Si C-L 2018 *International Journal of Polymer Science*

[6] Tursi A 2019 *Biofuel Research Journal* **22** 962

[7] Ribeiro L S, de Melo Órfão J J and Pereira M F R 2017 *Bioresource Technology*

[8] Muller N, Romero R, Grandon H and Segura C 2016 *Energy Fuels* **30** 10417

[9] Soares I B, Mendes K C S, Benachour M and Abreu C A M 2017 *Chemical Engineering Communications* **204** 1369

[10] Dietrich K, Mejia C H, Verschuren P, Rothenberg G and Shiju N R 2016 *Org Process Rev Dev* **21** 165
[11] Novitasari I, Rahayu D U C and Krisnadi Y K 2017 **496** *IOP Conference Series: Materials Science and Engineering*
[12] Zahara Z F 2018 *Evergreen Joint Journal of Novel Carbon Resource Sciences & Green Asia Strategy* **5** 67
[13] Signoretto M, Taghavi S, Ghedini E and Menegazzo F 2019 *Molecules* **24** 2760
[14] Scapin E, Rambo M G K, Viana G C C, Marasca N, Lacerda G E, Rambo M C D and Fernandes R d M N 2019 *Food Sci Technol* **40**
[15] Mamleeva N A, Autlov S A, Bazarnova N G and Lunin V V 2016 *Russian Journal of Bioorganic Chemistry* **42** 694
[16] Wang A G, Austin D and Song H 2017 *Biomass Volume Estimation and Valorization for Energy*
[17] Li S, Deng W, Li Y, Zhang Q and Wang Y 2018 *Journal of Energy Chemistry* **32** 138
[18] Yang W, Du X, Liu W, Wang Z, Dai H and Deng Y 2019 *Ind Eng Chem Res* **28** 22996
[19] Nogueira J S Mm Silva J P A, Mussatto S I and Carneiro L M 2020 Energies **13** 655
[20] Carminati S d A, Arantes A C C, Oliveira A C S d and Bianchi M *Matéria (Rio de Janeiro)* **23**
[21] Kupiainen L 2012 *Dilute Acid Catalysed Hydrolysis of Cellulose - Extension to Formic Acid.* Finland: University of Oulu
[22] Jia S, Cox B J, Guo X, Zhang Z C and Ekerdt J G 2010 *Chem Sus Chem* **3** 1078
[23] Takagaki A 2019 *Catalyst* **9** 907
[24] Oh H, Choi Y, Shin C, Nguyen T V T, Han Y, Kim H, Kim Y H, Lee J W, Jang J-W and Ryu J 2020 *ACS Catal* **10** 2060
[25] Li C, Zhao X, Wang A, Huber G W and Zhang T 2015 *Chem Rev* **115** 1159
[26] Girisuta B, Janssen L P B M and Heeres H J 2007 *Ind Eng Chem Res* **46** 1696
[27] Deepa A K and Dhepe P L 2014 *RSC Adv* **4** 12625
[28] Zhang Z, Liu B and Zhao Z K 2012 *Polymer Degradation and Stability* **97** 573
[29] Li X, Jiang Y, Wang L, Meng L, Wang W and Mu X 2012 *RSC Adv* **2** 6921
[30] Unkelbach G, Rößiger B and Pufky-Heinrich D 2018 *Lignin- Trends and Application*
[31] Otromke M, White R J and Sauerb J 2019 Carbon Resources **2** 59
[32] Pineda A and Lee A F 2016 Appl Petrochem Res **6** 243
[33] Albuquerque E M, Borges L E P and Fraga M A 2015 Green Chem **17** 3889
[34] Wattanapaphawong P, Sato O, Sato K, Mimura N, Reubroycharoen P and Yamaguchi A 2017 *Catalyst* **7** 221
[35] Luigi D V, Recchi C, Gonzalez-Ya˜nez E O, Davies T E, Lopez-Sanchez J A 2017 **202** 281
[36] Granone L I, Sieland F, Zheng N, Dillert R and Bahnemanna D W 2018 *Green Chem*
[37] Sun Y, Shi L, Wang H, Miao G, Kong L, Li S and Sun Y 2019 *Sustainable Energy and Fuels* **3** 1163
[38] Gumerova N I and Rompell A 2018 *Nature Reviews* **2**
[39] Silva M J and Teixeira M G 2017 *RSC Adv* **7** 8192
[40] Morales M D, Infantes-Molina A, Lazaro-Martinez J M, Romanellia G P, Pizzio L R and Rodriguez-Castellon E 2020 *Molecular Catalyst* **485** 110842
[41] Arantes A C C, Oliveira A C S, Borges W M d S, Bianchi M L and Resende E C 2017 *Revista Materia* **22**
[42] Li H-X, Zhuang X, Wang Q, Yang D, Cao Q and Jin L 2019 *Cellulose* **27** 285
[43] Timofeeva M N 2003 *Applied Catalysis* **256** 19
[44] Cottaa R F, Martins R A, Pereira M M, Rocha K A d S, Kozhevnikov E F, Kozhevnikov I V and Gusevskaya E V 2019 *Applied Catalysis* **584** 117173
[45] Hanif M A, Nisar S and Rashid U 2017 *Catalysis Reviews* **59** 165
[46] Gromov N V, taran O P, Delidovich I V, Pestunov A V, Rodikova Y A, Tatsenko D A, Zhizhina E G and V N 2016 *Catalysis Today* **278** 74

[47] Conceicao L R V, Reis C E R, Lima R, Cortez D V and Castro H F 2019 *RSC Adv* **9** 23450
[48] de Oliveira A d N, de Lima M A B, Pires L H d O, da Silva M R, da Luz P T S, Angelica R S, Filho G N d R, da Costa C E E, Luque R and Nascimento L A S 2019 *Materials* **12** 1431
[49] Guan W, Tsang C-W, Lin C S K, Len C, Hu H and Liang C 2019 *Bioresource Technology* **298**
[50] Chen Q, Shen C and He L 2018 *Acta Cryst* **74** 1182
[51] Demesa A G, Laari A, Sillanpaa M and Koiranen T 2017 *Molecules* **22** 1625
[52] Nayak A, Pulidindi I N and Rao C S 2020 *Renewable Energy* **190** 215
[53] Lee H V, Hamid S B A and Zain S K 2014 *The Scientific World Journal* **2014**
[54] Zhou L, Wang L, Diao Y, Yan R and Zhang S 2017 *Molecular Catalysis* **433** 153
[55] Marianou A A, Michailof C C, Ipsakis D, Triantafylidid K and Lappas A A 2019 *Green Chem* **21** 6161
[56] Gromov N V, Medvedeva T B, Taran O P, Timofeva M N, Said-Aizpuru O, Panchenko V N, Gerasimov E Y, Kozhevnikov I V and Pamon V N 2020 *Applied Catalysis A General* **595** 117489
[57] Sun Z, Wang S, Wang X and Jiang Z 2016 *Fuel* **164** 262
[58] Zhang J, Liu X, Hedhili M N< Zhu Y and Han Y 2011 *ChemCatchem* **3** 1294
[59] Zhao P, Zhang Y, Wang Y, Cul H, Song F, Sun X and Zhang L 2018 *Green Chem* **20** 1551
[60] Anderson E M, Crisci A, Murugappan K and Roman-Leshkov Y 2017 *ChemSuschem* **10** 10
[61] Chowdhury S R, Witte P T, Blank D H A, Alsters P L and Elshof J E 2006 *Chem Eur J* **12** 3061
[62] Tian J, Wang J, Zhao S, Jiang C, Zhang X and Wang X 2010 *Cellulose* **17** 587
[63] Zhou S, Yang X, Zhang Y, Jiang L, Zhou L, Lu T and Su Y 2019 *Cellulose* **26** 9135
[64] Albert J 2017 *Faraday Discuss* **202** 99
[65] Ogasawara Y, Itagaki S, Yamaguchi K and Mizuno N 2011 *ChemSusChem* **4** 519
[66] Dolmetsch T R 2017 *Phosphomolybdic Acid Catalysis of Cellulose Hydrolysis*. United States: East Tennessee State University
[67] Rafiee E and Kahrizi M 2013 *S Afr J Chem* **66** 145
[68] Wang S-S and Yang G-Y 2015 *Chem Rev* **115** 4893
[69] Eferadi-Kerenkan A, Do T and Kallaguine S 2018 *Catal Sci Technol* **8** 2257
[70] Batalha D C, Ferreira S O, da Silva R C and da Silva M J 2020 *ChemistrySelect* **5** 1976
[71] Tao M, Zhang D, Guan H, Huang G and Wang X 2016 *Scientific Reports* **6** 29840
[72] Du F, Li Y, Xian X, Li B, Tang P, Lu B and Lai J 2019 *Energy Fuels* **33** 7366
[73] Zhang Z, Wang B, Zhang Y, Zhang G and Wang Y 2019 *Advances in Polymer Technology*
[74] Guo T, Qiu M and Qi X 2019 *Applied Catalysis A General* **572** 168
[75] Zhou L, Wang L, Wang H, Cao Y, Yan R and Zhang S 2016 *J Thermodyn Catal* **7** 2160
[76] Alcaniz-Monge J, Bakkali B E, Trautwein G and Reinoso S 2017 *Applied Catalysis B: Environmental* **224** 194

AUTHOR INDEX